Springer Monographs in Math

Editors-in-Chief

Isabelle Gallagher, UFR de Mathématiques, Université Paris-Diderot, Paris, France

Minhyong Kim, School of Mathematics, Korea Institute for Advanced Study, Seoul, South Korea; Mathematical Institute, University of Warwick, Coventry, UK

Series Editors

Sheldon Axler, Department of Mathematics, San Francisco State University, San Francisco, CA, USA

Mark Braverman, Department of Mathematics, Princeton University, Princeton, NJ, USA

Maria Chudnovsky, Department of Mathematics, Princeton University, Princeton, NJ, USA

Tadahisa Funaki, Department of Mathematics, University of Tokyo, Tokyo, Japan

Sinan C. Güntürk, Department of Mathematics, Courant Institute of Mathematical Science, New York, NY, USA

Claude Le Bris, Cite Descartes, Champs-sur-Marne, CERMICS-ENPC, Marne la Vallée, France

Pascal Massart, Département de Mathématiques, Université de Paris-Sud, Orsay, France

Alberto A. Pinto, Department of Mathematics, University of Porto, Porto, Portugal

Gabriella Pinzari, Department of Mathematics, University of Padova, Padova, Italy

Ken Ribet, Department of Mathematics, University of California, Berkeley, CA, USA

René Schilling, Institut für Mathematische Stochastik, TU Dresden, Dresden, Germany

Panagiotis Souganidis, Department of Mathematics, University of Chicago, Chicago, IL, USA

Endre Süli, Mathematical Institute, University of Oxford, Oxford, UK

Shmuel Weinberger, Department of Mathematics, University of Chicago, Chicago, IL, USA

Boris Zilber, Department of Mathematics, Oxford University, Oxford, UK

This series publishes advanced monographs giving well-written presentations of the "state-of-the-art" in fields of mathematical research that have acquired the maturity needed for such a treatment. They are sufficiently self-contained to be accessible to more than just the intimate specialists of the subject, and sufficiently comprehensive to remain valuable references for many years. Besides the current state of knowledge in its field, an SMM volume should ideally describe its relevance to and interaction with neighbouring fields of mathematics, and give pointers to future directions of research.

More information about this series at http://www.springer.com/series/3733

Rudolf Gorenflo · Anatoly A. Kilbas · Francesco Mainardi · Sergei Rogosin

Mittag-Leffler Functions, Related Topics and Applications

Second Edition

Rudolf Gorenflo (1930–2017)
Mathematical Institute
Free University Berlin
Berlin, Germany

Francesco Mainardi
Department of Physics
and Astronomy
University of Bologna
Bologna, Italy

Anatoly A. Kilbas (1948–2010)
Department of Mathematics
and Mechanics
Belarusian State University
Minsk, Belarus

Sergei Rogosin
Department of Economics
Belarusian State University
Minsk, Belarus

ISSN 1439-7382   ISSN 2196-9922 (electronic)
Springer Monographs in Mathematics
ISBN 978-3-662-61552-2   ISBN 978-3-662-61550-8 (eBook)
https://doi.org/10.1007/978-3-662-61550-8

Mathematics Subject Classification: 33E12, 26A33, 34A08, 45K05, 44Axx, 60G22

1st edition: © Springer-Verlag Berlin Heidelberg 2014
2nd edition: © Springer-Verlag GmbH Germany, part of Springer Nature 2020
This work is subject to copyright. All rights are reserved by the Publisher, whether the whole or part of the material is concerned, specifically the rights of translation, reprinting, reuse of illustrations, recitation, broadcasting, reproduction on microfilms or in any other physical way, and transmission or information storage and retrieval, electronic adaptation, computer software, or by similar or dissimilar methodology now known or hereafter developed.
The use of general descriptive names, registered names, trademarks, service marks, etc. in this publication does not imply, even in the absence of a specific statement, that such names are exempt from the relevant protective laws and regulations and therefore free for general use.
The publisher, the authors and the editors are safe to assume that the advice and information in this book are believed to be true and accurate at the date of publication. Neither the publisher nor the authors or the editors give a warranty, expressed or implied, with respect to the material contained herein or for any errors or omissions that may have been made. The publisher remains neutral with regard to jurisdictional claims in published maps and institutional affiliations.

This Springer imprint is published by the registered company Springer-Verlag GmbH, DE part of Springer Nature.
The registered company address is: Heidelberger Platz 3, 14197 Berlin, Germany

To the memory of our colleagues and friends

*Anatoly Kilbas (1948–2010) and
Rudolf Gorenflo (1930–2017)*

Preface to the Second Edition

After the appearance of the first edition of our book "Mittag-Leffler Functions: Related Topics and Applications", we have observed a growing interest in the subject. Many new research articles and books have appeared. This is mainly due to the central role of the Mittag-Leffler functions in Fractional Calculus and Fractional Modeling. With this interest in mind, we decided to prepare the second edition of our book on Mittag-Leffler functions, presenting new ideas and results related to the theory and applications of this family of functions.

New results have been added to practically all sections of the book. In Chap. 3 "The Classical Mittag-Leffler Function", results on Mittag-Leffler summation as well as the notion of the Mittag-Leffler reproducing kernel Hilbert space are discussed. We present results applying the distribution of the zeros of the Mittag-Leffler function to the study of inverse problems for differential equations in Banach spaces (Chap. 4 "The Two-Parametric Mittag-Leffler Function"). Chapter 5 "Mittag-Leffler Functions With Three Parameters" discusses recent results on Le Roy type functions related to the Mittag-Leffler function but having a different nature. New applications related to all functions in this chapter have been added. Essentially enlarged is the next chapter (Chap. 6) concerning Mittag-Leffler functions depending on several parameters. Such functions have become important from both a theoretical and an applied point of view. We also discuss the properties of the Mittag-Leffler functions of several variables and with matrix argument. Numerical methods for these classes of functions are discussed too. We have completely rewritten the chapters dealing with applications (Chap. 8 "Applications to Fractional Order Equations", Chap. 9 "Applications to Deterministic Models", and Chap. 10 "Applications to Stochastic Models", which are essentially enlarged versions of Chaps. 7–9 from the first edition). Thus, we briefly discuss in Chap. 8 the main ideas of fractional control theory and present some numerical methods applied to the study of fractional models, including those related to the calculation of the values of the Mittag-Leffler functions. We also added a new chapter (Chap. 7), which describes the main properties of the classical Wright function, closely related to the Mittag-Leffler function. Consequently, the structure of Appendix F "Higher Transcendental Functions" has been changed, since now in App. F. 2 we deal mainly with the generalized Wright function, focussing not only on the general function of this type (the so-called Fox–Wright function $_pW_q$) but also on the most applicable special cases $_1W_1$ and $_0W_2$. Essential changes were also made to Appendix E "Elements of Fractional Calculus" in order to outline the role of less popular fractional constructions and to show which specific properties of these constructions give potential further applications of Grünwald–Letnikov, Marchaud, Hadamard, Erdélyi–Kober, and Riesz fractional derivatives.

Our book project could not have been realized without the constant support of our colleagues and friends. We are grateful to Roberto Garrappa for preparing short reviews of his results and allowing us to include them in the book. Additional thanks are due to Alexander Apelblat, Roberto Garra, Andrea Giusti, George Karneadakis, Virginia Kiryakova, Yuri Luchko, Arak Mathai, Edmundo Capelas de Oliveira, Gianni Pagnini, Enrico Scalas, José Tenreiro Machado, and Vladimir Uchaikin. Our wives, Giovanna and Maryna, were so polite to allow us to spend so much time on the book. The 2nd edition was discussed in Bologna and Berlin by three of us, but as of October 20, 2017, Professor Rudolf Gorenflo is no longer with us. We took the liberty to dedicate this second edition to our missed colleagues and friends, Anatoly Kilbas and Rudolf Gorenflo, keeping them as co-authors because of their essential role in realizing this project.

Bologna, Italy Francesco Mainardi
Minsk, Belarus Sergei Rogosin
March 2020

Preface to the First Edition

The study of the Mittag-Leffler function and its various generalizations has become a very popular topic in Mathematics and its Applications. However, during the twentieth century, this function was practically unknown to the majority of scientists, since it was ignored in most common books on special functions. As a noteworthy exception the handbook "Higher Transcendental Functions", vol. 3, by A. Erdelyi et al. deserves to be mentioned.

Now the Mittag-Leffler function is leaving its isolated role as *Cinderella* (using the term coined by F.G. Tricomi for the *incomplete gamma* function).

The recent growing interest in this function is mainly due to its close relation to the *Fractional Calculus* and especially to fractional problems which come from applications.

Our decision to write this book was motivated by the need to fill the gap in the literature concerning this function, to explain its role in modern pure and applied mathematics, and to give the reader an idea of how one can use such a function in the investigation of modern problems from different scientific disciplines.

This book is a fruit of collaboration between researchers in Berlin, Bologna and Minsk. It has highly profited from visits of SR to the Department of Physics at the University of Bologna and from several visits of RG to Bologna and FM to the Department of Mathematics and Computer Science at Berlin Free University under the European ERASMUS exchange. RG and SR appreciate the deep scientific atmosphere at the University of Bologna and the perfect conditions they met there for intensive research.

We are saddened that our esteemed and always enthusiastic co-author Anatoly A. Kilbas is no longer with us, having lost his life in a tragic accident on 28 June 2010 in the South of Russia. We will keep him, and our inspiring joint work with him, in living memory.

Berlin, Germany Rudolf Gorenflo
Bologna, Italy Francesco Mainardi
Minsk, Belarus Sergei Rogosin
March 2014

Contents

1	**Introduction**	1
2	**Historical Overview of the Mittag-Leffler Functions**	7
	2.1 A Few Biographical Notes On Gösta Magnus Mittag-Leffler	7
	2.2 The Contents of the Five Papers by Mittag-Leffler on New Functions	10
	2.3 Further History of Mittag-Leffler Functions	12
3	**The Classical Mittag-Leffler Function**	19
	3.1 Definition and Basic Properties	19
	3.2 Relations to Elementary and Special Functions	21
	3.3 Recurrence and Differential Relations	23
	3.4 Integral Representations and Asymptotics	24
	3.5 Distribution of Zeros	31
	3.6 Further Analytic Properties	37
	3.6.1 Additional Integral Properties	37
	3.6.2 Mittag-Leffler Summation of Power Series	41
	3.6.3 Mittag-Leffler Reproducing Kernel Hilbert Spaces	44
	3.7 The Mittag-Leffler Function of a Real Variable	46
	3.7.1 Integral Transforms	46
	3.7.2 The Complete Monotonicity Property	53
	3.7.3 Relation to Fractional Calculus	55
	3.8 Historical and Bibliographical Notes	57
	3.9 Exercises	60
4	**The Two-Parametric Mittag-Leffler Function**	63
	4.1 Series Representation and Properties of Coefficients	64
	4.2 Explicit Formulas. Relations to Elementary and Special Functions	65
	4.3 Differential and Recurrence Relations	66

4.4	Integral Relations and Asymptotics		69
4.5	The Two-Parametric Mittag-Leffler Function as an Entire Function		74
4.6	Distribution of Zeros		76
	4.6.1	Distributions of Zeros and Inverse Problems for Differential Equations in Banach Spaces	84
4.7	Computations With the Two-Parametric Mittag-Leffler Function		87
4.8	Further Analytic Properties		93
	4.8.1	Additional Integral and Differential Formulas	93
	4.8.2	Geometric Properties of the Mittag-Leffler Function	95
	4.8.3	An Extension for Negative Values of the First Parameter	97
4.9	The Two-Parametric Mittag-Leffler Function of a Real Variable		100
	4.9.1	Integral Transforms of the Two-Parametric Mittag-Leffler Function	100
	4.9.2	The Complete Monotonicity Property	101
	4.9.3	Relations to the Fractional Calculus	102
4.10	Historical and Bibliographical Notes		103
4.11	Exercises		107
5	**Mittag-Leffler Functions with Three Parameters**		**115**
5.1	The Prabhakar (Three-Parametric Mittag-Leffler) Function		115
	5.1.1	Definition and Basic Properties	115
	5.1.2	Integral Representations and Asymptotics	118
	5.1.3	Expansion on the Negative Semi-axes	121
	5.1.4	Integral Transforms of the Prabhakar Function	122
	5.1.5	Complete Monotonicity of the Prabhakar Function	123
	5.1.6	Fractional Integrals and Derivatives of the Prabhakar Function	125
	5.1.7	Relations to the Fox–Wright Function, H-function and Other Special Functions	127
5.2	The Kilbas–Saigo (Three-Parametric Mittag-Leffler) Function		128
	5.2.1	Definition and Basic Properties	128
	5.2.2	The Order and Type of the Entire Function $E_{\alpha,m,l}(z)$	129
	5.2.3	Recurrence Relations for $E_{\alpha,m,l}(z)$	132
	5.2.4	Connection of $E_{n,m,l}(z)$ with Functions of Hypergeometric Type	134
	5.2.5	Differentiation Properties of $E_{n,m,l}(z)$	136
	5.2.6	Complete Monotonicity of the Kilbas–Saigo Function	140

		5.2.7	Fractional Integration of the Kilbas–Saigo Function	141
		5.2.8	Fractional Differentiation of the Kilbas–Saigo Function	143
	5.3	The Le Roy Type Function		147
		5.3.1	Definition and Main Analytic Properties	147
		5.3.2	Integral Representations of the Le Roy Type Function	149
		5.3.3	Laplace Transforms of the Le Roy Type Function	152
		5.3.4	The Asymptotic Expansion on the Negative Semi-axis	153
		5.3.5	Extension to Negative Values of the Parameter α	157
	5.4	Historical and Bibliographical Notes		157
	5.5	Exercises		159
6	**Multi-index and Multi-variable Mittag-Leffler Functions**			**163**
	6.1	The Four-Parametric Mittag-Leffler Function: The Luchko–Kilbas–Kiryakova Approach		163
		6.1.1	Definition and Special Cases	163
		6.1.2	Basic Properties	164
		6.1.3	Integral Representations and Asymptotics	166
		6.1.4	Extended Four-Parametric Mittag-Leffler Functions	168
		6.1.5	Relations to the Wright Function and the H-Function	168
		6.1.6	Integral Transforms of the Four-Parametric Mittag-Leffler Function	170
		6.1.7	Integral Transforms with the Four-Parametric Mittag-Leffler Function in the Kernel	172
		6.1.8	Relations to the Fractional Calculus	175
	6.2	The Four-Parametric Mittag-Leffler Function: A Generalization of the Prabhakar Function		179
		6.2.1	Definition and General Properties	179
		6.2.2	The Four-Parametric Mittag-Leffler Function of a Real Variable	180
	6.3	Mittag-Leffler Functions with $2n$ Parameters		181
		6.3.1	Definition and Basic Properties	181
		6.3.2	Representations in Terms of Hypergeometric Functions	184
		6.3.3	Integral Representations and Asymptotics	186
		6.3.4	Extension of the $2n$-Parametric Mittag-Leffler Function	186
		6.3.5	Relations to the Wright Function and to the H-Function	188

		6.3.6	Integral Transforms with the Multi-parametric Mittag-Leffler Functions	189
		6.3.7	Relations to the Fractional Calculus.............	193
	6.4	Mittag-Leffler Functions of Several Variables		194
		6.4.1	Integral Representations	195
		6.4.2	Asymptotic Behavior for Large Values of Arguments...............................	198
	6.5	Mittag-Leffler Functions with Matrix Arguments		200
	6.6	Historical and Bibliographical Notes		201
	6.7	Exercises		207
7	**The Classical Wright Function**			209
	7.1	Definition and Basic Properties		209
	7.2	Relations to Elementary and Special Functions.............		211
	7.3	Integral Representations and Asymptotics.................		215
	7.4	Distribution of Zeros................................		218
	7.5	Further Analytic Properties		220
		7.5.1	Additional Properties of the Wright Function in the Complex Plane.........................	220
		7.5.2	Geometric Properties of the Wright Function	221
		7.5.3	Auxiliary Functions of the Wright Type	222
	7.6	The Wright Function of a Real Variable..................		225
		7.6.1	Relation to Fractional Calculus	225
		7.6.2	Laplace Transforms of the Mittag-Leffler and the Wright Functions	226
		7.6.3	Mainardi's Approach to the Wright Functions of the Second Kind	227
	7.7	Historical and Bibliographical Notes		231
	7.8	Exercises		233
8	**Applications to Fractional Order Equations**			235
	8.1	Fractional Order Integral Equations		235
		8.1.1	The Abel Integral Equation.....................	235
		8.1.2	Other Integral Equations Whose Solutions Are Represented Via Generalized Mittag-Leffler Functions....................................	239
	8.2	Fractional Ordinary Differential Equations		241
		8.2.1	Fractional Ordinary Differential Equations with Constant Coefficients	241
		8.2.2	Ordinary FDEs with Variable Coefficients	248
		8.2.3	Other Types of Ordinary Fractional Differential Equations....................................	251

	8.3	Optimal Control for Equations with Fractional Derivatives and Integrals...................................	253
		8.3.1 Linear Fractional-Order Controllers................	253
		8.3.2 Nonlinear Fractional-Order Controllers.............	254
		8.3.3 Modification of the Control Actions in Fractional-Order PID Controllers.................	255
		8.3.4 Further Possible Modifications of the Fractional-Order PID Controllers.............	256
	8.4	Differential Equations with Fractional Partial Derivatives.....	257
		8.4.1 Cauchy-Type Problems for Differential Equations with Riemann–Liouville Fractional Partial Derivatives.................................	258
		8.4.2 The Cauchy Problem for Differential Equations with Caputo Fractional Partial Derivatives..........	259
	8.5	Numerical Methods for the Solution of Fractional Differential Equations.................................	261
		8.5.1 Direct Numerical Methods......................	262
		8.5.2 Indirect Numerical Methods.....................	263
		8.5.3 Other Numerical Methods......................	264
	8.6	Historical and Bibliographical Notes.....................	264
	8.7	Exercises...	275
9	**Applications to Deterministic Models**........................		281
	9.1	Fractional Relaxation and Oscillations...................	281
		9.1.1 Simple Fractional Relaxation and Oscillation........	282
		9.1.2 The Composite Fractional Relaxation and Oscillations..............................	291
	9.2	Examples of Applications of the Fractional Calculus in Physical Models...................................	299
		9.2.1 Linear Visco-Elasticity.........................	299
		9.2.2 The Use of Fractional Calculus in Linear Viscoelasticity................................	300
		9.2.3 The General Fractional Operator Equation..........	303
	9.3	The Fractional Dielectric Models........................	303
		9.3.1 The Main Models for Anomalous Dielectric Relaxation...................................	307
		9.3.2 The Cole–Cole Model.........................	307
		9.3.3 The Davidson–Cole Model......................	310
		9.3.4 The Havriliak–Negami Model....................	313
	9.4	The Fractional Calculus in the Basset Problem.............	317
		9.4.1 The Equation of Motion for the Basset Problem......	318
		9.4.2 The Generalized Basset Problem..................	320
	9.5	Other Deterministic Fractional Models...................	325

	9.6	Historical and Bibliographical Notes 329
	9.7	Exercises 334

10 Applications to Stochastic Models 339
    10.1 Introduction 339
    10.2 The Mittag-Leffler Process According to Pillai 340
    10.3 Elements of Renewal Theory and Continuous Time
          Random Walks (CTRWs) 342
          10.3.1 Renewal Processes 342
          10.3.2 Continuous Time Random Walks (CTRWs) 343
          10.3.3 The Renewal Process as a Special CTRW 347
    10.4 The Poisson Process and Its Fractional Generalization
          (The Renewal Process of Mittag-Leffler Type) 349
          10.4.1 The Mittag-Leffler Waiting Time Density 349
          10.4.2 The Poisson Process 350
          10.4.3 The Renewal Process of Mittag-Leffler Type 351
          10.4.4 Thinning of a Renewal Process 355
    10.5 Fractional Diffusion and Subordination Processes 357
          10.5.1 Renewal Process with Reward 357
          10.5.2 Limit of the Mittag-Leffler Renewal Process 357
          10.5.3 Subordination in the Space-Time Fractional
                  Diffusion Equation 362
          10.5.4 The Rescaling and Respeeding Concept Revisited.
                  Universality of the Mittag-Leffler Density 365
    10.6 The Wright M-Functions in Probability 367
          10.6.1 The Absolute Moments of Order δ 368
          10.6.2 The Characteristic Function 369
          10.6.3 Relations with Lévy Stable Distributions 369
          10.6.4 The Wright $\mathbb{M}$-Function in Two Variables 373
    10.7 Historical and Bibliographical Notes 374
    10.8 Exercises 378

Appendix A: The Eulerian Functions 381

Appendix B: The Basics of Entire Functions 397

Appendix C: Integral Transforms 409

Appendix D: The Mellin–Barnes Integral 431

Appendix E: Elements of Fractional Calculus 439

Appendix F: Higher Transcendental Functions 463

References 499

Index 537

Chapter 1
Introduction

This book is devoted to an extended description of the properties of the Mittag-Leffler function, its numerous generalizations and their applications in different areas of modern science.

The function $E_\alpha(z)$ is named after the great Swedish mathematician **Gösta Magnus Mittag-Leffler** (1846–1927) who defined it by a power series

$$E_\alpha(z) = \sum_{k=0}^{\infty} \frac{z^k}{\Gamma(\alpha k + 1)}, \quad \alpha \in \mathbb{C}, \ \mathrm{Re}\,\alpha > 0, \tag{1.0.1}$$

and studied its properties in 1902–1905 in five subsequent notes [ML1, ML2, ML3, ML4, ML5-5] in connection with his summation method for divergent series.

This function provides a simple generalization of the exponential function because of the replacement of $k! = \Gamma(k+1)$ by $(\alpha k)! = \Gamma(\alpha k + 1)$ in the denominator of the power terms of the exponential series.

During the first half of the twentieth century the Mittag-Leffler function remained almost unknown to the majority of scientists. They unjustly ignored it in many treatises on special functions, including the most popular (Abramowitz and Stegun [AbrSte72] and its new version "NIST Handbook of Mathematical Functions" [NIST]). Furthermore, there appeared some relevant works where the authors arrived at series or integral representations of this function without recognizing it, e.g., (Gnedenko and Kovalenko [GneKov68]), (Balakrishnan [BalV85]) and (Sanz-Serna [San88]). A description of the most important properties of this function is present in the third volume [ErdBat-3] of the Handbook on Higher Transcendental Functions of the Bateman Project, (Erdelyi et al.). In it, the authors have included the Mittag-Leffler functions in their Chapter XVIII devoted to the so-called miscellaneous functions. The attribution of 'miscellaneous' to the Mittag-Leffler function is due to the fact that it was only later, in the sixties, that it was recognized to belong to

© Springer-Verlag GmbH Germany, part of Springer Nature 2020
R. Gorenflo et al., *Mittag-Leffler Functions, Related Topics and Applications*,
Springer Monographs in Mathematics,
https://doi.org/10.1007/978-3-662-61550-8_1

a more general class of higher transcendental functions, known as Fox H-functions (see, e.g., [MatSax78, KilSai04, MaSaHa10]). In fact, this class was well-established only after the seminal paper by Fox [Fox61]. A more detailed account of the Mittag-Leffler function is given in the treatise on complex functions by Sansone and Gerretsen [SanGer60]. However, the most specialized treatise, where more details on the functions of Mittag-Leffler type are given, is surely the book by Dzherbashyan [Dzh66], in Russian. Unfortunately, no official English translation of this book is presently available. Nevertheless, Dzherbashyan has done a lot to popularize the Mittag-Leffler function from the point of view of its special role among entire functions of a complex variable, where this function can be considered as the simplest non-trivial generalization of the exponential function.

Successful applications of the Mittag-Leffler function and its generalizations, and their direct involvement in problems of physics, biology, chemistry, engineering and other applied sciences in recent decades has made them better known among scientists. A considerable literature is devoted to the investigation of the analyticity properties of these functions; in the references we quote several authors who, after Mittag-Leffler, have investigated such functions from a mathematical point of view. At last, the 2000 Mathematics Subject Classification has included these functions in item 33E12: "Mittag-Leffler functions and generalizations".

Starting from the classical paper of Hille and Tamarkin [HilTam30] in which the solution of Abel integral equation of the second kind

$$\phi(x) - \frac{\lambda}{\Gamma(\alpha)} \int_0^x \frac{\phi(t)}{(x-t)^{1-\alpha}} dt = f(x), \ 0 < \alpha < 1, \ 0 < x < 1, \qquad (1.0.2)$$

is presented in terms of the Mittag-Leffler function, this function has become very important in the study of different types of integral equations. We should also mention the 1954 paper by Barret [Barr54], which was concerned with the general solution of the linear fractional differential equation with constant coefficients.

But the real importance of this function was recognized when its special role in fractional calculus was discovered (see, e.g., [SaKiMa93]). In recent times the attention of mathematicians and applied scientists towards the functions of Mittag-Leffler type has increased, overall because of their relation to the Fractional Calculus and its applications. Because the Fractional Calculus has attracted wide interest in different areas of applied sciences, we think that the Mittag-Leffler function is now beginning to leave behind its isolated life as Cinderella. We like to refer to the classical Mittag-Leffler function as the Queen Function of Fractional Calculus, and to consider all the related functions as her court.

A considerable literature is devoted to the investigation of the analytical properties of this function. In the references, in addition to purely mathematical investigations, we also mention several monographs, surveys and research articles dealing with different kinds of applications of the higher transcendental functions related to the Mittag-Leffler function. However, we have to point out once more that there exists no treatise specially devoted to the Mittag-Leffler function itself. In our opinion, it is now time for a book aimed at a wide audience. This book has to serve both as a textbook for beginners, describing the basic ideas and results in the area, and as a table-book for applied scientists in which they can find the most important facts for applications, and it should also be a good source for experts in Analysis and Applications, collecting deep results widely spread in the special literature. These ideas have been implemented into our plan for the present book. Because of the relevance of the Mittag-Leffler function to the theory and applications of Fractional Calculus we were invited to write a survey chapter for the first volume of the Handbook of Fractional Calculus with Applications [HAND1]. This chapter [GoMaRo19] presents in condensed form the main ideas of our book.

The book has the following structure. It can be formally considered as consisting of four main parts. The first part (INTRODUCTION AND HISTORY) consists of two chapters. The second part (THEORY) presents different aspects of the theory of the Mittag-Leffler function and its generalizations, in particular those arising in applied models. This part is divided into five chapters. The third part (APPLICATIONS) deals with different kinds of applications involving the Mittag-Leffler function and its generalizations. This part is divided into three chapters. Since the variety of models related to the Mittag-Leffler function is very large and rapidly growing, we mainly focus on how to use this function in different situations. We also separate theoretical applications dealing mainly with the solution of certain equations in terms of the Mittag-Leffler function from the more "practical" applications related to its use in modelling. Most of the auxiliary facts are collected in the fourth part consisting of six APPENDICES. The role of the appendices is multi-fold. First, we present those results which are helpful in reading the main text. Secondly, we discuss in part the machinery which can be omitted at the first reading of the corresponding chapter. Lastly, the appendices partly play the role of a handbook on some auxiliary subjects related to the Mittag-Leffler function. In this sense these appendices can be used to further develop the ideas contained in our book and in the references mentioned in it.

Each structural part of the book (either chapter or appendix) ends with a special section "Historical and Bibliographical Notes". We hope that these sections will help the readers to understand the features of the Mittag-Leffler function more deeply. We also hope that acquaintance with the book will give the readers new practical instruments for their research. In addition, since one of the aims of the book is to attract students, we present at the end of each chapter and each appendix a collection of exercises connected with different aspects of the theory and applications. Special attention is paid to the list of references which we have tried to make as complete as possible. Only seldom does the main text give references to the literature, the references are mainly deferred to the notes sections at the end of chapters and appendices. The bibliography contains a remarkably large number of references to articles and books not mentioned in the text, since they have attracted the author's attention over the last few decades and cover topics more or less related to this monograph. In the second edition we have significantly updated the bibliography. The interested reader will hopefully take advantage of this bibliography, enlarging and improving the scope of the monograph itself and developing new results.

Chapter 2 has in a sense a historical nature. We present here a few bibliographical notes about the creator of this book's subject, G.M. Mittag-Leffler. The contents of his pioneering works on the considered function is given here together with a brief description of the further development of the theory of the Mittag-Leffler function and its generalizations.

Chapter 3 is devoted to the classical Mittag-Leffler function (1.0.1). We collect here the main results on the function which were discovered during the century following Mittag-Leffler's definition. These are of an analytic nature, comprising rules of composition and asymptotic properties, and its character as an entire function of a complex variable. Special attention is paid to integral transforms related to the Mittag-Leffler function because of their importance in the solution of integral and differential equations. We point out its role in the Fractional Calculus and its place among the whole collection of higher transcendental functions.

In Chap. 4 we discuss questions similar to those of Chap. 3. This chapter deals with the simplest (and for applications most important) generalizations of the Mittag-Leffler function, namely the two-parametric Mittag-Leffler function

$$E_{\alpha,\beta}(z) = \sum_{k=0}^{\infty} \frac{z^k}{\Gamma(\alpha k + \beta)}, \quad \alpha, \beta \in \mathbb{C}, \ \mathrm{Re}\,\alpha > 0, \tag{1.0.3}$$

which was deeply investigated independently by Humbert and Agarwal in 1953 [Hum53, Aga53, HumAga53] and by Dzherbashyan in 1954 [Dzh54a, Dzh54b, Dzh54c] (but formally appeared first in the paper by Wiman [Wim05a]).

Chapter 4 presents the theory of two types of three-parametric Mittag-Leffler function. First of all it is the three-parametric Mittag-Leffler function (or Prabhakar function) introduced by Prabhakar [Pra71]

$$E_{\alpha,\beta}^{\gamma}(z) = \sum_{k=0}^{\infty} \frac{(\gamma)_k}{k!\,\Gamma(\alpha k + \beta)} z^k, \ \alpha, \beta, \gamma \in \mathbb{C}, \ \mathrm{Re}\,\alpha, \gamma > 0, \tag{1.0.4}$$

where $(\gamma)_k = \gamma(\gamma+1)\ldots(\gamma+k-1) = \frac{\Gamma(\gamma+k)}{\Gamma(\gamma)}$ is the Pochhammer symbol (see (A.17) in Appendix A). This function is now widely used for different applied problems. Another type of three-parametric Mittag-Leffler function is not as well-known as the Prabhakar function (1.0.4). It was introduced and studied by Kilbas and Saigo [KilSai95b] in connection with the solution of a new type of fractional differential equation. This function (the Kilbas–Saigo function) is defined as follows

$$E_{\alpha,m,l}(z) = \sum_{k=0}^{\infty} c_k z^k \ (z \in \mathbb{C}), \tag{1.0.5}$$

where

1 Introduction

$$c_0 = 1, \quad c_k = \prod_{i=1}^{k-1} \frac{\Gamma(\alpha[im+l]+1)}{\Gamma(\alpha[im+l+1]+1)} \quad (k=1,2,\cdots), \ \alpha \in \mathbb{C}, \ \operatorname{Re}\alpha > 0. \tag{1.0.6}$$

Some basic results on this function are also included in Chap. 5. In the second edition we also include in this chapter another three-parametric generalization of the Mittag-Leffler function, namely, the Le Roy type function

$$F_{\alpha,\beta}^{(\gamma)} := \sum_{k=1}^{\infty} \frac{z^k}{(\Gamma(\alpha k + \beta))^\gamma}, \tag{1.0.7}$$

which is a function of a different nature than the other functions in this chapter. The Le Roy type function is a simple generalization of the Le Roy function [LeR00], which appeared as a competitor of the Mittag-Leffler function in the study of divergent series.

By introducing additional parameters one can discover new interesting properties of these functions (discussed in Chaps. 3–5) and extend their range of applicability. This is exactly the case with the generalizations described in this chapter. Together with some appendices, the above mentioned chapters constitute a short course on the Mittag-Leffler function and its generalizations. This course is self-contained and requires only a basic knowledge of Real and Complex Analysis.

Chapter 6 is rooted deeper mathematically. The reader can find here a number of modern generalizations. The ideas leading to them are described in detail. The main focus is on four-parametric Mittag-Leffler functions (Dzherbashyan [Dzh60]) and $2n$-parametric Mittag-Leffler functions (Al-Bassam and Luchko [Al-BLuc95] and Kiryakova [Kir99]). Experts in higher transcendental functions and their applications will find here many interesting results, obtained recently by various authors. These generalizations will all be labelled by the name Mittag-Leffler, in spite of the fact that some of them can be considered for many values of parameters as particular cases of the general class of Fox H-functions. These H-functions offer a powerful tool for formally solving many problems, however by inserting relevant parameters one often arrives at functions whose behavior is easier to handle. This is the case for the Mittag-Leffler functions, and so these functions are often more appropriate for applied scientists who prefer direct work to a detour through a wide field of generalities.

In the second edition we have decided to give a much wider presentation (new Chap. 7) of the classical Wright function

$$\phi(\alpha, \beta; z) = \sum_{k=0}^{\infty} \frac{z^k}{k!\Gamma(\alpha k + \beta)}, \ \alpha > -1, \beta \in \mathbb{C}. \tag{1.0.8}$$

This function is closely related to the Mittag-Leffler function (especially to the four-parametric Mittag-Leffler function, see, e.g., [GoLuMa99, RogKor10]) and is of great importance for Fractional Calculus too. In spite of this similarity, some properties of the Wright function are not completely analogous to those of the Mittag-Leffler function. Thus, we found it important to discuss here the properties of the Wright function in detail.

The last three chapters deal with applications of the functions treated in the preceding chapters. We start (Chap. 8) with the "formal" (or mathematical) applications of Mittag-Leffler functions. The title of the chapter is "Applications to Fractional Order Equations". By fractional order equations we mean either integral equations with weak singularities or differential equations with ordinary or partial fractional derivatives. The collection of such equations involving Mittag-Leffler functions in their analysis or in their explicit solution is fairly big. Of course, we should note that a large number of fractional order equations arise in certain applied problems. We would like to separate the questions of mathematical analysis (solvability, asymptotics of solutions, their explicit presentation etc.) from the motivation and description of those models in which such equations arise. In Chap. 8 we focus on the development of a special "fractional" technique and give the reader an idea of how this technique can be applied in practice.

Further applications are presented in the two subsequent chapters devoted to mathematical modelling of special processes of interest in the applied sciences. Chapter 9 deals mainly with the role of Mittag-Leffler functions in discovering and analyzing deterministic models based on certain equations of fractional order. Special attention is paid to fractional relaxation and oscillation phenomena, to fractional diffusion and diffusive wave phenomena, to fractional models in dielectrics, models of particle motion in a viscous fluid, and to hereditary phenomena in visco-elasticity and hydrodynamics. These are models in physics, chemistry, biology etc., which by adopting a macroscopic viewpoint can be described without using probabilistic ideas and machinery.

In contrast, in Chap. 10 we describe the role of Mittag-Leffler functions in models involving randomness. We explain here the key role of probability distributions of Mittag-Leffler type which enter into a variety of stochastic processes, including fractional Poisson processes and the transition from continuous time random walk to fractional diffusion.

Our six appendices can be divided into two groups. First of all we present here some basic facts from certain areas of analysis. Such appendices are useful additions to the course of lectures which can be extracted from Chaps. 3–5. The second type of appendices constitute those which can help the reader to understand modern results in the areas in which the Mittag-Leffler function is essential and important. They serve to make the book self-contained.

The book is addressed to a wide audience. Special attention is paid to those topics which are accessible for students in Mathematics, Physics, Chemistry, Biology and Mathematical Economics. Also in our audience are experts in the theory of the Mittag-Leffler function and its applications. We hope that they will find the technical parts of the book and the historical and bibliographical remarks to be a source of new ideas. Lastly, we have to note that our main goal, which we always had in mind during the writing of the book, was to make it useful for people working in different areas of applications (even those far from pure mathematics).

Chapter 2
Historical Overview of the Mittag-Leffler Functions

2.1 A Few Biographical Notes On Gösta Magnus Mittag-Leffler

Gösta Magnus Mittag-Leffler was born on March 16, 1846, in Stockholm, Sweden. His father, John Olof Leffler, was a school teacher, and was also elected as a member of the Swedish Parliament. His mother, Gustava Vilhelmina Mittag, was a daughter of a pastor, who was a person of great scientific abilities. At his birth Gösta was given the name Leffler and later (when he was a student) he added his mother's name "Mittag" as a tribute to this family, which was very important in Sweden in the nineteenth century. Both sides of his family were of German origin.[1]

At the Gymnasium in Stockholm Gösta was training as an actuary but later changed to mathematics. He studied at the University of Uppsala, entering it in 1865. In 1872 he defended his thesis on applications of the argument principle and in the same year was appointed as a Docent (Associate Professor) at the University of Uppsala.

In the following year he was awarded a scholarship to study and work abroad as a researcher for three years. In October 1873 he left for Paris.

In Paris Mittag-Leffler met many mathematicians, such as Bouquet, Briot, Chasles, Darboux, and Liouville, but his main goal was to learn from Hermite. However, he found the lectures by Hermite on elliptic functions difficult to understand.

In Spring 1875 he moved to Berlin to attend lectures by Weierstrass, whose research and teaching style was very close to his own. From Weierstrass' lectures Mittag-Leffler learned many ideas and concepts which would later become the core of his scientific interests.

In Berlin Mittag-Leffler received news that professor Lorenz Lindelöf (Ernst Lindelöf's father) had decided to leave a chair at the University of Helsingfors (now Helsinki). At the same time Weierstrass requested from the ministry of education

---

[1]Many interesting aspects of Mittag-Leffler's life can be found in [Noe27, Har28a, Har28b].

the installation of a new position at his institute and suggested Mittag-Leffler for the position. In spite of this, Mittag-Leffler applied for the chair at Helsingfors. He got the chair in 1876 and remained at the University of Helsingfors for the next five years.

In 1881 the new University of Stockholm was founded, and Gösta Mittag-Leffler was the first to hold a chair in Mathematics there. Soon afterwards he began to organize the setting up of the new international journal *Acta Mathematica*. In 1882 Mittag-Leffler founded *Acta Mathematica* and served as the Editor-in-Chief of the journal for 45 years. The original idea for such a journal came from Sophus Lie in 1881, but it was Mittag-Leffler's understanding of the European scene, together with his political skills, that ensured the success of the journal. Later he invited many well-known mathematicians (Cantor, Poincaré and many others) to submit papers to this journal. Mittag-Leffler was always a good judge of the quality of the work submitted to him for publication.

The role of G. Mittag-Leffler as a founder of Acta Mathematica was more than simply an organizer of the mathematical Journal. We can cite from [Dau80, p. 261–263]: *"Gösta Mittag-Leffler was the founding editor of the journal Acta Mathematica. In the early 1870s it was meant, in part, to bring the mathematicians of Germany and France together in the aftermath of the France-Prussian War, and the political neutrality of Sweden made it possible for Mittag-Leffler to realize this goal by publishing articles in German and French, side by side. Even before the end of the First World War, Mittag-Leffler again saw his role as mediator, and began to work for a reconciliation between German and Allied mathematicians through the auspices of his journal. Similarly, G.H. Hardy was particularly concerned about the reluctance of many scientists in England to attempt any sort of rapprochement with the Central European countries and he sought to do all he could to bring English and German mathematicians together after the War.... Nearly half a century earlier, Mittag-Leffler saw himself in much the same position as mediator between belligerent mathematicians on both sides. In fact, he believed that he was in an especially suitable position to bring European scientific interests together after World War I, and he saw his Acta Mathematics as the perfect instrument for promoting a lasting rapprochement."*

In 1882 Gösta Mittag-Leffler married Signe af Linfors and they lived together until the end of his life.

Mittag-Leffler made numerous contributions to mathematical analysis, particularly in the areas concerned with limits, including calculus, analytic geometry and probability theory. He worked on the general theory of functions, studying relationships between independent and dependent variables.

His best known work deals with the analytic representation of a single-valued complex function, culminating in the Mittag-Leffler theorem. This study began as an attempt to generalize results from Weierstrass's lectures, where Weierstrass had described his theorem on the existence of an entire function with prescribed zeros each with a specified multiplicity. Mittag-Leffler tried to generalize this result to meromorphic functions while he was studying in Berlin. He eventually assembled his findings on generalizing Weierstrass' theorem to meromorphic functions in a paper which he published (in French) in 1884 in *Acta Mathematica*. In this paper

Mittag-Leffler proposed a series of general topological notions on infinite point sets based on Cantor's new set theory.

With this paper Mittag-Leffler became the sole proprietor of a theorem that later became widely known and so he took his place in the circle of internationally known mathematicians. Mittag-Leffler was one of the first mathematicians to support Cantor's theory of sets but, one has to remark, a consequence of this was that Kronecker refused to publish in *Acta Mathematica*. Between 1899 and 1905 Mittag-Leffler published a series of papers which he called "Notes" on the summation of divergent series. The aim of these notes was to construct the analytical continuation of a power series outside its circle of convergence. The region in which he was able to do this is now called Mittag-Leffler's star. Andre Weyl in his memorial [Weil82] says: "*A well-known anecdote has Oscar Wilde saying that he had put his genius into his life; into his writings he had put merely his talent. With at least equal justice it may be said of Mittag-Leffler that the Acta Mathematica were the product of his genius, while nothing more than talent went into his mathematical contributions. Genius transcends and defies analysis; but this may be a fitting occasion for examining some of the qualities involved in the creating and in the editing of a great mathematical journal.*"

In the same period Mittag-Leffler introduced and investigated in five subsequent papers a new special function, which is now very popular and useful for many applications. This function, as well as many of its generalizations, is now called the "Mittag-Leffler" function.[2]

His contribution is nicely summed up by Hardy [Har28a]: "*Mittag-Leffler was a remarkable man in many ways. He was a mathematician of the front rank, whose contributions to analysis had become classical, and had played a great part in the inspiration of later research; he was a man of strong personality, fired by an intense devotion to his chosen study; and he had the persistence, the position, and the means to make his enthusiasm count.*"

Gösta Mittag-Leffler passed away on July 7, 1927. During his life he received many honours. He was an honorary member or corresponding member of almost every mathematical society in the world including the Accademia Reale dei Lincei, the Cambridge Philosophical Society, the Finnish Academy of Sciences, the London Mathematical Society, the Moscow Mathematical Society, the Netherlands Academy of Sciences, the St. Petersburg Imperial Academy, the Royal Institution, the Royal Belgium Academy of Sciences and Arts, the Royal Irish Academy, the Swedish Academy of Sciences, and the Institute of France. He was elected a Fellow of the Royal Society of London in 1896. He was awarded honorary degrees from the Universities of Oxford, Cambridge, Aberdeen, St. Andrews, Bologna and Christiania (now Oslo).

---

[2]Since it is the subject of this book, we will give below a wider discussion of these five papers and of the role of the Mittag-Leffler functions.

2.2 The Contents of the Five Papers by Mittag-Leffler on New Functions

Let us begin with a description of the ideas which led to the introduction by Mittag-Leffler of a new transcendental function.

In 1899 Mittag-Leffler began the publication of a series of articles under the common title "*Sur la représentation analytique d'une branche uniforme d'une fonction monogène*" ("On the analytic representation of a single-valued branch of a monogenic function") published mainly in *Acta Mathematica* [ML5-1, ML5-2, ML5-3, ML5-4, ML5-5, ML5-6]. The first articles of this series were based on three reports presented by him in 1898 at the Swedish Academy of Sciences in Stockholm.

His research was connected with the following question:
Let $k_0, k_1, \ldots$ be a sequence of complex numbers for which

$$\lim_{\nu \to \infty} |k_\nu|^{1/\nu} = \frac{1}{r} \in \mathbb{R}_+$$

is finite. Then the series

$$FC(z) := k_0 + k_1 z + k_2 z^2 + \ldots$$

is convergent in the disk $D_r = \{z \in \mathbb{C} : |z| < r\}$ and divergent at any point with $|z| > r$. It determines a single-valued analytic function in the disk D_r.[3]

The questions discussed were:

(1) to determine the maximal domain on which the function $FC(z)$ possesses a single-valued analytic continuation;
(2) to find an analytic representation of the corresponding single-valued branch.

Abel [Abe26a] had proposed (see also [Lev56]) to associate with the function $FC(z)$ the entire function

$$F_1(z) := k_0 + \frac{k_1 z}{1!} + \frac{k_2 z^2}{2!} + \ldots + \frac{k_\nu z^\nu}{\nu!} + \ldots = \sum_{\nu=0}^{\infty} \frac{k_\nu z^\nu}{\nu!}.$$

This function was used by Borel (see, e.g., [Bor01]) to discover that the answer to the above question is closely related to the properties of the following integral (now called the *Laplace–Abel integral*):

$$\int_0^\infty e^{-\omega} F_1(\omega z) d\omega. \tag{2.2.1}$$

---

[3]The notation $FC(z)$ is not defined in Mittag-Leffler's paper. The letter "C" probably indicates the word 'convergent' in order to distinguish this function from its analytic continuation $FA(z)$ (see discussion below).

2.2 The Contents of the Five Papers by Mittag-Leffler on New Functions

An intensive study of these properties was carried out at the beginning of the twentieth century by many mathematicians (see, e.g., [ML5-3, ML5-5] and references therein).

Mittag-Leffler introduced instead of $F_1(z)$ a one-parametric family of (entire) functions

$$F_\alpha(z) := k_0 + \frac{k_1 z}{\Gamma(1 \cdot \alpha + 1)} + \frac{k_2 z^2}{\Gamma(2 \cdot \alpha + 1)} + \ldots = \sum_{\nu=0}^{\infty} \frac{k_\nu z^\nu}{\Gamma(\nu \cdot \alpha + 1)}, \quad (\alpha > 0),$$

and studied its properties as well as the properties of the generalized Laplace–Abel integral

$$\int_0^\infty e^{-\omega^{1/\alpha}} F_\alpha(\omega z) d\omega^{1/\alpha} = \int_0^\infty e^{-\omega} F_\alpha(\omega^\alpha z) d\omega. \tag{2.2.2}$$

The main result of his study was: in a maximal domain A (star-like with respect to origin) the analytic representation of the single-valued analytic continuation $FA(z)$ of the function $FC(z)$ can be represented in the following form

$$FA(z) = \lim_{\alpha \to 1} \int_0^\infty e^{-\omega} F_\alpha(\omega^\alpha z) d\omega. \tag{2.2.3}$$

For this reason analytic properties of the functions $F_\alpha(z)$ become highly important.

Due to this construction Mittag-Leffler decided to study the most simple function of the type $F_\alpha(z)$, namely, the function corresponding to the unit sequence k_ν. This function

$$E_\alpha(z) := 1 + \frac{z}{\Gamma(1 \cdot \alpha + 1)} + \frac{z^2}{\Gamma(2 \cdot \alpha + 1)} + \ldots = \sum_{\nu=0}^{\infty} \frac{z^\nu}{\Gamma(\nu \cdot \alpha + 1)}, \tag{2.2.4}$$

was introduced and investigated by G. Mittag-Leffler in five subsequent papers [ML1, ML2, ML3, ML4, ML5-5] (in particular, in connection with the above formulated questions). This function is known now as the *Mittag-Leffler function*.

In the *first paper* [ML1], devoted to his new function, Mittag-Leffler discussed the relation of the function $F_\alpha(z)$ with the above problem on analytic continuation. In particular, he posed the question of whether the domains of analyticity of the function

$$\lim_{\alpha \downarrow 1} \int_0^\infty e^{-\omega} F_\alpha(\omega^\alpha z) d\omega$$

and the function (introduced and studied by Le Roy [LeR00], see also [GaRoMa17, GoHoGa19] in relation to complete monotonicity of this function)

$$\lim_{\alpha \downarrow 1} \sum_{\nu=0}^{\infty} \frac{\Gamma(\nu\alpha + 1)}{\Gamma(\nu + 1)} k_\nu z^\nu$$

coincide.

In the *second paper* [ML2] the new function (i.e., the Mittag-Leffler function) appeared. Its asymptotic properties were formulated. In particular, Mittag-Leffler showed that $E_\alpha(z)$ behaves as $e^{z^{1/\alpha}}$ in the angle $-\frac{\pi\alpha}{2} < \arg z < \frac{\pi\alpha}{2}$ and is bounded for values of z with $\frac{\pi\alpha}{2} < |\arg z| \leq \pi$.[4]

In the *third paper* [ML3] the asymptotic properties of $E_\alpha(z)$ were discussed more carefully. Mittag-Leffler compared his results with those of Malmquist [Mal03], Phragmén [Phr04] and Lindelöf [Lin03], which they obtained for similar functions (the results form the background of the classical Phragmén–Lindelöf theorem [PhrLin08]).

The *fourth paper* [ML4] was completely devoted to the extension of the function $E_\alpha(z)$ (as well as the function $F_\alpha(z)$) to complex values of the parameter α.

Mittag-Leffler's most creative paper on the new function $E_\alpha(z)$ is his *fifth paper* [ML5-5]. In this article, he:

(a) found an integral representation for the function $E_\alpha(z)$;
(b) described the asymptotic behavior of $E_\alpha(z)$ in different angle domains;
(c) gave the formulas connecting $E_\alpha(z)$ with known elementary functions;
(d) provided the asymptotic formulas for

$$E_\alpha(z) = \frac{1}{2\pi i} \int_L \frac{1}{\alpha} e^{\omega^{1/\alpha}} \frac{d\omega}{\omega - z}$$

by using the so-called *Hankel integration path*;
(e) obtained detailed asymptotics of $E_\alpha(z)$ for negative values of the variable, i.e. for $z = -r$;
(f) compared in detail his asymptotic results for $E_\alpha(z)$ with the results obtained by Malmquist;
(g) found domains which are free of zeros of $E_\alpha(z)$ in the case of "small" positive values of parameter, i.e. for $0 < \alpha < 2$, $\alpha \neq 1$;
(h) applied his results on $E_\alpha(z)$ to answer the question of the domain of analyticity of the function $FA(z)$ and its analytic representation (see formula (2.2.3)).

2.3 Further History of Mittag-Leffler Functions

The importance of the new function was understood as soon as the first analytic results for it appeared. First of all, it is a very simple function playing a key role in the solution of a general problem of the theory of analytic functions. Secondly, the Mittag-Leffler function can be considered as a direct generalization of the exponential function, preserving some of its properties. Furthermore, $E_\alpha(z)$ has some interesting properties which later became essential for the description of many problems arising in applications.

---

[4]The behavior of $E_\alpha(z)$ on critical rays $|\arg z| = \pm\frac{\pi\alpha}{2}$ was not described.

2.3 Further History of Mittag-Leffler Functions

After Mittag-Leffler's introduction of the new function, one of the first results on it was obtained by Wiman [Wim05a]. He used Borel's method of summation of divergent series (which Borel applied to the special case of the Mittag-Leffler function, namely, for $\alpha = 1$, see [Bor01]). Using this method, Wiman gave a new proof of the asymptotic representation of $E_\alpha(z)$ in different angle domains. This representation was obtained for positive rational values of the parameter α. He also noted[5] that analogous asymptotic results hold for the two-parametric generalization $E_{\alpha,\beta}(z)$ of the Mittag-Leffler function (see (1.0.3)). Applying the obtained representation Wiman described in [Wim05b] the distribution of zeros of the Mittag-Leffler function $E_\alpha(z)$. The main focus was on two cases – to the case of real values of the parameter $\alpha \in (0, 2]$, $\alpha \neq 1$, and to the case of complex values of α, $\mathrm{Re}\,\alpha > 0$.

In [Phr04] Phragmén proved the generalization of the Maximum Modulus Principle for the case of functions analytic in an angle. For this general theorem the Mittag-Leffler function plays the role of the key example. It satisfies the inequality $|E_\alpha(z)| < C_1 \mathrm{e}^{|z|^\rho}$, $\rho = 1/\{\mathrm{Re}\,\alpha\}$, in an angular domain z, $|\arg z| \leq \frac{\pi}{2\rho}$, but although it is bounded on the boundary rays it is not constant in the whole angular domain. This means that the Mittag-Leffler function possesses a maximal angular domain (in the sense of the *Phragmén* or *Phragmén–Lindelöf theorem*, see [PhrLin08]) in which the above stated property holds.

One more paper devoted to the development of the asymptotic method of Mittag-Leffler appeared in 1905. Malmquist (a student of G. Mittag-Leffler) applied this method to obtain the asymptotics of a function similar to $E_\alpha(z)$, namely

$$\sum_\nu \frac{z^\nu}{\Gamma(1 + \nu a_\nu)},$$

where the sequence a_ν tends to zero as $\nu \to \infty$. The particular goal was to construct a simple example of an entire function which tends to zero along almost all rays when $|z| \to \infty$. Such an example

$$G(z) = \sum_\nu \frac{z^\nu}{\Gamma(1 + \frac{\nu}{(\log \nu)^\alpha})}, \quad 0 < \alpha < 1, \tag{2.3.1}$$

was constructed [Mal05] and carefully examined by using the calculus of residues for the integral representation of $G(z)$ (which is also analogous to that for E_α).

At the beginning of the twentieth century many mathematicians paid great attention to obtaining asymptotic expansions of special functions, in particular, those of hypergeometric type. The main reason for this was that these functions play an important role in the study of differential equations, which describe different phenomena. In the fundamental paper [Barn06] Barnes proposed a unified approach to the investigation of asymptotic expansions of entire functions defined by Taylor series. This

---

[5]But did not discuss in detail.

approach was based on the previous results of Barnes [Barn02] and Mellin [Mel02]. The essence of this approach is to use the representation of the quotient of the products of Gamma functions in the form of a contour integral which is handled by using the method of residues. This representation is now known as the *Mellin–Barnes integral formula* (see Appendix D). Among the functions which were treated in [Barn06] was the Mittag-Leffler function. The results of Barnes were further developed in his articles, including applications to the theory of differential equations, as well as in the articles by Mellin (see, e.g., [Mel10]). In fact, the idea of employing contour integrals involving Gamma functions of the variable in the subject of integration is due to Pincherle, whose suggestive paper [Pin88] was the starting point of Mellin's investigations (1895), although the type of contour and its use can be traced back to Riemann, as Barnes wrote in [Barn07b, p. 63].

Generalizations of the Mittag-Leffler function are proposed among other generalizations of the hypergeometric functions. For them similar approaches were used. Among these generalizations we should point out the collection of *Wright functions*, first introduced in 1935, see [Wri35a],

$$\phi(z; \rho, \beta) := \sum_{n=0}^{\infty} \frac{z^n}{\Gamma(n+1)\Gamma(\rho n + \beta)} = \sum_{n=0}^{\infty} \frac{z^n}{n!\Gamma(\rho n + \beta)}; \quad (2.3.2)$$

the collection of generalized hypergeometric functions, first introduced in 1928, see [Fox28],

$$_pF_q(z) = {}_pF_q\left(\alpha_1, \alpha_2, \ldots, \alpha_p; \beta_1, \beta_2, \ldots, \beta_q; z\right) = \sum_{k=0}^{\infty} \frac{(\alpha_1)_k \cdot (\alpha_2)_k \cdots (\alpha_p)_k}{(\beta_1)_k \cdot (\beta_2)_k \cdots (\beta_q)_k} \frac{z^k}{k!}; \quad (2.3.3)$$

[6]the collection of *Meijer G-functions* introduced in 1936, see [Mei36], and intensively treated in 1946, see [Mei46],

$$G_{p,q}^{m,n}\left(z \left| \begin{matrix} a_1, \ldots, a_p \\ b_1, \ldots, b_q \end{matrix} \right. \right)$$

$$= \frac{1}{2\pi i} \int_T \frac{\prod_{i=1}^{m} \Gamma(b_i + s) \prod_{i=1}^{n} \Gamma(1 - a_i - s)}{\prod_{i=m+1}^{q} \Gamma(1 - b_i - s) \prod_{i=n+1}^{p} \Gamma(a_i + s)} z^{-s} \mathrm{d}s, \quad (2.3.4)$$

---

[6]Here $(\cdot)_k$ is the Pochhammer symbol, see (A.17) in Appendix A.

2.3 Further History of Mittag-Leffler Functions

and the collection of more general *Fox H-functions*

$$H_{p,q}^{m,n}\left(z \left| \begin{array}{c} (a_1, \alpha_1), \ldots, (a_p, \alpha_p) \\ (b_1, \beta_1), \ldots, (b_q, \beta_q) \end{array} \right. \right)$$

$$= \frac{1}{2\pi i} \int_T \frac{\prod_{i=1}^{m} \Gamma(b_i + \beta_i s) \prod_{i=1}^{n} \Gamma(1 - a_i - \alpha_i s)}{\prod_{i=m+1}^{q} \Gamma(1 - b_i - \beta_i s) \prod_{i=n+1}^{p} \Gamma(a_i + \alpha_i s)} z^{-s} ds. \qquad (2.3.5)$$

Some generalizations of the Mittag-Leffler function appeared as a result of developments in integral transform theory. In this connection in 1953 Agarwal and Humbert (see [Hum53, Aga53, HumAga53]) and independently in 1954 Djrbashian (see [Dzh54a, Dzh54b, Dzh54c]) introduced and studied the *two-parametric Mittag-Leffler function* (or *Mittag-Leffler type function*)

$$E_{\alpha,\beta}(z) := \sum_{\nu=0}^{\infty} \frac{z^\nu}{\Gamma(\nu \cdot \alpha + \beta)}. \qquad (2.3.6)$$

We note once more that, formally, the function (2.3.6) first appeared in the paper of Wiman [Wim05a], who did not pay much attention to its extended study.

In 1971 Prabhakar [Pra71] introduced the *three-parametric Mittag-Leffler function* (or *generalized Mittag-Leffler function*, or *Prabhakar function*)

$$E_{\alpha,\beta}^{\rho}(z) := \sum_{\nu=0}^{\infty} \frac{(\rho)_\nu z^\nu}{\Gamma(\nu \cdot \alpha + \beta)}. \qquad (2.3.7)$$

This function appeared in the kernel of a first-order integral equation which Prabhakar treated by using Fractional Calculus.

Other *three-parametric Mittag-Leffler functions* (also called *generalized Mittag-Leffler functions* or *Mittag-Leffler type functions*, or *Kilbas–Saigo functions*) were introduced by Kilbas and Saigo (see, e.g., [KilSai95a])

$$E_{\alpha,m,l}(z) := \sum_{n=0}^{\infty} c_n z^n, \qquad (2.3.8)$$

where

$$c_0 = 1, \quad c_n = \prod_{i=0}^{n-1} \frac{\Gamma[\alpha(im + l) + 1]}{\Gamma[\alpha(im + l + 1) + 1]}.$$

These functions appeared in connection with the solution of new types of integral and differential equations and with the development of the Fractional Calculus. Nowadays they are referred to as Kilbas–Saigo functions.

One more generalization of the Mittag-Leffler function depending on three parameters was studied recently in [Ger12, GarPol13, GaRoMa17, GoHoGa19]

$$F_{\alpha,\beta}^{(\gamma)}(z) = \sum_{k=0}^{\infty} \frac{z^k}{[\Gamma(\alpha k + \beta)]^\gamma}, \quad z \in \mathbb{C}, \quad \alpha, \beta, \gamma \in \mathbb{C}. \tag{2.3.9}$$

It is related to the so-called *Le Roy function*

$$R_\gamma(z) = \sum_{k=0}^{\infty} \frac{z^k}{[(k+1)!]^\gamma}, \quad z \in \mathbb{C}. \tag{2.3.10}$$

The function (2.3.9) plays an important role in Probability Theory.

For real $\alpha_1, \alpha_2 \in \mathbb{R}$ ($\alpha_1^2 + \alpha_2^2 \neq 0$) and complex $\beta_1, \beta_2 \in \mathbb{C}$ the following function was introduced by Dzherbashian (=Djrbashian) [Dzh60] in the form of the series (in fact only for $\alpha_1, \alpha_2 > 0$)

$$E_{\alpha_1,\beta_1;\alpha_2,\beta_2}(z) \equiv \sum_{k=0}^{\infty} \frac{z^k}{\Gamma(\alpha_1 k + \beta_1)\Gamma(\alpha_2 k + \beta_2)} \quad (z \in \mathbb{C}). \tag{2.3.11}$$

Generalizing the four-parametric Mittag-Leffler function (2.3.11) Al-Bassam and Luchko [Al-BLuc95] introduced the following Mittag-Leffler type function

$$E((\alpha, \beta)_m; z) = \sum_{k=0}^{\infty} \frac{z^k}{\prod_{j=1}^{m} \Gamma(\alpha_j k + \beta_j)} \quad (m \in \mathbb{N}) \tag{2.3.12}$$

with $2m$ real parameters $\alpha_j > 0$; $\beta_j \in \mathbb{R}$ ($j = 1, ..., m$) and with complex $z \in \mathbb{C}$. In [Al-BLuc95] an explicit solution to a Cauchy type problem for a fractional differential equation is given in terms of (2.3.12). The theory of this class of functions was developed in the series of articles by Kiryakova et al. [Kir99, Kir00, Kir08, Kir10a, Kir10b].

In the last several decades the study of the Mittag-Leffler function has become a very important branch of Special Function Theory. Many important results have been obtained by applying integral transforms to different types of functions from the Mittag-Leffler collection. Conversely, Mittag-Leffler functions generate new kinds of integral transforms with properties making them applicable to various mathematical models.

A number of more general functions related to the Mittag-Leffler function will be discussed in Chap. 6 below.

Nowadays the Mittag-Leffler function and its numerous generalizations have acquired a new life. The recent notable increased interest in the study of their relevant properties is due to the close connection of the Mittag-Leffler function to the Fractional Calculus and its application to the study of Differential and Integral Equations (in particular, of fractional order). Many modern models of fractional type have recently been proposed in Probability Theory, Mechanics, Mathematical Physics, Chemistry, Biology, Mathematical Economics etc. Historical remarks concerning these subjects will be presented at the end of the corresponding chapters of this book.

Chapter 3
The Classical Mittag-Leffler Function

In this chapter we present the basic properties of the classical Mittag-Leffler function $E_\alpha(z)$ (see (1.0.1)). The material can be formally divided into two parts. Starting from the basic definition of the Mittag-Leffler function in terms of a power series, we discover that for parameter α with positive real part the function $E_\alpha(z)$ is an entire function of the complex variable z. Therefore we discuss in the first part the (analytic) properties of the Mittag-Leffler function as an entire function. Namely, we calculate its order and type, present a number of formulas relating it to elementary and special functions as well as recurrence relations and differential formulas, introduce some useful integral representations and discuss the asymptotics and distribution of zeros of the classical Mittag-Leffler function.

It is well-known that current applications mostly use the properties of the Mittag-Leffler function with real argument. Thus, in the second part (Sect. 3.7), we collect results of this type. They concern integral representations and integral transforms of the Mittag-Leffler function of a real variable, the complete monotonicity property and relations to Fractional Calculus. On first reading, people working in applications can partly omit some of the deeper mathematical material (say, that from Sects. 3.4–3.6).

3.1 Definition and Basic Properties

Following Mittag-Leffler's classical definition we consider the one-parametric Mittag-Leffler function as defined by the power series

$$E_\alpha(z) = \sum_{k=0}^{\infty} \frac{z^k}{\Gamma(\alpha k + 1)} \quad (\alpha \in \mathbb{C}). \tag{3.1.1}$$

Although information on this function is widely spread in the literature (see, e.g., [Dzh66, GupDeb07, MatHau08, HaMaSa11]), we think that a fairly complete presentation here will help the reader to understand the ideas and results presented later this book.

Applying to the coefficients $c_k := \dfrac{1}{\Gamma(\alpha k + 1)}$ of the series (3.1.1) the Cauchy–Hadamard formula for the radius of convergence

$$R = \limsup_{k \to \infty} \frac{|c_k|}{|c_{k+1}|}, \tag{3.1.2}$$

and the asymptotic formula [ErdBat-1, 1.18(4)]

$$\frac{\Gamma(z+a)}{\Gamma(z+b)} = z^{a-b}\left[1 + \frac{(a-b)(a-b-1)}{2z} + O\left(\frac{1}{z^2}\right)\right] \quad (z \to \infty, |\arg z| < \pi), \tag{3.1.3}$$

one can see that the series (3.1.1) converges in the whole complex plane for all $\operatorname{Re}\alpha > 0$. For all $\operatorname{Re}\alpha < 0$ it diverges everywhere on $\mathbb{C} \setminus \{0\}$. For $\operatorname{Re}\alpha = 0$ the radius of convergence is equal to

$$R = e^{\frac{\pi}{2}|\operatorname{Im}\alpha|}.$$

In particular, for $\alpha \in \mathbb{R}_+$ tending to 0 one obtains the following relation:

$$E_0(\pm z) = \sum_{k=0}^{\infty} (\pm 1)^k z^k = \frac{1}{1 \mp z}, \quad |z| < 1. \tag{3.1.4}$$

In the most interesting case, $\operatorname{Re}\alpha > 0$, the Mittag-Leffler function is an entire function. Moreover, it follows from the Cauchy inequality for the Taylor coefficients and simple properties of the Gamma function that there exists a number $k \geq 0$ and a positive number $r(k)$ such that

$$M_{E_\alpha}(r) := \max_{|z|=r} |E_\alpha(z)| < e^{r^k}, \quad \forall r > r(k). \tag{3.1.5}$$

This means that $E_\alpha(z)$ is an entire function of *finite order* (see, e.g., [Lev56]).

For $\alpha > 0$ by Stirling's asymptotic formula [ErdBat-1, 1.18(3)]

$$\Gamma(\alpha k + 1) = \sqrt{2\pi}\,(\alpha k)^{\alpha k + \frac{1}{2}}\, e^{-\alpha k}\,(1 + o(1)), \quad k \to \infty, \tag{3.1.6}$$

one can see that the Mittag-Leffler function satisfies for $\alpha > 0$ the relations

$$\limsup_{k \to \infty} \frac{k \log k}{\log \frac{1}{|c_k|}} = \lim_{k \to \infty} \frac{k \log k}{\log|\Gamma(\alpha k + 1)|} = \frac{1}{\alpha},$$

3.1 Definition and Basic Properties

and

$$\limsup_{k\to\infty} \left(k^{\frac{1}{\rho}}\sqrt[k]{|c_k|}\right) = \lim_{k\to\infty}\left(k^{\frac{1}{\rho}}\sqrt[k]{\frac{1}{|\Gamma(\alpha k+1)|}}\right) = \left(\frac{e}{\alpha}\right)^\alpha.$$

If $\operatorname{Re}\alpha > 0$, and $\operatorname{Im}\alpha \neq 0$, the corresponding result is valid too. This follows from formula (3.1.3), which in particular means

$$0 < C_1 < \left|\frac{\Gamma(\alpha k+1)}{\Gamma(\alpha_0 k+1)}\right| < C_2 < \infty$$

for certain positive constants C_1, C_2 and sufficiently large k. Thus one can define the order and type of the Mittag-Leffler function as an entire function (see the definitions of order and type in formulas (B.5), (B.6) of Appendix B).

Proposition 3.1 (Order and type.) *For each α, $\operatorname{Re}\alpha > 0$, the Mittag-Leffler function (3.1.1) is an entire function of order $\rho = \dfrac{1}{\operatorname{Re}\alpha}$ and type $\sigma = 1$.*

In a certain sense each $E_\alpha(z)$ is the simplest entire function among those having the same order (see, e.g., [Phr04, GoLuRo97]). The Mittag-Leffler function also furnishes examples and counter-examples for the growth and other properties of entire functions of finite order (see, e.g., [Buh25a]).

One can also observe that from the above Proposition 3.1 it follows that the function

$$E_\alpha(\sigma^\alpha z) = \sum_{k=0}^\infty \frac{(\sigma^\alpha z)^k}{\Gamma(\alpha k+1)}, \quad \sigma > 0,$$

has order $\rho = \dfrac{1}{\operatorname{Re}\alpha}$ and type σ.

3.2 Relations to Elementary and Special Functions

The Mittag-Leffler function plays an important role among special functions. First of all it is not difficult to obtain a number of its relations to elementary and special functions. The simplest relation is formula (3.1.4) representing $E_0(z)$ as the sum of a geometric series.

We collect in the following proposition other relations of this type.

Proposition 3.2 (Special cases.) *For all $z \in \mathbb{C}$ the Mittag-Leffler function satisfies the following relations*

$$E_1(\pm z) = \sum_{k=0}^{\infty} \frac{(\pm 1)^k z^k}{\Gamma(k+1)} = e^{\pm z}, \qquad (3.2.1)$$

$$E_2(-z^2) = \sum_{k=0}^{\infty} \frac{(-1)^k z^{2k}}{\Gamma(2k+1)} = \cos z, \qquad (3.2.2)$$

$$E_2(z^2) = \sum_{k=0}^{\infty} \frac{z^{2k}}{\Gamma(2k+1)} = \cosh z, \qquad (3.2.3)$$

$$E_{\frac{1}{2}}(\pm z^{\frac{1}{2}}) = \sum_{k=0}^{\infty} \frac{(\pm 1)^k z^{\frac{k}{2}}}{\Gamma(\frac{1}{2}k+1)} = e^z \left[1 + \mathrm{erf}(\pm z^{\frac{1}{2}})\right] = e^z \, \mathrm{erfc}(\mp z^{\frac{1}{2}}), \qquad (3.2.4)$$

where erf (erfc) *denotes the error function (complementary error function)*

$$\mathrm{erf}(z) := \frac{2}{\sqrt{\pi}} \int_0^z e^{-u^2} du, \quad \mathrm{erfc}(z) := 1 - \mathrm{erf}(z), \; z \in \mathbb{C},$$

and $z^{\frac{1}{2}}$ means the principal branch of the corresponding multi-valued function defined in the whole complex plane cut along the negative real semi-axis.

A more general formula for the function with half-integer parameter is valid

$$E_{\frac{p}{2}}(z) = {}_0F_{p-1}\left(; \frac{1}{p}, \frac{2}{p}, \ldots, \frac{p-1}{p}; \frac{z^2}{p^p}\right) \qquad (3.2.5)$$

$$+ \frac{2^{\frac{p+1}{2}} z}{p! \sqrt{\pi}} {}_1F_{2p-1}\left(1; \frac{p+2}{2p}, \frac{p+3}{2p}, \ldots, \frac{3p}{2p}; \frac{z^2}{p^p}\right),$$

where $_pF_q$ is the (p,q)-hypergeometric function

$$_pF_q(z) = {}_pF_q(a_1, a_2, \ldots, a_p; b_1, b_2, \ldots, b_q; z) = \sum_{k=0}^{\infty} \frac{(a_1)_k (a_2)_k \ldots (a_p)_k}{(b_1)_k (b_2)_k \ldots (b_q)_k} \frac{z^k}{k!}. \qquad (3.2.6)$$

◁ The formulas (3.2.1)–(3.2.3) follow immediately from the definition (3.1.1).

Let us prove formula (3.2.4). We first rewrite the series representation (3.1.1) assuming that $z^{\frac{1}{2}}$ is the principal branch of the corresponding multi-valued function and substituting z in place of $z^{\frac{1}{2}}$:

$$E_{\frac{1}{2}}(z) = \sum_{m=0}^{\infty} \frac{z^{2m}}{\Gamma(m+1)} + \sum_{m=0}^{\infty} \frac{z^{2m+1}}{\Gamma(m+\frac{3}{2})} = u(z) + v(z). \qquad (3.2.7)$$

The sum $u(z)$ is equal to e^{z^2}. To obtain the formula for the remaining function v one can use the series representation of the error function as in [ErdBat-1]

3.2 Relations to Elementary and Special Functions

$$\operatorname{erf}(z) = \frac{2}{\sqrt{\pi}} \mathrm{e}^{-z^2} \sum_{m=0}^{\infty} \frac{2^m}{(2m+1)!!} z^{2m+1}, \quad z \in \mathbb{C}. \tag{3.2.8}$$

An alternative proof can be obtained by a term-wise differentiation of the second series in (3.2.7). It follows that $v(z)$ satisfies the Cauchy problem for the first-order differential equation in $\mathbb{C}$.

$$v'(z) = 2\left[\frac{1}{\sqrt{\pi}} + zv(z)\right], \quad v(0) = 0.$$

Representation (3.2.4) follows from the solution of this problem

$$v(z) = \mathrm{e}^{z^2} \frac{2}{\sqrt{\pi}} \int_0^z \mathrm{e}^{-u^2} du = \mathrm{e}^{z^2} \operatorname{erf}(z).$$

To prove (3.2.5) one can simply use the definitions of the Mittag-Leffler function (3.1.1) with $\alpha = p/2$ and the generalized hypergeometric function (3.2.6) and compare the coefficients at the same powers in both sides of (3.2.5). ▷

For an interesting application of the function $E_{1/2}$, see [Gor98, Gor02].

3.3 Recurrence and Differential Relations

Proposition 3.3 (Recurrence relations.) *The following recurrence formulas relating the Mittag-Leffler function for different values of parameters hold:*

$$E_{p/q}(z) = \frac{1}{q} \sum_{l=0}^{q-1} E_{1/p}(z^{1/q} \mathrm{e}^{\frac{2\pi l i}{q}}), \quad q \in \mathbb{N}. \tag{3.3.1}$$

$$E_{\frac{1}{q}}(z^{\frac{1}{q}}) = \mathrm{e}^z \left[1 + \sum_{m=0}^{q-1} \frac{\gamma(1-\frac{m}{q}, z)}{\Gamma(1-\frac{m}{q})}\right], q = 2, 3, \ldots, \tag{3.3.2}$$

where $\gamma(a, z) := \int_0^z \mathrm{e}^{-u} u^{a-1} du$ *denotes the incomplete gamma function, and* $z^{1/q}$ *means the principal branch of the corresponding multi-valued function.*

◁ To prove relation (3.3.1) we use the well-known identity (discrete orthogonality relation)

$$\sum_{l=0}^{p-1} \mathrm{e}^{\frac{2\pi l k i}{p}} = \begin{cases} p, & \text{if } k \equiv 0 \pmod{p}, \\ 0, & \text{if } k \not\equiv 0 \pmod{p}. \end{cases} \tag{3.3.3}$$

This together with definition (3.1.1) of the Mittag-Leffler function gives

$$\sum_{l=0}^{p-1} E_\alpha(ze^{\frac{2\pi l i}{p}}) = p E_{\alpha \cdot p}(z^p), \quad p \geq 1. \tag{3.3.4}$$

Substituting $\dfrac{\alpha}{p}$ for α and $z^{\frac{1}{p}}$ for z we arrive at the desired relation (3.3.1) after setting $\alpha = p/q$.

We mention the following "symmetric" variant of (3.3.4):

$$E_\alpha(z) = \frac{1}{2m+1} \sum_{l=-m}^{m} E_{\alpha/(2m+1)}\left(z^{\frac{1}{2m+1}} e^{\frac{2\pi l i}{2m+1}}\right), \quad m \geq 0. \tag{3.3.5}$$

Relation (3.3.2) follows by differentiation, valid for all $p, q \in \mathbb{N}$ (see below). ▷

Proposition 3.4 (Differential relations.)

$$\left(\frac{d}{dz}\right)^p E_p(z^p) = E_p(z^p), \tag{3.3.6}$$

$$\frac{d^p}{dz^p} E_{p/q}\left(z^{p/q}\right) = E_{p/q}\left(z^{p/q}\right) + \sum_{k=1}^{q-1} \frac{z^{-kp/q}}{\Gamma(1-kp/q)}, \quad q = 2, 3, \ldots. \tag{3.3.7}$$

◁ These formulas are simple consequences of definition (3.1.1). ▷

Let $p = 1$ in (3.3.7). Multiplying both sides of the corresponding relation by e^{-z} we get

$$\frac{d}{dz}\left[e^{-z} E_{1/q}\left(z^{1/q}\right)\right] = e^{-z} \sum_{k=1}^{q-1} \frac{z^{-k/q}}{\Gamma(1-k/q)}.$$

By integrating and using the definition of the incomplete gamma function we arrive at the relation (3.3.2). The relation (3.3.2) shows that the Mittag-Leffler functions of rational order can be expressed in terms of exponentials and the incomplete gamma function. In particular, for $q = 2$ we obtain the relation

$$E_{1/2}(z^{1/2}) = e^z \left[1 + \frac{1}{\sqrt{\pi}} \gamma(1/2, z)\right]. \tag{3.3.8}$$

This is equivalent to relation (3.2.4) by the formula $\mathrm{erf}(z) = \dfrac{\gamma(1/2, z^2)}{\sqrt{\pi}}$.

3.4 Integral Representations and Asymptotics

Many important properties of the Mittag-Leffler function follow from its integral representations. Let us denote by $\gamma(\varepsilon; a)$ ($\varepsilon > 0, 0 < a \leq \pi$) a contour oriented by non-decreasing $\arg \zeta$ consisting of the following parts: the ray $\arg \zeta = -a, |\zeta| \geq \varepsilon$,

3.4 Integral Representations and Asymptotics

the arc $-a \leq \arg \zeta \leq a$, $|\zeta| = \varepsilon$, and the ray $\arg \zeta = a$, $|\zeta| \geq \varepsilon$. If $0 < a < \pi$, then the contour $\gamma(\varepsilon; a)$ divides the complex ζ-plane into two unbounded parts, namely $G^{(-)}(\varepsilon; a)$ to the left of $\gamma(\varepsilon; a)$ by orientation, and $G^{(+)}(\varepsilon; a)$ to the right of it. If $a = \pi$, then the contour consists of the circle $|\zeta| = \varepsilon$ and the twice passable ray $-\infty < \zeta \leq -\varepsilon$. In both cases the contour $\gamma(\varepsilon; a)$ is called the *Hankel path* (as it is used in the representation of the reciprocal of the Gamma function (see, e.g., [ErdBat-1, ErdBat-2, ErdBat-3])).

Lemma 3.1 • Let $0 < \alpha < 2$ and

$$\frac{\pi \alpha}{2} < \beta \leq \min\{\pi, \pi \alpha\}. \tag{3.4.1}$$

Then the Mittag-Leffler function can be represented in the form

$$E_\alpha(z) = \frac{1}{2\pi \alpha i} \int_{\gamma(\varepsilon;\beta)} \frac{e^{\zeta^{1/\alpha}}}{\zeta - z} d\zeta, \quad z \in G^{(-)}(\varepsilon; \beta); \tag{3.4.2}$$

$$E_\alpha(z) = \frac{1}{\alpha} e^{z^{1/\alpha}} + \frac{1}{2\pi \alpha i} \int_{\gamma(\varepsilon;\beta)} \frac{e^{\zeta^{1/\alpha}}}{\zeta - z} d\zeta, \quad z \in G^{(+)}(\varepsilon; \beta). \tag{3.4.3}$$

• Let $\alpha = 2$. Then the Mittag-Leffler function E_2 can be represented in the form

$$E_2(z) = \frac{1}{4\pi i} \int_{\gamma(\varepsilon;\pi)} \frac{e^{\zeta^{\frac{1}{2}}}}{\zeta - z} d\zeta, \quad z \in G^{(-)}(\varepsilon; \pi); \tag{3.4.4}$$

$$E_2(z) = \frac{1}{2} e^{z^{\frac{1}{2}}} + \frac{1}{4\pi i} \int_{\gamma(\varepsilon;\pi)} \frac{e^{\zeta^{\frac{1}{2}}}}{\zeta - z} d\zeta, \quad z \in G^{(+)}(\varepsilon; \pi). \tag{3.4.5}$$

In (3.4.2)–(3.4.5) the function $z^{\frac{1}{\alpha}}$ (or $\zeta^{\frac{1}{\alpha}}$) means the principal branch of the corresponding multi-valued function determined in the complex plane $\mathbb{C}$ cut along the negative semi-axis which is positive for positive z (respectively, ζ).

◁ We use in the proof the *Hankel integral representation* (or Hankel's formula) for the reciprocal of the Euler Gamma function (see formula (A.19a) in Appendix A)

$$\frac{1}{\Gamma(s)} = \frac{1}{2\pi i} \int_{\gamma(\varepsilon;a)} e^u u^{-s} du, \quad \varepsilon > 0, \quad \frac{\pi}{2} < a < \pi, \quad s \in \mathbb{C}. \tag{3.4.6}$$

Formula (3.4.6) is also valid for $a = \frac{\pi}{2}$, $\operatorname{Re} s > 0$, i.e.

$$\frac{1}{\Gamma(s)} = \frac{1}{2\pi i} \int_{\gamma(\varepsilon;\frac{\pi}{2})} e^u u^{-s} du, \quad \varepsilon > 0, \quad \operatorname{Re} s > 0. \tag{3.4.7}$$

We now rewrite formulas (3.4.6) and (3.4.7) in a slightly modified form. Let us begin with the integral representation (3.4.6). After the change of variables $u = \zeta^{1/\alpha}$ (in the case $1 \leq \alpha < 2$ we only consider the contours $\gamma(\varepsilon; \theta)$ with $\theta \in (\pi/2, \pi/\alpha)$) we arrive at

$$\frac{1}{\Gamma(s)} = \frac{1}{2\pi i \alpha} \int_{\gamma(\varepsilon;\beta)} e^{\zeta^{1/\alpha}} \zeta^{\frac{-s+1}{\alpha}-1} d\zeta, \quad \pi\alpha/2 < \beta \leq \min\{\pi, \alpha\pi\}. \quad (3.4.8)$$

Similarly, using the change of variables $u = \zeta^{1/2}$ in (3.4.7)[1] we have

$$\frac{1}{\Gamma(s)} = \frac{1}{4\pi i} \int_{\gamma(\varepsilon;\pi)} e^{\zeta^{1/2}} \zeta^{-\frac{s+1}{2}} d\zeta, \quad \mathrm{Re}\, s > 0. \quad (3.4.9)$$

Let us begin with the case $\alpha < 2$. First let $|z| < \varepsilon$. In this case

$$\sup_{\zeta \in \gamma(\varepsilon;\beta)} |z\zeta^{-1}| < 1.$$

It now follows from the integral representation (3.4.9) and the definition (3.1.1) of the function $E_\alpha(z)$ that for $0 < \alpha < 2$, $|z| < \varepsilon$,

$$E_\alpha(z) = \sum_{k=0}^{\infty} \frac{1}{2\pi i \alpha} \left\{ \int_{\gamma(\varepsilon;\beta)} e^{\zeta^{1/\alpha}} \zeta^{\frac{-\alpha k-1+1}{\alpha}-1} d\zeta \right\} z^k$$

$$= \frac{1}{2\pi i \alpha} \int_{\gamma(\varepsilon;\beta)} e^{\zeta^{1/\alpha}} \frac{1}{\zeta} \left\{ \sum_{k=0}^{\infty} (z\zeta^{-1})^k \right\} d\zeta$$

$$= \frac{1}{2\pi i \alpha} \int_{\gamma(\varepsilon;\beta)} \frac{e^{\zeta^{1/\alpha}}}{\zeta - z} d\zeta.$$

The last integral converges absolutely under condition (3.4.1) and represents an analytic function of z in each of the two domains: $G^{(-)}(\varepsilon; \beta)$ and $G^{(+)}(\varepsilon; \beta)$. On the other hand, the disk $|z| < \varepsilon$ is contained in the domain $G^{(-)}(\varepsilon; \beta)$ for any β. It follows from the Analytic Continuation Principle that the integral representation (3.4.2) holds for the whole domain $G^{(-)}(\varepsilon; \beta)$.

Let now $z \in G^{(+)}(\varepsilon; \beta)$. Then for any $\varepsilon_1 > |z|$ we have $z \in G^{(-)}(\varepsilon_1; \beta)$, and using formula (3.4.2) we arrive at

$$E_\alpha(z) = \frac{1}{2\pi i \alpha} \int_{\gamma(\varepsilon_1;\beta)} \frac{e^{\zeta^{1/\alpha}}}{\zeta - z} d\zeta. \quad (3.4.10)$$

---

[1] Since ε is assumed to tend to zero, in the following we will retain the same notation $\gamma(\varepsilon; \pi)$ for the path which appears after the change of variable.

3.4 Integral Representations and Asymptotics

On the other hand, for $\varepsilon < |z| < \varepsilon_1$, $|\arg z| < \beta$, it follows from the Cauchy integral theorem that

$$\frac{1}{2\pi i \alpha} \int_{\gamma(\varepsilon_1;\beta) - \gamma(\varepsilon;\beta)} \frac{e^{\zeta^{1/\alpha}}}{\zeta - z} \, d\zeta = \frac{1}{\alpha} e^{z^{1/\alpha}}. \tag{3.4.11}$$

The representation (3.4.3) of the function $E_\alpha(z)$ in the domain $G^{(+)}(\epsilon;\beta)$ now follows from (3.4.10) and (3.4.11).

To prove the integral representations (3.4.4) and (3.4.5) for $\alpha = 2$ we argue analogously to the case $0 < \alpha < 2$ using the representation (3.4.9). Recall that there is no need to revise formula (3.4.5) for $\alpha = 2$ since we have exact representations (3.2.2) and (3.2.3) in this case. ▷

It should be noted that integral representations (3.4.2)–(3.4.3) can be used for the representation of the function $E_\alpha(z)$, $0 < \alpha < 2$, at any point z of the complex plane. To obtain such a representation it is sufficient to consider contours $\gamma(\varepsilon;\beta)$ and $\gamma(\varepsilon;\pi)$ with parameter $\varepsilon < |z|$.

The above given representations (3.4.2)–(3.4.5) can be rewritten in a unique form, namely in the form of the classical *Mittag-Leffler integral representation*

$$E_\alpha(z) = \frac{1}{2\pi i} \int_{\text{Ha}_-} \frac{\zeta^{\alpha-1} e^\zeta}{\zeta^\alpha - z} \, d\zeta, \tag{3.4.12}$$

where the path of integration Ha_- is a loop which starts and ends at $-\infty$ approaching along the negative semi-axis and encircles the disk $|\zeta| \leq |z|^{1/\alpha}$ in the positive sense: $-\pi \leq \arg \zeta \leq \pi$ on Ha_- (this curve is also called the *Hankel path*, see Fig. A.3 in Appendix A).

The most interesting properties of the Mittag-Leffler function are associated with its asymptotic expansions as $z \to \infty$ in various sectors of the complex plane. These properties can be summarized as follows.

Proposition 3.5 *Let $0 < \alpha < 2$ and*

$$\frac{\pi \alpha}{2} < \theta < \min\{\pi, \alpha\pi\}. \tag{3.4.13}$$

Then we have the following asymptotics for formulas in which p is an arbitrary positive integer[2]:

---

[2] We adopt here and in what follows the *empty sum convention*: if the upper limit is smaller than the lower limit in a sum, then this sum is empty, i.e. has to be omitted. In particular, it can said that (3.4.14) and (3.4.15) also hold for $p = 0$.

$$E_\alpha(z) = \frac{1}{\alpha} \exp(z^{1/\alpha}) - \sum_{k=1}^{p} \frac{z^{-k}}{\Gamma(1-\alpha k)} + O\left(|z|^{-1-p}\right), \quad |z| \to \infty, \ |\arg z| \leq \theta,$$
(3.4.14)

$$E_\alpha(z) = -\sum_{k=1}^{p} \frac{z^{-k}}{\Gamma(1-\alpha k)} + O\left(|z|^{-1-p}\right), \quad |z| \to \infty, \ \theta \leq |\arg z| \leq \pi.$$
(3.4.15)

For the case $\alpha \geq 2$, we have

$$E_\alpha(z) = \frac{1}{\alpha} \sum_\nu \exp\left(z^{\frac{1}{\alpha}} e^{\frac{2\pi \nu i}{\alpha}}\right) - \sum_{k=1}^{p} \frac{z^{-k}}{\Gamma(1-\alpha k)} + O\left(|z|^{-1-p}\right), \quad (3.4.16)$$
$$|z| \to \infty, \ |\arg z| < \pi,$$

where the first sum is taken over all integers ν such that $|2\pi \nu + \arg z| \leq \frac{\pi\alpha}{2}$, i.e.,

$$\nu \in A(z) = \{n : n \in \mathbb{Z}, \ |\arg z + 2\pi n| \leq \frac{\pi\alpha}{2}\} \quad (3.4.17)$$

and where $\arg z$ can take any value between $-\pi$ and $+\pi$ inclusively.

◁ 1. Let us first prove asymptotic formula (3.4.14). Let β be chosen so that

$$\frac{\pi\alpha}{2} < \theta < \beta \leq \min\{\pi, \pi\alpha\}. \quad (3.4.18)$$

Substituting the expansion

$$\frac{1}{\zeta - z} = -\sum_{k=1}^{p} \frac{\zeta^{k-1}}{z^k} + \frac{\zeta^p}{z^p(\zeta - z)}, \quad p \geq 1, \quad (3.4.19)$$

into formula (3.4.3) with $\varepsilon = 1$, we get the representation of the function $E_\alpha(z)$ in the domain $G^{(+)}(1; \beta)$:

$$E_\alpha(z) = \frac{1}{\alpha} e^{z^{1/\alpha}} - \sum_{k=1}^{p} \left(\frac{1}{2\pi i \alpha} \int_{\gamma(1;\beta)} e^{\zeta^{1/\alpha}} \zeta^{k-1} \, d\zeta\right) z^{-k}$$
$$+ \frac{1}{2\pi i \alpha z^p} \int_{\gamma(1;\beta)} \frac{e^{\zeta^{1/\alpha}} \zeta^p}{\zeta - z} \, d\zeta. \quad (3.4.20)$$

Now Hankel's formula (3.4.6) yields

$$\frac{1}{2\pi i \alpha} \int_{\gamma(1;\beta)} e^{\zeta^{1/\alpha}} \zeta^{k-1} \, d\zeta = \frac{1}{\Gamma(1-k\alpha)}, \quad k \geq 1.$$

3.4 Integral Representations and Asymptotics

Using this formula and (3.4.20) we arrive under conditions (3.4.18) at

$$E_\alpha(z) = \frac{1}{\alpha} e^{z^{1/\alpha}} - \sum_{k=1}^{p} \frac{z^{-k}}{\Gamma(1-k\alpha)} \qquad (3.4.21)$$

$$+ \frac{1}{2\pi i \, \alpha z^p} \int_{\gamma(1;\beta)} \frac{e^{\zeta^{1/\alpha}} \zeta^p}{\zeta - z} \, d\zeta, \quad |\arg z| \le \theta, \quad |z| > 1.$$

We denote the last term in formula (3.4.21) by $I_p(z)$ and estimate it for sufficiently large $|z|$ and $|\arg z| \le \theta$. In this case we have

$$\min_{\zeta \in \gamma(1;\beta)} |\zeta - z| = |z| \sin(\beta - \theta)$$

and, consequently,

$$|I_p(z)| \le \frac{|z|^{-1-p}}{2\pi \, \alpha \, \sin(\beta - \theta)} \int_{\gamma(1;\beta)} |e^{\zeta^{1/\alpha}}| \, |\zeta^p| \, |d\zeta|. \qquad (3.4.22)$$

Note that the integral in the right-hand side of (3.4.22) converges since the contour $\gamma(1;\beta)$ consists of two rays $\arg \zeta = \pm\beta$, $|\zeta| \ge 1$, on which we have

$$|\exp\{\zeta^{1/\alpha}\}| = \exp\left\{\cos\frac{\beta}{\alpha} |\zeta|^{1/\alpha}\right\}, \quad \arg \zeta = \pm\beta, \quad |\zeta| \ge 1,$$

and $\cos\beta/\alpha < 0$ due to condition (3.4.18). The asymptotic formula (3.4.14) now follows from the representation (3.4.21) and the estimate (3.4.22).

2. To prove (3.4.15) let us choose a number β satisfying

$$\frac{\pi \alpha}{2} < \beta < \theta < \min\{\pi, \alpha\pi\}, \qquad (3.4.23)$$

and substitute representation (3.4.19) of the Cauchy kernel into formula (3.4.2) with $\varepsilon = 1$. It follows that

$$E_\alpha(z) = -\sum_{k=1}^{p} \frac{z^{-k}}{\Gamma(1-k\alpha)} + \frac{1}{2\pi i \, \alpha \, z^p} \int_{\gamma(1;\beta)} \frac{e^{\zeta^{1/\alpha}} \zeta^p}{\zeta - z} \, d\zeta \quad z \in G^{(-)}(1;\beta).$$

$$(3.4.24)$$

If $\theta \le |\arg z| \le \pi$ condition (3.4.23) gives, for sufficiently large $|z|$,

$$\min_{\zeta \in \gamma(1;\beta)} |\zeta - z| = |z| \sin(\theta - \beta).$$

For β chosen as in (3.4.23) the domain $\theta \le |\arg z| \le \pi$ is contained in the domain $G^{(-)}(1;\beta)$, and thus the result follows from representation (3.4.24) and the estimate

$$\left| E_\alpha(z) + \sum_{k=1}^{p} \frac{z^{-k}}{\Gamma(1-k\alpha)} \right| \leq \frac{|z|^{-1-p}}{2\pi \alpha \sin(\theta-\beta)} \int_{\gamma(1;\beta)} |e^{\zeta^{1/\alpha}}||\zeta^p| |\mathrm{d}\zeta|, \quad (3.4.25)$$

is valid for sufficiently large $|z|$ and $\theta \leq |\arg z| \leq \pi$.

3. To prove (3.4.16) we note first that formula (3.3.5) is true for any $\alpha > 0$ and $p \geq 0$. Fixing α, $\alpha \geq 2$, we can always choose an integer $m \geq 1$ such that $\alpha_1 = \dfrac{\alpha}{2m+1} < 2$ and, consequently, we can use (3.4.14)–(3.4.15) for any term of the sum in the right-hand side of formula (3.3.5).

Let $\dfrac{\pi \alpha_1}{2} < \theta < \min\{\pi, \pi \alpha_1\}$, $\alpha_1 = \dfrac{\alpha}{(2m+1)}$. Then it follows from the proven part of the Proposition and from (3.3.5) that

$$E_\alpha(z) = \frac{1}{2m+1} \sum_{\nu \in B(z)} \frac{2m+1}{\alpha} e^{z^{\frac{1}{\alpha}} e^{\frac{2\pi \nu i}{\alpha}}}$$

$$- \frac{1}{2m+1} \sum_{\nu=-m}^{m} \left\{ \sum_{k=1}^{q} \frac{z^{-\frac{k}{2m+1}} e^{-\frac{2\pi k \nu i}{2m+1}}}{\Gamma\left(1 - \frac{k\alpha}{2m+1}\right)} \right\} + O\left(|z|^{-\frac{q+1}{2m+1}} \right), \quad (3.4.26)$$

where

$$B(z) = \left\{ n : n \in \mathbb{Z}, \left| \arg\left(z^{\frac{1}{2m+1}} e^{\frac{2\pi n i}{2m+1}} \right) \right| \leq \theta \right\}. \quad (3.4.27)$$

The last inequality can be rewritten in the form $|\arg z + 2\pi n| \leq (2m+1)\theta$. Let us fix some z. If $\theta' > \dfrac{\pi \alpha}{2}$ and the difference $\theta' - \dfrac{\pi \alpha}{2}$ is small enough, then the inequalities $|\arg z + 2\pi n| \leq \dfrac{\pi \alpha}{2}$ and $|\arg z + 2\pi n| \leq \theta'$ have the same set of solutions with respect to $n \in \mathbb{Z}$.

Since the number $(2m+1)\theta > \dfrac{\pi \alpha}{2}$ can be chosen in an arbitrary small neighborhood of $\dfrac{\pi \alpha}{2}$, formula (3.4.27) can be rewritten in the form

$$E_\alpha(z) = \frac{1}{\alpha} \sum_{\nu \in A(z)} e^{z^{\frac{1}{\alpha}} e^{\frac{2\pi \nu i}{\alpha}}}$$

$$- \frac{1}{2m+1} \sum_{k=1}^{q} \frac{z^{-\frac{k}{2m+1}}}{\Gamma\left(1 - \frac{k\alpha}{2m+1}\right)} \left\{ \sum_{\nu=-m}^{m} e^{-\frac{2\pi k \nu i}{2m+1}} \right\} + O\left(|z|^{-\frac{q+1}{2m+1}} \right), \quad (3.4.28)$$

where the summation in the first sum is taken over the set $A(z)$ described in (3.4.17).

Formula (3.4.28) has been proved for any integer $q \geq 1$. To get from here the representation of the form (3.4.16) let us fix any $p \geq 1$ and choose $q = (2m+$

3.4 Integral Representations and Asymptotics

$1)(p+1) - 1$. Using formula (3.4.28) and the discrete orthogonality relation

$$\sum_{v=-m}^{m} e^{-\frac{2\pi k v i}{2m+1}} = \begin{cases} 2m+1, & \text{if } k \equiv 0 \pmod{(2m+1)}, \\ 0, & \text{if } k \not\equiv 0 \pmod{(2m+1)}, \end{cases}$$

we finally arrive at formula (3.4.16). ▷

As a simple consequence of Proposition 3.5 we have

Corollary 3.1 *Let* $0 < \alpha < 2$ *and* $\frac{\pi \alpha}{2} < \theta < \min\{\pi, \pi\alpha\}$. *Then we have the following estimates:*

1. *If* $|\arg z| \leq \theta$ *and* $|z| > 0$:

$$|E_\alpha(z)| \leq M_1 e^{\text{Re } z^{\frac{1}{\alpha}}} + \frac{M_2}{1+|z|}. \tag{3.4.29}$$

2. *If* $\theta \leq |\arg z| \leq \pi$ *and* $|z| \geq 0$:

$$|E_\alpha(z)| \leq \frac{M_2}{1+|z|}. \tag{3.4.30}$$

Here M_1 and M_2 are constants not depending on z.

Corollary 3.2 *Let* $0 < \alpha < 2$ *and* $|z| = r > 0$. *Then the following relations hold:*

1.
$$\lim_{r \to +\infty} e^{-r^{\frac{1}{\alpha}}} |E_\alpha(re^{i\theta})| = 0, \quad 0 < |\theta| < \min\{\pi, \pi\alpha\}, \tag{3.4.31}$$

and the limit in (3.4.31) is uniform with respect to θ.

2.
$$\lim_{r \to +\infty} e^{-r^{\frac{1}{\alpha}}} |E_\alpha(r)| = \frac{1}{\alpha}. \tag{3.4.32}$$

The last results were obtained by G. Mittag-Leffler [ML3].

3.5 Distribution of Zeros

In this section we consider the problem of the distribution of zeros of the Mittag-Leffler function $E_\alpha(z)$. The following lemma presents two situations in which this distribution is easily described.

Lemma 3.2 *(i) The Mittag-Leffler function $E_1(z)$ has no zero in $\mathbb{C}$.*
(ii) All zeros of the function $E_2(z)$ are simple and are situated on the real negative semi-axis. They are given by the formula

$$z_k = -\left(\frac{\pi}{2} + \pi k\right)^2, \quad k \in \mathbb{N}_0 = \{0, 1, 2, \ldots\}. \tag{3.5.1}$$

◁ The first statement follows immediately from formula (3.2.1)

$$E_1(z) = e^z, \quad z \in \mathbb{C},$$

and properties of the exponential function.

As for zeros of the function $E_2(z)$ one can use one of the representations (3.2.2) or (3.2.3). Hence the zeros are described by formula (3.5.1). They are simple (i.e. are of first order) by the differentiation formulas for the function $\cos\sqrt{-z}$. ▷

The following fact is commonly used:

Corollary 3.3 *The exponential function $E_1(z)$ is the only Mittag-Leffler function which has no zeros in the whole complex plane. All other functions $E_\alpha(z)$, $\operatorname{Re}\alpha > 0$, $\alpha \neq 1$, have infinitely many zeros in $\mathbb{C}$.*

This fact is partly a simple consequence of Proposition 3.1, which states that for $\operatorname{Re}\alpha > 0$ the function $E_\alpha(z)$ is an entire function of order $\rho = \dfrac{1}{\operatorname{Re}\alpha}$. Then for each α, $\operatorname{Re}\alpha \neq \dfrac{1}{n}$, $n \in \mathbb{N}$, the order of $E_\alpha(z)$ is a positive non-integer. Then, for these values of the parameter α, the statement follows from the general theory of entire functions (see, e.g., Levin [Lev56]). The statement is still valid for all $\alpha \neq 1$ including when α is the reciprocal of a natural number, but in this case the argument is much more delicate. We return to the proof later. Next we consider another simple case, $\operatorname{Re}\alpha \geq 2$.

Lemma 3.3 *For each value α, $\operatorname{Re}\alpha \geq 2$, the Mittag-Leffler function $E_\alpha(z)$ has infinitely many zeros lying on the real negative semi-axis.*

There exists only finitely many zeros of the function $E_\alpha(z)$, $\operatorname{Re}\alpha \geq 2$, which are not negative real (i.e., zeros belonging to $\mathbb{C}\setminus(-\infty, 0]$).

◁ If $\operatorname{Re}\alpha > 2$ then by Proposition 3.1 the function $E_\alpha(z)$ is an entire function of order $\rho = \dfrac{1}{\operatorname{Re}\alpha} < \dfrac{1}{2}$. The general theory of entire functions says (see, e.g.., [Lev56]) that in this case the set of zeros of such a function is denumerable. Therefore it suffices to determine only the location of these zeros.

Let us study for simplicity only the case of positive $\alpha > 2$. Consider the asymptotic formula (3.4.16). For each fixed $z \neq -x$, $x \geq 0$, the index set $A(z)$ in the sum on the right-hand side of (3.4.16) consists of a finite number of elements (see definition (3.4.17)). Moreover, straightforward calculations show that the modulus of each term in this sum is equal to

$$\varphi_\nu(z) = \left|\exp\left(z^{\frac{1}{\alpha}} e^{\frac{2\pi \nu i}{\alpha}}\right)\right| = e^{|z|^{\frac{1}{\alpha}} \cos\left(\frac{\arg z + 2\pi \nu}{\alpha}\right)}.$$

Comparing the values of the functions φ_ν for a fixed z but for different $\nu \in A(z)$ one can conclude that the function φ_0 is the maximal one. More precisely, if we consider

3.5 Distribution of Zeros

any angle $|\arg z| < \pi - \varepsilon$ with sufficiently small $\varepsilon > 0$, then there exists $\delta > 0$ and $r_0 > 0$ such that

$$|E_\alpha(z)| > e^{r^{\frac{1}{\alpha}} \left(\cos \frac{\pi - \varepsilon}{\alpha} - \delta\right)}, \quad |z| = r > r_0, \; |\arg z| < \pi - \varepsilon.$$

Both statements of the lemma follow since $\varepsilon > 0$ is an arbitrary small number. ▷

We also mention a result which was stated in [Wim05b], but was only proved in [OstPer97] (see the comments in [PopSed11, p. 56]).

Proposition 3.6 *All zeros of the classical Mittag-Leffler function $E_\alpha(z)$ with $\alpha > 2$ are simple and negative. These zeros z_n, $n = 1, 2, \ldots$, satisfy the following inequalities:*

$$-\left(\frac{\pi n}{\sin \frac{\pi}{\alpha}}\right)^\alpha < z_n < -\left(\frac{\pi(n-1)}{\sin \frac{\pi}{\alpha}}\right)^\alpha. \tag{3.5.2}$$

The most interesting case of the distribution of zeros of $E_\alpha(z)$ is that for $0 < \alpha < 2$, $\alpha \neq 1$. Let us first introduce some notation and a few simple facts. It follows from Proposition 3.5 (see formulas (3.4.14)–(3.4.15)) that the zeros of the function $E_\alpha(z)$ (if any) with sufficiently large modulus are situated in two angular domains

$$\Omega_\delta^{(\pm)} = \left\{z \in \mathbb{C} : \left|\arg z \mp \frac{\pi \alpha}{2}\right| < \delta\right\},$$

where δ is an arbitrary positive number, $\delta \in \left(0, \min\left\{\frac{\pi \alpha}{2}, \pi - \frac{\pi \alpha}{2}\right\}\right)$. Let us denote those zeros of $E_\alpha(z)$ which belong to the upper half-plane (lower half-plane) by $z_k^{(+)}$ (respectively by $z_k^{(-)}$) ordering each of these collections in increasing order by modulus, i.e. $|z_k^{(+)}| \leq |z_{k+1}^{(+)}|$ ($|z_k^{(-)}| \leq |z_{k+1}^{(-)}|$).

We also have to note that since for a real value of the parameter α the function $E_\alpha(z)$ has the symmetry property $E_\alpha(\bar{z}) = \overline{E_\alpha(z)}$, we have the following connection between "upper" and "lower" zeros:

$$z_k^{(-)} = \overline{z_k^{(+)}}, \quad k = 1, 2, \ldots. \tag{3.5.3}$$

The distribution of zeros of the function $E_\alpha(z)$ in the considered case $0 < \alpha < 2$, $\alpha \neq 1$, can be described in the following form:

Proposition 3.7 *All zeros of the function E_α, $0 < \alpha < 2$, $\alpha \neq 1$, with sufficiently large modulus are simple, and the following asymptotic formula for its zeros holds:*

$$z_k^{(\pm)} = e^{\pm i \frac{\pi \alpha}{2}} (2\pi k)^\alpha \left\{1 + O\left(\frac{\log k}{k}\right)\right\}, \quad k \to \infty. \tag{3.5.4}$$

◁ Due to the symmetry property (3.5.3) it suffices to consider only zeros lying in the upper half-plane. To do this we use the asymptotic formula (3.4.14). The reason for using an asymptotic description of the zeros of the function $E_\alpha(z)$ is two-fold. First, we select a part of the right-hand side of the asymptotic formula (3.4.14) and find its zeros with sufficiently large modulus. We also derive the asymptotic behavior of these auxiliary zeros. This helps us in the second step to approximate the zeros of the right-hand side of (3.4.14) i.e., of the function $E_\alpha(z)$), and to find the asymptotics for them.

Denote by

$$c_\alpha := \frac{\alpha}{\Gamma(1-\alpha)} = \frac{\sin \pi \alpha}{\pi} \Gamma(1+\alpha) \qquad (3.5.5)$$

the real constant which is positive for $0 < \alpha < 1$ and negative for $1 < \alpha < 2$. We also introduce the following function:

$$e_\alpha := z \, e^{z^{\frac{1}{\alpha}}} - c_\alpha. \qquad (3.5.6)$$

Then formula (3.4.14) can be rewritten in the form

$$\alpha z \, E_\alpha(z) = e_\alpha(z) + O\left(\frac{1}{z}\right), \quad |z| \to \infty. \qquad (3.5.7)$$

Let us first solve the equation

$$z \, e^{z^{\frac{1}{\alpha}}} = c_\alpha. \qquad (3.5.8)$$

Let us denote by L_0 the set of all z for which the modulus of the left- and right-hand sides of (3.5.8) coincide

$$L_0 := \left\{ z \in \mathbb{C} : |z \, e^{z^{\frac{1}{\alpha}}}| = |c_\alpha| \right\}.$$

It is equivalent to say either $z = r \, e^{i\phi} \in L_0$ or

$$r^{\frac{1}{\alpha}} \cos\left(\frac{\varphi}{\alpha}\right) = -\log r + \log |c_\alpha|. \qquad (3.5.9)$$

Hence L_0 is a (continuous) curve consisting of two branches $L_0^{(+)}$, $L_0^{(-)}$ (symmetric with respect to the real axis) whose equations can be written for large enough r in the form

$$L_0^{(\pm)} : \arg z := \phi = \pm \left\{ \frac{\pi \alpha}{2} + \alpha \frac{\log r}{r^{\frac{1}{\alpha}}} + \alpha \frac{\log |c_\alpha|}{r^{\frac{1}{\alpha}}} + O\left(\frac{\log^3 r}{r^{\frac{1}{\alpha}}}\right) \right\}. \qquad (3.5.10)$$

In order to compare the arguments of the left- and right-hand sides of (3.5.8) we calculate the argument of $z \, e^{z^{\frac{1}{\alpha}}}$

$$\arg\left(z \, e^{z^{\frac{1}{\alpha}}}\right) = \phi + r^{\frac{1}{\alpha}} \sin\left(\frac{\phi}{\alpha}\right).$$

3.5 Distribution of Zeros

Then for all $z \in L_0^{(+)}$ with sufficiently large $r = |z|$ we have from (3.5.10)

$$r^{\frac{1}{\alpha}} \sin\left(\frac{\phi}{\alpha}\right) = r^{\frac{1}{\alpha}} + O\left(\frac{\log^2 r}{r^{\frac{1}{\alpha}}}\right), \quad r \to \infty. \tag{3.5.11}$$

The zeros of the function e_α in the upper half-plane are those points $r\,e^{i\phi}$ on the curve $L_0^{(+)}$ which satisfy for certain $k \in \mathbb{N}$ the corresponding equality for arguments

$$\arg z\,e^{z^{\frac{1}{\alpha}}} = \arg c_\alpha + 2\pi k,$$

or

$$\phi + r^{\frac{1}{\alpha}} \sin\left(\frac{\phi}{\alpha}\right) = \arg c_\alpha + 2\pi k. \tag{3.5.12}$$

Since the left-hand side of the last equation satisfies the asymptotic formula (3.5.11), then for sufficiently large r there exists a denumerable collection of natural numbers k for which equality (3.5.12) holds. These values of k determine zeros of the function e_α with sufficiently large modulus. Denote these zeroes by $\zeta_k := \lambda_k\,e^{i\theta_k}$ and recall the equations for them

$$\begin{cases} \lambda_k^{\frac{1}{\alpha}} \cos\left(\frac{\theta_k}{\alpha}\right) = -\log \lambda_k + \log |c_\alpha|, \\ \theta_k + \lambda_k^{\frac{1}{\alpha}} \sin\left(\frac{\theta_k}{\alpha}\right) = \arg c_\alpha + 2\pi k. \end{cases} \tag{3.5.13}$$

It follows from the second relation that $\lambda_k \to \infty$ for $k \to \infty$. More precisely

$$\lambda_k^{\frac{1}{\alpha}} = 2\pi k + O\left(\frac{1}{\lambda_k^{\frac{1}{\alpha}}}\right).$$

Therefore the asymptotic formulas (3.5.10)–(3.5.11) lead us to the following asymptotics for λ_k and θ_k with respect to k:

$$\begin{cases} \theta_k = \frac{\pi\alpha}{2} + O\left(\frac{\log k}{k}\right), \quad k \to \infty, \\ \lambda_k = (2\pi k)^\alpha \left[1 + O\left(\frac{\log^2 k}{k^2}\right)\right], \quad k \to \infty. \end{cases} \tag{3.5.14}$$

These relations determine the behavior of the zeros ζ_k of the function $e_\alpha(z)$ lying outside a certain disk. The formulas (3.5.14) show in particular that such zeros are simple.

The continuity of the function $e_\alpha(z)$ on $L_0^{(+)}$ implies that for each pair of successive zeros ζ_k, ζ_{k+1} there exists a point $w_k = \omega_k\,e^{i\psi_k}$ on $L_0^{(+)}$ such that

$$\arg w_k \, \mathrm{e}^{w_k^{\frac{1}{\alpha}}} = \arg c_\alpha + 2\pi k + \pi,$$

and accordingly

$$e_\alpha(w_k) = -c_\alpha.$$

Let us introduce the domain $\Delta_k(\gamma)$ encircled by two curves $L_0^{(+)}(\pm\gamma)$ determined by the equations:

$$L_0^{(+)}(\pm\gamma) : r^{\frac{1}{\alpha}} \cos\left(\frac{\phi}{\alpha}\right) = -\log r + \log |c_\alpha| \pm \gamma, \qquad (3.5.15)$$

with a fixed sufficiently small $\gamma > 0$ and two circular arcs $l_j := \{z \in \mathbb{C} : |z| = \omega_j\}$, $j = k-1, k$. From the definition of these curves we obtain

$$|e_\alpha(z)| > |c_\alpha|, \quad z \in l_k, \quad k > N_0, \qquad (3.5.16)$$

$$|e_\alpha(z)| \geq |c_\alpha||\mathrm{e}^{\pm\gamma} - 1|, \quad z \in L_0^{(+)}(\pm\gamma). \qquad (3.5.17)$$

The right-hand sides of inequalities (3.5.16)–(3.5.17) are constants not depending on k. Therefore one can apply to (3.5.7) Rouché's theorem. This implies that the function $E_\alpha(z)$ has the same number of zeros in the domains Δ_k with $k \geq N_1 \geq N_0$ as the function $e_\alpha(z)$ has. On the other hand, according to the construction of the domain Δ_k, the function $e_\alpha(z)$ has in this domain exactly one zero ζ_k. Consequently, the function $E_\alpha(z)$ has exactly one simple zero $z_k^{(+)}$ inside of Δ_k, $k \geq N_1$, and

$$z_k^{(+)} = \zeta_k + \alpha_k, \quad \alpha_k = O(d_k), \qquad (3.5.18)$$

where d_k is the diameter of the domain Δ_k.

It can easily be shown that the perimeter of the contour $\partial \Delta_k$ has order $O\left(k^{\alpha-1}\right)$ and, consequently, $d_k = O\left(k^{\alpha-1}\right)$. This, together with the asymptotics of ζ_k in (3.5.14) and the representation (3.5.18), gives the following asymptotical formula for $z_k^{(+)}$:

$$z_k^{(+)} = \mathrm{e}^{i\frac{\pi\alpha}{2}}(2\pi k)^\alpha \left[1 + O\left(\frac{\log k}{k}\right)\right], \quad k \to \infty. \qquad (3.5.19)$$

If z_0 is a zero of the function $E_\alpha(z)$ in the upper half-plane with large enough modulus then by using formula (3.5.7) we arrive at the inequalities

$$-\log |z_0| + \log |c_\alpha| - \frac{\gamma}{2} < |z_0|^{\frac{1}{\alpha}} \cos\left(\frac{\arg z_0}{\alpha}\right) < -\log |z_0| + \log |c_\alpha| + \frac{\gamma}{2}.$$

This means that all zeros of the function $E_\alpha(z)$ in the upper half-plane with sufficiently large modulus are contained in the curvilinear strip between the curves $L_0^{(+)}(\pm\gamma)$. The latter inequalities also give the asymptotic representation (3.5.4) for the zeros $z_k^{(+)}$ of the function $E_\alpha(z)$. ▷

3.6 Further Analytic Properties

In this section we present some additional analytic results for the Mittag-Leffler function. We deal with integral properties of the Mittag-Leffler function and also describe its relation to some special functions. Mittag-Leffler's primary idea, namely summation of power series, is discussed too. Finally, we present a few results on the Mittag-Leffler function related to geometric function theory.

3.6.1 Additional Integral Properties

The first property describes the relation between the Mittag-Leffler function and the generalized Wright function (this relation is also known as the *Euler transform of the Mittag-Leffler function*, see, e.g., [MatHau08, p. 84]). Let $\alpha, \rho, \sigma \in \mathbb{C}$, $\gamma > 0$ and $\operatorname{Re}\alpha > 0$, $\operatorname{Re}\sigma > 0$, then the following representation holds:

$$\int_0^1 t^{\rho-1}(1-t)^{\sigma-1} E_\alpha(xt^\gamma)\,dt = \Gamma(\sigma) {}_2\Psi_2\left[\begin{array}{c}(\rho,\gamma),\,(1,1)\\(1,\alpha),\,(\sigma+\rho,\gamma)\end{array}\bigg|\,x\right], \quad (3.6.1)$$

where ${}_2\Psi_2$ is a special case of the generalized Wright function ${}_p\Psi_q$ (see formula (F.2.12) in Appendix F):

$${}_2\Psi_2(z) := {}_2\Psi_2\left[\begin{array}{c}(\rho,\gamma),\,(1,1)\\(1,\alpha),\,(\sigma+\rho,\gamma)\end{array}\bigg|\,z\right] = \sum_{k=0}^\infty \frac{\Gamma(\rho+\gamma k)\Gamma(1+k)}{\Gamma(1+\alpha k)\Gamma(\sigma+\rho+\gamma k)} \frac{x^k}{k!}.$$

Formula (3.6.1) follows from the series representation of the Mittag-Leffler function (3.1.1) and simple calculations involving properties of the Beta function (see formula (A.34) in Appendix A).

Next, we obtain the Mellin–Barnes integral representation for the Mittag-Leffler function (see, e.g., [MatHau08, p. 88] and also Appendix D).

Lemma 3.4 *Let $\alpha > 0$. Then the Mittag-Leffler function $E_\alpha(z)$ has the Mellin–Barnes integral representation*

$$E_\alpha(z) = \frac{1}{2\pi i}\int_{\mathcal{L}_{ic}} \frac{\Gamma(s)\Gamma(1-s)}{\Gamma(1-\alpha s)}(-z)^{-s}\,ds, \quad |\arg z| < \pi, \quad (3.6.2)$$

where the contour of integration $\mathcal{L}_{ic}$ is a straight line which starts at $c - i\infty$ and ends at $c + i\infty$, $(0 < c < 1)$ and thus leaves all poles $s = 0, -1, -2, \ldots$ of $\Gamma(s)$ to the left and all poles $s = 1, 2, 3, \ldots$ of $\Gamma(1 - s)$ to the right.

◁ As is standard for Mellin–Barnes representations, we can calculate the integral on the right-hand side of (3.6.2) by using Residue Theory:

$$\frac{1}{2\pi i} \int_{\mathcal{L}_{ic}} \frac{\Gamma(s)\Gamma(1-s)}{\Gamma(1-\alpha s)}(-z)^{-s} ds = \sum_{k=0}^{\infty} \operatorname{Res}_{s=-k} \frac{\Gamma(s)\Gamma(1-s)}{\Gamma(1-\alpha s)}(-z)^{-s}$$

$$= \sum_{k=0}^{\infty} \lim_{s \to -k} \frac{(s+k)\Gamma(s)\Gamma(1-s)}{\Gamma(1-\alpha s)}(-z)^{-s} = \sum_{k=0}^{\infty} \frac{(-1)^k \Gamma(1+k)}{k!\Gamma(1+\alpha k)}(-z)^k = E_\alpha(z).$$

▷

Two simple consequences of the representation (3.6.2) are the relations of the Mittag-Leffler function $E_\alpha(z)$ to the generalized Wright function (see formula (F.2.12) in Appendix F)

$$E_\alpha(z) = {}_1\Psi_1 \left[\begin{matrix} (1, 1) \\ (1, \alpha) \end{matrix} \bigg| z \right], \tag{3.6.3}$$

and to the Fox H-function (see formulae (F.4.1)–(F.4.4) in Appendix F)

$$E_\alpha(z) = H_{1,2}^{1,1} \left[-z \bigg| \begin{matrix} (0, 1) \\ (0, 1), (0, \alpha) \end{matrix} \right]. \tag{3.6.4}$$

These formulas follow immediately from the Mellin–Barnes representations of the generalized Wright function and Fox H-function (see representations (F.2.14) and (F.4.1), respectively, in Appendix F).

Further, we also obtain an integral representation for the Cauchy kernel $\dfrac{1}{\zeta - z}$. Let us first introduce some definitions and notation. Let $-\pi < \theta \leq \pi$ and $\alpha > 0$. By $(e^{-i\theta}\zeta)^{\frac{1}{\alpha}}$ we denote the branch of the corresponding multi-valued function having positive values $\exp\{\frac{1}{\alpha}\log|\zeta|\}$ when $\arg \zeta = \theta$ and by $L_\alpha(\theta; \nu)$, $\alpha > 0$, $\nu \geq 0$, we denote the curve

$$\operatorname{Re}(e^{-i\theta}\zeta)^{\frac{1}{\alpha}} = \nu, \quad |\arg \zeta - \theta| \leq \begin{cases} \pi, & \text{if } \alpha \geq 2, \\ \dfrac{\pi\alpha}{2}, & \text{if } 0 < \alpha \leq 2. \end{cases} \tag{3.6.5}$$

The equation of the curve $L_\alpha(\theta; \nu)$ can be written in polar coordinates as

$$r \cos^\alpha \frac{1}{\alpha}(\phi - \theta) = \nu^\alpha, \quad |\phi - \theta| \leq \begin{cases} \pi, & \text{if } \alpha \geq 2, \\ \frac{\pi\alpha}{2}, & \text{if } 0 < \alpha \leq 2. \end{cases} \tag{3.6.6}$$

3.6 Further Analytic Properties

It follows from this equation that the curve $L_\rho(\theta; \nu)$ is bounded and closed when $\alpha > 2$ and has two unbounded branches when $0 < \alpha \leq 2$.

Consequently, the ζ-plane is divided by the curve $L_\alpha(\theta; \nu)$ into two complementary simply-connected domains $D_\alpha^*(\theta; \nu)$ and $D_\alpha(\theta; \nu)$ containing, respectively, intervals $0 < |\zeta| < \nu^\alpha$ and $\nu^\alpha < |\zeta| < \infty$ of the ray $\arg \zeta = \theta$ (with the exception of the case $\alpha > 2$, $\nu = 0$ when the domain $D_\alpha(\theta; \nu)$ becomes the whole ζ-plane without the point $\zeta = 0$). The following properties of these domains are easily verified:

1. If $\zeta \in D_\alpha(\theta; \nu)$, then
$$\mathrm{Re}\,(e^{-i\theta}\zeta)^{\frac{1}{\alpha}} > \nu. \tag{3.6.7}$$

2. The domain $D_\alpha(\theta; \nu)$, $\nu > 0$, $\alpha \leq 2$, is contained in the angular domain
$$\Delta(\theta; \alpha) = \left\{ \zeta : |\arg \zeta - \theta| < \frac{\pi\alpha}{2} \right\}, \tag{3.6.8}$$

and $D_\alpha(\theta; 0) = \Delta(\theta; \alpha)$. The domain $D_\alpha^*(\theta; 0)$ coincides with the angular domain complementary to $\Delta(\theta; \alpha)$:
$$D_\alpha^*(\theta; 0) = \Delta^*(\theta; \alpha) = \left\{ \zeta : \frac{\pi\alpha}{2} < |\arg \zeta - \theta| \leq \pi \right\}. \tag{3.6.9}$$

The following lemma gives an integral representation of the Cauchy kernel using the Mittag-Leffler function.

Lemma 3.5 *Let $\alpha > 0$, $\nu > 0$ and $-\pi < \theta \leq \pi$ be some fixed parameters.*

1. If $z \in D_\alpha^*(\theta; \nu)$ and $\zeta \in \overline{D_\alpha(\theta; \nu)}$ then
$$\int_0^{+\infty} e^{-(e^{-i\theta}\zeta)^{\frac{1}{\alpha}}t} E_\alpha(e^{-i\theta}zt^\alpha)\,dt = e^{i\frac{\theta}{\alpha}} \frac{\zeta^{1-\frac{1}{\alpha}}}{\zeta - z}. \tag{3.6.10}$$

2. The integral (3.6.10) converges absolutely and uniformly with respect to the two variables z and ζ if
$$z \in G_\alpha^*(\theta; \nu), \; \zeta \in \overline{D_\alpha(\theta; \nu)}, \tag{3.6.11}$$

where $G_\alpha^*(\theta; \nu)$ is any bounded and closed subdomain of the domain $D_\alpha^*(\theta; \nu)$.

◁ It suffices to prove the statements of the lemma under the assumption
$$|z| \leq \nu_1^\alpha, \; \zeta \in \overline{D_\alpha(\theta; \nu)}, \tag{3.6.12}$$

where ν_1 is any number from the interval $(0, \nu)$.

Let us choose for a given value of ν_1, $0 < \nu_1 < \nu$, a number ϵ, $0 < \epsilon < \nu$,
$$q = \left(\frac{\nu_1}{\nu - \epsilon}\right)^\alpha < 1. \tag{3.6.13}$$

Then we arrive at the formula

$$\max_{0\leq t<+\infty} \{t^{k\alpha}e^{-(\nu-\epsilon)t}\} = \left(\frac{k\alpha}{\nu-\epsilon}\right)^{k\alpha} e^{-k\alpha},$$

and the estimate

$$\Gamma(1+k\alpha) > (k\alpha)^{k\alpha+1-\frac{1}{2}}e^{-k\alpha}$$

follows easily from Stirling's formula. Hence, when q is chosen as in (3.6.13) and $|z| \leq \nu_1^\alpha$, we have

$$\max_{0\leq t<+\infty} \left|\frac{(e^{-i\theta}z)^k t^{k\alpha}}{\Gamma(1+k\alpha)} e^{-(\nu-\epsilon)t}\right| \leq (k\alpha)^{-\frac{1}{2}} q^k, \quad k \geq k_0.$$

It follows that the expansion

$$e^{-(\nu-\epsilon)t} E_\alpha(e^{-i\theta}zt^\alpha) = \sum_{k=0}^{\infty} \frac{(e^{-i\theta}z)^k t^{k\alpha}}{\Gamma(1+k\alpha)} e^{-(\nu-\epsilon)t} \quad (3.6.14)$$

converges uniformly with respect to z and t when $|z| \leq \nu_1^\alpha$ and $0 \leq t < +\infty$.

Due to the relation $\mathrm{Re}\,(e^{-i\theta}\zeta)^{\frac{1}{\alpha}} > \nu$ and assumption (3.6.12) the series in (3.6.14) can be integrated term-by-term with respect to t along the semi-axis $[0, +\infty)$. Using the known formula

$$\int_0^{+\infty} e^{-(e^{-i\theta}\zeta)^{\frac{1}{\alpha}} t} t^{k\alpha} \, dt = \frac{\Gamma(1+k\alpha)}{(e^{-i\theta}\zeta)^{k+\frac{1}{\alpha}}}, \quad \mathrm{Re}\,(e^{-i\theta}\zeta)^{\frac{1}{\alpha}} > 0, \; k \geq 0,$$

we arrive at

$$\int_0^{+\infty} e^{-(e^{-i\theta}\zeta)^{\frac{1}{\alpha}} t} E_\alpha(e^{-i\theta}zt^\alpha) \, dt$$

$$= \sum_{k=0}^{\infty} \frac{(e^{-i\theta}z)^k}{\Gamma(1+k\alpha)} \int_0^{+\infty} e^{-(e^{-i\theta}\zeta)^{\frac{1}{\alpha}} t} t^{k\alpha} \, dt$$

$$= (e^{-i\theta}\zeta)^{-\frac{1}{\alpha}} \sum_{k=0}^{\infty} \left(\frac{z}{\zeta}\right)^k = e^{i\frac{\theta}{\alpha}} \frac{\zeta^{1-\frac{1}{\alpha}}}{\zeta - z},$$

since

$$|\zeta| \geq |\mathrm{Re}(e^{-i\theta}\zeta)^{\frac{1}{\alpha}}|^\alpha \geq \nu^\alpha > \nu_1^\alpha \geq |z|.$$

Thus we have proved formula (3.6.10) under the condition (3.6.12). ▷

3.6.2 Mittag-Leffler Summation of Power Series

An approach to generalizing the Borel summation method of power series (see [SanGer60, Chap. 8]) is due to the following simplification of the integral representation of the Cauchy kernel (3.6.10),

$$\frac{1}{1-z} = \int_0^\infty e^{-u} E_\alpha(zu^\alpha)du, \qquad (3.6.15)$$

which is valid throughout any region $\operatorname{Re} z^{1/\alpha} < 1$. Based on this, Mittag-Leffler considered a similar generalization for summation of power series

$$\sum_{\nu=0}^\infty a_\nu z^\nu. \qquad (3.6.16)$$

Let us briefly describe Mittag-Leffler's method following [SanGer60] in the case $0 < \alpha \leq 2$. The region of convergence for the integral (3.6.15) is bounded by the curve

$$r \cos^\alpha \frac{\theta}{\alpha} = 1, \quad -\frac{\pi\alpha}{2} < \arg z = \theta < \frac{\pi\alpha}{2}.$$

For $0 < \alpha < 1$ it looks similar to hyperbola, for $\alpha = 1$ it is a vertical straight line, and for $1 < \alpha < 2$ it looks similar to parabola, being a parabola for $\alpha = 2$. Let us consider the following series depending on z as a parameter

$$\sum_{\nu=0}^\infty a_\nu \frac{(zu^\alpha)^\nu}{\Gamma(\alpha\nu + 1)}. \qquad (3.6.17)$$

This series is convergent for each finite u and suitably chosen parameter z. Denote its sum by $A(zu^\alpha)$ and introduce the following integral

$$F_\alpha A(z) := \int_{\nu=0}^\infty e^{-u} A(zu^\alpha)du. \qquad (3.6.18)$$

If the series (3.6.17) and the integral (3.6.18) are convergent for the same value of parameter z, then $F_\alpha A(z)$ is called the *Mittag-Leffler sum* or B_α-sum. For $\alpha = 1$ it coincides with the Borel sum (see, e.g.., [Bor01, Har92, ReeSim78], and also the paper by N. Obrechkoff reprinted in the Journal *Fract. Calc. Appl. Anal.* [Obr16]).

The following theorem is due to G. Mittag-Leffler [ML08].

Theorem 3.1 *Let (3.6.16) be a power series which may or may not have a circle of convergence.*

Suppose that

(a) *there is a value $z_0 \neq 0$ for which an associated series (3.6.17) converges for all positive u;*
(b) *the integral (3.6.18) converges at $z = z_0$.*

Then (3.6.18) converges at any point z of the segment $(0, z_0]$.

◁ The integral (3.6.18) can be represented in the form of the Laplace integral

$$F_\alpha A(z) := s \int_{\nu=0}^{\infty} e^{-su} A(z_0 t^\alpha) dt \qquad (3.6.19)$$

with $t^\alpha = \vartheta u^\alpha$, $s = \vartheta^{-1/\alpha}$, $0 < \vartheta \leq 1$. Since the above Laplace integral converges for $s = 1$, it also converges for all $s \geq 1$. This gives the desired result. ▷

If variable s in the Laplace integral (3.6.19) is considered as a complex variable, then this integral is convergent (and thus is analytic) in the half-plane $\operatorname{Re} s > 1$ or, equivalently, $F_\alpha A(z)$ is analytic in the domain

$$\operatorname{Re} \left(\frac{z_0}{z} \right)^{1/\alpha} > 1.$$

This domain consists of several parts. We restrict ourselves to the part corresponding to

$$-\frac{\pi \alpha}{2} < \arg \frac{z_0}{z} < \frac{\pi \alpha}{2}.$$

This part of the domain is encircled by the curve

$$r = r_0 \cos^\alpha \frac{\psi}{\alpha}, \quad -\frac{\pi \alpha}{2} < \psi < \frac{\pi \alpha}{2},$$

where ψ is an angle between Oz_0 and Oz, $r = |z|$, and $r_0 = |z_0|$.

Corollary 3.4 *If the series (3.6.16) is B_α-summable at $z = z_0$, then it is B_α-summable at any point $z = \vartheta z_0$, $0 \leq \vartheta \leq 1$.*
Moreover, its sum is analytic in the domain

$$\operatorname{Re} \left(\frac{z_0}{z} \right)^{1/\alpha} > 1, \quad -\frac{\pi \alpha}{2} < \arg \frac{z_0}{z} < \frac{\pi \alpha}{2}. \qquad (3.6.20)$$

If the series (3.6.17) has a finite radius of convergence, then the sum of the following series

$$\sum_{\nu=0}^{\infty} a_\nu \frac{z^\nu}{\Gamma(\alpha \nu + 1)} \qquad (3.6.21)$$

3.6 Further Analytic Properties

is an analytic function of the complex variable z. Then the series (3.6.17) is uniformly convergent in u, $0 \leq u \leq \omega$.

Hence

$$\int_0^\omega e^{-u} \sum_{\nu=0}^\infty a_\nu \frac{(zu^\alpha)^\nu}{\Gamma(\alpha\nu+1)} du = \sum_{\nu=0}^\infty a_\nu z^\nu \int_0^\omega e^{-u} \frac{u^{\alpha\nu}}{\Gamma(\alpha\nu+1)} du. \quad (3.6.22)$$

The integral in the right-hand side is uniformly convergent in ω, $0 < \omega < \infty$, since

$$\int_0^\omega e^{-u} \frac{u^{\alpha\nu}}{\Gamma(\alpha\nu+1)} du < \int_0^\infty e^{-u} \frac{u^{\alpha\nu}}{\Gamma(\alpha\nu+1)} du.$$

Therefore, we can pass to the limit in (3.6.22) as $\omega \to \infty$ and by definition of the Gamma-function we obtain the following equality for all z in the domain of convergence

$$\int_0^\infty e^{-u} \sum_{\nu=0}^\infty a_\nu \frac{(zu^\alpha)^\nu}{\Gamma(\alpha\nu+1)} du = \sum_{\nu=0}^\infty a_\nu z^\nu. \quad (3.6.23)$$

This can be summarized in the form of the following

Theorem 3.2 *Let the integral*

$$\int_0^\infty e^{-u} \sum_{\nu=0}^\infty a_\nu \frac{(zu^\alpha)^\nu}{\Gamma(\alpha\nu+1)} du \quad (3.6.24)$$

be convergent at a certain point $z = z_0$.

Then it represents an analytic function in the domain (3.6.20) and provides an analytic continuation of the series (3.6.16) along the segment $(0, z_0]$. At every point of this segment the series is B_α-summable.

Let us recall that the *principle star* of the power series (see e.g., [KolYus96]) is the set of all rays originating at the center of the circle of convergence and extending to the first singular point of the analytic continuation of its sum. This set is a star-shaped domain (or *Mittag-Leffler star*) denoted $\mathcal{B}_\alpha$. It is known (see, e.g., [SanGer60]) that the integral

$$F_\alpha A(z) = \int_{\nu=0}^\infty e^{-u} A(zu^\alpha) du$$

is convergent at any inner point of the Mittag-Leffler star $\mathcal{B}_\alpha$.

3.6.3 Mittag-Leffler Reproducing Kernel Hilbert Spaces

In this subsection we mainly follow the paper [RoRuDi18] (see also [RosDix17]). The corresponding definitions of the Mittag-Leffler reproducing kernel Hilbert spaces are important due to their connection to classical problems of quantum mechanics (see [RoRuDi18] and references therein for details).

We start with the definition of the real-valued Mittag-Leffler reproducing kernel Hilbert space (RKHS), which was introduced in [RosDix17] in order to get a numerical method to estimate the Caputo fractional derivative (for the definition and properties of this kind of fractional derivative we refer to Appendix E, Sect. E.2).

Definition 3.1 The real-valued Mittag-Leffler reproducing kernel Hilbert space of order $q > 0$ is that associated with the Mittag-Leffler functions $\mathcal{K}_q(t, \lambda) = E_q(\lambda^q t^q)$ as the kernel, namely

$$ML^2(\mathbb{R}_+; q) := \left\{ f(t) = \sum_{n=0}^{\infty} a_n t^{qn} \,\Big|\, \sum_{n=0}^{\infty} |a_n|^2 \Gamma(qn+1) < \infty \right\}. \quad (3.6.25)$$

In [RoRuDi18] the complexification of the real-variable Mittag-Leffler RKHS is proposed as a way of generalizing the so-called Bargmann–Fock space (see, e.g., [ZhK12]) which is defined as the space of those entire functions which are L^2 under the Gaussian measure, i.e.

$$F^2(\mathbb{C}) := \left\{ f(z) = \sum_{n=0}^{\infty} a_n z^n \,\Big|\, \sum_{n=0}^{\infty} |a_n|^2 n! < \infty \right\}. \quad (3.6.26)$$

Definition 3.2 The qth order $(q > 0)$ Mittag-Leffler reproducing kernel Hilbert space of entire functions is that associated with the Mittag-Leffler functions $K_q(w, z) = E_q(\overline{w}z)$ as the kernel, namely

$$ML^2(\mathbb{C}; q) := \left\{ f(t) = \sum_{n=0}^{\infty} a_n z^n \,\Big|\, \sum_{n=0}^{\infty} |a_n|^2 \Gamma(qn+1) < \infty \right\}. \quad (3.6.27)$$

A direct verification gives the following property: the set of functions

$$\{g_n(z)\}_{n=0}^{\infty} = \left\{ \frac{z^n}{\sqrt{\Gamma(qn+1)}} \right\}_{n=0}^{\infty} \quad (3.6.28)$$

forms an orthonormal basis in $ML^2(\mathbb{C}; q)$.

Further, for $q = 1$ the Mittag-Leffler space ML^2 coincides with Bargmann-Fock space F^2. The norm in F^2 is given by the following

3.6 Further Analytic Properties

$$\|f\|_{F^2}^2 = \frac{1}{\pi} \int_{\mathbb{C}} |f(\zeta)|^2 e^{-|\zeta|^2} dA(\zeta), \qquad (3.6.29)$$

where the integration is performed with respect to the Lebesgue area measure $dA(\zeta)$. Hence, one can expect in which form the norm in ML^2 can be represented. The result is the following ([RoRuDi18, Theorem 3.2]):

$$\|f\|_{ML^2(\mathbb{C};q)}^2 = \frac{1}{q\pi} \int_{\mathbb{C}} |f(\zeta)|^2 |\zeta|^{\frac{2}{q}-2} e^{-|\zeta|^{\frac{2}{q}}} d\zeta. \qquad (3.6.30)$$

This immediately gives the monotonicity property of the spaces $ML^2(\mathbb{C}; q)$ with respect to the index q, namely, if $0 < q \le p$ and $f \in ML^2(\mathbb{C}; p)$, then $f \in ML^2(\mathbb{C}; q)$. Moreover

$$\lim_{q \to p-0} ML^2(\mathbb{C}; q) = ML^2(\mathbb{C}; p).$$

In particular, $F^2(\mathbb{C}) \subset ML^2(\mathbb{C}; q)$ for all $0 < q \le 1$.

From the standard point-wise estimate for the function $f \in ML^2(\mathbb{C}; q)$:

$$|f(z)|^2 \le E_q(|z|^2) \|f\|_{ML^2(\mathbb{C};q)}^2$$

and the growth of the Mittag-Leffler function (see Appendix B) follows the characterization of the growth of the functions $f \in ML^2(\mathbb{C}; q)$.

Proposition 3.8 ([RoRuDi18, Proposition 4.3]) *Let $q > 0$ be fixed. If $f \in ML^2(\mathbb{C}; q)$, then f is of order at most $\frac{2}{q}$, and if f has order $\frac{2}{q}$, then its type is not greater than $1/2$.*

This statement is in a sense invertible (see [RoRuDi18, Proof 4.4]).

Another result characterizes the asymptotical distribution of zeros of the entire functions belonging to $ML^2(\mathbb{C}; q)$. Let $q > 0$ and $\frac{2}{q} > \varepsilon > 0$. Then for any sequence $\{z_n\}$ of complex numbers satisfying

$$\sum_{n=1}^{\infty} \frac{1}{|z_n|^{\frac{2}{q}-\varepsilon}} < \infty,$$

there exists a function $f \in ML^2(\mathbb{C}; q)$ having these points as zeros.

The following result is an analog of the corresponding statement valid for the Bargmann–Fock space.

Theorem 3.3 *(a) for $0 < q < 1$ every square lattice is a zero set for $ML^2(\mathbb{C}; q)$;*
(b) for $q > 1$ no square lattice is a zero set for $ML^2(\mathbb{C}; q)$;
(c) for $q = 1$ some square lattice can be a zero set for $ML^2(\mathbb{C}; 1) = F^2(\mathbb{C})$.

The following definition generalizing the above definition of the RKHP of entire functions is useful when characterizing the Caputo fractional derivative.

Definition 3.3 Let $q > 0$. The Mittag-Leffler reproducing kernel Hilbert space of functions analytic in the slit complex plane is defined as

$$ML^2\left(\mathbb{C}\setminus\mathbb{R}_-; q\right) := \left\{f : \mathbb{C}\setminus\mathbb{R}_- \to \mathbb{C} \,\Big|\, f(z^{1/q}) \in ML^2\left(\mathbb{C}; q\right)\right\}. \quad (3.6.31)$$

This space is equipped with the following inner product (similar to the corresponding inner result in the Bargmann–Fock space)

$$\langle f(z), g(z)\rangle_{ML^2(\mathbb{C}\setminus\mathbb{R}_-;q)} = \langle f(z^{1/q}), g(z^{1/q})\rangle_{ML^2(\mathbb{C};q)}.$$

Finally, we obtain the following characterization of the Caputo fractional derivative D_*^q defined for any function $f \in ML^2\left(\mathbb{C}\setminus\mathbb{R}_-; q\right)$, $f(z) = \sum_{n=0}^{\infty} a_n z^{qn}$, by the following relation

$$\left(D_*^q f\right)(z) := \sum_{n=0}^{\infty} a_n \frac{\Gamma(qn+1)}{\Gamma(q(n-1)+1)} z^{q(n-1)}. \quad (3.6.32)$$

If v is an entire function with the power series $v(z) = \sum_{n=0}^{\infty} a_n z^n$, and $f(z) = v(z^q)$, then for $z = x \geq 0$ the left-hand side of (3.6.32) coincides with the standard Caputo fractional derivative $\left({}^C D_{0+}^q f\right)(x)$ defined on the restriction of the functions from $ML^2\left(\mathbb{C}\setminus\mathbb{R}_-; q\right)$ to the nonnegative real axis.

3.7 The Mittag-Leffler Function of a Real Variable

3.7.1 *Integral Transforms*

Let us recall a few basic facts about the Laplace transform (a more detailed discussion can be found in Appendix C, see also [BatErd54a, Wid46, DebBha15]). The classical Laplace transform is defined by the following integral formula

$$(\mathcal{L}f)(s) = \int_0^{\infty} e^{-st} f(t)\,dt, \quad (3.7.1)$$

provided that the function f (the *Laplace original*) is absolutely integrable on the semi-axis $(0, +\infty)$. In this case the image of the Laplace transform (also called the *Laplace image*), i.e., the function

$$F(s) = (\mathcal{L}f)(s) \quad (3.7.2)$$

3.7 The Mittag-Leffler Function of a Real Variable

(sometimes denoted as $F(s) = \tilde{f}(s)$) is defined and analytic in the half-plane $\operatorname{Re} s > 0$.

It may happen that the Laplace image can be analytically continued to the left of the imaginary axis $\operatorname{Re} s = 0$ into a larger domain, i.e., there exists a non-positive real number σ_s (called *the Laplace abscissa of convergence*) such that $F(s) = \tilde{f}(s)$ is analytic in the half-plane $\operatorname{Re} s \geq \sigma_s$. Then the inverse Laplace transform can be introduced by the so-called *Bromwich formula*

$$\left(\mathcal{L}^{-1} F\right)(t) = \frac{1}{2\pi i} \int_{\mathcal{L}_{ic}} e^{st} F(s) ds, \qquad (3.7.3)$$

where $\mathcal{L}_{ic} = (c - i\infty, c + i\infty)$, $c > \sigma_s$, and the integral is usually understood in the sense of the Cauchy principal value, i.e.,

$$\int_{\mathcal{L}_{ic}} e^{st} F(s) ds = \lim_{T \to +\infty} \int_{c-iT}^{c+iT} e^{st} F(s) ds.$$

If the Laplace transform (3.7.2) possesses an analytic continuation into the half-plane $\operatorname{Re} s \geq \sigma_s$ and the integral (3.7.3) converges absolutely on the line $\operatorname{Re} s = c > \sigma_s$, then at any continuity point t_0 of the original f the integral (3.7.3) gives the value of f at this point, i.e.,

$$\frac{1}{2\pi i} \int_{\mathcal{L}_{ic}} e^{st_0} \tilde{f}(s) ds = f(t_0). \qquad (3.7.4)$$

Thus, under these conditions, the operators $\mathcal{L}$ and $\mathcal{L}^{-1}$ constitute an inverse pair of operators. Correspondingly, the functions f and $F = \tilde{f}$ constitute a Laplace transform pair.[3] The following notation is used to denote this fact

$$f(t) \div \tilde{f}(s) = \int_0^\infty e^{-st} f(t) \, dt, \quad \operatorname{Re} s > \sigma_s, \qquad (3.7.5)$$

where σ_s is the abscissa of convergence. Here the sign $\div$ denotes the juxtaposition of a function (depending on $t \in \mathbb{R}^+$) with its Laplace transform (depending on $s \in \mathbb{C}$). In the following the conjugate variables $\{t, s\}$ may be given in another notation, e.g.. $\{r, s\}$, and the abscissa of the convergence may sometimes be omitted. Furthermore, throughout our analysis, we assume that the Laplace transforms obtained by our formal manipulations are invertible by using the Bromwich formula (3.7.3).

For the Mittag-Leffler function we have the Laplace integral relation

$$\int_0^\infty e^{-x} E_\alpha(x^\alpha z) \, dx = \frac{1}{1-z}, \quad \alpha \geq 0. \qquad (3.7.6)$$

---

[3] It is easily seen that these properties do not depend on the choice of the real number c.

This integral was evaluated by Mittag-Leffler, who showed that the region of convergence of the integral contains the unit disk and is bounded by the curve $\operatorname{Re} z^{1/\alpha} = 1$.

The *Laplace transform* of $E_\alpha(\pm t^\alpha)$ can be obtained from (3.7.6) by putting $x = st$ and $x^\alpha z = \pm t^\alpha$; we get

$$\mathcal{L}\left[E_\alpha(\pm t^\alpha)\right] := \int_0^\infty e^{-st} E_\alpha(\pm t^\alpha)\, dt = \frac{s^{\alpha-1}}{s^\alpha \mp 1}. \tag{3.7.7}$$

This result was used by Humbert [Hum53] to obtain a number of functional relations satisfied by $E_\alpha(z)$. Formula (3.7.7) can also be obtained by Laplace transforming the series (3.1.1) term-by-term, and summing the resulting series.

Recalling the scale property of the Laplace transform

$$q f(qt) \div \bar{F}(s/q), \quad \forall q > 0,$$

we have

$$E_\alpha\left[\pm(qt)^\alpha\right] \div \frac{s^{\alpha-1}}{s^\alpha \mp q^\alpha}, \quad \forall q > 0. \tag{3.7.8}$$

As an exercise we can invert the r.h.s. of (3.7.8) either by means of the expansion method (find the series expansion of the Laplace transform and then invert term-by-term to get the series representation of the l.h.s.) or by the *Bromwich inversion formula* (deform the Bromwich path into the Hankel path and, by means of an appropriate change of variable, obtain the integral representation of the left-hand side, see, e.g.. [CapMai71a]).

Using the asymptotic behavior of the function $E_\alpha(z)$ we will investigate further the Mellin integral transform of the function $E_\alpha(z)$ and its properties. The Mellin integral transform is defined by the formula

$$(\mathcal{M}f)(p) = f^*(p) := \int_0^\infty f(t) t^{p-1} dt \quad (p \in \mathbb{C}) \tag{3.7.9}$$

provided that the integral on the right-hand side exists.

In many cases, the function $E_\alpha(z)$ does not satisfy the convergence condition for the standard Mellin transform (see Theorem C.5 in Appendix C). Therefore we find the Mellin transform of a slightly different function. A good candidate for applying the general theory is the function

$$e_\alpha(x; \lambda) = \frac{1}{\lambda x}\{E_\alpha(\lambda x^\alpha) - 1\}, \tag{3.7.10}$$

where $x > 0$ and $\lambda \neq 0$ is any complex number. It happens that the function $\epsilon_\alpha(x; \lambda)$ satisfies, up to rotation, the above given convergence condition for certain values of parameters.

3.7 The Mittag-Leffler Function of a Real Variable

Lemma 3.6 *Fix an α in the interval $(\frac{1}{2}, 2]$. Then for each φ, $\frac{\pi\alpha}{2} \leq \varphi \leq 2\pi - \frac{\pi\alpha}{2}$, the function $e_\alpha(x; e^{i\varphi})$ is square-integrable on the positive semi-axis:*

$$e_\alpha(x; e^{i\varphi}) \in L_2(0, +\infty). \qquad (3.7.11)$$

◁ If $0 < x \leq 1$ one can choose a constant $C_1 > 0$ such that

$$|e_\alpha(x; e^{i\varphi})| \leq C_1 x^{\alpha-1}, \ 0 < x \leq 1, \ 0 \leq \varphi \leq 2\pi.$$

Thus with the condition $\alpha > \frac{1}{2}$ we have

$$e_\alpha(x; e^{i\varphi}) \in L_2(0, 1), \ 0 \leq \varphi \leq 2\pi. \qquad (3.7.12)$$

Consider now the behavior of the function $e_\alpha(x; e^{i\varphi})$ with $\frac{\pi\alpha}{2} \leq \varphi \leq 2\pi - \frac{\pi\alpha}{2}$ for $x \in (1, +\infty)$.

In the case $\alpha = 2$ we have $\varphi = \pi$. It follows then from (3.2.2) that

$$e_2(x; -1) = -\frac{1}{x}\left(E_2(-x^2) - 1\right) = \frac{1 - \cos x}{x} = O\left(\frac{1}{x}\right), \ x \to \infty.$$

Therefore

$$e_2(x; -1) \in L_2(1, \infty). \qquad (3.7.13)$$

If $\frac{1}{2} < \alpha < 2$ and $\frac{\pi\alpha}{2} \leq \varphi \leq 2\pi - \frac{\pi\alpha}{2}$ then one can use a variant of the asymptotic estimate (3.4.30) for the Mittag-Leffler function $E_\alpha(z)$, namely the estimate

$$|E_\alpha(e^{i\varphi} x^\alpha)| \leq C_2 x^{-\alpha}, \ \gamma \leq \varphi \leq 2\pi - \gamma, \ 1 \leq x < +\infty,$$

where $C_2 > 0$ is a constant not depending on φ, and $\frac{\pi\alpha}{2} < \gamma < \min\{\pi, \pi\alpha\}$. Thus

$$|e_\alpha(x; e^{i\alpha})| \leq C_2 x^{-1-\alpha} + \frac{1}{x}, \ \gamma \leq \varphi \leq 2\pi - \gamma, \ 1 \leq x < +\infty.$$

Therefore

$$e_\alpha(x; e^{i\alpha}) \in L_2(1, +\infty), \ \gamma \leq \alpha \leq 2\pi - \gamma. \qquad (3.7.14)$$

The same result follows for $\frac{\pi\alpha}{2} \leq \varphi \leq \gamma$ or $2\pi - \gamma \leq \varphi \leq 2\pi - \frac{\pi\alpha}{2}$ from the asymptotic formula (3.4.14). The lemma is proven. ▷

For further considerations we need the Mellin integral transforms of some elementary functions, related to the Mittag-Leffler function (cf., e.g., [AbrSte72, Mari83, NIST]).

Lemma 3.7 *If* $0 < \operatorname{Re} s < 1$, *then*

$$\int_0^{+\infty} \frac{e^{\pm ix} - 1}{\pm ix} x^{s-1} dx = \frac{\Gamma(s)}{1-s} e^{\pm i\frac{\pi}{2}s}, \quad (3.7.15)$$

$$\int_0^{+\infty} \frac{1 - \cos x}{x} x^{s-1} dx = \frac{\Gamma(s)}{1-s} \sin\frac{\pi s}{2}, \quad (3.7.16)$$

$$\int_0^{+\infty} \frac{\sin x}{x} x^{s-1} dx = \frac{\Gamma(s)}{1-s} \cos\frac{\pi s}{2}. \quad (3.7.17)$$

The following lemma describes asymptotic properties of the function

$$U_p(s) = \int_{-\frac{\pi}{2}}^{\frac{\pi}{2}} e^{pe^{i\psi} + i(s-1)\psi} d\psi, \quad 0 < \operatorname{Re} s < 1, \quad (3.7.18)$$

for $p \to +\infty$.

Lemma 3.8 *If* $0 < \operatorname{Re} s < 1$, *then*

$$\lim_{p \to +\infty} p^{s-1} U_p(s) = \frac{2\pi}{\Gamma(2-s)}. \quad (3.7.19)$$

◁ We can represent the function $U_p(s)$ as a Taylor series in p in a neighborhood of $p = 0$:

$$U_p(s) = \sum_{k=0}^{\infty} \frac{p^k}{k!} \int_{-\frac{\pi}{2}}^{\frac{\pi}{2}} e^{i(k-1+s)\psi} d\psi = 2 \sum_{k=0}^{\infty} \frac{p^k}{k!} \frac{\sin(k-1+s)\frac{\pi}{2}}{k-1+s}.$$

Here the series on the right-hand side can be rewritten as the sum of two series with even and odd indices of summation, respectively. Thus we arrive at

$$U_p(s) = 2\frac{\cos\frac{\pi s}{2}}{1-s} + 2\sum_{k=1}^{\infty} \frac{p^{2k}}{(2k)!} \frac{\sin\left(\pi k - \frac{\pi}{2} + \frac{\pi s}{2}\right)}{2k-1+s}$$

$$+ 2\sum_{k=0}^{\infty} \frac{p^{2k+1}}{(2k+1)!} \frac{\sin\left(\pi k + \frac{\pi s}{2}\right)}{2k+s}$$

$$= 2\frac{\cos\frac{\pi s}{2}}{1-s} - 2\cos\frac{\pi s}{2} \sum_{k=1}^{\infty} \frac{(-1)^k p^{2k}}{(2k)!(2k-1+s)}$$

$$+ 2\sin\frac{\pi s}{2} \sum_{k=0}^{\infty} \frac{(-1)^k p^{2k+1}}{(2k+1)!(2k+s)}$$

$$= 2\frac{\cos\frac{\pi s}{2}}{1-s} - 2p^{1-s} \cos\frac{\pi s}{2} \int_0^p \frac{\cos x - 1}{x} x^{s-1} dx$$

3.7 The Mittag-Leffler Function of a Real Variable

$$+ 2p^{1-s} \sin \frac{\pi s}{2} \int_0^p \frac{\sin x}{x} x^{s-1} \, dx. \tag{3.7.20}$$

Since $0 < \operatorname{Re} s < 1$, we can use the formulae (3.7.15) and (3.7.16) to obtain

$$\lim_{p \to +\infty} p^{s-1} U_p(s) = 2\cos \frac{\pi s}{2} \int_0^{+\infty} \frac{1 - \cos x}{x} x^{s-1} \, dx$$
$$+ 2\sin \frac{\pi s}{2} \int_0^{+\infty} \frac{\sin x}{x} x^{s-1} \, dx = 2\frac{\Gamma(s)}{1-s} \sin(\pi s).$$

The formula (3.7.19) and the statement of the lemma now follow from the last formula and the well-known formula for the Euler Gamma function (formula (A.13), Appendix A):

$$\Gamma(s)\Gamma(1-s) = \frac{\pi}{\sin(\pi s)}. \triangleright$$

In order to obtain the Mellin transform of the function connected with the Mittag-Leffler function $E_\alpha(z)$ we need one more auxiliary result. It is in a certain sense a special refinement of the Jordan lemma (see, e.g., [AblFok97]).

Let us draw a cut in the z-plane along the ray $\arg z = \varphi$, $\frac{\pi\alpha}{2} \leq \varphi \leq 2\pi - \frac{\pi\alpha}{2}$, $\frac{1}{2} < \alpha \leq 2$. Consider in the cut z-plane that branch of the function $z^{\frac{(s+\alpha-1)}{\alpha}-1}$, $\operatorname{Re} s = \frac{1}{2}$, which takes on the semi-axis $0 < x < +\infty$ the values $\exp\{(\frac{(s+\alpha-1)}{\alpha} - 1) \log x\}$. By l_R we denote a part of the circle $|z| = R$ on the z-plane with the cut along $\arg z = \varphi$.

Lemma 3.9 *Let $\frac{1}{2} < \alpha \leq 2$ be a fixed value of the parameter. Then we have for any s with $\operatorname{Re} s = \frac{1}{2}$*

$$\lim_{R \to +\infty} \int_{l_R} \frac{E_\alpha(z) - 1}{z} z^{\frac{(s+\alpha-1)}{\alpha}-1} \, dz = \frac{2\pi i}{\Gamma(2-s)}. \tag{3.7.21}$$

The proof follows from the asymptotic representations of the Mittag-Leffler function and the above proved Lemmas 3.7 and 3.8.

Finally we have the following:

Proposition 3.9 *Let $\frac{1}{2} < \alpha \leq 2$ be a fixed value of the parameter. Then the formula*

$$\int_0^{+\infty} \frac{E_\alpha(re^{i\varphi}) - 1}{re^{i\varphi}} r^{\frac{(s+\alpha-1)}{\alpha}-1} \, dr = \frac{\pi}{\Gamma(2-s)} \frac{e^{i\frac{(\pi-\varphi)}{\alpha}(s+\alpha-1)}}{\sin(\frac{\pi(s+\alpha-1)}{\alpha})}, \quad \operatorname{Re} s = \frac{1}{2}, \tag{3.7.22}$$

holds for any $\varphi \in \left[\frac{\pi\alpha}{2}, 2\pi - \frac{\pi\alpha}{2}\right]$.

◁ We consider the function $\frac{E_\alpha(z)-1}{z} z^{\frac{(s+\alpha-1)}{\alpha}-1}$, Re $s = \frac{1}{2}$, in the z-plane which is cut along the ray $\arg z = \varphi$, $\varphi \in \left[\frac{\pi\alpha}{2}, 2\pi - \frac{\pi\alpha}{2}\right]$. We denote by $L(R;\varepsilon)$ a closed positive oriented contour consisting of two circles $l_\varepsilon = \{z : |z| = \varepsilon\}$, $l_R = \{z : |z| = R\}$, $0 < \varepsilon < R$, and of the two sides of the cut $\arg z = \varphi$, $\varepsilon \leq |z| \leq R$.

By the Cauchy theorem,

$$\int_{L(R;\varepsilon)} \frac{E_\alpha(z) - 1}{z} z^{\frac{(s+\alpha-1)}{\alpha}-1} dz = 0.$$

This formula can be rewritten as

$$e^{-i\frac{(2\pi-\varphi)}{\alpha}(s+\alpha-1)} \int_\varepsilon^R \frac{E_\alpha(re^{i\varphi}) - 1}{re^{i\varphi}} r^{\frac{(s+\alpha-1)}{\alpha}-1} dr$$
$$+ \int_{l_R} \frac{E_\alpha(z) - 1}{z} z^{\frac{(s+\alpha-1)}{\alpha}-1} dz$$
$$- e^{i\frac{\varphi}{\alpha}(s+\alpha-1)} \int_\varepsilon^R \frac{E_\alpha(re^{i\varphi}) - 1}{re^{i\varphi}} r^{\frac{(s+\alpha-1)}{\alpha}-1} dr$$
$$+ \int_{l_\varepsilon} \frac{E_\alpha(z) - 1}{z} z^{\frac{(s+\alpha-1)}{\alpha}-1} dz = 0. \quad (3.7.23)$$

If $\varepsilon > 0$ is small enough, we have

$$\max_{|z|=\varepsilon} \left|\frac{E_\alpha(z)-1}{z}\right| \leq \frac{2}{\Gamma(\alpha+1)}.$$

For such ε and Re $s = \frac{1}{2}$ we get the estimate

$$\left|\int_{l_\varepsilon} \frac{E_\alpha(z)-1}{z} z^{\frac{(s+\alpha-1)}{\alpha}-1} dz\right| \leq \frac{2}{\Gamma(\alpha+1)} 2\pi\varepsilon^{\frac{\alpha-1/2}{\alpha}}.$$

Since $\alpha > \frac{1}{2}$, we conclude that

$$\lim_{\varepsilon \to 0} \int_{l_\varepsilon} \frac{E_\alpha(z)-1}{z} z^{\frac{(s+\alpha-1)}{\alpha}-1} dz = 0. \quad (3.7.24)$$

Passing to the limit in the identity (3.7.23) with $\varepsilon \to 0$ and using the formula (3.7.24) we arrive at the identity

$$e^{i\frac{\varphi}{\alpha}(s+\alpha-1)} \{e^{-i\frac{2\pi}{\alpha}(s+\alpha-1)} - 1\} \int_0^R \frac{E_\alpha(re^{i\varphi}) - 1}{re^{i\varphi}} r^{\frac{(s+\alpha-1)}{\alpha}-1} dr$$
$$= -\int_{l_R} \frac{E_\alpha(z) - 1}{z} z^{\frac{(s+\alpha-1)}{\alpha}-1} dz, \quad (3.7.25)$$

3.7 The Mittag-Leffler Function of a Real Variable

where the integral in the left-hand side converges absolutely for any finite R, since $\operatorname{Re} s = \frac{1}{2}$ and $\alpha > \frac{1}{2}$.

If $\operatorname{Re} s = \frac{1}{2}$ then $\operatorname{Re}\{\frac{(s+\alpha-1)}{\alpha}\} = 1 - \frac{1}{2\alpha} \in (0, 3/4]$. Consequently, $e^{-i 2\pi \frac{(s+\alpha-1)}{\alpha}} - 1 \neq 0$ in this case. Thus the identity (3.7.25) can be rewritten in the form

$$\int_0^R \frac{E_\alpha(re^{i\varphi}) - 1}{re^{i\varphi}} r^{\frac{(s+\alpha-1)}{\alpha} - 1} dr$$

$$= -\frac{e^{i\frac{(\pi-\varphi)}{\alpha}(s+\alpha-1)}}{2i \sin(\frac{\pi(s+\alpha-1)}{\alpha})} \int_{l_R} \frac{E_\alpha(z) - 1}{z} z^{\frac{(s+\alpha-1)}{\alpha} - 1} dz, \quad \operatorname{Re} s = \frac{1}{2}. \qquad (3.7.26)$$

Passing to the limit as $R \to \infty$ in the last identity and using the formula (3.7.21) we finally obtain the formula (3.7.22). ▷

Corollary 3.5 *Let $\frac{1}{2} < \alpha \leq 2$. Then the Mellin transform of the function $e_\alpha(x; e^{i\varphi})$ (see (3.7.10)) exists and the following representation holds*

$$\mathcal{M}\left(\frac{E_\alpha(t^\alpha e^{i\varphi}) - 1}{te^{i\varphi}}\right)(p) = \int_0^\infty e_\alpha(t; e^{i\varphi}) t^{p-1} dt \qquad (3.7.27)$$

$$= \frac{\pi}{\alpha \Gamma(1 - \alpha(p-1))} \frac{e^{i(\pi-\varphi)p}}{\sin(\pi p)}, \quad \operatorname{Re} s = \frac{1}{2},$$

where $\varphi \in \left[\frac{\pi\alpha}{2}, 2\pi - \frac{\pi\alpha}{2}\right]$.

3.7.2 The Complete Monotonicity Property

Definition 3.4 A function $f : (0, \infty) \to \mathbb{R}$ is called *completely monotonic* if it possesses derivatives $f^{(n)}(x)$ of any order $n = 0, 1, \ldots$, and the derivatives are alternating in sign, i.e.

$$(-1)^n f^{(n)}(x) \geq 0, \quad \forall x \in (0, \infty). \qquad (3.7.28)$$

The above property (see, e.g., [Wid46, p. 161]) is equivalent to the existence of a representation of the function f in the form of a Laplace–Stieltjes integral with non-decreasing density and non-negative measure $d\mu$

$$f(x) = \int_0^\infty e^{-xt} d\mu(t). \qquad (3.7.29)$$

Proposition 3.10 ([Poll48]) *The Mittag-Leffler function of negative argument $E_\alpha(-x)$ is completely monotonic for all $0 \leq \alpha \leq 1$.*

◁ Since $E_0(-x) = 1/(1+x)$ and $E_1(-x) = e^{-x}$ there is nothing to be proved in these cases. Let $0 < a < 1$. By a standard representation [Bie31]

$$E_\alpha(-x) = \frac{1}{2\pi i a} \int_L \frac{e^{t^{1/\alpha}}}{t+x} dt, \qquad (3.7.30)$$

where L consists of three parts as follows:

C_1: the line $y = (\tan \psi) x$ from $x = +\infty$ to $x = \varrho, \varrho > 0$;
C_2: an arc of the circle $|z| = \varrho \sec \psi$, $-\psi \leq \arg z \leq \psi$;
C_3: the reflection of C_1 in the x-axis.

We assume $\pi > \psi/\alpha > \pi/2$ while ϱ is arbitrary but fixed.

Let us replace $(x+t)^{-1}$ by $\int_0^\infty e^{-(x+t)} du$ in (3.7.30). The resulting double integral converges absolutely, so that one can interchange the order of integration to obtain

$$E_\alpha(-x) = \frac{1}{2\pi i \alpha} \int_0^\infty e^{-xu} du \int_L e^{t^{1/\alpha}} e^{-tu} dt. \qquad (3.7.31)$$

It remains to compute the function

$$F_\alpha(u) = \frac{1}{2\pi i a} \int_L e^{t^{1/\alpha}} e^{-tu} dt \qquad (3.7.32)$$

and to prove it is non-negative when $u \geq 0$ (see the remark concerning representation (3.7.29)). An integration by parts in (3.7.32) yields

$$F_\alpha(u) = \frac{1}{2\pi i \alpha u} \int_L e^{-tu} \left(\frac{1}{\alpha} t^{1/\alpha - 1}\right) e^{t^{1/\alpha}} dt. \qquad (3.7.33)$$

Now let $tu = z^\alpha$. Then

$$F_\alpha(u) = \frac{u^{-1-1/\alpha}}{\alpha} \frac{1}{2\pi i} \int_{L'} e^{-z^\alpha} e^{zu^{-1/\alpha}} dz, \qquad (3.7.34)$$

where L' is the image of L under the mapping.

Now consider the function

$$\Phi_\alpha(t) = \frac{1}{2\pi i} \int_{L'} e^{-z^\alpha} e^{zt} dz. \qquad (3.7.35)$$

This is known to be the inverse Laplace transform of

$$e^{-z^\alpha} = \int_0^\infty e^{-zt} \Phi_\alpha(t) dt,$$

3.7 The Mittag-Leffler Function of a Real Variable

which is completely monotonic [Poll48]. Hence

$$F_\alpha(u) = \frac{u^{-1-1/\alpha}}{\alpha} \Phi_\alpha(u^{-1/\alpha}) \geq 0.$$

The proof will be completed if we can show the existence of any derivative of $F_\alpha(u)$ for all $u \geq 0$. From the explicit series representation [Poll48] for the function $\Phi_\alpha(t)$ we deduce that

$$F_\alpha(u) = \frac{1}{\pi\alpha} \sum_{1}^{\infty} \frac{(-1)^{k-1}}{k!} \sin(\pi\alpha k) \, \Gamma(\alpha k + 1) \, u^{k-1}, \quad (3.7.36)$$

so that $F_\alpha(u)$ is an entire function. ▷

Note that it is also possible to obtain (3.7.34) directly from (3.7.36).

From Proposition 3.10 it follows, in particular, that $E_\alpha(-x)$ has no real zeros when $0 \leq \alpha \leq 1$.

3.7.3 Relation to Fractional Calculus

Let us recall a few definitions concerning fractional integrals and derivatives. The left- and right-sided Riemann–Liouville fractional integrals on any finite interval (a, b) are given by the formulas (see, e.g., [SaKiMa93, p. 33])

$$\left(I_{a+}^\alpha \varphi\right)(x) = \frac{1}{\Gamma(\alpha)} \int_a^x \frac{\varphi(t)}{(x-t)^{1-\alpha}} dt, \quad x > a, \quad (3.7.37)$$

$$\left(I_{b-}^\alpha \varphi\right)(x) = \frac{1}{\Gamma(\alpha)} \int_x^b \frac{\varphi(t)}{(t-x)^{1-\alpha}} dt, \quad x < b. \quad (3.7.38)$$

The right-sided fractional integral on a semi-axis (also called the right-sided Liouville fractional integral) is defined via the formula (see, e.g., [SaKiMa93, p. 94])

$$\left(I_-^\alpha \varphi\right)(x) = \frac{1}{\Gamma(\alpha)} \int_x^\infty \frac{\varphi(t)}{(t-x)^{1-\alpha}} dt, \quad -\infty < x < +\infty. \quad (3.7.39)$$

By simple calculation one can obtain the following values for the above integrals of power-type functions (see [SaKiMa93, p. 40])

$$\left(I_{a+}^{\alpha}(t-a)^{\beta-1}\right)(x) = \frac{\Gamma(\beta)}{\Gamma(\alpha+\beta)}(x-a)^{\alpha+\beta-1}, \quad x > a, \qquad (3.7.40)$$

$$\left(I_{b-}^{\alpha}(b-t)^{\beta-1}\right)(x) = \frac{\Gamma(\beta)}{\Gamma(\alpha+\beta)}(b-x)^{\alpha+\beta-1}, \quad x < b, \qquad (3.7.41)$$

$$\left(I_{-}^{\alpha}t^{-\beta-1}\right)(x) = \frac{\Gamma(1-\alpha+\beta)}{\Gamma(1+\beta)}x^{\alpha-\beta-1}, \quad -\infty < x < +\infty. \qquad (3.7.42)$$

The last integral can be calculated using the following property of the fractional integrals

$$\left(I_{-}^{\alpha}\varphi\left(\frac{1}{t}\right)\right)(x) = x^{\alpha-1}\left(t^{-\alpha-1}I_{0+}^{\alpha}\varphi(t)\right)\left(\frac{1}{x}\right). \qquad (3.7.43)$$

The values of fractional integrals of the Mittag-Leffler function (or, in other words, the composition of the fractional integrals with the Mittag-Leffler function) can be calculated by using the following auxiliary result.

Lemma 3.10 *Let $\alpha > 0$ and suppose $\lambda \in \mathbb{C}$ is not an eigenvalue of the Abel integral operator, i.e. $\nexists \varphi$ such that $\left(I_{0+}^{\alpha}\right)(x) = \lambda \varphi(x)$, $0 < x < \infty$. Then*

$$\frac{\lambda}{\Gamma(\alpha)} \int_0^x \frac{E_{\alpha}(\lambda t^{\alpha})}{(x-t)^{1-\alpha}} dt = E_{\alpha}(\lambda x^{\alpha}) - 1. \qquad (3.7.44)$$

◁ The proof follows from the Taylor expansion of E_{α} and by term-by-term integration using formula (3.7.40). ▷

We present here the above-mentioned composition formulas only for the left- and right-sided Riemann–Liouville fractional integrals and right-sided Liouville fractional integral. The formulas for other types of fractional integrals and derivatives (see Appendix E) can be obtained in a similar way.

Proposition 3.11 *Let $\operatorname{Re}\alpha > 0$, then the following formulas are satisfied*

$$I_{a+}^{\alpha}\left(E_{\alpha}(\lambda(t-a)^{\alpha})\right)(x) = \frac{1}{\lambda}\left\{E_{\alpha}(\lambda(x-a)^{\alpha}) - 1\right\}, \quad \lambda \neq 0. \qquad (3.7.45)$$

$$I_{b-}^{\alpha}\left(E_{\alpha}(\lambda(b-t)^{\alpha})\right)(x) = \frac{1}{\lambda}\left\{E_{\alpha}(\lambda(b-x)^{\alpha}) - 1\right\}, \quad \lambda \neq 0. \qquad (3.7.46)$$

$$I_{-}^{\alpha}\left(t^{-\alpha-1}E_{\alpha}(\lambda t^{-\alpha})\right)(x) = \frac{x^{\alpha-1}}{\lambda}\left\{E_{\alpha}(\lambda x^{-\alpha}) - 1\right\}, \quad \lambda \neq 0. \qquad (3.7.47)$$

The result follows immediately from Lemma 3.10.

The left- and right-sided fractional derivative of a non-integer order α ($m-1 < \alpha < m$) are defined by the formulas (see Appendix E)

$$D_{a+}^{\alpha}\phi(x) = \frac{1}{\Gamma(m-\alpha)}\frac{d^m}{dx^m}\int_a^x (x-\xi)^{m-\alpha-1}\phi(\xi)\,d\xi, \quad a < x < b, \qquad (3.7.48)$$

$$D_{b-}^{\alpha}\phi(x) = \frac{(-1)^m}{\Gamma(m-\alpha)}\frac{d^m}{dx^m}\int_x^b (\xi - x)^{m-\alpha-1}\phi(\xi)\,d\xi\,,\quad a < x < b\,. \quad (3.7.49)$$

Proposition 3.12 *Let $0 < \operatorname{Re}\alpha < 1$, then the following formulas hold*

$$D_{a+}^{\alpha}\left(E_{\alpha}(\lambda(t-a)^{\alpha})\right)(x) = \frac{(x-a)^{-\alpha}}{\Gamma(1-\alpha)} + \lambda E_{\alpha}(\lambda(x-a)^{\alpha}),\ a \neq 0. \quad (3.7.50)$$

$$D_{b-}^{\alpha}\left(E_{\alpha}(\lambda(b-t)^{\alpha})\right)(x) = \frac{(b-x)^{-\alpha}}{\Gamma(1-\alpha)} + \lambda E_{\alpha}(\lambda(b-x)^{\alpha}),\ a \neq 0. \quad (3.7.51)$$

Since the Mittag-Leffler function is infinitely differentiable, one can use in the considered case ($0 < \operatorname{Re}\alpha < 1$, $f(t) = E_{\alpha}(\lambda(t-a)^{\alpha})$) known relation formulas for fractional integrals and derivatives (see, e.g., [SaKiMa93, p. 35–36]):

$$\left(D_{a+}^{\alpha}f(t)\right)(x) = \frac{1}{\Gamma(1-\alpha)}\left[\frac{f(a)}{(x-a)^{\alpha}} + \int_a^x \frac{f'(t)}{(x-t)^{\alpha}}dt\right],$$

$$\left(D_{b-}^{\alpha}f(t)\right)(x) = \frac{1}{\Gamma(1-\alpha)}\left[\frac{f(b)}{(b-x)^{\alpha}} - \int_x^b \frac{f'(t)}{(x-t)^{\alpha}}dt\right].$$

3.8 Historical and Bibliographical Notes

Historical remarks regarding early results on the classical Mittag-Leffler function as well as certain direct generalizations have already been presented in the Introduction and in Sect. 2.3. We also mention here the treatise on complex functions by Sansone and Gerretsen [SanGer60], where a detailed account of these functions is given. However, the most specialized treatise, where more details on the functions of Mittag-Leffler type are given, is surely that by Dzherbashyan [Dzh66], in Russian. We also mention another book by this author from 1993 [Djr93], in English, where a brief description of the theory of Mittag-Leffler functions is given. For the basic facts on the Mittag-Leffler function presented in Sect. 3.4–3.6 we have followed the monograph [Dzh66] and the survey paper [PopSed11], see also [Per00].

In this section we have also mentioned more recent results related to the development of the theory of the Mittag-Leffler function $E_{\alpha}(z)$. We note that many mathematicians, not only in Mittag-Leffler's time, see, e.g.. Phragmén [Phr04], Malmquist [Mal03], Wiman [Wim05a, Wim05b], [LeR00], have recognized its importance, providing interesting results and applications, which unfortunately are not as well known as they deserve to be, see, e.g., Hille and Tamarkin [HilTam30], Pollard [Poll48], Humbert and Agarwal [Aga53], [Hum53]–[HumAga53], or the short reviews in books such as Buhl [Buh25b], Davis [Dav36], Evgrafov [Evg78], see also [Wong89]. We briefly outline some important results. In particular, in [Dav36] the Volterra function was mentioned

$$\nu(z) = \int_0^{+\infty} \frac{z^u}{\Gamma(\alpha u + 1)} \mathrm{d}t,$$

which can be considered as the continuous counterpart of the classical Mittag-Leffler function. It is treated in the Bateman handbook [ErdBat-3, Chap. 18]. The Volterra function and its generalizations are treated in the book [Ape08] and in the paper [GarMai16] (in relation to Ramanujan integrals).

The first asymptotic result which was obtained using Mittag-Leffler's method was that of Malmquist [Mal05]. Phragmén [Phr04] studied the Mittag-Leffler function as an example of an entire function of finite order satisfying his theorem generalizing the maximum modulus principle for analytic functions. In the same direction Buhl [Buh25a] showed that the Mittag-Leffler function furnishes examples and counterexamples for the growth and other properties of entire functions of finite order (see also [Evg78, GoLuRo97, WongZh02]).

The asymptotic expansions of $E_\alpha(z)$ were generalized by Wiman [Wim05a] (see also the papers by Barnes [Barn06] and Mellin [Mel10], in which the complete theory of the Mellin–Barnes integral representation is developed). Wright [Wri35a] introduced and discussed the integral representation of the function $\phi(\alpha, \beta; z)$ (called the *Wright function*). He also found [Wri40a] the connection of $\phi(\alpha, \beta; z)$ with the so-called generalized Bessel functions of order greater than one, which was a prototype of the two-parametric Mittag-Leffler function $E_{\alpha,\beta}(z)$.

The representation of the Cauchy kernel in the form (3.6.15) was found by Mittag-Leffler [ML08]. Using that as a basis, he developed a generalization of the Borel summation method for power series as presented in [SanGer60].

Humbert and Agarwal [Aga53], [Hum53]–[HumAga53] introduced the two-parametric Mittag-Leffler function $E_{\alpha,\beta}(z)$ (mentioned for the first time in the article by Wiman [Wim05a]) and investigated its properties (see also [HumDel53] for the generalized Mittag-Leffler function of two variables).

The Laplace transform of $E_\alpha(t^\alpha)$ was used by Humbert [Hum53] to obtain a number of functional relations satisfied by the Mittag-Leffler function.

Feller (see, e.g.., [Fel71], see also [GrLoSt90, Chap. 5]) conjectured the complete monotonicity of the Mittag-Leffler function $E_\alpha(-x)$ and proved it for $0 \leq \alpha \leq 1$ by using methods of Probability Theory. An analytic proof of this result was obtained by Pollard [Pol48]. The result presented in Sect. 3.7.2 is due to him. More recently, further aspects of the complete monotonicity were investigated by Mainardi [Mai14] (see also [Sim14]). In particular, some bounds for the function $E_\alpha(-x^\alpha)$ were found and illustrated by respective plots.

For pioneering works on the mathematical applications of the Mittag-Leffler function we refer to Hille and Tamarkin [HilTam30] and Barrett [Barr54]. The 1930 paper by Hille and Tamarkin deals with the solution of the Abel integral equation of the second kind (a particular fractional integral equation). The 1954 paper by Barrett con-

3.8 Historical and Bibliographical Notes

cerns the general solution of the linear fractional differential equation with constant coefficients.

Concerning earlier applications of the Mittag-Leffler function in physics, we refer to the contributions by K.S. Cole, see [Col33] (mentioned in the book by Davis [Dav36, p. 287]), in connection with nerve conduction, and by de Oliveira Castro [Oli39], and Gross [Gro47], in connection with dielectrical and mechanical relaxation, respectively. Subsequently, Caputo and Mainardi [CapMai71a], [CapMai71b], have proved that the Mittag-Leffler type functions appear whenever derivatives of fractional order are introduced in the constitutive equations of a linear viscoelastic body. Since then, several other authors have pointed out the relevance of such functions for fractional viscoelastic models.

In recent times the attention of mathematicians towards the Mittag-Leffler function has increased from both the analytical and numerical point of view, overall because of its relation with the Fractional Calculus and its applications, see Diethelm et al. [Die10], Hilfer and Seybold [HilSey06], Luchko et al. [LucGor99], [LucSri95], Gorenflo et al. [GoLoLu02], Samko et al. [SaKiMa93], Gorenflo and Vessella [GorVes91], Miller and Ross [MilRos93], Kiryakova [Kir94], Gorenflo and Rutman [GorRut94], Mainardi [Mai96b], [Mai97], Podlubny [Pod99], Kilbas and Saigo [KilSai04], Blank [Bla97], Gorenflo and Mainardi [GorMai97], Schneider [Sch90], Tarasov [Tar10, Tar13], Uchaikin [Uch13a, Uch13b], and Baleanu et al. [Bal-et-al17].

In fact there has been a revived interest in the Fractional Calculus because of its applications in different areas of physics and engineering. In addition to the books and papers already quoted, here we would like to draw the reader's attention to some relevant papers, in alphabetical order of the first author, Al Saqabi and Tuan [Al-STua96], Brankov and Tonchev [BraTon92], Berberan-Santos [Ber-S05a, Ber-S05b, Ber-S05c], and others (see Mainardi's book [Mai10] and the references therein).

This list, however, is not exhaustive. Details on Mittag-Leffler functions can also be found in some treatises devoted to the theory and/or applications of special functions, integral transforms and fractional calculus, e.g.. Davis [Dav36], Marichev [Mari83], Gorenflo and Vessella [GorVes91], Samko et al. [SaKiMa93], Kiryakova [Kir94], Carpinteri and Mainardi [CarMai97], Podlubny [Pod99], Hilfer [Hil00], West et al. [WeBoGr03], Kilbas et al. [KiSrTr06], Magin [Mag06], Debnath-Bhatta [DebBha15], Mathai and Haubold [MatHau08].

To the best of the authors' knowledge, earlier plots of the Mittag-Leffler functions can be found (presumably for the first time in the literature of fractional calculus and special functions) in the 1971 paper by Caputo and Mainardi [CapMai71b]. More precisely, these authors provided plots of the function $E_\nu(-t^\nu)$ for some values of $\nu \in (0, 1]$, adopting linear-logarithmic scales, in the framework of fractional relaxation for viscoelastic media. At the time, not only were such functions still almost ignored, but also the fractional calculus was not yet well accepted by the community of physicists.

Recently, numerical routines for functions of the Mittag-Leffler type have been provided, see e.g.. Gorenflo et al. [GoLoLu02] (with *MATHEMATICA*), Podlubny

[Pod06] (with *MATLAB*) and Seybold and Hilfer [SeyHil05]. Furthermore, in the NASA report by Freed et al. [FrDiLu02], an appendix is devoted to the table of Padè approximants for the Mittag-Leffler function $E_\alpha(-x)$.

Since the fractional calculus has actually attracted wide interest in different areas of the applied sciences, we think that the Mittag-Leffler function is now leaving its isolated life as *Cinderella* (using the term coined by F.G. Tricomi in the 1950s for the incomplete gamma function). We like to refer to the classical Mittag-Leffler function as the *Queen function of fractional calculus*, and to consider all the related functions as her court, see [MaiGor07].

For the first reading we recommend Chaps. 2, 3, 4, 5 in which the basic theory of the Mittag-Leffler function in one variable with one, two and three parameters is discussed. A short account of the further mathematical development of this theory is presented in Chap. 6.

3.9 Exercises

Example 3.1 Prove that the following relations hold:

$$E_3(z) = \frac{1}{2}\left[e^{z^{\frac{1}{3}}} + 2e^{-\frac{1}{2}z^{\frac{1}{3}}} \cdot \cos\left(\frac{\sqrt{3}}{2}z^{\frac{1}{3}}\right)\right], \quad z \in \mathbb{C},$$

$$E_4(z) = \frac{1}{2}\left[\cos\left(z^{\frac{1}{4}}\right) + \cosh\left(z^{\frac{1}{4}}\right)\right], \quad z \in \mathbb{C}.$$

Example 3.2 Deduce the following duplication formula

$$E_{2\alpha}(z^2) = \frac{1}{2}[E_\alpha(z) + E_\alpha(-z)], \quad z \in \mathbb{C} \ (\operatorname{Re}\alpha > 0).$$

Example 3.3 Evaluate the Laplace transform of the Mittag-Leffler function

$$E_\alpha(z) = \sum_{k=0}^{\infty} \frac{z^k}{\Gamma(\alpha k + 1)}.$$

Example 3.4 ([MatHau08, p. 90]) Prove that

$$\frac{\lambda}{\Gamma(\alpha)} \int_0^x \frac{E_\alpha(\lambda t^\alpha)}{(x-t)^{1-\alpha}} dt = E_\alpha(\lambda x^\alpha) - 1, \quad \operatorname{Re}\alpha > 0.$$

3.9 Exercises

Example 3.5 ([HaMaSa11]) Prove the following integral representations of the Mittag-Leffler function for $x \in \mathbb{R}$

$$E_\alpha(-x^\alpha) = \frac{2}{\pi} \sin\frac{\alpha\pi}{2} \int_0^\infty \frac{t^{\alpha-1}\cos xt}{1 + 2t^\alpha \cos\frac{\alpha\pi}{2} + t^{2\alpha}} dt, \quad \operatorname{Re}\alpha > 0;$$

$$E_\alpha(-x) = \frac{1}{\pi}\sin\alpha\pi \int_0^\infty \frac{t^{\alpha-1}}{1 + 2t^\alpha \cos\alpha\pi + t^{2\alpha}} e^{-tx^{\frac{1}{\alpha}}} dt, \quad \operatorname{Re}\alpha > 0;$$

$$E_\alpha(-x) = 1 - \frac{1}{2\alpha} + \frac{x^{\frac{1}{\alpha}}}{\pi} \int_0^\infty \arctan\left[\frac{t^\alpha + \cos\alpha\pi}{\sin\alpha\pi}\right] e^{-tx^{\frac{1}{\alpha}}} t\, dt, \quad \operatorname{Re}\alpha > 0.$$

Example 3.6 ([GorMai97, Ber-S05b]) Prove the following integro-functional relation for the Mittag-Leffler function

$$E_\alpha(-x) = \frac{2x}{\pi} \int_0^\infty \frac{E_{2\alpha}(-t^2)}{x^2 + t^2} dt, \quad 0 \leq \operatorname{Re}\alpha \leq 1.$$

Example 3.7 ([PengLi10]) Prove the following integral form of the semi-group property of the Mittag-Leffler function

$$\int_0^{t+s} \frac{E_\alpha(a\tau^\alpha)}{(t+s-\tau)^\alpha} d\tau - \int_0^t \frac{E_\alpha(a\tau^\alpha)}{(t+s-\tau)^\alpha} d\tau - \int_0^s \frac{E_\alpha(a\tau^\alpha)}{(t+s-\tau)^\alpha} d\tau$$

$$= \alpha \int_0^t \int_0^s \frac{E_\alpha(ar_1^\alpha) E_\alpha(ar_2^\alpha)}{(t+s-r_1-r_2)^\alpha} dr_1 dr_2.$$

Example 3.8 ([SanGer60]) Determine the Mittag-Leffler star (see Sect. 3.6.2) for the geometric series

$$\sum_{n=0}^\infty z^n.$$

Example 3.9 ([Tua17]) Let $\gamma(\varepsilon, \theta)$, $\frac{\pi\alpha}{2} < \theta < \pi\alpha$, be a contour defined in Sect. 4.7 (see formulas (4.7.1)–(4.7.2));

$$\lambda \in \left\{\mathbb{C} \setminus \{0\} : \arg\lambda < \frac{\pi\alpha}{2}, |\theta \quad \arg\lambda| \geq \theta_0, \theta_0 \subset (0, 0 \quad \frac{\pi\alpha}{2})\right\}.$$

Prove the following inequality

$$\left| E_\alpha(\lambda t^{\frac{1}{\alpha}}) - \frac{1}{\alpha} \exp(\lambda^{\frac{1}{\alpha}} t) \right| < \frac{M_1(\alpha, \lambda)}{t^\alpha}, \quad \forall t \geq t_0,$$

where

$$M_1(\alpha, \lambda) = \frac{\int_{\gamma(1,\theta)} \left| \exp(\zeta^{\frac{1}{\alpha}}) \right| d\zeta}{2\pi \alpha |\lambda| \sin \theta_0}, \quad t_0 = \frac{1}{|\lambda|^{\frac{1}{\alpha}} (1 - \sin \theta_0)^{\frac{1}{\alpha}}}.$$

Chapter 4
The Two-Parametric Mittag-Leffler Function

In this chapter we present the basic properties of the two-parametric Mittag-Leffler function $E_{\alpha,\beta}(z)$ (see (1.0.3)), which is the most straightforward generalization of the classical Mittag-Leffler function $E_\alpha(z)$ (see (3.1.1)). As in the previous chapter, the material can be formally divided into two parts. Starting from the basic definition of the Mittag-Leffler function as a power series, we discover that, for the first parameter α with positive real part and any complex value of the second parameter β, the function $E_{\alpha,\beta}(z)$ is an entire function of the complex variable z. Therefore we discuss in the first part the (analytic) properties of the two-parametric Mittag-Leffler function as an entire function. Namely, we calculate its order and type, present a number of formulas relating the two-parametric Mittag-Leffler function to elementary and special functions as well as recurrence relations and differentiation formulas, introduce some useful integral representations and discuss its asymptotics and the distribution of zeros of the considered function. An extension of the two-parametric Mittag-Leffler function to values of the first parameter α with non-positive real part is given here too.

It is well-known that in current applications the properties of the two-parametric Mittag-Leffler function of a real variable are often used. Thus, we collect in the second part (Sect. 4.9) results of this type. They concern integral representations and integral transforms of the two-parametric Mittag-Leffler function of a real variable, the complete monotonicity property, and relations to the fractional calculus. People working in applications can, at first reading, omit some of the deeper mathematical material (that from Sects. 4.4–4.8, say).

4.1 Series Representation and Properties of Coefficients

The Mittag-Leffler type function (or the *two-parametric Mittag-Leffler function*)

$$E_{\alpha,\beta}(z) = \sum_{k=0}^{\infty} \frac{z^k}{\Gamma(\alpha k + \beta)} \quad (\operatorname{Re}\alpha > 0, \ \beta \in \mathbb{C}) \qquad (4.1.1)$$

generalizes the classical Mittag-Leffler function

$$E_{\alpha}(z) = \sum_{k=0}^{\infty} \frac{z^k}{\Gamma(\alpha k + 1)} \quad (\operatorname{Re}\alpha > 0).$$

We have $E_{\alpha,1}(z) = E_{\alpha}(z)$. For any $\alpha, \beta \in \mathbb{C}$, $\operatorname{Re}\alpha > 0$, the function (4.1.1) is an entire function of order $\rho = 1/(\operatorname{Re}\alpha)$ and type $\sigma = 1$.

Indeed, let us consider the slightly more general function

$$E_{\alpha,\beta}(\sigma^{\alpha} z) = \sum_{k=0}^{\infty} \frac{(\sigma^{\alpha} z)^k}{\Gamma(\alpha k + \beta)}, \qquad (4.1.2)$$

i.e. take the coefficients in the form

$$c_k = \frac{\sigma^{\alpha k}}{\Gamma(\alpha k + \beta)} \quad (k = 0, 1, 2, \ldots), \qquad (4.1.3)$$

where $0 < \operatorname{Re}\alpha < +\infty$, $0 < \sigma < +\infty$ is an arbitrary real constant and β is a complex parameter. By using Stirling's formula (see, e.g., [ErdBat-1, 1.18 (3)]) we have

$$\Gamma(\alpha k + \beta) = \sqrt{2\pi}\, (\alpha k)^{\alpha k + \beta - \frac{1}{2}}\, e^{-\alpha k}\, [1 + o(1)], \quad k \to \infty \qquad (4.1.4)$$

and, consequently, for the sequence $\{c_k\}_0^{\infty}$ we immediately obtain

$$\lim_{k \to \infty} \frac{k \log k}{\log \frac{1}{|c_k|}} = \frac{1}{(\operatorname{Re}\alpha)} = \rho, \quad \lim_{k \to \infty} k |c_k|^{\rho/k} = e\rho\sigma. \qquad (4.1.5)$$

According to a well-known theorem in the theory of entire functions (see, e.g., formulas (B.5) and (B.6) in Appendix B), the function (4.1.2) has order $\rho = 1/\alpha$ and type σ for any β. Therefore, the two-parametric Mittag-Leffler function (4.1.1) has order $\rho = 1/(\operatorname{Re}\alpha)$ and type 1 for any $\operatorname{Re}\alpha > 0$ and any value of the parameter $\beta \in \mathbb{C}$.

4.2 Explicit Formulas. Relations to Elementary and Special Functions

Using definition (4.1.1) we obtain a number of formulas relating the two-parametric Mittag-Leffler function $E_{\alpha,\beta}$ to elementary functions (see, e.g. [HaMaSa11])

$$E_{1,1}(z) = e^z, \quad E_{1,2}(z) = \frac{e^z - 1}{z}, \qquad (4.2.1)$$

$$E_{2,1}(z) = \cosh \sqrt{z}, \quad E_{2,2}(z) = \frac{\sinh \sqrt{z}}{\sqrt{z}}. \qquad (4.2.2)$$

Some extra formulas, for other special values of parameters, are presented as exercises at the end of the chapter.

Here we also mention two recurrence relations for the function (4.1.1).

$$E_{\alpha,\beta}(z) = \frac{1}{\Gamma(\beta)} + z E_{\alpha,\beta+\alpha}(z), \qquad (4.2.3)$$

$$E_{\alpha,\beta}(z) = \beta E_{\alpha,\beta+1}(z) + \alpha z \frac{d}{dz} E_{\alpha,\beta+1}(z). \qquad (4.2.4)$$

Now we present other relations of the two-parametric Mittag-Leffler function to certain special functions introduced by different authors.

The hyperbolic functions of order n, denoted by $h_r(z, n)$, are defined, e.g., in [ErdBat-3, 18.2 (2)]. Their series representations relate these functions to the two-parametric Mittag-Leffler function (see [ErdBat-3, 18.2 (16)]):

$$h_r(z, n) = \sum_{k=0}^{\infty} \frac{z^{nk+r-1}}{(nk+r-1)!} = z^{r-1} E_{n,r}(z^n), \quad r = 1, 2, \ldots \qquad (4.2.5)$$

The trigonometric functions of order n, denoted by $k_r(z, n)$, are defined, e.g., in [ErdBat-3, 18.2 (18)]. With $\lambda = \exp\left\{\frac{\pi i}{n}\right\}$ the functions $k_r(z, n)$ and $h_r(z, n)$ are related by:

$$k_r(z, n) = \lambda^{1-r} h_r(\lambda z, n),$$

from which it follows

$$k_r(z, n) = \sum_{j=0}^{\infty} \frac{(-1)^j z^{nj+r-1}}{(nj+r-1)!} = z^{r-1} E_{n,r}(-z^n), \quad r = 1, 2, \ldots \qquad (4.2.6)$$

The relation to the *complementary error function* is also well known (see [MatHau08, pp. 80–81], [Dzh66, p. 297]):

$$E_{\frac{1}{2},1}(z) = \sum_{k=0}^{\infty} \frac{z^k}{\Gamma\left(\frac{k}{2}+1\right)} = e^{z^2}\operatorname{erfc}(-z), \quad (4.2.7)$$

where erfc is complementary to the error function erf:

$$\operatorname{erfc}(z) := \frac{2}{\sqrt{\pi}} \int_z^{\infty} e^{-u^2}\,du = 1 - \operatorname{erf}(z), \ z \in \mathbb{C}.$$

The Miller–Ross function is defined as follows (see [MilRos93]):

$$E_t(\nu, a) = t^{\nu} \sum_{k=0}^{\infty} \frac{(at)^k}{\Gamma(\nu+k+1)} = t^{\nu} E_{1,\nu+1}(at). \quad (4.2.8)$$

The Rabotnov function is represented as follows (see [RaPaZv69]):

$$R_{\alpha}(\beta, a) = t^{\alpha} \sum_{k=0}^{\infty} \frac{\beta^k t^{k(\alpha+1)}}{\Gamma((1+\alpha)(k+1))} = t^{\alpha} E_{\alpha+1,\alpha+1}(\beta t^{\alpha+1}). \quad (4.2.9)$$

4.3 Differential and Recurrence Relations

A term-by-term differentiation allows us to verify in an easy way that

$$\frac{d^m}{dz^m} E_{\alpha,\beta}(z) = \sum_{k=0}^{\infty} \frac{\Gamma(k+m+1)z^k}{k!\Gamma(\alpha k + \alpha m + \beta)} = m! E_{\alpha,\alpha m+\beta}^{m+1}(z) \ (m \geq 1), \quad (4.3.1)$$

where $E_{\alpha,\beta}^{\gamma}(z)$ denotes the 3-parameter Mittag-Leffler function (also known as the Prabhakar function), which will be introduced later on in Chap. 5.

The following formula [Djr93] expresses the first derivative of the ML function in terms of the difference of two instances of the same function

$$\frac{d}{dz} E_{\alpha,\beta}(z) = \frac{E_{\alpha,\beta-1}(z) + (1-\beta) E_{\alpha,\beta}(z)}{\alpha z}, \quad z \neq 0. \quad (4.3.2)$$

It can be generalized to derivatives of any integer order [GarPop18], and thus provides a summation formula of Djrbashian type: Let $\alpha > 0, \beta \in \mathbb{R}$ and $z \neq 0$. For any $m \in \mathbb{N}$

$$\frac{d^m}{dz^m} E_{\alpha,\beta}(z) = \frac{1}{\alpha^m z^m} \sum_{j=0}^{m} c_j^{(m)} E_{\alpha,\beta-j}(z), \quad (4.3.3)$$

where $c_0^{(0)} = 1$ and the remaining coefficients $c_j^{(m)}$, $j = 0, 1, \ldots, m$, are recursively evaluated as

4.3 Differential and Recurrence Relations

$$c_j^{(m)} = \begin{cases} (1-\beta-\alpha(m-1))c_0^{(m-1)} & j=0, \\ c_{j-1}^{(m-1)} + (1-\beta-\alpha(m-1)+j)c_j^{(m-1)} & 1 \le j \le m-1, \\ 1 & j=m. \end{cases} \quad (4.3.4)$$

An explicit closed form for the coefficients $c_j^{(m)}$ is not available. However, they can be computed [GarPop18] as the solution of a linear system of m equations in the m unknowns $c_j^{(m)}$, $j = 0, 1, \ldots, m-1$, where each equation is

$$\sum_{j=0}^{m-1} \frac{1}{\Gamma(\alpha l + \beta - j)} c_j^{(m)} = -\frac{1}{\Gamma(\alpha l + \beta - m)}, \quad l = 0, \ldots m-1.$$

By exploiting the recurrence relation in [Pra71, Eq. (2.4)] together with (4.3.1), the alternative Prabhakar type summation formula for derivatives of the ML function is obtained [GarPop18]: *Let $\alpha > 0$ and $\beta \in \mathbb{R}$. For any $m \in \mathbb{N}$*

$$\frac{d^m}{dz^m} E_{\alpha,\beta}(z) = \frac{1}{\alpha^m} \sum_{j=0}^{m} c_j^{(m)} E_{\alpha,\alpha m+\beta-j}(z), \quad (4.3.5)$$

where $c_j^{(m)}$, $j = 0, 1, \ldots, m$, are the same coefficients given in (4.3.4).

Formula (4.3.5) is mathematically equivalent to (4.3.3) but must be preferred for its better stability properties when computing derivatives of the ML function with z close to the origin.

Several applications involve the more general function $z^{\beta-1}E_{\alpha,\beta}(z^\alpha)$, for which differentiation formulas can easily be derived. The following differentiation formula is an immediate consequence of the definition of the two-parametric Mittag-Leffler function (4.1.1)

$$\left(\frac{d}{dz}\right)^m [z^{\beta-1}E_{\alpha,\beta}(z^\alpha)] = z^{\beta-m-1}E_{\alpha,\beta-m}(z^\alpha) \quad (m \ge 1). \quad (4.3.6)$$

We consider now some corollaries of formula (4.3.6). Let $\alpha = m/n$, $(m, n = 1, 2, \ldots)$ in (4.3.6). Then

$$\left(\frac{d}{dz}\right)^m [z^{\beta-1}E_{m/n,\beta}(z^{m/n})] \quad (4.3.7)$$

$$= z^{\beta-1}E_{m/n,\beta}(z^{m/n}) + z^{\beta-1}\sum_{k=1}^{n} \frac{z^{-\frac{m}{n}k}}{\Gamma\left(\beta - \frac{m}{n}k\right)} \quad (m, n \ge 1).$$

Since

$$\frac{1}{\Gamma(-s)} = 0 \quad (s = 0, 1, 2, \ldots),$$

it follows from (4.3.7) with $n=1$ and any $\beta = 0, 1, \ldots, m$ that

$$\left(\frac{d}{dz}\right)^m [z^{\beta-1} E_{m,\beta}(z^m)] = z^{\beta-1} E_{m,\beta}(z^m) \quad (m \geq 1). \tag{4.3.8}$$

Substituting $z^{n/m}$ in place of z in (4.3.7) we get

$$\left(\frac{m}{n} z^{1-\frac{n}{m}} \frac{d}{dz}\right)^m \left[z^{(\beta-1)\frac{n}{m}} E_{m,\beta}(z)\right] \tag{4.3.9}$$

$$= z^{(\beta-1)\frac{n}{m}} E_{m,\beta}(z) + z^{(\beta-1)\frac{n}{m}} \sum_{k=1}^{n} \frac{z^{-k}}{\Gamma\left(\beta - \frac{m}{n} k\right)} \quad (m, n = 1, 2, \ldots).$$

Let $m = 1$ in this formula. We then obtain the first-order differential equation for the function $z^{(\beta-1)n} E_{1/n,\beta}(z)$:

$$\frac{1}{n} \frac{d}{dz}\left[z^{(\beta-1)n} E_{1/n,\beta}(z)\right] - z^{n-1} [z^{(\beta-1)n} E_{1/n,\beta}(z)] \tag{4.3.10}$$

$$= z^{\beta n-1} \sum_{k=1}^{n} \frac{z^{-k}}{\Gamma\left(\beta - \frac{k}{n}\right)}, \quad (n = 1, 2, \ldots).$$

Solving this equation we obtain for any $z_0 \neq 0$,

$$E_{1/n,\beta}(z) = z^{(1-\beta)n} e^{z^n} \left\{ z_0^{(\beta-1)n} e^{-z_0^n} E_{1/n,\beta}(z_0) \right.$$

$$\left. + n \int_{z_0}^{z} e^{-\tau^n} \left(\sum_{k=1}^{n} \frac{\tau^{-k}}{\Gamma\left(\beta - \frac{k}{n}\right)} \tau^{\beta n-1}\right) d\tau \right\} \quad (n = 1, 2, \ldots). \tag{4.3.11}$$

Formula (4.3.11) is true with $z_0 = 0$ if $\beta = 1$. In this case we have

$$E_{1/n,1}(z) = e^{z^n} \left\{ 1 + n \int_0^z e^{-\tau^n} \left(\sum_{k=1}^{n-1} \frac{\tau^{k-1}}{\Gamma\left(\frac{k}{n}\right)}\right) d\tau \right\} \quad (n \geq 2). \tag{4.3.12}$$

In particular,

$$E_{1/2,1}(z) = e^{z^2} \left\{ 1 + \frac{2}{\sqrt{\pi}} \int_0^z e^{-\tau^2} d\tau \right\} = e^{z^2} \{1 + \mathrm{erf}\, z\} = e^{z^2} \mathrm{erfc}\,(-z), \tag{4.3.13}$$

and, consequently,

$$E_{1/2,1}(z) \sim 2 e^{z^2}, \quad |\arg z| < \frac{\pi}{4}, \quad |z| \to \infty.$$

4.3 Differential and Recurrence Relations

In the following lemma we collect together a number of known recurrence relations for the two-parametric Mittag-Leffler function.

Lemma 4.1 ([GupDeb07]) *For all $\alpha > 0$, $\beta > 0$ the following relations hold*

$$z^2 E_{\alpha,\beta+2\alpha}(z) = E_{\alpha,\beta}(z) - \frac{1}{\Gamma(\beta)} - \frac{z}{\Gamma(\beta+\alpha)}, \qquad (4.3.14)$$

$$z^3 E_{\alpha,\beta+3\alpha}(z) = E_{\alpha,\beta}(z) - \frac{1}{\Gamma(\beta)} - \frac{z}{\Gamma(\beta+\alpha)} - \frac{z^2}{\Gamma(\beta+2\alpha)}, \qquad (4.3.15)$$

$$z^4 E_{\alpha,\beta+3\alpha}(z) = E_{\alpha,\beta}(z) - \frac{1}{\Gamma(\beta)} - \frac{z}{\Gamma(\beta+\alpha)} - \frac{z^2}{\Gamma(\beta+2\alpha)} - \frac{z^3}{\Gamma(\beta+3\alpha)}. \qquad (4.3.16)$$

4.4 Integral Relations and Asymptotics

Using the well-known discrete orthogonality relation

$$\sum_{h=0}^{m-1} e^{i2\pi hk/m} = \begin{cases} m, & \text{if } k \equiv 0 \pmod{m} \\ 0, & \text{if } k \not\equiv 0 \pmod{m} \end{cases}$$

and definition (4.1.1) of the function $E_{\alpha,\beta}(z)$ we have

$$\sum_{h=0}^{m-1} E_{\alpha,\beta}(z\, e^{i2\pi h/m}) = m\, E_{\alpha/m,\beta}(z^m) \quad (m \geq 1). \qquad (4.4.1)$$

Substituting here $m\alpha$ for α and $z^{1/m}$ for z we obtain

$$E_{\alpha,\beta}(z) = \frac{1}{m} \sum_{h=0}^{m-1} E_{m\alpha,\beta}(z^{1/m} e^{i2\pi h/m}) \quad (m \geq 1). \qquad (4.4.2)$$

Similarly, the formula

$$E_{\alpha,\beta}(z) = \frac{1}{2m+1} \sum_{h=-m}^{m} E_{(2m+1)\alpha,\beta}(z^{1/(2m+1)} e^{i2\pi h/(2m+1)}) \quad (m \geq 0) \qquad (4.4.3)$$

can be obtained via the relation

$$\sum_{h=-m}^{m} e e^{i2\pi hk/(2m+1)} = \begin{cases} 2m+1, & \text{if } k \equiv 0 \pmod{2m+1}, \\ 0, & \text{if } k \not\equiv 0 \pmod{2m+1}. \end{cases}$$

Using (4.1.1) and term-by-term integration we arrive at

$$\int_0^z E_{\alpha,\beta}(\lambda t^\alpha) t^{\beta-1} dt = z^\beta E_{\alpha,\beta+1}(\lambda z^\alpha) \quad (\beta > 0), \qquad (4.4.4)$$

and furthermore, at the more general relation

$$\frac{1}{\Gamma(\alpha)} \int_0^z (z-t)^{\mu-1} E_{\alpha,\beta}(\lambda t^\alpha) t^{\beta-1} dt \qquad (4.4.5)$$
$$= z^{\mu+\beta-1} E_{\alpha,\mu+\beta}(\lambda z^\alpha) \quad (\mu > 0, \ \beta > 0),$$

where the integration is performed along the straight line connecting the points 0 and z.

It follows from formulas (4.4.5), (4.4.2) and (4.4.3) that

$$\frac{1}{\Gamma(\beta)} \int_0^z (z-t)^{\beta-1} e^{\lambda t} dt = z^\beta E_{1,\beta+1}(\lambda z) \quad (\beta > 0), \qquad (4.4.6)$$

$$\frac{1}{\Gamma(\beta)} \int_0^z (z-t)^{\beta-1} \cosh\sqrt{\lambda} t \, dt = z^\beta E_{2,\beta+1}(\lambda z^2) \quad (\beta > 0), \qquad (4.4.7)$$

$$\frac{1}{\Gamma(\beta)} \int_0^z (z-t)^{\beta-1} \frac{\sinh\sqrt{\lambda} t}{\sqrt{\lambda}} dt = z^{\beta+1} E_{2,\beta+2}(\lambda z^2) \quad (\beta > 0). \qquad (4.4.8)$$

Let us prove the relation

$$z^{\beta-1} E_{\alpha,\beta}(z^\alpha) = z^{\beta-1} E_{2\alpha,\beta}(z^{2\alpha}) \qquad (4.4.9)$$
$$+ \frac{1}{\Gamma(\alpha)} \int_0^z (z-t)^{\alpha-1} E_{2\alpha,\beta}(t^{2\alpha}) t^{\beta-1} dt \quad (\beta > 0).$$

First of all, we have by direct evaluations

$$\int_0^z E_{2\alpha,\beta}(t^{2\alpha}) t^{\beta-1} \left\{1 + \frac{(z-t)^\alpha}{\Gamma(\alpha+1)}\right\} dt$$

$$= \sum_{k=0}^\infty \frac{1}{\Gamma(2k\alpha+\beta)} \int_0^z t^{2k\alpha+\beta-1} \left\{1 + \frac{(z-t)^\alpha}{\Gamma(\alpha+1)}\right\} dt$$

$$= z^\beta \sum_{k=0}^\infty \frac{z^{2k\alpha}}{\Gamma(2k\alpha+\beta+1)} + z^\beta \sum_{k=0}^\infty \frac{z^{(2k+1)\alpha}}{\Gamma((2k+1)\alpha+\beta+1)}$$

$$= z^\beta \sum_{k=0}^\infty \frac{z^{k\alpha}}{\Gamma(k\alpha+\beta+1)} = z^\beta E_{\alpha,\beta+1}(z^\alpha).$$

4.4 Integral Relations and Asymptotics

This relation and formula (4.4.4) imply

$$\int_0^z E_{2\alpha,\beta}(t^{2\alpha}) t^{\beta-1} \left\{1 + \frac{(z-t)^\alpha}{\Gamma(\alpha+1)}\right\} dt$$

$$= \int_0^z E_{\alpha,\beta}(t^\alpha) t^{\beta-1} dt \quad (\beta > 0).$$

Differentiation of this formula with respect to z gives us formula (4.4.9).

Let us prove the formula

$$\int_0^l x^{\beta-1} E_{\alpha,\beta}(\lambda x^\alpha)(l-x)^{\nu-1} E_{\alpha,\nu}(\lambda^*(l-x)^\alpha) dx \qquad (4.4.10)$$

$$= \frac{\lambda E_{\alpha,\beta+\nu}(l^\alpha \lambda) - \lambda^* E_{\alpha,\beta+\nu}(l^\alpha \lambda^*)}{\lambda - \lambda^*} l^{\beta+\nu-1} \quad (\beta > 0, \ \nu > 0),$$

where λ and λ^* ($\lambda \neq \lambda^*$) are any complex parameters.

Indeed, using (4.1.1) for any λ and λ^* ($\lambda \neq \lambda^*$) and $\beta > 0$, $\nu > 0$ we find

$$\int_0^l x^{\beta-1} E_{\alpha,\beta}(\lambda x^\alpha)(l-x)^{\nu-1} E_{\alpha,\nu}(\lambda^*(l-x)^\alpha) dx$$

$$= \sum_{n=0}^\infty \sum_{m=0}^\infty \frac{\lambda^n (\lambda^*)^m}{\Gamma(n\alpha+\beta)\Gamma(m\alpha+\nu)} \int_0^l x^{n\alpha+\beta-1}(l-x)^{m\alpha+\nu-1} dx$$

$$= \sum_{n=0}^\infty \sum_{m=0}^\infty \frac{\lambda^n (\lambda^*)^m l^{(n+m)\alpha+\beta+\nu-1}}{\Gamma((m+n)\alpha+\beta+\nu)} = l^{\beta+\nu-1} \sum_{n=0}^\infty \sum_{k=n}^\infty \frac{\lambda^n (\lambda^*)^{k-n} l^{k\alpha}}{\Gamma(k\alpha+\beta+\nu)}$$

$$= l^{\beta+\nu-1} \sum_{k=0}^\infty \frac{(\lambda^*)^k l^{k\alpha}}{\Gamma(k\alpha+\beta+\nu)} \sum_{n=0}^k \left(\frac{\lambda}{\lambda^*}\right)^n = \frac{l^{\beta+\nu-1}}{\lambda-\lambda^*} \sum_{k=0}^\infty \frac{l^{k\alpha}(\lambda^{k+1} - (\lambda^*)^{k+1})}{\Gamma(k\alpha+\beta+\nu)}.$$

Using formula (4.1.1) once more, we arrive at (4.4.10).

Finally, we obtain two integral relations:

$$\int_0^{+\infty} e^{-t} E_{\alpha,\beta}(zt^\alpha) t^{\beta-1} dt = \frac{1}{1-z} \quad (\beta > 0, \ |z| < 1), \qquad (4.4.11)$$

$$\int_0^{+\infty} e^{-t^2/(4x)} E_{\alpha,\beta}(t^\alpha) t^{\beta-1} dt = \sqrt{\pi} x^{\beta/2} E_{2\alpha,\frac{1+\beta}{2}}(x^{2\alpha}) \quad (\beta > 0, \ x > 0).$$

$$(4.4.12)$$

First of all, since the Mittag-Leffler type function (4.1.1) is an entire function of order $\rho = 1/(\operatorname{Re}\alpha)$ and type 1 (see Sect. 4.1), we have for any $\sigma > 1$ the estimate:

$$|E_{\alpha,\beta}(z)| \leq C\exp\{\sigma|z|^\rho\}, \ |z| \geq |z_\sigma|.$$

Consequently, the integrals in the formulae (4.4.11) and (4.4.12) are convergent.

It is easy to check that term-by-term integration of the expansion

$$e^{-t} E_{\alpha,\beta}(zt^{\alpha}) t^{\beta-1} = \sum_{k=0}^{\infty} \frac{e^{-t} t^{k\alpha+\beta-1}}{\Gamma(k\alpha+\beta)} z^k$$

can be performed with respect to t along the interval $(0, +\infty)$ if $|z| < 1$. As a result, we arrive at formula (4.4.11).

Similarly, we have using term-by-term integration of the following expansion with respect to t:

$$e^{-t^2/(4x)} E_{\alpha,\beta}(t^{\alpha}) t^{\beta-1} = \sum_{k=0}^{\infty} \frac{t^{k\alpha+\beta-1}}{\Gamma(k\alpha+\beta)} e^{-t^2/(4x)} \quad (\beta > 0)$$

along the same interval $(0, +\infty)$ (integration is performable with any fixed $x > 0$!),

$$\int_0^{+\infty} e^{-t^2/(4x)} E_{\alpha,\beta}(t^{\alpha}) t^{\beta-1} \, dt = \sum_{k=0}^{\infty} \frac{\Gamma\left(\frac{k\alpha+\beta}{2}\right)}{2\, \Gamma(k\alpha+\beta)} (2\sqrt{x})^{k\alpha+\beta}.$$

Rewriting the right-hand side of the last relation by using the Lagrange formula

$$\Gamma(s)\Gamma(s+1/2) = \sqrt{\pi}\, 2^{1-2s}\, \Gamma(2s)$$

we obtain formula (4.4.12).

The Mittag-Leffler type function $E_{\alpha,\alpha}(z)$ plays an essential role in the linear Abel integral equation of the second kind (see, e.g., [GorVes91, GorMai97]).

Theorem 4.2 *Let a function $f(t)$ be in the function space $L_1(0, l)$. Let $\alpha > 0$ and λ be an arbitrary complex parameter. Then the integral equation*

$$u(t) = f(t) + \frac{\lambda}{\Gamma(\alpha)} \int_0^t (t-\tau)^{\alpha-1} u(\tau) \, d\tau, \quad t \in (0, l), \tag{4.4.13}$$

has a unique solution

$$u(t) = f(t) + \lambda \int_0^t (t-\tau)^{\alpha-1} E_{\alpha,\alpha}\left[\lambda (t-\tau)^{\alpha}\right] f(\tau) \, d\tau, \quad t \in (0, l), \tag{4.4.14}$$

in the space $L_1(0, l)$.

This result was discovered in the pioneering work of Hille and Tamarkin [HilTam30]. We give the proof of this theorem in Chap. 8.

We consider now a simple application of Theorem 4.2.

4.4 Integral Relations and Asymptotics

Let $z = t > 0$ and $t = \tau$ be the integration variable in (4.4.9). Then this representation can be considered as an integral equation of type (4.4.13) with

$$f(t) = t^{\beta-1} E_{\alpha,\beta}(t^\alpha), \quad \lambda = 1;$$

and solution

$$u(t) = t^{\beta-1} E_{2\alpha,\beta}(x^{2\alpha}).$$

Using formula (4.4.14) we obtain for the solution $u(t)$ the representation

$$u(t) = t^{\beta-1} E_{\alpha,\beta}(t^\alpha) \qquad (4.4.15)$$
$$- \int_0^t (t-\tau)^{\alpha-1} E_{\alpha,\beta}(\tau^\alpha) \tau^{\beta-1} E_{\alpha,\alpha}\left[-(t-\tau)^\alpha\right] d\tau.$$

We also mention asymptotic results for the Mittag-Leffler function $E_{\alpha,\beta}(z)$ which are essentially a refinement of results of Dzhrbashian (see [Dzh66]).

Theorem 4.3 *For all $0 < \alpha < 2$, $\beta \in \mathbb{C}$, $m \in \mathbb{N}$, the following asymptotic formulas hold:*

If $|\arg z| < \min\{\pi, \pi\alpha\}$, then

$$E_{\alpha,\beta}(z) = \frac{1}{\alpha} z^{(1-\beta)/\alpha} e^{z^{1/\alpha}} - \sum_{k=1}^m \frac{z^{-k}}{\Gamma(\beta-k\alpha)} + O\left(|z|^{-m-1}\right), \quad |z| \to \infty. \quad (4.4.16)$$

If $0 < \alpha < 1$, $\pi\alpha < |\arg z| < \pi$, then

$$E_{\alpha,\beta}(z) = -\sum_{k=1}^m \frac{z^{-k}}{\Gamma(\beta-k\alpha)} + O\left(|z|^{-m-1}\right), \quad |z| \to \infty. \quad (4.4.17)$$

Theorem 4.4 *For all $\alpha \geq 2$, $\beta \in \mathbb{C}$, $m \in \mathbb{N}$, the following asymptotic formula holds:*

$$E_{\alpha,\beta}(z) = \frac{1}{\alpha} \sum_{|\arg z + 2\pi n| < \frac{3\pi\alpha}{4}} \left(z^{1/\alpha} e^{2\pi i n/\alpha}\right)^{1-\beta} e^{z^{1/\alpha} e^{2\pi i n/\alpha}}$$
$$- \sum_{k=1}^m \frac{z^{-k}}{\Gamma(\beta-k\alpha)} + O\left(|z|^{-m-1}\right), \quad |z| \to \infty. \quad (4.4.18)$$

A more exact description of the remainders in formulas (4.4.16)–(4.4.18) is presented, e.g., in [PopSed11].

4.5 The Two-Parametric Mittag-Leffler Function as an Entire Function

One more immediate consequence of the results in [Sed94, Sed00, Sed04, Sed07] (see also [Pop02, Pop06, PopSed03]) is the fact that $E_{\alpha,\beta}$ with $\alpha > 0$, $\beta \in \mathbb{R}$, is a function of completely regular growth in the sense of Levin–Pfluger.

For an entire function $F(z)$ of finite order ρ this means that the following limit exists

$$h_F(\theta) = \lim_{r \to \infty} \frac{\log |F(re^{i\theta})|}{r^\rho}, \qquad (4.5.1)$$

when $r \to \infty$ takes all positive values avoiding an exceptional set C^0 of relatively small measure common for all rays $\arg z = \theta$ (see the formal definition in Sect. B.4 of Appendix B). Sometimes the limit in (4.5.1) is called the weak limit and is denoted by

$$\lim_{r \to \infty}{}^* .$$

The above property is known to be equivalent to the regularity of the distribution of zeros of an entire function (see [Lev56, Ron92]). The corresponding result is presented in Sect. B.4 of Appendix B.

To prove that $E_{\alpha,\beta}$ possesses this property let us consider the Weierstrass product representation (see Theorem B.1 in Appendix B). For the two-parametric Mittag-Leffler function this representation has the form

$$E_{\alpha,\beta} = \prod_{n=1}^{\infty} \left(1 - \frac{z}{z_n}\right) e^{\frac{z}{z_n} + \frac{z^2}{2z_n^2} + \ldots + \frac{z^{[1/\alpha]}}{[1/\alpha]z_n^{[1/\alpha]}}} \qquad (4.5.2)$$

$$\times \prod_{n=1}^{\infty} \left(1 - \frac{z}{z_{-n}}\right) e^{\frac{z}{z_{-n}} + \frac{z^2}{2z_{-n}^2} + \ldots + \frac{z^{[1/\alpha]}}{[1/\alpha]z_{-n}^{[1/\alpha]}}} \times e^{q_0 + q_1 z + \ldots + q_{[1/\alpha]} z^{[1/\alpha]}},$$

where z_n, z_{-n} are the zeros of $E_{\alpha,\beta}$ in the neighborhoods of the rays $\arg z = \frac{\pi\alpha}{2}$, $\arg z = -\frac{\pi\alpha}{2}$, respectively.

This formula can be written in terms of simple Weierstrass factors:

$$G(\zeta, p) = (1 - \zeta) e^{\zeta + \frac{\zeta^2}{2} + \ldots + \frac{\zeta^p}{p}}, \quad p \in \mathbb{N}. \qquad (4.5.3)$$

Then (4.5.2) becomes

$$E_{\alpha,\beta}(z) = \prod_{n=1}^{\infty} G\left(\frac{z}{z_n}, [1/\alpha]\right) \times \prod_{n=1}^{\infty} G\left(\frac{z}{z_{-n}}, [1/\alpha]\right) \times e^{q_0 + q_1 z + \ldots + q_{[1/\alpha]} z^{[1/\alpha]}}.$$

(4.5.4)

Formula (4.5.4) (or (4.5.2)) is not too useful for obtaining asymptotic results. By using the technique developed for the entire functions of completely regular growth

4.5 The Two-Parametric Mittag-Leffler Function as an Entire Function

(see, e.g., [GoLeOs91, Lev56, Ron92]), we can compare the asymptotic behavior of $E_{\alpha,\beta}$ with that of more simple functions. In order to formulate it in a more rigorous form let us introduce two pairs of sequences:

$$w_n = |z_n|e^{i\frac{\pi\alpha}{2}}, \ n = 1, 2, \ldots, \ w_{-n} = |z_{-n}|e^{-i\frac{\pi\alpha}{2}}, \ n = 1, 2, \ldots; \quad (4.5.5)$$

$$\omega_n = (2\pi n)^\alpha e^{i\frac{\pi\alpha}{2}}, \ n = 1, 2, \ldots, \ \omega_{-n} = (2\pi n)^\alpha e^{-i\frac{\pi\alpha}{2}}, \ n = 1, 2, \ldots; \quad (4.5.6)$$

and construct the corresponding Weierstrass products for these sequences

$$W(z) = \prod_{n=1}^{\infty} G\left(\frac{z}{w_n}, [1/\alpha]\right) \times \prod_{n=1}^{\infty} G\left(\frac{z}{w_{-n}}, [1/\alpha]\right) \times e^{q_0+q_1 z+\ldots+q_{[1/\alpha]}z^{[1/\alpha]}},$$
(4.5.7)

$$\Omega(z) = \prod_{n=1}^{\infty} G\left(\frac{z}{\omega_n}, [1/\alpha]\right) \times \prod_{n=1}^{\infty} G\left(\frac{z}{\omega_{-n}}, [1/\alpha]\right) \times e^{q_0+q_1 z+\ldots+q_{[1/\alpha]}z^{[1/\alpha]}}.$$
(4.5.8)

Lemma 4.5 *The functions $E_{\alpha,\beta}(z)$, $W(z)$, $\Omega(z)$ are functions of completely regular growth with the same characteristics, i.e. with the same angular density (see formula (B.14a) in Appendix B) and in the case of integer $\rho = [1/\alpha]$ also with the same coefficients of angular symmetry (see formula (B.14b) in Appendix B).*

To prove this result it suffices to examine the corresponding properties of the sequences $(z_{\pm n})$, $(w_{\pm n})$, $(\omega_{\pm n})$.

Lemma 4.6 *The functions $E_{\alpha,\beta}(z)$, $W(z)$, $\Omega(z)$ have the same asymptotic behavior, i.e. the following weak limits exist:*[1]

$$\lim_{r\to\infty}{}^* \frac{|E_{\alpha,\beta}(re^{i\theta})|}{|W(re^{i\theta})|} = \lim_{r\to\infty}{}^* \frac{|W(re^{i\theta})|}{|\Omega(re^{i\theta})|} = 1. \quad (4.5.9)$$

◁ Comparing representations (4.5.4) and (4.5.7) we arrive at the conclusion that the quotient under this limit is in fact a *Blaschke type product* for two rays. Such products were studied in [Gov94]. Repeating calculations of [Gov94] we get the existence of the first limit (4.5.9).

The second result of the lemma also follows from quite general considerations. Comparing sequences $(w_{\pm n})$ and $(\omega_{\pm n})$ we can see that there exists a *proximate order* (cf. [Lev56, GoLeOs91]) (i.e. a non-negative, non-decreasing function such that $\lim_{r\to\infty} \rho(r) = \rho$, $\lim_{r\to\infty} \rho'(r)r \log r = 0$) for which

$$|w_{\pm n}|^{\rho(r)} = |\omega_{\pm n}|^\rho. \quad (4.5.10)$$

---

[1] The meaning of the weak limit $\lim_{r\to\infty}^*$ is the same as in the definition of entire functions of completely regular growth, see (4.5.1).

A possible way to see this is to use an interpolating C^1-function $\rho(r)$ obeying the interpolation conditions

$$\rho(n) = \rho \frac{\log |\omega_{\pm n}|}{\log |w_{\pm n}|}, \quad n = 1, 2, \ldots \quad (4.5.11)$$

Since we need to obtain only an asymptotic result, the interpolation problem can be simplified and even solved in closed form. After constructing the proximate order it remains to apply the Valiron–Levin theory of entire functions of a proximate order (see, e.g., [Lev56]). This completes the proof. ▷

We are now in a position to discuss the global behavior of the Mittag-Leffler type function $E_{\alpha,\beta}(z)$. This behavior is described in terms of the indicator function for $E_{\alpha,\beta}(z)$ in the case $0 < \alpha < 2$ (we assume additionally that $\beta \neq 1, 0, -1, \ldots$, when $\alpha = 1$, see (B.7)):

$$h_{E_{\alpha,\beta}}(\theta) = \begin{cases} \cos \frac{\theta}{\alpha}, & 0 \leq |\theta| < \frac{\pi\alpha}{2}, \\ 0, & \frac{\pi\alpha}{2} \leq |\theta| \leq \pi. \end{cases} \quad (4.5.12)$$

4.6 Distribution of Zeros

In this section we mainly follow the article [Sed94] (see also [PopSed11]).

First of all we recall some relations of the Mittag-Leffler function to elementary functions for special values of the parameters α, β. In these cases the zeros of the Mittag-Leffler function can be found explicitly.

$$E_{1,1}(z) = \exp\{z\}, \quad E_{1,-m}(z) = z^{m+1}\exp\{z\}, \quad m \in \mathbb{Z}_+; \quad (4.6.1)$$

$$E_{2,1}(z) = \cosh\sqrt{z}, \quad E_{2,2}(z) = \frac{\sinh\sqrt{z}}{\sqrt{z}}, \quad E_{2,3}(z) = \frac{\cosh\sqrt{z}-1}{\sqrt{z}}, \quad (4.6.2)$$

$$E_{2,-2m}(z) = z^{m+1/2}\sinh\sqrt{z}, \quad m \in \mathbb{Z}_+, \quad E_{2,-(2m-1)}(z) = z^m\cosh\sqrt{z}, \quad m \in \mathbb{N}. \quad (4.6.3)$$

Thus the function $E_{1,1}(z)$ has no zero, while the function $E_{1,-m}(z)$ has its only zero at $z = 0$ of order $m + 1$. Below we can see that for all other values of parameters α, β (including those described in (4.6.2), (4.6.3)) the Mittag-Leffler function $E_{\alpha,\beta}(z)$ has an infinite number of zeros. We should also mention that the function $E_{2,3}(z)$ has double zeros at the points $z_n = -(2\pi n)^2$, $n \in \mathbb{N}$. This is the only case when the Mittag-Leffler function $E_{\alpha,\beta}(z)$ has an infinite number of multiple zeros.

In order to describe the distribution of zeros of the Mittag-Leffler function $E_{\alpha,\beta}(z)$ we introduce the following constants:

4.6 Distribution of Zeros

$$c_\beta = \frac{\alpha}{\Gamma(\beta - \alpha)}, \; d_\beta = \frac{\alpha}{\Gamma(\beta - 2\alpha)}, \; \tau_\beta = 1 + \frac{1-\beta}{\alpha}, \; \beta \neq \alpha - l, \; l \in \mathbb{Z}_+, \quad (4.6.4)$$

$$c_\beta = \frac{\alpha}{\Gamma(\beta - 2\alpha)}, \; d_\beta = \frac{\alpha}{\Gamma(\beta - 3\alpha)}, \; \tau_\beta = 2 + \frac{1-\beta}{\alpha}, \; \beta = \alpha - l, \; l \in \mathbb{Z}_+, \; \alpha \notin \mathbb{N}.$$
$$(4.6.5)$$

By construction $c_\beta \neq 0$ for all considered α and β. The values of parameters α and β, which are not mentioned in (4.6.4) and (4.6.5), are called exceptional values.

Theorem 4.7 *Let the values of parameters α, β satisfy one of the following relations:*
(1) $0 < \alpha < 2$, $\beta \in \mathbb{C}$, and $\beta \neq 0, -1, -2, \ldots$ when $\alpha = 1$;
(2) $\alpha = 2$, $\operatorname{Re} \beta > 3$.
Then all zeros z_n of the Mittag-Leffler function $E_{\alpha,\beta}(z)$ with sufficiently large modulus are simple and the asymptotic relation holds for $n \to \pm\infty$

$$(z_n)^{1/\alpha} = 2\pi i n - \tau_\beta \alpha \log 2\pi i n + \frac{d_\beta/c_\beta}{(2\pi i n)^\alpha} + (\tau_\beta \alpha)^2 \frac{\log 2\pi i n}{2\pi i n} - (\tau_\beta \alpha)^2 \frac{\log c_\beta}{2\pi i n} + \alpha_n,$$
$$(4.6.6)$$

where for $\alpha < 2$

$$\alpha_n = O\left(\frac{\log |n|}{|n|^{1+\alpha}}\right) + O\left(\frac{1}{|n|^{2\alpha}}\right) + O\left(\frac{\log^2 |n|}{|n|^2}\right), \quad (4.6.7)$$

but for $\alpha = 2$

$$\alpha_n = \frac{e^{\pm i\pi\beta}}{c_\beta^2 (2\pi n)^{-4\tau_\beta}} + O\left(\frac{1}{|n|^{-8\tau_\beta}}\right) + O\left(\frac{\log |n|}{|n|^{1-4\tau_\beta}}\right) + O\left(\frac{\log^2 |n|}{|n|^2}\right). \quad (4.6.8)$$

The proof of the theorem is based on the integral representation of the Mittag-Leffler function $E_{\alpha,\beta}(z)$ and on the following lemma.

Lemma 4.8 *Let $A \in \mathbb{C}$ and $\delta \in (0, \pi/2)$ be fixed numbers and let the sets Z, W be defined for all $R > 0$ by the relations*

$$Z = Z_{\delta,R} = \{z : |\arg z| < \pi - \delta, \; |z| > R\},$$
$$W = W_{\delta,R} = \{z : |\arg z| < \pi - 2\delta, \; |z| > 2R\}.$$

Then for sufficiently large $R > 0$ the equation

$$z - A \log z = w, \; w \in W,$$

has a unique zero $z \in Z$. This zero is simple and we have the asymptotics

$$z = w + A \log w + A^2 \frac{\log w}{w} + O\left(\frac{\log^2 w}{w^2}\right), \; w \to \infty. \quad (4.6.9)$$

◁ Proof of Theorem 4.7 ([PopSed11, pp. 35–37]). Let us put in the formulas (4.4.16), (4.4.17) $m = 1$ if β is defined as in (4.6.4), and $m = 2$ if β is defined as in (4.6.5). Then $c_\beta \neq 0$. Hence one can conclude from (4.4.16), (4.4.17) that for any $\varepsilon > 0$ all zeros z_n of the Mittag-Leffler function $E_{\alpha,\beta}(z)$ with sufficiently large modulus are situated in the angle $|\arg z| < \frac{\pi\alpha}{2} + \varepsilon$. In this angle $E_{\alpha,\beta}(z)$ satisfies the asymptotic relation

$$\alpha z^m E_{\alpha,\beta}(z) = z^{\tau_\beta} \exp\{z^{1/\alpha}\} - c_\beta - \frac{d_\beta}{z} + O\left(\frac{1}{z^2}\right), \quad |\arg z| < \frac{\pi\alpha}{2} + \varepsilon. \quad (4.6.10)$$

Therefore, there exists a sufficiently large r_0 such that all zeros z_n, $|z_n| > r_0$, can be found from the equation

$$\exp\{z^{1/\alpha} + \tau_\beta \log z\} = c_\beta + \frac{d_\beta}{z} + O\left(\frac{1}{z^2}\right). \quad (4.6.11)$$

Let us put

$$w = z^{1/\alpha} + \alpha\tau_\beta \log z^{1/\alpha}. \quad (4.6.12)$$

Then, by Lemma 4.8 we obtain the solution to (4.6.12) with respect to $z^{1/\alpha}$ in the form

$$z^{1/\alpha} = w + O(\log w).$$

Hence

$$\frac{1}{z} = \frac{1}{w^\alpha}\left(1 + O\left(\frac{\log w}{w}\right)\right) = \frac{1}{w^\alpha} + O\left(\frac{\log w}{w^{1+\alpha}}\right).$$

Substituting this relation into (4.6.11) we obtain the equation

$$\exp\{w\} = c_\beta + \frac{d_\beta}{w^\alpha} + O\left(\frac{\log w}{w^{1+\alpha}}\right) + O\left(\frac{1}{w^{2\alpha}}\right). \quad (4.6.13)$$

In particular,

$$\exp\{w\} = c_\beta + o(1), \quad w \to \infty. \quad (4.6.14)$$

Since all zeros of the function $\exp\{w\} - c_\beta$ are simple and are given by the formula $2\pi i n + \log c_\beta$, $n \in \mathbb{Z}$, by Rouché's theorem all zeros w_n of Eq. (4.6.14) with sufficiently large modulus are simple too and can be described by the formula

$$w_n = 2\pi i n + \log c_\beta + \epsilon_n, \quad \epsilon_n \to 0, \quad n \to \pm\infty. \quad (4.6.15)$$

Thus

$$\frac{1}{w_n} = \frac{1}{2\pi i n} + \left(1 + O\left(\frac{1}{n}\right)\right), \quad \log w_n = \log |n| + O(1), \quad n \to \pm\infty. \quad (4.6.16)$$

4.6 Distribution of Zeros

Therefore, if $w = w_n$ in (4.6.14), then

$$c_\beta \exp\{\epsilon_n\} = c_\beta + \frac{d_\beta}{(2\pi i n)^\alpha} + O\left(\frac{\log|n|}{|n|^{1+\alpha}}\right) + O\left(\frac{1}{|n|^{2\alpha}}\right).$$

Since the left-hand side of this relation is equal to $c_\beta + c_\beta \epsilon_n + O(\epsilon_n^2)$, we have $\epsilon_n = O\left(\frac{1}{|n|^\alpha}\right)$ and hence

$$\epsilon_n = \frac{d_\beta/c_\beta}{(2\pi i n)^\alpha} + O\left(\frac{\log|n|}{|n|^{1+\alpha}}\right) + O\left(\frac{1}{|n|^{2\alpha}}\right). \quad (4.6.17)$$

Substituting this relation into (4.6.15) we get

$$w_n = 2\pi i n + \log c_\beta + \frac{d_\beta/c_\beta}{(2\pi i n)^\alpha} + O\left(\frac{\log|n|}{|n|^{1+\alpha}}\right) + O\left(\frac{1}{|n|^{2\alpha}}\right), \quad n \to \pm\infty, \quad (4.6.18)$$

$$\log w_n = \log 2\pi i n + \frac{\log c_\beta}{2\pi i n} + O\left(\frac{1}{|n|^{1+\alpha}}\right), \quad n \to \pm\infty. \quad (4.6.19)$$

Now we note that the pre-images z_n of w_n satisfy $|\arg z_n| < \frac{\pi\alpha}{2} + \varepsilon$ and $\frac{1}{\alpha}\left(\frac{\pi\alpha}{2} + \varepsilon\right) < \pi$. Thus the conditions of Lemma 4.8 are satisfied and we obtain from this lemma and from (4.6.12) and (4.6.18)

$$z_n^{1/\alpha} = w_n - \alpha\tau_\beta \log w_n + (\alpha\tau_\beta)^2 \frac{\log w_n}{w_n} + O\left(\frac{\log^2 w_n}{w_n^2}\right). \quad (4.6.20)$$

Then the proof of the theorem in case (1) follows from (4.6.18) and (4.6.19).

In case (2) one can use a similar argument (see [PopSed11, pp. 36–37]) based on the following asymptotic formula

$$E_{2,\beta}(z) = \frac{1}{2}z^{(1-\beta)/2}\left(e^{\sqrt{z}} + e^{\mp i\pi(1-\beta)}e^{-\sqrt{z}}\right) - \sum_{k=1}^{m} \frac{1}{z^k \Gamma(\beta - 2k)} + O\left(\frac{1}{z^{m+1}}\right), \quad (4.6.21)$$

which is valid for $|z| \to \infty$ in the angles $0 \leq \arg z \leq \pi$ and $-\pi \leq \arg z \leq 0$, respectively. ▷

The most attractive result (see, e.g., [PopSed11, p. 37]) concerning the distribution of zeros of the Mittag-Leffler function is the following:

Theorem 4.9 *Let* $\operatorname{Re}\beta < 3$, $\beta \neq 2 - l, l \in \mathbb{Z}_+$. *Then all zeros* z_n *of the Mittag-Leffler function* $E_{2,\beta}$ *with sufficiently large modulus are simple and the following asymptotic formula holds* ($n \to \pm\infty$)

$$\sqrt{z_n} = \pi i \, (n - 1 + \beta/2) + (-1)^n \frac{c_\beta e^{-i\pi\beta/2}}{2(i\pi n)^{2\tau_\beta}} + O\left(\frac{1}{n^{6\mathrm{Re}\,\tau_\beta}}\right) + O\left(\frac{1}{n^{1+2\mathrm{Re}\,\tau_\beta}}\right),$$
(4.6.22)

where the single-valued branch of the function $\sqrt{z}$ is chosen by the relation $0 \leq \arg z < 2\pi$.

If β is real then all zeros z_n of $E_{2,\beta}$ with sufficiently large modulus are real.

Now we present a result on the distribution of zeros of the function $E_{2,3+i\gamma}(z)$, $\gamma \neq 0$, $\gamma \in \mathbb{R}$. For the proof of the following theorem we refer to [PopSed11, pp. 39–44].

Theorem 4.10 *(1) The set of multiple zeros of the function $E_{2,3+i\gamma}(z)$ is at most finite.*
(2) The sequence of all zeros z_n consists of two subsequences $z_n^+, n > n^+$, and $z_n^-, n < -n^-$, for which the following asymptotic relation holds

$$\sqrt{z_n^\pm} = 2\pi i n - \frac{\pi\gamma}{2} + \delta_n + O\left(\frac{1}{n}\right), \quad n \to \pm\infty,$$
(4.6.23)

where the sequence δ_n is defined by

$$\delta_n = \delta_n(\gamma) = \log\left(\eta e^{i\gamma \log 2\pi n} + \sqrt{\left(\eta e^{i\gamma \log 2\pi n}\right)^2 - 1}\right), \quad \eta = \frac{1}{\Gamma(1+i\gamma)},$$
(4.6.24)

where the principal branch is used for the values of the logarithmic function.
(3) The sequence $\zeta_n = \sqrt{z_n}$ asymptotically belongs to the semi-strips

$$\log \rho_1 < \left|\mathrm{Re}\,\zeta + \frac{\pi\gamma}{2}\right| < \log \rho_2, \quad \mathrm{Im}\,\zeta > 0,$$
(4.6.25)

where

$$\rho_1 = |\eta| + \sqrt{|\eta|^2 - 1}, \quad \rho_2 = |\eta| + \sqrt{|\eta|^2 + 1}.$$

(4) Every point of the interval $[0, \pi]$ is a limit point of the sequence $\mathrm{Im}\,\delta_n$, and every point of the intervals $[\log \rho_1, \log \rho_2]$, $[\log 1/\rho_2, \log 1/\rho_1]$ is a limit point of the sequence $\mathrm{Re}\,\delta_n$.
(5) There exist $R = R(\gamma) > 0$ and $N = N(\gamma) \in \mathbb{N}$ such that there is no point of the sequence $\zeta_k = \sqrt{z_k}$ in the disks

$$|\zeta - \pi i n| < R, \quad n > N.$$
(4.6.26)

It remains to consider the distribution of zeros of the function $E_{\alpha,\beta}$ in the case $\alpha > 2$.

Theorem 4.11 *([PopSed11, p. 45]) Let $\alpha > 2$. Then all zeros z_n of the Mittag-Leffler function $E_{\alpha,\beta}$ with sufficiently large modulus are simple and the following asymptotic formula holds:*

4.6 Distribution of Zeros

$$z_n = \left(\frac{\pi}{\sin \pi/\alpha}\left(n - \frac{1}{2} - \frac{\beta-1}{\alpha}\right) + \alpha_n\right)^\alpha, \tag{4.6.27}$$

where the sequence α_n is defined as described below:

(1) If the pair (α, β) is not mentioned in (4.6.4), (4.6.5), then

$$\alpha_n = O\left(e^{-\pi n(\cos\frac{\pi}{\alpha} - \cos\frac{3\pi}{\alpha})/\sin\frac{\pi}{\alpha}}\right).$$

(2) If $2 < \alpha < 4$, then

$$\alpha_n = O\left(n^{-\alpha \mathrm{Re}\, \tau_\beta} e^{-\pi n \cot\frac{\pi}{\alpha}}\right).$$

(3) If $\alpha \geq 4$, then

$$\alpha_n = e^{-\pi n \cot\frac{\pi}{\alpha}}\left(O\left(e^{\pi n \cos\frac{3\pi}{\alpha}/\sin\frac{\pi}{\alpha}}\right) + O\left(n^{-\alpha \mathrm{Re}\, \tau_\beta}\right)\right).$$

If β is real then all zeros z_n with sufficiently large modulus are real too.

In a series of articles by different authors the following question, which is very important for applications, was discussed: "Are all zeros of the Mittag-Leffler function $E_{\alpha,\beta}$ with $\alpha > 2$ simple and negative?" This question goes back to an article by Wiman [Wim05b]. Several attempts to answer this question have shown its non-triviality. In [OstPer97] this question was reformulated as the following problem: "For any $\alpha \geq 2$, find a set W_α consisting of those values of the positive parameter β such that all zeros of the Mittag-Leffler function $E_{\alpha,\beta}$ are simple and negative."

Let us give some answers to the above question, following [PopSed11].

Theorem 4.12 *For any $\alpha > 2$, $\beta \in (0, 2\alpha - 1]$, all complex zeros $(z_n(\alpha, \beta))_{n \in \mathbb{N}}$ of the Mittag-Leffler function $E_{\alpha,\beta}$ are simple and negative and satisfy the following inequalities*

$$-\xi_1^\alpha(\alpha, \beta) < z_1(\alpha, \beta) < -\frac{\Gamma(\alpha+\beta)}{\Gamma(\beta)}, \tag{4.6.28}$$

$$-\xi_n^\alpha(\alpha, \beta) < z_n(\alpha, \beta) < -\xi_{n-1}^\alpha(\alpha, \beta), \quad n \geq 2, \tag{4.6.29}$$

where

$$\xi_n^\alpha(\alpha, \beta) = \frac{\pi\left(n + \frac{\beta-1}{\alpha}\right)}{\sin\frac{\pi}{\alpha}}.$$

If $\alpha \geq 4$, then all zeros are simple and negative for any $\beta \in (0, 2\alpha]$.

Theorem 4.13 *Let $\alpha \geq 6$, $0 < \beta \leq 2\alpha$. Then for all n, $1 \leq n \leq \left[\frac{\alpha}{3}\right] - 1$, the zeros $z_n(\alpha, \beta)$ of the Mittag-Leffler function satisfy the following inequalities*

$$-\sqrt{2}\frac{\Gamma(\alpha n+\beta)}{\Gamma(\alpha(n-1)+\beta)} < z_n(\alpha,\beta) < -\frac{\Gamma(\alpha n+\beta)}{\Gamma(\alpha(n-1)+\beta)}. \quad (4.6.30)$$

Theorem 4.14 *For any $N \in \mathbb{N}$, $N \geq 3$, the zeros $z_n(N, N+1)$ of the Mittag-Leffler function $E_{N,N+1}(z)$ satisfy the relation*

$$z_n(N, N+1) = -\left[\frac{\pi n + \pi/2 + \alpha_n(N)}{\sin \pi/\alpha}\right]^\alpha, \quad n \in \mathbb{N}, \ n \geq [N/3], \quad (4.6.31)$$

where $\alpha_n(N) \in \mathbb{R}$, $|\alpha_n(N)| \leq x_n(N)$,

$$x_n(N) = \begin{cases} \exp\{-\pi n \cot \pi/\alpha\}, & 3 \leq N \leq 6, \\ \exp\{-2\pi n \sin 2\pi/\alpha\}, & 7 \leq N \leq 1400, \\ 1.01\exp\{-2\pi n \sin 2\pi/\alpha\}, & N > 1400. \end{cases}$$

If $N \geq 6$, $1 \leq n \leq [N/3] - 1$, then

$$-\frac{((n+1)N)!}{(nN)!}\left(1 + \frac{3/2[((n+1)N)!]^2}{(nN)!((n+2)N)!}\min\{1, Nn^{-2}\}\right) \quad (4.6.32)$$

$$< z_n(N, N+1) < -\frac{((n+1)N)!}{(nN)!}.$$

Next we present a few non-asymptotic results on the distribution of zeros of the Mittag-Leffler function.

Theorem 4.15 *Let $0 < \alpha < 1$. Then*

(1) *for $\beta \in \left(\bigcup_{n=0}^{\infty}[-n+\alpha, -n+1]\right)\bigcup[1, +\infty)$ the function $E_{\alpha,\beta}(z)$ has no negative zero;*

(2) *for $\beta \in \bigcup_{n=0}^{\infty}(-n, -n+\alpha)$ the function $E_{\alpha,\beta}(z)$ has one negative zero and it is simple.*

Theorem 4.16 *(I) Let $0 < \alpha < 1$, $\beta < 0$. Then*

(1) *for $\beta \in [-2n-1, -2n)$, $n \in \mathbb{Z}_+$, the function $E_{\alpha,\beta}(z)$ has one positive zero and it is simple;*

(2) *for $\beta \in [-2n, -2n+1)$, $n \in \mathbb{N}$, the set of zeros of the function $E_{\alpha,\beta}(z)$ is either empty, or consists of two simple points, or consists of one double point.*

(II) The function $E_{1,\beta}(z)$ has a unique simple positive zero, whenever $\beta \in (-2n-1, -2n)$, $n \in \mathbb{Z}_+$, and has no positive zero, whenever $\beta \in (-2n, -2n+1)$, $n \in \mathbb{N}$.

4.6 Distribution of Zeros

Theorem 4.17 *(I) Let one of the following conditions be satisfied:*

(1) $0 < \alpha < 1$, $\beta \in [1, 1+\alpha]$,
(2) $1 < \alpha < 2$, $\beta \in [\alpha - 1, 1] \cup [\alpha, 2]$.

Then all zeros of the function $E_{\alpha,\beta}(z)$ are located outside of the angle $|\arg z| \leq \frac{\pi\alpha}{2}$.

(II) Let $1 < \alpha < 2$, $\beta = 0$. Then all zeros $(\neq 0)$ are located outside of the angle $|\arg z| \leq \frac{\pi\alpha}{2}$.

Distribution of zeros of the Mittag-Leffler function for certain special values of parameters is important in applications. Below we present a result of such type which is related to the study of inverse problems for abstract differential equations in Banach spaces (see Sect. 4.6.1).

Theorem 4.18 ([KarTik17, Theorem 1.]) *Let the function $L(z)$ be defined for all $z \in \mathbb{C}$ by the formula*

$$L(z) = \int_0^1 dt \int_0^t e^{z(t-s)} ds = E_{1,3}(z). \qquad (4.6.33)$$

All zeros of this function are simple and form an infinite denumerable set of the type

$$z_k = x_k + iy_k, \quad k \in \mathbb{Z} \setminus \{0\}; \quad z_{-k} = \overline{z_k}, \; k \in \mathbb{N}.$$

Zeros with positive indices ($z_k = x_k + iy_k, k \in \mathbb{N}$) are situated in the upper half plane, their real and imaginary parts are strictly monotone, tend to $+\infty$ as $k \to +\infty$ and satisfy the following inequalities

$$x_k = \operatorname{Re} z_k > 2, \quad \frac{\pi}{3} + 2\pi k < y_k = \operatorname{Im} z_k < \frac{\pi}{2} + 2\pi k. \qquad (4.6.34)$$

Defining

$$b_k := \frac{\pi}{2} + 2\pi k, \qquad (4.6.35)$$

one can get the following representation of zeros of the function $L(z)$:

$$z_k = \ln b_k + ib_k + \alpha_k - i\beta_k, \quad k \in \mathbb{N}, \qquad (4.6.36)$$

where $\alpha_k > 0, \beta_k > 0$ for all $k \in \mathbb{N}$ and the following estimates hold

$$0 < \alpha_k < \frac{1}{2}\left(\frac{\ln b_k + 2}{b_k}\right)^2, \quad \frac{\ln b_k}{b_k} < \beta_k < \frac{\ln b_k + 2}{b_k}, \quad k \in \mathbb{N}. \qquad (4.6.37)$$

Note that the zeros of the functions $L(z)$ are related to the values of the so-called *Lambert function* $w(\zeta)$, which solves the equation

$$w(\zeta)e^{w(\zeta)} = \zeta.$$

The relation takes the form

$$z_k = -1 - w_{-k-1}(-e^{-1}), \quad k \in \mathbb{N},$$

where $w_j(\zeta)$ is the j-th single-valued branch of the multi-valued function $w(\zeta)$ (see for details [KarTik17]).

4.6.1 Distributions of Zeros and Inverse Problems for Differential Equations in Banach Spaces

The inverse problem is formulated here as in [PrOrVa00] (see also [TikEid02]). Here, following [TikEid02] (see also [TikEid94]), we describe some results concerning the subject of the title of this subsection. Let E be a complex Banach space, $u : [0, T] \to E[0, T]$ and A be a closed operator with domain $D(A) \subset E$. For a positive integer $N \geq 1$ we consider the following abstract differential equation in the Banach space E

$$\frac{d^N u(t)}{dt^N} = Au(t) + p, \ t \in [0, T], \quad (4.6.38)$$

with unknown parameter $p \in E$, subject to the initial conditions

$$u(0) = u_0, \ u'(0) = u_1, \ldots, u^{(N-1)}(0) = u_{N-1}, \quad u_j \in E, \ j = 0, 1, \ldots, N-1, \quad (4.6.39)$$

and the so-called terminal overdetermined condition

$$u(T) = u_N, \quad u_N \in E. \quad (4.6.40)$$

The inverse problem for the abstract differential equation (4.6.38) is to determine for given $u_0, u_1, \ldots, u_N \in E$ a pair of elements $(u(t), p)$ satisfying (4.6.38)–(4.6.40).

The relation of such an inverse problem to the distribution of zeros Λ_N of special cases of the Mittag-Leffler function

$$X_N(z) = \frac{1}{N!} + \frac{z}{(2N)!} + \frac{z^2}{(3N)!} + \ldots = \sum_{m=0}^{\infty} \frac{z^m}{\Gamma(mN + N + 1)} = E_{N,N+1}(z) \quad (4.6.41)$$

4.6 Distribution of Zeros

was discovered in a series of articles by Tikhonov and Eidelman (see e.g. [TikEid02], cf. [KarTik17]).

We present here the uniqueness theorem for the above inverse problem (following the proof from [TikEid02]) and formulate an existence theorem in a special case as it was stated in [KarTik17].

The main result in [TikEid02] reads: *Let the inverse problem (4.6.38)–(4.6.40) with given elements $u_0, u_1, \ldots, u_N \in E$ have a solution $(u(t), p)$. This solution is unique if and only if no number $\frac{\lambda_k}{T^N}$, $\lambda_k \in \Lambda_N$, is an eigenvalue of the operator A.*

For unique solvability of the inverse problem it suffices to prove that the corresponding problem subject to the homogeneous condition has only the trivial solution. Furthermore, without loss of generality we can replace the interval $[0, T]$ by $[0, 1]$.

Theorem 4.19 ([TikEid02, Theorem 2]) *The inverse problem*

$$\frac{d^N u(t)}{dt^N} = Au(t) + p, \quad t \in [0, 1], \tag{4.6.42}$$

$$u(0) = 0, \; u'(0) = 0, \ldots, u^{(N-1)}(0) = 0, \tag{4.6.43}$$

$$u(1) = 0 \tag{4.6.44}$$

has only the trivial solution $u(t) \equiv 0$, $p = 0$ if and only if no zero λ_k of the Mittag-Leffler function $X_N(z) = E_{N,N+1}(z)$ is an eigenvalue of the operator A.

◁ To find the relation between the zeros of the function (4.6.41) and the eigenvalues of the operator A we consider an auxiliary Cauchy problem for the differential equation

$$\frac{d^N x(t)}{dt^N} = \lambda u(t) + 1, \quad t \in [0, 1], \tag{4.6.45}$$

$$x(0) = x'(0) = \ldots = x^{(N-1)}(0) = 0, \tag{4.6.46}$$

$$x(1) = 0, \tag{4.6.47}$$

with a spectral parameter $\lambda \in \mathbb{C}$.

It is straightforward to check that the uniquely determined solution to this problem for each $\lambda \in \mathbb{C}$ (which will be denoted $x(t; \lambda)$) is

$$x(t; \lambda) = \frac{t^N}{N!} + \lambda \frac{t^{2N}}{(2N)!} + \ldots + \lambda^{m-1} \frac{t^{mN}}{(mN)!} + \ldots = t^N X_N(\lambda t^N). \tag{4.6.48}$$

Since

$$x(1; \lambda) = X_N(\lambda),$$

condition (4.6.47) is satisfied if and only if $\lambda \in \Lambda_N$, i.e. λ is a zero of the function $X_N(z)$.

First, we prove the necessity of the conditions in this theorem. Let $\lambda_k \in \Lambda_N$ be an eigenvalue of the operator A and $f_k \neq 0$ be the corresponding eigenvector. Then the pair $u(t) = x(t; \lambda_k) f_k$, $p_k = f_k$ satisfies all the conditions (4.6.42)–(4.6.44), i.e. it is a nontrivial solution to the homogeneous inverse problem.

Let us now prove the sufficiency of the conditions in Theorem 4.19. Assume that no zero λ_k of the function $X_N(z\lambda)$ is an eigenvalue of the operator A. Let $(u(t); p)$ be an arbitrary solution to the problem (4.6.42)–(4.6.44). Our aim is to show that it is the trivial one, i.e. $u(t) \equiv 0$, $p = 0$.

We introduce for arbitrary $\lambda \in \mathbb{C}$ and sufficiently small $\varepsilon > 0$ the integral containing $x(t; \lambda)$ and the above-mentioned solution $u(t)$:

$$I_\varepsilon := \int_\varepsilon^{1-\varepsilon} x^{(N+1)}(t; \lambda) u(1-t) dt.$$

Integrating I_ε by parts gives

$$\left(x^{(N)}(t; \lambda) u(1-t) + x^{(N-1)}(t; \lambda) u'(1-t) + \ldots + x'(t; \lambda) u^{(N-1)}(1-t) \right) \Big|_\varepsilon^{1-\varepsilon}$$

$$+ A \int_\varepsilon^{1-\varepsilon} x'(t; \lambda) u(1-t) dt + (x(1-\varepsilon; \lambda) - x(\varepsilon; \lambda)) p.$$

By taking the limit as $\varepsilon \to 0$, using the boundary conditions, the fact that the operator A is closed and the following notation

$$f(\lambda) := \int_0^1 x'(t; \lambda) u(1-t) dt, \qquad (4.6.49)$$

we get $I_0 = A f(\lambda) + X_N(\lambda) p$. Since by (4.6.45) the following identity holds true $x^{(N+1)}(t; \lambda) = \lambda x'(t; \lambda)$, we get the final form of the previous relation

$$(\lambda - A) f(\lambda) = X_N(\lambda) p, \quad \forall \lambda \in \mathbb{C}. \qquad (4.6.50)$$

From the properties of the function $x(t; \lambda)$ it follows that $f(\lambda)$ is an entire vector function of the variable $\lambda \in \mathbb{C}$ and ranges in the Banach space E. Since A is a closed operator, one can differentiate relation (4.6.50) in λ:

$$(\lambda - A) f^{(n)}(\lambda) + n f^{(n-1)}(\lambda) = X_N^{(n)}(\lambda) p, \quad n = 1, 2, \ldots. \qquad (4.6.51)$$

Since by assumption no zero λ_k of $X_N(\lambda)$ is an eigenvalue of the operator A, it follows from (4.6.51) that $f(\lambda_k) = 0$ for all $\lambda_k \in \Lambda_N$. Moreover, by (4.6.51) we obtain that if λ_k is a zero of $X_N(\lambda)$ of a certain multiplicity, then it will also be a zero of $f(\lambda)$ of a not smaller multiplicity.

To complete the proof it is instructive to consider the case $N \geq 3$ (the cases $N = 1$ and $N = 2$ are straightforward (see [TikEid02]) since in both cases the zeros of $X_N(\lambda)$ are determined explicitly). From representations (4.6.49) and (4.6.48) we get

$$f(\lambda) = \int_0^1 X_N(\lambda t^N) \left(N^{-1} t u'(1-t)\right) dt^N = \int_0^1 X_N(\lambda s) v(s) ds,$$

where $v(s) = N^{-1} s^{1/N} u'(1 - s^{1/N})$ is a continuous vector function of $s \in [0; 1]$ uniquely determined by the function $u(t)$. Let us now apply the functional f^* from E^* to $f(\lambda)$. Defining $F(\lambda) := f^*(f(\lambda))$ and $h(s) := f^*(v(s))$, we obtain that the entire function $F(\lambda)$ of the variable $\lambda \in \mathbb{C}$ and the continuous scalar function $h(s), s \in [0, 1]$, are related by the formula

$$F(\lambda) = \int_0^1 X_N(\lambda s) h(s) ds.$$

Above we proved that all zeros of the function $X_N(\lambda)$ are zeros of the function $f(\lambda)$ (retaining their multiplicity). Hence, the same is true for the function $F(\lambda)$ and thus

$$Q(\lambda) = \frac{1}{X_N(\lambda)} \int_0^1 X_N(\lambda s) h(s) ds$$

is an entire function of the complex variable $\lambda \in \mathbb{C}$. Hence, using Carleman's reasoning, we obtain $h(s) \equiv 0$ for $0 \leq s \leq 1$ (see [PolSze76, Sect. 4, the solution of problem 199]) since the order of the function $X_N(\lambda)$ is equal to $\rho = \frac{1}{N} \leq \frac{1}{3} < \frac{1}{2}$ and this function has nonzero Taylor coefficients.

Thus, $h(s) = f^*(v(s)) \equiv 0$ for all $0 \leq s \leq 1$. Therefore $v(s) \equiv 0$ for $0 \leq s \leq 1$ by the Hahn–Banach theorem. It readily follows from the definition of $v(s)$ that $u'(t) \equiv 0$ for $0 \leq t \leq 1$, and since $u(0) = u(1) = 0$, we have $u(t) \equiv 0$; consequently, $p = 0$. The proof is complete. ▷

4.7 Computations With the Two-Parametric Mittag-Leffler Function

Mittag-Leffler type functions play a basic role in the solution of fractional differential equations and integral equations of Abel type. Therefore, it seems important as a first step to develop their theory and stable methods for their numerical computation.

Integral representations play a prominent role in the analysis of entire functions. For the two-parametric Mittag-Leffler function (4.1.1) such representations in the form of an improper integral along the Hankel loop have been treated in the case $\beta = 1$ and in the general case with arbitrary β by Erdélyi et al. [ErdBat 3] and Dzherbashyan [Dzh54a, Dzh66]. They considered the representations

$$E_{\alpha,\beta}(z) = \frac{1}{2\pi i \alpha} \int_{\gamma(\epsilon;\delta)} \frac{e^{\zeta^{1/\alpha}} \zeta^{(1-\beta)/\alpha}}{\zeta - z} d\zeta, \quad z \in G^{(-)}(\epsilon;\delta), \quad (4.7.1)$$

$$E_{\alpha,\beta}(z) = \frac{1}{\alpha} z^{(1-\beta)/\alpha} e^{z^{1/\alpha}} + \frac{1}{2\pi i \alpha} \int_{\gamma(\epsilon;\delta)} \frac{e^{\zeta^{1/\alpha}} \zeta^{(1-\beta)/\alpha}}{\zeta - z} d\zeta, \quad z \in G^{(+)}(\epsilon;\delta), \quad (4.7.2)$$

under the conditions

$$0 < \alpha < 2, \, \pi\alpha/2 < \delta < \min\{\pi, \pi\alpha\}. \quad (4.7.3)$$

The contour $\gamma(\epsilon;\delta)$ consists of two rays $S_{-\delta}$ ($\arg \zeta = -\delta$, $|\zeta| \geq \epsilon$) and S_{δ} ($\arg \zeta = \delta$, $|\zeta| \geq \epsilon$) and a circular arc $C_{\delta}(0;\epsilon)$ ($|\zeta| = \epsilon$, $-\delta \leq \arg \leq \delta$). On its left side there is a region $G^{(-)}(\epsilon,\delta)$, on its right side a region $G^{(+)}(\epsilon,\delta)$.

Using the integral representations in (4.7.1) and (4.7.2) it is not difficult to obtain asymptotic expansions for the Mittag-Leffler function in the complex plane (see Theorems 4.3, 4.4). Let $0 < \alpha < 2$, β be an arbitrary number, and δ be chosen to satisfy the condition (4.7.3). Then we have, for any $p \in \mathbb{N}$ (and for $p = 0$ if the "empty sum convention" is adopted) and $|z| \to \infty$

$$E_{\alpha,\beta}(z) = \frac{1}{\alpha} z^{(1-\beta)/\alpha} e^{z^{1/\alpha}} - \sum_{k=1}^{p} \frac{z^{-k}}{\Gamma(\beta - \alpha k)} + O\left(|z|^{-1-p}\right), \, \forall z, |\arg z| \leq \delta. \quad (4.7.4)$$

Analogously, for all z, $\delta \leq |\arg z| \leq \pi$, we have

$$E_{\alpha,\beta}(z) = -\sum_{k=1}^{p} \frac{z^{-k}}{\Gamma(\beta - \alpha k)} + O\left(|z|^{-1-p}\right). \quad (4.7.5)$$

These formulas are used in the numerical algorithm presented in this section (proposed in [GoLoLu02]). In what follows attention is restricted to the case $\beta \in \mathbb{R}$, the most important one in the applications. For the purpose of numerical computation we look for integral representations better suited than (4.7.1) and (4.7.2). Defining

$$\phi(\zeta, z) = \frac{e^{\zeta^{1/\alpha}} \zeta^{(1-\beta)/\alpha}}{\zeta - z}$$

4.7 Computations With the Two-Parametric Mittag-Leffler Function

we represent the integral in formulas (4.7.1) and (4.7.2) in the form

$$I = \frac{1}{2\pi i \alpha} \int_{\gamma(\epsilon;\delta)} \phi(\zeta, z)d\zeta = \frac{1}{2\pi i \alpha} \int_{S_{-\delta}} \phi(\zeta, z)d\zeta \qquad (4.7.6)$$
$$+ \frac{1}{2\pi i \alpha} \int_{C_\delta(0;\epsilon)} \phi(\zeta, z)d\zeta + \frac{1}{2\pi i \alpha} \int_{S_\delta} \phi(\zeta, z)d\zeta = I_1 + I_2 + I_3.$$

The integrals I_1, I_2 and I_3 have to be transformed. For I_1 we take $\zeta = re^{-i\delta}$, $\epsilon \leq r < \infty$, and get

$$I_1 = \frac{1}{2\pi i \alpha} \int_{S_{-\delta}} \phi(\zeta, z)d\zeta = \frac{1}{2\pi i \alpha} \int_{+\infty}^{\epsilon} \frac{e^{(re^{-i\delta})^{1/\alpha}} (re^{-i\delta})^{(1-\beta)/\alpha}}{(re^{-i\delta}) - z} e^{-i\delta} dr. \qquad (4.7.7)$$

Analogously, by using $\zeta = re^{i\delta}$, $\epsilon \leq r < \infty$,

$$I_3 = \frac{1}{2\pi i \alpha} \int_{S_\delta} \phi(\zeta, z)d\zeta = \frac{1}{2\pi i \alpha} \int_{\epsilon}^{+\infty} \frac{e^{(re^{i\delta})^{1/\alpha}} (re^{i\delta})^{(1-\beta)/\alpha}}{(re^{i\delta}) - z} e^{i\delta} dr. \qquad (4.7.8)$$

For I_2 with $\zeta = \epsilon e^{i\varphi}$, $-\delta \leq \varphi \leq \delta$

$$I_2 = \frac{1}{2\pi i \alpha} \int_{C_\delta(0;\epsilon)} \phi(\zeta, z)d\zeta = \frac{1}{2\pi i \alpha} \int_{-\delta}^{\delta} \frac{e^{(\epsilon e^{i\varphi})^{1/\alpha}} (\epsilon e^{i\varphi})^{(1-\beta)/\alpha}}{(\epsilon e^{i\varphi}) - z} \epsilon i e^{i\varphi} d\varphi \qquad (4.7.9)$$
$$= \frac{\epsilon^{1+(1-\beta)/\alpha}}{2\pi \alpha} \int_{-\delta}^{\delta} \frac{e^{\epsilon^{1/\alpha}(e^{i\varphi/\alpha})} e^{i\varphi(1-\beta)/\alpha+1)}}{\epsilon e^{i\varphi} - z} d\varphi = \int_{-\delta}^{\delta} P[\alpha, \beta, \epsilon, \varphi, z] d\varphi,$$

where

$$P[\alpha, \beta, \epsilon, \varphi, z] = \frac{\epsilon^{1+(1-\beta)/\alpha}}{2\pi\alpha} \frac{e^{\epsilon^{1/\alpha} \cos(\varphi/\alpha)} (\cos(\omega) + i \sin(\omega))}{\epsilon e^{i\varphi} - z}, \qquad (4.7.10)$$

$$\omega = \epsilon^{1/\alpha} \sin(\varphi/\alpha) + \varphi(1 + (1-\beta)/\alpha).$$

The sum I_1 and I_3 can be rewritten as

$$I_1 + I_3 = \int_\epsilon^{+\infty} K[\alpha, \beta, \delta, r, z] dr, \qquad (4.7.11)$$

where

$$K[\alpha, \beta, \delta, r, z] = \frac{1}{2\pi\alpha} r^{(1-\beta)/\alpha} e^{r^{1/\alpha} \cos(\delta/\alpha)} \frac{r \sin(\psi - \delta) - z \sin(\psi)}{r^2 - 2rz \cos(\delta) + z^2}, \qquad (4.7.12)$$

$$\psi = r^{1/\alpha} \sin(\delta/\alpha) + \varphi(1 + (1-\beta)/\alpha).$$

Using the above notation formulas (4.7.1) and (4.7.2) can be rewritten in the form

$$E_{\alpha,\beta}(z) = \int_{\epsilon}^{+\infty} K[\alpha, \beta, \epsilon, \varphi, z] dr + \int_{-\delta}^{\delta} P[\alpha, \beta, \epsilon, \varphi, z] d\varphi, \quad z \in G^{(-)}(\epsilon; \delta),$$
(4.7.13)

$$E_{\alpha,\beta}(z) = \int_{\epsilon}^{+\infty} K[\alpha, \beta, \epsilon, \varphi, z] dr + \int_{-\delta}^{\delta} P[\alpha, \beta, \epsilon, \varphi, z] d\varphi \quad (4.7.14)$$

$$+ \frac{1}{\alpha} z^{(1-\beta)/\alpha} e^{z^{1/\alpha}}, \quad z \in G^{(+)}(\epsilon; \delta).$$

Let us now consider the case $0 < \alpha \leq 1, z \neq 0$. By condition (4.7.3) we can choose $\delta = \min\{\pi, \pi\alpha\} = \pi\alpha$. Then the kernel function (4.7.12) looks simpler:

$$K[\alpha, \beta, \pi\alpha, r, z] = \widetilde{K}[\alpha, \beta, r, z] \quad (4.7.15)$$

$$= \frac{1}{2\pi\alpha} r^{(1-\beta)/\alpha} e^{-r^{1/\alpha}} \frac{r \sin(\pi(1-\beta)) - z \sin(\pi(1-\beta+\alpha))}{r^2 - 2rz\cos(\pi\alpha) + z^2}.$$

We distinguish three possibilities for $\arg z$ in the formulas (4.7.13)–(4.7.15) for the computation of the function $E_{\alpha,\beta}(z)$ at an arbitrary point $z \in \mathbb{C}, z \neq 0$, namely

(A) $|\arg z| > \pi\alpha$;
(B) $|\arg z| = \pi\alpha$;
(C) $|\arg z| < \pi\alpha$.

The following theorems give representation formulas suitable for further numerical calculations.

Theorem 4.20 *Under the conditions*

$$0 < \alpha \leq 1, \ \beta \in \mathbb{R}, \ |\arg z| > \pi\alpha, \ z \neq 0,$$

the function $E_{\alpha,\beta}(z)$ has the representations

$$E_{\alpha,\beta}(z) = \int_{\epsilon}^{+\infty} \widetilde{K}[\alpha, \beta, r, z] dr + \int_{-\pi\alpha}^{\pi\alpha} P[\alpha, \beta, \epsilon, \varphi, z] d\varphi, \ \epsilon > 0, \ \beta \in \mathbb{R},$$
(4.7.16)

$$E_{\alpha,\beta}(z) = \int_{0}^{+\infty} \widetilde{K}[\alpha, \beta, r, z] dr, \text{ if } \beta < 1 + \alpha, \quad (4.7.17)$$

$$E_{\alpha,\beta}(z) = -\frac{\sin(\pi\alpha)}{\pi\alpha} \int_{0}^{+\infty} \frac{e^{-r^{1/\alpha}}}{r^2 - 2rz\cos(\pi\alpha) + z^2} dr - \frac{1}{z}, \text{ if } \beta = 1 + \alpha.$$
(4.7.18)

4.7 Computations With the Two-Parametric Mittag-Leffler Function

Theorem 4.21 *Under the conditions*

$$0 < \alpha \leq 1, \ \beta \in \mathbb{R}, \ |\arg z| = \pi\alpha, \ z \neq 0,$$

the function $E_{\alpha,\beta}(z)$ has the representations

$$E_{\alpha,\beta}(z) = \int_{\epsilon}^{+\infty} \widetilde{K}[\alpha, \beta, r, z] dr + \int_{-\pi\alpha}^{\pi\alpha} P[\alpha, \beta, \epsilon, \varphi, z] d\varphi, \ \epsilon > |z|, \quad (4.7.19)$$

where the kernel functions $\widetilde{K}[\alpha, \beta, r, z]$ and $P[\alpha, \beta, \epsilon, \varphi, z]$ are given by the formulas (4.7.15) and (4.7.10), respectively.

Theorem 4.22 *Under the conditions*

$$0 < \alpha \leq 1, \ \beta \in \mathbb{R}, \ |\arg z| < \pi\alpha, \ z \neq 0,$$

the function $E_{\alpha,\beta}(z)$ has the representations

$$E_{\alpha,\beta}(z) = \int_{\epsilon}^{+\infty} \widetilde{K}[\alpha, \beta, r, z] dr + \int_{-\pi\alpha}^{\pi\alpha} P[\alpha, \beta, \epsilon, \varphi, z] d\varphi \quad (4.7.20)$$

$$+ \frac{1}{\alpha} z^{(1-\beta)/\alpha} e^{z^{1/\alpha}}, \ 0 < \epsilon < |z|, \ \beta \in \mathbb{R};$$

$$E_{\alpha,\beta}(z) = \int_{0}^{+\infty} \widetilde{K}[\alpha, \beta, r, z] dr + \frac{1}{\alpha} z^{(1-\beta)/\alpha} e^{z^{1/\alpha}}, \ \text{if } \beta < 1 + \alpha; \quad (4.7.21)$$

$$E_{\alpha,\beta}(z) = -\frac{\sin(\pi\alpha)}{\pi\alpha} \int_{0}^{+\infty} \frac{e^{-r^{1/\alpha}}}{r^2 - 2rz\cos(\pi\alpha) + z^2} dr \quad (4.7.22)$$

$$- \frac{1}{z} + \frac{1}{\alpha z} e^{z^{1/\alpha}}, \ \text{if } \beta = 1 + \alpha,$$

where the kernel functions $\widetilde{K}[\alpha, \beta, r, z]$ and $P[\alpha, \beta, \epsilon, \varphi, z]$ are given by formulas (4.7.15) and (4.7.10), respectively.

Therefore, for arbitrary $z \neq 0$ and $0 < \alpha \leq 1$ the Mittag-Leffler function $E_{\alpha,\beta}(z)$ can be represented by one of the formulas (4.7.16)–(4.7.22). These formulas are used for numerical computation if $q < |z|, 0 < q < 1$ and $0 < \alpha \leq 1$. In the case $|z| \leq q$, $0 < q < 1$, the values of the Mittag-Leffler function are computed for arbitrary $\alpha > 0$ by using series representation (4.1.1). The case $\alpha > 1$ is reduced to the case $0 < \alpha \leq 1$ by using recursion formulas. To compute the function $E_{\alpha,\beta}(z)$ for arbitrary $z \in \mathbb{C}$ with arbitrary indices $\alpha > 0, \beta \in \mathbb{R}$, three possibilities are distinguished:

(A) $|z| \leq q, 0 < q < 1$ (q is a fixed number), $\alpha > 0$;
(B) $|z| > q, 0 < \alpha \leq 1$;
(C) $|z| > q, \alpha > 1$.

In each case the Mittag-Leffler function can be computed with the prescribed accuracy $\rho > 0$. In case (A) the computations are based on the following result:

Theorem 4.23 *In case (A) the Mittag-Leffler function can be computed with the prescribed accuracy $\rho > 0$ by use of the formula*

$$E_{\alpha,\beta}(z) = \sum_{k=0}^{k_0} \frac{z^k}{\Gamma(\alpha k + \beta)} + \mu(z), \quad |\mu(z)| < \rho, \qquad (4.7.23)$$

where

$$k_0 = \max\{[(1-\beta)/\alpha] + 1; [\ln(\rho(1-|z|))/\ln(|z|)]\}.$$

In case (B) one can use the integral representations (4.7.16)–(4.7.22). For this it is necessary to compute numerically either the improper integral

$$I = \int_a^\infty \widetilde{K}[\alpha, \beta, r, z] dr, \quad a \in \{0; \epsilon\},$$

and/or the integral

$$J = \int_{-\pi\alpha}^{\pi\alpha} P[\alpha, \beta, \epsilon, \varphi, z] d\varphi, \quad \epsilon > 0.$$

To calculate the first (improper) integral I of the bounded function $\widetilde{K}[\alpha, \beta, r, z]$ the following theorem is used:

Theorem 4.24 *The representation*

$$I = \int_a^\infty \widetilde{K}[\alpha, \beta, r, z] dr = \int_a^{r_0} \widetilde{K}[\alpha, \beta, r, z] dr + \mu(r), \quad |\mu(r)| \leq \rho, \quad a \in \{0; \epsilon\}, \qquad (4.7.24)$$

is valid under the conditions

$$0 < \alpha \leq 1, \quad |z| > q > 0,$$

$$r_0 = \begin{cases} \max\{1, 2|z|, (-\ln(\pi\rho/6))^\alpha\}, & \text{if } \beta \geq 0; \\ \max\{(1+|\beta|)^\alpha, 2|z|, (-2\ln(\pi\rho/(6(|\beta|+2)(2|\beta|)^{|\beta|})))^\alpha\}, & \text{if } \beta < 0. \end{cases}$$

The second integral J (the integrand $P[\alpha, \beta, \epsilon, \varphi, z]$ being bounded and the limits of integration being finite) can be calculated with prescribed accuracy $\rho > 0$ by one of many product quadrature methods.

In case (C) the following recursion formula is used (see [Dzh66])

$$E_{\alpha,\beta}(z) = \frac{1}{m} \sum_{l=0}^{m-1} E_{\alpha/m,\beta}(z^{1/m} e^{2\pi i l/m}), \quad m \geq 1. \qquad (4.7.25)$$

4.7 Computations With the Two-Parametric Mittag-Leffler Function

In order to reduce case (C) to cases (B) and (A) one can take $m = [\alpha] + 1$ in formula (4.7.25). Then $0 < \alpha/m < 1$, and we calculate the functions $E_{\alpha/m,\beta}(z^{1/m}e^{2\pi i l/m})$ as in case (A) if $|z|^{1/m} \leq q < 1$, and as in case (B) if $|z|^{1/m} > q$.

Remark 4.25 The ideas and techniques employed for the Mittag-Leffler function can be used for the numerical calculation of other functions of hypergeometric type. In particular, the same method with some small modifications can be applied to the Wright function, which plays a very important role in the theory of partial differential equations of fractional order (see, e.g., [BucLuc98, GoLuMa00, Luc00, LucGor98, MaLuPa01]). To this end, the following representations of the Wright function (see [GoLuMa99]) can be used in place of the corresponding representations of the Mittag-Leffler function:

$$\phi(\rho, \beta; z) = \sum_{k=0}^{\infty} \frac{z^k}{\Gamma(\rho k + \beta)}, \quad \rho > -1, \quad \beta \in \mathbb{C},$$

$$\phi(\rho, \beta; z) = \frac{1}{2\pi i} \int_{\mathrm{Ha}} e^{\zeta + z\zeta^{-\rho}} \zeta^{-\beta} d\zeta, \quad \rho > -1, \quad \beta \in \mathbb{C},$$

where Ha denotes the Hankel path in the ζ-plane with a cut along the negative real semi-axis $\arg \zeta = \pi$.

4.8 Further Analytic Properties

4.8.1 Additional Integral and Differential Formulas

Here we present a number of integral and differential formulas for the Mittag-Leffler function. The integral relations below can be easily established by the application of classical formulas for Gamma and Beta functions (see Appendix A) and other techniques.

$$\int_0^\infty e^{-x} x^{\beta-1} E_{\alpha,\beta}(x^\alpha z) \, dx = \frac{1}{1-z} \quad (|z| < 1), \tag{4.8.1}$$

$$\int_0^x (x-\zeta)^{\beta-1} E_\alpha(\zeta^\alpha) \, d\zeta = \Gamma(\beta) x^\beta E_{\alpha,\beta+1}(x^\alpha), \tag{4.8.2}$$

$$\int_0^\infty e^{-sx} x^{m\alpha+\beta-1} E_{\alpha,\beta}^{(m)}(\pm \lambda x^\alpha) \, dx = \frac{m! s^{\alpha-\beta}}{s^\alpha \mp \lambda}, \tag{4.8.3}$$

where $\alpha, \beta \in \mathbb{C}$, $\mathrm{Re}\, \alpha > 0$, $\mathrm{Re}\, \beta > 0$ and in (4.8.3) $|\lambda s^{-\alpha}| < 1$.

$$\int_0^x \zeta^{\beta_1-1} E_{\alpha,\beta_1}(\lambda\zeta^\alpha)(x-\zeta)^{\beta_2-1} E_{\alpha,\beta_2}(\mu(x-\zeta)^\alpha)\,d\zeta$$

$$= \frac{x^{\beta_1+\beta_2-1}}{\lambda-\mu}\{E_{\alpha,\beta_1+\beta_2}(\lambda x^\alpha) - E_{\alpha,\beta_1+\beta_2}(\mu x^\alpha)\}. \tag{4.8.4}$$

The following differential relations can be found by direct calculation.

$$\left(\frac{\partial}{\partial z}\right)^n \left[z^{\beta-1} E_{\alpha,\beta}(\lambda z^\alpha)\right] = z^{\beta-n-1} E_{\alpha,\beta-n}(\lambda z^\alpha), \tag{4.8.5}$$

$$\left(\frac{\partial}{\partial \lambda}\right)^n \left[z^{\beta-1} E_{\alpha,\beta}(\lambda z^\alpha)\right] = n!z^{\alpha n+\beta-1} E_{\alpha,\alpha n+\beta}^{n+1}(\lambda z^\alpha), \tag{4.8.6}$$

where $E_{\alpha,\alpha n+\beta}^{n+1}(z)$ is the Prabhakar three-parametric function (see Sect. 5.1 below).

A special case of the two-parametric Mittag-Leffler function (the so-called α-*exponential function*) is of interest for many applications. It is defined in the following way:

$$e_\alpha^{\lambda z} := z^{\alpha-1} E_{\alpha,\alpha}(\lambda z^\alpha) \quad (z \in \mathbb{C} \setminus \{0\}, \lambda \in \mathbb{C}). \tag{4.8.7}$$

For all $\alpha \in \mathbb{C}$, $\operatorname{Re}\alpha > 0$, it can be represented in the form of a series

$$e_\alpha^{\lambda z} = z^{\alpha-1} \sum_{k=0}^\infty \lambda^k \frac{z^{\alpha k}}{\Gamma((k+1)\alpha)}, \tag{4.8.8}$$

which converges in $\mathbb{C} \setminus \{0\}$ and determines in this domain an analytic function. The simple properties of this function:

1) $\lim_{z\to 0} z^{1-\alpha} e_\alpha^{\lambda z} = \frac{1}{\Gamma(\alpha)} \quad (\operatorname{Re}\alpha > 0),$ \hfill (4.8.9)

2) $e_1^{\lambda z} = e^{\lambda z},$ \hfill (4.8.10)

justify its name. However, the α-exponential function does not satisfy the main property of the exponential function, i.e.,

$$e_\alpha^{\lambda z} e_\alpha^{\mu z} \neq e_\alpha^{(\lambda+\mu)z}. \tag{4.8.11}$$

For $0 < \alpha < 2$ the α-exponential function satisfies a simple asymptotic relation

$$e_\alpha^{\lambda z} = \frac{\lambda^{(1-\alpha)/\alpha}}{\alpha} \exp\{\lambda^{1/\alpha} z\} - \sum_{k=1}^{N-1} \frac{\lambda^{-k-1}}{\Gamma(-\alpha k)} \frac{1}{z^{\alpha k+1}} + O\left(\frac{1}{z^{\alpha N+1}}\right), \tag{4.8.12}$$

where $z \to \infty$, $N \in \mathbb{N} \setminus \{1\}$, $|\arg(\lambda z^\alpha)| \leq \mu$, $\frac{\pi\alpha}{2} < \mu < \min\{\pi, \pi\alpha\}$, and

$$e_\alpha^{\lambda z} = -\sum_{k=1}^{N-1} \frac{\lambda^{-k-1}}{\Gamma(-\alpha k)} \frac{1}{z^{\alpha k+1}} + O\left(\frac{1}{z^{\alpha N+1}}\right), \qquad (4.8.13)$$

where $z \to \infty$, $N \in \mathbb{N} \setminus \{1\}$, $\mu \leq |\arg(\lambda z^\alpha)| \leq \pi$.

When $\alpha \geq 2$ the asymptotic behavior at infinity of the α-exponential function is more complicated (see, e.g., [KiSrTr06, pp. 51–52]).

4.8.2 Geometric Properties of the Mittag-Leffler Function

In this subsection we discuss (following [BanPra16]) certain geometric properties of the Mittag-Leffler function $E_{\alpha,\beta}(z)$ of the complex variable z.

Let us recall some necessary definitions (see, e.g. [BanPra16, Goo83, Dur83]). Let $\mathcal{H}(\mathbb{U})$ be the class of analytic functions in the open unit disk $\mathbb{U} = z : |z| < 1$. A subclass $\mathcal{A} \subset \mathcal{H}(\mathbb{U})$ consists of all functions $f \in \mathcal{H}(\mathbb{U})$ which are normalized by $f(0) = 0;\ f'(0) = 1$:

$$f(z) = z + a_2 z^2 + \ldots + a_n z^n + \ldots, \quad z \in \mathbb{U}.$$

A function f is called a *univalent function* in a domain D if it is one-to-one in D. A function $f \in \mathcal{A}$ is called *starlike* (with respect to the origin 0), if $tw \in f(\mathbb{U})$ whenever $w \in f(\mathbb{U})$ and $t \in [0, 1]$, i.e. $f(\mathbb{U})$ is starlike with respect to the origin. A function $f \in \mathcal{A}$ is known to be a convex function if $f(\mathbb{U})$ is a convex domain. The subsets of $\mathcal{A}$ consisting of starlike and convex functions are denoted by S^* and $\mathcal{K}$, respectively. Given $0 \leq \eta < 1$, a function $f \in \mathcal{A}$ is called a *starlike function of order η*, denoted by $f \in S^*(\eta)$, if

$$\mathrm{Re}\left(\frac{zf'(z)}{f(z)}\right) > \eta,\ z \in \mathbb{U}.$$

Furthermore, for a given $0 \leq \eta < 1$, a function $f \in \mathcal{A}$ is called a convex function of order η, denoted by $f \in \mathcal{K}(\eta)$, if

$$\mathrm{Re}\left(1 + \frac{zf''(z)}{f'(z)}\right) > \eta,\ z \in \mathbb{U}.$$

In particular, $S^*(0) = S^*$, $\mathcal{K}(0) = \mathcal{K}$, and it is a well-known fact that $f \in \mathcal{A}$ is convex if and only if zf' is starlike.

A function $f \in \mathcal{A}$ is called close-to-convex in $\mathbb{U}$ if the range $f(\mathbb{U})$ is close-to-convex, i.e. the complement of $f(\mathbb{U})$ can be written as the union of non-intersecting half-lines. A function $f \in \mathcal{A}$ is close-to-convex in $\mathbb{U}$ if there exists a starlike function g (which need not be normalized) in $\mathbb{U}$ such that

$$\operatorname{Re}\left(\frac{zf'(z)}{g(z)}\right) > 0, \ z \in \mathbb{U}.$$

The class of all close-to-convex functions in $\mathbb{U}$ is denoted by $\mathcal{C}$. It is well known that close-to-convex functions are univalent in $\mathbb{U}$, but it is easy to verify that $\mathcal{K} \subset \mathcal{S}^* \subset \mathcal{C}$.

Observe that the Mittag-Leffler function $E_{\alpha,\beta}(z)$ does not belong to the family $\mathcal{A}$. Thus, it is natural to consider (following [BanPra16]) the normalization of the Mittag-Leffler function:

$$\mathbb{E}_{\alpha,\beta}(z) := \Gamma(\beta) z E_{\alpha,\beta}(z) \qquad (4.8.14)$$

$$= z + \sum_{k=2}^{\infty} \frac{\Gamma(\beta)}{\Gamma(\alpha(n-1)+\beta)} z^n, \ z \in \mathbb{C}; \ \operatorname{Re}\alpha > 0, \beta \in \mathbb{C}.$$

Below we present certain geometric results in the case of real values of parameters $\alpha > 0, \beta \in \mathbb{R}$, whilst the definition (4.8.14) holds for more general values of parameters.

Let us present two sufficient conditions for starlikeness of $\mathbb{E}_{\alpha,\beta}(z)$ in $\mathbb{U}$.

Theorem 4.26 ([BanPra16, Theorem 2.1]) *Let $\alpha \geq 1, \beta \geq 1$ and $\Gamma(\alpha+\beta) > 4\Gamma(\beta)$. Then $\mathbb{E}_{\alpha,\beta}(z)$ is starlike in $\mathbb{U}$, $\mathbb{E}_{\alpha,\beta} \in \mathcal{A}$.*

◁ To get the result, it suffices (see [Fej36]) to show that the following expressions are nonnegative:

$$\underline{\Delta}_n = na_n - (n+1)a_{n+1}, \quad \underline{\Delta}_n^2 = na_n - 2(n+1)a_{n+1} + (n+2)a_{n+2}.$$

This follows from the conditions on the parameters α, β and the properties of the Gamma function. ▷

Theorem 4.27 ([BanPra16, Theorem 2.2]) *Let $\alpha \geq 1, \beta \geq (3+\sqrt{17})/2$. Then $\mathbb{E}_{\alpha,\beta}(z)$ is starlike in $\mathbb{U}$, $\mathbb{E}_{\alpha,\beta} \in \mathcal{A}$.*

◁ Let us define the function $p(z) = \frac{z \mathbb{E}'_{\alpha,\beta}(z)}{\mathbb{E}_{\alpha,\beta}(z)}, z \in \mathbb{U}$. Since $\frac{\mathbb{E}_{\alpha,\beta}(z)}{z} \neq 0$ in $\mathbb{U}$, the function p is analytic in $\mathbb{U}$ and $p(0) = 1$. To prove the result, we need to show that $\operatorname{Re} p(z) > 0$, $z \in \mathbb{U}$. Since for $\alpha \geq 1, \beta \geq 1$ and any $n \in \mathbb{N}$ we have

$$\Gamma(\beta+n) \leq \Gamma(\alpha n + \beta),$$

then for any $n \in \mathbb{N} \setminus \{1\}$

$$\frac{n\Gamma(\beta)}{\Gamma(\alpha n+\beta)} \leq \frac{n}{\beta(\beta+1)\ldots(\beta+n-1)} < \frac{1}{\beta(\beta+1)^{n-2}}.$$

For any $z \in \mathbb{U}$ we then have two inequalities

4.8 Further Analytic Properties

$$\left|\mathbb{E}'_{\alpha,\beta}(z) - \frac{\mathbb{E}_{\alpha,\beta}(z)}{z}\right| = \left|\sum_{n=1}^{\infty} \frac{n\Gamma(\beta)}{\Gamma(\alpha n + \beta)} z^n\right| < \frac{1}{\beta} + \frac{1}{\beta}\sum_{n=0}^{\infty}\left(\frac{1}{\beta+1}\right)^n = \frac{2\beta+1}{\beta^2},$$

$$\left|\frac{\mathbb{E}_{\alpha,\beta}(z)}{z}\right| > 1 - \left|\sum_{n=1}^{\infty} \frac{\Gamma(\beta)}{\Gamma(\alpha n + \beta)} z^n\right| \geq 1 - \frac{1}{\beta}\sum_{n=0}^{\infty}\left(\frac{1}{\beta+1}\right)^n = \frac{\beta^2-\beta-1}{\beta^2}.$$

This gives the result of the theorem due to the assumption on β and the following inequality

$$|p(z) - 1| = \left|\frac{\mathbb{E}'_{\alpha,\beta}(z) - \frac{\mathbb{E}_{\alpha,\beta}(z)}{z}}{\frac{\mathbb{E}_{\alpha,\beta}(z)}{z}}\right| < \frac{2\beta+1}{\beta^2-\beta-1}. \triangleright$$

We also mention a weaker geometric result related to $\mathbb{E}_{\alpha,\beta}(z)$ restricted to $\mathbb{D}_{1/2} = \{z \in \mathbb{C} : |z| < 1/2\}$.

Theorem 4.28 ([BanPra16, Theorem 2.2]) *If*
(a) $\alpha \geq 1$ *and* $\beta \geq (1+\sqrt{5})/2$, *then* $\mathbb{E}_{\alpha,\beta}(z)$ *is univalent and starlike in* $\mathbb{D}_{1/2}$;
(b) $\alpha \geq 1$ *and* $\beta \geq (3+\sqrt{17})/2$, *then* $\mathbb{E}_{\alpha,\beta}(z)$ *is convex in* $\mathbb{D}_{1/2}$.

4.8.3 An Extension for Negative Values of the First Parameter

The two-parametric Mittag-Leffler function (4.1.1), defined in the form of a series, exists only for the values of parameters $\operatorname{Re}\alpha > 0$ and $\beta \in \mathbb{C}$. However, by using an existing integral representation formula for the two-parametric Mittag-Leffler function (see, e.g., [Dzh66, KiSrTr06]) it is possible to determine an extension of the two-parametric Mittag-Leffler function to other values of the first parameter.

In this section we present an analytic continuation of the Mittag-Leffler function depending on real parameters $\alpha, \beta \in \mathbb{R}$ by extending its domain to negative $\alpha < 0$. Here we follow the results of [Han-et-al09].

The following integral representation of the Mittag-Leffler function is known (see, e.g., [Dzh66, KiSrTr06])

$$E_{\alpha,\beta}(z) = \frac{1}{2\pi}\int_{\text{Ha}} \frac{t^{\alpha-\beta}e^t}{t^\alpha - z} dt, \quad z \in \mathbb{C}, \tag{4.8.15}$$

where the contour of integration Ha is the so-called Hankel path, a loop starting and ending at $-\infty$, and encircling the disk $|t| \leq |z|^{1/\alpha}$ counterclockwise.

To find an equation which can determine $E_{-\alpha,\beta}(z)$, we rewrite the integral representation of the Mittag-Leffler function (4.8.15) as

$$E_{\alpha,\beta}(z) = \frac{1}{2\pi} \int_{Ha} \frac{e^t}{t^\beta - zt^{-\alpha+\beta}} dt \qquad (4.8.16)$$

and expand part of the integrand in (4.8.16) in partial fractions as follows:

$$\frac{1}{t^\beta - zt^{-\alpha+\beta}} = \frac{1}{t^\beta} - \frac{1}{t^\beta - z^{-1}t^{\alpha+\beta}}. \qquad (4.8.17)$$

Substituting Eq. (4.8.17) into (4.8.16) yields

$$E_{\alpha,\beta}(z) = \frac{1}{2\pi} \int_{Ha} \frac{e^t}{t^\beta} dt - \frac{1}{2\pi} \int_{Ha} \frac{e^t}{t^\beta - z^{-1}t^{\alpha+\beta}} dt, \quad z \in \mathbb{C} \setminus \{0\}. \qquad (4.8.18)$$

This gives the following definition of the Mittag-Leffler function with negative value of the first parameter:

$$E_{-\alpha,\beta}(z) = \frac{1}{\Gamma(\beta)} - E_{\alpha,\beta}\left(\frac{1}{z}\right), \quad \alpha > 0, \ \beta \in \mathbb{R}; z \in \mathbb{C} \setminus \{0\}. \qquad (4.8.19)$$

In particular,

$$E_{-\alpha}(z) := E_{-\alpha,1}(z) = 1 - E_\alpha\left(\frac{1}{z}\right), \quad \alpha > 0; z \in \mathbb{C} \setminus \{0\}.$$

By using the known recurrence formula

$$E_{\alpha,\beta}(z) = \frac{1}{\Gamma(\beta)} + zE_{\alpha,\alpha+\beta}(z)$$

we obtain another variant of the definition (4.8.19)

$$E_{-\alpha,\beta}(z) = -\frac{1}{z} E_{\alpha,\alpha+\beta}\left(\frac{1}{z}\right), \quad \alpha > 0, \ \beta \in \mathbb{R}; z \in \mathbb{C} \setminus \{0\}. \qquad (4.8.20)$$

Direct calculations show that definitions (4.8.19) and (4.8.20) determine the same function, analytic in $\mathbb{C} \setminus \{0\}$.

By taking the limit in (4.8.19) as $\alpha \to +0$ we get the definition of $E_{0,\beta}(z)$

$$E_{0,\beta}(z) = \frac{1}{\Gamma(\beta)(1-z)}, \quad \beta \in \mathbb{R}; |z| < 1. \qquad (4.8.21)$$

Obviously, this function can be analytically continued in the domain $\mathbb{C} \setminus \{1\}$.

From the definition of the two-parametric Mittag-Leffler function (4.1.1) we obtain the following series representation of the extended Mittag-Leffler function (i.e. the function corresponding to negative values of the first parameter):

4.8 Further Analytic Properties

$$E_{-\alpha,\beta}(z) = -\sum_{k=1}^{\infty} \frac{1}{\Gamma(\alpha z + \beta)} \left(\frac{1}{z}\right)^k, \quad z \in \mathbb{C} \setminus \{0\}. \tag{4.8.22}$$

By using this representation and the above definitions of the extended Mittag-Leffler function (4.8.19) (or (4.8.20)) one can obtain functional, differential and recurrence relations which are analogous to corresponding relations for the two-parametric function with positive first parameter.

Proposition 4.29 *Let $\alpha > 0$, $\beta \in \mathbb{R}$. Then the following formulas are valid for all values of parameters for which all items are defined.*

A. *Recurrence relations.*

$$E_{-\alpha,\beta}(z) + E_{-\alpha,\beta}(-z) = 2E_{-2\alpha,\beta}(z^2); \tag{4.8.23}$$

$$E_{-n\alpha,\beta}(z) = \frac{1}{n} \sum_{k=0}^{n-1} E_{-\alpha,\beta}(ze^{-2\pi i k/n}); \tag{4.8.24}$$

$$E_{-\alpha,\beta}(z) = z^n E_{-\alpha,\beta-\alpha n}(z) + \sum_{k=0}^{n-1} \frac{z^k}{\Gamma(\beta - \alpha k)}; \tag{4.8.25}$$

$$E_{-\alpha}(-z) = E_{-2\alpha}(z^2) + E_{-2\alpha}(z^2) - zE_{-2\alpha,\alpha+1}(z^2). \tag{4.8.26}$$

B. *Differential relations.*

$$\frac{d}{dz}\left[z^{1-\beta}E_{-\alpha,\beta}(z^\alpha)\right] = -z^{-\beta}E_{-\alpha,\beta-1}(z^\alpha); \tag{4.8.27}$$

$$\frac{d}{dz}\left[E_{-\alpha}(z)\right] = -\frac{1}{\Gamma(\alpha+1)} + \frac{1}{\alpha}E_{-\alpha,\alpha}(z); \tag{4.8.28}$$

$$\frac{d^n}{dz^n}\left[E_{-n}(z^{-n})\right] = E_{-n}(z^{-n}). \tag{4.8.29}$$

C. *Functional relations.*

$$\int_0^z E_{-\alpha,\beta}(t^\alpha) t^{-\beta-1} dt = -z^{-\beta} E_{-\alpha,\beta+1}(z^\alpha); \tag{4.8.30}$$

$$\mathcal{L}\left[z^{\beta-1} E_{-\alpha,\beta}\left(\frac{1}{\pm az^\alpha}\right)\right] = \frac{\mp a}{s^\beta(s^\alpha \mp a)}. \tag{4.8.31}$$

4.9 The Two-Parametric Mittag-Leffler Function of a Real Variable

4.9.1 Integral Transforms of the Two-Parametric Mittag-Leffler Function

The following form of the Laplace transform of the two-parametric Mittag-Leffler function is most often used in applications:

$$\left(\mathcal{L} t^{\beta-1} E_{\alpha,\beta}(\lambda t^\alpha)\right)(s) = \frac{s^{\alpha-\beta}}{s^\alpha - \lambda} \quad (\operatorname{Re} s > 0, \ \lambda \in \mathbb{C}, \ |\lambda s^{-\alpha}| < 1). \quad (4.9.1)$$

It can be shown directly that the Laplace transform of the two-parametric Mittag-Leffler function $E_{\alpha,\beta}(t)$ is given in terms of the Wright function (see, e.g., [KiSrTr06, p. 44])

$$\left(\mathcal{L} E_{\alpha,\beta}(t)\right)(s) = \frac{1}{s} {}_2\Psi_1 \left[\begin{matrix} (1,1),(1,1) \\ (\alpha,\beta) \end{matrix} \bigg| \frac{1}{s} \right] \quad (\operatorname{Re} s > 0). \quad (4.9.2)$$

From the Mellin–Barnes integral representation of the two-parametric Mittag-Leffler function we arrive at the following formula for the Mellin transform of this function

$$\left(\mathcal{M} E_{\alpha,\beta}(-t)\right)(s) = \int_0^\infty E_{\alpha,\beta}(-t) t^{s-1} dt = \frac{\Gamma(s)\Gamma(1-s)}{\Gamma(\beta - \alpha s)} \quad (0 < \operatorname{Re} < 1). \quad (4.9.3)$$

To conclude this subsection, we consider the Fourier transform of the two-parametric Mittag-Leffler function $E_{\alpha,\beta}(|t|)$ with $\alpha > 1$. Performing a term-by-term integration of the series we get the formula ($\alpha > 1$):

$$\left(\mathcal{F} E_{\alpha,\beta}(|t|)\right)(x) := \int_{-\infty}^{+\infty} e^{ixt} E_{\alpha,\beta}(|t|) dt = \frac{\delta(x)}{\Gamma(\beta)} - \frac{2}{x^2} {}_2\Psi_1 \left[\begin{matrix} (2,2),(1,1) \\ (\alpha+\beta, 2\alpha) \end{matrix} \bigg| -\frac{1}{x^2} \right], \quad (4.9.4)$$

where $\delta(\cdot)$ is the Dirac delta function.

Since for all t

$$E_{\alpha,\beta}(|t|) - \frac{1}{\Gamma(\beta)} = |t| E_{\alpha,\alpha+\beta}(|t|), \quad (4.9.5)$$

formula (4.9.4) can be simplified

$$\left(\mathcal{F}\,|t|E_{\alpha,\alpha+\beta}(|t|)\right)(x) = -\frac{2}{x^2}{}_2\Psi_1\left[\begin{array}{c}(2,2),(1,1)\\(\alpha+\beta,2\alpha)\end{array}\Big|-\frac{1}{x^2}\right]\;(\alpha>1,\beta\in\mathbb{C}).$$
(4.9.6)

4.9.2 The Complete Monotonicity Property

Let us show that the generalized Mittag-Leffler function $E_{\alpha,\beta}(-x)$ possesses the complete monotonicity property for $0 \leq \alpha \leq 1$, $\beta \geq \alpha$. In fact, this result follows from the complete monotonicity of the classical Mittag-Leffler function $E_\alpha(-x)$ due to the following technical lemmas.

Lemma 4.30 *For all $\alpha \geq 0$*

$$E_{\alpha,\alpha}(-x) = -\alpha \frac{d}{dx}E_\alpha(-x).$$

◁ This follows from the standard properties of the integral depending on a parameter.
▷

Lemma 4.31 *Let $\beta > \alpha > 0$. Then the following identity holds:*

$$E_{\alpha,\beta}(-x) = \frac{1}{\alpha\Gamma(\beta-\alpha)}\int_0^1 \left(1-t^{1/\alpha}\right)^{\beta-\alpha-1} E_{\alpha,\alpha}(-tx)\,dt. \qquad (4.9.7)$$

◁ Let us take $E_{\alpha,\alpha}(-tx)$ in the form of a series and substitute it into the right-hand side of (4.9.7). By interchanging the order of integration and summation (which can be easily justified) we obtain that the right-hand side is equal to

$$\frac{1}{\alpha\Gamma(\beta-\alpha)}\sum_{k=0}^\infty \frac{(-x)^k}{\Gamma(\alpha k+\alpha)}\int_0^1 t^k\left(1-t^{1/\alpha}\right)^{\beta-\alpha-1}\,dt.$$

Calculating these integrals we arrive at the series representation for $E_{\alpha,\beta}(-x)$.
▷

Observe that

$$E_{0,\beta}(-x) = \frac{1}{\alpha\Gamma(\beta)}\frac{1}{1+x},\quad \beta>0,$$

$$E_{0,\beta}(-x) = 0,\quad \beta=0.$$

In both cases $E_{0,\beta}(-x)$ is completely monotonic.

The complete monotonicity of $E_{\alpha,\beta}(-x)$ then follows immediately from Pollard's result [Poll48], see Sect. 3.7.2 of this book.

4.9.3 Relations to the Fractional Calculus

Here we present a few formulas related to the values of the fractional integrals and derivatives of the two-parametric Mittag-Leffler function (see, e.g., [HaMaSa11, pp. 15–16]). Let us start with the left-sided Riemann–Liouville integral. Suppose that $\operatorname{Re}\alpha > 0$, $\operatorname{Re}\beta > 0$, $\operatorname{Re}\gamma > 0$, $a \in \mathbb{R}$. Then by using the series representation and the left-sided Riemann–Liouville integral of the power function we get

$$\left(I_{0+}^{\alpha} t^{\gamma-1} E_{\beta,\gamma}(at^{\beta})\right)(x) = x^{\alpha+\gamma-1}\left(E_{\beta,\alpha+\gamma}(ax^{\beta})\right), \qquad (4.9.8)$$

and, in particular, if $a \neq 0$, then (for $\beta = \alpha$)

$$\left(I_{0+}^{\alpha} t^{\gamma-1} E_{\alpha,\gamma}(at^{\alpha})\right)(x) = \frac{x^{\gamma-1}}{a}\left(E_{\alpha,\gamma}(ax^{\alpha}) - \frac{1}{\Gamma(\gamma)}\right). \qquad (4.9.9)$$

In the same manner one can obtain the formula

$$\left(I_{0+}^{\alpha} t^{\alpha-1} E_{\alpha,\beta}(at^{\alpha})\right)(x) = \frac{x^{\alpha-1}}{a}\left(E_{\alpha,\beta}(ax^{\alpha}) - \frac{1}{\Gamma(\beta)}\right). \qquad (4.9.10)$$

Analogously, one can calculate the right-sided fractional Riemann–Liouville integral of the two-parametric Mittag-Leffler function in the case $\operatorname{Re}, \alpha > 0$, $\operatorname{Re}\beta > 0$, $a \in \mathbb{R}$, $a \neq 0$

$$\left(I_{-}^{\alpha} t^{-\alpha-\gamma} E_{\beta,\gamma}(at^{-\beta})\right)(x) = x^{-\gamma}\left(E_{\beta,\alpha+\gamma}(ax^{-\beta})\right). \qquad (4.9.11)$$

If we suppose additionally that $\operatorname{Re}(\alpha + \gamma) > \operatorname{Re}\beta$, then the last formula can be rewritten as

$$\left(I_{-}^{\alpha} t^{-\alpha-\gamma} E_{\beta,\gamma}(at^{-\beta})\right)(x) = \frac{x^{\beta-\gamma}}{a}\left(E_{\beta,\alpha+\gamma-\beta}(ax^{-\beta}) - \frac{1}{\Gamma(\alpha+\gamma-\beta)}\right), \qquad (4.9.12)$$

and, in particular,

$$\left(I_{-}^{\alpha} t^{-\alpha-\beta} E_{\alpha,\beta}(at^{-\alpha})\right)(x) = \frac{x^{\alpha-\beta}}{a}\left(E_{\alpha,\beta}(ax^{-\alpha}) - \frac{1}{\Gamma(\beta)}\right). \qquad (4.9.13)$$

In the case of the fractional differentiation of the two-parametric Mittag-Leffler function we have

$$\left(D_{0+}^{\alpha} t^{\gamma-1} E_{\beta,\gamma}(at^{\beta})\right)(x) = x^{\gamma-\alpha-1}\left(E_{\beta,\gamma-\alpha}(ax^{\beta})\right), \qquad (4.9.14)$$
$$\operatorname{Re}\alpha > 0,\ \operatorname{Re}\beta > 0, a \in \mathbb{R}.$$

If we assume extra conditions on the parameters, namely $\operatorname{Re}\gamma > \operatorname{Re}\beta$, $\operatorname{Re}\gamma > \operatorname{Re}(\alpha+\beta)$, $a \neq 0$, then the following relations hold:

$$\left(D_{0+}^{\alpha} t^{\gamma-1} E_{\beta,\gamma}(at^{\beta})\right)(x) = \frac{x^{\gamma-\alpha-1}}{\Gamma(\gamma-\alpha)} + ax^{\gamma-\alpha+\beta-1} E_{\beta,\gamma-\alpha+\beta}(ax^{\beta}). \quad (4.9.15)$$

In particular (see [KilSai95b]), for $\operatorname{Re}\alpha > 0$, $\operatorname{Re}\beta > \operatorname{Re}\alpha + 1$, one can prove

$$\left(D_{0+}^{\alpha} t^{\beta-1} E_{\alpha,\beta}(at^{\alpha})\right)(x) = \frac{x^{\beta-\alpha-1}}{\Gamma(\beta-\alpha)} + ax^{\beta-1}\left(E_{\alpha,\beta}(ax^{\alpha})\right). \quad (4.9.16)$$

Finally, the right-sided (Liouville) fractional derivative of the two-parametric Mittag-Leffler function satisfies the relation (see, e.g., [KiSrTr06, p. 86])

$$\left(D_{-}^{\alpha} t^{\alpha-\beta} E_{\alpha,\beta}(at^{-\alpha})\right)(x) = \frac{x^{-\beta}}{\Gamma(\beta-\alpha)} + ax^{-\alpha-\beta}\left(E_{\alpha,\beta}(ax^{-\alpha})\right), \quad (4.9.17)$$

valid for all $\operatorname{Re}\alpha > 0$, $\operatorname{Re}\beta > \operatorname{Re}\alpha + 1$.

We also mention two extra integral relations for the two-parametric Mittag-Leffler function which are useful for applications.

Lemma 4.32 *Let $\alpha > 0$ and $\beta > 0$. Then the following formula holds*

$$\frac{1}{\Gamma(\alpha)} \int_0^x \frac{t^{\beta-1} E_{2\alpha,\beta}(t^{2\alpha})}{(x-t)^{1-\alpha}} dt = x^{\beta-1}\left[E_{\alpha,\beta}(x^{\alpha}) - E_{2\alpha,\beta}(x^{2\alpha})\right]. \quad (4.9.18)$$

Corollary 4.33 *Formula (4.9.18) means*

$$I_{0+}^{\alpha}\left(t^{\beta-1} E_{2\alpha,\beta}(t^{2\alpha})\right)(x) = x^{\beta-1}\left[E_{\alpha,\beta}(x^{\alpha}) - E_{2\alpha,\beta}(x^{2\alpha})\right]. \quad (4.9.19)$$

4.10 Historical and Bibliographical Notes

The two-parametric Mittag-Leffler function first appeared in the paper by Wiman 1905 [Wim05a], but he did not pay too much attention to it. Much later this function was rediscovered by Humbert and Agarwal, who studied it in detail in 1953 [Aga53] (see also [Hum53, HumAga53]). A new function was obtained by replacing the additive constant 1 in the argument of the Gamma function in (3.1.1) by an arbitrary complex parameter β. Later, when we deal with Laplace transform pairs, the parameter β will be required to be positive like α.

Using the integral representations for $E_{\alpha,\beta}(z)$ Dzherbashian [Dzh54a, Dzh54b], [Dzh66, Chap. III, Sect. 2] proved formulas for the asymptotic representation of $E_{\alpha,\beta}(z)$ at infinity, and in [Dzh66, Chap. III, Sect. 4] he gave applications of these to the construction of Fourier type integrals and to the proof of theorems on pointwise convergence of these integrals on functions defined and summable with exponential-power weight on a finite system of rays. Note that the developed technique is based on the representation of entire functions in the form of sums of integral transforms with kernels of the form $E_{\alpha,\beta}(z)$.

By using asymptotic properties of the function $E_{\alpha,\beta}(z)$, Dzherbashian (see [Dzh66, Chap. III, Sect. 2]) found its Mellin transform, established certain functional identities and proved the inversion formula for the following integral transform with the function $E_{\alpha,\beta}(z)$ in the kernel

$$\int_0^\infty E_{\alpha,\beta}(e^{i\varphi}x^\alpha t^\alpha)t^{\beta-1}f(t)dt \qquad (4.10.1)$$

in the space $L_2(\mathbb{R}_+)$.

In [Bon-et-al02] the properties of the integral transforms with Mittag-Leffler function in the kernel

$$\int_0^\infty E_{\alpha,\beta}(-xt)f(t)dt \quad (x > 0) \qquad (4.10.2)$$

are studied in weighted spaces of r-summable functions

$$\mathcal{L}_{\nu,r} = \left\{ f : \|f\|_{\nu,r} \equiv \left(\int_0^\infty |t^\nu f(t)| \frac{dt}{t}\right)^{1/r}, \quad 1 \leq r < \infty, \quad \nu \in \mathbb{R} \right\}. \qquad (4.10.3)$$

The conditions for the boundedness of such an operator as a mapping from one space to another were found, the images of these spaces under such a mapping were described, and inversion formulas were established. These results are based on the representation of (4.10.2) as a special case of the general **H**-transform (see Sect. F.3 in Appendix F).

In recent years mathematicians' attention towards the Mittag-Leffler type functions has increased, both from the analytical and numerical point of view, overall because of their relation to the fractional calculus. In addition to the books and papers already quoted in the text, here we would like to draw the reader's attention to some recent papers on the Mittag-Leffler type functions, e.g., Al Saqabi and Tuan [Al-STua96], Kilbas and Saigo [KilSai96], Gorenflo, Luchko and Rogosin [GoLuRo97] and Mainardi and Gorenflo [MaiGor00]. Since the fractional calculus has now received wide interest for its applications in different areas of physics and engineering, we expect that the Mittag-Leffler function will soon occupy its place as the Queen Function of Fractional Calculus.

The remarkable asymptotic properties of the Mittag-Leffler function have provoked an interest in the investigation of the distribution of the zeros of $E_{\alpha,\beta}(z)$. Several articles have been devoted to this problem (see [Dzh84, DzhNer68, OstPer97,

4.10 Historical and Bibliographical Notes

Poly21, Pop02, Psk05, Psk06, Sed94, Sed00, Wim05b]). An extended survey of the results is presented in [PopSed11]. Also studied is the related question of the distribution of zeros of sections and tails of the Mittag-Leffler function (see [Ost01, Zhe02]) and of some associated special functions (see [GraCso06, Luc00]).

The obtained results have found an application in the study of certain problems in spectral theory (see, e.g. [Dzh70, Djr93, Nak03]), approximation theory (see, e.g. [Sed98]), and in treating inverse problems for abstract differential equations (see, e.g., [TikEid02, TikEid05, KarTik17]).

Except in the case when $\alpha = 1$, $\beta = -m$, $m \in \{-1\} \cup \mathbb{Z}_+$, the function $E_{\alpha,\beta}(z)$ has an infinite set of zeros (see [Sed94]). In [Wim05b] it was shown that for $\alpha \geq 2$ all zeros of the classical Mittag-Leffler function $E_{\alpha,1}(z)$ are negative and simple (see also [Poly21], where the case $\alpha = N \in \mathbb{N}$, $N > 1$, is considered). In [Dzh84] it was proved that same result is valid for $E_{2,\beta}(z)$, $1 < \beta < 3$. Note that all zeros of the function $E_{2,3}(z) = \cosh \frac{\sqrt{z}-1}{z}$ are twofold and negative, but the function $E_{2,\beta}(z)$, $\beta > 3$, has no real zero.

Ostrovski and Pereselkova [OstPer97] formulated the problem to describe the set $\mathcal{W}$ of pairs (α, β) such that all zeros of $E_{\alpha,\beta}(z)$ are negative and simple. The authors conjectured that

$$\mathcal{W} = \{(\alpha, \beta) | \alpha \geq 2, 0 < \beta < 1 + \alpha\}.$$

It was shown, in particular, that $(\alpha, 1)$, $(\alpha, 2) \in \mathcal{W}$ for all real $\alpha \geq 2$, and $\{(2^m, \beta) | m \in \mathbb{N}, 0 < \beta < 1 + 2^m\} \subset \mathcal{W}$.

The asymptotic behavior of the zeros of the function $E_{\alpha,\beta}(z)$ is the subject of several investigations. In [Sed94] asymptotic formulas for the zeros $z_n(\alpha, \beta)$ of $E_{\alpha,\beta}(z)$ were found for all $\alpha > 0$ and $\beta \in \mathbb{C}$. For $0 < \alpha < 2$ this asymptotic representation as $n \to \pm\infty$ is more exact and has the form

$$(z_n(\alpha, \beta))^{1/\alpha} = 2\pi i n + a(\alpha, \beta) \left(\log |n| + \frac{\pi i}{2} \text{sign } n \right)$$
$$+ b(\alpha, \beta) + O\left(n^{-\alpha} + \frac{1}{n} \log |n|\right).$$

The values of $a(\alpha, \beta)$, $b(\alpha, \beta)$ are given in [Sed94]. A way of enumerating the zeros compatible with this asymptotical formula is proposed in [Sed00].

The material of Sect. 4.6.1 is due to the paper [TikEid02]. Note that another proof of the uniqueness theorem is given in [KarTik17] for the following inverse problem for an abstract evolution differential equation in a Banach space E

$$\frac{du(t)}{dt} = Au(t) + g,$$

$$u(0) = u_0, \quad \frac{1}{T} \int_0^T u(t) dt = u_1,$$

where A is a closed linear operator with domain $D(A) \subset E$ and the elements $u_0, u_1 \in E$ are given. It was shown that the uniqueness of such a problem is related to the distribution of zeros of the Mittag-Leffler function $E_{1,3}(z)$. The distribution of these zeros is studied in detail in [KarTik17].

Schneider [Sch96] has proved that the generalized Mittag-Leffler function $E_{\alpha,\beta}(-x)$ is completely monotonic for positive values of parameters α, β if and only if $0 < \alpha \leq 1$, $\beta \geq \alpha$. The proof was based on the use of the corresponding probability measures and the Hankel integration path.

An analytic proof presented in Sect. 4.1.5 is due to Miller and Samko [MilSam97]. Note that the main formula (4.9.7) used in this proof is a special case of a more general relation due to Dzherbashian [Dzh66, p. 120] which states that $x^{\beta+\gamma-1} E_{\alpha,\beta+\gamma}(-x^\alpha)$ is the fractional integral of order γ of the function $x^{\beta-1} E_{\alpha,\beta}(-x^\alpha)$. However, the result presented above is more simple and straightforward.

Later (see, [MilSam01]), the proof of the complete monotonicity of some other special functions was given by Miller and Samko, see also [Mai10]. One of the possible approaches to the proof of complete monotonicity is the use of the Volterra and Bernstein functions, see [Ape08, Bern28, Boc37].

As a challenging open problem related to the Special Functions of Fractional Calculus (such as the multi-index Mittag-Leffler functions), we mention the possibility of their numerical computation and graphical interpretation, plots and tables, and implementations in software packages such as Mathematica, Maple, Matlab, etc. As mentioned earlier, the Classical Special Functions are already implemented there. For their Fractional Calculus analogues, numerical algorithms and software packages have been developed only for the classical Mittag-Leffler function $E_{\alpha;\beta}(z)$ and the Wright function $\phi(\alpha, \beta; z)$! Numerical results and plots for the Mittag-Leffler functions for basic values of indices can be found in Caputo–Mainardi [CapMai71b] (one of the first attempts!) and Gorenflo–Mainardi [GorMai97]. Among the very recent achievements, we mention the following results: Podlubny [Pod06, Pod11, PodKac09] (a Matlab routine that calculates the Mittag-Leffler function with desired accuracy), Gorenflo et al. [GoLoLu02], Diethelm et al. [Die-et-al05] (algorithms for the numerical evaluation of the Mittag-Leffler function and a package for computation with Mathematica), Hilfer–Seybold [HilSey06] (an algorithm for extensive numerical calculations for the Mittag-Leffler function in the whole complex plane, based on its integral representations and exponential asymptotics), Luchko [Luc08] (algorithms for computation of the Wright function with prescribed accuracy), etc.

The results concerning calculation of the Mittag-Leffler function presented in Sect. 4.7 are based on the paper [GoLoLu02]. In [GoLoLu02] a numerical scheme for computation of the Mittag-Leffler function is given in pseudocode using a specially developed algorithm based on the above formulated results (see also the MatLab routine by Podlubny [Pod06, Pod11], and numerical computations of the Mittag-Leffler function performed by Hilfer and Seybold [SeyHil05, HilSey06, SeyHil08]).

Efficient techniques for the computation of the Mittag-Leffler functions with two parameters based on the numerical inversion of the Laplace transform have be studied in [GarPop13, Gar15]: essentially, a quadrature rule is applied on a suitably selected contour in the complex plane obtained after deforming the Bromwich line. Parabolic contours are usually chosen since their simple geometry allows an in-depth error analysis leading to the accurate tuning of the main parameters, thus obtaining high accuracy. A Matlab code for the evaluation of the Mittag-Leffler function based on the results in [Gar15] is freely available on the Mathworks website.[2]

4.11 Exercises

4.11.1 ([Ber-S05b]) Prove the following relations:

$$E_\alpha(-x) = E_{2\alpha}(x^2) - xE_{2\alpha,1+\alpha}(x^2), \quad x \in \mathbb{R}, \quad \operatorname{Re}\alpha > 0, \quad (4.11.1a)$$

$$E_\alpha(-ix) = E_{2\alpha}(-x^2) - ixE_{2\alpha,1+\alpha}(-x^2), \quad x \in \mathbb{R}, \quad \operatorname{Re}\alpha > 0. \quad (4.11.1b)$$

4.11.2 ([SaKaKi03]) Prove the following recurrence relation

$$z^m E_{\alpha,\beta+m\alpha}(z) = E_{\alpha,\beta}(z) - \sum_{n=0}^{m-1} \frac{z^n}{\Gamma(\beta+n\alpha)}, \quad \operatorname{Re}\alpha > 0, \ \operatorname{Re}\beta > 0, \ m \in \mathbb{N}.$$

4.11.3 ([Ber-S05b, p. 432]) Let the family of functions H_α be given by the formula

$$H_\alpha(k) = \frac{2}{\pi} \int_0^\infty E_{2\alpha}(-t^2) \cdot \cos(kt) dt, \quad k > 0, \ 0 \leq \alpha \leq 1,$$

where its power series in k have the form

$$H_\alpha(k) = \frac{1}{\pi} \sum_{n=0}^\infty b_n(\alpha) k^n, \quad 0 \leq \alpha < 1.$$

Deduce the following asymptotic formula for $E_\alpha(-x)$:

$$E_\alpha(-x) = \frac{1}{\pi} \sum_{n=0}^\infty \frac{b_n(\alpha)}{x^{n+1}}, \quad 0 \leq \alpha < 1.$$

Hint. Use the relation (4.11.1a).

4.11.4 Using the series representation of the two-parametric Mittag-Leffler function (4.1.1) prove the following recurrence relations

---
[2] www.mathworks.com/matlabcentral/fileexchange/48154-the-mittag-leffler-function.

$$E_{1,1}(z) + E_{1,1}(-z) = 2 E_{2,1}(z^2) \iff e^z + e^{-z} = 2 \cosh(z),$$
$$E_{1,1}(z) - E_{1,1}(-z) = 2z E_{2,2}(z^2) \iff e^z - e^{-z} = 2 \sinh(z),$$
(4.11.4a)

or in a more general form:

$$E_{\alpha,\beta}(z) + E_{\alpha,\beta}(-z) = 2 E_{2\alpha,\beta}(z^2),$$
$$E_{\alpha,\beta}(z) - E_{\alpha,\beta}(-z) = 2z E_{2\alpha,\alpha+\beta}(z^2);$$
(4.11.4b)

$$E_{1,3}(z) = \frac{e^z - 1 - z}{z^2}$$
(4.11.4c)

or in a more general form (for any $m \in \mathbb{N}$):

$$E_{1,m}(z) = \frac{1}{z^{m-1}} \left\{ e^z - \sum_{k=0}^{m-2} \frac{z^k}{k!} \right\}.$$
(4.11.4d)

4.11.5 With $\alpha > 0$ show that

$$t^{\alpha-1} E_{\alpha,\alpha}(-t^\alpha) = -\frac{d}{dt} E_\alpha(-t^\alpha).$$

4.11.6 Prove the following integral representations ([Bra96, p. 58])

$$\frac{1}{1+z^\alpha} = \int_0^\infty e^{-zx} E_{\alpha,\alpha}(-x^\alpha) dx;$$
(4.11.6a)

$$\log(1 + z^\alpha) = \int_0^\infty (1 - e^{-zx}) \frac{\alpha E_{\alpha,1}(-x^\alpha)}{x} dx.$$
(4.11.6b)

4.11.7 Prove the following differential relations for the two-parametric Mittag-Leffler function ([GupDeb07])

$$3 E_{1,4}(z) + 5z E'_{1,4} + z^2 E''_{1,4} = E_{1,2} - E_{1,3}.$$
(4.11.7a)

$$n(n+2) E_{\alpha,n+3}(z) + z\alpha[2n + \alpha + 2] E'_{\alpha,n+3} + z^2 E''_{\alpha,n+3} = E_{\alpha,n+1} - E_{\alpha,n+2},$$
(4.11.7b)

which hold for any $\alpha > 0$ and any $n = 1, 2, \ldots$.

4.11.8 Prove the following Laplace transform pair for the auxiliary functions of Mittag-Leffler type defined below

$$e_{\alpha,\beta}(t;\lambda) := t^{\beta-1} E_{\alpha,\beta}(-\lambda t^\alpha) \div \frac{s^{\alpha-\beta}}{s^\alpha + \lambda} = \frac{s^{-\beta}}{1 + \lambda s^{-\alpha}}.$$

4.11 Exercises

4.11.9 ([HaMaSa11]) Evaluate the following integrals:

$$\int_0^x \frac{E_\alpha(t^\alpha)}{(x-t)^{1-\beta}} dt \tag{4.11.9a}$$

for $\operatorname{Re}\alpha > 0$, $\operatorname{Re}\beta > 0$.

$$\int_0^\infty e^{-st} t^{m\alpha+\beta-1} E_{\alpha,\beta}^{(m)}(\pm at^\alpha) dt \tag{4.11.9b}$$

for $\operatorname{Re} s > 0$, $\operatorname{Re}\alpha > 0$, $\operatorname{Re}\beta > 0$, where

$$E_{\alpha,\beta}^{(m)}(z) = \frac{d^m}{dz^m} E_{\alpha,\beta}(z).$$

Answers.
(4.11.9a)

$$\Gamma(\beta) x^\beta E_{\alpha,\beta+1}(x^\alpha).$$

(4.11.9b)

$$\frac{m! s^{\alpha-\beta}}{(s^\alpha \mp a)^{m+1}}.$$

4.11.10 Prove the following formulas for half-integer values of parameters [Han-et-al09]:

$$E_{1/2,1/2}(\pm x) = \frac{1}{\sqrt{x}} \pm x e^{x^2} [1 \pm \operatorname{erf}(x)]. \tag{4.11.10a}$$

$$E_{1/2,1}(\pm x) = e^{x^2}[1 \pm \operatorname{erf}(x)]. \tag{4.11.10b}$$

$$E_{1,1/2}(+x) = \frac{1}{\sqrt{x}} + \sqrt{x} e^{+x} \operatorname{erf}(\sqrt{x}). \tag{4.11.10c}$$

$$E_{1,1/2}(-x) = \frac{1}{\sqrt{x}} + i\sqrt{x} e^{-x} \operatorname{erf}(i\sqrt{x}). \tag{4.11.10d}$$

$$E_{1,3/2}(+x) = e^{+x} \frac{\operatorname{erf}(\sqrt{x})}{\sqrt{x}}. \tag{4.11.10e}$$

$$E_{1,3/2}(-x) = -i e^{-x} \frac{\operatorname{erf}(i\sqrt{x})}{\sqrt{x}}. \tag{4.11.10f}$$

4.11.11 Prove the following formulas for negative integer values of parameters [Han-et-al09]:

$$E_{-1,2}\left(\pm\frac{1}{x}\right) = 1 \pm \frac{1-e^{\pm x}}{x}. \quad (4.11.11a)$$

$$E_{-2,1}\left(+\frac{1}{x^2}\right) = 1 - \cosh x. \quad (4.11.11b)$$

$$E_{-2,1}\left(-\frac{1}{x^2}\right) = 1 - \cos x. \quad (4.11.11c)$$

$$E_{-2,2}\left(+\frac{1}{x^2}\right) = 1 - \frac{\sinh x}{x}. \quad (4.11.11d)$$

$$E_{-2,2}\left(-\frac{1}{x^2}\right) = 1 - \frac{\sin x}{x}. \quad (4.11.11e)$$

4.11.12 Prove the following formulas for negative semi-integer values of parameters [Han-et-al09]:

$$E_{-1/2,1/2}\left(\pm\frac{1}{x}\right) = \mp x e^{x^2}[1 \pm \mathrm{erf}(x)]. \quad (4.11.12a)$$

$$E_{-1/2,1}\left(\pm\frac{1}{x}\right) = 1 - e^{x^2}[1 \pm \mathrm{erf}(x)]. \quad (4.11.12b)$$

$$E_{-1,1/2}\left(+\frac{1}{x}\right) = \sqrt{x}e^{+x}\mathrm{erf}(\sqrt{x}). \quad (4.11.12c)$$

$$E_{-1,1/2}\left(-\frac{1}{x}\right) = -i\sqrt{x}e^{-x}\mathrm{erf}(i\sqrt{x}). \quad (4.11.12c)$$

$$E_{-1,3/2}\left(+\frac{1}{x}\right) = \frac{2}{\sqrt{x}} - e^{+x}\frac{\mathrm{erf}(\sqrt{x})}{\sqrt{x}}. \quad (4.11.12d)$$

$$E_{-1,3/2}\left(-\frac{1}{x}\right) = \frac{2}{\sqrt{x}} + ie^{-x}\frac{\mathrm{erf}(i\sqrt{x})}{\sqrt{x}}. \quad (4.11.12e)$$

4.11.13 Prove the following formula for the Laplace transform of the derivatives of the Mittag-Leffler function ([KiSrTr06, p. 50]):

$$\left(\mathcal{L}t^{\alpha n+\beta-1}\left(\frac{\partial}{\partial\lambda}\right)^n E_{\alpha,\beta}(\lambda t^\alpha)\right)(s) = \frac{n!s^{\alpha-\beta}}{(s^\alpha-\lambda)^{n+1}} \quad (|\lambda s^{-\alpha}|<1).$$

4.11 Exercises

4.11.14 ([Cap13]) Prove the following relations for the Mittag-Leffler functions with positive integer values of parameters (Capelas relations).

$$\sum_{k=1}^{m} z^{k-1} E_{m,k}(z^m) = e^z, \quad m \in \mathbb{N}, \tag{4.11.14a}$$

$$E_{1,m}(z) = \frac{1}{z^{m-1}} \left(e^z - \sum_{k=0}^{m-2} \frac{z^k}{k!} \right), \quad m \in \mathbb{N}. \tag{4.11.14b}$$

4.11.15 ([BanPra16]). Let

$$\mathbb{E}_{\alpha,\beta}(z) := \Gamma(\beta) z E_{\alpha,\beta}(z).$$

Prove the following relations of this function to some elementary functions

$$\begin{cases} \mathbb{E}_{0,1}(z) = \frac{z}{1-z}; \ \mathbb{E}_{1,1}(z) = ze^z; \ \mathbb{E}_{2,1}(z) = z\cosh(\sqrt{z}); \\ \mathbb{E}_{1,2}(z) = e^z - 1; \ \mathbb{E}_{1,3}(z) = \frac{2(e^z - 1 - z)}{z}; \\ \mathbb{E}_{1,4}(z) = \frac{6(e^z - 1 - z) - 3z^2}{z^2}; \ \mathbb{E}_{2,2}(z) = \sqrt{z}\sinh(\sqrt{z}); \\ \mathbb{E}_{3,1}(z) = \frac{z}{3} \left[e^{\sqrt[3]{z}} + 2e^{-\frac{\sqrt[3]{z}}{2}} \cos\left(\frac{\sqrt{3}}{2}\sqrt[3]{z}\right) \right]. \end{cases}$$

4.11.16 ([AnsShe14]) Prove the following (inverse Laplace transform of the Mittag-Leffler function)

$$t^{\beta-1} E_{\alpha,\beta}(-\lambda t^\alpha) = \frac{1}{\pi} \int_0^\infty e^{-rt} \frac{r^{2\alpha-\beta} \sin(\beta\pi) - \lambda r^{\alpha-\beta} \sin((\alpha-\beta)\pi)}{\lambda^2 + 2\lambda r^\alpha \cos(\alpha\pi) + r^{2\alpha}} dr.$$

4.11.17 ([Tua17]) In the notation of Exercise 3.9.9 prove the following inequality

$$\left| t^{\alpha-1} E_{\alpha,\alpha}(\lambda t^{\frac{1}{\alpha}}) - \frac{1}{\alpha} \lambda^{\frac{1-\alpha}{\alpha}} \exp(\lambda^{\frac{1}{\alpha}} t) \right| < \frac{M_2(\alpha, \lambda)}{t^{\alpha+1}}, \quad \forall t \geq t_0,$$

where

$$M_2(\alpha, \lambda) = \frac{\int_{\gamma(1,\theta)} \left| \exp(\zeta^{\frac{1}{\alpha}}) \zeta^{\frac{1}{\alpha}} \right| d\zeta}{2\pi \alpha |\lambda|^2 \sin \theta_0}, \quad t_0 = \frac{1}{|\lambda|^{\frac{1}{\alpha}}(1 - \sin \theta_0)^{\frac{1}{\alpha}}}.$$

4.11.18 ([WaZhOR18])

(a) Let $\alpha \in (0, 1]$, $\beta \in \mathbb{R}$. Prove that the following inequality holds for all $t > 0$:

$$\left| t^{\beta-1} E_{\alpha,\beta}(\lambda t^\alpha) - \frac{1}{\alpha} \lambda^{(1-\beta)/\alpha} \exp\left(\lambda^{1/\alpha} t\right) \right| \leq \frac{m_1(\alpha, \beta, \lambda)}{t^{2\alpha-\beta+1}} + \frac{m_2(\alpha, \beta, \lambda)}{t^{\alpha-\beta+1}},$$

where

$$m_1(\alpha, \beta, \lambda) = \frac{|\sin(\pi\beta)| \int_0^\infty r^{(1-\beta+\alpha)/\alpha} \exp\left(-r^{1/\alpha}\right) dr}{\sin^2(\pi\alpha)\pi\alpha\lambda^2},$$

$$m_2(\alpha, \beta, \lambda) = \frac{|\sin(\pi(\beta-\alpha))| \int_0^\infty r^{(1-\beta)/\alpha} \exp\left(-r^{1/\alpha}\right) dr}{\sin^2(\pi\alpha)\pi\alpha\lambda}.$$

(b) Let $\alpha \in (1, 2]$, $\beta > 0$. Prove that the following inequality holds for all $t > 0$:

$$\left| t^{\beta-1} E_{\alpha,\beta}(\lambda t^\alpha) - \frac{1}{\alpha} \lambda^{(1-\beta)/\alpha} \exp\left(\lambda^{1/\alpha} t\right) \right|$$

$$\leq m(\alpha, \beta, \lambda) \left(\frac{1}{t^{2\alpha-\beta+1}} + \frac{1}{t^{\alpha-\beta+1}} \right) + \frac{2}{\alpha\lambda^{1-1/\alpha}} \exp\left(t\lambda^{1/\alpha} \cos\left(\frac{\pi}{\alpha}\right) \right),$$

where

$$m(\alpha, \beta, \lambda) = \max\{m_1(\alpha, \beta, \lambda), m_2(\alpha, \beta, \lambda)\}.$$

4.11.19 ([WaZhOR18]) Let α, β, λ be arbitrary positive numbers. Prove that the following inequality holds for all $t > 0$:

$$-\frac{1}{\Gamma(\beta+1)} + \beta E_{\alpha,\beta+1}(-\lambda t^\alpha) \leq E_{\alpha,\beta}(-\lambda t^\alpha) \leq \beta E_{\alpha,\beta+1}(-\lambda t^\alpha).$$

4.11.20 ([PrMaBa18]) Let $0 \leq \gamma < 1$, $\alpha \geq 1$. Prove that for all $\beta > \beta_2$ the normalized Mittag-Leffler function $\mathbb{E}_{\alpha,\beta}(z) = \Gamma(\beta) z E_{\alpha,\beta}(z)$ belongs to the following Hardy spaces in the unit disc $\mathbb{D}$:

$$\mathbb{E}_{\alpha,\beta}(z) = \begin{cases} H^{\frac{1}{1-2\gamma}}(\mathbb{D}), & \gamma \in [0, 1/2), \\ H^\infty(\mathbb{D}), & \gamma \geq 1/2, \end{cases}$$

where β_2 is the largest root of the equation

$$(1-\gamma)(\beta^2 - \beta - 1)(\beta^2 - 4\beta + 3) - (2-\gamma)\beta(\beta^- 2\beta - 3) - \beta(\beta^2 - 1) = 0.$$

4.11.21 ([Rad16]) Let $\mathbb{E}_{\alpha,\beta}(z) = \Gamma(\beta) z E_{\alpha,\beta}(z)$ be the normalized Mittag-Leffler function. Denote by $(\mathbb{E}_{\alpha,\beta})_m(z)$ its $m+1$-th partial sums:

$$(\mathbb{E}_{\alpha,\beta})_0(z) = z, \quad (\mathbb{E}_{\alpha,\beta})_m(z) = z + \sum_{n=1}^m A_n z^{n+1}, \; m \in \mathbb{N}.$$

1. Prove that for all $\alpha \geq 1$, $\beta \geq \frac{1+\sqrt{5}}{2}$ the following inequalities hold for all z in the unit disc $\mathbb{D}$:

4.11 Exercises

$$\operatorname{Re}\left\{\frac{\mathbb{E}_{\alpha,\beta}(z)}{(\mathbb{E}_{\alpha,\beta})_m(z)}\right\} \geq \frac{\beta^2 - \beta - 1}{\beta^2}, \qquad (4.11.21.1a)$$

$$\operatorname{Re}\left\{\frac{(\mathbb{E}_{\alpha,\beta})_m(z)}{\mathbb{E}_{\alpha,\beta}(z)}\right\} \geq \frac{\beta^2}{\beta^2 + \beta + 1}. \qquad (4.11.21.1b)$$

2. Prove that for all $\alpha \geq 1$, $\beta \geq \frac{3+\sqrt{17}}{2}$ the following inequalities hold for all z in the unit disc $\mathbb{D}$:

$$\operatorname{Re}\left\{\frac{\mathbb{E}'_{\alpha,\beta}(z)}{(\mathbb{E}_{\alpha,\beta})'_m(z)}\right\} \geq \frac{\beta^2 - 3\beta - 2}{\beta^2}, \qquad (4.11.21.2a)$$

$$\operatorname{Re}\left\{\frac{(\mathbb{E}_{\alpha,\beta})'_m(z)}{\mathbb{E}'_{\alpha,\beta}(z)}\right\} \geq \frac{\beta^2}{\beta^2 + 3\beta + 2}. \qquad (4.11.21.2b)$$

Chapter 5
Mittag-Leffler Functions with Three Parameters

5.1 The Prabhakar (Three-Parametric Mittag-Leffler) Function

5.1.1 Definition and Basic Properties

The Prabhakar generalized Mittag-Leffler function [Pra71] is defined as

$$E_{\alpha,\beta}^{\gamma}(z) := \sum_{n=0}^{\infty} \frac{(\gamma)_n}{n!\,\Gamma(\alpha n + \beta)} z^n, \quad Re\,(\alpha) > 0,\ Re\,(\beta) > 0,\ \gamma > 0, \quad (5.1.1)$$

where $(\gamma)_n = \gamma(\gamma+1)\ldots(\gamma+n-1)$ (see formula (A.17) in Appendix A).

For $\gamma = 1$ we recover the two-parametric Mittag-Leffler function

$$E_{\alpha,\beta}(z) := \sum_{n=0}^{\infty} \frac{z^n}{\Gamma(\alpha n + \beta)}, \quad (5.1.2)$$

and for $\gamma = \beta = 1$ we recover the classical Mittag-Leffler function

$$E_{\alpha}(z) := \sum_{n=0}^{\infty} \frac{z^n}{\Gamma(\alpha n + 1)}. \quad (5.1.3)$$

Let $\alpha, \beta > 0$. Then termwise Laplace transformation of series (5.1.1) yields

$$\int_0^\infty e^{-st}\, t^{\beta-1}\, E_{\alpha,\beta}^{\gamma}(at^\alpha)\,\mathrm{d}t = s^{-\beta} \sum_{n=0}^{\infty} \frac{\Gamma(\gamma+n)}{\Gamma(\gamma)} \left(\frac{a}{s}\right)^n. \quad (5.1.4)$$

On the other hand (binomial series!)

$$(1+z)^{-\gamma} = \sum_{n=0}^{\infty} \frac{\Gamma(1-\gamma)}{\Gamma(1-\gamma-n)n!} z^n = \sum_{n=0}^{\infty} (-1)^n \frac{\Gamma(\gamma+n)}{\Gamma(\gamma)n!} z^n. \qquad (5.1.5)$$

Comparison of (5.1.4) and (5.1.5) yields the Laplace transform pair

$$t^{\beta-1} E_{\alpha,\beta}^{\gamma}(at^{\alpha}) \div \frac{s^{-\beta}}{(1-as^{-\alpha})^{\gamma}}. \qquad (5.1.6)$$

Equation (5.1.6) holds (by analytic continuation) for $\operatorname{Re} \alpha > 0$, $\operatorname{Re} \beta > 0$.
In particular we get the known Laplace transform pairs

$$t^{\beta-1} E_{\alpha,\beta}(at^{\alpha}) \div \frac{s^{\alpha-\beta}}{s^{\alpha}-a}, \qquad (5.1.7)$$

$$E_{\alpha}(at^{\alpha}) \div \frac{s^{\alpha-1}}{s^{\alpha}-a}. \qquad (5.1.8)$$

Note that the pre-factor $t^{\beta-1}$ is essential for the above Laplace transform pairs.

From the above Laplace transform pair one can obtain the complete monotonicity (CM) of the function

$$E_{\alpha,1}^{\gamma}(-t^{\alpha}) \div \frac{1}{s(1+s^{-\alpha})^{\gamma}}. \qquad (5.1.9)$$

The proof is based on the following theorem (see [GrLoSt90, Theorem 2.6]): if the real-valued function $F(x)$, $x > 0$; $\lim_{x\to+\infty} F(x) = 0$, possesses an analytic continuation in $\mathbb{C} \setminus \mathbb{R}_-$ and satisfies the inequalities $\operatorname{Im} zF(z) \geq 0$ for $\operatorname{Im} z > 0$ and $\operatorname{Im} F(x) \geq 0$ for $0 < x <= +\infty$, then F is the Laplace transform of a function, locally integrable on $(0, +\infty)$ and completely monotone on this interval.

The conditions of this theorem for the function $\frac{1}{s(1+s^{-\alpha})^{\gamma}}$ can be verified directly (cf., e.g.. [HanSer08, p. 292], [OrsPol09]). A more general result holds for all $\lambda > 0$ (see, e.g.. [CdOMai11]) for the function $e_{\alpha,\beta}^{\gamma}(t;\lambda) := t^{\beta-1} E_{\alpha,\beta}^{\gamma}(-\lambda t^{\alpha})$:

$$e_{\alpha,\beta}^{\gamma}(t;\lambda) = t^{\beta-1} E_{\alpha,\beta}^{\gamma}(-\lambda t^{\alpha}) \quad \text{CM} \quad \text{iff} \quad \begin{cases} 0 < \alpha, \beta \leq 1, \\ 0 < \gamma \leq \beta/\alpha. \end{cases} \qquad (5.1.10)$$

In the same manner we recover the known result

$$E_{\alpha}(-t^{\alpha}) \quad \text{CM} \quad \text{if} \quad 0 < \alpha \leq 1. \qquad (5.1.11)$$

Cases of Reducibility

Here we present some formulas connecting the values of three-parametric (Prabhakar) Mittag-Leffler functions with different values of parameters (see, e.g.. [MatHau08]).

5.1 The Prabhakar (Three-Parametric Mittag-Leffler) Function

(i) If $\alpha, \beta, \gamma \in \mathbb{C}$ are such that $\mathrm{Re}\,\alpha > 0$, $\mathrm{Re}\,\beta > 0$, $\mathrm{Re}\,(\beta - \alpha) > 0$, then

$$zE^\gamma_{\alpha,\beta} = E^\gamma_{\alpha,\beta-\alpha} - E^{\gamma-1}_{\alpha,\beta-\alpha}. \tag{5.1.12}$$

(ii) If $\alpha, \beta \in \mathbb{C}$ are such that $\mathrm{Re}\,\alpha > 0$, $\mathrm{Re}\,\beta > 0$, $(\alpha - \beta) \notin \mathbb{N}_0$, then

$$zE^1_{\alpha,\beta} = E_{\alpha,\beta-\alpha} - \frac{1}{\Gamma(\beta-\alpha)}. \tag{5.1.13}$$

(iii) If $\alpha, \beta \in \mathbb{C}$ are such that $\mathrm{Re}\,\alpha > 0$, $\mathrm{Re}\,\beta > 1$, then

$$\alpha E^2_{\alpha,\beta} = E_{\alpha,\beta-1} - (1 + \alpha - \beta)E_{\alpha,\beta}. \tag{5.1.14}$$

Differentiation of the Three-Parametric Mittag-Leffler Function

If $\alpha, \beta, \gamma, z, w \in \mathbb{C}$, then for any $n = 1, 2, \ldots$, and any β, $\mathrm{Re}\,\beta > n$, the following formula holds:

$$\left(\frac{d}{dz}\right)^n \left[z^{\beta-1} E^\gamma_{\alpha,\beta}(wz^\alpha)\right] = z^{\beta-n-1} E^\gamma_{\alpha,\beta-n}(wz^\alpha). \tag{5.1.15}$$

In particular, for any $n = 1, 2, \ldots$, and any β, $\mathrm{Re}\,\beta > n$,

$$\left(\frac{d}{dz}\right)^n \left[z^{\beta-1} E_{\alpha,\beta}(wz^\alpha)\right] = z^{\beta-n-1} E_{\alpha,\beta-n}(wz^\alpha) \tag{5.1.16}$$

and for any $n = 1, 2, \ldots$, and any β, $\mathrm{Re}\,\beta > n$,

$$\left(\frac{d}{dz}\right)^n \left[z^{\beta-1} \phi(\gamma, \beta; wz)\right] = \frac{\Gamma(\beta)}{\Gamma(\beta-n)} z^{\beta-n-1} \phi(\gamma, \beta-n; wz), \tag{5.1.17}$$

where

$$\phi(\gamma, \beta; z) := {}_1F_1(\gamma, \beta; z) = \Gamma(\beta) E^\gamma_{1,\beta}. \tag{5.1.18}$$

◁ To prove formula (5.1.15) one can use term-by-term differentiation of the power series representation of the three-parametric Mittag-Leffler function. Thus we get

$$\left(\frac{d}{dz}\right)^n \left[z^{\beta-1} E^\gamma_{\alpha,\beta}(wz^\alpha)\right] = \sum_{k=0}^\infty \frac{(\gamma)_k}{\Gamma(\alpha k + \beta)} \left(\frac{d}{dz}\right)^n \left[\frac{w^k z^{\alpha k + \beta - 1}}{k!}\right]$$

$$= z^{\beta-n-1} E^\gamma_{\alpha,\beta-n}(wz^\alpha), \quad \mathrm{Re}\,\beta > n,$$

and the result follows. ▷

Integrals of the Three-Parametric Mittag-Leffler Function

By integration of series (5.1.1) we get the following.

If $\alpha, \beta, \gamma, z, w \in \mathbb{C}$, $\operatorname{Re}\alpha > 0$, $\operatorname{Re}\beta > 0$, $\operatorname{Re}\gamma > 0$, then

$$\int_0^z t^{\beta-1} E_{\alpha,\beta}^\gamma(wt^\alpha) dt = z^\beta E_{\alpha,\beta+1}^\gamma(wz^\alpha). \tag{5.1.19}$$

In particular,

$$\int_0^z t^{\beta-1} E_{\alpha,\beta}(wt^\alpha) dt = z^\beta E_{\alpha,\beta+1}(wz^\alpha), \tag{5.1.20}$$

and

$$\int_0^z t^{\beta-1} \phi(\gamma, \beta; wz) dt = \frac{1}{\beta} z^\beta \phi(\gamma, \beta+1; wz). \tag{5.1.21}$$

5.1.2 Integral Representations and Asymptotics

As for any function of the Mittag-Leffler type, the three parametric Mittag-Leffler function can be represented via the Mellin–Barnes integral.

Let $\alpha \in \mathbb{R}_+$, $\beta, \gamma \in \mathbb{C}$, $\beta \neq 0$, $\operatorname{Re}\gamma > 0$. Then we have the representation

$$E_{\alpha,\beta}^\gamma(z) = \frac{1}{\Gamma(\gamma)} \frac{1}{2\pi i} \int_L \frac{\Gamma(s)\Gamma(\gamma - s)}{\Gamma(\beta - \alpha s)} (-z)^{-s} ds, \tag{5.1.22}$$

where $|\arg z| < \pi$, the contour of integration begins at $c - i\infty$, ends at $c + i\infty$, $0 < c < \operatorname{Re}\gamma$, and separates all poles of the integrand at $s = -k$, $k = 0, 1, 2, \ldots$ to the left and all poles at $s = n + \gamma$, $n = 0, 1, \ldots$ to the right.

◁ The integral in the R.H.S. of (5.1.22) is equal to the sum of residues at the poles $s = 0, -1, -2, \ldots$. Hence

$$\int_L \frac{\Gamma(s)\Gamma(\gamma - s)}{\Gamma(\beta - \alpha s)}(-z)^{-s} ds = \sum_{k=0}^\infty \lim_{s \to -k}\left[\frac{(s+k)\Gamma(s)\Gamma(\gamma - s)(-z)^{-s}}{\Gamma(\beta - \alpha s)}\right]$$

$$= \sum_{k=0}^\infty \frac{(-1)^k}{k!} \frac{\Gamma(\gamma + k)}{\Gamma(\beta + \alpha k)}(-z)^k$$

$$= \Gamma(\gamma) \sum_{k=0}^\infty \frac{(\gamma)_k}{\Gamma(\beta + \alpha k)} \frac{z^k}{k!} = \Gamma(\gamma) E_{\alpha,\beta}^\gamma(z),$$

and thus (5.1.22) follows. ▷

5.1 The Prabhakar (Three-Parametric Mittag-Leffler) Function

If β is a sufficiently large real number, one can use Stirling's formula, valid for any fixed a

$$\Gamma(z+a) \approx \sqrt{2\pi} z^{z+a-1/2} e^{-z}, \text{ as } |z| \to \infty, \tag{5.1.23}$$

in order to get the following asymptotic formula ($a > 0, \alpha > 0, \beta > 0, \gamma > 0$)

$$\Gamma(\alpha) E_{\alpha,\beta}^{\gamma}(a(\alpha x)^{\gamma}) \approx \sum_{k=0}^{\infty} \frac{(\beta)_k a^k x^{\gamma k}}{k!} \frac{\sqrt{2\pi} \alpha^{\alpha-1/2} e^{-\alpha}}{\sqrt{2\pi} \alpha^{\alpha-1/2+\gamma k} e^{-\alpha}} \tag{5.1.24}$$

$$= \sum_{k=0}^{\infty} \frac{(\beta)_k}{k!} \left(a \left(\frac{x}{\alpha}\right)^{\gamma}\right)^k = \frac{1}{\left(1 + a\left(\frac{x}{\alpha}\right)^{\gamma}\right)^{\beta}}, \text{ as } x \to +\infty.$$

As in the case of the Mittag-Leffler function with two parameters, the asymptotic behavior of the three parametric function critically depends on the values of the parameters α, β, γ and cannot easily be described. In principle, an asymptotic expansion of the Prabhakar function can be found from its representation via a generalized Wright function or H-function (see Sect. 5.1.5 below) by using an approach of Braaksma [Bra62] (cf. [KiSrTr06]).

Asymptotic expansions of the Prabhakar function for large values of argument were given, e.g.., in [GarGar18]. On the basis of this work, the asymptotic expansion of the Fox–Wright functions for large arguments has been studied by Paris [Par10] in a paper describing an efficient algorithm for the derivation of the coefficients in the asymptotic expansion of Fox–Wright functions.

An asymptotic expansion will be considered for α a real positive number and β, γ arbitrary complex numbers with $\gamma \neq -1, -2, \ldots$. Following the approach proposed in [Par10], we introduce two functions

$$H(z) := \frac{z^{-\gamma}}{\Gamma(\gamma)} \sum_{k=0}^{\infty} \frac{(-1)^k \Gamma(k+\gamma)}{k! \Gamma(\beta - \alpha(k+\gamma))} z^{-k}, \tag{5.1.25}$$

$$F(z) := \frac{e^{z^{1/\alpha}} z^{\frac{\gamma-\beta}{\alpha}}}{\Gamma(\gamma) \alpha^{\gamma}} \sum_{k=0}^{\infty} c_k z^{-\frac{k}{\alpha}}, \tag{5.1.26}$$

where the coefficients $c_k = c_k(\alpha, \beta, \gamma)$ are obtained as coefficients in the following expansion

$$F_{\alpha,\beta}^{\gamma}(z) := \frac{\Gamma(s+\gamma)\Gamma(\alpha s + \psi)}{\Gamma(s+1)\Gamma(\alpha s + \beta)} = \alpha^{1-\gamma} \left(c_0 + \sum_{j=1}^{\infty} \frac{c_j}{(\alpha s + \psi)_j} \right) \tag{5.1.27}$$

for $|s| \to \infty$ in $|\arg s| \leq \pi - \varepsilon$ and any arbitrarily small $\varepsilon > 0$. Here $(x)_j = x(x+1)\ldots(x+j-1)$ denotes the Pochhammer symbol and $\psi := 1 - \gamma + \beta$.

The main results for the asymptotic expansions of the Prabhakar function can be given thanks to the following theorems related to the asymptotic expansion of the Wright function (for the proofs see [Wri35b], [ParKam01, Sect. 2.3] or [Par10, Theorems 1, 2 and 3]).

Theorem 5.1 ([GarGar18, Theorem 3]) *Let $0 < \alpha < 2$. Then*

$$E_{\alpha,\beta}^{\gamma}(z) \sim \begin{cases} F(z) + H(ze^{\mp pii}), & \text{if } |\arg z| < \frac{\pi\alpha}{2}, \\ H(ze^{\mp pii}), & \text{if } |\arg(-z)| < \frac{\pi(2-\alpha)}{2}, \end{cases} \quad (5.1.28)$$

as $|z| \to \infty$ with the sign in $H(ze^{\mp pii})$ being chosen according to whether z lies in the upper or lower half-plane, respectively.

Theorem 5.2 ([GarGar18, Theorem 4]) *Let $\alpha = 2$ and $|\arg z| \leq \pi$. Then*

$$E_{\alpha,\beta}^{\gamma}(z) \sim F(z) + F(ze^{\mp 2\pi i}) + H(ze^{\mp \pi i}), \text{ as } |z| \to \infty, \quad (5.1.29)$$

with the sign in $F(ze^{\mp 2\pi i})$ and $H(ze^{\mp \pi i})$ being chosen according to whether z lies in the upper or lower half-plane, respectively.

Theorem 5.3 ([GarGar18, Theorem 5]) *Let $\alpha > 2$ and $|\arg z| \leq \pi$. Then*

$$E_{\alpha,\beta}^{\gamma}(z) \sim \sum_{l=-P}^{P} F(ze^{2\pi i l}), \text{ as } |z| \to \infty, \quad (5.1.30)$$

where P is the integer such that $2P+1$ is the smallest odd integer satisfying $2P+1 > \frac{\alpha}{2}$.

The evaluation of the coefficients c_k in $F(z)$ is not an easy task and, indeed, each coefficient is a function (of increasing complexity as k increases) of the three parameters α, β and γ. An algorithm for their computation is proposed in [Par10] and in [GarGar18, Appendix A] the main steps for the application to the Prabhakar function are described. This algorithm makes it possible to numerically evaluate any number of c_k's. The first few coefficients are however explicitly listed here

$$c_0 = 1,$$
$$c_1 = \frac{(\gamma-1)(\alpha\gamma+\gamma-2\beta)}{2},$$
$$c_2 = \frac{(\gamma-1)(\gamma-2)\left(3(\alpha+1)^2(\gamma+1)^2 - (\alpha+1)(\alpha+12\beta+5)\gamma + 12\beta(\beta+1)\right)}{24}.$$

5.1.3 Expansion on the Negative Semi-axes

Due to its essential importance for applications (see, e.g.. [Xu17]), special attention is paid in [GarGar18] to the asymptotic expansion of the Prabhakar function for real negative arguments.

Theorem 5.4 *Let $\alpha > 0$ and $t > 0$. Then the following asymptotic expansion holds for $t \to +\infty$:*

$$E^\gamma_{\alpha,\beta}(-t) \sim \begin{cases} H(t), & \text{if } 0 < \alpha < 2, \\ C_0(t) + H(t), & \text{if } \alpha = 2, \\ \sum_{l=0}^{P-1} C_l(t), & \text{if } \alpha > 2, \end{cases} \quad (5.1.31)$$

where $P = \left[\frac{\alpha/2+1}{2}\right]$, $[x]$ means the largest integer smaller than x and

$$C_l(t) = \frac{2 \exp\{t^{1/\alpha} \cos \frac{(2l+1)\pi}{\alpha}\}}{\alpha^\gamma \Gamma(\gamma)} \times$$

$$\sum_{k=0}^\infty c_k t^{\frac{\gamma-\beta-k}{\alpha}} \cos\left(\frac{(2l+1)\pi(\gamma-\beta-k)}{\alpha} + t^{1/\alpha} \sin\frac{(2l+1)\pi}{\alpha}\right).$$

◁ From the above Theorems 5.1–5.3 it follows that (since $-t = te^{\pi i}$) as $t \to \infty$

$$E^\gamma_{\alpha,\beta}(-t) \sim \begin{cases} H(t), & \text{if } 0 < \alpha < 2, \\ F(te^{\pi i}) + F(te^{-\pi i}) + H(t), & \text{if } \alpha = 2, \\ \sum_{l=-P}^{P} F(te^{\pi i(2l+1)}), & \text{if } \alpha > 2. \end{cases}$$

Since

$$\exp\{(te^{\pi i l})^{1/\alpha}\} = \exp\left\{t^{1/\alpha} \cos \frac{\pi l}{\alpha}\right\} \left(\cos\left(t^{1/\alpha} \sin \frac{\pi l}{\alpha}\right) + i \sin\left(t^{1/\alpha} \sin \frac{\pi l}{\alpha}\right)\right),$$

$$(te^{\pi i l})^{\frac{\gamma-\beta-l}{\alpha}} = t^{\frac{\gamma-\beta-l}{\alpha}} \left(\cos \frac{\pi l(\gamma-\beta-l)}{\alpha} + i \sin \frac{\pi l(\gamma-\beta-l)}{\alpha}\right),$$

defining $C_l(t) := F(te^{\pi i l}) + F(te^{-\pi i l})$, after standard manipulations we find that expressions $C_l(t)$ have the form predicted in the theorem, and for $C_l(t) = C_{2l+1}(t)$ all $l = 0, 1, \ldots, P-1$. Furthermore,

$$F(te^{\pi i(2P+1)}) = \frac{2\exp\{t^{1/\alpha}\cos\frac{(2P+1)\pi}{\alpha}\}}{\alpha^\gamma \Gamma(\gamma)} \sum_{k=0}^\infty c_k t^{\frac{\gamma-\beta-k}{\alpha}} \times$$

$$\left(\cos\left(\frac{(2P+1)\pi(\gamma-\beta-k)}{\alpha} + t^{1/\alpha}\sin\frac{(2P+1)\pi}{\alpha}\right) + \right.$$

$$\left. i\sin\left(\frac{(2P+1)\pi(\gamma-\beta-k)}{\alpha} + t^{1/\alpha}\sin\frac{(2P+1)\pi}{\alpha}\right) \right)$$

$$= A_P(t) + iB_P(t).$$

If $\alpha > 2$ then $2P + 1 > \alpha/2$ and hence $\cos\frac{(2P+1)\pi}{\alpha} < 0$. Thus $A_P(t) \to 0$ and $B_P(t) \to 0$ exponentially as $t \to \infty$ and therefore all terms $C_l(t)$ with $l \geq P$ in the expansion of $E^\gamma_{\alpha,\beta}(-t)$ can be neglected. ▷

5.1.4 Integral Transforms of the Prabhakar Function

In a similar way as for the two-parametric Mittag-Leffler function one can calculate the *Laplace transform of the Prabhakar function* (the three-parametric Mittag-Leffler function)

$$\left(\mathcal{L} E^\gamma_{\alpha,\beta}(t)\right)(s) = \int_0^\infty e^{-st} E^\gamma_{\alpha,\beta}(t) dt = \frac{1}{s} {}_2\Psi_1\left[\begin{matrix}(\gamma,1),(1,1)\\(\beta,\alpha)\end{matrix}\bigg|\frac{1}{s}\right] \quad (\text{Re}\,s > 0). \tag{5.1.32}$$

The most useful variant of the Laplace transform of the Prabhakar function is the following formula (see, e.g.., [KiSrTr06, p. 47]):

$$\left(\mathcal{L} t^{\beta-1} E^\gamma_{\alpha,\beta}(\lambda t^\alpha)\right)(s) = \frac{s^{\alpha\gamma-\beta}}{(s^\alpha - \lambda)^\gamma}, \tag{5.1.33}$$

which is valid for all Re $s > 0$, Re $\beta > 0$, $\lambda \in \mathbb{C}$ such that $|\lambda s^{-\alpha}| < 1$.

Applying the Mellin inversion formula to (5.1.22) we obtain the *Mellin transform of three-parametric Mittag-Leffler function*

$$\left(\mathcal{M} E^\gamma_{\alpha,\beta}(-wt)\right)(s) = \int_0^\infty t^{s-1} E^\gamma_{\alpha,\beta}(-wt) dt = \frac{\Gamma(s)\Gamma(\gamma-s)}{\Gamma(\gamma)\Gamma(\beta-\alpha s)} w^{-s}. \tag{5.1.34}$$

Further we take into account the integral relation for the *Whittaker function*

$$\int_0^\infty t^{\nu-1} e^{-t/2} W_{\lambda,\mu}(t) dt = \frac{\Gamma(1/2+\mu+\nu)\Gamma(1/2-\mu+\nu)}{\Gamma(1-\lambda+\nu)}, \quad \text{Re}\,(\nu\pm\mu) > -1/2, \tag{5.1.35}$$

5.1 The Prabhakar (Three-Parametric Mittag-Leffler) Function

where

$$\mathcal{W}_{\lambda,\mu}(x) = e^{-x/2} x^{c/2} U(a, c, x), \quad a = 1/2 - \mu + \nu, \ c = 2\mu + 1,$$

and $U(a, c, x)$ is the *Tricomi function* (or *confluent hypergeometric function*, or *degenerate hypergeometric function*) defined, e.g.., by the integral

$$U(a, c, x) = \frac{1}{\Gamma(a)} \int_0^\infty \frac{t^{a-1}}{t+b} e^{-xt} dt, \quad b = 1 + a - c.$$

By using (5.1.35) we obtain the so-called *Whittaker integral transform* of the three-parametric Mittag-Leffler function

$$\int_0^\infty t^{\rho-1} e^{-\frac{pt}{2}} \mathcal{W}_{\lambda,\mu}(pt) E^\gamma_{\alpha,\beta}(wt^\delta) dt = \frac{p^{-\rho}}{\Gamma(\gamma)} {}_3\Psi_2 \left[\frac{w}{p^\delta} \left| \begin{array}{c} (\gamma, 1), \ (\frac{1}{2} \pm \mu + \rho, \delta) \\ (\beta, \alpha), \ (1 - \lambda + \rho, \delta) \end{array} \right. \right],$$
(5.1.36)

where ${}_3\Psi_2$ is the generalized Wright function, and $|\text{Re } \mu| < 1/2, \text{Re } \rho > 0, \left|\frac{w}{p^\delta}\right| < 1$. As a particular case of this formula we can obtain the Laplace transform of the three-parametric Mittag-Leffler function. Indeed, since

$$\mathcal{W}_{\pm 1/2, 0}(t) = e^{-t/2},$$

the *Laplace transform* of $E^\gamma_{\alpha,\beta}$ can be represented by the relation

$$\int_0^\infty t^{\rho-1} e^{-pt} E^\gamma_{\alpha,\beta}(wt^\delta) dt = \frac{p^{-\rho}}{\Gamma(\gamma)} {}_2\Psi_1 \left[\frac{w}{p^\delta} \left| \begin{array}{c} (\gamma, 1), \ (\rho, \delta) \\ (\beta, \alpha) \end{array} \right. \right], \qquad (5.1.37)$$

where $\text{Re } \alpha > 0, \text{Re } \beta > 0, \text{Re } \rho > 0, \text{Re } p > 0, p > |w|^{\frac{1}{\text{Re } \alpha}}$. In particular, for $\rho = \beta$ and $\delta = \alpha$ this result coincides with that obtained in [Pra71, Eq. 2.5]

$$\int_0^\infty t^{\beta-1} e^{-pt} E^\gamma_{\alpha,\beta}(wt^\alpha) dt = p^{-\beta} \left(1 - wp^{-\alpha}\right)^{-\gamma}, \qquad (5.1.38)$$

where $\text{Re } \alpha > 0, \text{Re } \beta > 0, \text{Re } p > 0, p > |w|^{\frac{1}{\text{Re } \alpha}}$.

5.1.5 Complete Monotonicity of the Prabhakar Function

Here we present the result on the complete monotonicity of the ("small") Prabhakar function

$$e^\gamma_{\alpha,\beta}(t) = t^{\beta-1} E^\gamma_{\alpha,\beta}(-t^\alpha), \quad t \geq 0, \qquad (5.1.39)$$

following [MaiGar15]. For some particular values of the parameters this function and its Laplace transform

$$\mathcal{E}_{\alpha,\beta}^{\gamma}(s) := \mathcal{L}\left(e_{\alpha,\beta}^{\gamma}(t)\right)(s) = \frac{s^{\alpha\gamma-\beta}}{(s^{\alpha}+1)^{\gamma}}, \quad \mathrm{Re}\,s > 0, \ |s^{\alpha}| > 1, \quad (5.1.40)$$

provide the response function and the complex susceptibility ($s = -i\omega$), respectively, found in the most common models for non-Debye (or anomalous) relaxation in dielectrics.

In [MaiGar15] a simplified proof is given of the complete monotonicity of $e_{\alpha,\beta}^{\gamma}(t)$ for the following values of parameters

$$0 < \alpha \leq 1, \quad 0 < \alpha\gamma \leq \beta \leq 1. \quad (5.1.41)$$

Note first that for $0 < \alpha < 1$ the Laplace transform in (5.1.40) exhibits a branch cut on the negative real semi-axis but has no poles. Therefore, the inversion of the Laplace transform through the Bromwich integral reduces to the evaluation of the integral on an equivalent Hankel path which starts from $-\infty$ along the lower negative real axis, encircles the small circle $|s| = \varepsilon$ in the positive sense and returns to $-\infty$ along the upper negative real axis. It can be shown (see, e.g.., [CdOMai11]) by calculating the limit as $\varepsilon \to 0$ that

$$e_{\alpha,\beta}^{\gamma}(t) = \int_0^{+\infty} e^{-rt} K_{\alpha,\beta}^{\gamma}(r)\,dr, \quad (5.1.42)$$

where

$$K_{\alpha,\beta}^{\gamma}(r) = \mp\frac{1}{\pi}\mathrm{Im}\left[\mathcal{E}_{\alpha,\beta}^{\gamma}(s)\Big|_{s=re^{\pm i\pi}}\right] \quad (5.1.43)$$

denotes the spectral distribution of $e_{\alpha,\beta}^{\gamma}(t)$. In other words, since $\mathcal{E}_{\alpha,\beta}^{\gamma}(s)$ is required to be the iterated Laplace transform of $K_{\alpha,\beta}^{\gamma}(r)$, we recognize that it is the Stieltjes transform of the spectral distribution. As a consequence, the spectral distribution can be determined as the inverse Stieltjes transform of $\mathcal{E}_{\alpha,\beta}^{\gamma}(s)$ via the so-called *Titchmarsh inversion formula* (see, e.g.., [Tit86, Wid46]).

By virtue of the Bernstein theorem (see, e.g.. [SchSoVo12]), to ensure the complete monotonicity of $e_{\alpha,\beta}^{\gamma}(t)$ the spectral distribution $K_{\alpha,\beta}^{\gamma}(r)$ has to be shown to be nonnegative for all $r \geq 0$. In [MaiGar15], the spectral distribution is computed explicitly from the Titchmarsh formula and thus the conditions of non-negativity are derived. Indeed,

$$K_{\alpha,\beta}^{\gamma}(r) = \frac{r^{-\beta}}{\pi}\mathrm{Im}\left[e^{i\beta\pi}\left(\frac{r^{\alpha}+e^{-i\alpha\pi}}{r^{\alpha}+2\cos(\alpha\pi)+r^{-\alpha}}\right)^{\gamma}\right]$$

$$= -\frac{r^{\alpha\gamma-\beta}}{\pi}\mathrm{Im}\left[\frac{e^{i(\alpha\gamma-\beta)\pi}}{(r^{\alpha}e^{i\alpha\pi}+1)^{\gamma}}\right],$$

5.1 The Prabhakar (Three-Parametric Mittag-Leffler) Function

which gives us directly the following representation of the spectral distribution

$$K_{\alpha,\beta}^{\gamma}(r) = \frac{r^{\alpha\gamma-\beta}}{\pi} \frac{\sin(\gamma\theta_{\alpha}(r) + (\beta - \alpha\gamma)\pi)}{(r^{2\alpha} + 2r^{\alpha}\cos(\alpha\pi) + 1)^{\gamma/2}}, \tag{5.1.44}$$

where the value of $\theta_{\alpha}(r)$ is chosen as follows

$$\theta_{\alpha}(r) = \arctan\left[\frac{r^{\alpha}\sin(\alpha\pi)}{r^{\alpha}\cos(\alpha\pi) + 1}\right] \in [0, \pi]. \tag{5.1.45}$$

It can be easily checked that $\theta_{\alpha}(r)$ is a non-negative and increasing function of r bounded by $\alpha\pi \leq \pi$. In fact, for $r \gg 1$ we have

$$\frac{r^{\alpha}\sin(\alpha\pi)}{r^{\alpha}\cos(\alpha\pi) + 1} = \frac{\sin(\alpha\pi)}{\cos(\alpha\pi) + \frac{1}{r^{\alpha}}} \leq \frac{\sin(\alpha\pi)}{\cos(\alpha\pi)} = \tan(\alpha\pi). \tag{5.1.46}$$

The above consideration leads us immediately to the validity of the complete monotonicity of the function $e_{\alpha,\beta}^{\gamma}(t)$ in the case when its parameters satisfy the inequalities (5.1.41).

Note that the obtained result is illustrated by numerical calculations in [MaiGar15].

5.1.6 Fractional Integrals and Derivatives of the Prabhakar Function

Theorem 5.5 *Let $\mu, \alpha, \beta > 0, a \in \mathbb{R}$. Then the following formulas for the Riemann–Liouville and the Liouville fractional integration and differentiation of the Prabhakar function hold:*

(i)

$$\left\{I_{0+}^{\mu}\left[t^{\beta-1}E_{\alpha,\beta}^{\gamma}(at^{\alpha})\right]\right\}(x) = x^{\beta+\mu-1}E_{\alpha,\beta+\mu}^{\gamma}(ax^{\alpha}), \tag{5.1.47}$$

where

$$\left(I_{0+}^{\mu}\varphi(t)\right)(x) = \frac{1}{\Gamma(\mu)}\int_{0}^{x}\frac{\varphi(t)}{(x-t)^{1-\mu}}dt, \quad \operatorname{Re}\mu > 0,$$

is the left-sided Riemann–Liouville fractional integral (see, e.g.., [SaKiMa93, p. 33]).

(ii)

$$\left\{I_{-}^{\mu}\left[t^{-\mu-\beta}E_{\alpha,\beta}^{\gamma}(at^{-\alpha})\right]\right\}(x) = x^{-\beta}E_{\alpha,\beta+\mu}^{\gamma}(ax^{-\alpha}), \tag{5.1.48}$$

where

$$\left(I_-^\mu \varphi(t)\right)(x) = \frac{1}{\Gamma(\mu)} \int_x^\infty \frac{\varphi(t)}{(t-x)^{1-\mu}} dt, \ \operatorname{Re}\mu > 0,$$

is the right-sided Liouville fractional integral (see, e.g.., [SaKiMa93, p. 94]).

(iii)
$$\left\{D_{0+}^\mu \left[t^{\beta-1} E_{\alpha,\beta}^\gamma(at^\alpha)\right]\right\}(x) = x^{\beta-\mu-1} E_{\alpha,\beta-\mu}^\gamma(ax^\alpha), \tag{5.1.49}$$

where

$$\left(D_{0+}^\mu \varphi(t)\right)(x) = \frac{1}{\Gamma(n-\mu)} \left(\frac{d}{dx}\right)^n \int_0^x \frac{\varphi(t)}{(x-t)^{\mu-n+1}} dt, \ \operatorname{Re}\mu > 0, \ n = [\mu]+1,$$

is the left-sided Riemann–Liouville fractional derivative (see, e.g.., [SaKiMa93, p. 37]).

(iv) If $\beta - \mu + \{\mu\} > 1$, then

$$\left\{D_-^\mu \left[t^{\mu-\beta} E_{\alpha,\beta}^\gamma(at^{-\alpha})\right]\right\}(x) = x^{-\beta} E_{\alpha,\beta-\mu}^\gamma(ax^{-\alpha}), \tag{5.1.50}$$

where

$$\left(D_-^\mu \varphi(t)\right)(x) = \frac{(-1)^n}{\Gamma(n-\mu)} \left(\frac{d}{dx}\right)^n \int_x^\infty \frac{\varphi(t)}{(t-x)^{\mu-n+1}} dt, \ \operatorname{Re}\mu > 0, \ n = [\mu]+1,$$

is the right-sided Liouville fractional derivative (see, e.g.., [SaKiMa93, p. 95]).

◁ The proof follows from the definitions of the corresponding fractional integrals and derivatives. Thus, to prove relation (5.1.47) we put

$$K \equiv \left\{I_{0+}^\mu \left[t^{\beta-1} E_{\alpha,\beta}^\gamma(at^\alpha)\right]\right\}(x) = \frac{1}{\Gamma(\mu)} \int_0^x (x-t)^{\mu-1} \sum_{n=0}^\infty \frac{(\gamma)_n a^n t^{n\alpha+\beta-1}}{\Gamma(n\alpha+\beta)n!} dt.$$

Since the series converges for any $t > 0$, interchanging the order of integration and summation and evaluating the inner integral by means of the Beta function yields

$$K \equiv x^{\mu+\beta-1} \sum_{n=0}^\infty \frac{(\gamma)_n (ax^\alpha)^n}{\Gamma(\mu+n\alpha+\beta)n!} = x^{\beta+\mu-1} E_{\alpha,\beta+\mu}^\gamma(ax^\alpha).$$

The proof is complete. ▷

5.1.7 Relations to the Fox–Wright Function, H-function and Other Special Functions

Due to the integral representation (5.1.22) the three-parametric Mittag-Leffler function can be considered as a special case of the H-function (see, e.g.., [KilSai04])

$$E^\gamma_{\alpha,\beta}(z) = \frac{1}{\Gamma(\gamma)} H^{1,1}_{1,2}\left[-z \left| \begin{array}{c} (1-\gamma, 1) \\ (0,1), \ (1-\beta, \alpha) \end{array}\right.\right] \qquad (5.1.51)$$

as well as a special case of the Fox–Wright generalized hypergeometric function $_pW_q$ (see, e.g.. Slater [Sla66], Mathai and Saxena [MatSax73])

$$E^\gamma_{\alpha,\beta}(z) = \frac{1}{\Gamma(\gamma)} {}_1W_1\left[z \left| \begin{array}{c} (\gamma, 1) \\ (\beta, \alpha) \end{array}\right.\right]. \qquad (5.1.52)$$

In particular, when $\alpha = 1$ the Prabhakar function $E^\gamma_{1,\beta}(z)$ coincides with the Kummer confluent hypergeometric function $\Phi(\gamma; \beta; z)$, apart from the constant factor $(\Gamma(\beta))^{-1}$

$$E^\gamma_{\alpha,\beta}(z) = \frac{1}{\Gamma(\beta)} \Phi(\gamma; \beta; z), \qquad (5.1.53)$$

and when $\alpha = m \in \mathbb{N}$ is a positive integer then $E^\gamma_{m,\beta}(z)$ is related to the generalized hypergeometric function

$$E^\gamma_{m,\beta}(z) = \frac{1}{\Gamma(\beta)} {}_1F_m\left(\gamma; \frac{\beta}{m}, \frac{\beta+1}{m}, \ldots, \frac{\beta+m-1}{m}; \frac{z}{m^m}\right). \qquad (5.1.54)$$

Here we present other special functions, which are connected with the Prabhakar function (5.1.1) (see [KiSaSa04]). The following relation holds

$$E^{-k}_{m,\beta+1}(z) = \frac{\Gamma(k+1)}{\Gamma(km+\beta+1)} Z^{(\beta)}_k(z; m) \quad (k,m \in \mathbb{N}; \beta \in \mathbb{C}), \qquad (5.1.55)$$

where $Z^{(\beta)}_k(z; m)$ is a polynomial of degree k in z^m studied in [Kon67]. In particular,

$$Z^{(\beta)}_k(z; 1) = L^\beta_k, \quad (k \in \mathbb{N}; \beta \in \mathbb{C}), \qquad (5.1.56)$$

where L^β_k is the Laguerre polynomial (see, e.g.., [ErdBat-2, Sect. 10.12]), and hence

$$E^{-k}_{1,\beta+1}(z) = \frac{\Gamma(k+1)}{\Gamma(k+\beta+1)} L^\beta_k, \quad (k \in \mathbb{N}; \beta \in \mathbb{C}). \qquad (5.1.57)$$

The Laguerre function L^β_ν (see, e.g.., [ErdBat-1, 6.9(37)]) is also a special case of the Prabhakar function (5.1.1):

$$E^{-\nu}_{1,\beta+1}(z) = \frac{\Gamma(\nu+1)}{\Gamma(\beta+1)} L^\beta_\nu, \quad (\nu, \beta \in \mathbb{C}). \tag{5.1.58}$$

The following relation in terms of the Kummer confluent hypergeometric function ($m \in \mathbb{N}$; $\beta, \gamma \in \mathbb{C}$)

$$E^\gamma_{m,\beta}(z) = \frac{(2\pi)^{(m-1)/2}}{(m)^{\beta-1/2}} \prod_{k=0}^{m-1} \frac{1}{\Gamma((\beta+k)/m)} \Phi\left(\gamma, \frac{\beta+k}{m}; \frac{z}{m^m}\right) \tag{5.1.59}$$

is deduced from the definition of the Prabhakar function and from the multiplication formula of Gauss and Legendre for the Gamma function [ErdBat-1, 1.2(11)] (see Sect. A.1.4 in Appendix A).

5.2 The Kilbas–Saigo (Three-Parametric Mittag-Leffler) Function

5.2.1 Definition and Basic Properties

Another form of the three-parametric Mittag-Leffler function is defined by Kilbas and Saigo [KilSai95b] (see also [KilSai95a]) (now called the *Kilbas–Saigo function*)

$$E_{\alpha,m,l}(z) = \sum_{k=0}^\infty c_k z^k \quad (z \in \mathbb{C}; \alpha, m \in \mathbb{R}, l \in \mathbb{C}), \tag{5.2.1}$$

where

$$c_0 = 1, \quad c_k = \prod_{j=0}^{k-1} \frac{\Gamma(\alpha[jm+l]+1)}{\Gamma(\alpha[jm+l+1]+1)} \quad (k = 1, 2, \cdots). \tag{5.2.2}$$

In (5.2.2) an empty product is defined to be equal to one,[1] α, m are real numbers and $l \in \mathbb{C}$ such that

$$\alpha > 0, \ m > 0, \ \alpha(jm+l)+1 \neq -1, -2, -3, \cdots \ (j = 0, 1, 2, \cdots). \tag{5.2.3}$$

The function (5.2.1) was introduced in [KilSai95b] (see also [KilSai95a]).

In particular, if $m = 1$, the conditions in (5.2.3) take the form

$$\alpha > 0, \ \alpha(j+l)+1 \neq -1, -2, -3, \cdots \ (j = 0, 1, 2, \cdots) \tag{5.2.4}$$

---

[1] In what follows we will call this assumption the "Empty Product Convention".

5.2 The Kilbas–Saigo (Three-Parametric Mittag-Leffler) Function

and (5.2.1) is reduced to the two-parametric Mittag-Leffler function given in [Chap. 4, formula (4.1.1)]:
$$E_{\alpha,1,l}(z) = \Gamma(\alpha l + 1)E_{\alpha,\alpha l+1}(z). \tag{5.2.5}$$

Therefore we call the three-parametric Mittag-Leffler function $E_{\alpha,m,l}(z)$ a Kilbas–Saigo function. As we shall see later, the Kilbas–Saigo function is used to solve in closed form new classes of integral and differential equations of fractional order.

When $\alpha = n \in \mathbb{N} = \{1, 2, \cdots\}$, $E_{n,m,l}(z)$ takes the form

$$E_{n,m,l}(z) = 1 + \sum_{k=1}^{\infty} \prod_{j=0}^{k-1} \left(\frac{1}{[n(jm+l)+1]\cdots[n(jm+l)+n]} \right) z^k, \tag{5.2.6}$$

where n, m and l are real numbers such that

$$n \in \mathbb{N}, \ m > 0, \ n(jm+l) \neq -1, -2, \cdots, -n \ (j = 0, 1, 2, \cdots). \tag{5.2.7}$$

5.2.2 The Order and Type of the Entire Function $E_{\alpha,m,l}(z)$

In this subsection we give a few characteristics of $E_{\alpha,m,l}(z)$. First of all we show that the Kilbas–Saigo function is an entire function.

Lemma 5.6 *If α, m and l are real numbers such that the conditions (5.2.3) are satisfied, then $E_{\alpha,m,l}(z)$ is an entire function of the variable z.*

◁ According to (5.2.2) and the relation [Appendix A, formula (A.27)] with $z = \alpha nm$, $a = \alpha l + \alpha + 1$ and $b = \alpha l + 1$, we have the asymptotic estimate

$$\frac{c_k}{c_{k+1}} = \frac{\Gamma[\alpha(km+l+1)+1]}{\Gamma[\alpha(km+l)+1]} \sim (\alpha mk)^\alpha \to \infty \ (k \to \infty).$$

Therefore the radius of convergence R of the series (5.2.1) is equal to $+\infty$, i.e. $E_{\alpha,m,l}(z)$ is an entire function. ▷

Corollary 5.7 *For $\alpha > 0$, $m > 0$ and $\operatorname{Re} l > -1/\alpha$ the function $E_{\alpha,m,l}(z)$ is an entire function of z.*

Corollary 5.8 *If $\alpha = n \in \mathbb{N}$, $m > 0$ and l are real numbers such that the conditions (5.2.7) are satisfied, then the Kilbas–Saigo function $E_{n,m,l}(z)$ given by (5.2.6) is an entire function of z.*

The order and type of the Kilbas–Saigo function (5.2.1) is given by the following statement.

Theorem 5.9 *If α, m and l are real numbers such that the conditions (5.2.3) are satisfied, then $E_{\alpha,m,l}(z)$ is an entire function of order $\rho = 1/\alpha$ and type $\sigma = 1/m$. Moreover we have the asymptotic estimate*

$$|E_{\alpha,m,l}(z)| < \exp\left(\left[\frac{1}{m} + \epsilon\right]|z|^{1/\alpha}\right), \quad |z| \geq r_0 > 0, \qquad (5.2.8)$$

whenever $\epsilon > 0$ is sufficiently small.

◁ Applying formula (B.5) (Appendix B) we first find the order ρ of $E_{\alpha,m,l}(z)$. According to (5.2.2) we have

$$c_k = \frac{\Gamma(\alpha l + 1)\Gamma(\alpha l + \alpha m + 1)\cdots\Gamma(\alpha l + \alpha m[k-1] + 1)}{\Gamma(\alpha l + \alpha + 1)\Gamma(\alpha l + \alpha + \alpha m + 1)\cdots\Gamma(\alpha l + \alpha + \alpha m[k-1] + 1)}.$$

Let

$$z_n = \alpha l + \alpha m n + 1 \ (n \in \mathbb{N} = \{1, 2, \cdots\}). \qquad (5.2.9)$$

Using (5.2.8) with $z = z_n$, $a = 0$ and $b = \alpha$ we obtain that for any $d > 0$ there exists an $n_0 \in \mathbb{N}$ such that

$$(1-d)z_n^\alpha \leq \left|\frac{\Gamma(z_n + \alpha)}{\Gamma(z_n)}\right| \leq (1+d)z_n^\alpha \quad \forall n > n_0. \qquad (5.2.10)$$

Therefore for $k > n_0$ we have

$$\log\left(\frac{1}{|c_k|}\right) = \sum_{n=0}^{k-1} \log\left|\frac{\Gamma(z_n + \alpha)}{\Gamma(z_n)}\right|$$

$$= \sum_{n=0}^{n_0} \log\left|\frac{\Gamma(z_n + \alpha)}{\Gamma(z_n)}\right| + \sum_{n=n_0+1}^{k-1} \log\left|\frac{\Gamma(z_n + \alpha)}{\Gamma(z_n)}\right|$$

$$\leq d_1 + \sum_{n=n_0+1}^{k-1} \log(1+d) + \sum_{n=n_0+1}^{k-1} \log(z_n^\alpha)$$

$$= d_1 + (k - n_0 - 1)\log(1+d)$$

$$+ \alpha \sum_{n=n_0+1}^{k-1}\left[\log(n) + \log(\alpha m) + \log\left(1 + \frac{\alpha l + 1}{n\alpha m}\right)\right]$$

and hence

$$\log\left(\frac{1}{|c_k|}\right) \leq d_3 + kd_4 + \alpha k \log(k), \qquad (5.2.11)$$

5.2 The Kilbas–Saigo (Three-Parametric Mittag-Leffler) Function

where d_3 and d_4 are certain positive constants. Similarly

$$\log\left(\frac{1}{|c_k|}\right) \geq d_4 + (k - n_0 - 1)\log(1 - d)$$

$$+ \alpha \sum_{n=n_0+1}^{k-1} \left[\log(n) + \log(\alpha m) + \log\left(1 + \frac{\alpha l + 1}{nlm}\right)\right]$$

and

$$\log\left(\frac{1}{|c_k|}\right) \geq d_5 + kd_6 + \alpha k \log(k) \tag{5.2.12}$$

for some real constants d_5, d_6.

It follows from (5.2.11) and (5.2.12) that the usual limit

$$\lim_{k \to \infty} \frac{k \log(k)}{\log(1/|c_k|)} = \frac{1}{\alpha} \tag{5.2.13}$$

exists and hence in accordance with formula (B.5) (Appendix B) the order ρ of $E_{\alpha,m,l}(z)$ is given by

$$\rho = \frac{1}{\alpha}. \tag{5.2.14}$$

Next we use formula (B.6) (Appendix B) to find the type σ of the function $E_{\alpha,m,l}(z)$. Applying (5.2.10), (5.2.9) and (5.2.2) we have

$$\prod_{n=0}^{n_0} \left|\frac{\Gamma(z_n + \alpha)}{\Gamma(z_n)}\right| \left(\frac{1}{1+d}\right)^{k-n_0-1} \prod_{n=n_0+1}^{k-1} z_n^\alpha \leq |c_k|$$

$$\leq \prod_{n=0}^{n_0} \left|\frac{\Gamma(z_n + \alpha)}{\Gamma(z_n)}\right| \left(\frac{1}{1-d}\right)^{k-n_0-1} \prod_{n=n_0+1}^{k-1} z_n^\alpha. \tag{5.2.15}$$

Using this formula and the asymptotic relation

$$\prod_{n=n_0}^{k-1} \left(\frac{1}{\alpha nm}\right)^\alpha \sim \left(\frac{1}{k!}\right)^\alpha \left(\frac{1}{\alpha m}\right)^{\alpha k} \sim (2\pi k)^{-\alpha} \left(\frac{e}{\alpha km}\right)^{\alpha k} \quad (k \to \infty)$$

from formula (B.6) (Appendix B), we obtain $\sigma^\alpha = m^{-\alpha}$ and hence the type σ of $E_{\alpha,m,l}(z)$ is given by

$$\sigma = \frac{1}{m}. \tag{5.2.16}$$

The asymptotic estimate (5.2.8) follows from (5.2.13)–(5.2.14) and the definitions of the order and type of an entire function given in Appendix B. This completes the proof of the theorem. ▷

Corollary 5.10 *If $\alpha = n \in \mathbb{N}$, $m > 0$ and l are real numbers such that the conditions (5.2.7) are satisfied, then the Kilbas–Saigo function $E_{n,m,l}(z)$ given by (5.2.6) is an entire function of z with order $\rho = 1/n$ and type $\sigma = 1/m$.*

Corollary 5.11 *The classical Mittag-Leffler function $E_\alpha(z)$ and the two-parametric Mittag-Leffler function $E_{\alpha,\beta}(z)$, given respectively in [Chap. 3, formula (3.1.1)] and in [Chap. 4, formula (4.1.1)], have the same order and type:*

$$\rho = \frac{1}{\alpha}, \quad \sigma = 1. \tag{5.2.17}$$

Remark 5.12 The assertions of Corollary 5.11 coincide with those in [Chap. 3, Proposition 3.1] and [Chap. 4, Sect. 4.1].

Remark 5.13 Theorem 5.9 shows that the Kilbas–Saigo function (5.2.1) has the same order as the Mittag-Leffler functions [Chap. 3, formula (3.1.1)] and [Chap. 4, formula (4.1.1)]. But the type of $E_{\alpha,m,l}(z)$ depends on m.

5.2.3 Recurrence Relations for $E_{\alpha,m,l}(z)$

In this subsection we give recurrence relations for $E_{\alpha,m,l}(z)$.

Theorem 5.14 *Let α, m and l be real numbers such that the condition (5.2.3) is satisfied and let $n \in \mathbb{N}$. Then the following recurrence relation holds*

$$z^n \left[E_{\alpha,m,l+nm}(z) - 1 \right] = \prod_{j=0}^{n-1} \frac{\Gamma[\alpha(jm+l+1)+1]}{\Gamma[\alpha(jm+l)+1]} \times$$

$$\left[E_{\alpha,m,l}(z) - 1 - \sum_{k=1}^{n} \left(\prod_{j=0}^{k-1} \frac{\Gamma[\alpha(jm+l)+1]}{\Gamma[\alpha(jm+l+1)+1]} \right) z^k \right]. \tag{5.2.18}$$

◁ By (5.2.1) we have

$$E_{\alpha,m,l+nm}(z) = 1 + \sum_{k=1}^{\infty} \left(\prod_{j=0}^{k-1} \frac{\Gamma[\alpha(jm+nm+l)+1]}{\Gamma[\alpha(jm+nm+l+1)+1]} \right) z^k.$$

5.2 The Kilbas–Saigo (Three-Parametric Mittag-Leffler) Function

Changing the summation indices $s = j + n$ and $p = k + n$, we obtain

$$E_{\alpha,m,l+nm}(z) = 1 + \sum_{k=1}^{\infty} \left(\prod_{s=n}^{n+k-1} \frac{\Gamma[\alpha(sm+l)+1]}{\Gamma[\alpha(sm+l+1)+1]} \right) z^k$$

$$= 1 + \sum_{p=n+1}^{\infty} \left(\prod_{s=n}^{p-1} \frac{\Gamma[\alpha(sm+l)+1]}{\Gamma[\alpha(sm+l+1)+1]} \right) z^{p-n}$$

$$= 1 + \prod_{s=0}^{n-1} \frac{\Gamma[\alpha(sm+l+1)+1]}{\Gamma[\alpha(sm+l)+1]} \sum_{p=n+1}^{\infty} \left(\prod_{s=0}^{p-1} \frac{\Gamma[\alpha(sm+l)+1]}{\Gamma[\alpha(sm+l+1)+1]} \right) z^{p-n}$$

$$= 1 + \frac{1}{z^n} \prod_{s=0}^{n-1} \frac{\Gamma[\alpha(sm+l+1)+1]}{\Gamma[\alpha(sm+l)+1]}$$

$$\times \left[1 + \sum_{p=1}^{\infty} \left(\prod_{s=0}^{p-1} \frac{\Gamma[\alpha(sm+l)+1]}{\Gamma[\alpha(sm+l+1)+1]} \right) z^p \right.$$

$$\left. - \sum_{p=1}^{n} \left(\prod_{s=0}^{p-1} \frac{\Gamma[\alpha(sm+l)+1]}{\Gamma[\alpha(sm+l+1)+1]} \right) z^p - 1 \right]$$

$$= 1 + \frac{1}{z^n} \prod_{s=0}^{n-1} \frac{\Gamma[\alpha(sm+l)+1]}{\Gamma[\alpha(sm+l+1)+1]}$$

$$\times \left[E_{\alpha,m,l}(z) - \sum_{p=1}^{n} \left(\prod_{s=0}^{p-1} \frac{\Gamma[\alpha(sm+l)+1]}{\Gamma[\alpha(sm+l+1)+1]} \right) z^p - 1 \right]$$

and (5.2.18) is proved. ▷

Corollary 5.15 *If the conditions of Theorem 5.14 are satisfied, then*

$$zE_{\alpha,m,l+m}(z) = \frac{\Gamma(\alpha l + \alpha + 1)}{\Gamma(\alpha l + 1)} \left[E_{\alpha,m,l}(z) - 1 \right] \qquad (5.2.19)$$

and

$$z^n E_{\alpha,m,l+nm}(z) = \prod_{j=0}^{n-1} \frac{\Gamma[\alpha(jm+l+1)+1]}{\Gamma[\alpha(jm+l)+1]}$$

$$\times \left[E_{\alpha,m,l}(z) - 1 - \sum_{k=1}^{n-1} \left(\prod_{j=0}^{k-1} \frac{\Gamma[\alpha(jm+l)+1]}{\Gamma[\alpha(jm+l+1)+1]} \right) z^k \right] \qquad (5.2.20)$$

for $n = 2, 3, \cdots$.

The following two corollaries show what the above properties look like in special cases.

Corollary 5.16 *If $\alpha > 0$, $\beta > 0$ and $n \in \mathbb{N}$, then for the two-parametric Mittag-Leffler function $E_{\alpha,\alpha n+\beta}(z)$ we have the recurrence relations*

$$zE_{\alpha,\alpha+\beta}(z) = E_{\alpha,\beta}(z) - \frac{1}{\Gamma(\beta)} \qquad (5.2.21)$$

and

$$z^n \Gamma(\alpha n + \beta) E_{\alpha,\alpha n+\beta}(z)$$

$$= \prod_{j=0}^{n-1} \frac{\Gamma(\alpha j + \alpha + \beta)}{\Gamma(\alpha j + \beta)} \left[\Gamma(\beta) E_{\alpha,\beta}(z) - 1 - \sum_{k=1}^{n-1} \left(\prod_{j=0}^{k-1} \frac{\Gamma(\alpha j + \beta)}{\Gamma(\alpha j + \alpha + \beta)} \right) z^k \right] \qquad (5.2.22)$$

for $n = 2, 3, \cdots$.

Corollary 5.17 *If $\alpha > 0$ and $n \in \mathbb{N}$, then the Mittag-Leffler function $E_{\alpha,\alpha+1}(z)$ is expressed via the Mittag-Leffler function (3.1.1) by*

$$zE_{\alpha,\alpha+1}(z) = E_\alpha(z) - 1 \qquad (5.2.23)$$

and

$$z^n E_{\alpha,\alpha n+1}(z) = E_\alpha(z) - \sum_{j=1}^{n} \frac{z^j}{\Gamma(\alpha j + 1)} \qquad (5.2.24)$$

for $n = 2, 3, \cdots$.

5.2.4 Connection of $E_{n,m,l}(z)$ with Functions of Hypergeometric Type

As we have mentioned in Sect. 5.2.1, the Kilbas–Saigo function $E_{\alpha,m,l}(z)$ is a generalization of the two-parametric Mittag-Leffler function $E_{\alpha,\beta}(z)$ presented in Chap. 4, formula (4.1.1), in particular, of the Mittag-Leffler function $E_\alpha(z)$ presented in Chap. 3, formula (3.1.1). When $\alpha = n \in \mathbb{N}$, $E_{n,m,l}(z)$ in (5.2.6) becomes a function of hypergeometric type. Such a function ${}_pF_q(a_1, a_2, \cdots, a_p; b_1, b_2, \cdots, b_q; z)$ for $p \in \mathbb{N}_0 = \mathbb{N} \cup \{0\}$, $q \in \mathbb{N}_0$, $a_1, a_2, \cdots, a_p \in \mathbb{C}$, $b_1, b_2, \cdots, b_q \in \mathbb{C}$ and $z \in \mathbb{C}$, $|z| < 1$, is defined by the hypergeometric series [ErdBat-1]

$$_pF_q(a_1, a_2, \cdots, a_p; b_1, b_2, \cdots, b_q; z) = \sum_{k=0}^{\infty} \frac{(a_1)_k (a_2)_k \cdots (a_p)_k}{(b_1)_k (b_2)_k \cdots (b_q)_k} \frac{z^k}{k!}, \qquad (5.2.25)$$

where $(\cdot)_k$ is the Pochhammer symbol defined in [Appendix A, formula (A.17)].

5.2 The Kilbas–Saigo (Three-Parametric Mittag-Leffler) Function

Theorem 5.18 *Let the conditions (5.2.7) be satisfied. Then the function $E_{n,m,l}(z)$ is given via the hypergeometric function by*

$$E_{n,m,l}(z) = {}_1F_n\left(1; \frac{nl+1}{nm}, \frac{nl+2}{nm} \cdots \frac{nl+n}{nm}; \frac{z}{(nm)^n}\right) \quad (5.2.26)$$

and is an entire function of z with order $\rho = 1/n$ and type $\sigma = 1/m$.

◁ According to (5.2.1), (5.2.25) and (5.2.26) we have

$$E_{n,m,l}(z) = 1 + \sum_{k=1}^{\infty}\left(\prod_{j=0}^{k-1}\frac{\Gamma(n[jm+l]+1)}{\Gamma(n[jm+l+1]+1)}\right)z^k$$

$$= 1 + \sum_{k=1}^{\infty}\frac{k!}{(nl+1)\cdots(nl+1+(k-1)nm)\cdots(nl+n)\cdots(nl+n+(k-1)nm)}\frac{z^k}{k!}$$

$$= 1 + \sum_{k=1}^{\infty}\frac{(1)_k}{([nl+1])/[nm])_k \cdots ([nl+n]/[nm])_k}\frac{1}{(nm)^{nk}}\frac{z^k}{k!}$$

$$= {}_1F_n\left(1; \frac{nl+1}{nm}, \frac{nl+2}{nm} \cdots \frac{nl+n}{nm}; \frac{z}{(nm)^n}\right)$$

and (5.2.26) is proved. The last assertion follows from Theorem 5.9. This completes the proof. ▷

Corollary 5.19 *If $n \in \mathbb{N}$ and $\beta > 0$, then the Mittag-Leffler function $E_{n,\beta}(z)$ is given by*

$$E_{n,1,(\beta-1)/n}(z) = \Gamma(\beta)E_{n,\beta}(z) \quad (5.2.27)$$

$$= {}_1F_n\left(1; \frac{\beta}{n}, \frac{\beta+1}{n} \cdots \frac{\beta+n-1}{n}; \frac{z}{(n)^n}\right)$$

and is an entire function of order $\rho = 1/n$ and type $\sigma = 1$.

Corollary 5.20 *If*

$$m > 0, \ l \in \mathbb{R}, \ jm+l \neq -1, -2, -3, \cdots \ (j = 0, 1, 2, \cdots), \quad (5.2.28)$$

then $E_{1,m,l}(z)$ is given by

$$E_{1,m,l}(z) = {}_1F_1\left(1; \frac{l+1}{m}; \frac{z}{m}\right) = \Gamma\left(\frac{l+1}{m}\right)E_{1,(l+1)/m}\left(\frac{z}{m}\right) \quad (5.2.29)$$

and is an entire function of z with order $\rho = 1$ and type $\sigma = 1/m$.

Corollary 5.21 *If* $l \neq -1, -2, -3, \cdots$, *then* $E_{1,1,l}(z)$ *is given by*

$$E_{1,1,l}(z) = {}_1F_1(1; l+1; z) = \Gamma(l+1)E_{1,l+1}(z) \qquad (5.2.30)$$

and is an entire function of z with order $\rho = 1$ and type $\sigma = 1$. In particular, if $l \in \mathbb{N}_0$,

$$E_{1,1,l}(z) = l! E_{1,l+1}(z) = \frac{l!}{z^l}\left[e^z - \sum_{k=0}^{l-1}\frac{z^k}{k!}\right] \qquad (5.2.31)$$

and

$$E_{1,1,0}(z) = E_1(z) = e^z. \qquad (5.2.32)$$

5.2.5 Differentiation Properties of $E_{n,m,l}(z)$

In this subsection we give two differentiation formulas for the Kilbas–Saigo function (5.2.1) with an integer first parameter $\alpha = n$. The first of these is given by the following statement.

Theorem 5.22 *Let $n \in \mathbb{N}$, $m > 0$ and $l \in \mathbb{R}$ be such that the conditions in (5.2.7) are satisfied and let $\lambda \in \mathbb{C}$. Then the following differentiation formula*

$$\left(\frac{d}{dz}\right)^n \left[z^{n(l-m+1)} E_{n,m,l}(\lambda z^{nm})\right] = \prod_{j=1}^{n}[n(l-m)+j]z^{n(l-m)} + \lambda z^{nl} E_{n,m,l}(\lambda z^{nm}) \qquad (5.2.33)$$

holds. In particular, if

$$n(l-m) = -j \text{ for some } j = 1, 2, \cdots, n, \qquad (5.2.34)$$

then

$$\left(\frac{d}{dz}\right)^n \left[z^{n(l-m+1)} E_{n,m,l}(\lambda z^{nm})\right] = \lambda z^{nl} E_{n,m,l}(\lambda z^{nm}). \qquad (5.2.35)$$

◁ If $\lambda = 0$, then by (5.2.1) $E_{n,m,l}(z) = 1$, and formula (5.2.33) takes the well-known form

$$\left(\frac{d}{dz}\right)^n \left[z^{n(l-m+1)}\right] = \prod_{j=1}^{n}[n(l-m)+j]z^{n(l-m)}.$$

If $\lambda \neq 0$, then by (5.2.1) we have

5.2 The Kilbas–Saigo (Three-Parametric Mittag-Leffler) Function

$$\left(\frac{d}{dz}\right)^n \left[z^{n(l-m+1)} E_{n,m,l}(\lambda z^{nm})\right]$$

$$= \left(\frac{d}{dz}\right)^n \left[z^{n(l-m+1)} + \sum_{k=1}^{\infty} c_k \lambda^k z^{n(l-m+1)+nmk}\right]$$

$$= [n(l-m+1)][n(l-m+1)-1]\cdots[n(l-m+1)-n+1]z^{n(l-m)}$$

$$+ \sum_{k=1}^{\infty} \left(\prod_{j=0}^{k-1} \frac{\Gamma(n[jm+l]+1)}{\Gamma(n[jm+l+1]+1)}\right) \frac{\Gamma[n\{(k-1)m+l+1\}+1]}{\Gamma[n(\{(k-1)m+l\}+1]} \lambda^k z^{n(l-m)+nmk}$$

$$= \prod_{k=1}^{n}[n(l-m)+k] z^{n(l-m)} + \lambda z^{nl}$$

$$+ \sum_{k=2}^{\infty} \left(\prod_{j=0}^{k-2} \frac{\Gamma(n[jm+l]+1)}{\Gamma(n[jm+l+1]+1)}\right) \lambda^k z^{n(l-m)+nmk}.$$

By index substitution $s = k-1$ we obtain

$$\left(\frac{d}{dz}\right)^n \left[z^{n(l-m+1)} E_{n,m,l}(\lambda z^{nm})\right]$$

$$= \prod_{k=1}^{n}[n(l-m)+j] z^{n(l-m)} + \lambda z^{nl}$$

$$+ \sum_{s=1}^{\infty} \left(\prod_{j=0}^{s-1} \frac{\Gamma(n[jm+l]+1)}{\Gamma(n[jm+l+1]+1)}\right) \lambda^{s+1} z^{nl+nms}$$

$$= \prod_{k=1}^{n}[n(l-m)+k] z^{n(l-m)}$$

$$+ \lambda z^{nl} \left[1 + \sum_{s=1}^{\infty} \left(\prod_{j=0}^{s-1} \frac{\Gamma(n[jm+l]+1)}{\Gamma(n[jm+l+1]+1)}\right) (\lambda z^{nm})^s\right]$$

and (5.2.33) is proved in accordance with (5.2.1). (5.2.35) follows from (5.2.33). The theorem is proved. ▷

Corollary 5.23 *If $\alpha = n = 1, 2, \cdots$, $\beta > 0$ and $\lambda \in \mathbb{C}$, then for the Mittag-Leffler type function $E_{n,\beta}(az^n)$ in [Chap. 4, formula (4.1.1)] we have*

$$\left(\frac{d}{dz}\right)^n \left[z^{\beta-1} E_{n,\beta}(\lambda z^n)\right] = \frac{1}{\Gamma(\beta-n)} z^{\beta-n-1} + \lambda z^{\beta-1} E_{n,\beta}(\lambda z^n). \quad (5.2.36)$$

Special cases of the above property have the form:

Corollary 5.24 *If $\alpha = n = 1, 2, \cdots$, $\beta = k \in \mathbb{N}$ ($1 \le k \le n$) and $\lambda \in \mathbb{C}$, then for the Mittag-Leffler type function $E_{n,k}(\lambda z^n)$ we have*

$$\left(\frac{d}{dz}\right)^n \left[z^{k-1} E_{n,k}(\lambda z^n)\right] = \lambda z^{k-1} E_{n,k}(\lambda z^n). \tag{5.2.37}$$

In particular, when $k = 1$, for the Mittag-Leffler function $E_n(\lambda z^n)$ in [Chap. 3, formula (3.1.1)] we have

$$\left(\frac{d}{dz}\right)^n \left[E_n(\lambda z^n)\right] = \lambda E_n(\lambda z^n). \tag{5.2.38}$$

Remark 5.25 By [Chap. 4, formula (4.1.1)], the relation (5.2.36) can be represented in the form

$$\left(\frac{d}{dz}\right)^n \left[z^{\beta-1} E_{n,\beta}(\lambda z^n)\right] = \lambda z^{\beta-n-1} E_{n,\beta-n}(\lambda z^n), \tag{5.2.39}$$

which coincides with [Chap. 4, formula (4.3.6)].

Remark 5.26 When $a = 1$, the relation (5.2.38) coincides with [Chap. 3, formula (3.1.1)].

Another differentiation relation for the Kilbas–Saigo function (5.2.1) is given by the following:

Theorem 5.27 *Let $\alpha = n \in \mathbb{N}$, $m > 0$ and $l \in \mathbb{R}$ be such that the conditions in (5.2.7) are satisfied and let $\lambda \in \mathbb{C}$. Then for $E_{n,m,l}(az^{-nm})$ the differentiation formula*

$$\left(\frac{d}{dz}\right)^n \left[z^{n(m-l)-1} E_{n,m,l}(\lambda z^{-nm})\right] \tag{5.2.40}$$

$$= \prod_{j=1}^{n} [n(m-l) - j] z^{n(m-l-1)-1} + (-1)^n \lambda z^{-n(l+1)-1} E_{n,m,l}(\lambda z^{-nm})$$

holds. In particular, if the conditions

$$n(l - m) = j \text{ for some } j = 1, 2, \cdots, n, \tag{5.2.41}$$

are satisfied, then

$$\left(\frac{d}{dz}\right)^n \left[z^{n(m-l)-1} E_{n,m,l}(\lambda z^{-nm})\right] = (-1)^n \lambda z^{-n(l+1)-1} E_{n,m,l}(\lambda z^{-nm}). \tag{5.2.42}$$

5.2 The Kilbas–Saigo (Three-Parametric Mittag-Leffler) Function

◁ If $\lambda = 0$, then by (5.2.1) $E_{n,m,l}(z) = 1$, and formula (5.2.40) takes the well-known form

$$\left(\frac{d}{dz}\right)^n \left[z^{n(m-l)-1}\right] = \prod_{j=1}^{n} [n(m-l) - j] z^{n(m-l-1)-1}.$$

When $\lambda \neq 0$, according to (5.2.1) we have

$$\left(\frac{d}{dz}\right)^n \left[z^{n(m-l)-1} E_{n,m,l}(\lambda z^{-nm})\right]$$

$$= \left(\frac{d}{dz}\right)^n z^{n(m-l)-1} \left[1 + \sum_{k=1}^{\infty} c_k \lambda^k z^{n(m-l)-nmk-1}\right]$$

$$= [n(m-l) - 1] \cdots [n(m-l) - n] z^{n(m-l)-1-n}$$

$$+ \sum_{k=1}^{\infty} \lambda^k \left(\prod_{j=0}^{k-1} \frac{\Gamma(n[jm+l]+1)}{\Gamma(n[jm+l+1]+1)}\right)$$

$$\times [n(m-l) - nmk - 1] \cdots [n(m-l) - nmk - n] z^{n(m-l)-nmk-n-1}$$

$$= \prod_{k=1}^{n} [n(m-l) - k] z^{n(m-l-1)-1}$$

$$+ (-1)^n \sum_{k=1}^{\infty} \left(\prod_{j=0}^{k-1} \frac{\Gamma(\alpha[jm+l]+1)}{\Gamma(\alpha[jm+l+1]+1)}\right)$$

$$\times \frac{\Gamma(n[\{k-1\}m+l+1]+1)}{\Gamma(n[\{k-1\}m+l]+1)} \lambda^j z^{n(m-l)-nmk-n-1}$$

$$= \prod_{k=1}^{n} [n(m-l) - k] z^{n(m-l-1)-1}$$

$$+ (-1)^n a z^{-n(l+1)-1} \left[1 + \sum_{k=2}^{\infty} \lambda^k \left(\prod_{j=0}^{k-2} \frac{\Gamma(n[jm+l]+1)}{\Gamma(n[jm+l+1]+1)}\right) \lambda^k z^{-nm(k-1)}\right].$$

By changing the index $s = k - 1$ we obtain

$$\left(\frac{d}{dz}\right)^n \left[z^{n(m-l)-1} E_{n,m,l}(\lambda z^{-nm})\right]$$

$$= \prod_{k=1}^{n} [n(m-l) - k] z^{n(m-l-1)-1}$$

$$+ (-1)^n \lambda z^{-n(l+1)-1} \left[1 + \sum_{s=1}^{\infty} \left(\prod_{j=0}^{s-1} \frac{\Gamma(n[jm+l]+1)}{\Gamma(n[jm+l+1]+1)}\right) (\lambda z^{-nm})^s\right]$$

and (5.2.40) is proved in accordance with (5.2.1). (5.2.42) follows from (5.2.40). The theorem is proved. ▷

Special cases of the above property have the form:

Corollary 5.28 *If $\alpha = n = 1, 2, \cdots, \beta > 0$ and $\lambda \in \mathbb{C}$, then for the two-parametric Mittag-Leffler function $E_{n,\beta}(az^{-n})$ in [Chap. 4, formula (4.1.1)] we have*

$$\left(\frac{d}{dz}\right)^n \left[z^{n-\beta} E_{n,\beta}(\lambda z^{-n})\right] = \frac{(-1)^n}{\Gamma(\beta - n)} z^{-\beta} + (-1)^n \lambda z^{-n-\beta} E_{n,\beta}(\lambda z^{-n}). \quad (5.2.43)$$

Corollary 5.29 *If $\alpha = n = 1, 2, \cdots, \beta = k \in \mathbb{N}$ ($1 \leq k \leq n$) and $\lambda \in \mathbb{C}$, then for the two-parametric Mittag-Leffler function $E_{n,k}(az^{-n})$ we have*

$$\left(\frac{d}{dz}\right)^n \left[z^{n-k} E_{n,k}(\lambda z^{-n})\right] = (-1)^n \lambda z^{-n-k} E_{n,k}(\lambda z^{-n}). \quad (5.2.44)$$

In particular, when $k = 1$, for the Mittag-Leffler function $E_n(\lambda z^{-n})$ we have

$$\left(\frac{d}{dz}\right)^n \left[z^{n-1} E_n(\lambda z^{-n})\right] = (-1)^n \lambda z^{-n-1} E_n(\lambda z^{-n}). \quad (5.2.45)$$

Remark 5.30 The relations (5.2.33), (5.2.40) and (5.2.35), (5.2.42) can be considered as inhomogeneous and homogeneous differential equations of order n for the functions $z^{n(l-m+1)} E_{n,m,l}(\lambda z^{nm})$ and $z^{n(m-l)-1} E_{n,m,l}(\lambda z^{-nm})$, respectively. In this way explicit solutions of new classes of ordinary differential equations were obtained in [KilSai95b, SaiKil98], [SaiKil00]. In particular, (5.2.36), (5.2.43) and (5.2.37), (5.2.44) are inhomogeneous and homogeneous differential equations for the functions $z^{\beta-1} E_{n,\beta}(\lambda z^n)$ and $z^{n-j} E_{n,\beta}(\lambda z^{-n})$. The function $z^{j-1} E_{n,\beta}(\lambda z^n)$ as an explicit solution to a differential equation was found earlier by reduction of the differential equation to the corresponding Volterra integral equation [SaKiMa93, Section 42.1].

5.2.6 Complete Monotonicity of the Kilbas–Saigo Function

Complete monotonicity of the Kilbas–Saigo function $E_{\alpha,m,l}(z)$ was treated in connection with relaxation phenomena in dielectrics (see [CdOMai14, GaMaMa16]). In [CdOMai14], this question was studied in the case $0 < \alpha \leq 1$. The authors also made a conjecture about the domain of parameters α, m, l for which the function $E_{\alpha,m,l}(z)$ is completely monotonic. The conjecture is based on physical and numerical considerations. This conjecture was partially solved in [BoSiVa19]. The result reads.

5.2 The Kilbas–Saigo (Three-Parametric Mittag-Leffler) Function

Proposition 5.31 ([BoSiVa19, Prop. 4.3]) *Let $\alpha, m > 0$ and $l > -1/\alpha$. The Kilbas–Saigo function*

$$x \mapsto E_{\alpha,m,l}(-x)$$

is completely monotonic on $(0, +\infty)$ if and only if $\alpha \leq 1$ and $l \geq m - 1/\alpha$.

5.2.7 Fractional Integration of the Kilbas–Saigo Function

In this subsection we present applications of the Riemann–Liouville and Liouville fractional integrals $(I_{0+}^\alpha \varphi)(x)$ and $(I_-^\alpha \varphi)(x)$ of order $\alpha > 0$, defined by the respective formulas [SaKiMa93, formulas (5.1) and (5.3)]

$$(I_{0+}^\alpha \varphi)(x) = \frac{1}{\Gamma(\alpha)} \int_0^x \frac{\varphi(t)}{(x-t)^{1-\alpha}} dt \quad (x > 0) \tag{5.2.46}$$

and

$$(I_-^\alpha \varphi)(x) = \frac{1}{\Gamma(\alpha)} \int_x^\infty \frac{\varphi(t)}{(t-x)^{1-\alpha}} dt \quad (x > 0), \tag{5.2.47}$$

to the Kilbas–Saigo function (5.2.1).

The first statement shows the effect of I_{0+}^α on $E_{\alpha,m,l}(z)$.

Theorem 5.32 *Let $\alpha > 0$, $m > 0$, $l > -1/\alpha$ and $\lambda \in \mathbb{C}$. Then*

$$\lambda \left(I_{0+}^\alpha \left[t^{\alpha l} E_{\alpha,m,l} \left(\lambda t^{\alpha m} \right) \right] \right)(x) = x^{\alpha(l-m+1)} \left[E_{\alpha,m,l} \left(\lambda x^{\alpha m} \right) - 1 \right]. \tag{5.2.48}$$

◁ If $\lambda = 0$, then (5.2.48) takes the form $0 = 0$. Let $\lambda \neq 0$. In accordance with (5.2.46) and (5.2.1) we have

$$J \equiv \lambda \left(I_{0+}^\alpha \left[t^{\alpha l} E_{\alpha,m,l} \left(\lambda t^{\alpha m} \right) \right] \right)(x)$$

$$= \frac{\lambda}{\Gamma(\alpha)} \int_0^x (x-t)^{\alpha-1} \left[t^{\alpha l} + \sum_{k=1}^\infty \lambda^k \prod_{j=0}^{k-1} \frac{\Gamma(\alpha[jm+l]+1)}{\Gamma(\alpha[jm+l+1]+1)} t^{\alpha(mk+l)} \right] dt.$$

Interchanging integration and summation and evaluating the inner integrals by using the well-known formula [SaKiMa93, formula (2.44)]

$$\left(I_{0+}^\alpha \left[t^{\beta-1} \right] \right)(x) = \frac{\Gamma(\beta)}{\Gamma(\alpha+\beta)} x^{\alpha+\beta-1} \quad (\alpha > 0, \ \beta > 0), \tag{5.2.49}$$

we find

Where the interchange is possible since all integrals converge under the conditions of the theorem. Shifting the summation index $k+1$ to k we obtain

$$J = x^{\alpha(l-m+1)} \sum_{k=1}^{\infty} \lambda^k \left(\prod_{j=0}^{k-1} \frac{\Gamma(\alpha[jm+l]+1)}{\Gamma(\alpha[jm+l+1]+1)} \right) x^{\alpha m k}$$

$$= x^{\alpha(l-m+1)} \left[E_{\alpha,m,l} \left(\lambda x^{\alpha m} \right) - 1 \right].$$

This completes the proof of the theorem. ▷

The following corollary shows what the above properties look like in special cases.

Corollary 5.33 *For $\alpha > 0$, $\beta > 0$ and $\lambda \in \mathbb{C}$ the following formulas hold:*

$$\lambda \left(I_{0+}^{\alpha} \left[t^{\beta-1} E_{\alpha,\beta} (\lambda t^{\alpha}) \right] \right)(x) = x^{\beta-1} \left[E_{\alpha,\beta} (\lambda x^{\alpha}) - 1 \right], \quad (5.2.50)$$

$$\lambda \left(I_{0+}^{\alpha} \left[E_{\alpha} (\lambda t^{\alpha}) \right] \right)(x) = E_{\alpha} (\lambda x^{\alpha}) - 1. \quad (5.2.51)$$

Remark 5.34 In view of [Chap. 4, formula (4.1.1)] (5.2.50) can be written as

$$\lambda \left(I_{0+}^{\alpha} \left[t^{\beta-1} E_{\alpha,\beta} (\lambda t^{\alpha}) \right] \right)(x) = x^{\alpha+\beta-1} E_{\alpha,\alpha+\beta} (\lambda x^{\alpha}), \quad (5.2.52)$$

which coincides with the formula [SaKiMa93, Table 9.1, formula 23], if $\lambda = 1$. In particular, (5.2.51) takes the form

$$\lambda \left(I_{0+}^{\alpha} \left[E_{\alpha} (\lambda t^{\alpha}) \right] \right)(x) = x^{\alpha} E_{\alpha,\alpha+1} (\lambda x^{\alpha}), \quad (5.2.53)$$

by putting $\beta = 1$ in (5.2.52).

Next, we calculate the right-sided Liouville fractional operator I_{-}^{α} of the generalized Mittag-Leffler function.

Theorem 5.35 *Let $\alpha > 0, m > 0, l > -1/\alpha$ and $\lambda \in \mathbb{C}$. Then the following formula holds:*

$$\lambda \left(I_{-}^{\alpha} \left[t^{-\alpha(l+1)-1} E_{\alpha,m,l} \left(\lambda t^{-\alpha m} \right) \right] \right)(x) = x^{-\alpha(l-m)-1)} \left[E_{\alpha,m,l} \left(\lambda x^{-\alpha m} \right) - 1 \right]. \quad (5.2.54)$$

◁ If $\lambda = 0$, then (5.2.54) takes the form $0 = 0$. If $\lambda \neq 0$ we have in accordance with (5.2.47) and (5.2.1)

$$J = \lambda \left[\left(I_{0+}^{\alpha} \left[t^{\alpha l} \right] \right)(x) + \sum_{k=1}^{\infty} \lambda^{k+1} \left(\prod_{j=0}^{k-1} \frac{\Gamma(\alpha[jm+l]+1)}{\Gamma(\alpha[jm+l+1]+1)} \right) \left(I_{0+}^{\alpha} \left[t^{\alpha(mk+l)} \right] \right)(x) \right]$$

$$= \sum_{k=0}^{\infty} \lambda^{k+1} \left(\prod_{j=0}^{k} \frac{\Gamma(\alpha[jm+l]+1)}{\Gamma(\alpha[jm+l+1]+1)} \right) x^{\alpha(mk+l+1)},$$

5.2 The Kilbas–Saigo (Three-Parametric Mittag-Leffler) Function

Interchanging integration and summation and evaluating the inner integrals be using the formula [SaKiMa93, Table 9.3, formula 1]

$$\left(I_{-}^{\alpha}\left[t^{-\gamma}\right]\right)(x) = \frac{\Gamma(\gamma - \alpha)}{\Gamma(\gamma)} x^{\alpha - \gamma} \quad (\gamma > \alpha > 0), \tag{5.2.55}$$

and using the same arguments as in the proof of Theorem 5.32 we have

$$J = \lambda \Bigg[\left(I_{-}^{\alpha}\left[t^{-\alpha(l+1)-1}\right]\right)(x)$$
$$+ \sum_{k=1}^{\infty} \lambda^{k+1} \left(\prod_{j=0}^{k-1} \frac{\Gamma(\alpha[jm+l]+1)}{\Gamma(\alpha[jm+l+1]+1)} \right) \left(I_{-}^{\alpha}\left[t^{-\alpha(mk+l+1)-1}\right]\right)(x) \Bigg]$$
$$= \sum_{k=0}^{\infty} \lambda^{k+1} \left(\prod_{j=0}^{k} \frac{\Gamma(\alpha[jm+l]+1)}{\Gamma(\alpha[jm+l+1]+1)} \right) x^{-\alpha(mk+l)-1}$$
$$= x^{-\alpha(l-m)-1} \sum_{k=1}^{\infty} \lambda^k \left(\prod_{j=0}^{k-1} \frac{\Gamma(\alpha[jm+l]+1)}{\Gamma(\alpha[jm+l+1]+1)} \right) x^{-\alpha mk}$$
$$= x^{-\alpha(l-m)-1} \left[E_{\alpha,m,l}\left(\lambda x^{-\alpha m}\right) - 1 \right],$$

and the theorem is proved. ▷

The following corollary shows what the above properties look like in special cases.

Corollary 5.36 *For $\alpha > 0$, $\beta > 0$ and $\lambda \in \mathbb{C}$ the following formulas hold:*

$$\lambda \left(I_{-}^{\alpha}\left[t^{-\alpha-\beta} E_{\alpha,\beta}\left(\lambda t^{-\alpha}\right)\right]\right)(x) = x^{\alpha-\beta} \left[E_{\alpha,\beta}\left(\lambda x^{-\alpha}\right) - \frac{1}{\Gamma(\beta)} \right], \tag{5.2.56}$$

$$\lambda \left(I_{-}^{\alpha}\left[t^{-\alpha-1} E_{\alpha}\left(\lambda t^{-\alpha}\right)\right]\right)(x) = x^{\alpha-1} \left[E_{\alpha}\left(\lambda x^{-\alpha}\right) - 1 \right]. \tag{5.2.57}$$

5.2.8 Fractional Differentiation of the Kilbas–Saigo Function

In this subsection we present applications of the Riemann–Liouville and Liouville fractional derivatives D_{0+}^{α} and D_{-}^{α}, defined by the respective formulas [SaKiMa93, formulas (5.8)]

$$(D_{0+}^{\alpha} f)(x) = \left(\frac{d}{dx}\right)^n (I_0^{n-\alpha} f)(x) \tag{5.2.58}$$

$$= \left(\frac{d}{dx}\right)^n \frac{1}{\Gamma(n-\alpha)} \int_0^x \frac{f(t)}{(x-t)^{1-n+\alpha}} dt \quad (x > 0; \ n = [\alpha]+1)$$

and

$$(D_-^\alpha f)(x) = \left(-\frac{d}{dx}\right)^n (I_-^{n-\alpha} f)(x) \tag{5.2.59}$$

$$= \left(-\frac{d}{dx}\right)^n \frac{1}{\Gamma(n-\alpha)} \int_x^\infty \frac{f(t)}{(t-x)^{1-n+\alpha}} dt \quad (x > 0; \ n = [\alpha] + 1)),$$

to the Kilbas–Saigo function (5.2.1).

The application of D_{0+}^α to $E_{\alpha,m,l}(z)$ is given by the following statement.

Theorem 5.37 Let $\alpha > 0$ and $m > 0$ be such that

$$l > m - 1 - \frac{1}{\alpha}, \quad \alpha(jm + l) \neq 0, -1, -2, \cdots \ (j \in \neq \mathbb{N}_0),$$

and let $\lambda \in \mathbb{C}$. Then the following formula holds:

$$\left(D_{0+}^\alpha \left[t^{\alpha(l-m+1)} E_{\alpha,m,l}\left(\lambda t^{\alpha m}\right)\right]\right)(x)$$
$$= \frac{\Gamma[\alpha(l-m+1)+1]}{\Gamma[\alpha(l-m)+1]} x^{\alpha(l-m)} + \lambda x^{\alpha l} E_{\alpha,m,l}\left(\lambda x^{\alpha m}\right). \tag{5.2.60}$$

In particular, if $\alpha(l-m) = -j$ for some $j = 1, \cdots, -[-\alpha]$, then

$$\left(D_{0+}^\alpha \left[t^{\alpha(l-m+1)} E_{\alpha,m,l}\left(\lambda t^{\alpha m}\right)\right]\right)(x) = \lambda x^{\alpha l} E_{\alpha,m,l}\left(\lambda x^{\alpha m}\right). \tag{5.2.61}$$

◁ If $\lambda = 0$, then $E_{\alpha,m,l}(z) = 1$, and applying the formula [SaKiMa93, formula (2.44)]

$$\left(D_{0+}^\alpha \left[t^{\beta-1}\right]\right)(x) = \frac{\Gamma(\beta)}{\Gamma(\beta-\alpha)} x^{\beta-\alpha-1} \quad (\alpha > 0, \ \beta > 0), \tag{5.2.62}$$

with $\beta = \alpha(l - m + 1) + 1$, we have

$$\left(D_{0+}^\alpha \left[t^{\alpha(l-m+1)}\right]\right)(x) = \frac{\Gamma[\alpha(l-m+1)+1]}{\Gamma[\alpha(l-m)+1]} x^{\alpha(l-m)},$$

which proves (5.2.60) for $\lambda = 0$.

If $\lambda \neq 0$, then setting $n = [\alpha] + 1$, using (5.2.1) and (5.2.58), interchanging the order of summation and integration and applying (5.2.62), we have

$$J \equiv \left(D_{0+}^\alpha \left[t^{\alpha(l-m+1)} E_{\alpha,m,l}\left(\lambda t^{\alpha m}\right)\right]\right)(x)$$
$$= \left(D_{0+}^\alpha \left[t^{\alpha(l-m+1)}\right]\right)(x)$$
$$+ \left(\frac{d}{dx}\right)^n \left(I_{0+}^{n-\alpha} \left[\sum_{k=1}^\infty \lambda^k \left(\prod_{j=0}^{k-1} \frac{\Gamma(\alpha[jm+l]+1)}{\Gamma(\alpha[jm+l+1]+1)}\right) t^{\alpha(km+l-m+1)}\right]\right)(x)$$

5.2 The Kilbas–Saigo (Three-Parametric Mittag-Leffler) Function

$$= \frac{\Gamma[\alpha(l-m+1)+1]}{\Gamma[\alpha(l-m)+1]} x^{\alpha(l-m)}$$

$$+ \sum_{k=1}^{\infty} \lambda^k \left(\prod_{j=0}^{k-1} \frac{\Gamma(\alpha[jm+l]+1)}{\Gamma(\alpha[jm+l+1]+1)} \right) \left(D_{0+}^{\alpha} \left[t^{\alpha(km+l-m+1)} \right] \right)(x)$$

$$= \frac{\Gamma[\alpha(l-m+1)+1]}{\Gamma[\alpha(l-m)+1]} x^{\alpha(l-m)}$$

$$+ \sum_{k=1}^{\infty} \lambda^k \left(\prod_{j=0}^{k-1} \frac{\Gamma(\alpha[jm+l]+1)}{\Gamma(\alpha[jm+l+1]+1)} \right) \frac{\Gamma[\alpha(km+l-m+1)+1]}{\Gamma[\alpha(km+l-m)+1]} x^{\alpha(km+l-m)}$$

$$= \frac{\Gamma[\alpha(l-m+1)+1]}{\Gamma[\alpha(l-m)+1]} x^{\alpha(l-m)}$$

$$+ \lambda x^{\alpha l} \left[1 + \sum_{k=2}^{\infty} \lambda^{k-1} \left(\prod_{j=0}^{k-2} \frac{\Gamma(\alpha[jm+l]+1)}{\Gamma(\alpha[jm+l+1]+1)} \right) x^{\alpha(km-m)} \right]$$

$$= \frac{\Gamma[\alpha(l-m+1)+1]}{\Gamma[\alpha(l-m)+1]} x^{\alpha(l-m)} + x^{\alpha l} E_{\alpha,m,l} \left(\lambda x^{\alpha m} \right).$$

This yields (5.2.60).

Formula (5.2.61) then follows from the fact that the Gamma function has poles at every non-positive integer. This completes the proof of the theorem. ▷

Corollary 5.38 *For $\alpha > 0$, $\beta > 0$ and $\lambda \in \mathbb{C}$ the following formula holds:*

$$\left(D_{0+}^{\alpha} \left[t^{\beta-1} E_{\alpha,\beta} \left(\lambda t^{\alpha} \right) \right] \right)(x) = \frac{x^{\beta-\alpha-1}}{\Gamma(\beta-\alpha)} + \lambda x^{\beta-1} E_{\alpha,\beta} \left(\lambda x^{\alpha} \right). \quad (5.2.63)$$

If further $\beta - \alpha = 0, -1, -2, \cdots$, then

$$\left(D_{0+}^{\alpha} \left[t^{\beta-1} E_{\alpha,\beta} \left(\lambda t^{\alpha} \right) \right] \right)(x) = \lambda x^{\beta-1} E_{\alpha,\beta} \left(\lambda x^{\alpha} \right). \quad (5.2.64)$$

In particular, for $\beta = 1$ we have

$$\left(D_{0+}^{\alpha} \left[E_{\alpha} \left(\lambda t^{\alpha} \right) \right] \right)(x) = \frac{x^{-\alpha}}{\Gamma(1-\alpha)} + \lambda E_{\alpha} \left(\lambda x^{\alpha} \right). \quad (5.2.65)$$

Remark 5.39 If $\alpha = n \in \mathbb{N}$, then

$$\left(D_{0+}^n f \right)(x) = y^{(n)}(x) \quad (n \in \mathbb{N}), \quad (5.2.66)$$

and therefore formula (5.2.60) is reduced to relation (5.2.33), proved in Theorem 5.22 under weaker conditions. Similarly (5.2.65) is reduced to (5.2.38).

The next statement holds for the operator D_-^{α} given by (5.2.59).

Theorem 5.40 Let $\lambda \in \mathbb{C}$ and let $\alpha > 0$ and $m > 0$ be such that $l > m - \{\alpha\}/\alpha$, where $\{\alpha\}$ is the fractional part of α. Then

$$\left(D_-^\alpha \left[t^{\alpha(m-l)-1} E_{\alpha,m,l}\left(\lambda t^{-\alpha m}\right)\right]\right)(x) \tag{5.2.67}$$
$$= \frac{\Gamma[\alpha(l-m+1)+1]}{\Gamma[\alpha(l-m)+1]} x^{\alpha(m-l-1)-1} + \lambda x^{-\alpha(l+1)-1} E_{\alpha,m,l}\left(\lambda x^{-\alpha m}\right).$$

◁ If $\lambda = 1$, then $E_{\alpha,m,l}(z) = 1$ and applying the formula [KiSrTr06, formula (2.2.13)]

$$\left(D_-^\alpha \left[t^{-\gamma}\right]\right)(x) = \frac{\Gamma(\gamma+\alpha)}{\Gamma(\gamma)} x^{-\gamma-\alpha} \quad (\alpha > 0, \ \gamma > 1 - \{\alpha\}) \tag{5.2.68}$$

with $\gamma = \alpha(l-m)$ we get

$$\left(D_-^\alpha \left[t^{\alpha(m-l)-1}\right]\right)(x) = \frac{\Gamma[\alpha(l-m+1)+1]}{\Gamma[\alpha(l-m)+1]} x^{\alpha(m-l-1)},$$

which proves (5.2.67) for $\lambda = 0$.

If $\lambda \neq 0$, then we note that, in accordance with the condition $l > m - \{\alpha\}/\alpha$, $l > -\{\alpha\}/\alpha > -1/\alpha$. Therefore condition (5.2.3) is satisfied and the Kilbas–Saigo function $E_{\alpha,m,l}(z)$ is properly defined. In view of (5.2.59) and (5.2.1) we have

$$J \equiv \left(D_-^\alpha \left[t^{\alpha(m-l)-1} E_{\alpha,m,l}\left(\lambda t^{-\alpha m}\right)\right]\right)(x) = \left(D_-^\alpha \left[t^{-\alpha(l-m)-1}\right]\right)(x)$$
$$+ \sum_{k=1}^{\infty} \lambda^k \left(\prod_{j=0}^{k-1} \frac{\Gamma(\alpha[jm+l]+1)}{\Gamma(\alpha[jm+l+1]+1)} \right) \left(D_-^\alpha \left[t^{-\alpha(km+l-m)-1}\right]\right)(x).$$

Applying (5.2.68) with $\gamma = \alpha(km+l-m)+1$ $(k \in \mathbb{N}_0)$, we obtain

$$J = \frac{\Gamma[\alpha(l-m+1)+1]}{\Gamma[\alpha(l-m)+1]} x^{\alpha(m-l-1)-1}$$
$$+ \sum_{k=1}^{\infty} \lambda^k \left(\prod_{j=0}^{k-1} \frac{\Gamma(\alpha[jm+l]+1)}{\Gamma(\alpha[jm+l+1]+1)} \right) \frac{\Gamma[\alpha(km+l-m+1)+1]}{\Gamma[\alpha(km+l-m)+1]} x^{-\alpha(km+l-m-1)-1}$$
$$= \frac{\Gamma[\alpha(l-m+1)+1]}{\Gamma[\alpha(l-m)+1]} x^{\alpha(m-l-1)-1} + \lambda x^{-\alpha(l-1)-1}$$
$$+ \sum_{k=2}^{\infty} \lambda^k \left(\prod_{j=0}^{k-2} \frac{\Gamma(\alpha[jm+l]+1)}{\Gamma(\alpha[jm+l+1]+1)} \right) x^{-\alpha(km+l-m-1)-1}$$
$$= \frac{\Gamma[\alpha(l-m+1)+1]}{\Gamma[\alpha(l-m)+1]} x^{\alpha(m-l-1)-1} + \lambda x^{-\alpha(l-1)-1}$$
$$+ \lambda x^{-\alpha(l-1)-1} \left[1 + \sum_{k=1}^{\infty} \lambda^k \left(\prod_{j=0}^{k-1} \frac{\Gamma(\alpha[jm+l]+1)}{\Gamma(\alpha[jm+l+1]+1)} \right) x^{-\alpha km} \right].$$

5.2 The Kilbas–Saigo (Three-Parametric Mittag-Leffler) Function

This, in accordance with (5.2.1), yields (5.2.67), and the theorem is proved. ▷

Corollary 5.41 *For $\alpha > 0$, $\beta > [\alpha] + 1$ and $\lambda \in \mathbb{C}$ the following formula holds:*

$$\left(\mathcal{D}_-^\alpha \left[t^{\alpha-\beta} E_{\alpha,\beta}\left(\lambda t^{-\alpha}\right)\right]\right)(x) = \frac{x^{-\beta}}{\Gamma(\beta-\alpha)} + \lambda x^{-\alpha-\beta} E_{\alpha,\beta}\left(\lambda x^{-\alpha}\right). \quad (5.2.69)$$

Remark 5.42 If $\alpha = n \in \mathbb{N}$, then according to (5.2.66), formula (5.2.67) is reduced to relation (5.2.40), proved in Theorem 5.27 under weaker conditions. Similarly (5.2.69) is reduced to (5.2.43).

5.3 The Le Roy Type Function

Recently S. Gerhold [Ger12] and R. Garra and F. Polito [GarPol13] independently introduced a new function related to the special functions of the Mittag-Leffler family. This function is a generalization of the function studied by É. Le Roy [LeR99, LeR00] in the period 1895–1905 in connection with the problem of analytic continuation of power series with a finite radius of convergence. The study of this function was continued in [GaRoMa17]. In this section we mostly follow the results in [GaRoMa17].

5.3.1 Definition and Main Analytic Properties

In [Ger12, GarPol13] a new function was introduced

$$F_{\alpha,\beta}^{(\gamma)}(z) = \sum_{k=0}^{\infty} \frac{z^k}{[\Gamma(\alpha k + \beta)]^\gamma}, \quad z \in \mathbb{C}, \quad \alpha, \beta, \gamma \in \mathbb{C}, \quad (5.3.1)$$

which is related to the so-called *Le Roy function* (see, e.g.. [GaRoMa17])

$$R_\gamma(z) = \sum_{k=0}^{\infty} \frac{z^k}{[(k+1)!]^\gamma}, \quad z \in \mathbb{C}. \quad (5.3.2)$$

Here, for short, we use the name *Le Roy type function* for $F_{\alpha,\beta}^{(\gamma)}(z)$ defined by (5.3.1). It can also be considered as a generalization of the Wright function, since $F_{\alpha,\beta}^{(\gamma)}(z)$ has some properties similar to those of the *Wright function* (or better to say, of the *Fox–Wright function* $_p\Psi_q(z)$, see below).

By applying the properties of the Gamma function one can see that (5.3.1) is an entire function of the complex variable z for all values of the parameters such that $\operatorname{Re}\alpha > 0$, $\beta \in \mathbb{R}$ and $\gamma > 0$.

The order ρ and the type σ of the Le Roy type function can be found directly from the series representation (5.3.1) by using standard formulas for $\rho := \rho_F$ and $\sigma := \sigma_F$ valid for any entire function of the form (see e.g.. Appendix B below in this book)

$$F(z) = \sum_{n=0}^{\infty} c_n z^n,$$

namely

$$\rho = \limsup_{n\to\infty} \frac{n \log n}{\log \frac{1}{|c_n|}}, \qquad (5.3.3)$$

$$(\sigma e \rho)^{\frac{1}{\rho}} = \limsup_{n\to\infty} \left(n^{\frac{1}{\rho}} |c_n|^{\frac{1}{n}} \right). \qquad (5.3.4)$$

By using the Stirling formula for the Gamma function (see e.g.. Appendix A below in this book)

$$\Gamma(\alpha z + \beta) \approx \sqrt{2\pi} e^{-\alpha z} (\alpha z)^{\alpha z + \beta - 1/2} \left(1 + O\left(\frac{1}{z}\right) \right) \qquad (5.3.5)$$

we get the following result, which helps us to predict the maximal possible growth of the function $F_{\alpha,\beta}^{(\gamma)}(z)$.

Lemma 5.43 *Let $\alpha, \beta, \gamma > 0$. The order and type of the entire Le Roy type function $F_{\alpha,\beta}^{(\gamma)}(z)$ are*

$$\rho_{F_{\alpha,\beta}^{(\gamma)}} = \frac{1}{\alpha\gamma}, \quad \sigma_{F_{\alpha,\beta}^{(\gamma)}} = \alpha. \qquad (5.3.6)$$

These formulas still hold for any α, β, γ such that $\mathrm{Re}\,\alpha > 0$, $\beta \in \mathbb{C}$, $\gamma > 0$ if the parameter α is replaced with $\mathrm{Re}\,\alpha$ in (5.3.6).

Note that the above results agree well with the corresponding ones for the order and type of the Mittag-Leffler function (1.0.1) and its multi-index extension (see Chaps. 3, 4, 6 in this book and [Al-BLuc95, Kir99, Kir10b, KiKoRo13]).

The study of the asymptotic behavior of the Le Roy type function is of special interest due to existing and perspective applications. The main result in [Ger12] reads that (5.3.1) has the following asymptotic

$$F_{\alpha,\beta}^{(\gamma)}(z) \sim \frac{1}{\alpha\sqrt{\gamma}} (2\pi)^{(1-\gamma)/2} z^{(\gamma-2\beta\gamma+1)/2\alpha\gamma} e^{\gamma z^{1/\alpha\gamma}}, \quad |z| \to \infty, \qquad (5.3.7)$$

in the sector

5.3 The Le Roy Type Function

$$|\arg z| \leq \begin{cases} \frac{1}{2}\alpha\gamma\pi - \varepsilon, & 0 < \alpha\gamma < 2, \\ (2 - \frac{1}{2}\alpha\gamma)\pi - \varepsilon, & 2 \leq \alpha\gamma < 4, \\ 0, & 4 \leq \alpha\gamma, \end{cases} \quad (5.3.8)$$

where ε is an arbitrary small number. This result was obtained by using the saddle point method as described in [Evg78] and the purpose of the analysis in [Ger12] was to apply asymptotics in order to deliver certain holonomicity results for power series.

5.3.2 Integral Representations of the Le Roy Type Function

One of the important tools used to study the behavior of Mittag-Leffler type functions is their Mellin–Barnes integral representation (see e.g.. [GKMR, ParKam01]). Below we establish two integral representations for our function $E_{\alpha,\beta}^{(\gamma)}$ which use a technique similar to that in the Mellin–Barnes formulas. However, we should note that our integral representations cannot always be called Mellin–Barnes type representations since in the case of non-integer γ the integrands in these formulas contain a function $[\Gamma(\alpha s + \beta)]^\gamma$ which is multi-valued in s.

For simplicity we consider here and in what follows the function $F_{\alpha,\beta}^{(\gamma)}(z)$ with positive values of all parameters ($\alpha, \beta, \gamma > 0$). In this case the function $\Gamma(\alpha s + \beta)$ is a meromorphic function of the complex variable s with just simple poles at the points $s = -\frac{\beta+k}{\alpha}$, $k = 0, 1, 2, \ldots$. We fix the principal branch of the multi-valued function $[\Gamma(\alpha s + \beta)]^\gamma$ by drawing the cut along the negative semi-axes starting from $-\frac{\beta}{\alpha}$, ending at $-\infty$ and by supposing that $[\Gamma(\alpha x + \beta)]^\gamma$ is positive for all positive x. In addition, let the function $(-z)^s$ be defined in the complex plane cut along the negative semi-axis and

$$(-z)^s = \exp\{s[\log|z| + i \arg(-z)]\},$$

where $\arg(-z)$ is any arbitrary chosen branch of $\mathrm{Arg}(-z)$.

Theorem 5.44 *Let $\alpha, \beta, \gamma > 0$ and $[\Gamma(\alpha s + \beta)]^\gamma$, $(-z)^s$ be the described branches of the corresponding multi-valued functions. Then the Le Roy type function has the following $\mathcal{L}_{+\infty}$-integral representation*

$$F_{\alpha,\beta}^{(\gamma)}(z) = \frac{1}{2\pi i} \int_{\mathcal{L}_{+\infty}} \frac{\Gamma(-s)\Gamma(1+s)}{[\Gamma(\alpha s + \beta)]^\gamma} (-z)^s \, ds + \frac{1}{[\Gamma(\beta)]^\gamma}, \quad z \in \mathbb{C} \setminus (-\infty, 0],$$

(5.3.9)

where $\mathcal{L}_{+\infty}$ is a right loop situated in a horizontal strip starting at the point $+\infty + i\varphi_1$ and terminating at the point $+\infty + i\varphi_2$, $-\infty < \varphi_1 < 0 < \varphi_2 < +\infty$, crossing the real line at a point c, $0 < c < 1$.

◁ The chosen contour $\mathcal{L}_{+\infty}$ separate the poles $s = 1, 2, \ldots$ of the function $\Gamma(-s)$ and $s = -1, -2, \ldots$ of the function $\Gamma(1+s)$, together with the pole at $s = 0$ of the function $\Gamma(-s)$. So, the integral locally exists (see, e.g.., [KilSai04, p. 1], [ParKam01, p. 66]).

Now we prove the convergence of the integral in (5.3.9). To this end we use the reflection formula for the Gamma function [GKMR, p. 250]

$$\Gamma(z)\Gamma(1-z) = \frac{\pi}{\sin \pi z}, \quad z \notin \mathbb{Z}, \tag{5.3.10}$$

and the Stirling formula (5.3.5), which holds for any $\alpha, \beta > 0$.

First we note that on each ray $s = x + i\varphi_j$, $j = 1, 2$, $\varphi_j > 0$, we have

$$\Gamma(-s)\Gamma(1+s) = -\frac{1}{s}\Gamma(1-s)s\Gamma(s) = \frac{-\pi}{\sin \pi s}$$
$$= \frac{-2\pi i}{\cos \pi x \left(e^{-\pi \varphi_j} - e^{\pi \varphi_j}\right) + i \sin \pi x \left(e^{-\pi \varphi_j} + e^{\pi \varphi_j}\right)}$$

and hence,

$$|\Gamma(-s)\Gamma(1+s)| = \frac{\pi}{\sqrt{\sinh^2 \pi \varphi_j + \sin^2 \pi x}}.$$

Since

$$\sinh^2 \pi \varphi_j + \sin^2 \pi x > \sinh^2 \pi \varphi_j > 0,$$

this gives

$$|\Gamma(-s)\Gamma(1+s)| \leq C_1, \quad s \in \mathcal{L}_{+\infty}. \tag{5.3.11}$$

Next, it follows from (5.3.5) that

$$\log [\Gamma(\alpha s + \beta)]^\gamma$$
$$= \gamma \left[\frac{1}{2} \log 2\pi + (\alpha s + \beta - 1/2) \log \alpha s - \alpha s + \log\left(1 + \mathcal{O}\left(z^{-1}\right)\right)\right]$$
$$= \gamma \frac{1}{2} \log 2\pi + \gamma(\alpha x + i\alpha\varphi_j + \beta - 1/2) \log (\alpha x + i\alpha\varphi_j)$$
$$\quad - \gamma\alpha(x + i\varphi_j) + \gamma \log \left(1 + \mathcal{O}\left(z^{-1}\right)\right).$$

Hence

$$\log \|[\Gamma(\alpha s + \beta)]^\gamma| = \operatorname{Re} \log [\Gamma(\alpha s + \beta)]^\gamma$$
$$= \gamma \frac{1}{2} \log 2\pi - \gamma \alpha x + \gamma(\alpha x + \beta - 1/2)(\log \alpha + \log |x + i\varphi_j|)$$
$$\quad - \gamma \alpha \varphi_j \arg (x + i\varphi_j) + \gamma \operatorname{Re} \log \left(1 + \mathcal{O}\left(z^{-1}\right)\right)$$

5.3 The Le Roy Type Function

and therefore,

$$|[\Gamma(\alpha s + \beta)]^\gamma| = C_2 e^{-\gamma \alpha x} \alpha^{\gamma x} |x + i\varphi_j|^{\gamma(\alpha x + \beta - 1/2)}. \tag{5.3.12}$$

At last,

$$|(-z)^s| = |z|^x e^{-\varphi_j \arg(-z)}, \quad z = x + i\varphi_j. \tag{5.3.13}$$

The obtained asymptotic relations (5.3.11)–(5.3.13) give us the convergence of the integral in (5.3.9) for each fixed $z \in \mathbb{C} \setminus (-\infty, 0]$.

Finally, we evaluate the integral by using the residue theorem (since the poles $s = 1, 2, \ldots$ are bypassed by the contour $\mathcal{L}_{+\infty}$):

$$\frac{1}{2\pi i} \int_{\mathcal{L}_{+\infty}} \frac{\Gamma(-s)\Gamma(1+s)}{[\Gamma(\alpha s + \beta)]^\gamma} (-z)^s ds = -\sum_{k=1}^{\infty} \operatorname{Res}_{s=k} \left[\frac{\Gamma(-s)\Gamma(1+s)}{[\Gamma(\alpha s + \beta)]^\gamma} (-z)^s \right].$$

Since

$$\operatorname{Res}_{s=k} \Gamma(-s) = -\frac{(-1)^k}{k!}, \quad \Gamma(1+k) = k!,$$

then we obtain the final relation

$$\frac{1}{2\pi i} \int_{\mathcal{L}_{+\infty}} \frac{\Gamma(-s)\Gamma(1+s)(-z)^s ds}{[\Gamma(\alpha s + \beta)]^\gamma} = \sum_{k=1}^{\infty} \frac{z^k}{[\Gamma(\alpha k + \beta)]^\gamma} = F_{\alpha, \beta}^{(\gamma)}(z) - \frac{1}{[\Gamma(\beta)]^\gamma}. \triangleright$$

Now we get another form of the representation of the Le Roy type function via a generalization of the Mellin–Barnes integral. We consider the multi-valued function $[\Gamma(\alpha(-s) + \beta)]^\gamma$ and fix its principal branch by drawing the cut along the positive semi-axis starting from $\frac{\beta}{\alpha}$ and ending at $+\infty$ and supposing that $[\Gamma(\alpha(-x) + \beta)]^\gamma$ is positive for all negative x. We also define the function z^{-s} in the complex plane cut along positive semi-axis as

$$z^{-s} = \exp\{(-s)[\log|z| + i \arg z]\},$$

where $\arg z$ is any arbitrary chosen branch of $\operatorname{Arg} z$.

Theorem 5.45 *Let $\alpha, \beta, \gamma > 0$ and $[\Gamma(\alpha(-s) + \beta)]^\gamma$, z^{-s} be the described branches of the corresponding multi-valued functions. Then the Le Roy type function has the following $\mathcal{L}_{-\infty}$-integral representation*

$$F_{\alpha, \beta}^{(\gamma)}(z) = \frac{1}{2\pi i} \int_{\mathcal{L}_{-\infty}} \frac{\Gamma(s)\Gamma(1-s)}{[\Gamma(\alpha(-s) + \beta)]^\gamma} z^{-s} ds + \frac{1}{[\Gamma(\beta)]^\gamma}, \tag{5.3.14}$$

where $\mathcal{L}_{-\infty}$ is a left loop situated in a horizontal strip starting at the point $-\infty + i\varphi_1$ and terminating at the point $-\infty + i\varphi_2$, $-\infty < \varphi_1 < 0 < \varphi_2 < +\infty$, crossing the real line at a point c, $-1 < c < 0$.

The proof repeats all the arguments of the proof of Theorem 5.44 by using the behavior of the integrand on the contour $\mathcal{L}_{-\infty}$ and calculating the residue at the poles $s = -1, -2, \ldots$.

Remark 5.46 Note that in both representations (5.3.9) and (5.3.14) we cannot include the term corresponding to the pole at $s = 0$ in the integral term, since in this case either $\mathcal{L}_{+\infty}$ or $\mathcal{L}_{-\infty}$ should cross the branch cut of the corresponding multi-valued function.

5.3.3 Laplace Transforms of the Le Roy Type Function

Let us consider the case $\gamma > 1$ and evaluate the Laplace transform pair related to the Le Roy type function by means of an expression which is similar to that used to obtain the Laplace transform of the Mittag-Leffler function

Lemma 5.47 *Let $\alpha, \beta > 0$, $\gamma > 1$ be positive numbers, and $\lambda \in \mathbb{C}$. The Laplace transform of the Le Roy type function is*

$$\mathcal{L}\left\{t^{\beta-1} F_{\alpha,\beta}^{(\gamma)}(\lambda t^\alpha)\right\}(s) = \frac{1}{s^\beta} F_{\alpha,\beta}^{(\gamma-1)}(\lambda s^{-\alpha}). \tag{5.3.15}$$

◁ For the above mentioned values of its parameters $F_{\alpha,\beta}^{(\gamma)}(\cdot)$ is an entire function of its argument. Therefore the below interchanging of the integral and the sum is valid

$$\mathcal{L}\left\{t^{\beta-1} F_{\alpha,\beta}^{(\gamma)}(\lambda t^\alpha)\right\}(s) = \int_0^\infty e^{-st} \sum_{k=0}^\infty t^{\beta-1} \frac{\lambda^k t^{\alpha k}}{[\Gamma(\alpha k + \beta)]^\gamma} dt$$

$$= \sum_{k=0}^\infty \frac{\lambda^k}{[\Gamma(\alpha k + \beta)]^\gamma} \int_0^\infty e^{-st} t^{\beta-1} t^{\alpha k} dt$$

$$= \sum_{k=0}^\infty \frac{\lambda^k}{[\Gamma(\alpha k + \beta)]^\gamma} \frac{\Gamma(\alpha k + \beta)}{s^{\alpha k + \beta}} = \frac{1}{s^\beta} F_{\alpha,\beta}^{(\gamma-1)}(\lambda s^{-\alpha}),$$

which allows us to conclude the proof. ▷

Corollary 5.48 *For particular values of the parameter γ formula (5.3.15) allows us to establish the following simple relationships between the Laplace transform of the Le Roy type function and the Mittag-Leffler function:*

5.3 The Le Roy Type Function

$$\gamma = 2: \quad \mathcal{L}\left\{t^{\beta-1} F_{\alpha,\beta}^{(2)}(\lambda t^\alpha)\right\}(s) = \frac{1}{s^\beta} E_{\alpha,\beta}(\lambda s^{-\alpha}), \tag{5.3.16}$$

$$\gamma = 3: \quad \mathcal{L}\left\{t^{\beta-1} F_{\alpha,\beta}^{(3)}(\lambda t^\alpha)\right\}(s) = \frac{1}{s^\beta} E_{\alpha,\beta;\alpha,\beta}(\lambda s^{-\alpha}), \tag{5.3.17}$$

where $E_{\alpha,\beta}$ and $E_{\alpha,\beta;\alpha,\beta}$ are respectively the 2-parameter and 4-parameter Mittag-Leffler functions in the sense of Luchko–Kilbas–Kiryakova (see [GKMR, Chs. 4, 6]).

For any arbitrary positive integer value of the parameter γ the Laplace transform of the Le Roy type-function can be represented in terms of the *generalized Wright function*, known also as the *Fox–Wright function* (see e.g.. [GKMR, Appendix F]):

$$_pW_q(z) \equiv {}_pW_q(z)\left[\begin{matrix}(\rho_1, a_1), \ldots, (\rho_p, a_p) \\ (\sigma_1, b_1), \ldots, (\sigma_q, b_q)\end{matrix}; z\right] = \sum_{k=0}^\infty \frac{z^k}{k!} \frac{\prod_{r=1}^p \Gamma(\rho_r k + a_r)}{\prod_{r=1}^q \Gamma(\sigma_r k + b_r)},$$
(5.3.18)

where p and q are integers and $\rho_r, a_r, \sigma_r, b_r$ are real or complex parameters.

Lemma 5.49 *Let $\alpha, \beta > 0$ and $\gamma = m \in \mathbb{N}$. The Laplace transform of the Le Roy type function is given by*

$$\mathcal{L}\left\{F_{\alpha,\beta}^{(m)}(t)\right\}(s) = \frac{1}{s} {}_2W_m\left(\left[\begin{matrix}(1,1),(1,1) \\ \underbrace{(\beta,\alpha), \ldots, (\beta,\alpha)}_{m-\text{times}}\end{matrix}; \frac{1}{s}\right]\right). \tag{5.3.19}$$

Formula (5.3.19) is obtained directly by using the definitions of the Laplace transform and the generalized Wright function (cf. [KiSrTr06, p. 44]).

5.3.4 The Asymptotic Expansion on the Negative Semi-axis

In this section we study the asymptotic expansion of the Le Roy type function for large arguments. In particular, we pay attention to the case of a positive integer parameter $\gamma = m$ and, with major emphasis, we discuss the behavior of the function along the negative real semi-axis.

Since in this case (positive integer $\gamma = m$), the Le Roy type function is a particular instance of the generalized Wright function (5.3.18), namely

$$F_{\alpha,\beta}^{(m)}(z) = {}_1\Psi_m(z),$$

with $\rho_1 = 1$, $a_1 = 1$, $\sigma_1 = \sigma_2 = \cdots = \sigma_m = \alpha$ and $b_1 = b_2 = \cdots = b_m = \beta$, some of the results on the expansion of the Wright function, discussed first in

[Wri40a, Wri40c] and successively in [Bra62, Par10], can be exploited to derive suitable expansions of the Le Roy type function.

In particular, by applying to $F_{\alpha,\beta}^{(m)}(z)$ the reasoning proposed in [Par10], we introduce the functions

$$H(z) = \sum_{k=0}^{\infty} \frac{(-1)^k z^{-(k+1)}}{[\Gamma(\beta - \alpha(k+1))]^m} = -\sum_{k=1}^{\infty} \frac{(-1)^k z^{-k}}{[\Gamma(\beta - \alpha k)]^m} \qquad (5.3.20)$$

and

$$E(z) = m^{\frac{1}{2}(m+1)-m\beta} z^{\frac{m+1-2m\beta}{2\alpha m}} e^{mz^{\frac{1}{\alpha m}}} \sum_{j=0}^{\infty} A_j m^{-j} z^{-\frac{j}{\alpha m}},$$

where the A_j are the coefficients in the inverse factorial expansion of

$$\frac{\Gamma(\alpha m s + \theta')}{[\Gamma(\alpha s + \beta)]^m} = \alpha m \sum_{j=0}^{M-1} \frac{A_j}{(\alpha m s + \theta')_j} + \frac{\mathcal{O}(1)}{(\alpha m s + \theta')_M}, \quad \theta' = m\beta - \frac{m-1}{2},$$

with $(x)_j = x(x+1)\cdots(x+j-1)$ denoting the Pochhammer symbol. The following results directly descend from Theorem 1, 2 and 3 in [Par10].

Theorem 5.50 *Let $m \in \mathbb{N}$ and $0 < \alpha m < 2$. Then*

$$F_{\alpha,\beta}^{(m)}(z) \sim \begin{cases} E(z) + H(ze^{\mp\pi i}), & \text{if } |\arg z| \leq \frac{1}{2}\pi\alpha m, \\ H(ze^{\mp\pi i}), & \text{otherwise,} \end{cases} \quad \text{as } |z| \to \infty,$$

with the upper or lower signs chosen according as $\arg z > 0$ or $\arg z < 0$, respectively.

Theorem 5.51 *Let $m \in \mathbb{N}$, $\alpha m = 2$ and $|\arg z| \leq \pi$. Then*

$$F_{\alpha,\beta}^{(m)}(z) \sim E(z) + E(ze^{\mp 2\pi i}) + H(ze^{\mp\pi i}), \quad \text{as } |z| \to \infty,$$

with the upper or lower signs chosen according as $\arg z > 0$ or $\arg z < 0$, respectively.

Theorem 5.52 *Let $m \in \mathbb{N}$, $\alpha m > 2$ and $|\arg z| \leq \pi$. Then*

$$F_{\alpha,\beta}^{(m)}(z) \sim \sum_{r=-P}^{P} E(ze^{2\pi i r}), \quad \text{as } |z| \to \infty,$$

with P the integer such that $2P + 1$ is the smallest odd integer satisfying $2P + 1 > \frac{1}{2}m\alpha$.

5.3 The Le Roy Type Function

Deriving the coefficients A_j in $E(z)$ is quite a cumbersome process (a sophisticated algorithm is however described in [Par10]). Nevertheless, the first coefficient

$$A_0 = \frac{1}{\alpha}(2\pi)^{(1-m)/2} m^{-1-\frac{1}{2}m+m\beta}$$

is explicitly available, thus allowing us to write

$$E(z) = a_0 z^{\frac{m+1-2m\beta}{2\alpha m}} e^{mz^{\frac{1}{\alpha m}}} \left(1 + \mathcal{O}(z^{-\frac{1}{\alpha m}})\right), \qquad (5.3.21)$$

where

$$a_0 = \frac{1}{\alpha\sqrt{m}}(2\pi)^{(1-m)/2}.$$

We are then able to represent the asymptotic behavior of the Le Roy type function on the real negative semi axis by means of the following theorem.

Theorem 5.53 *Let $\alpha > 0$, $m \in \mathbb{N}$ and $t > 0$. Then*

$$F_{\alpha,\beta}^{(m)}(-t) \sim \begin{cases} H(t), & 0 < \alpha m < 2, \\ G(t) + H(t), & \alpha m = 2, \\ G(t), & 2 < \alpha m, \end{cases} \qquad t \to \infty,$$

where $H(t)$ is the same function introduced in (5.3.20) and

$$G(t) = 2a_0 t^{\frac{m+1-2m\beta}{2\alpha m}} \exp\left(mt^{\frac{1}{\alpha m}} \cos\frac{\pi}{\alpha m}\right) \cos\left(\frac{\pi(m+1-2m\beta)}{2\alpha m} + mt^{\frac{1}{\alpha m}} \sin\frac{\pi}{\alpha m}\right).$$

◁ Since for real and negative values $z = -t$, with $t > 0$, we can write $z = te^{i\pi}$, the use of Theorems 5.50–5.52 allows us to describe the asymptotic behavior of the Le Roy type function along the negative semi-axis according to

$$F_{\alpha,\beta}^{(m)}(-t) \sim \begin{cases} H(t), & 0 < \alpha m < 2, \\ E(te^{i\pi}) + E(te^{-i\pi}) + H(t), & \alpha m = 2, \end{cases} \qquad t \to \infty,$$

and when $\alpha m > 2$ we have, for an integer $P \geq 1$,

$$F_{\alpha,\beta}^{(m)}(-t) \sim \sum_{r=-P}^{P} E(te^{i(2r+1)\pi}), \quad 2(2P-1) \leq \alpha m < 2(2P+1). \qquad (5.3.22)$$

We define

$$\phi_r(t) := \frac{r\pi(m+1-2m\beta)}{2\alpha m} + mt^{\frac{1}{\alpha m}} \sin\frac{r\pi}{\alpha m},$$

and, by means of some standard trigonometric identities, we observe that

$$E(te^{ir\pi}) = a_0 t^{\frac{m+1-2m\beta}{2\alpha m}} \exp\left(mt^{\frac{1}{\alpha m}} \cos \frac{r\pi}{\alpha m}\right)\left[\cos\phi_r(t) + i\sin\phi_r(t)\right],$$

from which it is immediate that

$$E(te^{ir\pi}) + E(te^{-ir\pi}) = 2a_0 t^{\frac{m+1-2m\beta}{2\alpha m}} \exp\left(mt^{\frac{1}{\alpha m}} \cos \frac{r\pi}{\alpha m}\right)\cos\phi_r(t),$$

and, clearly, for $\alpha m < 2(2P+1)$ we have

$$\lim_{t \to \infty} E(te^{i(2P+1)\pi}) = 0.$$

Therefore, after introducing the functions

$$G_r(t) = 2a_0 t^{\frac{m+1-2m\beta}{2\alpha m}} \exp\left(mt^{\frac{1}{\alpha m}} \cos \frac{r\pi}{\alpha m}\right)\cos\left(\frac{r\pi(m+1-2m\beta)}{2\alpha m} + mt^{\frac{1}{\alpha m}} \sin \frac{r\pi}{\alpha m}\right)$$

for $r = 1, 2, \ldots, P$, with

$$P = \left\lfloor \frac{1}{2}\left(\frac{\alpha m}{2} + 1\right) \right\rfloor$$

and $\lfloor x \rfloor$ the greatest integer smaller than x, we can summarize the asymptotic behavior of the Le Roy type function as $t \to \infty$ by means of

$$F_{\alpha,\beta}^{(m)}(-t) \sim \begin{cases} H(t), & 0 < \alpha m < 2, \\ G_1(t) + H(t), & \alpha m = 2, \\ \sum_{r=1}^{P} G_r(t), & \alpha m > 2. \end{cases} \quad (5.3.23)$$

Observe now that since $\cos \frac{\pi}{\alpha m} > \cos \frac{2\pi}{\alpha m} > \cdots > \cos \frac{P\pi}{\alpha m} > 0$ the exponential in $G_1(t)$ dominates the exponential in the other functions $G_r(t)$, $r \geq 2$, which can therefore be neglected for $t \to \infty$ and hence the proof follows after putting $G(t) = G_1(t)$. ▷

We note that the asymptotic representation for $\alpha m < 2$ is similar to a well-known representation for the 2-parametric Mittag-Leffler function used in [GoLoLu02], also for computational purposes.

As we can clearly observe, $\alpha m = 2$ is a threshold value (compare with (5.3.6) for the order of this entire function) for the asymptotic behavior of $F_{\alpha,\beta}^{(m)}(-t)$ as $t \to \infty$. Whenever $\alpha m < 2$ the function is expected to decay in an algebraic way, while for $\alpha m > 2$ an increasing but oscillating behavior is instead expected.

5.3 The Le Roy Type Function

5.3.5 Extension to Negative Values of the Parameter α

The $\mathcal{L}_{-\infty}$-integral representation can be used to extend the function $F_{\alpha,\beta}^{(\gamma)}(z)$ to negative values of the parameter α (we follow here the approach described in [KiKoRo13]). To clearly distinguish the two cases we denote this extended Le Roy type function by $\mathcal{F}_{-\alpha,\beta}^{(\gamma)}(z)$.

Definition 5.54 The function $\mathcal{F}_{-\alpha,\beta}^{(\gamma)}(z)$, α, β, γ is defined by the following

$$\mathcal{F}_{-\alpha,\beta}^{(\gamma)}(z) = -\frac{1}{2\pi i} \int_{\mathcal{L}_{-\infty}} \frac{\Gamma(-s)\Gamma(1+s)}{[\Gamma(-\alpha s + \beta)]^\gamma}(-z)^s ds, \tag{5.3.24}$$

where $\mathcal{L}_{-\infty}$ is a right loop situated in a horizontal strip starting at the point $-\infty + i\varphi_1$ and terminating at the point $-\infty + i\varphi_2$, $-\infty < \varphi_1 < 0 < \varphi_2 < +\infty$, crossing the real line at a point c, $-1 < c < 0$, values of $(-z)^s$ are calculated as described above, and the branch of the multi-valued function $[\Gamma(-\alpha s + \beta)]^\gamma$ is defined in the complex plane cut along the positive semi-axes starting from $\frac{\beta}{\alpha}$ and ending at $+\infty$, with $[\Gamma(-\alpha x + \beta)]^\gamma$ being positive for all negative x.

Using the slight correction of the proof of Theorem 5.44 we get the following result.

Theorem 5.55 Let $\alpha, \beta, \gamma > 0$, then the extended Le Roy type function (5.3.24) satisfies the following series representation

$$\mathcal{F}_{-\alpha,\beta}^{(\gamma)}(z) = -\sum_{k=1}^{\infty} \frac{1}{[\Gamma(\alpha k + \beta)]^\gamma} \frac{1}{z^k}, \quad z \in \mathbb{C} \setminus \{0\}. \tag{5.3.25}$$

Corollary 5.56 The Le Roy type function (5.3.1) and its extension (5.3.24) are connected via the following relation

$$\mathcal{F}_{-\alpha,\beta}^{(\gamma)}(z) = \frac{1}{[\Gamma(\beta)]^\gamma} - F_{\alpha,\beta}^{(\gamma)}\left(\frac{1}{z}\right). \tag{5.3.26}$$

Observe that the relation (5.3.26) is similar to the ones presented in [Han-et-al09, KiKoRo13].

5.4 Historical and Bibliographical Notes

By means of the series representation, a generalization of the Mittag-Leffler function $E_{\alpha,\beta}^\gamma(z)$ was introduced by Prabhakar in [Pra71] (see also [MatHau08]). This function is a special case of the Wright generalized hypergeometric function [Wri35a, Wri35b] as well as the H-function [MaSaHa10]. For various properties of this function with applications, see [Pra71]. Like any function of the Mittag-Leffler type, the three parametric Mittag-Leffler function $E_{\alpha,\beta}^{\gamma}(z)$ can be represented via the Mellin–Barnes integral.

Differentiation and integration formulas for the three-parametric Mittag-Leffler function (Prabhakar function) were obtained in [KiSaSa04]. Some of these formulas were generalized and given in the form of the Laplace transform (see [Sax02]). Relations connecting the function $E_{\alpha,\beta}^{\gamma}(z)$ and the Riemann–Liouville fractional integrals and derivatives are given in [SaxSai05].

Most of the interest in the Prabhakar function is related to the description of relaxation and response in anomalous dielectrics of Havriliak–Negami type (e.g.., see [GarGar18, GaMaMa16, Pan18, StaWer16]), a model of complex susceptibility introduced to keep into account the simultaneous nonlocality and nonlinearity observed in the response of disordered materials and heterogeneous systems [Mis09]. Further applications of the Prabhakar function are however encountered in probability theory [Go-et-al16, Jam10, PogTom16], in the study of stochastic processes [D'OPol17, PolSca16] and of systems with strong anisotropy [ChaTon06, Ton07], in fractional viscoelasticity [GiuCol18], in the solution of some fractional boundary-value problems [BazDim13, BazDim14, EshAns16, FiCaVa12, LucSri95], in the description of dynamical models of spherical stellar systems [AnVHBa12] and in connection with other fractional or integral differential equations [AskAns16, KiSaSa02, LiSaKa17]. We should also mention a survey of the key results and applications emerging from the Prabhakar function [Giu-et-al20].

Another three-parametric generalization of the Mittag-Leffler function $E_{\alpha,m,l}(z)$ was proposed in the form of a power series by Kilbas and Saigo as the solution of a certain Abel–Volterra type integral equation (see [KilSai95b]). In particular, in [KilSai95b], a number of differential relations involving the Kilbas–Saigo function $E_{\alpha,m,l}(z)$ were obtained. The corresponding formulas were considered as inhomogeneous and homogeneous differential equations of order n for the functions $z^{n(l-m+1)}E_{n,m,l}(\lambda z^{nm})$ and $z^{n(m-l)-1}E_{n,m,l}(\lambda z^{-nm})$, respectively. In this way explicit solutions of new classes of ordinary differential equations were obtained in [KilSai95a, SaiKil98, SaiKil00] (see also [KiSrTr06]). Analytic properties of this function are studied in [GoKiRo98]. One of the first applications of the Kilbas–Saigo function is given in the paper by Orsinger and Polito [OrsPol09], where the birth-death stochastic process was discussed. A theoretical application is presented by Kilbas and Repin [KilRep10]. They studied an analog of the Tricomi problem for mixed type equations with fractional partial derivatives. In [HanSer12] the solution of the Bloch–Torrey equation for space-time fractional anisotropic diffusion is expressed in terms of the three-parametric Mittag-Leffler function $E_{\alpha,m,l}(z)$. Recently, an interest in the use of the Kilbas–Saigo function has been revisited, see e.g.. [CdOMai14], where relaxation phenomena in dielectrics are discussed by using the Kilbas–Saigo function, and [Mag-et-al19] presents recent results on ultraslow diffusion in heterogeneous media. A numerical routine for the Kilbas–Saigo function is presented in [Mag19].

5.4 Historical and Bibliographical Notes

Le Roy introduced his function

$$R_\gamma(z) = \sum_{k=0}^\infty \frac{z^k}{[(k+1)!]^\gamma}, \quad z \in \mathbb{C}, \tag{5.4.1}$$

and used it in [LeR99] to study the analytic continuation of the sum of power series. This reason for the origin of (5.4.1) sounds close to Mittag-Leffler's idea to introduce the function $E_\alpha(z)$ for the aims of analytic continuation (we have to note that Mittag-Leffler and Le Roy were working on this idea in competition). The Le Roy function is involved in the solution of problems of various types; in particular it has recently been used in the construction of a Convey–Maxwell–Poisson distribution [ConMax62], which is important due to its ability to model count data with different degrees of over- and under-dispersion [Pog16, SLSPL].

An interest in the Le Roy type function has been revisited with its generalization proposed in [Ger12, GarPol13, GaRoMa17]. The work in [Ger12] was devoted to the study of asymptotic properties of $F^{(\gamma)}_{\alpha,\beta}(z)$ as an analytic function in some sectors of the complex plane (it was implicitly shown that this function has order $\rho = 1/\alpha\gamma$ and type $\sigma = \gamma$). In [GarPol13] the function $F^{(\gamma)}_{\alpha,\beta}(z)$ is considered from an operational point of view. More specifically, the properties of this function are studied in relation to some integro-differential operators involving the Hadamard fractional derivatives (e.g.., see [SaKiMa93, Sect. 18.3]) or hyper-Bessel-type operators. By using these properties, the operational (or formal) solutions to certain boundary and initial value problems for fractional differential equations are derived. An application of the developed technique to a modified Lamb–Bateman integral equation is also presented. Further properties of the Le Roy type function are studied in [GaRoMa17].

Among the most studied properties of the above discussed three-parametric Mittag-Leffler functions is their complete monotonicity, see, e.g.., [GarGar18, GiuCol18], cf. [AnhMcV]. We also recall the paper [BoSiVa19], where the relations of the Le Roy function and the Kilbas–Saigo function to certain probability distributions is mentioned and the complete monotonicity of the respective cases of these functions is discussed.

5.5 Exercises

5.5.1 Prove the following reducibility formulas for the three-parametric Mittag-Leffler function $E^\gamma_{\alpha,\beta}(z)$:

(a) If $\beta, \gamma \in \mathbb{C}$ are such that $\operatorname{Re}\beta > 0$, $\operatorname{Re}(\gamma - \beta) > 2$, then

$$zE^3_{\beta,\gamma} = \frac{1}{2\beta^2}\left[E_{\beta,\gamma-\beta-2)} - (2\gamma - 3\beta - 3)E_{\beta,\gamma-\beta-1} \right.$$
$$\left. + (2\beta^2 + \gamma^2 - 3\beta\gamma + 3\beta - 2\gamma + 1)E_{\beta,\gamma-\beta}\right].$$

(b) If $\beta, \gamma \in \mathbb{C}$ are such that $\operatorname{Re}\beta > 0$, $\operatorname{Re}\gamma > 2$, then

$$E^3_{\beta,\gamma} = \frac{1}{2\beta^2}\left[E_{\beta,\gamma-2)} - (2\gamma - 3\beta - 3)E_{\beta,\gamma-1} \right.$$
$$\left. + (2\beta^2 + \gamma^2 - 3\beta\gamma + 3\beta - 2\gamma + 1)E_{\beta,\gamma-\beta}\right].$$

5.5.2 ([Cap19, p. 112]) Let $\operatorname{Re}\alpha > 0$, $\operatorname{Re}\beta > 0$, $\gamma > 0$. Prove the following

$$(\beta - \alpha\gamma)E^\gamma_{\alpha,\beta+1} = E^\gamma_{\alpha,\beta} - \alpha\gamma E^{\gamma+1}_{\alpha,\beta+1}.$$

5.5.3 ([MatHau08, p. 96]) Prove the following integral relations for the three-parametric Mittag-Leffler function $E^\gamma_{\alpha,\beta}(z)$ valid for all $\operatorname{Re}\alpha > 0$, $\operatorname{Re}\beta > 0$, $\gamma > 0$, $\operatorname{Re}\delta > 0$:

(a) $\frac{1}{\Gamma(\delta)}\int_0^1 u^{\beta-1}(1-u)^{\delta-1} E^\gamma_{\alpha,\beta}(zu^\alpha)du = E^\gamma_{\alpha,\beta+\delta}(z);$

(b) $\frac{1}{\Gamma(\delta)}\int_t^x (x-u)^{\delta-1}(u-t)^{\beta-1} E^\gamma_{\alpha,\beta}(\lambda(u-t)^\alpha)du$

$= (x-t)^{\beta+\delta-1} E^\gamma_{\alpha,\beta+\delta}(\lambda(x-t)^\alpha).$

5.5.4 ([KiSaSa02, p. 383–384]) Let

$$\left(\mathbf{E}^\gamma_{\rho,\mu,\omega;a+}\varphi\right)(x) = \int_a^x (x-t)^{\mu-1} E^\gamma_{\rho,\mu}[\omega(x-t)^\rho]\varphi(t)dt, \quad x > a,$$

be the integral transform with the Prabhakar function in the kernel.

(a) Find the value of this transform of the power-type function $(t-a)^{\beta-1}$.
(b) Calculate the composition of the operator $\mathbf{E}^\gamma_{\rho,\mu,\omega;a+}$ and the left-sided Riemann–Liouville fractional integration operator I^α_{a+}.
(c) Prove the semigroup property of the integral transform with the Prabhakar function in the kernel:

$$\mathbf{E}^\gamma_{\rho,\mu,\omega;a+}\mathbf{E}^\sigma_{\rho,\mu,\omega;a+}\varphi = \mathbf{E}^{\gamma+\sigma}_{\rho,\mu,\omega;a+}\varphi.$$

5.5.5 ([KilSai00, p. 194]) Show that the linear homogeneous differential equation

$$y^{(n)}(x) = ax^\beta y(x) \quad (0 < x \leq d < +\infty)$$

$(a \neq 0,\ \beta\mathbb{R},\ \beta > -n,\ (n+\beta)(i+1) \neq 1, 2, \ldots, n-1,\ i \in \mathbb{N})$ has n solutions of the form

$$y_j(x) = x^{j-1} E_{n,1+\beta/n,(\beta+j-)/n}(ax^{\beta+n}) \quad (j = 1, 2, \ldots, n).$$

Prove that if $\beta \geq 0$, then these solutions are linearly independent.

5.5 Exercises

5.5.6 ([KilSai00, p. 197]) Let $n \in \mathbb{N}$, $\beta \in \mathbb{R}$, $\beta > -n$, $f_k, \mu_k \in \mathbb{R}$, $k = 0, 1, \ldots, p$, and $i(n + \beta) + \mu_k \neq -j$ ($i \in \mathbb{N}_0$, $k = 0, 1, \ldots, p$, $j = 1, 2, \ldots, n$). Show that the inhomogeneous differential equation

$$y^{(n)}(x) = ax^\beta y(x) + \sum_{k=0}^{p} f_k x^{\mu_k} \quad (0 < x \leq d < +\infty)$$

has a particular solution of the form

$$y_0(x) = \sum_{k=0}^{p} \left[\prod_{j=1}^{n} \frac{1}{\mu_k + j} \right] f_k x^{\mu_k + n} E_{n, 1 + \beta/n, 1 + (\beta + \mu_k)/n}(ax^{n+\beta}).$$

5.5.7 ([KilSai96, p. 365]) Let $\alpha > 0$, $m > 0$, $\mu_k > -1$, $f_k \in \mathbb{R}$ ($k0, 1, \ldots, l$). Prove that the Abel–Volterra equation with quasi-polynomial free term

$$\varphi(x) = \frac{ax^{\alpha(m-1)}}{\Gamma(\alpha)} \int_0^x \frac{\varphi(t)}{(x-t)^{1-\alpha}} dt + \sum_{k=0}^{l} f_k x^{\mu_k} \quad (0 < x < d \leq +\infty)$$

has a unique solution of the form

$$\varphi(x) = \sum_{k=0}^{l} f_k x^{\mu_k} E_{\alpha, m, \mu_k/\alpha}(ax^{\alpha m}).$$

Chapter 6
Multi-index and Multi-variable Mittag-Leffler Functions

6.1 The Four-Parametric Mittag-Leffler Function: The Luchko–Kilbas–Kiryakova Approach

6.1.1 Definition and Special Cases

Consider the function defined for $\alpha_1, \alpha_2 \in \mathbb{R}$ ($\alpha_1^2 + \alpha_2^2 \neq 0$) and $\beta_1, \beta_2 \in \mathbb{C}$ by the series

$$E_{\alpha_1,\beta_1;\alpha_2,\beta_2}(z) \equiv \sum_{k=0}^{\infty} \frac{z^k}{\Gamma(\alpha_1 k + \beta_1)\Gamma(\alpha_2 k + \beta_2)} \quad (z \in \mathbb{C}). \tag{6.1.1}$$

Such a function with positive $\alpha_1 > 0$, $\alpha_2 > 0$ and real $\beta_1, \beta_2 \in \mathbb{R}$ was introduced by Dzherbashian [Dzh60]. When $\alpha_1 = \alpha$, $\beta_1 = \beta$ and $\alpha_2 = 0, \beta_2 = 1$, this function coincides with the Mittag-Leffler function (4.1.1):

$$E_{\alpha,\beta;0,1}(z) = E_{\alpha,\beta}(z) \equiv \sum_{k=0}^{\infty} \frac{z^k}{\Gamma(\alpha k + \beta)} \quad (z \in \mathbb{C}). \tag{6.1.2}$$

Therefore (6.1.1) is sometimes called the *generalized Mittag-Leffler function* or *four-parametric Mittag-Leffler function*.

Certain special functions of Bessel type are expressed in terms of $E_{\alpha_1,\beta_1;\alpha_2,\beta_2}(z)$:

The Bessel function of the first kind (see e.g., [ErdBat-2, n. 7.2.1-2], [NIST, p. 217, 219])

$$J_\nu(z) = \left(\frac{z}{2}\right)^\nu E_{1,\nu+1;1,1}\left(-\frac{z^2}{4}\right). \tag{6.1.3}$$

The Struve function (see e.g., [ErdBat-2, n. 7.5.4], [NIST, p. 288])

$$\mathbf{H}_\nu(z) = \left(\frac{z}{2}\right)^{\nu+1} E_{1,\nu+3/2;1,3/2}\left(-\frac{z^2}{4}\right). \tag{6.1.4}$$

The Lommel function (see e.g., [ErdBat-2, n. 7.5.5])

$$S_{\mu,\nu}(z) = \frac{z^{\mu+1}}{4}\Gamma\left(\frac{\mu-\nu+1}{2}\right)\Gamma\left(\frac{\mu+\nu+1}{2}\right)E_{1,\frac{\mu-\nu+1}{2};1,\frac{\mu+\nu+1}{2}}\left(-\frac{z^2}{4}\right). \tag{6.1.5}$$

The Bessel–Maitland function (see e.g., [Kir94, App. E, ii])

$$J_\nu^\mu(z) = E_{\mu,\nu+1;1,1}(-z). \tag{6.1.6}$$

The generalized Bessel–Maitland function (see e.g., [Kir94, App. E, ii])

$$J_{\nu,\lambda}^\mu(z) = \left(\frac{z}{2}\right)^{\nu+2\lambda} E_{\mu,\lambda+\nu+1;1,\lambda+1}(-z). \tag{6.1.7}$$

6.1.2 Basic Properties

First of all we prove that (6.1.1) is an entire function if $\alpha_1 + \alpha_2 > 0$.

Theorem 6.1 *Let $\alpha_1, \alpha_2 \in \mathbb{R}$ and $\beta_1, \beta_2 \in \mathbb{C}$ be such that $\alpha_1^2 + \alpha_2^2 \neq 0$ and $\alpha_1 + \alpha_2 > 0$. Then $E_{\alpha_1,\beta_1;\alpha_2,\beta_2}(z)$ is an entire function of $z \in \mathbb{C}$ of order*

$$\rho = \frac{1}{\alpha_1 + \alpha_2} \tag{6.1.8}$$

and type

$$\sigma = \left(\frac{\alpha_1+\alpha_2}{|\alpha_1|}\right)^{\frac{\alpha_1}{\alpha_1+\alpha_2}} \left(\frac{\alpha_1+\alpha_2}{|\alpha_2|}\right)^{\frac{\alpha_2}{\alpha_1+\alpha_2}}. \tag{6.1.9}$$

◁ Rewrite (6.1.1) as the power series

$$E_{\alpha_1,\beta_1;\alpha_2,\beta_2}(z) = \sum_{k=0}^\infty c_k z^k, \quad c_k = \frac{1}{\Gamma(\alpha_1 k + \beta_1)\Gamma(\alpha_2 k + \beta_2)}. \tag{6.1.10}$$

Using Stirling's formula for the Gamma function we obtain

$$\frac{|c_k|}{|c_{k+1}|} \sim |\alpha_1|^{\alpha_1}|\alpha_2|^{\alpha_2} k^{\alpha_1+\alpha_2} \to +\infty \ (k \to \infty).$$

Thus, $E_{\alpha_1,\beta_1;\alpha_2,\beta_2}(z)$ is an entire function of z when $\alpha_1 + \alpha_2 > 0$.

We use [Appendix B, formulas (B.5) and (B.6)] to evaluate the order ρ and the type σ of (6.3.1). For this we apply the asymptotic formula for the logarithm of the

6.1 The Four-Parametric Mittag-Leffler Function: The Luchko ...

Gamma function $\Gamma(z)$ at infinity [ErdBat-1, 1.18(1)]:

$$\log \Gamma(z) = \left(z - \frac{1}{2}\right) \log z - z + \frac{1}{2} \log(2z) + O\left(\frac{1}{z}\right) \quad (|z| \to \infty, \ |\arg z| < \pi). \tag{6.1.11}$$

Applying this formula and taking (6.1.10) into account, we deduce the asymptotic estimate

$$\log\left(\frac{1}{c_k}\right) \sim k \log(k)(\alpha_1 + \alpha_2) \quad (k \to \infty)$$

from which, in accordance with [Appendix B, (B.5)], we obtain (6.1.8).

Further, according to [Appendix A, (A.24)], we have

$$\Gamma(\alpha_j k + \beta_j) \tag{6.1.12}$$

$$= (2\pi)^{1/2} \left(\alpha_j k + \beta_j\right)^{\alpha_j k + \beta_j - \frac{1}{2}} e^{-(\alpha_j k + \beta_j)} \left[1 + O\left(\frac{1}{k}\right)\right] \quad (k \to \infty)$$

for $j = 1, 2$, and we obtain the asymptotic estimate

$$\Gamma(\alpha_1 k + \beta_1)\Gamma(\alpha_2 k + \beta_2) \sim 2\pi \prod_{j=1}^{2} (\alpha_j k)^{\alpha_j k + \beta_j - \frac{1}{2}} e^{-\alpha_j k} \quad (k \to \infty). \tag{6.1.13}$$

From (6.1.10) and (6.1.13) we have

$$\limsup_{k \to \infty} \left(k^{1/\rho} |c_k|^{1/k}\right) = \limsup_{k \to \infty} k^{1/\rho} \prod_{j=1}^{2} \left[(|\alpha_j| k)^{-\alpha_j} e^{\alpha_j}\right]$$

$$= e^{\alpha_1 + \alpha_2} \prod_{j=1}^{2} |\alpha_j|^{-\alpha_j} = e^{1/\rho} \prod_{j=1}^{2} |\alpha_j|^{-\alpha_j}.$$

Substituting this relation into [Appendix B, (B.6)] we obtain

$$\sigma = \frac{1}{\rho} \left(\prod_{j=1}^{2} |\alpha_j|^{-\alpha_j}\right)^{\rho} = (\alpha_1 + \alpha_2) \left(|\alpha_1|^{-\alpha_1} |\alpha_2|^{-\alpha_2}\right)^{\frac{1}{\alpha_1 + \alpha_2}}$$

$$= \left(\frac{\alpha_1 + \alpha_2}{|\alpha_1|}\right)^{\frac{\alpha_1}{\alpha_1 + \alpha_2}} \left(\frac{\alpha_1 + \alpha_2}{|\alpha_2|}\right)^{\frac{\alpha_2}{\alpha_1 + \alpha_2}},$$

which proves (6.1.9). ▷

Remark 6.2 For $\alpha_1 > 0$ and $\alpha_2 > 0$, relations (6.1.8) and (6.1.9) were proved by Dzherbashian [Dzh60].

6.1.3 Integral Representations and Asymptotics

The four-parametric Mittag-Leffler function has the Mellin–Barnes integral representation

$$E_{\alpha_1,\beta_1;\alpha_2,\beta_2}(z) = \frac{1}{2\pi i} \int_{\mathcal{L}} \frac{\Gamma(s)\Gamma(1-s)}{\Gamma(\beta_1 - \alpha_1 s)\Gamma(\beta_2 - \alpha_2 s)}(-z)^{-s} ds, \qquad (6.1.14)$$

where $\mathcal{L} = \mathcal{L}_{-\infty}$ is a left loop, i.e. the contour which is situated in a horizontal strip, starting at $-\infty + i\varphi_1$ and ending at $-\infty + i\varphi_2$, with $-\infty < \varphi_1 < 0 < \varphi_2 < +\infty$. This contour separates poles of the Gamma functions $\Gamma(s)$ and $\Gamma(1-s)$.

By using (6.1.14) the function $E_{\alpha_1,\beta_1;\alpha_2,\beta_2}$ can be extended to non-real values of the parameters. If the parameters $\alpha_1, \beta_1; \alpha_2, \beta_2$ are such that $\mathrm{Re}\,(\alpha_1 + \alpha_2) > 0$, then the integral (6.1.14) converges for all $z \neq 0$. This is a consequence of the following asymptotic formulas for the function $H(s) = \frac{\Gamma(s)\Gamma(1-s)}{\Gamma(\beta_1 - \alpha_1 s)\Gamma(\beta_2 - \alpha_2 s)}$ in the integrand of (6.1.14), where $s = t + i\sigma, (t \to -\infty)$, and the properties of the Mellin–Barnes integral:

– for $\mathrm{Re}\,\alpha_1 > 0,\ \mathrm{Re}\,\alpha_2 > 0$

$$|H(s)| \sim M_1 \left(\frac{|t|}{e}\right)^{\mathrm{Re}(\alpha_1+\alpha_2)t} \frac{[\mathrm{Re}(\alpha_1)^{\mathrm{Re}(\alpha_1)}\mathrm{Re}(\alpha_2)^{\mathrm{Re}(\alpha_2)}]^t}{|t|^{\sum_{j=1}^{2}[\mathrm{Re}(\beta_j)+\sigma\mathrm{Im}(\alpha_i)]-1}}; \qquad (6.1.15)$$

– for $\mathrm{Re}\,\alpha_1 < 0,\ \mathrm{Re}\,\alpha_2 > 0$

$$|H(s)| \sim M_2 \left(\frac{|t|}{e}\right)^{\mathrm{Re}(\alpha_1+\alpha_2)t} \frac{[|\mathrm{Re}(\alpha_1)|^{\mathrm{Re}(\alpha_1)}\mathrm{Re}(\alpha_2)^{\mathrm{Re}(\alpha_2)}]^t}{|t|^{\sum_{i=1}^{2}[\mathrm{Re}(\beta_i)+\sigma\mathrm{Im}(\alpha_i)]-1}} e^{-\pi\mathrm{Im}(\alpha_1)t}; \qquad (6.1.16)$$

– for $\mathrm{Re}\,\alpha_1 > 0,\ \mathrm{Re}\,\alpha_2 < 0$

$$|H(s)| \sim M_3 \left(\frac{|t|}{e}\right)^{\mathrm{Re}(\alpha_1+\alpha_2)t} \frac{[\mathrm{Re}(\alpha_1)^{\mathrm{Re}(\alpha_1)}|\mathrm{Re}(\alpha_2)|^{\mathrm{Re}(\alpha_2)}]^t}{|t|^{\sum_{i=1}^{2}[\mathrm{Re}(\beta_i)+\sigma\mathrm{Im}(\alpha_i)]-1}} e^{-\pi\mathrm{Im}(\alpha_2)t}. \qquad (6.1.17)$$

We do not present here exact asymptotic formulas for $E_{\alpha_1,\beta_1;\alpha_2,\beta_2}(z)$ as $z \to \infty$. They can be considered as formulas for a special case of the generalized Wright function and H-function (see Sect. 6.1.5 below).

From the series representation of the four-parametric Mittag-Leffler function we derive a simple asymptotics at zero, valid in the case $\mathrm{Re}\,\{\alpha_1 + \alpha_2\} > 0$ for all $N \in \mathbb{N}$:

$$E_{\alpha_1,\beta_1;\alpha_2,\beta_2}(z) = \sum_{k=0}^{N} \frac{z^k}{\Gamma(\alpha_1 k + \beta_1)\Gamma(\alpha_2 k + \beta_2)} + O\left(|z|^{N+1}\right),\ z \to 0. \quad (6.1.18)$$

6.1 The Four-Parametric Mittag-Leffler Function: The Luchko ...

The following integral representation of the four-parametric Mittag-Leffler function (see [RogKor10]) shows its tight connection to the generalized Wright function (see Chap. 7).

Let $0 < \alpha_j < 2$, $\beta_j \in \mathbb{C}$, $j = 1, 2$. Then the following representation of the four-parametric generalized Mittag-Leffler function $E_{\alpha_1,\beta_1;\alpha_2,\beta_2}(z)$ holds ([RogKor10]).

$$E_{\alpha_1,\beta_1;\alpha_2,\beta_2}(z) = \quad (6.1.19)$$

$$\begin{cases} I_0(z), & z \in G^{(-)}(\epsilon, \mu_2), \\[1em] I_0(z) + \dfrac{z^{\frac{-\beta_2+1}{\alpha_2}}}{2\pi i \alpha_2} \phi\left(\dfrac{\alpha_1}{\alpha_2}, \dfrac{\beta_1\alpha_2 - \beta_2\alpha_1 + 1}{\alpha_2}; z^{\frac{1}{\alpha_2}}\right), & z \in G^{(\mp)}(\epsilon, \mu_1, \mu_2), \\[1em] I_0(z) + \dfrac{z^{\frac{-\beta_2+1}{\alpha_2}}}{2\pi i \alpha_2} \phi\left(\dfrac{\alpha_1}{\alpha_2}, \dfrac{\beta_1\alpha_2 - \beta_2\alpha_1 + 1}{\alpha_2}; z^{\frac{1}{\alpha_2}}\right) \\[1em] \quad + \dfrac{z^{\frac{-\beta_1+1}{\alpha_1}}}{2\pi i \alpha_1} \phi\left(\dfrac{\alpha_2}{\alpha_1}, \dfrac{\beta_2\alpha_1 - \beta_1\alpha_2 + 1}{\alpha_1}; z^{\frac{1}{\alpha_1}}\right), & z \in G^{(+)}(\epsilon, \mu_1), \end{cases}$$

$$(6.1.20)$$

with

$$I_0(z) = \dfrac{-1}{4\pi^2 \alpha_1 \alpha_2} \left\{ \int_{\gamma(\epsilon;\mu_1)} e^{\zeta_1^{1/\alpha_1}} \zeta_1^{\frac{(-\beta_1+1)}{\alpha_1}} d\zeta_1 \int_{\gamma(\epsilon;\mu_2)} \dfrac{e^{\zeta_2^{1/\alpha_2}} \zeta_2^{\frac{(-\beta_2+1)}{\alpha_2}} d\zeta_2}{\zeta_1\zeta_2 - z} \right\},$$

$$(6.1.21)$$

$$\phi(\alpha, \beta; z) := \sum_{k=0}^{\infty} \dfrac{z^k}{k!\Gamma(\alpha k + \beta)}, \quad (6.1.22)$$

where $\phi(\alpha, \beta; z)$ is the classical Wright function (see Appendix F), $\mu_j \in \left(\frac{\pi\alpha_j}{2}, \min\{\pi\alpha_j, \pi\}\right)$, $0 < \mu_1 < \mu_2 < 2$, and $\epsilon > 0$ is an arbitrary positive number.

Here $\gamma(\epsilon; \theta)$ ($\epsilon > 0$, $0 < \theta \leq \pi$) is a contour with non-decreasing $\arg \zeta$ consisting of the following parts:

(1) the ray $\arg \zeta = -\theta$, $|\zeta| \geq \epsilon$;
(2) the arc $-\theta \leq \arg \zeta \leq \theta$ of the circle $|\zeta| = \epsilon$;
(3) the ray $\arg \zeta = \theta$, $|\zeta| \geq \epsilon$.

In the case $0 < \theta < \pi$ the complex ζ-plane is divided by the contour $\gamma(\epsilon; \theta)$ into two unbounded parts: the domain $G^{(-)}(\epsilon; \theta)$ to the left of the contour and the domain $G^{(+)}(\epsilon; \theta)$ to the right. If $\theta = \pi$, the contour $\gamma(\epsilon; \theta)$ consists of the circle $|\zeta| = \epsilon$ and of the cut $-\infty < \zeta \leq -\epsilon$. In this case the domain $G^{(-)}(\epsilon; \theta)$ becomes the circle $|\zeta| < \epsilon$ and the domain $G^{(+)}(\epsilon; \theta)$ becomes the domain $\{\zeta : |\arg \zeta| < \pi, |\zeta| > \epsilon\}$.

For two different values of $\theta_1, \theta_2, 0 < \theta_1 < \theta_2 < \pi$ the union of the two unbounded domains between the curves $\gamma(\epsilon; \theta_1)$ and $\gamma(\epsilon; \theta_2)$ is denoted by $G^{(\mp)}(\epsilon; \theta_1, \theta_2)$.

6.1.4 Extended Four-Parametric Mittag-Leffler Functions

Let the contour $\mathcal{L}$ in the Mellin–Barnes integral

$$\frac{1}{2\pi i} \int_{\mathcal{L}} \frac{\Gamma(s)\Gamma(1-s)}{\Gamma(\beta_1 - \alpha_1 s)\Gamma(\beta_2 - \alpha_2 s)} (-z)^{-s} ds, \qquad (6.1.23)$$

now coincide with the right loop $\mathcal{L}_{+\infty}$, i.e. with a curve starting at $+\infty + i\varphi_1$ and ending at $+\infty + i\varphi_2$ ($-\infty < \varphi_1 < \varphi_2 < +\infty$), leaving the poles of $\Gamma(s)$ at the left and the poles of $\Gamma(1-s)$ at the right. Then this integral exists for all $z \neq 0$ whenever Re $\{\alpha_1 + \alpha_2\} < 0$.

Thus the integral (6.1.23) possesses an extension to another set of parameters. It defines a new function which is called the *extended generalized Mittag-Leffler function* and is denoted $\mathcal{E}_{\alpha_1,\beta_1;\alpha_2,\beta_2}(z)$ (see [KilKor05], [KilKor06a]).

Using the same approach as before, i.e. calculating the integral (6.1.23) by the Residue Theorem, one can obtain the following Laurent series representation of $\mathcal{E}_{\alpha_1,\beta_1;\alpha_2,\beta_2}(z)$:

$$\mathcal{E}_{\alpha_1,\beta_1;\alpha_2,\beta_2}(z) = \sum_{k=0}^{\infty} \frac{d_k}{z^{k+1}}, \qquad (6.1.24)$$

where

$$d_k = -\frac{1}{\Gamma(-\alpha_1(k+1) - \beta_1)\Gamma(-\alpha_2(k+1) - \beta_2)}.$$

In the case Re $\{\alpha_1 + \alpha_2\} < 0$ the series (6.1.24) is convergent for all $z \in \mathbb{C}, z \neq 0$. The function $\mathcal{E}_{\alpha_1,\beta_1;\alpha_2,\beta_2}(z)$ has an asymptotics at $z \to 0$ similar to that of the standard four-parametric Mittag-Leffler function $E_{\alpha_1,\beta_1;\alpha_2,\beta_2}(z)$, Re $\{\alpha_1 + \alpha_2\} > 0$ at $z \to \infty$. The asymptotics of $\mathcal{E}_{\alpha_1,\beta_1;\alpha_2,\beta_2}(z)$ at $z \to \infty$ can be displayed in the form

$$\mathcal{E}_{\alpha_1,\beta_1;\alpha_2,\beta_2}(z) = \sum_{k=0}^{N} \frac{d_k}{z^{k+1}} + O\left(\frac{1}{|z|^{N+1}}\right), \quad z \to \infty. \qquad (6.1.25)$$

6.1.5 Relations to the Wright Function and the H-Function

For short, let us use the common notation $\mathcal{E}_{\alpha_1,\beta_1;\alpha_2,\beta_2}$ for the usual four-parametric Mittag-Leffler function and for its extension in this subsection. For real values of

6.1 The Four-Parametric Mittag-Leffler Function: The Luchko ...

the parameters $\alpha_1, \alpha_2 \in \mathbb{R}$ and complex values of $\beta_1, \beta_2 \in \mathbb{C}$ the four-parametric Mittag-Leffler function $\mathcal{E}_{\alpha_1,\beta_1;\alpha_2,\beta_2}$ can be represented in terms of the generalized Wright function and the H-function.

These representations follow immediately from the Mellin–Barnes integral representation of the function $\mathcal{E}_{\alpha_1,\beta_1;\alpha_2,\beta_2}$ and the properties of the corresponding integrals.

Let us present some formulas relating $\mathcal{E}_{\alpha_1,\beta_1;\alpha_2,\beta_2}$ to the generalized Wright function $_p\Psi_q$:

(1) If $\alpha_1 + \alpha_2 > 0$ and the contour of integration in (6.1.14) is chosen as $\mathcal{L} = \mathcal{L}_{-\infty}$, then

$$\mathcal{E}_{\alpha_1,\beta_1;\alpha_2,\beta_2}(z) = {}_1\Psi_2 \left[\begin{matrix} (1,1) \\ (\beta_1,\alpha_1), (\beta_2,\alpha_2) \end{matrix} \bigg| z \right]. \tag{6.1.26}$$

(2) If $\alpha_1 + \alpha_2 < 0$ and the contour of integration in (6.1.14) is chosen as $\mathcal{L} = \mathcal{L}_{+\infty}$, then

$$\mathcal{E}_{\alpha_1,\beta_1;\alpha_2,\beta_2}(z) = \frac{1}{z} {}_1\Psi_2 \left[\begin{matrix} (1,1) \\ (\beta_1 - \alpha_1, -\alpha_1), (\beta_2 - \alpha_2, -\alpha_2) \end{matrix} \bigg| \frac{1}{z} \right]. \tag{6.1.27}$$

Analogously, one can obtain the following representation of $\mathcal{E}_{\alpha_1,\beta_1;\alpha_2,\beta_2}$ in terms of the H-function:

(1) If $\alpha_1 > 0, \alpha_2 > 0$ and the contour of integration in (6.1.14) is chosen as $\mathcal{L} = \mathcal{L}_{-\infty}$, then

$$\mathcal{E}_{\alpha_1,\beta_1;\alpha_2,\beta_2}(z) = H^{1,1}_{1,3} \left[\begin{matrix} (0,1) \\ (0,1), (1-\beta_1,\alpha_1), (1-\beta_2,\alpha_2) \end{matrix} \bigg| z \right]. \tag{6.1.28}$$

(2) If $\alpha_1 > 0, \alpha_2 < 0$ and the contour of integration in (6.1.14) is chosen as $\mathcal{L} = \mathcal{L}_{-\infty}$ when $\alpha_1 + \alpha_2 > 0$ or $\mathcal{L} = \mathcal{L}_{+\infty}$ when $\alpha_1 + \alpha_2 < 0$, then

$$\mathcal{E}_{\alpha_1,\beta_1;\alpha_2,\beta_2}(z) = H^{1,1}_{2,2} \left[\begin{matrix} (0,1), (\beta_2, -\alpha_2) \\ (0,1), (1-\beta_1,\alpha_1) \end{matrix} \bigg| x \right]. \tag{6.1.29}$$

(3) If $\alpha_1 < 0, \alpha_2 > 0$ and the contour of integration in (6.1.14) is chosen as $\mathcal{L} = \mathcal{L}_{-\infty}$ when $\alpha_1 + \alpha_2 > 0$ or $\mathcal{L} = \mathcal{L}_{+\infty}$ when $\alpha_1 + \alpha_2 < 0$, then

$$\mathcal{E}_{\alpha_1,\beta_1;\alpha_2,\beta_2}(z) = H^{1,1}_{2,2} \left[\begin{matrix} (0,1), (\beta_1, -\alpha_1) \\ (0,1), (1-\beta_2,\alpha_2) \end{matrix} \bigg| x \right]. \tag{6.1.30}$$

(4) If $\alpha_1 < 0, \alpha_2 < 0$ and the contour of integration in (6.1.14) is chosen as $\mathcal{L} = \mathcal{L}_{+\infty}$, then

$$\mathcal{E}_{\alpha_1,\beta_1;\alpha_2,\beta_2}(z) = H^{1,1}_{3,1} \left[\begin{matrix} (0,1), (\beta_1, -\alpha_1), (\beta_2, -\alpha_2) \\ (0,1) \end{matrix} \bigg| x \right]. \tag{6.1.31}$$

6.1.6 Integral Transforms of the Four-Parametric Mittag-Leffler Function

In order to present elements of the theory of integral transforms of the extended four-parametric Mittag-Leffler function we introduce a set of weighted Lebesgue spaces $\mathcal{L}_{\nu,r}(\mathbb{R}_+)$. These spaces are suitable for the above mentioned integral transforms since the latter are connected with the classical Mellin transform (see, e.g., [Mari83, p. 36–39]).

Let us denote by $\mathcal{L}_{\nu,r}(\mathbb{R}_+)$ ($1 \leq r \leq \infty$, $\nu \in \mathbb{R}$) the space of all Lebesgue measurable functions f such that $\|f\|_{\nu,r} < \infty$, where

$$\|f\|_{\nu,r} \equiv \left(\int_0^\infty |t^\nu f(t)|^r \, \frac{dt}{t} \right)^{1/r} < \infty \ (1 \leq r < \infty); \ \|f\|_{\nu,\infty} \equiv \operatorname*{ess\,sup}_{t>0} \|t^\nu f(t)\|.$$
(6.1.32)

In particular, for $\nu = 1/r$ the spaces $\mathcal{L}_{\nu,r}$ coincide with the classical spaces of r-summable functions: $\mathcal{L}_{1/r,r} = \mathcal{L}_r(\mathbb{R}_+)$ endowed with the norm

$$\|f\|_r = \left\{ \int_0^\infty |f(t)|^r \, dt \right\}^{1/r} < \infty \ (1 \leq r < \infty).$$

For any function $f \in \mathcal{L}_{\nu,r}(\mathbb{R}_+)$ ($1 \leq r \leq 2$) its Mellin transform $\mathcal{M}f$ is defined (see, e.g., [KilSai04, (3.2.5)]) by the equality

$$(\mathcal{M}f)(s) = \int_{-\infty}^{+\infty} f(e^\tau) e^{s\tau} \, d\tau \quad (s = \nu + it; \ \nu, t \in \mathbb{R}).$$
(6.1.33)

If $f \in \mathcal{L}_{\nu,r} \cap \mathcal{L}_{\nu,1}$, then the transform (6.1.33) can be written in the form of the classical Mellin transform with $\operatorname{Re} s = \nu$ (see Appendix C):

$$(\mathcal{M}f)(s) = \int_0^{+\infty} f(t) t^{s-1} \, dt.$$
(6.1.34)

An inverse Mellin transform in this case can be determined by the formula

$$f(t) = \frac{1}{2\pi i} \int_{\nu - i\infty}^{\nu + i\infty} (\mathcal{M}f)(s) t^{-s} ds \quad (\nu = \operatorname{Re} s).$$

We have for the Mellin transform of the generalized hypergeometric Wright function

6.1 The Four-Parametric Mittag-Leffler Function: The Luchko ...

$$\mathcal{M}\left[{}_p\Psi_q\left[\begin{matrix}(a_i,\alpha_i)_{1,p}\\(b_j,\beta_j)_{1,q}\end{matrix}\bigg|t\right]\right](s) = \frac{\Gamma(s)\prod_{i=1}^{p}\Gamma(a_i-\alpha_i s)}{\prod_{j=1}^{q}\Gamma(b_j-\beta_j s)}, \qquad (6.1.35)$$

$$\left(\alpha_i > 0,\ \beta_j > 0;\ i = 1,\ldots,p;\ j = 1,\ldots,q;\ 0 < \operatorname{Re} s < \min_{1\le i\le p}\left[\frac{\operatorname{Re}(a_i)}{\alpha_i}\right]\right),$$

and, in particular, for the Mellin transform of the classical Wright function

$$\mathcal{M}[\phi(\alpha,\beta;t)](s) = \frac{\Gamma(s)}{\Gamma(\beta-\alpha s)} \quad (\operatorname{Re} s > 0). \qquad (6.1.36)$$

The Mellin transform of the H-function under certain assumptions on its parameters coincides with the function $\mathcal{H}_{p,q}^{m,n}(s)$ in the Mellin–Barnes integral representation of the H-function (see [PrBrMa-V3, 8.4.51.11], [KilSai04, Theorem 2.2]).

Let us introduce the following parameters characterizing the behavior of the H-function (see Appendix F)

$$\mathcal{H}_{p,q}^{m,n}(z) = \mathcal{H}_{p,q}^{m,n}\left[z\bigg|\begin{matrix}(a_i,\alpha_i)_{1,p}\\(b_j,\beta_j)_{1,q}\end{matrix}\right]$$

$$a^* = \sum_{i=1}^{n}\alpha_i - \sum_{i=n+1}^{p}\alpha_i + \sum_{j=1}^{m}\beta_j - \sum_{j=m+1}^{q}\beta_j,$$

$$\mu = \sum_{j=1}^{q}\beta_j - \sum_{i=1}^{p}\alpha_i + \frac{p-q}{2},\quad \Delta = \sum_{j=1}^{q}\beta_j - \sum_{i=1}^{p}\alpha_i,$$

$$\alpha = -\min_{1\le j\le m}\left[\frac{\operatorname{Re} b_j}{\beta_j}\right],\ \beta = \min_{1\le i\le n}\left[\frac{1-\operatorname{Re} a_i}{\alpha_i}\right]. \qquad (6.1.37)$$

Let $a^* \ge 0$, $s \in \mathbb{C}$ be such that

$$\alpha < \operatorname{Re} s < \beta \qquad (6.1.38)$$

and for $a^* = 0$ assume the following additional inequality holds:

$$\Delta\operatorname{Re} s + \operatorname{Re}\mu < -1. \qquad (6.1.39)$$

Then the Mellin transform of the H-function exists and satisfies the relation

$$\left(\mathcal{M}H_{p,q}^{m,n}\left[z\bigg|\begin{matrix}(a_i,\alpha_i)_{1,p}\\(b_j,\beta_j)_{1,q}\end{matrix}\right]\right)(s) = \mathcal{H}_{p,q}^{m,n}\left[\begin{matrix}(a_i,\alpha_i)_{1,p}\\(b_j,\beta_j)_{1,q}\end{matrix}\bigg|s\right]. \qquad (6.1.40)$$

Since the four-parametric Mittag-Leffler function is related to the generalized Wright function and to the H-function (see Sect. 6.1.5), then one can use (6.1.35) or (6.1.40) to define the Mellin transform of the function $E_{\alpha_1,\beta_1;\alpha_2,\beta_2}(z)$ and of its extension $\mathcal{E}_{\alpha_1,\beta_1;\alpha_2,\beta_2}(z)$.

6.1.7 Integral Transforms with the Four-Parametric Mittag-Leffler Function in the Kernel

Integral transforms with the four-parametric Mittag-Leffler function in the kernel can be considered as a special case of the more general **H**-transform. Let us recall a few facts from the theory of the **H**-transform following [KilSai04]). The **H**-transform is introduced as a Mellin-type convolution with the H-function in the kernel:

$$(\mathbf{H}f)(x) = \int_0^\infty H^{m,n}_{p,q}\left[xt \left| \begin{matrix} (a_i, \alpha_i)_{1,p} \\ (b_j, \beta_j)_{1,q} \end{matrix}\right.\right] f(t)\, dt \quad (x > 0). \tag{6.1.41}$$

Let us recall some results on the **H**-transform in $\mathcal{L}_{\nu,2}$-type spaces following [KilSai04, Chap. 3] (elements of the so-called $\mathcal{L}_{\nu,2}$-theory of **H**-transforms). Here we use the notation (6.1.37) for the parameters a^*, μ, Δ, α, β. We also introduce a so-called exceptional set $\mathcal{E}_{\mathcal{H}}$ for the function $\mathcal{H}(s)$:

$$\mathcal{E}_{\mathcal{H}} = \{\nu \in \mathbb{R} : \alpha < 1 - \nu < \beta \text{ and } \mathcal{H}(s) \text{ has zeros on } \operatorname{Re} s = 1 - \nu\}. \tag{6.1.42}$$

Let

(i) $\alpha < 1 - \nu < \beta$ and suppose one of the following conditions holds:
(ii) $a^* > 0$, or
(iii) $a^* = 0$, $\Delta(1 - \nu) + \operatorname{Re}\mu \leq 0$.

Then the following statements are satisfied:

(a) There exists an injective transform $\mathbf{H}^* \in [\mathcal{L}_{\nu,2}, \mathcal{L}_{1-\nu,2}]$ such that for any $f \in \mathcal{L}_{\nu,2}$ the Mellin transform satisfies the relation

$$(\mathcal{M}\mathbf{H}^* f)(s) = \mathcal{H}^{m,n}_{p,q}\left[\left. \begin{matrix} (a_i, \alpha_i)_{1,p} \\ (b_j, \beta_j)_{1,q} \end{matrix}\right| s \right] (\mathcal{M}f)(1 - s) \quad (\operatorname{Re} s = 1 - \nu). \tag{6.1.43}$$

If $a^* = 0$, $\Delta(1 - \nu) + \operatorname{Re}\mu = 0$, $\nu \notin \mathcal{E}_{\mathcal{H}}$, then $\mathbf{H}^*$ is bijective from $\mathcal{L}_{\nu,2}$ onto $\mathcal{L}_{1-\nu,2}$.

(b) For any $f, g \in \mathcal{L}_{\nu,2}$ the following equality holds:

$$\int_0^\infty f(x)(\mathbf{H}^* g)(x)\, dx = \int_0^\infty (\mathbf{H}^* f)(x) g(x)\, dx. \tag{6.1.44}$$

6.1 The Four-Parametric Mittag-Leffler Function: The Luchko ...

(c) Let $f \in \mathcal{L}_{\nu,2}$, $\lambda \in \mathbb{C}$ and $h > 0$. If $\mathrm{Re}\,\lambda > (1-\nu)h - 1$, then for almost all $x > 0$ the transform $\mathbf{H}^*$ can be represented in the form:

$$(\mathbf{H}^* f)(x) = hx^{1-(\lambda+1)/h} \frac{\mathrm{d}}{\mathrm{d}x} x^{(\lambda+1)/h}$$

$$\times \int_0^\infty H^{m,n+1}_{p+1,q+1}\left[xt \left| \begin{matrix} (-\lambda,h), (a_i,\alpha_i)_{1,p} \\ (b_j,\beta_j)_{1,q}, (-\lambda-1,h) \end{matrix} \right. \right] f(t)\,\mathrm{d}t. \qquad (6.1.45)$$

If $\mathrm{Re}\,\lambda < (1-\nu)h - 1$, then

$$(\mathbf{H}^* f)(x) = -hx^{1-(\lambda+1)/h} \frac{\mathrm{d}}{\mathrm{d}x} x^{(\lambda+1)/h}$$

$$\times \int_0^\infty H^{m+1,n}_{p+1,q+1}\left[xt \left| \begin{matrix} (a_i,\alpha_i)_{1,p}, (-\lambda,h) \\ (-\lambda-1,h), (b_j,\beta_j)_{1,q} \end{matrix} \right. \right] f(t)\,\mathrm{d}t. \qquad (6.1.46)$$

(d) The $\mathbf{H}^*$-transform does not depend on ν in the following sense: if two values of the parameter, say ν and $\tilde{\nu}$, satisfy condition (i) and one of the conditions (ii) or (iii), and if the transforms $\mathbf{H}^*$ and $\tilde{\mathbf{H}}^*$ are defined by the relation (6.1.43) in $\mathcal{L}_{\nu,2}$ and $\mathcal{L}_{\tilde{\nu},2}$, respectively, then $\mathbf{H}^* f = \tilde{\mathbf{H}}^* f$ for any $f \in \mathcal{L}_{\nu,2} \cap \mathcal{L}_{\tilde{\nu},2}$.
(e) If either $a^* > 0$ or $a^* = 0$, and $\Delta(1-\nu) + \mathrm{Re}\,\mu < 0$, then for any $f \in \mathcal{L}_{\nu,2}$ we have $\mathbf{H}^* f = \mathbf{H} f$, i.e. $\mathbf{H}^*$ is defined by the equality (6.1.41).

An extended $\mathcal{L}_{\nu,r}$-theory (for any $1 \le r \le +\infty$) of the $\mathbf{H}$-transform is presented in [KilSai04].

The integral transform with the four-parametric Mittag-Leffler function in the kernel is defined for $\alpha_1, \alpha_2 \in \mathbb{R}$, $\beta_1, \beta_2 \in \mathbb{C}$ by the formula:

$$\left(\mathbf{E}_{\alpha_1,\beta_1;\alpha_2,\beta_2} f\right)(x) = \int_0^\infty \mathcal{E}_{\alpha_1,\beta_1;\alpha_2,\beta_2}(-xt) f(t)\,\mathrm{d}t \quad (x > 0), \qquad (6.1.47)$$

where for $\alpha_1 + \alpha_2 > 0$ the kernel $\mathcal{E}_{\alpha_1,\beta_1;\alpha_2,\beta_2} = E_{\alpha_1,\beta_1;\alpha_2,\beta_2}$ (i.e. it is the four-parametric generalized Mittag-Leffler function defined by (6.1.1)), and for $\alpha_1 + \alpha_2 < 0$ the kernel $\mathcal{E}_{\alpha_1,\beta_1;\alpha_2,\beta_2}$ is the extended four-parametric generalized Mittag-Leffler function defined by (6.1.23).

The properties of this transform follow from its representation as a special case of the $\mathbf{H}$-transform.

(1) If $\alpha_1 > 0$, $\alpha_2 > 0$, then

$$\left(\mathbf{E}_{\alpha_1,\beta_1;\alpha_2,\beta_2} f\right)(x) = \int_0^\infty H^{1,1}_{1,3}\left[xt \left| \begin{matrix} (0,1) \\ (0,1), (1-\beta_1,\alpha_1), (1-\beta_2,\alpha_2) \end{matrix} \right. \right] f(t)\,\mathrm{d}t. \qquad (6.1.48)$$

(2) If $\alpha_1 > 0$, $\alpha_2 < 0$, then

$$\left(\mathbf{E}_{\alpha_1,\beta_1;\alpha_2,\beta_2} f\right)(x) = \int_0^\infty H_{2,2}^{1,1}\left[xt \left|\begin{array}{l}(0,1),(\beta_2,-\alpha_2)\\(0,1),(1-\beta_1,\alpha_1)\end{array}\right.\right] f(t)dt. \quad (6.1.49)$$

(3) If $\alpha_1 < 0$, $\alpha_2 > 0$, then

$$\left(\mathbf{E}_{\alpha_1,\beta_1;\alpha_2,\beta_2} f\right)(x) = \int_0^\infty H_{2,2}^{1,1}\left[xt \left|\begin{array}{l}(0,1),(\beta_1,-\alpha_1)\\(0,1),(1-\beta_2,\alpha_2)\end{array}\right.\right] f(t)dt. \quad (6.1.50)$$

(4) If $\alpha_1 < 0$, $\alpha_2 < 0$, then

$$\left(\mathbf{E}_{\alpha_1,\beta_1;\alpha_2,\beta_2} f\right)(x) = \int_0^\infty H_{3,1}^{1,1}\left[xt \left|\begin{array}{l}(0,1),(\beta_1,-\alpha_1),(\beta_2,-\alpha_2)\\(0,1)\end{array}\right.\right] f(t)dt. \quad (6.1.51)$$

Based on (6.1.48)–(6.1.51) and on the above presented elements of the $\mathcal{L}_{\nu,2}$-theory of the **H**-transform one can formulate the following results for the integral transforms with the four-parametric generalized Mittag-Leffler function in the kernel. Let us present these only in the case (1) (i.e. when $\alpha_1 > 0$, $\alpha_2 > 0$). All other cases can be considered analogously (see, e.g.. [KilKor06a], [KilKor06b]).

Let $\alpha_1 > 0$, $\alpha_2 > 0$. Then the parameters $a^*, \mu, \Delta, \alpha, \beta$ are related to the parameters of the four-parametric Mittag-Leffler function as follows:

$$a^* = 2 - \alpha_1 - \alpha_2, \ \Delta = \alpha_1 + \alpha_2, \ \mu = 1 - \beta_1 - \beta_2, \ \alpha = 0, \ \beta = 1.$$

Let $0 < \nu < 1$, $\alpha_1 > 0$, $\alpha_2 > 0$ and $\beta_1, \beta_2 \in \mathbb{C}$ be such that $\alpha_1 + \alpha_2 < 2$ or $\alpha_1 + \alpha_2 = 2$ and $3 - 2\nu \leq \operatorname{Re}(\beta_1 + \beta_2)$. Then:

(a) There exists an injective mapping $\mathbf{E}^*_{\alpha_1,\beta_1;\alpha_2,\beta_2} \in [\mathcal{L}_{\nu,2}, \mathcal{L}_{1-\nu,2}]$ such that for any $f \in \mathcal{L}_{\nu,2}$ the following relation holds:

$$\left(\mathcal{M}\mathbf{E}^*_{\alpha_1,\beta_1;\alpha_2,\beta_2} f\right)(s) = \frac{\Gamma(s)\Gamma(1-s)}{\Gamma(\beta_1 - \alpha_1 s)\Gamma(\beta_2 - \alpha_2 s)} (\mathcal{M}f)(1-s) \quad (\operatorname{Re} s = 1 - \nu). \quad (6.1.52)$$

If either $\alpha_1 + \alpha_2 < 2$ or $\alpha_1 + \alpha_2 = 2$ and $3 - 2\nu \leq \operatorname{Re}(\beta_1 + \beta_2)$ and the additional conditions

$$s \neq \frac{\beta_1 + k}{\alpha_1}, \ s \neq \frac{\beta_2 + l}{\alpha_2} \ (k, l = 0, 1, 2, \cdots), \ \text{for } \operatorname{Re} s = 1 - \nu, \quad (6.1.53)$$

are satisfied, then the operator $\mathbf{E}^*$ is bijective from $\mathcal{L}_{\nu,2}$ onto $\mathcal{L}_{1-\nu,2}$.

(b) For any $f, g \in \mathcal{L}_{\nu,2}$ we have the integration by parts formula

$$\int_0^\infty f(x) \mathbf{E}^*_{\alpha_1,\beta_1;\alpha_2,\beta_2} g(x) dx = \int_0^\infty \mathbf{E}^*_{\alpha_1,\beta_1;\alpha_2,\beta_2} f(x) g(x) dx. \qquad (6.1.54)$$

(c) If $f \in \mathcal{L}_{\nu,2}$, $\lambda \in \mathbb{C}$, $h > 0$, then $\mathbf{E}^*_{\alpha_1,\beta_1;\alpha_2,\beta_2} f$ is represented in the form:

$$\left(\mathbf{E}^*_{\alpha_1,\beta_1;\alpha_2,\beta_2} f\right)(x) = h x^{1-(\lambda+1)/h} \frac{d}{dx} x^{(\lambda+1)/h}$$

$$\times \int_0^\infty H^{1,2}_{2,4}\left[xt \left| \begin{matrix} (-\lambda, h), (0, 1) \\ (0, 1), (1-\beta_1, \alpha_1), (1-\beta_2, \alpha_2), (-\lambda-1, h) \end{matrix} \right. \right] f(t) dt \qquad (6.1.55)$$

when $\operatorname{Re} \lambda > (1-\nu)h - 1$, and in the form:

$$\left(\mathbf{E}^*_{\alpha_1,\beta_1;\alpha_2,\beta_2} f\right)(x) = -h x^{1-(\lambda+1)/h} \frac{d}{dx} x^{(\lambda+1)/h}$$

$$\times \int_0^\infty H^{2,1}_{2,4}\left[xt \left| \begin{matrix} (0, 1), (-\lambda, h) \\ (-\lambda-1, h), (0, 1), (1-\beta_1, \alpha_1), (1-\beta_2, \alpha_2) \end{matrix} \right. \right] f(t) dt \qquad (6.1.56)$$

when $\operatorname{Re} \lambda < (1-\nu)h - 1$.

(d) The mapping $\mathbf{E}^*_{\alpha_1,\beta_1;\alpha_2,\beta_2}$ does not depend on ν in the following sense: if $0 < \nu_1, \nu_2 < 1$ and the mappings $\mathbf{E}^*_{\alpha_1,\beta_1;\alpha_2,\beta_2;1}$, $\mathbf{E}^*_{\alpha_1,\beta_1;\alpha_2,\beta_2;2}$ are defined on the spaces $\mathcal{L}_{\nu_1,2}$, $\mathcal{L}_{\nu_2,2}$ respectively, then $\mathbf{E}^*_{\alpha_1,\beta_1;\alpha_2,\beta_2;1} f = \mathbf{E}^*_{\alpha_1,\beta_1;\alpha_2,\beta_2;2} f$ for all $f \in \mathcal{L}_{\nu_1,2} \cap \mathcal{L}_{\nu_2,2}$.

(e) If $f \in \mathcal{L}_{\nu,2}$ and either $\alpha_1 + \alpha_2 < 2$ or $\alpha_1 + \alpha_2 = 2$ and $3 - 2\nu \leq \operatorname{Re}(\beta_1 + \beta_2)$, then for all $f \in \mathcal{L}_{\nu,2}$ we have $\mathbf{E}^*_{\alpha_1,\beta_1;\alpha_2,\beta_2} f = \mathbf{E}_{\alpha_1,\beta_1;\alpha_2,\beta_2} f$, i.e. the mapping $\mathbf{E}^*_{\alpha_1,\beta_1;\alpha_2,\beta_2}$ is defined by the formula (6.1.48).

6.1.8 Relations to the Fractional Calculus

Let us present a number of (left- and right-sided) Riemann–Liouville fractional integration and differentiation formulas for the four-parametric Mittag-Leffler function. Both cases ($\alpha_1 + \alpha_2 > 0$ and $\alpha_1 + \alpha_2 < 0$) will be considered simultaneously (see [KiKoRo13]). For simplicity we use the notation $\mathcal{E}_{\alpha_1,\beta_1;\alpha_2,\beta_2}$ for the four-parametric Mittag-Leffler function in both cases.

Let $\alpha_1, \alpha_2 \in \mathbb{R}$, $\alpha_1 \neq 0$, $\alpha_2 \neq 0$, $\beta_1, \beta_2 \in \mathbb{C}$, and let the contour of integration in (6.1.14) be chosen as $\mathcal{L} = \mathcal{L}_{-\infty}$ when $\alpha_1 + \alpha_2 > 0$, and as $\mathcal{L} = \mathcal{L}_{+\infty}$ when $\alpha_1 + \alpha_2 < 0$. Let the additional parameters $\gamma, \sigma, \lambda \in \mathbb{C}$ be such that $\operatorname{Re} \gamma > 0$, $\operatorname{Re} \sigma > 0$ and $\omega \in \mathbb{R}$, ($\omega \neq 0$).

The left-sided Riemann–Liouville fractional integral of the four-parametric Mittag-Leffler function is given by the following formulas:

(a) If $\alpha_1 < 0$ and $\alpha_2 > 0$, then for $x > 0$

$$\left(I_{0+}^{\gamma} t^{\sigma-1} \mathcal{E}_{\alpha_1,\beta_1;\,\alpha_2,\beta_2}(\lambda t^{\omega})\right)(x) =$$

$$\begin{cases} x^{\sigma+\gamma-1} H_{3,3}^{1,2}\left[-\lambda x^{\omega} \,\bigg|\, \begin{matrix} (0,1), & (1-\sigma,\omega), & (\beta_1,-\alpha_1) \\ (0,1), & (1-\sigma-\gamma,\omega), & (1-\beta_2,\alpha_2) \end{matrix}\right] & (\omega > 0), \\[1em] x^{\sigma+\gamma-1} H_{3,3}^{2,1}\left[-\lambda x^{\omega} \,\bigg|\, \begin{matrix} (0,1), & (\sigma+\gamma,-\omega), & (\beta_1,-\alpha_1) \\ (0,1), & (\sigma,-\omega), & (1-\beta_2,\alpha_2) \end{matrix}\right] & (\omega < 0). \end{cases}$$

(b) If $\alpha_1 < 0$ and $\alpha_2 < 0$, then for $x > 0$

$$\left(I_{0+}^{\gamma} t^{\sigma-1} \mathcal{E}_{\alpha_1,\beta_1;\,\alpha_2,\beta_2}(\lambda t^{\omega})\right)(x) =$$

$$\begin{cases} x^{\sigma+\gamma-1} H_{4,2}^{1,2}\left[-\lambda x^{\omega} \,\bigg|\, \begin{matrix} (0,1), & (1-\sigma,\omega), & (\beta_1,-\alpha_1), & (\beta_2,-\alpha_2) \\ (0,1), & (1-\sigma-\gamma,\omega) & & \end{matrix}\right] \\ \hfill (\omega > 0), \\[1em] x^{\sigma+\gamma-1} H_{4,2}^{2,1}\left[-\lambda x^{\omega} \,\bigg|\, \begin{matrix} (0,1), & (\sigma+\gamma,-\omega), & (\beta_1,-\alpha_1), & (\beta_2,-\alpha_2) \\ (0,1), & (\sigma,-\omega) & & \end{matrix}\right] \\ \hfill (\omega < 0). \end{cases}$$

(c) If $\alpha_1 > 0$ and $\alpha_2 > 0$, then for $x > 0$

$$\left(I_{0+}^{\gamma} t^{\sigma-1} \mathcal{E}_{\alpha_1,\beta_1;\,\alpha_2,\beta_2}(\lambda t^{\omega})\right)(x) =$$

$$\begin{cases} x^{\sigma+\gamma-1} H_{2,4}^{1,2}\left[-\lambda x^{\omega} \,\bigg|\, \begin{matrix} (0,1), & (1-\sigma,\omega) \\ (0,1), & (1-\sigma-\gamma,\omega), & (1-\beta_1,\alpha_1), & (1-\beta_2,\alpha_2) \end{matrix}\right] \\ \hfill (\omega > 0), \\[1em] x^{\sigma+\gamma-1} H_{2,4}^{2,1}\left[-\lambda x^{\omega} \,\bigg|\, \begin{matrix} (0,1), & (\sigma+\gamma,-\omega) \\ (0,1), & (\sigma,-\omega), & (1-\beta_1,\alpha_1), & (1-\beta_2,\alpha_2) \end{matrix}\right] \\ \hfill (\omega < 0). \end{cases}$$

The right-sided Riemann–Liouville fractional integral of the four-parametric Mittag-Leffler function is given by the following formulas:

(a) If $\alpha_1 < 0$ and $\alpha_2 > 0$, then for $x > 0$

$$\left(I_{-}^{\gamma} t^{-\sigma} \mathcal{E}_{\alpha_1,\beta_1;\,\alpha_2,\beta_2}(\lambda t^{-\omega})\right)(x) =$$

$$\begin{cases} x^{\gamma-\sigma} H_{3,3}^{1,2}\left[-\lambda x^{-\omega} \,\bigg|\, \begin{matrix} (0,1), & (1-\sigma+\gamma,\omega), & (\beta_1,-\alpha_1) \\ (0,1), & (1-\sigma,\omega), & (1-\beta_2,\alpha_2) \end{matrix}\right] & (\omega > 0), \\[1em] x^{\gamma-\sigma} H_{3,3}^{2,1}\left[-\lambda x^{-\omega} \,\bigg|\, \begin{matrix} (0,1), & (\sigma,-\omega), & (\beta_1,-\alpha_1) \\ (0,1), & (\sigma-\gamma,-\omega), & (1-\beta_2,\alpha_2) \end{matrix}\right] & (\omega < 0). \end{cases}$$

6.1 The Four-Parametric Mittag-Leffler Function: The Luchko ...

(b) If $\alpha_1 < 0$ and $\alpha_2 < 0$, then for $x > 0$

$$\left(I_-^\gamma t^{-\sigma} \mathcal{E}_{\alpha_1,\beta_1;\,\alpha_2,\beta_2}(\lambda t^{-\omega})\right)(x) =$$

$$\begin{cases} x^{\gamma-\sigma} H_{4,2}^{1,2}\left[-\lambda x^{-\omega} \Big| \begin{matrix}(0,1),\ (1-\sigma+\gamma,\omega),\ (\beta_1,-\alpha_1),\ (\beta_2,-\alpha_2)\\ (0,1),\ (1-\sigma,\omega)\end{matrix}\right] \\ \hfill (\omega > 0), \\ \\ x^{\gamma-\sigma} H_{4,2}^{2,1}\left[-\lambda x^{-\omega} \Big| \begin{matrix}(0,1),\ (\sigma,-\omega),\quad (\beta_1,-\alpha_1),\ (\beta_2,-\alpha_2)\\ (0,1),\ (\sigma-\gamma,-\omega)\end{matrix}\right] \\ \hfill (\omega < 0). \end{cases}$$

(c) If $\alpha_1 > 0$ and $\alpha_2 > 0$, then for $x > 0$

$$\left(I_-^\gamma t^{-\sigma} \mathcal{E}_{\alpha_1,\beta_1;\,\alpha_2,\beta_2}(\lambda t^{-\omega})\right)(x) =$$

$$\begin{cases} x^{\gamma-\sigma} H_{2,4}^{1,2}\left[-\lambda x^{-\omega} \Big| \begin{matrix}(0,1),\ (1-\sigma+\gamma,\omega)\\ (0,1),\ (1-\sigma,\omega),\quad (1-\beta_1,\alpha_1),\ (1-\beta_2,\alpha_2)\end{matrix}\right] \\ \hfill (\omega > 0), \\ \\ x^{\gamma-\sigma} H_{2,4}^{2,1}\left[-\lambda x^{-\omega} \Big| \begin{matrix}(0,1),\ (\sigma,-\omega)\\ (0,1),\ (\sigma-\gamma,-\omega),\ (1-\beta_1,\alpha_1),\ (1-\beta_2,\alpha_2)\end{matrix}\right] \\ \hfill (\omega < 0). \end{cases}$$

The left-sided Riemann–Liouville fractional derivative of the four-parametric Mittag-Leffler function is given by the following formulas:

(a) If $\alpha_1 < 0$ and $\alpha_2 > 0$, then for $x > 0$

$$\left(D_{0+}^\gamma t^{\sigma-1} \mathcal{E}_{\alpha_1,\beta_1;\,\alpha_2,\beta_2}(\lambda t^{\omega})\right)(x) =$$

$$\begin{cases} x^{\sigma-\gamma-1} H_{3,3}^{2,1}\left[-\lambda x^{\omega} \Big| \begin{matrix}(0,1),\ (1-\sigma,\omega),\quad (\beta_1,-\alpha_1)\\ (0,1),\ (1-\sigma+\gamma,\omega),\ (1-\beta_2,\alpha_2)\end{matrix}\right] & (\omega > 0), \\ \\ x^{\sigma-\gamma-1} H_{3,3}^{1,2}\left[-\lambda x^{\omega} \Big| \begin{matrix}(0,1),\ (\sigma-\gamma,-\omega),\ (\beta_1,-\alpha_1)\\ (0,1),\ (\sigma,-\omega),\quad (1-\beta_2,\alpha_2)\end{matrix}\right] & (\omega < 0). \end{cases}$$

(b) If $\alpha_1 < 0$ and $\alpha_2 < 0$, then for $x > 0$

$$\left(D_{0+}^\gamma t^{\sigma-1} \mathcal{E}_{\alpha_1,\beta_1;\,\alpha_2,\beta_2}(\lambda t^{\omega})\right)(x) =$$

$$\begin{cases} x^{\sigma-\gamma-1} H_{4,2}^{1,2}\left[-\lambda x^{\omega} \Big| \begin{matrix}(0,1),\ (1-\sigma,\omega),\quad (\beta_1,-\alpha_1),\ (\beta_2,-\alpha_2)\\ (0,1),\ (1-\sigma+\gamma,\omega)\end{matrix}\right] \\ \hfill (\omega > 0), \\ \\ x^{\sigma-\gamma-1} H_{4,2}^{2,1}\left[-\lambda x^{\omega} \Big| \begin{matrix}(0,1),\ (\sigma-\gamma,-\omega),\ (\beta_1,-\alpha_1),\ (\beta_2,-\alpha_2)\\ (0,1),\ (\sigma,-\omega)\end{matrix}\right] \\ \hfill (\omega < 0). \end{cases}$$

(c) If $\alpha_1 > 0$ and $\alpha_2 > 0$, then for $x > 0$

$$\left(D_{0+}^{\gamma} t^{\sigma-1} \mathcal{E}_{\alpha_1,\beta_1;\,\alpha_2,\beta_2}(\lambda t^{\omega})\right)(x) =$$
$$\begin{cases} x^{\sigma-\gamma-1} H_{2,4}^{1,2}\left[-\lambda x^{\omega} \left| \begin{array}{l} (0,1),\ (1-\sigma,\omega) \\ (0,1),\ (1-\sigma+\gamma,\omega),\ (1-\beta_1,\alpha_1),\ (1-\beta_2,\alpha_2) \end{array}\right.\right] \\ \hfill (\omega > 0), \\[2ex] x^{\sigma-\gamma-1} H_{2,4}^{2,1}\left[-\lambda x^{\omega} \left| \begin{array}{l} (0,1),\ (\sigma-\gamma,-\omega) \\ (0,1),\ (\sigma,-\omega),\quad\ (1-\beta_1,\alpha_1),\ (1-\beta_2,\alpha_2) \end{array}\right.\right] \\ \hfill (\omega < 0). \end{cases}$$

The right-sided Riemann–Liouville fractional derivative of the four-parametric Mittag-Leffler function is given by the following formulas:

(a) If $\alpha_1 < 0$ and $\alpha_2 > 0$, then for $x > 0$

$$\left(D_{-}^{\gamma} t^{-\sigma} \mathcal{E}_{\alpha_1,\beta_1;\,\alpha_2,\beta_2}(\lambda t^{-\omega})\right)(x) =$$
$$\begin{cases} x^{-\sigma-\gamma} H_{3,3}^{2,1}\left[-\lambda x^{-\omega} \left| \begin{array}{ll} (0,1),\ (1-\sigma-\gamma,\omega),\ (\beta_1,-\alpha_1) \\ (0,1),\ (1-\sigma,\omega),\quad\ (1-\beta_2,\alpha_2) \end{array}\right.\right] & (\omega > 0), \\[2ex] x^{-\sigma-\gamma} H_{3,3}^{1,2}\left[-\lambda x^{-\omega} \left| \begin{array}{ll} (0,1),\ (\sigma,-\omega),\quad\ (\beta_1,-\alpha_1) \\ (0,1),\ (\sigma+\gamma,-\omega),\ (1-\beta_2,\alpha_2) \end{array}\right.\right] & (\omega < 0). \end{cases}$$

(b) If $\alpha_1 < 0$ and $\alpha_2 < 0$, then for $x > 0$

$$\left(D_{-}^{\gamma} t^{-\sigma} \mathcal{E}_{\alpha_1,\beta_1;\,\alpha_2,\beta_2}(-\lambda t^{-\omega})\right)(x) =$$
$$\begin{cases} x^{-\sigma-\gamma} H_{4,2}^{2,1}\left[-\lambda x^{-\omega} \left| \begin{array}{l} (0,1),\ (1-\sigma-\gamma,\omega),\ (\beta_1,-\alpha_1),\ (\beta_2,-\alpha_2) \\ (0,1),\ (1-\sigma,\omega) \end{array}\right.\right] \\ \hfill (\omega > 0), \\[2ex] x^{-\sigma-\gamma} H_{4,2}^{1,2}\left[-\lambda x^{-\omega} \left| \begin{array}{l} (0,1),\ (\sigma,-\omega),\quad\ (\beta_1,-\alpha_1),\ (\beta_2,-\alpha_2) \\ (0,1),\ (\sigma+\gamma,-\omega) \end{array}\right.\right] \\ \hfill (\omega < 0). \end{cases}$$

(c) If $\alpha_1 > 0$ and $\alpha_2 > 0$, then for $x > 0$

$$\left(D_{-}^{\gamma} t^{-\sigma} \mathcal{E}_{\alpha_1,\beta_1;\,\alpha_2,\beta_2}(-\lambda t^{-\omega})\right)(x) =$$
$$\begin{cases} x^{-\sigma-\gamma} H_{2,4}^{2,1}\left[-\lambda x^{-\omega} \left| \begin{array}{l} (0,1),\ (1-\sigma-\gamma,\omega) \\ (0,1),\ (1-\sigma,\omega),\quad\ (1-\beta_1,\alpha_1),\ (1-\beta_2,\alpha_2) \end{array}\right.\right] \\ \hfill (\omega > 0), \\[2ex] x^{-\sigma-\gamma} H_{2,4}^{1,2}\left[-\lambda x^{-\omega} \left| \begin{array}{l} (0,1),\ (\sigma,-\omega) \\ (0,1),\ (\sigma+\gamma,-\omega),\ (1-\beta_1,\alpha_1),\ (1-\beta_2,\alpha_2) \end{array}\right.\right] \\ \hfill (\omega < 0). \end{cases}$$

6.2 The Four-Parametric Mittag-Leffler Function: A Generalization of the Prabhakar Function

In this section we mainly follow the article [SriTom09].

6.2.1 Definition and General Properties

A generalization of the Prabhakar function (5.1.1) is proposed in [SriTom09] in the following form

$$E_{\alpha,\beta}^{\gamma,\kappa}(z) := \sum_{n=0}^{\infty} \frac{(\gamma)_{\kappa n}}{\Gamma(\alpha n + \beta)} \frac{z^n}{n!} \quad (z; \beta, \gamma \in \mathbb{C}; \operatorname{Re}\alpha > \max\{0, \operatorname{Re}\kappa - 1\}; \operatorname{Re}\kappa > 0), \tag{6.2.1}$$

where $(\gamma)_\delta$ with $\delta > 0$ is the generalized Pochhammer symbol $(\gamma)_\delta = \frac{\Gamma(\gamma+\delta)}{\Gamma(\gamma)}$ (cf. Appendix A, Sect. A.1.5). The function (6.2.1) is sometimes called the four-parametric Mittag-Leffler function (*a four-parametric generalization of the Prabhakar function*). With $\kappa = q \in \mathbb{N}_0$, $\min\{\operatorname{Re}\beta, \operatorname{Re}\gamma\} > 0$, this definition coincides with the definition proposed in [ShuPra07].

Theorem 6.3 ([SriTom09, Thm. 1]) *The four-parametric Mittag-Leffler function $E_{\alpha,\beta}^{\gamma,\kappa}(z)$ defined by (6.2.1) is an entire function in the complex z-plane of order ρ and type σ given by*

$$\rho = \frac{1}{\operatorname{Re}(\alpha - \kappa) + 1}, \quad \sigma = \frac{1}{\rho}\left(\frac{(\operatorname{Re}\kappa)^{\operatorname{Re}\kappa}}{(\operatorname{Re}\alpha)^{\operatorname{Re}\alpha}}\right)^\rho. \tag{6.2.2}$$

Moreover, the power series in the defining equation (6.2.1) converges absolutely in the disc $|z| < \frac{(\operatorname{Re}\alpha)^{\operatorname{Re}\alpha}}{(\operatorname{Re}\kappa)^{\operatorname{Re}\kappa}}$ whenever

$$\operatorname{Re}\alpha = \operatorname{Re}\kappa - 1 > 0.$$

◁ The proof follows from the asymptotic properties of the Gamma function

$$\Gamma(z) = z^z e^{-z}\sqrt{\frac{2\pi}{z}}\left[1 + \frac{1}{12z} + \frac{1}{288z^2} + O\left(\frac{1}{z^3}\right)\right],$$

where

$$(z \to \infty, \ |\arg z| \leq \pi - \varepsilon (0 < \varepsilon < \pi)),$$

and

$$\frac{\Gamma(z+a)}{\Gamma(z+b)} = z^{a-b}\left[1 + \frac{(a-b)(a+b-1)}{2z} + O\left(\frac{1}{z^2}\right)\right],$$

where $a, b \in \mathbb{C}$ and $z \to \infty$ along any curve joining $z = 0$ and $z = \infty$ provided $z \neq -a, -a - 1, \ldots$ and $z \neq -b, -b - 1, \ldots$.

To determine the radius of convergence R of the power series $\sum_{n=0}^{\infty} c_n z^n$ one can use the Cauchy–Hadamard formula

$$R = \limsup_{n \to \infty} \left| \frac{c_n}{c_{n+1}} \right|,$$

and for the order ρ and the type σ of an entire function the following standard formulas (see [Lev56])

$$\rho = \limsup_{n \to \infty} \frac{n \log n}{\log 1/|c_n|}, \quad e\rho\sigma = \limsup_{n \to \infty} n |c_n|^{\frac{\rho}{n}}.$$

▷

A number of further properties of the four-parametric Mittag-Leffler function follows from its relation with the Fox–Wright function

$$E_{\alpha,\beta}^{\gamma,\kappa}(z) = \frac{1}{\Gamma(\gamma)} {}_1W_1(z) \left[\begin{matrix} (\gamma, \kappa) \\ (\beta, \alpha) \end{matrix} \bigg| z \right], \qquad (6.2.3)$$

and with the Fox H-function (see [AgMiNi15])

$$E_{\alpha,\beta}^{\gamma,\kappa}(z) = \frac{1}{\Gamma(\gamma)} H_{2,2}^{1,2}(z) \left[z \bigg| \begin{matrix} (1 - \gamma, \kappa), (0, 1) \\ (0, 1), (1 - \beta, \alpha) \end{matrix} \right]. \qquad (6.2.4)$$

6.2.2 The Four-Parametric Mittag-Leffler Function of a Real Variable

Following [SriTom09] one can introduce an integral operator with the four-parametric Mittag-Leffler function in the kernel

$$\left(\mathcal{E}_{a+;\alpha,\beta}^{\omega;\gamma,\kappa} \varphi \right)(x) := \int_a^x (x-t)^{\beta-1} E_{\alpha,\beta}^{\gamma,\kappa}(\omega(x-t)^{\alpha-1}) \varphi(t) dt. \qquad (6.2.5)$$

It is well-defined for the following values of parameters:

$$\gamma, \omega \in \mathbb{C}; \ \operatorname{Re} \alpha > \max\{0, \operatorname{Re} \kappa - 1\}, \ \min\{\operatorname{Re} \beta, \operatorname{Re} \kappa\} > 0.$$

Moreover, this operator is bounded in the Lebesgue space L_1 on any finite interval $[a, b]$, $b > a$:

$$\| \mathcal{E}_{a+;\alpha,\beta}^{\omega;\gamma,\kappa} \varphi \|_1 \leq C_1 \| \varphi \|_1,$$

where

$$C_1 = (b-a)^{\operatorname{Re}\beta} \sum_{n=0}^{\infty} \frac{|(\gamma)_{\kappa n}|}{(n\operatorname{Re}\alpha + \operatorname{Re}\beta)|\Gamma(\alpha n + \beta)|} \frac{|\omega(b-a)^{\operatorname{Re}\alpha}|^n}{n!}.$$

The following theorem describes the action of the Riemann–Liouville fractional integral I_{a+}^{μ} and derivative D_{a+}^{μ} as well as the generalized Riemann–Liouville fractional derivative (the Hilfer fractional derivative)

$$\left(D_{a+}^{\mu,\nu} \phi \right)(x) := \left(I_{a+}^{\nu(1-\mu)} \frac{d}{dx} \left(I_{a+}^{(1-\nu)(1-\mu)} \phi \right) \right)(x) \qquad (6.2.6)$$

on the four-parametric Mittag-Leffler function $E_{\alpha,\beta}^{\gamma,\kappa}(t)$

Theorem 6.4 ([SriTom09, Thm 3]) *Let $x > a$, $a \in \mathbb{R}_+$, $0 < \mu < 1$, $0 \leq \nu \leq 1$, and*

$$\operatorname{Re}\alpha > \max\{0, \operatorname{Re}\kappa - 1\}, \ \min\{\operatorname{Re}\beta, \operatorname{Re}\kappa, \operatorname{Re}\lambda\} > 0, \ \gamma, \omega \in \mathbb{C}.$$

Then the following relations hold:

$$\left(I_{a+}^{\lambda} \left((t-a)^{\beta-1} E_{\alpha,\beta}^{\gamma,\kappa}(\omega(t-a)^{\alpha}) \right) \right)(x) = (x-a)^{\beta+\lambda-1} E_{\alpha,\beta+\lambda}^{\gamma,\kappa}(\omega(t-a)^{\alpha}), \qquad (6.2.7)$$

$$\left(D_{a+}^{\lambda} \left((t-a)^{\beta-1} E_{\alpha,\beta}^{\gamma,\kappa}(\omega(t-a)^{\alpha}) \right) \right)(x) = (x-a)^{\beta-\lambda-1} E_{\alpha,\beta-\lambda}^{\gamma,\kappa}(\omega(t-a)^{\alpha}), \qquad (6.2.8)$$

$$\left(D_{a+}^{\mu,\nu} \left((t-a)^{\beta-1} E_{\alpha,\beta}^{\gamma,\kappa}(\omega(t-a)^{\alpha}) \right) \right)(x) = (x-a)^{\beta-\mu-1} E_{\alpha,\beta-\mu}^{\gamma,\kappa}(\omega(t-a)^{\alpha}). \qquad (6.2.9)$$

The Laplace transform of the four-parametric Mittag-Leffler function is given by the following formula, which can be obtained using a term-by-term transformation of the corresponding power series

$$\mathcal{L}\left[x^{a-1} E_{\alpha,\beta}^{\gamma,\kappa}(\omega x^b) \right](s) = \frac{s^{-a}}{\Gamma(\gamma)} {}_2W_1 \left[\begin{matrix} (a,b), (\gamma,\kappa) \\ (\beta,\alpha) \end{matrix} \Big| \frac{\omega}{s^b} \right]. \qquad (6.2.10)$$

6.3 Mittag-Leffler Functions with $2n$ Parameters

6.3.1 Definition and Basic Properties

Consider the function defined for $\alpha_i \in \mathbb{R}$ ($\alpha_1^2 + \cdots + \alpha_n^2 \neq 0$) and $\beta_i \in \mathbb{C}$ ($i = 1, \cdots, n \in \mathbb{N}$) by

$$E((\alpha, \beta)_n; z) = \sum_{k=0}^{\infty} \frac{z^k}{\prod_{j=1}^{n} \Gamma(\alpha_j k + \beta_j)} \quad (z \in \mathbb{C}). \tag{6.3.1}$$

When $n = 1$, (6.3.1) coincides with the Mittag-Leffler function (4.1.1):

$$E((\alpha, \beta)_1; z) = E_{\alpha,\beta}(z) \equiv \sum_{k=0}^{\infty} \frac{z^k}{\Gamma(\alpha k + \beta)} \quad (z \in \mathbb{C}), \tag{6.3.2}$$

and, for $n = 2$, with the four-parametric function (6.1.1):

$$E((\alpha, \beta)_2; z) = E_{\alpha_1, \beta_1; \alpha_2, \beta_2}(z) \equiv \sum_{k=0}^{\infty} \frac{z^k}{\Gamma(\alpha_1 k + \beta_1)\Gamma(\alpha_2 k + \beta_2)} \quad (z \in \mathbb{C}). \tag{6.3.3}$$

First of all we prove that (6.3.1) under the condition $\alpha_1 + \alpha_2 + \cdots + \alpha_n > 0$ is an entire function.

Theorem 6.5 *Let $n \in \mathbb{N}$ and $\alpha_i \in \mathbb{R}$, $\beta_i \in \mathbb{C}$ $(i = 1, 2, \cdots, n)$ be such that*

$$\alpha_1^2 + \cdots + \alpha_n^2 \neq 0, \quad \alpha_1 + \alpha_2 + \cdots + \alpha_n > 0. \tag{6.3.4}$$

Then $E((\alpha, \beta)_n; z)$ is an entire function of $z \in \mathbb{C}$ of order

$$\rho = \frac{1}{(\alpha_1 + \alpha_2 + \cdots + \alpha_n)} \tag{6.3.5}$$

and type

$$\sigma = \prod_{i=1}^{n} \left(\frac{\alpha_1 + \cdots + \alpha_n}{|\alpha_i|}\right)^{\frac{\alpha_i}{\alpha_1 + \cdots + \alpha_n}}. \tag{6.3.6}$$

◁ Rewrite (6.3.1) as the power series

$$E((\alpha, \beta)_n; z) = \sum_{k=0}^{\infty} c_k z^k, \quad c_k = \left[\prod_{j=1}^{n} \Gamma(\alpha_j k + \beta_j)\right]^{-1}. \tag{6.3.7}$$

According to the asymptotic property (A.27) we have

$$\frac{|c_k|}{|c_{k+1}|} \sim \prod_{j=1}^{n} |\alpha_j k|^{\alpha_j} = \prod_{j=1}^{n} |\alpha_j|^{\alpha_j} k^{\alpha_1 + \alpha_2 + \cdots + \alpha_n} \to +\infty \ (k \to \infty).$$

Then, if $\alpha_1 + \alpha_2 \cdots + \alpha_n > 0$, we see that $R = \infty$, where R is the radius of convergence of the power series in (6.3.7). This means that $E((\alpha, \beta)_n; z)$ is an entire function of z.

6.3 Mittag-Leffler Functions with $2n$ Parameters

We use [Appendix B, (B.5) and (B.6)] to evaluate the order ρ and the type σ of (6.3.1). Applying Stirling's formula for the Gamma function $\Gamma(z)$ at infinity and taking (6.3.7) into account, we have

$$\log\left(\frac{1}{c_k}\right) = \log\left[\prod_{j=1}^{n}\Gamma(\alpha_j k + \beta_j)\right]$$

$$= \sum_{j=1}^{n}\left(\alpha_j k + \beta_j - \frac{1}{2}\right)\log(\alpha_j k) - \sum_{j=1}^{n}(\alpha_j k) + \frac{n}{2}\log(2\pi) + O\left(\frac{1}{k}\right) \quad (k \to \infty).$$

Hence the following asymptotic estimate holds:

$$\log\left(\frac{1}{c_k}\right) \sim k\log(k)(\alpha_1 + \alpha_2 + \cdots + \alpha_n) \quad (k \to \infty). \tag{6.3.8}$$

Thus, in accordance with [Appendix B, (B.5)], we obtain (6.3.5).

Further, according to (6.1.12) we obtain the asymptotic estimate

$$\prod_{j=1}^{n}\Gamma(\alpha_j k + \beta_j) \sim (2\pi)^{n/2}\prod_{j=1}^{n}(\alpha_j k)^{\alpha_j k + \beta_j - \frac{1}{2}}e^{-\alpha_j k} \quad (k \to \infty). \tag{6.3.9}$$

By (6.3.7) and (6.3.9) we have

$$\limsup_{k\to\infty}\left(k^{1/\rho}|c_k|^{1/k}\right) = \limsup_{k\to\infty} k^{1/\rho}\prod_{j=1}^{n}\left[(|\alpha_j|k)^{-\alpha_j}e^{\alpha_j}\right]$$

$$= e^{\alpha_1+\alpha_2+\cdots+\alpha_n}\prod_{j=1}^{n}|\alpha_j|^{-\alpha_j} = e^{1/\rho}\prod_{j=1}^{n}|\alpha_j|^{-\alpha_j}.$$

Substituting this relation into [Appendix B, (B.6)] we have

$$\sigma = \frac{1}{\rho}\left(\prod_{j=1}^{n}|\alpha_j|^{-\alpha_j}\right)^{\rho} = (\alpha_1 + \alpha_2 + \cdots + \alpha_n)\left(\prod_{j=1}^{n}|\alpha_j|^{-\alpha_j}\right)^{\frac{1}{\alpha_1+\cdots+\alpha_n}}$$

$$= \prod_{j=1}^{n}\left(\frac{\alpha_1 + \cdots + \alpha_n}{|\alpha_j|}\right)^{\frac{\alpha_j}{\alpha_1+\cdots+\alpha_n}},$$

which proves (6.3.6). ▷

Remark 6.6 In the general case $\alpha_1 + \cdots \alpha_n > 0$ the relations (6.3.5) and (6.3.6) have been proved by Kilbas and Koroleva [KilKor05] (and also in a paper by

Rogosin, Kilbas and Koroleva [KiKoRo13]), while in the particular case $\alpha_j > 0$ ($j = 1, \ldots, n$), in the works of Kiryakova, as [Kir99], [Kir00]. Note that if $n > 1$ the type σ in (6.3.6) is greater than 1 (Th.1, Kiryakova [Kir10b]).

Remark 6.7 When $n = 1, \alpha_1 = \alpha > 0$ and $\beta_1 = \beta \in \mathbb{C}$, relations (6.3.5) and (6.3.6) yield the known order and type of the Mittag-Leffler function $E_{\alpha,\beta}(z)$ in (4.1.1) [Sect. 4.1]:

$$\rho = \frac{1}{\alpha}, \quad \sigma = 1. \tag{6.3.10}$$

Remark 6.8 When $n = 2$, $\alpha_j \in \mathbb{R}$, $\beta_j \in \mathbb{C}$ ($j = 1, 2$) with $\alpha_1^2 + \alpha_2^2 \neq 0$ and $\alpha_1 + \alpha_2 > 0$, formulas (6.3.5) and (6.3.6) coincide with (6.1.8) and (6.1.9), respectively.

6.3.2 Representations in Terms of Hypergeometric Functions

We consider the generalized Mittag-Leffler function $E((\alpha, \beta)_n; z)$ in (6.3.1) under the conditions of Theorem 6.5. First we give a representation of $E((\alpha, \beta)_n; z)$ in terms of the generalized Wright hypergeometric function ${}_p\Psi_q(z)$ defined in Appendix F, (F.2.6)]. By (A.17), $(1)_k = k! = \Gamma(k+1)$ ($k \in \mathbb{N}_0$) and we can rewrite (6.3.1) in the form

$$E((\alpha, \beta)_n; z) = \sum_{k=0}^{\infty} \frac{\Gamma(k+1)}{\prod_{j=1}^{n} \Gamma(\beta_j + \alpha_j k)} \frac{z^k}{k!} \quad (z \in \mathbb{C}). \tag{6.3.11}$$

This yields the following representation of $E((\alpha, \beta)_n; z)$ via the generalized Wright hypergeometric function ${}_1\Psi_n(z)$:

$$E((\alpha, \beta)_n; z) = {}_1\Psi_n\left[\begin{array}{c}(1, 1) \\ (\beta_1, \alpha_1), \cdots, (\beta_n, \alpha_n)\end{array}\bigg| z\right] \quad (z \in \mathbb{C}). \tag{6.3.12}$$

Next we consider the generalized Mittag-Leffler function (6.3.1) with $n \geq 2$ and $\alpha_j = m_j \in \mathbb{N}$ ($j = 1, \cdots, n$):

$$E((m, \beta)_n; z) = \sum_{k=0}^{\infty} \frac{z^k}{\prod_{j=1}^{n} \Gamma(m_j k + \beta_j)}$$

$$= \sum_{k=0}^{\infty} \frac{(1)_k}{\prod_{j=1}^{n} \Gamma(m_j k + \beta_j)} \frac{z^k}{k!} \quad (z \in \mathbb{C}). \tag{6.3.13}$$

According to (A.14) with $z = k + \frac{\beta_j}{m_j}$, $m = m_j$ ($j = 1, \cdots, n$) and (A.17) we have

6.3 Mittag-Leffler Functions with 2n Parameters

$$\Gamma(m_j k + \beta_j) = \Gamma\left[m_j\left(k + \frac{\beta_j}{m_j}\right)\right]$$

$$= (2\pi)^{(1-m_j)/2} m_j^{m_j k + \beta_j - \frac{1}{2}} \prod_{s=0}^{m_j-1} \Gamma\left(\frac{\beta_j + s}{m_j} + k\right)$$

$$= (2\pi)^{(1-m_j)/2} m_j^{m_j k + \beta_j - \frac{1}{2}} \prod_{s=0}^{m_j-1} \Gamma\left(\frac{\beta_j + s}{m_j}\right) \left(\frac{\beta_j + s}{m_j}\right)_k$$

$$= m_j^{m_j k} \left[(2\pi)^{(1-m_j)/2} m_j^{\beta_j - \frac{1}{2}} \prod_{s=0}^{m_j-1} \Gamma\left(\frac{\beta_j + s}{m_j}\right)\right] \prod_{s=0}^{m_j-1} \left(\frac{\beta_j + s}{m_j}\right)_k.$$

Then applying (A.14) with $z = \frac{\beta_j}{m_j}$, $m = m_j$, we get

$$\Gamma(m_j k + \beta_j) = m_j^{m_j k} \Gamma(\beta_j) \prod_{s=0}^{m_j-1} \left(\frac{\beta_j + s}{m_j}\right)_k.$$

Hence

$$E\left((m, \beta)_n; z\right) = \frac{1}{\prod_{j=1}^{n} \Gamma(\beta_j)} \sum_{k=0}^{\infty} \frac{(1)_k}{\prod_{j=1}^{n} \prod_{s=0}^{m_j-1} \left(\frac{\beta_j + s}{m_j}\right)_k} \left(\frac{z}{\prod_{j=1}^{n} m_j^{m_j}}\right)^k \frac{1}{k!}.$$

Therefore, we obtain the following representation of the 2n-parametric Mittag-Leffler function via a generalized hypergeometric function in the case of positive integer first parameters $\alpha_j = m_j \in \mathbb{N}$ $(j = 1, \cdots, n)$

$$E\left((m, \beta)_n; z\right) = \frac{1}{\prod_{j=1}^{n} \Gamma(\beta_j)} \tag{6.3.14}$$

$$\times {}_1F_{m_1+\ldots+m_n}\left(1; \frac{\beta_1}{m_1}, \ldots, \frac{\beta_1 + m_1 - 1}{m_1}, \ldots, \frac{\beta_n}{m_n}, \ldots, \frac{\beta_n + m_n - 1}{m_n}; \frac{z}{\prod_{j=1}^{n} m_j^{m_j}}\right).$$

6.3.3 Integral Representations and Asymptotics

The $2n$-parametric Mittag-Leffler function can be introduced either in the form of a series (6.3.1) or in the form of a Mellin–Barnes integral

$$E_{(\alpha,\beta)_n}(z) = \frac{1}{2\pi i} \int_{\mathcal{L}} \frac{\Gamma(s)\Gamma(1-s)}{\prod_{j=1}^{n}\Gamma(\beta_j - \alpha_j s)} (-z)^{-s} ds \quad (z \neq 0). \tag{6.3.15}$$

For $\operatorname{Re}\alpha_1 + \ldots + \alpha_n > 0$ one can choose the left loop $\mathcal{L}_{-\infty}$ as a contour of integration in (6.3.15). Calculating this integral by using Residue Theory we immediately obtain the series representation (6.3.1).

If $\alpha_j > 0$; $\beta_j \in \mathbb{R}$ $(j = 1, \ldots, n)$, then the $2n$-parametric Mittag-Leffler function $E((\alpha, \beta)_n; z)$ is an entire function of the complex variable $z \in \mathbb{C}$ of finite order, see Theorem 6.5.

This result gives an upper bound for the growth of the $2n$-parametric Mittag-Leffler function at infinity, namely, for any positive $\varepsilon > 0$ there exists a positive r_ε such that

$$\left| E_{(\alpha,\beta)_n}(z) \right| < \exp\{(\sigma + \varepsilon)|z|^\rho\}, \quad \forall z, \, |z| > r_\varepsilon. \tag{6.3.16}$$

More precisely, the asymptotic behavior of the function $E_{(\alpha,\beta)_n}(z)$ can be described using the representation of the latter in terms of the H-function with special values of parameters (see Sect. 6.3.7 below) and asymptotic results for the H-function (see [KilSai04]).

6.3.4 Extension of the 2n-Parametric Mittag-Leffler Function

An extension of the $2n$-parametric Mittag-Leffler function is given by the representation

$$\mathcal{E}((\alpha, \beta)_n; z) = \frac{1}{2\pi i} \int_{\mathcal{L}} \frac{\Gamma(s)\Gamma(1-s)}{\prod_{j=1}^{n}\Gamma(\beta_j - \alpha_j s)} (-z)^{-s} ds \quad (z \neq 0), \tag{6.3.17}$$

where the right loop $\mathcal{L} = \mathcal{L}_{+\infty}$ is chosen as the contour of integration $\mathcal{L}$.

By using Stirling's asymptotic formula for the Gamma function

$$|\Gamma(x + iy)| = (2\pi)^{1/2} |x|^{x-1/2} e^{-x - \pi[1 - \operatorname{sign}(x)]y/2} \quad (x, y \in \mathbb{R}; \, |x| \to \infty), \tag{6.3.18}$$

one can show directly that with the above choice of the integration contour the integral (6.3.17) is convergent for all values of parameters $\alpha_1, \ldots, \alpha_n \in \mathbb{C}, \beta_1, \ldots, \beta_n \in \mathbb{C}$ such that $\operatorname{Re}\alpha_1 + \ldots + \alpha_n < 0$ (cf., e.g., [KilKor05]).

6.3 Mittag-Leffler Functions with $2n$ Parameters

Under these conditions (the choice of contour and assumption on the parameters) the integral (6.3.17) can be calculated by using Residue Theory. This gives the following Laurent series representation of the extended $2n$-parametric Mittag-Leffler function: let $\alpha_j, \beta_j \in \mathbb{C}$ ($j = 1...n$), $z \in \mathbb{C}$ ($z \neq 0$) with $\operatorname{Re} \alpha_1 + \ldots + \alpha_n < 0$ and $\mathcal{L} = \mathcal{L}_{+\infty}$, then the function $\mathcal{E}((\alpha, \beta)_n; z)$ has the Laurent series representation

$$\mathcal{E}((\alpha, \beta)_n; z) = \sum_{k=0}^{\infty} \frac{d_k}{z^{k+1}}, \quad d_k = \prod_{j=1}^{n} \frac{1}{\Gamma(-\alpha_j k - \alpha_j + \beta_j)}. \qquad (6.3.19)$$

The series in (6.3.19) is convergent for all $z \in \mathbb{C} \setminus \{0\}$. Convergence again follows from the asymptotic properties of the Gamma function, which yield the relation

$$\frac{|d_k|}{|d_{k+1}|} \sim \prod_{j=1}^{n} \left[|\alpha_j|^{-\operatorname{Re}(\alpha_j)} e^{\operatorname{Im}(\alpha_j) \arg(-\alpha_j k)} \right] k^{-\sum_{j=1}^{n} \operatorname{Re}(\alpha_j)} \quad (k \to \infty).$$

By using the series representation of the extended $2n$-parametric Mittag-Leffler function it is not hard to obtain an asymptotic formula for $z \to \infty$. Namely, if $\alpha_j, \beta_j \in \mathbb{C}$ ($j = 1, \ldots, n$), $z \in \mathbb{C}$ ($z \neq 0$) and $\operatorname{Re} \alpha_1 + \ldots + \alpha_n) < 0$, with contour of integration in (6.3.17) chosen as $\mathcal{L} = \mathcal{L}_{+\infty}$, then for any $N \in \mathbb{N}$ we have for $z \to \infty$ the asymptotic representation

$$\mathcal{E}((\alpha, \beta)_n; z) = \sum_{k=0}^{N} \frac{1}{\prod_{j=1}^{n} \Gamma(-\alpha_j k - \alpha_j + \beta_j) z^{k+1}} \left[1 + O\left(\frac{1}{z}\right) \right] \quad (z \to \infty).$$

The main term of this asymptotics is equal to

$$\mathcal{E}((\alpha, \beta)_n; z) = \prod_{j=1}^{n} \frac{1}{\Gamma(-\alpha_j + \beta_j)} \left[1 + O\left(\frac{1}{z}\right) \right] \quad (z \to \infty).$$

The asymptotics at $z \to 0$ is more complicated. It can be derived by using the relations of the extended $2n$-parametric Mittag-Leffler function with the generalized Wright function and the H-functions (see Sect. 6.3.5 below) and the asymptotics of the latter presented in [KilSai04].

Another possible way to get the asymptotics of $\mathcal{E}((\alpha, \beta)_n; z)$ for $z \to 0$ is to use the following. If $\alpha_j, \beta_j \in \mathbb{C}$ ($j = 1...n$), $z \in \mathbb{C}$ ($z \neq 0$), $\operatorname{Re} \alpha_1 + \ldots + \alpha_n < 0$, $\mathcal{L} = \mathcal{L}_{+\infty}$, then the extended $2n$-parametric Mittag-Leffler function can be presented in terms of the "usual" $2n$-parametric Mittag-Leffler function:

$$\mathcal{E}((\alpha, \beta)_n; z) = \frac{1}{z} E\left((-\alpha, \beta - \alpha)_n; \frac{1}{z}\right). \qquad (6.3.20)$$

6.3.5 Relations to the Wright Function and to the H-Function

In this section we present some formulas representing the $2n$-parametric Mitttag-Leffler function $E((\alpha, \beta)_n; z)$ and its extension $\mathcal{E}((\alpha, \beta)_n; z)$ in terms of the generalized Wright function $_p\Psi_q$ and the H-function.

For short, we use the same notation $\mathcal{E}((\alpha, \beta)_n; z)$ for the $2n$-parametric Mitttag-Leffler function and for its extension. These functions differ in values of the parameters α_j and in the choice of the contour of integration $\mathcal{L}$ in their Mellin–Barnes integral representation.

For real values of the parameters $\alpha_j \in \mathbb{R}$ and complex $\beta_j \in \mathbb{C}$ $(j = 1, \ldots, n)$ the following representations hold:

(1) if $\sum_{j=1}^n \alpha_j > 0$, $\mathcal{L} = \mathcal{L}_{-\infty}$, then

$$\mathcal{E}((\alpha, \beta)_n; z) = {}_1\Psi_n \left[\begin{array}{c} (1, 1) \\ (\beta_1, \alpha_1), \ldots, (\beta_n, \alpha_n) \end{array} \bigg| z \right]; \qquad (6.3.21)$$

(2) if $\sum_{j=1}^n \alpha_j < 0$, $\mathcal{L} = \mathcal{L}_{+\infty}$, then

$$\mathcal{E}((\alpha, \beta)_n; z) = \frac{1}{z} {}_1\Psi_2 \left[\begin{array}{c} (1, 1) \\ (\beta_1 - \alpha_1, -\alpha_1), \ldots, (\beta_n - \alpha_n, -\alpha_n) \end{array} \bigg| \frac{1}{z} \right]. \qquad (6.3.22)$$

The above representations can be obtained by comparing the series representation of the corresponding functions. In the case (6.3.22) one can also use the relation (6.3.20).

In the same manner one can obtain the following representations of the $2n$-parametric Mittag-Leffler function and its extension in terms of the H-function:

(1) if $\alpha_j > 0$ $(j = 1, \ldots, n)$, and $\mathcal{L} = \mathcal{L}_{-\infty}$, then

$$\mathcal{E}((\alpha, \beta)_n; z) = H^{1,1}_{1, n+1} \left[\begin{array}{c} (0, 1) \\ (0, 1)(1 - \beta_1, \alpha_1), \ldots, (1 - \beta_n, \alpha_n) \end{array} \bigg| z \right]; \qquad (6.3.23)$$

(2) if $\alpha_j > 0$ $(j = 1, \ldots, p, \ p < n)$, $\alpha_j < 0$ $(j = p+1, \ldots, n)$, and either $\sum_{j=1}^n \alpha_j > 0$, $\mathcal{L} = \mathcal{L}_{-\infty}$, or $\sum_{j=1}^n \alpha_j < 0$, $\mathcal{L} = \mathcal{L}_{+\infty}$, then

$$\mathcal{E}((\alpha, \beta)_n; z) = H^{1,1}_{n-p+1, p+1} \left[\begin{array}{c} (0, 1)(\beta_{p+1}, -\alpha_{p+1}) \ldots (\beta_n, -\alpha_n) \\ (0, 1)(1 - \beta_1, \alpha_1), \ldots, (1 - \beta_p, \alpha_p) \end{array} \bigg| z \right]; \qquad (6.3.24)$$

(3) if $\alpha_j < 0$ $(j = 1, \ldots, p, \ p < n)$, $\alpha_j > 0$ $(j = p+1, \ldots, n)$, and either $\sum_{j=1}^n \alpha_j > 0$, $\mathcal{L} = \mathcal{L}_{-\infty}$, or $\sum_{j=1}^n \alpha_j < 0$, $\mathcal{L} = \mathcal{L}_{+\infty}$, then

6.3 Mittag-Leffler Functions with 2n Parameters

$$\mathcal{E}((\alpha,\beta)_n;z) = H^{1,1}_{p+1,n-p+1}\left[\begin{array}{c}(0,1)(\beta_1,-\alpha_1)\ldots(\beta_p,-\alpha_p)\\(0,1)(1-\beta_{p+1},\alpha_{p+1}),\ldots,(1-\beta_n,\alpha_n)\end{array}\middle|z\right];$$
(6.3.25)

(4) if $\alpha_j < 0$, $(j = 1,\ldots,n)$ and $\mathcal{L} = \mathcal{L}_{+\infty}$, then

$$\mathcal{E}((\alpha,\beta)_n;z) = H^{1,1}_{n+1,1}\left[\begin{array}{c}(0,1)(\beta_1,-\alpha_1),\ldots,(\beta_n,-\alpha_n)\\(0,1)\end{array}\middle|z\right]. \quad (6.3.26)$$

6.3.6 Integral Transforms with the Multi-parametric Mittag-Leffler Functions

Here we consider only the case when the parameters α_i in the definition of the $2n$-parametric Mittag-Leffler function and its extension are real numbers.

Since the $2n$-parametric Mittag-Leffler function is related to the generalized Wright function and to the H-function with special values of parameters (see Sect. 6.3.5), one can use (6.1.35) or (6.1.40) to define the Mellin transform of the function $E((\alpha,\beta)_n;z)$ and of its extension $\mathcal{E}((\alpha,\beta)_n;z)$.

Now we present a few results on integral transforms with the $2n$-parametric function in the kernel. The transforms are defined by the formula

$$\left(\mathbf{E}(\alpha,\beta)_n f\right)(x) = \int_0^\infty \mathcal{E}((\alpha,\beta)_n;-xt)f(t)dt \quad (x>0), \quad (6.3.27)$$

with the $2n$-parametric Mittag-Leffler function in the kernel. These transforms are special cases of more general **H**-transforms (see Sect. 6.1.7). This can be seen from the definition of the **H**-transforms (6.1.32) and the following formulas which relate $\mathbf{E}(\alpha,\beta)_n$-transforms to **H**-transforms under different assumptions on the parameters.

(1) Let $\alpha_j > 0$ $(j = 1,\ldots,n)$, $\mathcal{L} = \mathcal{L}_{-\infty}$, then

$$(\mathbf{E}(\alpha,\beta)_n f)(x) = \quad (6.3.28)$$

$$\int_0^\infty H^{1,1}_{1,n+1}\left[xt\middle|\begin{array}{c}(0,1)\\(0,1),(1-\beta_1,\alpha_1),\ldots,(1-\beta_n,\alpha_n)\end{array}\right]f(t)dt.$$

(2) Let $\alpha_j > 0$ $(j = 1,\ldots,p,\ p<n)$, $\alpha_j < 0$ $(j = p+1,\ldots,n)$ and either $\sum_{j=1}^n \alpha_j > 0$, $\mathcal{L} = \mathcal{L}_{-\infty}$ or $\sum_{j=1}^n \alpha_j < 0$, $\mathcal{L} = \mathcal{L}_{+\infty}$, then

$$(\mathbf{E}(\alpha,\beta)_n f)(x) = \qquad (6.3.29)$$

$$\int_0^\infty H^{1,1}_{n-p+1,p+1}\left[xt \left| \begin{matrix} (0,1), (\beta_{p+1}, -\alpha_{p+1}), \ldots, (\beta_n, -\alpha_n) \\ (0,1), (1-\beta_1, \alpha_1), \ldots, (1-\beta_p, \alpha_p) \end{matrix} \right. \right] f(t)\mathrm{d}t.$$

(3) Let $\alpha_j < 0$ ($j = 1, \ldots, p$, $p < n$), $\alpha_j > 0$ ($j = p+1, \ldots, n$) and either $\sum_{j=1}^n \alpha_j > 0$, $\mathcal{L} = \mathcal{L}_{-\infty}$, or $\sum_{j=1}^n \alpha_j < 0$, $\mathcal{L} = \mathcal{L}_{+\infty}$, then

$$(\mathbf{E}(\alpha,\beta)_n f) = \qquad (6.3.30)$$

$$\int_0^\infty H^{1,1}_{p+1,n-p+1}\left[xt \left| \begin{matrix} (0,1), (\beta_1, -\alpha_1), \ldots, (\beta_p, -\alpha_p) \\ (0,1), (1-\beta_{p+1}, \alpha_{p+1}), \ldots, (1-\beta_n, \alpha_n) \end{matrix} \right. \right] f(t)\mathrm{d}t.$$

$$(6.3.31)$$

(4) Let $\alpha_j < 0$, ($j = 1, \ldots, n$) and $\mathcal{L} = \mathcal{L}_{+\infty}$, then

$$(\mathbf{E}(\alpha,\beta)_n f)(x) = \int_0^\infty H^{1,1}_{n+1,1}\left[xt \left| \begin{matrix} (0,1), (\beta_1, -\alpha_1), \ldots, (\beta_n, -\alpha_n) \\ (0,1) \end{matrix} \right. \right] f(t)\mathrm{d}t.$$

$$(6.3.32)$$

Convergence of the integrals depends on the values of some constants (as defined in formula (F.4.9), Appendix F). The constant a^* takes different values in the above cases:

1) $a^* = 2 - \sum_{j=1}^n \alpha_j$; (2) $a^* = 2 - \sum_{j=1}^p \alpha_j + \sum_{j=p+1}^n \alpha_j$;

3) $a^* = 2 + \sum_{j=1}^p \alpha_j - \sum_{j=p+1}^n \alpha_j$; 4) $a^* = 2 + \sum_{j=1}^n \alpha_j$;

and the constants $\Delta, \mu, \alpha, \beta$ take the same values in all four cases:

$$\Delta = \sum_{j=1}^n \alpha_j;\ \mu = \frac{n}{2} - \sum_{j=1}^n \beta_j;\ \alpha = 0;\ \beta = 1.$$

We present results on $\mathbf{E}(\alpha,\beta)_n$-transforms for two essentially different cases, namely for the case when all α_j are positive, and for the case when some of them are negative.

A. Let $0 < \nu < 1$, $\alpha_j > 0$ ($j = 1, \ldots, n$), $\beta_j \in \mathbb{C}$ ($j = 1, \ldots, n$) be such that either $0 < \sum_{j=1}^n \alpha_j < 2$ or $\sum_{j=1}^n \alpha_j = 2$ and $2\nu + \sum_{j=1}^n \mathrm{Re}\,\beta_j \geq 2 + \frac{n}{2}$.

(a) There exists an injective mapping (transform) $\mathbf{E}^*(\alpha,\beta)_n \in [\mathcal{L}_{\nu,2}, \mathcal{L}_{1-\nu,2}]$ such that the equality

6.3 Mittag-Leffler Functions with $2n$ Parameters

$$\left(\mathcal{M}\mathbf{E}^*(\alpha,\beta)_n f\right)(s) = \frac{\Gamma(s)\Gamma(1-s)}{\prod_{j=1}^n \Gamma(\beta_j - \alpha_j s)} (\mathcal{M}f)(1-s), \quad (\operatorname{Re} s = 1-\nu) \tag{6.3.33}$$

holds for any $f \in \mathcal{L}_{\nu,2}$.

If $\sum_{j=1}^n \alpha_j = 2$, $2\nu + \sum_{j=1}^n \operatorname{Re}\beta_j = 2 + \frac{n}{2}$ and

$$s \ne \frac{\beta_1 + k}{\alpha_1}, \ldots, s \ne \frac{\beta_n + l}{\alpha_n} \quad (k,l = 0, 1, 2, \cdots) \text{ for } \operatorname{Re} s = 1-\nu, \tag{6.3.34}$$

then the mapping $\mathbf{E}^*(\alpha, \beta)_n$ is bijective from $\mathcal{L}_{\nu,2}$ onto $\mathcal{L}_{1-\nu,2}$.

(b) For any $f, g \in \mathcal{L}_{\nu,2}$ the following integration by parts formula holds:

$$\int_0^\infty f(x)\left(\mathbf{E}^*(\alpha,\beta)_n g\right)(x)\mathrm{d}x = \int_0^\infty \left(\mathbf{E}^*(\alpha,\beta)_n f\right)(x)g(x)\mathrm{d}x. \tag{6.3.35}$$

(c) If $f \in \mathcal{L}_{\nu,2}$, $\lambda \in \mathbb{C}$, $h > 0$, then the value $\mathbf{E}^*(\alpha, \beta)_n f$ can be represented in the form:

$$\left(\mathbf{E}^*(\alpha, \beta)_n f\right)(x) = hx^{1-(\lambda+1)/h}\frac{\mathrm{d}}{\mathrm{d}x}x^{(\lambda+1)/h} \tag{6.3.36}$$

$$\times \int_0^\infty H_{2,n+2}^{1,2}\left[xt \middle| \begin{matrix}(-\lambda, h), (0, 1)\\(0, 1), (1-\beta_1, \alpha_1), \ldots, (1-\beta_n, \alpha_n), (-\lambda-1, h)\end{matrix}\right] f(t)\mathrm{d}t, \tag{6.3.37}$$

when $\operatorname{Re}(\lambda) > (1-\nu)h - 1$, or

$$\left(\mathbf{E}^*(\alpha, \beta)_n f\right)(x) = -hx^{1-(\lambda+1)/h}\frac{\mathrm{d}}{\mathrm{d}x}x^{(\lambda+1)/h} \tag{6.3.38}$$

$$\times \int_0^\infty H_{2,n+2}^{2,1}\left[xt \middle| \begin{matrix}(0, 1), (-\lambda, h)\\(-\lambda-1, h), (0, 1), (1-\beta_1, \alpha_1), \ldots, (1-\beta_n, \alpha_n)\end{matrix}\right] f(t)\mathrm{d}t, \tag{6.3.39}$$

when $\operatorname{Re}(\lambda) < (1-\nu)h - 1$.

(d) The mapping $\mathbf{E}^*(\alpha, \beta)_n$ does not depend on ν in the following sense: if two values of the parameter $0 < \nu_1, \nu_2 < 1$ and the corresponding mappings $\mathbf{E}^*(\alpha, \beta)_{n;1}$, $\mathbf{E}^*(\alpha, \beta)_{n;2}$ are defined on the spaces $\mathcal{L}_{\nu_1,2}$, $\mathcal{L}_{\nu_2,2}$, respectively, then $\mathbf{E}^*(\alpha, \beta)_{n;1} f = \mathbf{E}^*(\alpha, \beta)_{n;2} f$ for all $f \in \mathcal{L}_{\nu_1,2} \cap \mathcal{L}_{\nu_2,2}$.

(e) If $f \in \mathcal{L}_{\nu,2}$ and either $0 < \sum_{j=1}^{n} \alpha_j < 2$ or $\sum_{j=1}^{n} \alpha_j = 2$ and $2\nu + \sum_{j=1}^{n} \operatorname{Re} \beta_j \geq 2 + \frac{n}{2}$, then the mapping (transform) $\mathbf{E}^*(\alpha, \beta)_n$ coincides with the transform $\mathbf{E}(\alpha, \beta)_n$ given by the formula (6.3.27), i.e. $\mathbf{E}^*(\alpha, \beta)_n f = \mathbf{E}(\alpha, \beta)_n f, \forall f \in \mathcal{L}_{\nu,2}$.

B. Let $0 < \nu < 1$, $\alpha_j > 0$ $(j = 1, \ldots, jp, j\, p < n)$ and $\alpha_j < 0$ $(j = p+1, j \ldots, n)$, $\beta_j \in \mathbb{C}$ $(j = 1, \ldots, n)$, be such that either $2 - \sum_{j=1}^{p} \alpha_j + \sum_{j=p+1}^{n} \alpha_j > 0$ or $2 - \sum_{j=1}^{p} \alpha_i + \sum_{j=p+1}^{n} \alpha_j = 0$ and $(1 - \nu) \sum_{j=1}^{n} \alpha_j + \frac{n}{2} \leq \sum_{j=1}^{n} \beta_i$.

(a) There exists an injective mapping (transform) $\mathbf{E}^*(\alpha, \beta)_n \in [\mathcal{L}_{\nu,2}, \mathcal{L}_{1-\nu,2}]$ such that the equality (6.3.33) holds for any $f \in \mathcal{L}_{\nu,2}$.
If $2 - \sum_{j=1}^{p} \alpha_j + \sum_{j=p+1}^{n} \alpha_j = 0$, $(1 - \nu) \sum_{j=1}^{n} \alpha_j + \frac{n}{2} = \sum_{j=1}^{n} \beta_j$ and the parameter s (which determines the line of integration for the inverse Mellin transform in (6.3.33)) satisfies (6.3.34), then the mapping $\mathbf{E}^*(\alpha, \beta)_n$ is bijective from $\mathcal{L}_{\nu,2}$ onto $\mathcal{L}_{1-\nu,2}$.

(b) For any $f, g \in \mathcal{L}_{\nu,2}$ the integration by parts formula (6.3.35) is satisfied.

(c) If $f \in \mathcal{L}_{\nu,2}, \lambda \in \mathbb{C}, h > 0$, then the value $\mathbf{E}^*(\alpha, \beta)_n f$ can be represented in the form:

$$\left(\mathbf{E}^*(\alpha, \beta)_n f\right)(x) = h x^{1-(\lambda+1)/h} \frac{d}{dx} x^{(\lambda+1)/h} \quad (6.3.40)$$

$$\times \int_0^\infty H_{n-p+2,p+2}^{1,2}\left[xt \left|\begin{array}{c}(-\lambda, h), (0, 1), (\beta_{p+1}, -\alpha_{p+1}), \ldots, (\beta_n, -\alpha_n) \\ (0, 1), (1 - \beta_1, \alpha_1), \ldots, (1 - \beta_p, \alpha_p), (-\lambda - 1, h)\end{array}\right.\right] f(t) dt,$$

when $\operatorname{Re}(\lambda) > (1 - \nu)h - 1$, or

$$\left(\mathbf{E}^*(\alpha, \beta)_n f\right)(x) = -h x^{1-(\lambda+1)/h} \frac{d}{dx} x^{(\lambda+1)/h} \quad (6.3.41)$$

$$\times \int_0^\infty H_{n-p+2,p+2}^{2,1}\left[xt \left|\begin{array}{c}(0, 1), (\beta_{p+1}, -\alpha_{p+1}), \ldots, (\beta_n, -\alpha_n), (-\lambda, h) \\ (-\lambda - 1, h), (0, 1), (1 - \beta_1, \alpha_1), \ldots, (1 - \beta_p, \alpha_p)\end{array}\right.\right] f(t) dt,$$

when $\operatorname{Re}(\lambda) < (1 - \nu)h - 1$.

(d) The mapping $\mathbf{E}^*(\alpha, \beta)_n$ does not depend on ν in the following sense: if $0 < \nu_1, \nu_2 < 1$ and the mappings $\mathbf{E}^*(\alpha, \beta)_{n;1}, \mathbf{E}^*(\alpha, \beta)_{n;2}$ are defined on the spaces $\mathcal{L}_{\nu_1,2}, \mathcal{L}_{\nu_2,2}$, respectively, then $\mathbf{E}^*(\alpha, \beta)_{n;1} f = \mathbf{E}^*(\alpha, \beta)_{n;2} f$ for all $f \in \mathcal{L}_{\nu_1,2} \cap \mathcal{L}_{\nu_2,2}$.

6.3 Mittag-Leffler Functions with 2n Parameters

(e) If $f \in \mathcal{L}_{\nu,2}$ and either $2 - \sum_{i=1}^{p} \alpha_i + \sum_{i=p+1}^{n} \alpha_i > 0$ or $2 - \sum_{i=1}^{p} \alpha_i + \sum_{i=p+1}^{n} \alpha_i = 0$ and $(1-\nu)\sum_{i=1}^{n} \alpha_i + \frac{n}{2} \leq \sum_{i=1}^{n} \beta_i$, then the mapping (transform) $\mathbf{E}^*(\alpha, \beta)_n$ coincides with the transform $\mathbf{E}(\alpha, \beta)_n$ given by the formula (6.3.27), i.e. $\mathbf{E}^*(\alpha, \beta)_n f = \mathbf{E}(\alpha, \beta)_n f, \forall f \in \mathcal{L}_{\nu,2}$.

6.3.7 Relations to the Fractional Calculus

In this subsection we present a few formulas relating the $2n$-parametric Mittag-Leffler function (with different values of the parameters α_j) to the left- and right-sided Riemann–Liouville fractional integral and derivative. For short, we use the same notation $\mathcal{E}((\alpha, \beta)_n; z)$ for the $2n$-parametric Mitttag-Leffler function and for its extension. These functions differ in values of the parameters α_i and in the choice of the contour of integration $\mathcal{L}$ in their Mellin–Barnes integral representation. The results in this subsection are obtained (see [KiKoRo13]) by using known formulas for the fractional integration and differentiation of power-type functions (see [SaKiMa93, (2.44) and formula 1 in Table 9.3]).

Let $\alpha_j \in \mathbb{R}$, $\alpha_j \neq 0$ ($j = 1, \ldots, n$), $\alpha_1 < 0, \ldots, \alpha_l < 0, \alpha_{l+1} > 0, \ldots, \alpha_n > 0$ ($1 \leq l \leq n$) and let the contour $\mathcal{L}$ be given by one of the following:

$$\mathcal{L} = \mathcal{L}_{-\infty} \text{ if } \alpha_1 + \ldots + \alpha_n > 0 \text{ or } \mathcal{L} = \mathcal{L}_{+\infty} \text{ if } \alpha_1 + \ldots + \alpha_n < 0.$$

Let $\gamma, \sigma, \lambda \in \mathbb{C}$ be such that $\text{Re}(\gamma) > 0$, $\text{Re}(\sigma) > 0$ and $\omega \in \mathbb{R}$, $(\omega \neq 0)$. Then the following assertions are true.

A. Calculation of the left-sided Riemann–Liouville fractional integral.

(a) If $\omega > 0$, then for $x > 0$

$$\left(I_{0+}^{\gamma} t^{\sigma-1} \mathcal{E}((\alpha, \beta)_n; \lambda t^{\omega})\right)(x) \tag{6.3.42}$$

$$= x^{\sigma+\gamma-1} H^{1,2}_{2+l, 2+n-l}\left[-\lambda x^{\omega} \Bigg| \begin{array}{l} (0,1), (1-\sigma, \omega), (\beta_j, -\alpha_j)_{1,l} \\ (0,1), (1-\sigma-\gamma, \omega), (1-\beta_j, \alpha_j)_{l+1,n} \end{array}\right].$$

(b) If $\omega < 0$, then for $x > 0$

$$\left(I_{0+}^{\gamma} t^{\sigma-1} \mathcal{E}((\alpha, \beta)_n; \lambda t^{\omega})\right)(x) \tag{6.3.43}$$

$$= x^{\sigma+\gamma-1} H^{2,1}_{2+l, 2+n-l}\left[-\lambda x^{\omega} \Bigg| \begin{array}{l} (0,1), (\gamma+\sigma, -\omega), (\beta_j, -\alpha_j)_{1,l} \\ (0,1), (\sigma, -\omega), \quad (1-\beta_j, \alpha_j)_{l+1,n} \end{array}\right].$$

B. Calculation of the right-sided Liouville fractional integral.

(a) If $\omega > 0$, then for $x > 0$

$$\left(I_-^\gamma t^{-\sigma}\mathcal{E}((\alpha,\beta)_n;\lambda t^{-\omega})\right)(x) \tag{6.3.44}$$

$$= x^{\gamma-\sigma} H^{1,2}_{2+l,2+n-l}\left[-\lambda x^\omega \left|\begin{array}{l}(0,1),\ (1-\sigma+\gamma,\omega),\ (\beta_j,-\alpha_j)_{1,l}\\(0,1),\ (1-\sigma,\omega),\quad (1-\beta_j,\alpha_j)_{l+1,n}\end{array}\right.\right].$$

(b) If $\omega < 0$, then for $x > 0$

$$\left(I_-^\gamma t^{-\sigma}\mathcal{E}((\alpha,\beta)_n;\lambda t^{-\omega})\right)(x) \tag{6.3.45}$$

$$= x^{\gamma-\sigma} H^{2,1}_{2+l,2+n-l}\left[-\lambda x^{-\omega} \left|\begin{array}{l}(0,1),\ (\sigma,-\omega),\quad (\beta_j,-\alpha_j)_{1,l}\\(0,1),\ (\sigma-\gamma,-\omega),\ (1-\beta_j,\alpha_j)_{l+1,n}\end{array}\right.\right].$$

C. Calculation of the left-sided Riemann–Liouville fractional derivative.

(a) If $\omega > 0$, then for $x > 0$

$$\left(D_{0+}^\gamma t^{\sigma-1}\mathcal{E}((\alpha,\beta)_n;\lambda t^\omega)\right)(x) \tag{6.3.46}$$

$$= x^{\sigma-\gamma-1} H^{1,2}_{2+l,2+n-l}\left[-\lambda x^\omega \left|\begin{array}{l}(0,1),\ (1-\sigma,\omega),\quad (\beta_j,-\alpha_j)_{1,l}\\(0,1),\ (1-\sigma+\gamma,\omega),\ (1-\beta_j,\alpha_j)_{l+1,n}\end{array}\right.\right].$$

(b) If $\omega < 0$, then for $x > 0$

$$\left(D_{0+}^\gamma t^{\sigma-1}\mathcal{E}((\alpha,\beta)_n;\lambda t^\omega)\right)(x) \tag{6.3.47}$$

$$= x^{\sigma-\gamma-1} H^{2,1}_{2+l,2+n-l}\left[-\lambda x^\omega \left|\begin{array}{l}(0,1),\ (\gamma-\sigma,-\omega),\ (\beta_j,-\alpha_j)_{1,l}\\(0,1),\ (\sigma,-\omega),\quad (1-\beta_j,\alpha_j)_{l+1,n}\end{array}\right.\right].$$

D. Calculation of the right-sided Liouville fractional derivative.

(a) If $\omega > 0$, then for $x > 0$

$$\left(D_-^\gamma t^{-\sigma}\mathcal{E}((\alpha,\beta)_n;-\lambda t^{-\omega})\right)(x) \tag{6.3.48}$$

$$= x^{-\sigma-\gamma} H^{2,1}_{2+l,2+n-l}\left[-\lambda x^{-\omega} \left|\begin{array}{l}(0,1),\ (1-\sigma-\gamma,\omega),\ (\beta_j,-\alpha_j)_{1,l}\\(0,1),\ (1-\sigma,\omega),\quad (1-\beta_j,\alpha_j)_{l+1,n}\end{array}\right.\right].$$

(b) If $\omega < 0$ then for $x > 0$

$$\left(D_-^\gamma t^{-\sigma}\mathcal{E}((\alpha,\beta)_n;\lambda t^{-\omega})\right)(x) \tag{6.3.49}$$

$$= x^{-\sigma-\gamma} H^{1,2}_{2+l,2+n-l}\left[-\lambda x^{-\omega} \left|\begin{array}{l}(0,1),\ (\sigma,-\omega),\quad (\beta_j,-\alpha_j)_{1,l}\\(0,1),\ (\sigma+\gamma,-\omega),\ (1-\beta_j,\alpha_j)_{l+1,n}\end{array}\right.\right].$$

6.4 Mittag-Leffler Functions of Several Variables

In this section we present a few results on Mittag-Leffler function of several variables (see, e.g., [Lav18])

6.4 Mittag-Leffler Functions of Several Variables

$$E_{(\alpha)_n,\beta}(z_1, z_2, \ldots, z_n) = \sum_{m_1,m_2,\ldots,m_n \geq 0} \frac{z_1^{m_1} z_2^{m_2} \cdots z_n^{m_n}}{\Gamma(\alpha_1 m_1 + \alpha_2 m_2 + \ldots + \alpha_n m_n + \beta)}$$

$$=: \sum_{m=0}^{\infty} \frac{\mathbf{z}^{\mathbf{m}}}{\Gamma(\langle , \mathbf{m}\rangle + \beta)}, \quad = (\alpha_1, \alpha_2, \ldots, \alpha_n) \in \mathbb{C}^n, \operatorname{Re}\alpha_j > 0, \ \beta \in \mathbb{C}. \quad (6.4.1)$$

Other forms of the Mittag-Leffler type functions of several variables can be found too (see, e.g., [GaMaKa13], [Dua18], [Mam18] and references therein). There is a natural interest in studying the properties of this class of functions as it is related to the presentation of solutions of systems of linear fractional differential equations (in particular, of incommensurate orders).

In order to avoid additional technical details we focus here only on the case of two variables. The definition of the Mittag-Leffler function of two complex variables is similar the above presented for n variables (to avoid additional indexing we use the following variable names: $x = z_1$, $y = z_2$, $\alpha = \alpha_1$, $\beta = \alpha_2$, $\gamma = \beta$, $n = m_1$, $m = m_2$)

$$E_{\alpha,\beta;\gamma}(x, y) = \sum_{n,m \geq 0} \frac{x^n y^m}{\Gamma(\alpha n + \beta m + \gamma)}, \quad \alpha, \beta, \gamma \in \mathbb{C}, \ \operatorname{Re}\alpha, \operatorname{Re}\beta > 0. \quad (6.4.2)$$

It is straightforward to check that, under above conditions, $E_{\alpha,\beta;\gamma}(x, y)$ is an entire function of two complex variables $(x, y) \in \mathbb{C}^2$.

6.4.1 Integral Representations

For applications it is interesting to describe the behavior of the function (6.4.2) for large values of arguments. For this we use the known results in the case of the Mittag-Leffler function of one variable. First, we find the integral representations of the considered function $E_{\alpha,\beta;\gamma}(x, y)$. Let us recall the definition of the Hankel path (see Sect. 3.4). For fixed $\theta \in (0, \pi)$, $\varepsilon > 0$ it is denoted by $\omega(\varepsilon, \theta)$. The path oriented by non-decreasing $\arg \zeta$ consists of two rays $S_\theta := \{\zeta \in \mathbb{C} : \arg \zeta = \theta, |\zeta| > \varepsilon\}$, $S_{-\theta} := \{\arg \zeta = -\theta, |\zeta| > \varepsilon\}$ and a part of the circle $C_\varepsilon(\theta) := \{\zeta \in \mathbb{C} : |\zeta| = \varepsilon, -\theta \leq \arg \zeta \leq \theta\}$. When $\theta = \pi$ the rays $S_{\pm\theta}$ degenerate into parts of the sides of negative semi-axes. This path divides the complex plane into two domains $\Omega^{(-)}(\varepsilon; \theta)$ and $\Omega^{(+)}(\varepsilon; \theta)$ which are situated, respectively, to the left and to the right of $\omega(\theta, \varepsilon)$ with respect to the orientation on it.

Below we derive integral representations of $E_{\alpha,\beta;\gamma}(x, y)$ in four different domains in $\mathbb{C}^2$, namely $\Omega^{(-)}(\varepsilon_\alpha; \theta_\alpha) \times \Omega^{(-)}(\varepsilon_\beta; \theta_\beta)$, $\Omega^{(+)}(\varepsilon_\alpha; \theta_\alpha) \times \Omega^{(-)}(\varepsilon_\beta; \theta_\beta)$, $\Omega^{(-)}(\varepsilon_\alpha; \theta_\alpha) \times \Omega^{(+)}(\varepsilon_\beta; \theta_\beta)$, and $\Omega^{(+)}(\varepsilon_\alpha; \theta_\alpha) \times \Omega^{(+)}(\varepsilon_\beta; \theta_\beta)$. For this we use two representations of the reciprocal to the Gamma functions appearing in definition (6.4.2) (see Sect. 3.4 of this book).

$$\frac{1}{\Gamma(\alpha n + \beta m + \gamma)} = \frac{1}{2\pi i \alpha} \int_{\omega(\varepsilon,\theta_\alpha)} e^{\zeta^{\frac{1}{\alpha}}} \zeta^{\frac{-\alpha n - \beta m - \gamma + 1}{\alpha}} d\zeta, \qquad (6.4.3)$$

$$\frac{1}{\Gamma(\alpha n + \beta m + \gamma)} = \frac{1}{2\pi i \beta} \int_{\omega(\varepsilon,\theta_\beta)} e^{\zeta^{\frac{1}{\beta}}} \zeta^{\frac{-\alpha n - \beta m - \gamma + 1}{\alpha}} d\zeta. \qquad (6.4.4)$$

In the first integral (6.4.3) we have inequalities for θ_α (see, e.g., Sect. 3.4)

$$\frac{\pi\alpha}{2} < \theta_\alpha \le \min\{\pi; \pi\alpha\},$$

and in the second integral (6.4.4) we have analogous inequalities for θ_β (see, e.g., Sect. 3.4)

$$\frac{\pi\beta}{2} < \theta_\beta \le \min\{\pi; \pi\beta\}.$$

In order to satisfy both sets of inequalities we put $\theta_\alpha = \frac{\theta}{\beta}$, $\theta_\beta = \frac{\theta}{\alpha}$ and fix θ such that

$$\frac{\pi\alpha\beta}{2} < \theta \le \min\{\pi, , \pi\alpha, \pi\beta, \pi\alpha\beta\}, \qquad (6.4.5)$$

and write $\epsilon_\alpha := \varepsilon^{1/\beta}$, $\epsilon_\beta := \varepsilon^{1/\alpha}$. In all cases we also suppose that α, β are "small", i.e.

$$0 < \alpha, \ \beta < 2, \ \alpha\beta < 2. \qquad (6.4.6)$$

Note that it follows from (6.4.6) that the left-hand side is smaller than the right-hand side in (6.4.5).

Let us start with the derivation of the integral representation in the first domain. Let $y \in \Omega^{(-)}(\varepsilon_\beta; \theta_\beta)$, $x \in \mathbb{C}$, $|x| < \varepsilon_\alpha$. Then

$$\sup_{\zeta \in \omega(\theta_\beta, \varepsilon_\beta)} |x\zeta^{-\alpha/\beta}| < 1.$$

This allows us to reduce to the one-dimensional case, due to the identity

$$E_{\alpha,\beta;\gamma}(x, y) = \sum_{n=0}^\infty x^n \sum_{m=0}^\infty \frac{y^m}{\Gamma(\beta m + (\alpha n + \gamma))} = \sum_{n=0}^\infty x^n E_{\beta,\alpha n+\gamma}(y),$$

and the corresponding integral representation for $E_{\beta,\alpha n+\gamma}(y)$:

$$E_{\alpha,\beta;\gamma}(x, y) = \sum_{n=0}^\infty x^n \frac{1}{2\pi i \beta} \int_{\omega(\varepsilon_\beta,\theta_\beta)} \frac{e^{\zeta^{1/\beta}} \zeta^{\frac{1-\alpha n-\gamma}{\beta}}}{\zeta - y} d\zeta$$

6.4 Mittag-Leffler Functions of Several Variables

$$= \frac{1}{2\pi i \beta} \int\limits_{\omega(\varepsilon_\beta,\theta_\beta)} \frac{e^{\zeta^{1/\beta}} \zeta^{\frac{1-\gamma}{\beta}} d\zeta}{\zeta - y} \sum_{n=0}^{\infty} \left(x\zeta^{-\alpha/\beta}\right)^n = \frac{1}{2\pi i \beta} \int\limits_{\omega(\varepsilon_\beta,\theta_\beta)} \frac{e^{\zeta^{1/\beta}} \zeta^{\frac{1+\alpha-\gamma}{\beta}} d\zeta}{(\zeta - y)(\zeta^{\alpha/\beta} - x)}.$$

By changing variables $\zeta = \xi^{1/\alpha}$ we arrive at the following representation

$$E_{\alpha,\beta;\gamma}(x,y) = \frac{1}{2\pi i \alpha \beta} \int\limits_{\omega(\varepsilon,\theta)} \frac{e^{\xi^{\frac{1}{\alpha\beta}}} \xi^{\frac{\alpha+\beta-\gamma}{\alpha\beta}-1} d\xi}{(\xi^{1/\beta} - x)(\xi^{1/\alpha} - y)}, \quad |x| < \varepsilon_\alpha, \; y \in \Omega^{(-)}(\varepsilon_\beta;\theta_\beta).$$
(6.4.7)

Since the circle $x \in \mathbb{C}$, $|x| < \varepsilon_\alpha$, is contained in the domain of analyticity of the right-hand side of (6.4.7), by the Principle of Analytic Continuation formula (6.4.7) is valid in $\Omega^{(-)}(\varepsilon_\alpha;\theta_\alpha) \times \Omega^{(-)}(\varepsilon_\beta;\theta_\beta)$.

In order to prove the formula for the domain $\Omega^{(-)}(\varepsilon_\alpha;\theta_\alpha) \times \Omega^{(+)}(\varepsilon_\beta;\theta_\beta)$ we take $\varepsilon' > \varepsilon$. Then by the previous case we obtain for $y \in \Omega^{(-)}(\varepsilon'_\beta;\theta_\beta)$, $y < \varepsilon'_\beta$ and $x \in \Omega^{(-)}(\varepsilon_\alpha;\theta_\alpha)$ the following representation

$$E_{\alpha,\beta;\gamma}(x,y) = \frac{1}{2\pi i \beta} \int\limits_{\omega(\varepsilon'_\beta,\theta_\beta)} \frac{e^{\zeta^{1/\beta}} \zeta^{\frac{1+\alpha-\gamma}{\beta}} d\zeta}{(\zeta - y)(\zeta^{\alpha/\beta} - x)}. \qquad (6.4.8)$$

On the other hand, for each $\varepsilon_\beta < |y| < \varepsilon'_\beta$, $|\arg y| < \theta_\beta$, we have by the Cauchy theorem

$$E_{\alpha,\beta;\gamma}(x,y) = \frac{1}{2\pi i \beta} \int\limits_{\omega(\varepsilon'_\beta,\theta_\beta)-\omega(\varepsilon_\beta,\theta_\beta)} \frac{e^{\zeta^{1/\beta}} \zeta^{\frac{1+\alpha-\gamma}{\beta}} d\zeta}{(\zeta - y)(\zeta^{\alpha/\beta} - x)} = \frac{1}{\beta} \frac{e^{y^{1/\beta}} y^{\frac{1+\alpha-\gamma}{\beta}}}{y^{\alpha/\beta} - x}. \qquad (6.4.9)$$

By adding the difference between the right-hand side and the middle integral in the last formula to the right-hand side of (6.4.8) and performing the change of variables in the integral term we arrive at the following representation

$$E_{\alpha,\beta;\gamma}(x,y) = \frac{1}{\beta} \frac{e^{y^{1/\beta}} y^{\frac{1+\alpha-\gamma}{\beta}}}{y^{\alpha/\beta} - x} + \frac{1}{2\pi i \alpha \beta} \int\limits_{\omega(\varepsilon,\theta)} \frac{e^{\xi^{\frac{1}{\alpha\beta}}} \xi^{\frac{\alpha+\beta-\gamma}{\alpha\beta}-1} d\xi}{(\xi^{1/\beta} - x)(\xi^{1/\alpha} - y)} \qquad (6.4.10)$$

valid for all $(x,y) \in \Omega^{(-)}(\varepsilon_\alpha;\theta_\alpha) \times \Omega^{(+)}(\varepsilon_\beta;\theta_\beta)$. The result in the domain $\Omega^{(+)}(\varepsilon_\alpha;\theta_\alpha) \times \Omega^{(-)}(\varepsilon_\beta;\theta_\beta)$ has a symmetric form, namely

$$E_{\alpha,\beta;\gamma}(x,y) = \frac{1}{\alpha} \frac{e^{x^{1/\alpha}} x^{\frac{1+\beta-\gamma}{\alpha}}}{x^{\beta/\alpha} - y} + \frac{1}{2\pi i \alpha \beta} \int\limits_{\omega(\varepsilon,\theta)} \frac{e^{\xi^{\frac{1}{\alpha\beta}}} \xi^{\frac{\alpha+\beta-\gamma}{\alpha\beta}-1} d\xi}{(\xi^{1/\beta} - x)(\xi^{1/\alpha} - y)}. \qquad (6.4.11)$$

Finally, by unifying the argument of the previous cases we obtain an integral representation of the Mittag-Leffler function of two variables in the domain $\Omega^{(+)}(\varepsilon_\alpha; \theta_\alpha) \times \Omega^{(+)}(\varepsilon_\beta; \theta_\beta)$:

$$E_{\alpha,\beta;\gamma}(x,y) = \frac{1}{\alpha} \frac{e^{x^{1/\alpha}} x^{\frac{1+\beta-\gamma}{\alpha}}}{x^{\beta/\alpha} - y} + \frac{1}{\beta} \frac{e^{y^{1/\beta}} y^{\frac{1+\alpha-\gamma}{\beta}}}{y^{\alpha/\beta} - x} \qquad (6.4.12)$$

$$+ \frac{1}{2\pi i \alpha \beta} \int_{\omega(\varepsilon,\theta)} \frac{e^{\xi^{\frac{1}{\alpha\beta}}} \xi^{\frac{\alpha+\beta-\gamma}{\alpha\beta}-1} d\xi}{(\xi^{1/\beta} - x)(\xi^{1/\alpha} - y)}.$$

By assumption, each of the points x and y lies on the right-hand side of the Hankel contours $\omega(\varepsilon_\alpha; \theta_\alpha)$ and $\omega(\varepsilon_\beta; \theta_\beta)$, respectively. Note that the parameters in the definition of the above paths depend on a certain number $\varepsilon > 0$. Now choose ε^1 ($\varepsilon^1 > \varepsilon$) such that one of the coordinates is to the right of the contour (say y) and the other coordinate to its left (i.e. x). This means $(x,y) \in \Omega^{(-)}(\varepsilon_\alpha^1; \theta_\alpha) \times \Omega^{(+)}(\varepsilon_\beta^1; \theta_\beta)$. In this case we have representation (6.4.11) with ε replaced by ε^1 in the integral. This integral can be rewritten as

$$\frac{1}{2\pi i \alpha} \int_{\omega(\varepsilon_\alpha^1, \theta_\alpha)} \frac{e^{\zeta^{1/\alpha}} \zeta^{\frac{1+\beta-\gamma}{\alpha}} d\zeta}{(\zeta - x)(\zeta^{\beta/\alpha} - y)}.$$

For each $\varepsilon_\alpha < |x| < \varepsilon_\alpha^1$, $|\arg x| < \theta_\alpha$, we have by the Cauchy theorem

$$\frac{1}{2\pi i \beta} \int_{\omega(\varepsilon_\alpha^1, \theta_\alpha) - \omega(\varepsilon_\alpha, \theta_\alpha)} \frac{e^{\zeta^{1/\alpha}} \zeta^{\frac{1+\beta-\gamma}{\alpha}} d\zeta}{(\zeta - x)(\zeta^{\beta/\alpha} - x)} = \frac{1}{\alpha} \frac{e^{x^{1/\alpha}} x^{\frac{1+\beta-\gamma}{\alpha}}}{x^{\beta/\alpha} - y}. \qquad (6.4.13)$$

As before, this immediately yields the desired formula (6.4.12), valid in the domain $\Omega^{(+)}(\varepsilon_\alpha; \theta_\alpha) \times \Omega^{(+)}(\varepsilon_\beta; \theta_\beta)$.

6.4.2 Asymptotic Behavior for Large Values of Arguments

Here we describe the asymptotic behavior of the Mittag-Leffler function $E_{\alpha,\beta;\gamma}(x,y)$ of two complex variables x and y for large values of $|x|$ and $|y|$. The result follows from the above integral representations and the standard techniques for the description of the asymptotics of the corresponding integrals presented in Sects. 3.4 and 4.4.

Theorem 6.9 ([Lav18, Thm. 3.1]) *Let* $0 < \alpha, \beta < 2$, $\alpha\beta < 2$ *and the angle* θ *be chosen as*

$$\frac{\pi\alpha\beta}{2} < \theta \leq \min\{\pi, \pi\alpha, \pi\beta, \pi\alpha\beta\}.$$

6.4 Mittag-Leffler Functions of Several Variables

Then, for all pairs of positive integers $\mathbf{p} = (p_\alpha, p_\beta)$, $p_\alpha, p_\beta > 1$, the following asymptotic formulas for the function $E_{\alpha,\beta;\gamma}(x,y)$ hold as $|x| \to \infty$, $|y| \to \infty$.

(i) If $|\arg x| < \frac{\theta}{\beta}$, $|\arg y| < \frac{\theta}{\alpha}$, then

$$E_{\alpha,\beta;\gamma}(x,y) = \frac{1}{\alpha} \frac{e^{x^{1/\alpha}} x^{\frac{1+\beta-\gamma}{\alpha}}}{x^{\beta/\alpha} - y} + \frac{1}{\beta} \frac{e^{y^{1/\beta}} y^{\frac{1+\alpha-\gamma}{\beta}}}{y^{\alpha/\beta} - x} \quad (6.4.14)$$

$$+ \sum_{n=1}^{p_\alpha} \sum_{m=1}^{p_\beta} \frac{x^{-n} y^{-m}}{\Gamma(\gamma - \alpha n - \beta m)} + o\left(|xy|^{-1} |x|^{-p_\alpha}\right) + o\left(|xy|^{-1} |y|^{-p_\beta}\right);$$

(ii) If $|\arg x| < \frac{\theta}{\beta}$, $\frac{\theta}{\alpha} < |\arg y| \leq \pi$, then

$$E_{\alpha,\beta;\gamma}(x,y) = \frac{1}{\alpha} \frac{e^{x^{1/\alpha}} x^{\frac{1+\beta-\gamma}{\alpha}}}{x^{\beta/\alpha} - y} \quad (6.4.15)$$

$$+ \sum_{n=1}^{p_\alpha} \sum_{m=1}^{p_\beta} \frac{x^{-n} y^{-m}}{\Gamma(\gamma - \alpha n - \beta m)} + o\left(|xy|^{-1} |x|^{-p_\alpha}\right) + o\left(|xy|^{-1} |y|^{-p_\beta}\right);$$

(iii) If $\frac{\theta}{\beta} < |\arg x| \leq \pi$, $|\arg y| < \frac{\theta}{\alpha}$, then

$$E_{\alpha,\beta;\gamma}(x,y) = \frac{1}{\beta} \frac{e^{y^{1/\beta}} y^{\frac{1+\alpha-\gamma}{\beta}}}{y^{\alpha/\beta} - x} \quad (6.4.16)$$

$$+ \sum_{n=1}^{p_\alpha} \sum_{m=1}^{p_\beta} \frac{x^{-n} y^{-m}}{\Gamma(\gamma - \alpha n - \beta m)} + o\left(|xy|^{-1} |x|^{-p_\alpha}\right) + o\left(|xy|^{-1} |y|^{-p_\beta}\right);$$

(iv) If $\frac{\theta}{\beta} < |\arg x| \leq \pi$, $\frac{\theta}{\alpha} < |\arg y| \leq \pi$, then

$$E_{\alpha,\beta;\gamma}(x,y) = \sum_{n=1}^{p_\alpha} \sum_{m=1}^{p_\beta} \frac{x^{-n} y^{-m}}{\Gamma(\gamma - \alpha n - \beta m)} + o\left(|xy|^{-1} |x|^{-p_\alpha}\right) + o\left(|xy|^{-1} |y|^{-p_\beta}\right).$$

(6.4.17)

The result of the theorem is obtained by expanding and further estimating the kernel in the integral terms. The complete proof is presented in [Lav18] (cf. [GoLoLu02]).

6.5 Mittag-Leffler Functions with Matrix Arguments

In this section the problem of defining and evaluating Mittag-Leffler functions with matrix arguments is discussed. The idea to generalize a given function of a scalar variable to matrix arguments goes back to the work of Cayley (1858) and nowadays this topic attracts the attention of researchers due to its applications to numerical solutions of fractional multiterm differential equations and fractional partial differential equations, in control theory and so on.

Since Mittag-Leffler functions are entire, it is not a problem to introduce the formal definition

$$E_{\alpha,\beta}(A) = \sum_{j=0}^{\infty} \frac{A^j}{\Gamma(\alpha j + \beta)}, \qquad (6.5.1)$$

which is valid for any $n \times n$ square matrix A. This series representation is suitable for defining the value of a Mittag-Leffler function with matrix argument but not for practical and computational needs since the main issues related to the slow convergence of (6.5.1) and the possible numerical cancellation in summing terms with alternate signs are amplified by the presence of the matrix argument.

The *Jordan canonical form* provides an alternative way to introduce a function with matrix argument which (if suitably modified) can also be exploited for computational purposes.

If the $n \times n$ matrix A has s distinct eigenvalues $\lambda_k, k = 1, \ldots, s$, each with geometric multiplicity m_k (namely the smallest integer such that $(A - \lambda_k I)^{m_k} = 0$), the Jordan canonical form of A is

$$A = Z \begin{pmatrix} J_1 & & & \\ & J_2 & & \\ & & \ddots & \\ & & & J_s \end{pmatrix} Z^{-1}, \quad J_k = \begin{pmatrix} \lambda_k & 1 & & \\ & \lambda_k & \ddots & \\ & & \ddots & 1 \\ & & & \lambda_k \end{pmatrix} \in \mathbb{C}^{m_k \times m_k}.$$

Based on the Jordan canonical form it is possible to define the extension of a Mittag-Leffler function to a matrix argument according to

$$E_{\alpha,\beta}(A) = Z \begin{pmatrix} E_{\alpha,\beta}(J_1) & & & \\ & E_{\alpha,\beta}(J_2) & & \\ & & \ddots & \\ & & & E_{\alpha,\beta}(J_s) \end{pmatrix} Z^{-1}$$

with each Jordan block $J_k, k = 1, \ldots, s$, being mapped to

6.5 Mittag-Leffler Functions with Matrix Arguments

$$E_{\alpha,\beta}(J_k) = \begin{pmatrix} E_{\alpha,\beta}(\lambda_k) & E^{[1]}_{\alpha,\beta}(\lambda_k) & E^{[2]}_{\alpha,\beta}(\lambda_k) & \cdots & E^{[m_k-1]}_{\alpha,\beta}(\lambda_k) \\ & E_{\alpha,\beta}(\lambda_k) & E^{[2]}_{\alpha,\beta}(\lambda_k) & \cdots & E^{[m_k-2]}_{\alpha,\beta}(\lambda_k) \\ & & \ddots & \ddots & \vdots \\ & & & E_{\alpha,\beta}(\lambda_k) & E^{[1]}_{\alpha,\beta}(\lambda_k) \\ & & & & E_{\alpha,\beta}(\lambda_k) \end{pmatrix},$$

where, for compactness, we denote by $E^{[k]}_{\alpha,\beta}(z)$ the k-th term in the Taylor expansion of $E_{\alpha,\beta}(z)$, for which we incidentally note its relationship with the Prabhakar function $E^k_{\alpha,\beta}(z)$ since

$$E^{[k]}_{\alpha,\beta}(z) = \frac{1}{k!}\frac{d^k}{dz^k}E_{\alpha,\beta}(z) = E^{k+1}_{\alpha,\alpha k+\beta}(z).$$

It is clear that the evaluation of a Mittag-Leffler function with matrix arguments reduces to the evaluation of derivatives of the scalar function in the spectrum of the matrix. We refer to Sect. 4.3 for a more detailed discussion about derivatives of Mittag-Leffler functions.

From the practical point of view, however, evaluating the Jordan canonical form is an ill-conditioned problem and, except for matrices with favorable properties, in most cases it cannot be used in practice. A more efficient strategy considers the Schur–Parlett algorithm [DavHig03], which is based on the Schur decomposition of the matrix argument combined with Parlett recurrence to evaluate the matrix function of the triangular factors. In this case extensive computation of derivatives of scalar Mittag-Leffler functions is required. This problem has been extensively discussed in [GarPop18], where a series of applications to fractional calculus are also illustrated.

The numerical experiments presented in [GarPop18] have shown that combining the Schur–Parlett algorithm with techniques for the evaluation of derivatives of Mittag-Leffler functions makes it possible to evaluate the matrix Mittag-Leffler functions with high accuracy, in some cases very close to machine precision. A Matlab code for evaluating Mittag-Leffler functions with matrix arguments is freely available in the file exchange service of the Mathworks website.[1]

6.6 Historical and Bibliographical Notes

In recent decades, starting from the eighties in the last century, we have observed a rapidly increasing interest in the classical Mittag-Leffler function and its generalizations. This interest mainly stems from their use in the explicit solution of certain classes of fractional differential equations (especially those modelling processes of fractional relaxation, oscillation, diffusion and waves). This topic is under develop-

---

[1] www.mathworks.com/matlabcentral/fileexchange/66272-mittag-leffler-function-with-matrix-arguments.

ment and scientists are looking for further applications of the results presented in this chapter and their generalizations.

For $\alpha_1, \alpha_2 \in \mathbb{R}$ ($\alpha_1^2 + \alpha_2^2 \neq 0$) and $\beta_1, \beta_2 \in \mathbb{C}$ the four-parametric Mittag-Leffler function is defined by the series

$$E_{\alpha_1,\beta_1;\alpha_2,\beta_2}(z) \equiv \sum_{k=0}^{\infty} \frac{z^k}{\Gamma(\alpha_1 k + \beta_1)\Gamma(\alpha_2 k + \beta_2)} \quad (z \in \mathbb{C}). \qquad (6.6.1)$$

For positive $\alpha_1 > 0$, $\alpha_2 > 0$ and real $\beta_1, \beta_2 \in \mathbb{R}$ it was introduced by Djrbashian [Dzh60]. When $\alpha_1 = \alpha$, $\beta_1 = \beta$ and $\alpha_2 = 0$, $\beta_2 = 1$, it coincides with the Mittag-Leffler function $E_{\alpha,\beta}(z)$:

$$E_{\alpha,\beta;0,1}(z) = E_{\alpha,\beta}(z) \equiv \sum_{k=0}^{\infty} \frac{z^k}{\Gamma(\alpha k + \beta)} \quad (z \in \mathbb{C}). \qquad (6.6.2)$$

Generalizing the four-parametric Mittag-Leffler function, Al-Bassam and Luchko [Al-BLuc95] introduced the Mittag-Leffler type function

$$E((\alpha, \beta)_n; z) = \sum_{k=0}^{\infty} \frac{z^k}{\prod_{j=1}^{n} \Gamma(\alpha_j k + \beta_j)} \quad (n \in \mathbb{N}) \qquad (6.6.3)$$

with $2n$ real parameters $\alpha_j > 0$; $\beta_j \in \mathbb{R}$ ($j = 1, ..., n$) and with complex $z \in \mathbb{C}$. In [Al-BLuc95] an explicit solution to a Cauchy type problem for a fractional differential equation is given in terms of (6.6.3). The theory of this class of functions was developed in a series of articles by Kiryakova et al. [Kir99], [Kir00], [Kir08], [Kir10a], [Kir10b].

Among the results dealing with multi-index Mittag-Leffler functions we point out those which show their relation to a general class of special functions, namely to Fox's H-function. Representations of the multi-index Mittag-Leffler functions as special cases of the H-function and the generalized Wright function are obtained in [AlKiKa02], [Kir10b]. Relations of such multi-index functions to the Erdelyi–Kober (E-K) operators of fractional integration are discussed. The novel Mittag-Leffler functions are also used as generating functions of a class of so-called Gelfond–Leontiev (G-L) operators of generalized differentiation and integration. Laplace-type integral transforms corresponding to these G-L operators are considered too. The multi-index Mittag-Leffler functions (6.6.3) can be regarded as "fractional index" analogues of the hyper-Bessel functions, and the multiple Borel–Dzrbashian integral transforms (being H-transforms) as "fractional index" analogues of the Obrechkoff transforms (being G-transforms).

In a more precise terminology, these are Gelfond–Leontiev (G-L) operators of generalized differentiation and integration with respect to the entire function, a multi-

6.6 Historical and Bibliographical Notes

index generalization of the Mittag-Leffler function. Fractional multi-order integral equations

$$y(z) - \lambda \mathcal{L} y(z) = f(z) \tag{6.6.4}$$

and initial value problems for the corresponding fractional multi-order differential equations

$$\mathfrak{D} y(z) - \lambda y(z) = f(z) \tag{6.6.5}$$

are considered. From the known solution of the Volterra-type integral equation with m-fold integration, via a Poisson-type integral transformation $\mathcal{P}$ as a transformation (transmutation) operator, the corresponding solution of the integral equation (6.6.4) is found. Then a solution of the fractional multi-order differential equation (6.6.5) comes out, in an explicit form, as a series of integrals involving Fox's H-functions. For each particularly chosen right-hand side function $f(z)$, such a solution can be evaluated as an H-function. Special cases of the equations considered here lead to solutions in terms of the Mittag-Leffler, Bessel, Struve, Lommel and hyper-Bessel functions, and some other known generalized hypergeometric functions.

In [Kir10b] (see also [Kir10a]) a brief description of recent results by Kiryakova et al. on an important class of "Special Functions of Fractional Calculus" is presented. These functions became important in solutions of fractional order (or multi-order) differential and integral equations, control systems and refined mathematical models of various physical, chemical, economical, management and bioengineering phenomena. The notion "Special Functions of Fractional Calculus" essentially means the Wright generalized hypergeometric function ${}_p\Psi_q$, as a special case of the Fox H-function.

A generalization of the Prabhakar type function was given by Shukla and Prajapati [ShuPra07]:

$$E_{\alpha;\beta}^{\gamma,\kappa}(z) = E(\alpha, \beta; \gamma, \kappa; z) = \sum_{k=0}^{\infty} \frac{(\gamma)_{\kappa n} z^n}{\Gamma(\alpha n + \beta)} \quad (n \in \mathbb{N}), \tag{6.6.6}$$

where the generalized Pochhammer symbol is defined by

$$(\gamma)_{\kappa n} = \frac{\Gamma(\gamma + \kappa n)}{\Gamma(\gamma)}.$$

In [SriTom09] the existence of the function (6.6.6) for a wider set of parameters was shown, and its relation to the fractional calculus operators was described (see also [AgMiNi15], [GaShMa15]). Definition (6.6.6) was combined with (6.6.3) in [SaxNis10] (see also [Sax-et-al10]). As a result, the following definition of the generalized multi-index Mittag-Leffler function appears:

$$E^{\gamma,\kappa}_{(\alpha_j,\beta_j)_m}(z) = E_{\gamma,\kappa}((\alpha_j,\beta_j)_{j=1}^m;z) = \sum_{n=0}^{\infty} \frac{(\gamma)_{\kappa n} z^n}{\prod_{j=1}^{m} \Gamma(\alpha_j n + \beta_j)} \quad (m \in \mathbb{N}). \quad (6.6.7)$$

A four-parametric generalization of the Mittag-Leffler function similar to (6.6.6) (a so-called k-Mittag-Leffler function) was proposed in [DorCer12]

$$E^{\gamma}_{k,\alpha,\beta} := \sum_{n=0}^{\infty} \frac{(\gamma)_{n,k}}{\Gamma_k(\alpha n + \beta) n!} z^n, \quad (6.6.8)$$

with Pochhammer k-symbol

$$(z)_{n,k} := z(z+k)\ldots(z+(n-1)k)$$

and k-Gamma function

$$\Gamma_k(z) = \int_0^{\infty} t^{z-1} e^{-\frac{t^k}{k}} dt = k^{1-\frac{z}{k}} \Gamma\left(\frac{z}{k}\right), \quad (z)_{n,k} = \frac{\Gamma_k(z+nk)}{\Gamma_k(z)},$$

appearing in the definition. A generalization of the function (6.6.8) (a $(p-k)$-Mittag-Leffler function) was proposed and studied in [CeLuDo18]:

$$_p E^{\gamma}_{k,\alpha,\beta} := \sum_{n=0}^{\infty} \frac{_p(\gamma)_{n,k}}{_p\Gamma_k(\alpha n + \beta) n!} z^n, \quad (6.6.9)$$

$$_p(z)_{n,k} := \frac{zp}{k}\left(\frac{zp}{k}+p\right)\left(\frac{zp}{k}+2p\right)\ldots\left(\frac{zp}{k}+(n-1)p\right),$$

$$_p\Gamma_k(z) = \int_0^{\infty} t^{z-1} e^{-\frac{t^k}{p}} dt = \frac{p^{\frac{z}{k}}}{k}\Gamma\left(\frac{z}{k}\right), \quad _p(z)_{n,k} = \frac{_p\Gamma_k(z+nk)}{_p\Gamma_k(z)}.$$

Generalizations of the Mittag-Leffler function involving the Beta function and generalized Beta function were defined and studied in [OzaYlm14], [MiPaJo16].

Chudasama and Dave proposed a unification of the Mittag-Leffler and Wright functions in the following form, with conditions on the parameters (Re$(\alpha\delta) \geq 0$, Re$(\beta\delta + \sigma\gamma - \delta/2 - r + 1) > 0$, $\alpha, \sigma \neq 0$, $\mu \in \mathbb{C}$)

$$E^{\sigma,\nu,\gamma}_{\alpha,\beta,\delta}(\mu,r;z) = \sum_{k=0}^{\infty} \frac{(\mu)_{rk}}{\Gamma^{\delta k}(\alpha k + \beta)\Gamma^{\gamma}(\sigma k + \nu)} \frac{z^k}{k!}. \quad (6.6.10)$$

6.6 Historical and Bibliographical Notes

On the basis of the above described results a special H-transform was constructed in [Al-MKiVu02] (see also [KilSai04]). This transform turns out to exhibit many properties similar to the Laplace transform. Moreover, the inverse transform and the operational calculus, which is based on it, are related to the recently introduced multi-index Mittag-Leffler function. Some basic operational properties, complex and real inversion formulas, as well as a convolution theorem, have been derived.

Further generalizations of the Mittag-Leffler functions have been proposed recently in [Pan-K11], [Pan-K12], [Pan-K13].

The $3m$-parametric Mittag-Leffler functions generalizing the Prabhakar three parametric Mittag-Leffler function are introduced by the relation

$$E_{(\alpha_j),(\beta_j)}^{(\gamma_j),m} = \sum_{k=0}^{\infty} \frac{(\gamma_1)_k \ldots (\gamma_m)_k}{\Gamma(\alpha_1 k + \beta_1) \ldots \Gamma(\alpha_m k + \beta_m)} \frac{z^k}{k!}, \qquad (6.6.11)$$

where $(\gamma)_k$ is the Pochhammer symbol, $\alpha_j, \beta_j, \gamma_j \in \mathbb{C}$, $j = 1, \ldots, m$, $\text{Re}\,\alpha_j > 0$. These are entire functions for which the order and the type have been calculated. Representations of the $3m$-parametric Mittag-Leffler functions as generalized Wright functions and Fox H-functions have been obtained. Special cases of novel special functions have been discussed. Composition formulas with Riemann–Liouville fractional integrals and derivatives have been given. Analogues of the Cauchy–Hadamard, Abel, Tauber and Hardy–Littlewood theorems for the three multi-index Mittag-Leffler functions have also been presented.

Pathway type fractional integration of the $3m$-parametric Mittag-Leffler functions is performed in [JaAgKi17].

Two important families of special functions, namely the Bessel functions and Mittag-Leffler functions, and their multi-parametric generalizations are discussed in [Pan-K16]. The following main problems related to the classical and generalized functions of Bessel and Mittag-Leffler type are studied: integral representations and convergence, asymptotic behavior, Tauberian type theorems, completeness of systems of these functions, representations in terms of the generalized Wright function, the Meijer G- and the Fox H-functions with special values of parameters. Special attention is paid to the relations of these functions to the problems of Fractional Calculus.

The extension of the Mittag-Leffler function to a wider set of parameters by using Mellin–Barnes integrals was realized in a series of papers [KilKor05]–[KilKor06c] (see also the paper [Han-et-al09]). The method of extension of different special functions having a representation via a Mellin–Barnes integral has been developed recently.

First of all we have to mention the paper [Han-et-al09]. In this paper the Mittag-Leffler function $E_{\alpha,\beta}(z)$ for negative values of the parameter α is introduced. This definition is based on an analytic continuation of the integral representation

$$E_{\alpha,\beta}(z) = \frac{1}{2\pi i} \int_{\text{Ha}} \frac{t^{\alpha-\beta} e^t}{t^\alpha - z} dt, \quad z \in \mathbb{C}, \qquad (6.6.12)$$

where the path of integration Ha is the Hankel path, a loop starting and ending at $-\infty$, and encircling the disk $|t| \leq |z|^{1/\alpha}$ counterclockwise in the positive sense: $-\pi < \arg t \leq \pi$ on Ha. The integral representation of $E_{\alpha,\beta}(z)$ given in Eq. (6.6.12) can be shown to satisfy the criteria for analytic continuation by noting that for the domain $\alpha > 0$, Eq. (6.6.12) is equivalent to the infinite series representation for the Mittag-Leffler function. This is accomplished by expanding the integrand in Eq. (6.6.12) in powers of z and integrating term-by-term, making use of Hankel's contour integral for the reciprocal of the Gamma function (see, e.g., [NIST]).

To find a defining equation for $E_{-\alpha,\beta}(z)$, the integral representation of the Mittag-Leffler function is rewritten as

$$E_{\alpha,\beta}(z) = \frac{1}{2\pi i} \int_{Ha} \frac{e^t}{t^\beta - zt^{-\alpha+\beta}} dt, \quad z \in \mathbb{C}. \tag{6.6.13}$$

By expanding a part of the integrand in Eq. (6.6.13) into partial fractions

$$\frac{1}{t^\beta - zt^{-\alpha+\beta}} = \frac{1}{t^\beta} - \frac{1}{t^\beta - z^{-1}t^{\alpha+\beta}},$$

substituting it into (6.6.13) we get another representation

$$E_{\alpha,\beta}(z) = \frac{1}{2\pi i} \int_{Ha} \frac{e^t}{t^\beta} dt - \frac{1}{2\pi i} \int_{Ha} \frac{e^t}{t^\beta - z^{-1}t^{\alpha+\beta}} dt, \quad z \in \mathbb{C} \setminus \{0\}. \tag{6.6.14}$$

Thus we arrive at the following definition of the Mittag-Leffler function $E_{\alpha,\beta}(z)$ for negative values of the parameter α:

$$E_{-\alpha,\beta}(z) = \frac{1}{\Gamma(\beta)} - E_{\alpha,\beta}\left(\frac{1}{z}\right). \tag{6.6.15}$$

General properties of $E_{-\alpha,\beta}(z)$ were discussed and many of the common relationships between Mittag-Leffler functions of negative α were compared with their analogous relationships for positive α. A special case of (6.6.15), namely the function $E_{-\alpha}(z)$, has found application in the analysis of the transient kinetics of a two-state model for anomalous diffusion (see [Shu01]). The Mittag-Leffler functions with negative α and the results of this work are likely to become increasingly important as fractional-order differential equations find more applications.

This method of extension was also applied recently in [Kil-et-al12] for the generalized hypergeometric functions. This paper is devoted to the study of a certain function ${}_p\mathcal{F}_q[z] \equiv {}_p\mathcal{F}_q[a_1, \cdots, a_p; b_1, \cdots, b_q; z]$ (with complex $z \neq 0$ and complex parameters a_j ($j = 1, \cdots p$) and b_j ($j = 1, \cdots, q$)), represented by the Mellin–Barnes integral. Such a function is an extension of the classical generalized hypergeometric function ${}_pF_q[a_1, \cdots, a_p; b_1, \cdots, b_q; z]$ defined for all complex $z \in \mathbb{C}$ when $p < q + 1$ and for $|z| < 1$ when $p = q + 1$. Conditions are given for the existence

6.6 Historical and Bibliographical Notes

of $_p\mathcal{F}_q[z]$ and of its representations by the Meijer G-function and the H-function. Such an approach allows us to give meaning to the function $_p\mathcal{F}_q[z]$ for all ranges of parameters when $p < q + 1$, $p = q + 1$ and $p > q + 1$. The series representations and the asymptotic expansions of $_p\mathcal{F}_q[z]$ at infinity and at the origin are established. Special cases have been considered.

In Sect. 6.4 we mainly follow the article [Lav18]. Several other attempts to consider Mittag-Leffler functions and their generalizations as functions of several complex variables have to be mentioned: [SaKaSa11], [GaMaKa13], [Dua18], [Mam18] along with the book [SrGuGo82] devoted to the multivariable analog of the Fox H-function. We also have to mention here the article [YuZha06] in which an $(n+1)$-variable analog of the Mittag-Leffler function is introduced and studied

$$\varepsilon(t, y; \alpha, \beta, \gamma) := t^{\beta-1} E_{\alpha,\beta}(-D|\mathbf{y}|^\gamma t^\alpha),$$

where $t > 0$ is a time variable, $\mathbf{y} = (y_1, y_2, \ldots, y_n) \in \mathbb{R}^n$, α, β, γ are arbitrary real parameters and D is a physical constant. This function is used in the study of the diffusion-wave equation in $(n + 1)$ variables.

Section 6.5 presents in a condensed way the results from [GarPop18], which is devoted to the numerical evaluation of Mittag-Leffler functions with a matrix argument. The corresponding routine implemented in Matlab is also mentioned there. The evaluation of matrix Mittag-Leffler functions is closely related to the evaluation of exponential functions, a problem which has been deeply investigated due its applications to the solution of ordinary differential equations. Several methods have been proposed for matrix exponentials and a comparative discussion is available in the famous review paper by Moler and Van Loan [MolvLoa78] and in its 2003 extension [MolvLoa03]. Unfortunately, not all the methods presented in these two papers can be applied to Mittag-Leffler functions, often due to the absence of the semigroup property, which is exploited in several methods for the computation of the exponential.

The method described in Sect. 6.5 is however based on the work in [DavHig03] which is successive to the two reviews by Moler and Van Loan. Although it exploits some ideas (such as the Schur decomposition) already discussed in these papers, it exploits the more sophisticated Schur–Parlett algorithm, which is presently one of the most powerful methods for matrix computations.

6.7 Exercises

6.7.1 Let I_{0+}^γ be the left-sided Riemann–Liouville fractional integral and $\mathcal{E}_{\alpha_1,\beta_1;\alpha_2,\beta_2}(z)$ be either the four-parametric Mittag-Leffler function or its extension. In the case $\alpha_1 > 0$, $\alpha_2 < 0$ calculate the following compositions

a) $\left(I_{0+}^\gamma t^{\sigma-1} \mathcal{E}_{\alpha_1,\beta_1;\alpha_2,\beta_2}(\lambda t^\omega)\right)(x)$ $(\omega, \lambda > 0,\ 0 < x \leq d < +\infty)$;

b) $\left(I_{0+}^{\gamma} t^{-\sigma} \mathcal{E}_{\alpha_1,\beta_1;\alpha_2,\beta_2}(\lambda t^{-\omega})\right)(x)$ $(\omega, \lambda > 0, \ 0 < x \le d < +\infty)$.

6.7.2 Let D_{0+}^{γ} be the left-sided Riemann–Liouville fractional derivative and $\mathcal{E}_{\alpha_1,\beta_1;\alpha_2,\beta_2}(z)$ be either the four-parametric Mittag-Leffler function or its extension. In the case $\alpha_1 > 0$, $\alpha_2 < 0$ calculate the following compositions

(a) $\left(D_{0+}^{\gamma} t^{\sigma-1} \mathcal{E}_{\alpha_1,\beta_1;\alpha_2,\beta_2}(\lambda t^{\omega})\right)(x)$ $(\omega, \lambda > 0, \ 0 < x \le d < +\infty)$;

(b) $\left(D_{0+}^{\gamma} t^{-\sigma} \mathcal{E}_{\alpha_1,\beta_1;\alpha_2,\beta_2}(\lambda t^{-\omega})\right)(x)$ $(\omega, \lambda > 0, \ 0 < x \le d < +\infty)$.

6.7.3 In the case of positive integer $\alpha_1 = m_1$ and $\alpha_2 = m_2$ represent the four-parametric Mittag-Leffler function $E_{m_1,\beta_1;m_2,\beta_2}(z)$ in term of a generalized hypergeometric function $_pF_q$ with appropriate p, q.

6.7.4 [KirLuc10, p. 601]. Prove that the Laplace transform of a hyper-Bessel type generalized hypergeometric function $_0\Psi_m$ is related to the $2n$-parametric Mittag-Leffler function as follows

$$\left(\mathcal{L}_0 \Psi_m \left[\begin{matrix} -\ -\ - \\ (\beta_1, \alpha_1), \ldots, (\beta_n, \alpha_n) \end{matrix} \right]\right)(s) = \frac{1}{s} E_{(\alpha_1,\beta_1),\ldots,(\alpha_n,\beta_n)}\left(\frac{1}{s}\right).$$

6.7.5 [KirLuc10, p. 603–604]. Let $I_{(\beta_i),n}^{(\gamma_i),(\delta_i)} f(z) = \left[\prod_{i=1}^{n} I_{(\beta_i),n}^{(\gamma_i),(\delta_i)}\right] f(z)$ be the generalized fractional integral of multi-order, where

$$I_{\beta}^{\gamma,\delta} f(z) = \frac{1}{\Gamma(\delta)} \int_0^1 (1-\sigma)^{\delta-1} \sigma^{\gamma} f(z) \sigma^{1/\beta} d\sigma \ (\delta, \beta > 0, \gamma \in \mathbb{R})$$

is the Erdelyi–Kober fractional integral.

Prove the following formulas

$$(\lambda z) \left(I_{(1/\alpha_i),n}^{(\beta_i-1),(\alpha_i)} E_{(\alpha_i),(\beta_i)}\right)(\lambda z) = E_{(\alpha_i),(\beta_i)}(\lambda z) - \frac{1}{\prod_{i=1}^{n} \Gamma(\beta_i)},$$

$$\left(D_{(1/\alpha_i),n}^{(\beta_i-1-\alpha_i),(\alpha_i)} E_{(\alpha_i),(\beta_i)}\right)(\lambda z) = (\lambda z) E_{(\alpha_i),(\beta_i)}(\lambda z) + \frac{1}{\prod_{i=1}^{n} \Gamma(\beta_i - \alpha_i)}.$$

Chapter 7
The Classical Wright Function

7.1 Definition and Basic Properties

This chapter deals with the classical Wright function. Like the functions of Mittag-Leffler type, the functions of Wright type are known to play fundamental roles in various applications of the fractional calculus. This is mainly due to the fact that they are interrelated with the Mittag-Leffler functions through Laplace and Fourier transformations.

The Wright function is defined via a power series

$$\phi(\alpha, \beta; z) = \sum_{k=0}^{\infty} \frac{z^k}{k! \Gamma(\alpha k + \beta)}, \quad \alpha > -1, \beta \in \mathbb{C}. \tag{7.1.1}$$

Originally Wright assumed $\alpha \geq 0$, see [Wri33, Wri35a, Wri35b] and he only considered $-1 < \alpha < 0$ later, in 1940, see [Wri40b]. We note that in the handbook of the Bateman Project [ErdBat-3, Chap. 18], presumably a misprint, the first index α is restricted to be non-negative.

In the section on historical and bibliographical notes we will provide more information on the Wright functions for the readers' convenience.

We have to mention the tight relation between the family of the Mittag-Leffler functions and the Wright function. The latter can be considered as a special case of the Mittag-Leffler function with four parameters (see Chap. 6 of the present book, [RogKor10, ParVin16]):

$$E_{\alpha_1, \beta_1; \alpha_2, \beta_2}(z) = \sum_{k=0}^{\infty} \frac{z^k}{\Gamma(\alpha_1 k + \beta_1) \Gamma(\alpha_2 k + \beta_2)}$$

In recent times another notation has been introduced by Mainardi, who has revisited this function and pointed out its applications in certain partial differential equations of fractional order, see e.g. [Mai94a, Mai96a, Mai96b, Mai10]. In this notation, we write the Wright function as

$$W_{\lambda,\mu}(z) = \sum_{k=0}^{\infty} \frac{z^k}{k!\Gamma(\lambda k + \mu)}, \quad \lambda > -1, \mu \in \mathbb{C}. \tag{7.1.2}$$

As a consequence we have replaced ϕ by W and the parameters $\{\alpha, \beta\}$ by $\{\lambda, \mu\}$.

In this chapter, from now on, we prefer to use the Mainardi notation (7.1.2), which is also present in most recent papers and books. However, in other parts of our treatise the alternative notation (7.1.1) can be found. Following Mainardi, see e.g. [Mai10, Appendix F] we distinguish between the Wright functions of the *first kind* ($\lambda \geq 0$) and the *second kind* ($-1 < \lambda < 0$).

It follows from the Stirling asymptotic formula for the Gamma function,

$$\Gamma(z) = \sqrt{2\pi} z^{z-1/2} e^{-z} \left(1 + O(\frac{1}{z})\right) \quad (|\arg z| < \pi - \varepsilon, \varepsilon > 0, |z| \to \infty),$$

that the Wright function is an entire function.

To characterize the behavior of the Wright function at infinity we first calculate its order ρ and type σ. These characteristics for an entire function represented in the form of a power series $f(z) = \sum_{k=0}^{\infty} c_k z^k$ can be calculated by using standard formulas (see [Lev56] and Appendix B below)

$$\rho = \limsup_{k\to\infty} \frac{k \log k}{\log \frac{1}{|c_k|}}, \quad (\sigma e \rho)^{\frac{1}{\rho}} = \limsup_{k\to\infty} k^{\frac{1}{\rho}} \sqrt[k]{|c_k|}$$

and the Stirling asymptotic formula.

The corresponding result reads: *the Wright function $W_{\lambda,\mu}(z)$ ($\lambda > -1$; $\mu \neq -n(n \in \mathbb{N}_0)$ if $\lambda = 0$) is an entire function of finite order with the order ρ and the type σ given by the formulas*

$$\rho = \frac{1}{1+\lambda}, \quad \sigma = (1+\lambda)|\lambda|^{-\frac{1}{1+\lambda}}. \tag{7.1.3}$$

Remark 7.1 In the case $\lambda = 0$ the Wright function is reduced to the exponential function with the constant factor $\frac{1}{\Gamma(\mu)}$:

$$W_{0,\mu}(z) = \frac{\exp(z)}{\Gamma(\mu)}, \tag{7.1.4}$$

which turns out to vanish identically for $\mu = -n$ ($n \in \mathbb{N}_0$). For all other values of the parameter μ and $\lambda = 0$ formulas (7.1.3) (with $\sigma = \lim_{\lambda \to 0}(1+\lambda)|\lambda|^{-\frac{1}{1+\lambda}} = 1$) are still valid.

7.1 Definition and Basic Properties

The basic characteristic of the growth of an entire function $f = f(z)$ of finite order ρ in different directions is its indicator function $h = h_f(\theta)$ ($|\theta| \leq \pi$) defined by

$$h(\theta) = \limsup_{r \to +\infty} \frac{\log |f(re^{i\theta})|}{r^\rho}.$$

The corresponding result reads ([Luc00, Theorem 2]): *Let $\lambda > -1$; $\mu \neq -n$ ($n \in \mathbb{N}_0$) if $\lambda = 0$. Then the indicator function $h_W(\theta)$ of the Wright function $W_{\lambda,\mu}(z)$ is given by one of the following formulas:*

(a) *in the case $\lambda \geq 0$ by*

$$h_W(\theta) = \sigma \cos(\rho\theta) \quad (|\theta| \leq \pi); \tag{7.1.5}$$

(b) *in the cases*

  (i) $-\frac{1}{3} \leq \lambda < 0$,
  (ii) $\lambda = -\frac{1}{2}$, $\mu = -n$ ($n \in \mathbb{N}_0$),
  (iii) $\lambda = -\frac{1}{2}$, $\mu = \frac{1}{2} - n$ ($n \in \mathbb{N}_0$)

  by

$$h_W(\theta) = \begin{cases} -\sigma \cos(\theta + \pi), & \text{for } -\pi \leq \theta \leq 0, \\ -\sigma \cos(\theta - \pi), & \text{for } 0 \leq \theta \leq \pi, \end{cases} \tag{7.1.6}$$

(c) *in the case $-1 < \lambda < -\frac{1}{3}$ ($\mu \neq -n$ ($n \in \mathbb{N}_0$) and $\mu \neq \frac{1}{2} - n$ ($n \in \mathbb{N}_0$) if $\lambda = -\frac{1}{2}$) by*

$$h_W(\theta) = \begin{cases} -\sigma \cos(\theta + \pi), & \text{for } -\pi \leq \theta \leq \frac{3\pi}{2\rho} - \pi, \\ 0, & \text{for } |\theta| \leq \pi - \frac{3\pi}{2\rho}, \\ -\sigma \cos(\theta - \pi), & \text{for } \pi - \frac{3\pi}{2\rho} \leq \theta \leq \pi. \end{cases} \tag{7.1.7}$$

Here ρ and σ are the order and type of the Wright function, respectively, defined by (7.1.3).

7.2 Relations to Elementary and Special Functions

As was already mentioned, the degenerate case $\lambda = 0$ of the Wright function leads to the following representation via the exponential function (see (7.1.4)):

$$W_{0,\mu}(z) = \frac{\exp(z)}{\Gamma(\mu)},$$

which vanishes identically for $\mu = -n$, $n \in \mathbb{N}_0$.

For $\lambda = 1$ and $\mu = \nu + 1$ the Wright function turns out to be related to the well-known Bessel function of the first kind J_ν (see [NIST, p. 217]) and the modified Bessel function I_ν (see [NIST, p. 249]) by the following identity

$$(z/2)^\nu \, W_{1,\nu+1}(\mp z^2/4) := (z/2)^\nu \sum_{k=0}^\infty (\mp 1)^k \frac{\left(z^2/4\right)^k}{\Gamma(k+\nu+1)} = \begin{cases} J_\nu(z), \\ I_\nu(z). \end{cases} \quad (7.2.1)$$

The functions J_ν and I_ν are known to be solutions to the Bessel differential equation

$$z^2 \frac{d^2 w}{dz^2} + z \frac{dw}{dz} + (z^2 - \nu^2) = 0,$$

and to the modified Bessel differential equation

$$z^2 \frac{d^2 w}{dz^2} + z \frac{dw}{dz} - (z^2 + \nu^2) = 0,$$

respectively.

In view of this property some authors refer to the Wright function as the Wright generalized Bessel function (also misnamed as the Bessel–Maitland function after the second name of Edward Maitland Wright) and introduce the notation

$$J_\nu^{(\lambda)}(z) := (z/2)^\nu \sum_{k=0}^\infty (-1)^k \frac{(z/2)^{2k}}{\Gamma(\lambda k + \nu + 1)}, \quad J_\nu^{(1)}(z) = J_\nu(z). \quad (7.2.2)$$

As a matter of fact, the Wright function appears as the natural generalization of the entire function known as the Bessel–Clifford function, see e.g. [Kir94, p. 336], and referred to by Tricomi, see e.g. [Tri60], [Gat73, pp. 196–197], as the uniform Bessel function

$$T_\nu(z) := z^{-\nu/2} J_\nu(2\sqrt{z}) = \sum_{k=0}^\infty \frac{(-1)^k z^k}{k! \, \Gamma(k+\nu+1)} = W_{1,\nu+1}(-z).$$

Some of the properties which the Wright functions share with the popular Bessel functions were enumerated by Wright himself. Hereafter, we quote two relevant relations from the Bateman Project [ErdBat-3], which can easily be derived from the definitions of the Wright functions (7.1.1)–(7.1.2) or from its integral representation (see the next subsection)

$$\lambda z \, W_{\lambda,\lambda+\mu}(z) = W_{\lambda,\mu-1}(z) + (1-\mu) \, W_{\lambda,\mu}(z), \quad (7.2.3)$$

$$\frac{d}{dz} W_{\lambda,\mu}(z) = W_{\lambda,\lambda+\mu}(z). \quad (7.2.4)$$

7.2 Relations to Elementary and Special Functions

Another natural relation of the Wright function is to the function of hypergeometric type. In the case of the rational first parameter, the Wright function is represented via the following formula

$$W_{\frac{n}{m},\mu}(z) = \sum_{p=0}^{m-1} \frac{z^p}{p!\Gamma(\frac{n}{m}p+\beta)} {}_0F_{m+n-1}(-; \Delta(n, \frac{\beta}{n}+\frac{p}{m}), \Delta^*(m, \frac{p+1}{m}); \frac{z^m}{m^m n^n}), \quad (7.2.5)$$

where

$$_pF_q = {}_pF_q((\mathbf{a})_p; (\mathbf{b})_q; z) = \sum_{k=0}^{\infty} \frac{(a_1)_k, \ldots, (a_p)_k}{(b_1)_k, \ldots, (b_q)_k} \frac{z^k}{k!}$$

is the generalized hypergeometric function (see [NIST, Chap. 16]), and the vectors $\Delta(n, c)$, $\Delta^*(m, d)$ in (7.2.5) are defined by the formulas

$$\Delta(n, c) = \left((c)_n, (c+\frac{1}{n})_n, \ldots, (c+\frac{n-1}{n})_n\right), \quad c = \frac{\beta}{n}+\frac{p}{m},$$

$$\Delta^*(m, d) = \Delta(m, d) \setminus \{1\}, \quad d = \frac{p+1}{m}.$$

This expression means that element $(1)_m$ is removed from the collection $(d)_m$, $(d+\frac{1}{m})_m, \ldots, (d+\frac{m-1}{m})_m$ (note that such an element always exists since $0 \le p \le m-1$). Representation (7.2.5) can be obtained (see [GoLuMa99]) by calculation, via the residue theorem, of the integral in the Mellin–Barnes type integral representation of the Wright function with rational first parameter. In the above formulas $(a)_m$, $m = 0, 1, 2, \ldots$, denotes Pochhammer's symbol $(a)_m := a(a+1)\ldots(a+m-1)$ (for $a > 0$ we have $(a)_m = \frac{\Gamma(a+m)}{\Gamma(a)}$).

The same considerations can be applied in the case of negative rational λ but under the additional condition that the parameter μ is also a rational number. In particular, we obtain the formulas (see, e.g., [GoLuMa99])

$$W_{-\frac{1}{2},-n}(z) = \frac{(-1)^{n+1}}{\pi} \Gamma(\frac{3}{2}+n) {}_1F_1(\frac{3}{2}+n; \frac{3}{2}; -\frac{z^2}{4}), \quad n \in \mathbb{N}_0, \quad (7.2.6)$$

$$W_{-\frac{1}{2},\frac{1}{2}-n}(z) = \frac{(-1)^n}{\pi} \Gamma(\frac{1}{2}+n) {}_1F_1(\frac{1}{2}+n; \frac{1}{2}; -\frac{z^2}{4}), \quad n \in \mathbb{N}_0. \quad (7.2.7)$$

For $n = 0$ we have some interesting formulas related to the Gaussian, as pointed out by Mainardi, see e.g. [Mai10], in the case when z is replaced with $-z$

$$W_{-\frac{1}{2},0}(z) = -\frac{z}{2\sqrt{\pi}} e^{-z^2/4}, \quad (7.2.8)$$

$$W_{-\frac{1}{2},\frac{1}{2}}(z) = -\frac{1}{\sqrt{\pi}} e^{-z^2/4}. \quad (7.2.9)$$

By using the Kummer formula [NIST, p. 325]

$$_1F_1(a; c; z) = e^z {}_1F_1(c - a; c; -z)$$

we can represent (7.2.6) and (7.2.7) in the form

$$W_{-\frac{1}{2},-n}(z) = e^{-z^2/4} z P_n(z^2), \quad n \in \mathbb{N}_0, \tag{7.2.10}$$

$$W_{-\frac{1}{2},\frac{1}{2}-n}(z) = e^{-z^2/4} Q_n(z^2), \quad n \in \mathbb{N}_0, \tag{7.2.11}$$

where $P_n(z), Q_n(z)$ are polynomials of degree n defined as

$$P_n(z) = \frac{(-1)^{n+1}}{\pi} \Gamma(\frac{3}{2} + n) {}_1F_1(-n; \frac{3}{2}; z/4),$$

$$Q_n(z) = \frac{(-1)^n}{\pi} \Gamma(\frac{1}{2} + n) {}_1F_1(-n; \frac{1}{2}; z/4).$$

In a paper by Stankovic [Sta70], the following representation of the Wright function is obtained

$$W_{-\frac{2}{3},0}(-x^{-\frac{2}{3}}) = -\frac{1}{2\sqrt{3\pi}} \exp\left\{-\frac{2}{27x^2}\right\} \mathcal{W}_{-\frac{1}{2},\frac{1}{6}}\left(-\frac{4}{27x}\right) \tag{7.2.12}$$

via the Whittaker function $\mathcal{W}_{\mu,\nu}(x)$, which is defined as a solution to the following differential equation

$$\frac{d^2}{dx^2} \mathcal{W}_{\mu,\nu}(x) + \left(-\frac{1}{4} + \frac{\mu}{x} + \frac{\nu^2}{4x^2}\right) \mathcal{W}_{\mu,\nu}(x) = 0.$$

We note that in this book we have given priority to Mittag-Leffler and Wright functions in that we have adopted for them the notation E and W, respectively, formerly used for the Exponential integral and Whittaker function. As a consequence, for the latter functions we have adopted the calligraphic notation $\mathcal{E}$ and $\mathcal{W}$.

Another formula for the Wright function with rational parameters can be mentioned (see the paper by Mainardi and Tomirotti [MaiTom95])

$$W_{-\frac{1}{3},\frac{2}{3}}(z) = 3^{2/3} \text{Ai}(-z/3^{1/3}), \tag{7.2.13}$$

where Ai(z) is the Airy function (see [NIST, Chap. 9]).

Lastly, the Wright function can be represented in terms of the Fox H-function with special values of parameters

$$W_{\lambda,\mu}(z) = H_{0,2}^{1,0}\left[-z \,\bigg|\, \begin{matrix} \overline{} \\ (0,1), (1-\mu, \lambda) \end{matrix}\right]. \tag{7.2.14}$$

7.3 Integral Representations and Asymptotics

Probably the most important characteristic of a special function is its asymptotics. In the case of an entire function there are deep relations between its asymptotic behavior in the neighborhood of its only singular point – the essential singularity at $z = \infty$ – and other properties of this function, including the distribution of its zeros (see, for example, Evgrafov [Evg78], Levin [Lev56]). It follows from the Stirling asymptotic formula for the Gamma function that the Wright function

$$W_{\lambda,\mu}(z) = \sum_{k=0}^{\infty} \frac{z^k}{k!\Gamma(\lambda k + \mu)}, \ \lambda > -1, \ \mu \in \mathbb{C},$$

is an entire function of z for $\lambda > -1$ and, consequently, as we will see in the later parts of our survey, some elements of the general theory of entire functions can be applied.

The complete picture of the asymptotic behavior of the Wright function for large values of z was given by Wright [Wri35a] in the case $\lambda > 0$ and by Wright [Wri40b] in the case $-1 < \lambda < 0$. In both cases he used the method of steepest descent and the integral representation

$$W_{\lambda,\mu}(z) = \frac{1}{2\pi i} \int_{\text{Ha}} e^{\zeta + z\zeta^{-\lambda}} \zeta^{-\mu} d\zeta, \ \lambda > -1, \ \mu \in \mathbb{C}, \tag{7.3.1}$$

where Ha denotes the Hankel path in the ζ-plane with a cut along the negative real semi-axis $\arg\zeta = \pi$. Formula (7.3.1) is obtained by substituting the Hankel representation for the reciprocal of the Gamma function

$$\frac{1}{\Gamma(s)} = \frac{1}{2\pi i} \int_{\text{Ha}} e^{\zeta} \zeta^{-s} d\zeta, \ s \in \mathbb{C} \tag{7.3.2}$$

for $s = \lambda k + \mu$ into (7.1.2) and changing the order of integration and summation.

Let us first consider the case $\lambda > 0$.

Theorem 7.2 *If $\lambda > 0$, $\arg(-z) = \xi$, $|\xi| \leq \pi$, and*

$$Z_1 = (\lambda|z|)^{1/(\lambda+1)} e^{i(\xi+\pi)/(\lambda+1)}, \ Z_2 = (\lambda|z|)^{1/(\lambda+1)} e^{i(\xi-\pi)/(\lambda+1)},$$

then we have

$$W_{\lambda,\mu}(z) = H(Z_1) + H(Z_2), \tag{7.3.3}$$

where $H(Z)$ is given by

$$H(Z) = Z^{\frac{1}{2}-\mu} e^{\frac{1+\lambda}{\lambda} Z} \left\{ \sum_{m=0}^{M} \frac{(-1)^m a_m}{Z^m} + O\left(\frac{1}{|Z|^{M+1}}\right) \right\}, \ Z \to \infty \tag{7.3.4}$$

and the a_m, $m = 0, 1, \ldots$, are defined as the coefficients of v^{2m} in the expansion of

$$\frac{\Gamma(m+\tfrac{1}{2})}{2\pi}\left(\frac{2}{\lambda+1}\right)^{m+\tfrac{1}{2}}(1-v)^{-\beta}\{g(v)\}^{-2m-1}$$

with

$$g(v) = \left\{1 + \frac{\lambda+2}{3}v + \frac{(\lambda+2)(\lambda+3)}{3\cdot 4}v^2 + \ldots\right\}^{\tfrac{1}{2}}.$$

In particular, if $\mu \in \mathbb{R}$ we get the asymptotic expansion of the Wright function $W_{\lambda,\mu}(-x)$ for $x \to +\infty$ in the form

$$W_{\lambda,\mu}(-x) = x^{p(\tfrac{1}{2}-\beta)}e^{\sigma x^p \cos \pi p}\cos\left(\pi p(\tfrac{1}{2}-\mu) + \sigma x^p \sin \pi p\right)\{c_1 + O(x^{-p})\}, \tag{7.3.5}$$

where $p = \frac{1}{1+\lambda}$, $\sigma = (1+\lambda)\lambda^{-\tfrac{\lambda}{1+\lambda}}$ and the constant c_1 can be exactly evaluated.

If we exclude from the consideration an arbitrary small angle containing the negative real semi-axis, we get a simpler result.

Theorem 7.3 *If $\lambda > 0$, $\arg z = \theta$, $|\theta| \leq \pi - \epsilon$, $\epsilon > 0$, and*

$$Z = (\lambda|z|)^{1/(\lambda+1)}e^{i\theta/(\lambda+1)},$$

then we have

$$\phi(\lambda, \beta; z) = H(Z), \tag{7.3.6}$$

where $H(z)$ is given by (7.3.4).

In the case $\lambda = 0$ the Wright function is reduced to the exponential function with the constant factor $1/\Gamma(\beta)$:

$$\phi(0, \beta; z) = \exp(z)/\Gamma(\beta), \tag{7.3.7}$$

which turns out to vanish identically for $\beta = -n$, $n = 0, 1, \ldots$.

To formulate the results for the case $-1 < \lambda < 0$ we introduce some notation. Let

$$y = -z, \quad -\pi < \arg z \leq \pi, \quad -\pi < \arg y \leq \pi, \tag{7.3.8}$$

and let

$$Y = (1+\lambda)\left((-\lambda)^{-\lambda}y\right)^{1/(1+\lambda)}. \tag{7.3.9}$$

Theorem 7.4 *If $-1 < \lambda < 0$, $|\arg y| \leq \min\{\tfrac{3}{2}\pi(1+\lambda), \pi\} - \varepsilon$, $\varepsilon > 0$, then*

$$W_{\lambda,\mu}(z) = I(Y), \tag{7.3.10}$$

7.3 Integral Representations and Asymptotics

where

$$I(Y) = Y^{\frac{1}{2}-\mu} e^{-Y} \left\{ \sum_{m=0}^{M-1} A_m Y^{-m} + O(Y^{-M}) \right\}, \quad Y \to \infty, \qquad (7.3.11)$$

and the coefficients A_m, $m = 0, 1 \ldots$, are defined by the asymptotic expansion

$$\frac{\Gamma(1-\mu-\lambda t)}{2\pi(-\lambda)^{-\lambda t}(1+\lambda)^{(1+\lambda)(t+1)}\Gamma(t+1)} = \sum_{m=0}^{M-1} \frac{(-1)^m A_m}{\Gamma((1+\lambda)t + \mu + \frac{1}{2} + m)}$$

$$+ O\left(\frac{1}{\Gamma((1+\lambda)t + \mu + \frac{1}{2} + M)}\right),$$

valid for $\arg t$, $\arg(-\lambda t)$, and $\arg(1 - \mu - \lambda t)$ all lying between $-\pi$ and π and t tending to infinity.

If $-1/3 \leq \lambda < 0$, the only region not covered by Theorem 7.4 is the neighborhood of the positive real semi-axis. Here we have the following result.

Theorem 7.5 *If $-1/3 < \lambda < 0$, $|\arg z| \leq \pi(1+\lambda) - \varepsilon$, $\varepsilon > 0$, then*

$$W_{\lambda,\mu}(z) = I(Y_1) + I(Y_2), \qquad (7.3.12)$$

where $I(Y)$ is defined by (7.3.11),

$$Y_1 = (1+\lambda)\left((-\lambda)^{-\lambda} z e^{\pi i}\right)^{1/(1+\lambda)}, \quad Y_2 = (1+\lambda)\left((-\lambda)^{-\lambda} z e^{-\pi i}\right)^{1/(1+\lambda)}, \qquad (7.3.13)$$

hence

$$Y_1 = Y \text{ if } -\pi < \arg z \leq 0, \text{ and } Y_2 = Y \text{ if } 0 < \arg z \leq \pi.$$

As a consequence we get the asymptotic expansion of the Wright function $W_{\lambda,\mu}(x)$ for $x \to +\infty$ in the case $-1/3 < \lambda < 0$, $\mu \in \mathbb{R}$ in the form:

$$W_{\lambda,\mu}(x) = x^{p(\frac{1}{2}-\mu)} e^{-\sigma x^p \cos \pi p} \cos(\pi p(\frac{1}{2} - \mu) - \sigma x^p \sin \pi p) \{c_2 + O(x^{-p})\}, \qquad (7.3.14)$$

where $p = \frac{1}{1+\lambda}$, $\sigma = (1+\lambda)(-\lambda)^{-\frac{\lambda}{1+\lambda}}$ and the constant c_2 can be exactly evaluated. When $-1 < \lambda < -1/3$, there is a region of the plane in which the expansion is algebraic.

Theorem 7.6 *If $-1 < \lambda < -1/3$, $|\arg z| \leq \frac{1}{2}\pi(-1 - 3\lambda) - \varepsilon$, $\varepsilon > 0$, then*

$$W_{\lambda,\mu}(z) = J(z), \quad z \to \infty, \qquad (7.3.15)$$

where

$$J(z) = \sum_{m=0}^{M-1} \frac{z^{(\mu-1-m)/(-\lambda)}}{(-\lambda)\Gamma(m+1)\Gamma(1+(\beta-m-1)/(-\lambda))} + O(z^{\frac{\beta-1-M}{-\lambda}}). \quad (7.3.16)$$

Finally, the asymptotic expansions of the Wright function in the neighborhood of the positive real semi-axis in the case $\lambda = -1/3$ and in the neighborhood of the lines $\arg z = \pm\frac{1}{2}\pi(-1-3\lambda)$ when $-1 < \lambda < -1/3$ are given by the following results by Wright.

Theorem 7.7 *If $\lambda = -1/3$, $|\arg z| \leq \pi(1+\lambda) - \varepsilon$, $\varepsilon > 0$, then*

$$W_{\lambda,\mu}(z) = I(Y_1) + I(Y_2) + J(z), \quad (7.3.17)$$

where $I(Y)$ is defined by (7.3.11), Y_1, Y_2 by (7.3.13), and $J(z)$ by (7.3.16).

Theorem 7.8 *If $-1 < \lambda < -1/3$, $|\arg z \pm \frac{1}{2}\pi(-1-3\lambda)| \leq \pi(1+\lambda) - \varepsilon$, $\varepsilon > 0$, then*

$$W_{\lambda,\mu}(z) = I(Y) + J(z), \quad (7.3.18)$$

where $I(Y)$ is defined by (7.3.11) and $J(z)$ by (7.3.16).

The above given results contain the complete description of the asymptotic behavior of the Wright function for large values of z and for all values of the parameters $\lambda > -1$, $\mu \in \mathbb{C}$. We will use them repeatedly in our further discussions.

7.4 Distribution of Zeros

In the case $\lambda = 0$ the Wright function is an exponential function with a constant factor (equal to zero if $\beta = -n$, $n \in \mathbb{N}_0$) and it has no zeros. For $\lambda = -\frac{1}{2}$, $\mu = -n$ ($n \in \mathbb{N}_0$) and $\lambda = -\frac{1}{2}$, $\mu = \frac{1}{2} - n$ ($n \in \mathbb{N}_0$) the Wright function is reduced to a product of an exponential function and a polynomial of degree $2n + 1$ and $2n$, respectively (see formulas (7.2.10)–(7.2.11)), and it has exactly $2n + 1$ and $2n$ zeros in the complex plane, respectively. For all other values of parameters the Wright function has an infinite number of zeros. Following Luchko [Luc00] we present here the asymptotics of zeros of the Wright function in two cases, namely, $\lambda \geq -\frac{1}{3}$ and $-1 < \lambda < -\frac{1}{3}$ ($\mu \neq -n$, $n \in \mathbb{N}_0$ or $\mu \neq \frac{1}{2} - n$, $n \in \mathbb{N}_0$ if $\lambda = -\frac{1}{2}$).

Theorem 7.9 ([Luc00, Theorem 3]) *Let $(\gamma_k)_{k=1}^\infty$ be the sequence of zeros of the function $W_{\lambda,\mu}(z)$ ($\lambda \geq -\frac{1}{3}$, but $\lambda \neq 0$, $\beta \in \mathbb{R}$), where $\gamma_k \leq \gamma_{k+1}$ and each zero is counted according to its multiplicity. Then:*

(A) In the case $\lambda > 0$ all zeros with large enough k are simple and lie on the negative real semi-axis. The asymptotic formula

$$\gamma_k = -\left(\frac{\pi k + \pi(\rho\mu - \frac{\rho-1}{2})}{\sigma \sin \pi\rho}\right)^{1/\rho} \left(1 + O(k^{-2})\right) \quad (k \to +\infty) \quad (7.4.1)$$

7.4 Distribution of Zeros

holds. *Here and in the next formulas ρ and σ are the order and type of the Wright function given by (7.1.3), respectively.*

(B) In the case $-\frac{1}{3} \leq \lambda < 0$ all zeros with large enough k are simple, lie on the positive real semi-axis, and the asymptotic formula

$$\gamma_k = \left(\frac{\pi k + \pi(\rho\mu - \frac{\rho-1}{2})}{-\sigma \sin \pi\rho}\right)^{1/\rho} \left(1 + O(k^{-2})\right) \quad (k \to +\infty) \tag{7.4.2}$$

holds.

Remark 7.10 Combining the representation

$$J_\nu(z) = \left(\frac{z}{2}\right)^\nu \phi(1, \nu + 1; -\frac{z^2}{4})$$

with the asymptotic formula (7.4.1) we get the known formula (see, for example, [Wat66, p. 506]) for the asymptotic expansion of the large zeros r_k of the Bessel function $J_\nu(z)$:

$$r_k = \pi \left(k + \frac{\nu}{2} - \frac{1}{4}\right) + O(k^{-1}) \quad (k \to +\infty).$$

We consider now the case $-1 < \lambda < -\frac{1}{3}$. It follows from the asymptotic formula (7.4.2) that in this case all zeros of the function $W_{\lambda,\mu}(z)$ with large enough absolute value lie inside the angular domains

$$\Omega_\varepsilon^{(\pm)} = \{z \in \mathbb{C} : |\arg z \mp (\pi - \frac{3\pi}{2\rho})| < \varepsilon\},$$

where ε is any number in the interval $(0, \min\{\pi - \frac{3\pi}{2\rho}, \frac{3\pi}{2\rho}\})$. Consequently, the function $W_{\lambda,\mu}(z)$ has on the real axis only finitely many zeros. Let

$$\left(\gamma_k^{(+)}\right)_{k=1}^\infty \in \Pi^+ = \{z \in \mathbb{C} : \text{Im} > 0\},$$

$$\left(\gamma_k^{(-)}\right)_{k=1}^\infty \in \Pi^- = \{z \in \mathbb{C} : \text{Im} < 0\},$$

be sequences of zeros of the function $W_{\lambda,\mu}(z)$ in the upper and lower half-plane, respectively, such that $|\gamma_k^{(+)}| \leq |\gamma_{k+1}^{(+)}|$ and $|\gamma_k^{(-)}| \leq |\gamma_{k+1}^{(-)}|$ and each zero is counted according to its multiplicity. Then the following result holds.

Theorem 7.11 ([Luc00, Theorem 4]) *In the case $-1 < \lambda < -\frac{1}{3}$ ($\beta \neq -n$ ($n \in \mathbb{N}_0$) and $\beta \neq \frac{1}{2} - n$ ($n \in \mathbb{N}_0$)) all zeros of the function $W_{\lambda,\mu}(z)$ with large enough k are simple and the asymptotic formula*

$$\gamma_k^{(\pm)} = e^{\pm i(\pi - \frac{3\pi}{2\rho})} \left(\frac{2\pi k}{\sigma}\right)^{1/\rho} \left(1 + O(\frac{\log k}{k})\right) \quad (k \to +\infty) \tag{7.4.3}$$

holds.

7.5 Further Analytic Properties

7.5.1 Additional Properties of the Wright Function in the Complex Plane

Summarizing all the results concerning the asymptotic behavior of the Wright function, its indicator function and the distribution of its zeros, we get the following theorem.

Theorem 7.12 ([GoLuMa99, Theorem 2.4.5]) *The Wright function $W_{\lambda,\mu}(z)$, $\lambda > -1$, is an entire function of completely regular growth.*

We recall ([Lev56, Chap. 3]) that an entire function $f(z)$ of finite order ρ is called a function of completely regular growth (CRG-function) if for all θ, $|\theta| \leq \pi$, there exist a set $E_\theta \subset \mathbb{R}^+$ and the limit

$$\lim_{\substack{r \to +\infty \\ r \in E_\theta^*}} \frac{\log |f(re^{i\theta})|}{r^\rho}, \tag{7.5.1}$$

where

$$E_\theta^* = \mathbb{R}^+ \setminus E_\theta, \quad \lim_{r \to +\infty} \frac{\text{meas}(E_\theta \cap (0, r))}{r} = 0.$$

It is known ([Evg78, Chap. 2.6]) that the zeros of a CRG-function $f(z)$ are regularly distributed, that is, they possess a finite angular density

$$\lim_{r \to +\infty} \frac{n(r, \theta)}{r^\rho} = \nu(\theta), \tag{7.5.2}$$

where $n(r, \theta)$ is the number of zeros of $f(z)$ in the sector $0 < \arg z < \theta$, $|z| < r$ and ρ is the order of $f(z)$.

From the other side, the angular density $\nu(\theta)$ is connected with the indicator function $h(\theta)$ of a CRG-function. In particular (see [Evg78, Chap. 2.6]), the jump of $h'(\theta)$ at $\theta = \theta_0$ is equal to $2\pi p \Delta$, where Δ is the density of zeros of $f(z)$ in an arbitrarily small angle containing the ray $\arg z = \theta_0$.

In our case we get from Theorem 7.12, that the derivative of the indicator function of the Wright function has a jump $2\sigma p \sin \pi p$ at $\theta = \pi$ for $\lambda > 0$, the same jump at $\theta = 0$ for $-1/3 < \lambda < 0$, and a jump σp at $\theta = \pm(\pi - \frac{3\pi}{2\rho})$ for $-1 < \lambda < -1/3$

($\beta \neq -n$, $n = 0, 1, \ldots$ and $\mu \neq 1/2 - n$, $n = 0, 1, \ldots$ if $\lambda = -1/2$), where again ρ and σ are the order and type of the Wright function, respectively; if $\lambda = 0$ or $\lambda = -1/2$ and either $\mu = -n$, $n = 0, 1, \ldots$, or $\mu = 1/2 - n$, $n = 0, 1, \ldots$, the derivative of the indicator function has no jumps. As we see, the behavior of the derivative of the indicator function of the Wright function is in accordance with the distribution of its zeros given in Sect. 7.4, as predicted by the general theory of CRG-functions.

7.5.2 Geometric Properties of the Wright Function

The Wright function (more exactly, certain normalizations of this function) has certain geometric properties, including univalency, starlikeness, convexity and close-to-convexity in the open unit disk. For the above properties it is necessary to consider only functions belonging to the class $\mathcal{A}$ of analytic functions f in the unit disc normalized by the conditions $f(0) = 0$, $f'(0) = 1$. Let us describe some results of this type following Prajapat [Pra15]. The necessary definitions are presented in Sect. 4.8.2 (cf. [Goo83, Dur83]).

Observe that the restriction of the Wright function $W_{\lambda,\mu}(z)$ to the unit disc does not belong to the class $\mathcal{A}$. Thus, it is natural to consider the following two kinds of normalization of the Wright function:

$$\mathbb{W}^{(1)}_{\lambda,\mu}(z) = \Gamma(\mu) z W_{\lambda,\mu}(z) = \sum_{n=0}^{\infty} \frac{\Gamma(\mu) z^{n+1}}{n! \Gamma(\lambda n + \mu)}, \quad \lambda > -1, \mu > 0; z \in \mathbb{U};$$

$$\mathbb{V}_{\lambda,\mu}(z) := \frac{\mathbb{W}_{\lambda,\mu}(z)}{z};$$

$$\mathbb{W}^{(2)}_{\lambda,\mu}(z) = \Gamma(\lambda + \mu) \left[W_{\lambda,\mu}(z) - \frac{1}{\Gamma(\mu)} \right] = \sum_{n=0}^{\infty} \frac{\Gamma(\lambda + \mu) z^{n+1}}{(n+1)! \Gamma(\lambda n + \lambda + \mu)},$$

$$\lambda > -1, \lambda + \mu > 0; z \in \mathbb{U}.$$

Note that
$$\mathbb{W}^{(1)}_{1,\nu+1}(-z) = \mathbb{J}_\nu(z) := \Gamma(\nu+1) z^{1-\nu/2} J_\nu(2\sqrt{z}).$$

We present here a few of the main results of Prajapat [Pra15], which read

(a) for $\lambda \geq 1$, $\beta \geq 1 + \sqrt{3}$ the function $\mathbb{W}^{(1)}_{\lambda,\mu}(z)$ is starlike in $\mathbb{U}$;
(b) for $\lambda \geq 1$, $\lambda + \mu \geq 1 + \sqrt{3}$ the function $\mathbb{V}_{\lambda,\beta}(z)$ is convex in $\mathbb{U}$;
(c) for $\lambda \geq 1$, $\lambda + \mu \geq 1 + \sqrt{3}$ the function $\mathbb{W}^{(2)}_{\lambda,\mu}(z)$ is convex in $\mathbb{U}$.

These results can be obtained by estimating the corresponding magnitudes characterizing starlikeness or convexity.

Unfortunately Prajapat's notation for the second normalization of the classical Wright function coincides with that used for the Whittaker function. We have changed the notation for the two Prajapat functions to $\mathbb{W}^{(1)}$ and $\mathbb{W}^{(2)}$.

7.5.3 Auxiliary Functions of the Wright Type

In the earliest analysis of the time-fractional diffusion-wave equation Mainardi [Mai94a], see also [Mai10], introduced *two auxiliary functions of the Wright type* $F_\nu(z)$ and $M_\nu(z)$, where z is a complex variable and ν is a real parameter $0 < \nu < 1$. Both functions turn out to be analytic in the whole complex plane, i.e. they are entire functions. Their respective *integral representations* read,

$$F_\nu(z) := \frac{1}{2\pi i} \int_{\text{Ha}} e^{\zeta - z\zeta^\nu} \, d\zeta, \quad 0 < \nu < 1, \ z \in \mathbb{C}, \tag{7.5.3}$$

$$M_\nu(z) := \frac{1}{2\pi i} \int_{\text{Ha}} e^{\zeta - z\zeta^\nu} \frac{d\zeta}{\zeta^{1-\nu}}, \quad 0 < \nu < 1, \ z \in \mathbb{C}. \tag{7.5.4}$$

Clearly these functions are special cases of the classical Wright function of the second kind, namely:

$$F_\nu(z) = W_{-\nu,0}(-z), \quad M_\nu(z) = W_{-\nu,1-\nu}(-z), \tag{7.5.5}$$

and their connection follows immediately from (7.5.3)–(7.5.4) via an integration by parts in (7.5.4)

$$M_\nu(z) = \nu z M_\nu(z).$$

The *series representations* for these auxiliary functions can be obtained by using the well-known reflection formula for the Gamma function $\Gamma(\zeta)\Gamma(1-\zeta) = \pi/\sin\pi\zeta$:

$$F_\nu(z) := \sum_{n=1}^{\infty} \frac{(-z)^n}{n!\,\Gamma(-\nu n)} = -\frac{1}{\pi} \sum_{n=1}^{\infty} \frac{(-z)^n}{n!} \Gamma(\nu n + 1) \sin(\pi \nu n), \tag{7.5.6}$$

$$M_\nu(z) := \sum_{n=0}^{\infty} \frac{(-z)^n}{n!\,\Gamma[-\nu n + (1-\nu)]} = \frac{1}{\pi} \sum_{n=1}^{\infty} \frac{(-z)^{n-1}}{(n-1)!} \Gamma(\nu n) \sin(\pi \nu n). \tag{7.5.7}$$

Furthermore we note that $F_\nu(0) = 0$, $M_\nu(0) = 1/\Gamma(1-\nu)$ and that the relations (7.5.5) can also be derived from the series representations (7.5.6)–(7.5.7) and the definition of the Wright function.

Explicit expressions of $F_\nu(z)$ and $M_\nu(z)$ in terms of known functions are expected for some particular values of ν. It was shown in [MaiTom95] that for $\nu = 1/q$, with an integer $q \geq 2$, the auxiliary function $M_\nu(z)$ can be expressed as a sum of $(q-1)$

7.5 Further Analytic Properties

simpler entire functions,

$$M_{1/q}(z) = \frac{1}{\pi} \sum_{h=1}^{q-1} c(h,q) \, G(z;h,q) \tag{7.5.8}$$

with

$$c(h,q) = (-1)^{h-1} \, \Gamma(h/q) \, \sin(\pi h/q), \tag{7.5.9}$$

$$G(z;h,q) = \sum_{m=0}^{\infty} (-1)^{m(q+1)} \left(\frac{h}{q}\right)_m \frac{z^{qm+h-1}}{(qm+h-1)!}. \tag{7.5.10}$$

Here $(a)_m$, $m = 0, 1, 2, \ldots$, denotes Pochhammer's symbol

$$(a)_m := \frac{\Gamma(a+m)}{\Gamma(a)} = a(a+1)\ldots(a+m-1).$$

We note that $(-1)^{m(q+1)}$ is equal to $(-1)^m$ for q even and $+1$ for q odd. In the particular cases $q = 2$, $q = 3$ we find, respectively

$$M_{1/2}(z) = \frac{1}{\sqrt{\pi}} \exp\left(-z^2/4\right),$$

$$M_{1/3}(z) = 3^{2/3} \, \text{Ai}\left(z/3^{1/3}\right),$$

where Ai denotes the *Airy function*, see e.g. [NIST].

Furthermore it can be proved that $M_{1/q}(z)$ (for integer ≥ 2) satisfies the differential equation of order $q-1$,

$$\frac{d^{q-1}}{dz^{q-1}} M_{1/q}(z) + \frac{(-1)^q}{q} z \, M_{1/q}(z) = 0, \tag{7.5.11}$$

subjected to the $q-1$ initial conditions at $z = 0$, derived from the series expansion in (7.5.8)–(7.5.10),

$$M_{1/q}^{(h)}(0) = \frac{(-1)^h}{\pi} \Gamma[(h+1)/q] \sin[\pi(h+1)/q], \quad h = 0, 1, \ldots q-2. \tag{7.5.12}$$

We note that, for $q \geq 4$, Eq. (7.5.11) is akin to the *hyper-Airy* differential equation of order $q-1$, see e.g. Bender and Orszag [BenOrs87]. Consequently, the function $M_\nu(z)$ is a generalization of the hyper-Airy function. In the limiting case $\nu = 1$ we get $M_1(z) = \delta(z-1)$, i.e. the M function degenerates into a generalized function of Dirac type.

Since these functions are used to describe time-fractional diffusion processes, it is appropriate to consider the M_ν function for a positive (real) argument, which will be denoted by r.

The asymptotic representation of $M_\nu(r)$, as $r \to \infty$, can be obtained by using the ordinary saddle-point method. Choosing as a variable r/ν rather than r the computation is easier and yields, see [MaiTom95]

$$M_\nu(r/\nu) \sim a(\nu)\, r^{(\nu-1/2)/(1-\nu)} \exp\left[-b(\nu)\, r^{(1/(1-\nu))}\right], \quad r \to +\infty, \qquad (7.5.13)$$

where $a(\nu) = 1/\sqrt{2\pi(1-\nu)} > 0$, and $b(\nu) = (1-\nu)/\nu > 0$.

The above asymptotic representation is consistent with the first term of the asymptotic expansion

$$W_{-\nu,\mu}(z) = Y^{1/2-\mu} e^{-Y} \left(\sum_{m=0}^{M-1} A_m\, Y^{-m} + O(|Y|^{-M})\right), \quad |z| \to \infty,$$

with $Y = (1-\nu)(-\nu^\nu z)^{1/(1-\nu)}$, obtained by Wright for $W_{-\nu,\mu}(-r)$.

In fact, taking $\mu = 1 - \nu$ so $1/2 - \mu = \nu - 1/2$, we obtain

$$M_\nu(r) \sim A_0\, Y^{\nu-1/2} \exp(-Y), \quad r \to \infty, \qquad (7.5.14)$$

where

$$A_0 = \frac{1}{\sqrt{2\pi(1-\nu)^\nu\, \nu^{2\nu-1}}}, \quad Y = (1-\nu)(\nu^\nu r)^{1/(1-\nu)}. \qquad (7.5.15)$$

Because of the above exponential decay, any moment of order $\delta > -1$ for $M_\nu(r)$ is finite. In fact,

$$\int_0^\infty r^\delta\, M_\nu(r)\, dr = \frac{\Gamma(\delta+1)}{\Gamma(\nu\delta+1)}, \quad \delta > -1,\ 0 < \nu < 1. \qquad (7.5.16)$$

In particular, we get the normalization property in $\mathbb{R}^+$, $\int_0^\infty M_\nu(r)\, dr = 1$.

We can now obtain the Laplace transform pairs related to our auxiliary functions. Indeed, following Mainardi [Mai97] and using the integral representations (7.5.3)–(7.5.4), we get

$$\frac{1}{r} F_\nu\left(cr^{-\nu}\right) = \frac{c\nu}{r^{\nu+1}} M_\nu\left(cr^{-\nu}\right) \div \exp(-cs^\nu), \quad 0 < \nu < 1,\ c > 0. \qquad (7.5.17)$$

By applying the formula for differentiation of the image of the Laplace transform to Eq. (7.5.17), we get a Laplace transform pair useful for our further discussions, namely

7.5 Further Analytic Properties

$$\frac{1}{r^\nu} M_\nu(cr^{-\nu}) \div s^{\nu-1} \exp(-cs^\nu), \quad 0 < \nu < 1, \quad c > 0. \tag{7.5.18}$$

As particular cases of Eqs. (7.5.17)–(7.5.18), we recover the well-known pairs, see e.g. [Doe74],

$$\frac{1}{2r^{3/2}} M_{1/2}(1/r^{1/2})) = \frac{1}{2\sqrt{\pi}} r^{-3/2} \exp\left(-1/(4r^2)\right) \div \exp\left(-s^{1/2}\right) \tag{7.5.19}$$

$$\frac{1}{r^{1/2}} M_{1/2}(1/r^{1/2})) = \frac{1}{\sqrt{\pi}} r^{-1/2} \exp\left(-1/(4r^2)\right) \div s^{-1/2} \exp\left(-s^{1/2}\right). \tag{7.5.20}$$

More generally we can use the results of Stankovic in [Sta70] to state the following relevant formula for the Laplace transform related to Wright functions of the second kind:

$$r^{\mu-1} W_{-\nu,\mu}(-cr^{-\nu}) \div s^{-\mu} \exp(-cs^\nu), \quad 0 < \nu < 1, \ \mu > 0, \ c > 0. \tag{7.5.21}$$

We note that this formula can be *formally* derived by anti-transforming the Laplace transform term by term, as was done independently in a paper by Buchen and Mainardi in 1975 without being aware of the Wright functions.

7.6 The Wright Function of a Real Variable

7.6.1 Relation to Fractional Calculus

For the left-sided Riemann–Liouville fractional integration operator and for the right-sided Liouville fractional integration operator these formulas read:

Let $\lambda, \gamma, a \in \mathbb{C}, \mu > 0$.

(a) *If* $\operatorname{Re}\lambda > 0$ *and* $\operatorname{Re}\gamma > 0$, *then*

$$\left(I_{0+}^\lambda [t^{\gamma-1} \phi(\mu, \gamma, at^\mu)]\right)(x) = x^{\gamma+\lambda-1} \phi(\mu, \gamma + \lambda, ax^\mu). \tag{7.6.1}$$

(b) *If* $\operatorname{Re}\gamma > \operatorname{Re}\lambda > 0$, *then*

$$\left(I_-^\lambda [t^{-\gamma} \phi(\mu, \gamma - \lambda, at^{-\mu})]\right)(x) = x^{\lambda-\gamma} \phi(\mu, \gamma, ax^{-\mu}). \tag{7.6.2}$$

For the left-sided Riemann–Liouville fractional differentiation operator and for the right-sided Liouville fractional differentiation operator these formulas read:

Let $\lambda, \gamma, a \in \mathbb{C}, \mu > 0$.

(c) If $\operatorname{Re}\lambda > 0$ and $\operatorname{Re}\gamma > 0$, then

$$\left(D_{0+}^{\lambda}[t^{\gamma-1}\phi(\mu,\gamma,at^{\mu})]\right)(x) = x^{\gamma-\lambda-1}\phi(\mu,\gamma-\lambda,ax^{\mu}). \qquad (7.6.3)$$

(d) If $\operatorname{Re}\gamma > [\operatorname{Re}\lambda] + 1 - \operatorname{Re}\lambda$, then

$$\left(D_{-}^{\lambda}[t^{-\gamma}\phi(\mu,\gamma+\lambda,at^{-\mu})]\right)(x) = x^{-\lambda-\gamma}\phi(\mu,\gamma,ax^{-\mu}). \qquad (7.6.4)$$

Analogous formulas are valid for the Bessel–Maitland function (7.2.2):

Let $\lambda, \nu, a \in \mathbb{C}, \operatorname{Re}\lambda > 0, \operatorname{Re}\nu > -1$, and let $\mu > 0$. Then the following relations hold

$$\left(I_{0+}^{\lambda}[t^{\nu} J_{\nu}^{(\mu)}(at^{\mu})]\right)(x) = x^{\nu+\lambda} J_{\nu+1+\lambda}^{(\mu)}(ax^{\mu}), \qquad (7.6.5)$$

$$\left(I_{-}^{\lambda}[t^{-\lambda-\nu-1} J_{\nu}^{(\mu)}(at^{-\mu})]\right)(x) = x^{-\nu-1} J_{\nu+1+\lambda}^{(\mu)}(ax^{-\mu}), \qquad (7.6.6)$$

$$\left(D_{0+}^{\lambda}[t^{\nu} J_{\nu}^{(\mu)}(at^{\mu})]\right)(x) = x^{\nu-\lambda} J_{\nu+1-\lambda}^{(\mu)}(ax^{\mu}), \qquad (7.6.7)$$

if additionally $\operatorname{Re}\nu > [\operatorname{Re}\lambda]$, then

$$\left(D_{-}^{\lambda}[t^{\lambda-\nu-1} J_{\nu}^{(\mu)}(at^{-\mu})]\right)(x) = x^{-\nu-1} J_{\nu+1-\lambda}^{(\mu)}(ax^{-\mu}). \qquad (7.6.8)$$

Formulas (7.6.1)–(7.6.8) can be obtained by using formulas for fractional integration and differentiation of the power monomial and summation of the corresponding series.

7.6.2 Laplace Transforms of the Mittag-Leffler and the Wright Functions

In the case $\lambda > 0$ the Wright function is an entire function of order less than 1 and consequently its Laplace transform can be obtained by transforming term-by-term its Taylor expansion (7.1.1) at the origin. As a result we get ($0 \leq t < +\infty$, $s \in \mathbb{C}$, $0 < \varepsilon < |s|$, ε arbitrarily small)

$$W_{\lambda,\mu}(\pm t) \div \mathcal{L}\left[W_{\lambda,\mu}(\pm t); s\right] = \int_0^{\infty} e^{-st} W_{\lambda,\mu}(\pm t)\, dt \qquad (7.6.9)$$

$$= \int_0^{\infty} e^{-st} \sum_{k=0}^{\infty} \frac{(\pm t)^k}{k!\Gamma(\lambda k + \mu)}\, dt = \sum_{k=0}^{\infty} \frac{(\pm 1)^k}{k!\Gamma(\lambda k + \mu)} \int_0^{\infty} e^{-st} t^k\, dt$$

$$= \frac{1}{s}\sum_{k=0}^{\infty} \frac{(\pm s^{-1})^k}{\Gamma(\lambda k + \mu)} = \frac{1}{s} E_{\lambda,\mu}(\pm s^{-1}), \quad \lambda > 0, \ \mu \in \mathbb{C},$$

7.6 The Wright Function of a Real Variable

where $\div$ denotes the juxtaposition of a function $\varphi(t)$ with its Laplace transform $\tilde{\varphi}(s)$, and

$$E_{\lambda,\mu}(z) = \sum_{k=0}^{\infty} \frac{z^k}{\Gamma(\lambda k + \mu)}, \quad \lambda > 0, \; \mu \in \mathbb{C}, \tag{7.6.10}$$

is the two-parametric Mittag-Leffler function. In this case the resulting Laplace transform turns out to be analytic, vanishing at infinity and exhibiting an essential singularity at $s = 0$.

For $-1 < \lambda < 0$ the method just applied cannot be used since then the Wright function is an entire function of order greater than one. The existence of the Laplace transform of the function $W_{\lambda,\mu}(-t)$, $t > 0$, follows in this case from Theorem 7.4, which tells us that the function $W_{\lambda,\mu}(z)$ is exponentially small for large z in a sector of the plane containing the negative real semi-axis. To get the transform in this case we use an idea given in Mainardi [Mai97]. Recalling the integral representation (7.3.1) we have $(-1 < \lambda < 0)$

$$W_{\lambda,\mu}(-t) \div \int_0^{\infty} e^{-st} W_{\lambda,\mu}(-t) \, dt = \int_0^{\infty} e^{-st} \frac{1}{2\pi i} \int_{Ha} e^{\zeta - t\zeta^{-\lambda}} \zeta^{-\mu} \, d\zeta \, dt$$

$$= \frac{1}{2\pi i} \int_{Ha} e^{\zeta} \zeta^{-\mu} \int_0^{\infty} e^{-t(s+\zeta^{-\lambda})} \, dt \, d\zeta \tag{7.6.11}$$

$$= \frac{1}{2\pi i} \int_{Ha} \frac{e^{\zeta} \zeta^{-\mu}}{s + \zeta^{-\lambda}} \, d\zeta = E_{-\lambda, \mu - \lambda}(-s),$$

again with the two-parametric Mittag-Leffler function. We use here the integral representation (see Djrbashian [Dzh66] and Gorenflo and Mainardi [GorMai97])

$$E_{\lambda,\mu}(z) = \frac{1}{2\pi i} \int_{Ha} \frac{e^{\zeta} \zeta^{\lambda - \mu}}{\zeta^{\lambda} - z} \, d\zeta, \tag{7.6.12}$$

which is obtained by substituting the Hankel representation (7.3.2) for the reciprocal of the gamma function into the series representation (7.6.10).

7.6.3 Mainardi's Approach to the Wright Functions of the Second Kind

For the sake of convenience we first derive the Laplace transform for the special case of $M_\nu(r)$; the exponential decay as $r \to \infty$ of the *original* function provided by (7.5.13) ensures the existence of the *image* function. From the integral representation (7.5.7) of the M_ν function we obtain

$$M_\nu(r) \doteqdot \frac{1}{2\pi i} \int_0^\infty e^{-sr} \left[\int_{Ha} e^{\sigma - r\sigma^\nu} \frac{d\sigma}{\sigma^{1-\nu}} \right] dr$$

$$= \frac{1}{2\pi i} \int_{Ha} e^\sigma \sigma^{\nu-1} \left[\int_0^\infty e^{-r(s+\sigma^\nu)} dr \right] d\sigma = \frac{1}{2\pi i} \int_{Ha} \frac{e^\sigma \sigma^{\nu-1}}{\sigma^\nu + s} d\sigma.$$

Then, by recalling the integral representation (3.4.12) of the Mittag-Leffler function,

$$E_\lambda(z) = \frac{1}{2\pi i} \int_{Ha} \frac{\zeta^{\lambda-1} e^\zeta}{\zeta^\lambda - z} d\zeta, \quad \lambda > 0,$$

we obtain the Laplace transform pair

$$M_\nu(r) \doteqdot E_\nu(-s), \quad 0 < \nu < 1. \tag{7.6.13}$$

Although transforming the Taylor series of $M_\nu(r)$ term-by-term is not legitimate, this procedure yields a series of negative powers of s that represents the asymptotic expansion of the correct Laplace transform, $E_\nu(-s)$, as $s \to \infty$, in a sector around the positive real axis. Indeed we get

$$\sum_{n=0}^\infty \frac{\int_0^\infty e^{-sr}(-r)^n \, dr}{n! \Gamma(-\nu n + (1-\nu))} = \sum_{n=0}^\infty \frac{(-1)^n}{\Gamma(-\nu n + 1 - \nu)} \frac{1}{s^{n+1}}$$

$$= \sum_{m=1}^\infty \frac{(-1)^{m-1}}{\Gamma(-\nu m + 1)} \frac{1}{s^m} \sim E_\nu(-s), \quad s \to \infty,$$

which is consistent with the asymptotic expansion (3.4.15).

We note that (7.6.13) contains the well-known Laplace transform pair, see e.g. [Doe74],

$$M_{1/2}(r) := \frac{1}{\sqrt{\pi}} \exp\left(-r^2/4\right) \doteqdot E_{1/2}(-s) := \exp\left(s^2\right) \operatorname{erfc}(s),$$

which is valid for all $s \in \mathbb{C}$.

In the limit as $\lambda \to 0^-$ we formally obtain the Laplace transform pair

$$W_{0^-,\mu}(-r) := \frac{e^{-r}}{\Gamma(\mu)} \doteqdot \frac{1}{\Gamma(\mu)} \frac{1}{s+1}.$$

In order to be consistent with (7.6.13) we rewrite

$$W_{0^-,\mu}(-r) \doteqdot E_{0,\mu}(-s) = \frac{1}{\Gamma(\mu)} E_0(-s), \quad |s| < 1. \tag{7.6.14}$$

7.6 The Wright Function of a Real Variable

Therefore, as $\lambda \to 0^{\pm}$, we note a sort of continuity in the formal results [Mai10, (F.23)] and (7.6.14) because

$$\frac{1}{(s+1)} = \begin{cases} (1/s)\, E_0(-1/s), & |s| > 1; \\ E_0(-s), & |s| < 1. \end{cases} \qquad (7.6.15)$$

We now point out the relevant Laplace transform pair related to the *auxiliary* functions of argument $r^{-\nu}$ proved in [Mai94a, Mai96a, Mai96b]:

$$\frac{1}{r} F_\nu(1/r^\nu) = \frac{\nu}{r^{\nu+1}} M_\nu(1/r^\nu) \div e^{-s^\nu}, \quad 0 < \nu < 1, \qquad (7.6.16)$$

$$\frac{1}{\nu} F_\nu(1/r^\nu) = \frac{1}{r^\nu} M_\nu(1/r^\nu) \div \frac{e^{-s^\nu}}{s^{1-\nu}}, \quad 0 < \nu < 1. \qquad (7.6.17)$$

We recall that the Laplace transform pairs in (7.6.16) were formerly considered by [Poll46], who provided a rigorous proof based on a formal result by [Hum45]. Later [Mik59] achieved a similar result based on his theory of operational calculus, and finally, albeit unaware of the previous results, [BucMai75] derived the result in a formal way. We note, however, that none of these authors were informed about the Wright functions. To our actual knowledge, the first author to derive the Laplace transforms pairs (7.6.16)–(7.6.17) in terms of Wright functions of the second kind was Stankovič, see [Sta70].

Hereafter, we will provide two independent proofs of (7.6.16) by carrying out the inversion of $\exp(-s^\nu)$, either by the complex Bromwich integral formula, see [Mai94a, Mai96b], or by the formal series method, see [BucMai75]. We can proceed similarly for the Laplace transform pair (7.6.17).

For the complex integral approach we deform the Bromwich path Br into the Hankel path Ha, which is equivalent to the original path, and we set $\sigma = sr$. Recalling (7.5.3)–(7.5.4), we get

$$\mathcal{L}^{-1}\left[\exp(-s^\nu)\right] = \frac{1}{2\pi i} \int_{Br} e^{sr - s^\nu} ds = \frac{1}{2\pi i\, r} \int_{Ha} e^{\sigma - (\sigma/r)^\nu} d\sigma$$

$$= \frac{1}{r} F_\nu(1/r^\nu) = \frac{\nu}{r^{\nu+1}} M_\nu(1/r^\nu).$$

For the series approach, let us expand the Laplace transform in a series of positive powers of s and formally invert term by term. Then, after recalling (7.5.6)–(7.5.7), we obtain:

$$\mathcal{L}^{-1}\left[\exp(-s^\nu)\right] = \sum_{n=0}^{\infty} \frac{(-1)^n}{n!} \mathcal{L}^{-1}\left[s^{\nu n}\right] = \sum_{n=1}^{\infty} \frac{(-1)^n}{n!} \frac{r^{-\nu n - 1}}{\Gamma(-\nu n)}$$

$$= \frac{1}{r} F_\nu(1/r^\nu) = \frac{\nu}{r^{\nu+1}} M_\nu(1/r^\nu).$$

We note the relevance of Laplace transforms (7.6.13) and (7.6.16) in pointing out the non-negativity of the Wright function $M_\nu(x)$ and the complete monotonicity of the Mittag-Leffler functions $E_\nu(-x)$ for $x > 0$ and $0 < \nu < 1$. In fact, since $\exp(-s^\nu)$ denotes the Laplace transform of a probability density (precisely, the extremal Lévy stable density of index ν, see [Fel71]) the l.h.s. of (7.6.16) must be non-negative, and so also must the l.h.s. of (7.6.13). As a matter of fact the Laplace transform pair (7.6.13) shows that, replacing s by x, the spectral representation of the Mittag-Leffler function $E_\nu(-x)$ can be expressed in terms of the Wright M-function $M_\nu(r)$, that is:

$$E_\nu(-x) = \int_0^\infty e^{-rx} M_\nu(r)\, dr\,, \quad 0 < \nu < 1,\ x \geq 0. \quad (7.6.18)$$

We now recognize that Eq. (7.6.18) is consistent with the equations in [GKMR, Proposition 3.23] derived by [Pol148].

It is instructive to compare the spectral representation of $E_\nu(-x)$ with that of the function $E_\nu(-t^\nu)$. From the properties of the Mittag-Leffler function of a real argument (see, e.g. [GKMR, Sect. 3.7]) we can write

$$E_\nu(-t^\nu) = \int_0^\infty e^{-rt} K_\nu(r)\, dr\,, \quad 0 < \nu < 1,\ t \geq 0, \quad (7.6.19)$$

where the *spectral function* reads

$$K_\nu(r) = \frac{1}{\pi} \frac{r^{\nu-1} \sin(\nu\pi)}{r^{2\nu} + 2 r^\nu \cos(\nu\pi) + 1}. \quad (7.6.20)$$

The relationship between $M_\nu(r)$ and $K_\nu(r)$ is worth exploring. Both functions are non-negative, integrable and normalized in $\mathbb{R}^+$, so they can be adopted in probability theory as density functions. The normalization conditions derive from Eqs. (7.6.18) and (7.6.19) since

$$\int_0^{+\infty} M_\nu(r)\, dr = \int_0^{+\infty} K_\nu(r)\, dr = E_\nu(0) = 1\,.$$

In the following section we will discuss the probability interpretation of the M_ν function with support both in $\mathbb{R}^+$ and in $\mathbb{R}$ whereas for K_ν we note that it has been interpreted as the spectral distribution of relaxation/retardation times in the fractional Zener viscoelastic model, see [Mai10, Chap. 3, Sect. 3.2, Fig. 3.3].

We also note that for certain renewal processes, functions of Mittag-Leffler and Wright type can be adopted as probability distributions of waiting times, as shown in [MaGoVi05], where such distributions are compared. We refer the interested reader to that paper for details.

7.7 Historical and Bibliographical Notes

The *Wright* function, which we denote by $W_{\lambda,\mu}(z)$, $z \in \mathbb{C}$, with the parameters $\lambda > -1$ and $\mu \in \mathbb{C}$, is so named after the British mathematician E. Maitland Wright, who introduced and investigated it between 1935 and 1940 [Wri35a, Wri35b, Wri40a, Wri40b]. We note that originally Wright considered such a function restricted to $\lambda \geq 0$ in his paper [Wri35a] in connection with his investigations in the asymptotic theory of partitions. Only later, in 1940, did he extend his investigation to $-1 < \lambda < 0$ [Wri40b].

Like for the Mittag-Leffler functions, a description of the most important properties of the Wright functions (with relevant references up to the fifties) can be found in the third volume of the Bateman Project [ErdBat-3], in chapter XVIII on *Miscellaneous Functions*. However, probably a misprint, there λ is restricted to be positive.

Relevant investigations on the Wright functions have been carried out by Stanković, [Sta70, GajSta76], in Kiryakova's book [Kir94, p. 336], and, later, by Luchko and Gorenflo (1998) [LucGor98], Gorenflo, Luchko and Mainardi (1999, 2000) [GoLuMa99, GoLuMa00] and Luchko (2000) [Luc00]. Just recently Luchko [?] and Paris [Par19] have considered this function in two chapters of the Handbook of Fractional Calculus and Application (HFCA).

The order and type of the Wright function as well as its indicator function have been calculated by Luchko in [Luc00]. For $-1 < \lambda < 0$ this had already been done by Dzherbashian [=Djrbashian] in [Djr93], but in relation to the Mittag-Leffler function without knowing the existence of the Wright function. At least in the Russian edition of 1966 the name Wright is not mentioned.

The fact that the function $W_{\lambda,\mu}(z)$ is an entire function for all values of the parameters $\lambda > -1$ and $\mu \in \mathbb{C}$ was already known to Wright (see [Wri35a, Wri40b]).

In the paper by Djrbashian and Bagian [DjrBag75] (see also Djrbashian [Djr93]) the order and type of this function as well as an estimate of its indicator function were given for the case $-1 < \lambda < 0$. Wright [Wri40b] also remarked that the zeros of the function [Wri40b] lie near the positive real semi-axis if $-1/3 \leq \lambda < 0$ and near the two lines $\arg z = \pm \frac{1}{2}\pi(3\alpha + 1)$ if $-1 < \alpha < -1/3$. In [GoLuMa99] the investigation of the Wright function from the viewpoint of the theory of entire functions is continued. Exact formulas for the order, the type and the indicator function of the entire function $W_{\lambda,\mu}(z)$ for $\lambda > -1$, $\mu \in \mathbb{C}$ are given. On the basis of these results the problem of the distribution of the zeros of the Wright function is considered. In all cases this function is shown to be a function of completely regular growth.

It can be seen from the formulas (7.1.5)–(7.1.7) that the indicator function $h_W(\theta)$ of the Wright function $W_{\lambda,\mu}(z)$ reduces to the function $\cos \theta$ – the indicator function of the exponential function e^z – if $\lambda \to 0$.

This property does not hold for another generalization of the exponential function – the Mittag-Leffler function. Even though

$$E_1(z) = e^z,$$

the indicator function of the Mittag-Leffler function $E_{\alpha,\beta}(z)$, given for $0 < \alpha < 2$, $\alpha \neq 1$ by ([Evg78, Chap. 2.7])

$$h_{E_{\alpha,\beta}}(\theta) = \begin{cases} \cos\theta/\alpha, & |\theta| \leq \frac{\pi\alpha}{2}, \\ 0, & \frac{\pi\alpha}{2} \leq |\theta| \leq \pi \end{cases}$$

does not coincide with the indicator function of e^z if $\alpha \to 1$.

Geometric properties (such as starlikeness, convexity and close-to-convexity) of different normalizations of the classical Wright function are discussed by Prajapat in [Pra15] (see also [BanPra16]).

In Sect. 7.3 we mainly follow the papers of Gorenflo, Luchko and Mainardi [GoLuMa99, GoLuMa00]. Further asymptotic result for the classical Wright function as well as for the Fox–Wright function $_pW_q$ can be found in papers by Paris [Par10, Par17, ParVin16].

Composition formulas of fractional integrals and derivatives with the classical Wright function are presented in the paper by Kilbas [Kil05].

The Laplace inversion in Eq. (7.5.17) was properly carried out by Pollard (1948) [Pol48] (based on a formal result by Humbert (1945) [Hum45]) and by Mikusiński (1959) [Mik59]. A formal series inversion was carried out by Buchen and Mainardi (1975) [BucMai75], albeit unaware of the previous results.

The Wright function has appeared in papers related to partial differential equations of fractional order. Considering boundary-value problems for the fractional diffusion-wave equation, i.e., the linear partial integro-differential equation obtained from the classical diffusion or wave equation by replacing the first- or second-order time derivative by a fractional derivative of order α with $0 < \alpha \leq 2$, it was found that the corresponding Green functions can be represented in terms of the Wright function of the second kind. A very informative survey of these results can be found in the paper by Mainardi [Mai97] and in the survey paper by Mainardi, Luchko and Pagnini devoted to the 70th anniversary of the late Prof. Gorenflo [MaLuPa01].

Finally, in the papers [BucLuc98, GoLuMa00] the scale-invariant solutions of some partial differential equations of fractional order have been given in terms of the Wright and the generalized Wright functions (see also the present book).

The special cases $\lambda = -\nu$, $\mu = 0$ and $\lambda = -\nu$, $\mu = 1 - \nu$ with $0 < \nu < 1$ and z replaced by $-z$ provide the Wright type functions, $F_\nu(z)$ and $M_\nu(z)$, respectively, that have been so denoted and investigated by Mainardi (1994–1997), see [Mai97]. Since these functions are of special interest for us, we returned to them to present a detailed analysis, see also Gorenflo, Luchko and Mainardi (1999, 2000) [GoLuMa99, GoLuMa00]. We have referred to them as the auxiliary functions of the Wright type.

A long exposition on the properties of classical Wright function and its applications is presented in [Luc19].

7.8 Exercises

7.8.1 ([AnsShe14, Lemma 3.1]) Let $E_{(\alpha_1,\beta_1),(\alpha_2,\beta_2)}(z)$ be the four-parametric Mittag-Leffler function

$$E_{(\alpha_1,\beta_1),(\alpha_2,\beta_2)}(z) = \sum_{k=0}^{\infty} \frac{z^k}{\Gamma(\alpha_1 k + \beta_1)\Gamma(\alpha_2 k + \beta_2)}, \quad \alpha_j > 0, \ \beta_j > 0, \ j = 1, 2.$$

Prove the following identities ($t > 0$):

(a)
$$t^{\beta\gamma-2} E_{\alpha\gamma,\beta\gamma}(-\lambda t^{\alpha\gamma}) = \int_0^{\infty} E_{\alpha,\beta}(-\lambda\tau^{\alpha})\phi(-\gamma, 0; -\tau t^{-\gamma})\mathrm{d}\tau;$$

$$\alpha > 0, \beta > 0, 0 < \gamma < 1;$$

(b)
$$t^{\gamma-2} E_{(\gamma,\alpha),(\gamma,\beta)}(t^{\gamma}) = \int_0^{\infty} \phi(\alpha, \beta; \tau)\phi(-\gamma, 0; -\tau t^{-\gamma})\mathrm{d}\tau;$$

$$\alpha > 0, \beta > 0, 0 < \gamma < 1;$$

(c)
$$\phi(-\alpha_1\alpha_2, 0; -\lambda t^{-\alpha_1\alpha_2}) = \int_0^{\infty} \phi(-\alpha_1, 0; -\lambda\tau^{\alpha_1})\phi(-\alpha_2, 0; -\tau t^{-\alpha_2})\frac{\mathrm{d}\tau}{\tau};$$

$$0 < \alpha_1, \alpha_2 < 1.$$

7.8.2 ([AnsShe14, Lemma 3.4]) Let $\mathcal{H}_n, \mathcal{F}_s, \mathcal{F}_c$ be the Hankel, the Fourier-sine and the Fourier-cosine transforms, respectively:

$$\mathcal{H}_n\{f(t); s\} = \int_0^{\infty} t J_n(st) f(t) \mathrm{d}t,$$

$$\mathcal{F}_s\{f(t); s\} = \sqrt{\frac{2}{\pi}} \int_0^{\infty} \sin(st) f(t) \mathrm{d}t,$$

$$\mathcal{F}_c\{f(t); s\} = \sqrt{\frac{2}{\pi}} \int_0^{\infty} \cos(st) f(t) \mathrm{d}t.$$

Prove the following identities:

(a)
$$\mathcal{L}\left(t^{n/2}\mathcal{H}_n\{u^{-n}\phi(\alpha,\beta;u^2);2\sqrt{t}\};s\right) = \frac{1}{2s^n}E_{\alpha,\beta}(s);$$

(b)
$$\mathcal{L}\left(\mathcal{F}_s\{\phi(\alpha,\beta;u^2);2\sqrt{t}\};s\right) = \sqrt{\frac{1}{s}}E_{\alpha,\beta}(s);$$

(c)
$$\mathcal{L}\left(t^{-1/2}\mathcal{F}_c\{\phi(\alpha,\beta;u^2);2\sqrt{t}\};s\right) = \sqrt{s}E_{\alpha,\beta}(s).$$

7.8.3 ([Meh17, Lemma 1]) Prove that for any $\beta > \alpha > 0$ the following integral representation of the classical Wright function holds:

$$\phi(\alpha,\beta;t) = c_{\alpha,\beta}\int_0^1 (1-\tau^{1/\alpha})^{\beta-\alpha-1}\phi(\alpha,\alpha;t\tau)d\tau,\ t \in \mathbb{R}$$

with $c_{\alpha,\beta} = \frac{1}{\alpha\Gamma(\beta-\alpha)}$.

7.8.4 ([Pra15, Theorem 2.9]) Let $\alpha \geq 1$ and $\beta > \beta^*$, where β^* is a positive root of the equation $\beta^2 - \sqrt{5}\beta - \sqrt{5} = 0$. Show that the normalized Wright function $\mathbb{W}_{\alpha,\beta}(z) = \Gamma(\beta)z\phi(\alpha,\beta;z)$ is starlike in $\mathbb{U}$.

Chapter 8
Applications to Fractional Order Equations

In this chapter we consider a number of integral equations and differential equations (mainly of fractional order). In representations of their solution, the Mittag-Leffler function, its generalizations and some closely related functions are used.

8.1 Fractional Order Integral Equations

8.1.1 The Abel Integral Equation

Let us consider the Abel integral equation of the first kind in the classical setting (i.e. for $0 < \alpha < 1$)

$$\frac{1}{\Gamma(\alpha)} \int_0^t \frac{u(\tau)}{(t-\tau)^{1-\alpha}} \, d\tau = f(t), \ 0 < \alpha < 1, \tag{8.1.1}$$

where $f(t)$ is a given function. We easily recognize that this equation can be expressed in terms of a fractional integral, i.e.

$$\left(I_{0+}^{\alpha} u\right)(t) = f(t), \ 0 < \alpha < 1. \tag{8.1.2}$$

Consequently it is solved in terms of a fractional derivative:

$$u(t) = \left(D_{0+}^{\alpha} f\right)(t). \tag{8.1.3}$$

To this end we need to recall the definition of a fractional integral and derivative and the property $D_{0+}^{\alpha} I_{0+}^{\alpha} = \mathbb{I}$. Certainly, the solution (8.1.3) exists if the right-hand side satisfies certain conditions (see the discussion below).

A formal solution can be obtained for the Eq. (8.1.1) (or, what is equivalent, to the Eq. (8.1.2)) for any positive value of the parameter α.

Let us consider Eq. (8.1.2) with an arbitrary positive parameter α. Let $m = -[-\alpha]$, i.e. $m - 1 < \alpha \le m$, $m \in \mathbb{N}$. We apply the operator $I_{a+}^{m-\alpha}$ to both sides of Eq. (8.1.2)

$$I_{a+}^{m-\alpha} I_{0+}^{\alpha} u = I_{0+}^{m-\alpha} f. \tag{8.1.4}$$

Then using the semigroup property of the fractional integral operators we get

$$(I^m u)(t) = \left(I_{a+}^{m-\alpha} f\right)(t), \tag{8.1.5}$$

where I^m is the m-times repeated integral. Since m-times differentiation D^m is the left-inverse operator to I^m we obtain finally

$$u(t) = \left(D^m I_{0+}^{m-\alpha} f\right)(t) =: \left(D_{0+}^{\alpha} f\right)(t), \tag{8.1.6}$$

with

$$D_{0+}^{\alpha} = D^m I_{0+}^{m-\alpha}.$$

Thus, the solution to Eq. (8.1.1) has the form (8.1.6), if it exists. The problem is that Eqs. (8.1.2) and (8.1.4) are not equivalent since the operator D_{0+}^{α} is only left-inverse to I_{0+}^{α}, but not right-inverse. Solvability conditions for the Abel integral equation of the first kind are given in [SaKiMa93, pp. 31, 39]:

Let α be a positive non-integer (i.e. $m - 1 < \alpha < m$), then the Abel integral equation of the first kind has a solution in $L_1(a, b)$ iff $f_{m-\alpha}(x) \in AC^m([a, b])$, where

$$f_{m-\alpha}(x) = \frac{1}{\Gamma(m - \{\alpha\})} \int_a^x \frac{f(t)}{(x - t)^{\{\alpha\}}} dt \tag{8.1.7}$$

and

$$f_{m-\alpha}^{(k)}(a) = 0, \ k = 0, 1, \ldots, m - 1. \tag{8.1.8}$$

Condition (8.1.7) means that the function $f_{m-\alpha}$ is $(m - 1)$-times differentiable and the $(m - 1)$th derivative $f_{m-\alpha}^{(m-1)}$ is absolutely continuous on the interval $[a, b]$.

An alternative approach is to use the Laplace transform for the solution of the Abel integral equation (8.1.1). Note that the operator in the left-hand side of (8.1.2) can be written in the form of the Laplace convolution. Let us for simplicity consider only the case $m = 1$.

The relation $I^{\alpha} u(t) = \Phi_{\alpha}(t) * u(t) \div \tilde{u}(s)/s^{\alpha}$ holds with $\Phi_{\alpha}(t) = \frac{1}{\Gamma(\alpha)} \frac{1}{t^{1-\alpha}}$. This gives

$$\frac{\tilde{u}(s)}{s^{\alpha}} = \tilde{f}(s) \implies \tilde{u}(s) = s^{\alpha} \tilde{f}(s). \tag{8.1.9}$$

8.1 Fractional Order Integral Equations

Now one can choose two different ways to get the inverse Laplace transform from (8.1.9), according to the standard rules.

(a) Writing (8.1.9) as

$$\widetilde{u}(s) = s \left[\frac{\widetilde{f}(s)}{s^{1-\alpha}} \right], \qquad (8.1.10)$$

we obtain

$$u(t) = \frac{1}{\Gamma(1-\alpha)} \frac{d}{dt} \int_0^t \frac{f(\tau)}{(t-\tau)^\alpha} d\tau. \qquad (8.1.11)$$

(b) On the other hand, writing (8.1.9) as

$$\widetilde{u}(s) = \frac{1}{s^{1-\alpha}} [s \, \widetilde{f}(s) - f(0^+)] + \frac{f(0^+)}{s^{1-\alpha}}, \qquad (8.1.12)$$

we obtain

$$u(t) = \frac{1}{\Gamma(1-\alpha)} \int_0^t \frac{f'(\tau)}{(t-\tau)^\alpha} d\tau + f(0^+) \frac{t^{-\alpha}}{\Gamma(1-a)}. \qquad (8.1.13)$$

Thus, the solutions (8.1.11) and (8.1.13) are expressed in terms of the Riemann–Liouville and Caputo fractional derivatives D_{0+}^α and $^CD_{0+}^\alpha$ respectively, according to properties of fractional derivatives with $m = 1$.

Method (b) requires $f(t)$ to be differentiable with $\mathcal{L}$-transformable derivative; consequently $0 \leq |f(0+)| < \infty$. Then it turns out from (8.1.13) that $u(0^+)$ can be infinite if $f(0+) \neq 0$, being $u(t) = O(t^{-\alpha})$, as $t \to 0^+$. Method (a) requires weaker conditions in that the integral on the right-hand side of (8.1.11) must vanish as $t \to 0+$; consequently $f(0+)$ could be infinite but with $f(t) = O(t^{-\nu})$, $0 < \nu < 1 - \alpha$ as $t \to 0+$. Then it turns out from (8.1.11) that $u(0+)$ can be infinite if $f(0+)$ is infinite, being $u(t) = O(t^{-(\alpha+\nu)})$, as $t \to 0+$.

Finally, let us remark that the case of Eq. (8.1.1) with $0 < \alpha < 1$ replaced by $\alpha > 0$ can be treated analogously. If $m - 1 < \alpha \leq m$ with $m \in \mathbb{N}$, then again we have (8.1.2), now with $D_{0+}^\alpha f(t)$ given by the formula which can also be obtained by the Laplace transform method.

The Abel Integral Equation of the Second Kind

Let us now consider the Abel equation of the second kind

$$u(t) + \frac{\lambda}{\Gamma(\alpha)} \int_0^t \frac{u(\tau)}{(t-\tau)^{1-\alpha}} d\tau = f(t), \quad \alpha > 0, \ \lambda \in \mathbb{C}. \qquad (8.1.14)$$

In terms of the fractional integral operator this equation reads

$$\left(1 + \lambda I_{0+}^\alpha\right) u(t) = f(t), \qquad (8.1.15)$$

and consequently it can be formally solved as follows:

$$u(t) - \left(1 + \lambda I_{0+}^{\alpha}\right)^{-1} f(t) = \left(1 + \sum_{n=1}^{\infty} (-\lambda)^n I_{0+}^{\alpha n}\right) f(t). \qquad (8.1.16)$$

The formula is obtained by using the standard technique of successive approximation. Covergence of the Neumann series simply follows for any function $f \in \mathcal{C}[0, a]$ (cf., e.g., [GorVes91, p. 130]). Note that

$$I_{0+}^{\alpha n} f(t) = \Phi_{\alpha n}(t) * f(t) = \frac{t_{+}^{\alpha n - 1}}{\Gamma(\alpha n)} * f(t).$$

Thus the formal solution reads

$$u(t) = f(t) + \left(\sum_{n=1}^{\infty} (-\lambda)^n \frac{t_{+}^{\alpha n - 1}}{\Gamma(\alpha n)}\right) * f(t). \qquad (8.1.17)$$

Recalling

$$e_{\alpha}(t; \lambda) := E_{\alpha}(-\lambda t^{\alpha}) = \sum_{n=0}^{\infty} \frac{(-\lambda t^{\alpha})^n}{\Gamma(\alpha n + 1)}, \quad t > 0, \; \alpha > 0, \; \lambda \in \mathbb{C}, \qquad (8.1.18)$$

$$\sum_{n=1}^{\infty} (-\lambda)^n \frac{t_{+}^{\alpha n - 1}}{\Gamma(\alpha n)} = \frac{d}{dt} E_{\alpha}(-\lambda t^{\alpha}) = e'_{\alpha}(t; \lambda), \; t > 0. \qquad (8.1.19)$$

Finally, the solution reads

$$u(t) = f(t) + e'_{\alpha}(t; \lambda) * f(t). \qquad (8.1.20)$$

Of course the above formal proof can be made rigorous. Simply observe that because of the rapid growth of the Gamma function the infinite series in (8.1.17) and (8.1.19) are uniformly convergent in every bounded interval of the variable t so that term-wise integrations and differentiations are allowed.[1]

Alternatively one can use the Laplace transform, which will allow us to obtain the solution in different forms, including the result (8.1.20).

Applying the Laplace transform to (8.1.14) we obtain

$$\left[1 + \frac{\lambda}{s^{\alpha}}\right] \tilde{u}(s) = \tilde{f}(s) \implies \tilde{u}(s) = \frac{s^{\alpha}}{s^{\alpha} + \lambda} \tilde{f}(s) = s \frac{s^{\alpha - 1}}{s^{\alpha} + \lambda} \tilde{f}(s). \qquad (8.1.21)$$

---

[1] In other words, we use that the Mittag-Leffler function is an entire function for all $\alpha > 0$.

8.1 Fractional Order Integral Equations

Now, let us proceed to find the inverse Laplace transform of (8.1.21) using the Laplace transform pair (see Appendix C)

$$e_\alpha(t;\lambda) := E_\alpha(-\lambda t^\alpha) \div \frac{s^{\alpha-1}}{s^\alpha+\lambda}. \qquad (8.1.22)$$

We have

$$e'_\alpha(t;\lambda) \div s\frac{s^{\alpha-1}}{s^\alpha+\lambda}. \qquad (8.1.23)$$

Therefore the inverse Laplace transform of the right-hand side of (8.1.21) is the Laplace convolution of $e'_\alpha(t;\lambda)$ and f, i.e. the solution to the Abel integral equation of the second kind has the form

$$u(t) = f(t) + e'_\alpha(t;\lambda) * f(t) = f(t) + \int_0^t f(t-\tau)e'_\alpha(\tau;\lambda)d\tau. \qquad (8.1.24)$$

Formally, one can apply integration by parts and rewrite (8.1.24):

$$u(t) = f(t) + \int_0^t f'(t-\tau)e_\alpha(\tau;\lambda)d\tau + f(0+)e_\alpha(t;\lambda). \qquad (8.1.25)$$

Note that this formula is more restrictive with respect to conditions on the given function f (see, e.g. [GorVes91]).

If the function f is continuous on the interval $[0, a]$ then formula (8.1.20) (or, what is the same, (8.1.24)) gives the unique continuous solution to the Abel integral equation of the second kind (8.1.14) for any real λ. The existence of the integral in (8.1.20) follows for any absolutely integrable function f.

8.1.2 Other Integral Equations Whose Solutions Are Represented Via Generalized Mittag-Leffler Functions

In many physical applications the Abel integral equations of the first kind arise in more general forms:

$$\frac{1}{\Gamma(\alpha)}\int_a^x \frac{u(t)dt}{(h(x)-h(t))^{1-\alpha}} = f(x), \quad a < x < b, \qquad (8.1.26)$$

$$\frac{1}{\Gamma(\alpha)} \int_x^b \frac{u(t)dt}{(h(t) - h(x))^{1-\alpha}} = f(x), \quad a < x < b, \tag{8.1.27}$$

where h is a strictly increasing differentiable function in (a, b). An especially important case is $h(x) = x^2$, $a = 0$, $\alpha = 1/2$ (see, e.g., [GorVes91, Chap. 3]).

Following [GorVes91], we treat Eq. (8.1.26) using the substitutions $\xi = h(t)$, $a' = h(a)$, $b' = h(b)$ and introducing a new unknown function v and new right-hand side g:

$$v(\tau) = \frac{u\left(h^{-1}(\tau)\right)}{h'\left(h^{-1}(\tau)\right)}, \quad g(\xi) = f\left(h^{-1}(\tau)\right), \quad a' < \xi < b'.$$

Using this notation, Eq. (8.1.26) becomes the Abel integral equation of the first kind. Thus by inverse substitutions in (8.1.11) we obtain the formal solution to (8.1.26) in the following form

$$u(x) = \frac{1}{\Gamma(1-\alpha)} \frac{d}{dt} \int_a^x \frac{h'(t) f(t)}{(h(x) - h(t))^\alpha} dt, \quad a < x < b. \tag{8.1.28}$$

In a similar way one can obtain the formal solution to (8.1.27)

$$u(x) = -\frac{1}{\Gamma(1-\alpha)} \frac{d}{dt} \int_x^b \frac{h'(t) f(t)}{(h(t) - h(x))^\alpha} dt, \quad a < x < b. \tag{8.1.29}$$

Another method for solving (8.1.26) and (8.1.27) is given in [Sri63]. Thus, in the case of Eq. (8.1.26) we multiply both sides of the equation by

$$\frac{1}{\Gamma(1-\alpha)} \frac{h'(x)}{(h(y) - h(x))^\alpha}, \quad x \leq y \leq b,$$

and integrate with respect to x over the interval (a, y). Then after changing the order of integration we obtain the following relation

$$\frac{1}{\Gamma(\alpha)\Gamma(1-\alpha)} \int_a^y u(t) dt \int_t^y \frac{h'(x) dx}{(h(y) - h(x))^\alpha (h(x) - h(t))^{1-\alpha}} = \int_a^y \frac{h'(x) f(x) dx}{(h(y) - h(x))^\alpha}.$$

We observe that after making a suitable substitution one can show that the inner integral on the left-hand side is equal to $\Gamma(\alpha)$. Hence formula (8.1.28) follows.

Solvability conditions of Eqs. (8.1.26) and (8.1.27) in different functional spaces can be derived by using arguments similar to that for the classical Abel integral equation of the first kind (see, e.g., [SaKiMa93, p. 31]).

8.1 Fractional Order Integral Equations

A number of integral equations which reduce to the Abel integral equation of the first or the second kind are presented in the Handbook of Integral Equations [PolMan08]. Among them we single out the following:

1.
$$\int_0^x \left(1 + b\sqrt{x-t}\right) y(t) dt = f(x), \quad b = \text{const}, \tag{8.1.30}$$

which reduces to the Abel integral equation of the second kind by differentiating with respect to x;

2.
$$\int_0^x \left(b + \frac{1}{\sqrt{x-t}}\right) y(t) dt = f(x), \quad b = \text{const}, \tag{8.1.31}$$

which can be solved as a combination of the Abel integral equations of the first and the second kind.

8.2 Fractional Ordinary Differential Equations

8.2.1 Fractional Ordinary Differential Equations with Constant Coefficients

Here we focus on results concerning ordinary fractional differential equations (FDEs) which are solved in an explicit form via the Mittag-Leffler function, its generalizations, and related special functions. We choose the most simple equations in order to reach the main aim of this subsection, namely, to demonstrate the role of the Mittag-Leffler function in the solution of ordinary FDEs.

For a wider exposition presenting a classification of equations and initial boundary value problems for FDEs, and different methods of solution, we refer to the monograph [KiSrTr06] (see also [Bal-et-al17, Die10, Die19, Pod99] and references therein).

A Cauchy Type Problem for One-Term Equations

Let us start with a simple linear ordinary differential equation with one fractional derivative of Riemann–Liouville type

$$\left(D_{a+}^{\alpha} y\right)(x) - \lambda y(x) = f(x) \quad (a < x \leq b; \; \alpha > 0; \; \lambda \in \mathbb{R}). \tag{8.2.1}$$

Standard initial conditions for such an equation are the so-called *Cauchy type initial conditions*

$$\left(D_{a+}^{\alpha-k}y\right)(a+) = b_k \ (b_k \in \mathbb{R}, \ k=1,\ldots,n = -[-\alpha]), \qquad (8.2.2)$$

where $[\cdot]$ indicates the integral part of a number.

If we suppose that the right-hand side in (8.2.1) is Hölder-continuous, i.e. $f \in \mathcal{C}_\gamma[a,b]$, $0 \le \gamma < 1, \gamma < \alpha$, then (see [KiSrTr06, p. 172]) the Cauchy type problem (8.2.1)–(8.2.2) is equivalent in the space $\mathcal{C}_{n-\alpha}[a,b]$ to the Volterra integral equation

$$y(x) = \sum_{j=1}^{n} \frac{b_j(x-a)^{\alpha-j}}{\Gamma(\alpha-j+1)} + \frac{\lambda}{\Gamma(\alpha)} \int_a^x \frac{y(t)}{(x-t)^{1-\alpha}} dt + \frac{1}{\Gamma(\alpha)} \int_a^x \frac{f(t)}{(x-t)^{1-\alpha}} dt. \qquad (8.2.3)$$

Here and in what follows $\mathcal{C}_0[a,b]$ means simply the set of continuous functions on $[a,b]$, i.e. $\mathcal{C}[a,b]$.

One can solve Eq. (8.2.3) by the method of successive approximation (for the justification of this method in the present case, see, e.g. [KiSrTr06, pp. 172, 222]). If we set

$$y_0(x) = \sum_{j=1}^{n} \frac{b_j}{\Gamma(\alpha-j+1)}(x-a)^{\alpha-j}, \qquad (8.2.4)$$

then we get the recurrent relation

$$y_m(x) = y_0(x) + \frac{\lambda}{\Gamma(\alpha)} \int_a^x \frac{y_{m-1}(t)}{(x-t)^{1-\alpha}} dt + \frac{1}{\Gamma(\alpha)} \int_a^x \frac{f(t)}{(x-t)^{1-\alpha}} dt. \qquad (8.2.5)$$

This can be rewritten in terms of the Riemann–Liouville fractional integrals

$$y_m(x) = y_0(x) + \lambda \left(I_{a+}^{\alpha} y_{m-1}\right)(x) + \left(I_{a+}^{\alpha} f\right)(x). \qquad (8.2.6)$$

Performing successive substitutions one can obtain from (8.2.6) the following formula for the mth approximation y_m to the solution of (8.2.3)

$$y_m(x) = \sum_{j=1}^{n} b_j \sum_{k=1}^{m+1} \frac{\lambda^{k-1}(x-a)^{\alpha k-j}}{\Gamma(\alpha k-j+1)} + \int_a^x \left[\sum_{k=1}^{m} \frac{\lambda^{k-1}}{\Gamma(\alpha k)}(x-t)^{\alpha k-1}\right] f(t) dt. \qquad (8.2.7)$$

Taking the limit as $m \to \infty$ we get the solution of the integral equation (8.2.3) (and thus of the Cauchy type problem (8.2.1)–(8.2.2))

$$y(x) = \sum_{j=1}^{n} b_j \sum_{k=1}^{\infty} \frac{\lambda^{k-1}(x-a)^{\alpha k-j}}{\Gamma(\alpha k-j+1)} + \int_a^x \left[\sum_{k=1}^{\infty} \frac{\lambda^{k-1}}{\Gamma(\alpha k)}(x-t)^{\alpha k-1}\right] f(t) dt, \qquad (8.2.8)$$

or

8.2 Fractional Ordinary Differential Equations

$$y(x) = \sum_{j=1}^{n} b_j \sum_{k=0}^{\infty} \frac{\lambda^k (x-a)^{\alpha k + \alpha - j}}{\Gamma(\alpha k + \alpha - j + 1)} \quad (8.2.9)$$

$$+ \int_a^x \left[\sum_{k=0}^{\infty} \frac{\lambda^k}{\Gamma(\alpha k + \alpha)} (x-t)^{\alpha k + \alpha - 1} \right] f(t) dt.$$

The latter yields the following representation of the solution to (8.2.1)–(8.2.2) in terms of the Mittag-Leffler function:

$$y(x) = \sum_{j=1}^{n} b_j (x-a)^{\alpha - j} E_{\alpha, \alpha - j + 1} \left[\lambda (x-a)^{\alpha} \right] \quad (8.2.10)$$

$$+ \int_a^x (x-t)^{\alpha - 1} E_{\alpha, \alpha} \left[\lambda (x-t)^{\alpha} \right] f(t) dt.$$

The Cauchy Problem for One-Term Equations

Another important problem for linear ordinary differential equations is the Cauchy problem for FDEs with one fractional derivative of Caputo type

$$\left({}^C D_{a+}^{\alpha} y \right)(x) - \lambda y(x) = f(x) \quad (a \leq x \leq b; \; n-1 \leq \alpha < n; \; n \in \mathbb{N}; \; \lambda \in \mathbb{R}), \quad (8.2.11)$$

$$y^{(k)}(a) = c_k \quad (c_k \in \mathbb{R}, \; k = 1, \ldots, n-1). \quad (8.2.12)$$

In this case it is possible to pose initial conditions in the same form as that for ordinary differential equations (i.e. by involving usual derivatives) due to the properties of the Caputo derivative (see the corresponding discussion in Appendix E).

Under the assumption $f \in C_\gamma[a, b]$, $0 \leq \gamma < 1$, $\gamma < \alpha$, the Cauchy problem (8.2.11)–(8.2.12) is equivalent (see, e.g., [KiSrTr06, pp. 172, 230]) to the Volterra integral equation

$$y(x) = \sum_{j=0}^{n-1} \frac{c_j}{j!} (x-a)^j + \frac{\lambda}{\Gamma(\alpha)} \int_a^x \frac{y(t)}{(x-t)^{1-\alpha}} dt + \frac{1}{\Gamma(\alpha)} \int_a^x \frac{f(t)}{(x-t)^{1-\alpha}} dt.$$
(8.2.13)

By applying the method of successive approximation with initial approximation

$$y_0(x) = \sum_{j=0}^{n-1} \frac{c_j}{j!} (x-a)^j \quad (8.2.14)$$

we get the solution to the Volterra equation (8.2.13) (and thus to the Cauchy problem (8.2.11)–(8.2.12)) in the form

$$y(x) = \sum_{j=0}^{n-1} c_j \sum_{k=0}^{\infty} \frac{\lambda^k (x-a)^{\alpha k+j}}{\Gamma(\alpha k+j+1)} + \int_a^x \left[\sum_{k=1}^{\infty} \frac{\lambda^{k-1}}{\Gamma(\alpha k)} (x-t)^{\alpha k-1} \right] f(t) dt.$$
(8.2.15)

The latter yields the following representation of the solution to (8.2.11)–(8.2.12) in terms of the Mittag-Leffler function:

$$y(x) = \sum_{j=1}^{n} b_j (x-a)^j E_{\alpha, j+1}\left[\lambda (x-a)^\alpha\right] + \int_a^x (x-t)^{\alpha-1} E_{\alpha,\alpha}\left[\lambda(x-t)^\alpha\right] f(t) dt.$$
(8.2.16)

8.2.1.1 Solution Methods for Multi-term Ordinary FDEs

The Operational Method

The operational calculus for fractional differential equations has been developed in series of articles (see, e.g. [HiLuTo09] and the references therein). The idea of this approach goes back to the work by Mikusiński [Mik59] in which the Laplace convolution

$$(f * g)(x) = \int_0^x f(x-t) g(t) dt \qquad (8.2.17)$$

is interpreted as an algebraic multiplication in a ring of continuous functions on the real half-axis.

We briefly describe this approach (for a more detailed exposition we refer to the book [KiSrTr06, Sect. 4.3]). For any $\lambda \geq 1$ the mapping $\circ_\lambda$,

$$(f \circ_\lambda g)(x) = \left(I_{0+}^{\lambda-1} f * g\right)(x) = \int_0^x I_{0+}^{\lambda-1} f(x-t) g(t) dt \qquad (8.2.18)$$

becomes the convolution (without zero divisors) of the Liouville fractional integral operator I_{0+}^α ($\alpha > 0$) in the space

$$C_{-1} := \left\{ y \in C(0, +\infty) : \exists p > -1, y(x) = x^p y_1(x), y_1(x) \in C[0, +\infty) \right\}.$$
(8.2.19)

If $\alpha > 0$ and $1 \leq \lambda < \alpha + 1$, then the fractional integral I_{0+}^α has the following convolutional representation

$$\left(I_{0+}^\alpha f\right)(x) = (h \circ_\lambda f)(x), \quad h(x) = \frac{x^{\alpha-\lambda}}{\Gamma(\alpha-\lambda-1)}. \qquad (8.2.20)$$

8.2 Fractional Ordinary Differential Equations

By the semigroup property and linearity of the fractional integral I_{0+}^α, the space $\mathcal{C}_{-1}$ with the operations $\circ_\lambda$ and "+" becomes a commutative ring without zero divisors. As in the Mikusiński approach, one can see that this ring can be extended to the quotient field

$$\mathcal{P} = \mathcal{C}_{-1} \times (\mathcal{C}_{-1} \setminus \{0\}) / \sim \qquad (8.2.21)$$

with the standard equivalence relation determined by the convolution. Then the algebraic inverse to I_{0+}^α in the quotient field $\mathcal{P}$ can be defined as an element S of $\mathcal{P}$ which is reciprocal to the element $h(x)$ in the field $\mathcal{P}$:

$$S = \frac{I}{h} := \frac{h}{(h \circ_\lambda h)} = \frac{h}{h^2}, \qquad (8.2.22)$$

where I is an identity in $\mathcal{P}$ with respect to convolution.

For any $\alpha > 0$ and $m \in \mathbb{N}$ we introduce the space

$$\Omega_\alpha^m(\mathcal{C}_{-1}) := \left\{ y \in \mathcal{C}_{-1} : \left(D_{0+}^\alpha\right)^k y \in \mathcal{C}_{-1}, \ k = 1, \ldots, m \right\}, \qquad (8.2.23)$$

where $\left(D_{0+}^\alpha\right)^k$ is the kth power of the operator $\left(D_{0+}^\alpha\right)$:

$$\left(D_{0+}^\alpha\right)^k (\cdot) = \left(D_{0+}^\alpha \ldots D_{0+}^\alpha\right)(\cdot).$$

If $\alpha > 0$, $m \in \mathbb{N}$ and $f \in \Omega_\alpha^m(\mathcal{C}_{-1})$, then the following relation

$$\left(\left(D_{0+}^\alpha\right)^m f\right) = S^m - \sum_{k=0}^{m-1} S^{m-k} F\left(\left(D_{0+}^\alpha\right)^k f\right) \qquad (8.2.24)$$

holds in the field $\mathcal{P}$. Here $F = E - I_{0+}^\alpha D_{0+}^\alpha$ is a projector of I_{0+}^α determined in $\Omega_\alpha^1(\mathcal{C}_{-1})$ by the formula

$$\left(E - I_{0+}^\alpha D_{0+}^\alpha y\right)(x) = \sum_{k=1}^n \frac{\left(D_{0+}^{\alpha-k} y\right)(0+)}{\Gamma(\alpha - k - 1)} x^{k-\alpha}, \ y \in \Omega_\alpha^1(\mathcal{C}_{-1}), \ n - 1 < \alpha \leq n,$$

and $E : \Omega_\alpha^1(\mathcal{C}_{-1}) \to \Omega_\alpha^1(\mathcal{C}_{-1})$ is the identity operator on $\Omega_\alpha^1(\mathcal{C}_{-1})$.

If $\alpha > 0$, $1 \leq \lambda < \alpha + 1$, $m \in \mathbb{N}$, $\omega \in \mathbb{C}$, then convolution relations follow from the analytical property in the field $\mathcal{P}$ (see, e.g., [KiSrTr06, p. 265])

$$\frac{I}{S - \omega} = \frac{h}{E - \omega h} = h\left(E + \omega h + \omega^2 h^2 + \ldots\right) = x^{\alpha - \lambda} E_{\alpha, \alpha - \lambda + 1}(\omega x^\alpha),$$
$$(8.2.25)$$

$$\frac{I}{(S - \omega)^m} = x^{m\alpha - \lambda} E_{\alpha, m\alpha - \lambda + 1}^m(\omega x^\alpha), \qquad (8.2.26)$$

where $E_{\alpha,\beta}$, $E_{\alpha,\beta}^\gamma$ are two- and three-parametric Mittag-Leffler functions, respectively.

By using these relations one can represent solutions to the Cauchy type problem for certain multi-term fractional differential equations in terms of the Mittag-Leffler function. Since the application of this method critically depends on the analytical properties of the corresponding differential operator, we have taken the liberty of reproducing an example of such a problem from [KiSrTr06]. For this problem the method gives the solution in a final form.

Example ([KiSrTr06, p. 269]) Let $\omega \in \mathbb{C}$, $\lambda \in \mathbb{R}$ be arbitrary numbers such that $1 \leq \lambda < 3/2$. We consider in the space $\Omega_{1/2}^3 (\mathcal{C}_{-1})$ the following Cauchy type problem:

$$\left(D_{0+}^{3/2} y\right)(x) - \omega y'(x) + \lambda^2 \left(D_{0+}^{1/2} y\right)(x) - \omega \lambda^2 y(x) = f(x), \quad f \in \mathcal{C}_{-1} \quad (8.2.27)$$

$$\lim_{x \to 0+} \left(D_{0+}^{1/2} y\right)(x) = 0, \quad y(0) = 0, \quad \lim_{x \to 0+} \left(I_{0+}^{1/2} y\right)(x) = 0. \quad (8.2.28)$$

Here $\alpha = 1/2$ and the problem is reduced in the field $\mathcal{P}$ to the algebraic equation

$$S^3 y - \omega S^2 y + \lambda^2 S y - \omega \lambda^2 y = f. \quad (8.2.29)$$

Since the polynomial on the left-hand side possesses a simple decomposition

$$P(S) = S^3 - \omega S^2 + \lambda^2 S - \omega \lambda^2 = (S^2 + \lambda^2)(S - \omega)$$

one can find a representation of an algebraic inverse to the differential operator in (8.2.27) as the reciprocal to $P(S)$ in the field $\mathcal{P}$:

$$\frac{I}{P(S)} = \frac{1}{\omega^2 + \lambda^2} \left\{ -\frac{S}{S^2 + \lambda^2} - \frac{\omega}{S^2 + \lambda^2} + \frac{I}{S - \omega} \right\}. \quad (8.2.30)$$

Using the identity (8.2.25) we have the solution of the considered problem in the form

$$y(x) = \int_0^x K(x - t) f(t) dt, \quad (8.2.31)$$

where

$$K(x) = \frac{1}{\omega^2 + \lambda^2} \quad (8.2.32)$$

$$\times \left\{ -x^{\frac{1}{2} - \lambda} E_{1,2-\lambda}(-\lambda^2 x) - \omega x^{1-\lambda} E_{1,2-\lambda}(-\lambda^2 x) + x^{\frac{1}{2} - \lambda} E_{\frac{1}{2}, \frac{3}{2} - \lambda}(\omega x^{\frac{1}{2}}) \right\}.$$

8.2 Fractional Ordinary Differential Equations

The Laplace Transform Method

Another useful method which leads to the explicit solution of ordinary fractional differential equations is the Laplace transform method. The direct application of the Laplace transform to the homogeneous multi-term fractional differential equation

$$\sum_{k=1}^{m} A_k \left({}^C D_{0+}^{\alpha_k} y\right)(x) + A_0 y(x) = 0 \ (x > 0, \ m \in \mathbb{N}, \ 0 < \alpha_1 < \ldots < \alpha_m) \tag{8.2.33}$$

yields an explicit representation of the fundamental system $\{y_i(x)\}, i = 0, \ldots, l-1, l = -[-\alpha_m]$, for this equation, i.e. of the solutions to the Cauchy problems

$$y_i^{(j)}(0) = \delta_i^j, \ i, j = 0, 1, \ldots, l-1, \tag{8.2.34}$$

for Eq. (8.2.33). This representation critically depends on the properties of the so-called quasi-polynomial

$$P(s) = A_0 + \sum_{k=1}^{m} A_k s^{\alpha_k}, \ s \in \mathbb{C}.$$

Thus, the system of fundamental solutions for the one-term fractional differential equation

$$\left({}^C D_{0+}^{\alpha} y\right)(x) - \lambda y(x) = 0 \ (x > 0, \ l-1 < \alpha \le l, \ l \in \mathbb{N}) \tag{8.2.35}$$

has the form

$$y_i(x) = x^i E_{\alpha, i+1}(\lambda x^\alpha), \quad i = 0, \ldots, l-1; \tag{8.2.36}$$

and the system of fundamental solutions for two-term fractional differential equations with derivatives of orders α and β ($\alpha > \beta > 0, l-1 < \alpha \le l, l \in \mathbb{N}, n-1 < \beta \le n, n \in \mathbb{N}$)

$$\left({}^C D_{0+}^{\alpha} y\right)(x) - \lambda \left({}^C D_{0+}^{\beta} y\right)(x) = 0 \ (x > 0) \tag{8.2.37}$$

consists of two groups of functions (see [KiSrTr06, p. 314])

$$y_i(x) = x^i E_{\alpha-\beta, i+1}(\lambda x^{\alpha-\beta}) - \lambda x^{\alpha-\beta+i} E_{\alpha-\beta, \alpha-\beta+i+1}(\lambda x^{\alpha-\beta}), \ i = 0, \ldots, n-1; \tag{8.2.38}$$

$$y_i(x) = x^i E_{\alpha-\beta, i+1}(\lambda x^{\alpha-\beta}), \quad i = n, \ldots, l-1. \tag{8.2.39}$$

For multi-term equations with a greater number of derivatives, the system of fundamental solutions has also been found (see [KiSrTr06, pp. 319–321]). The corresponding formulas are rather cumbersome. They have the form of series with coefficients represented in terms of the Wright function ${}_1\Psi_1$.

The inhomogeneous equation, corresponding to (8.2.33), can also be solved by using the Laplace transform method. The solution is represented in the form

$$y(x) = \int_0^x G_{\alpha_1,\ldots,\alpha_m}(x-t) f(t) dt + \sum_{i=0}^{l-1} c_i y_i(x), \qquad (8.2.40)$$

where $\{y_i(x)\}, i = 0, \ldots, l-1$, is the fundamental system of solutions of the homogeneous equation and

$$G_{\alpha_1,\ldots,\alpha_m}(x) = \left(\mathcal{L}^{-1}\left[\frac{1}{P(s)}\right]\right)(x) \qquad (8.2.41)$$

is the function determined by the inverse transform of the reciprocal to the quasi-polynomial (an analogue of Green's function). In some cases the function $G_{\alpha_1,\ldots,\alpha_m}(x)$ can be represented in an explicit form (see [KiSrTr06]).

In a similar way the Cauchy type problem for multi-term equations can be studied using the Laplace transform method.

8.2.2 Ordinary FDEs with Variable Coefficients

Here we describe certain approaches concerning the solution of ordinary differential equations with variable coefficients. All of them are related to the case when the coefficients in the equations are power-type functions.

Let us consider first the following Cauchy type problem for the one-term differential equation with Riemann–Liouville fractional derivative:

$$\left(D_{0+}^\alpha y\right)(x) - \lambda (x-a)^\beta y(x) = 0 \ (a < x \leq b; \alpha > 0, \ \beta > -\{\alpha\}; \ \lambda \in \mathbb{R}), \qquad (8.2.42)$$

$$\left(D_{0+}^{\alpha-k} y\right)(a+) = b_k \ (b_k \in \mathbb{R}; \ k = 1, \ldots, n, \ n = -[-\alpha]). \qquad (8.2.43)$$

As in the above considered case of the equations with constant coefficients, one can prove that problem (8.2.42)–(8.2.43) is equivalent in the space $C_{n-\alpha}[a, b]$ to the Volterra integral equation

$$y(x) = \sum_{j=1}^n \frac{b_j (x-a)^{\alpha-j}}{\Gamma(\alpha - j + 1)} + \frac{\lambda}{\Gamma(\alpha)} \int_a^x \frac{(x-a)^\beta y(t)}{(x-t)^{1-\alpha}} dt. \qquad (8.2.44)$$

By applying the successive approximation method and using formulas for fractional integrals of power-type functions we can derive the following series representation of the solution to (8.2.42)–(8.2.43):

8.2 Fractional Ordinary Differential Equations

$$y(x) = \sum_{j=1}^{n} \frac{b_j (x-a)^{\alpha-j}}{\Gamma(\alpha-j+1)} \left[1 + \sum_{k=1}^{\infty} c_{k,j} \left(\lambda(x-a)^{\alpha+\beta}\right)^k \right], \quad (8.2.45)$$

where

$$c_{k,j} = \prod_{r=1}^{k} \frac{\Gamma(r(\alpha+\beta) - j + 1)}{\Gamma(r(\alpha+\beta) + \alpha - j + 1)}.$$

Changing in the last product the index r to $i = r - 1$ we arrive at the representation of the solution to (8.2.42)–(8.2.43) in terms of the three-parametric Mittag-Leffler function (Kilbas–Saigo function) $E_{\alpha,m,l}(z)$ (see Sect. 5.2)

$$y(x) = \sum_{j=1}^{n} \frac{b_j (x-a)^{\alpha-j}}{\Gamma(\alpha-j+1)} E_{\alpha, 1+\beta/\alpha, 1+(\beta-j)/\alpha} \left(\lambda(x-a)^{\alpha+\beta}\right). \quad (8.2.46)$$

In a similar way we can consider the Cauchy problem for the one-term fractional differential equation with Caputo fractional derivative

$$\left(^C D_{0+}^{\alpha} y\right)(x) - \lambda(x-a)^{\beta} y(x) = 0 \ (a \leq x \leq b; \alpha > 0, \ \beta > -\alpha; \ \lambda \in \mathbb{R}), \quad (8.2.47)$$

$$y^{(k)}(a) = d_k \ (d_k \in \mathbb{R}; \ k = 0, \ldots, n-1, \ n-1 < \alpha < n). \quad (8.2.48)$$

In this case the Cauchy problem (8.2.47)–(8.2.48) is equivalent in the space $\mathcal{C}^{n-1}[a,b]$ to the Volterra integral equation

$$y(x) = \sum_{j=0}^{n-1} \frac{d_j(x-a)^j}{j!} + \frac{\lambda}{\Gamma(\alpha)} \int_a^x \frac{(x-a)^{\beta} y(t)}{(x-t)^{1-\alpha}} dt. \quad (8.2.49)$$

Here, the successive approximation method gives the following representation (see [KiSrTr06, p. 233]) of the solution to (8.2.46)–(8.2.47) in terms of the three-parametric Mittag-Leffler function (Kilbas–Saigo function):

$$y(x) = \sum_{j=0}^{n-1} \frac{d_j}{j!} (x-a)^j E_{\alpha, 1+\beta/\alpha, (\beta+j)/\alpha} \left(\lambda(x-a)^{\alpha+\beta}\right). \quad (8.2.50)$$

The Cauchy type problem for the inhomogeneous equation corresponding to Eq. (8.2.42) is treated via the differentiation formulas (of integer and fractional order) for the three-parametric Mittag-Leffler function derived in [KilSai95b, GoKiRo98]. Let

$$\alpha > 0, \ n-1 < \alpha < n, n \in \mathbb{N}; \ \beta > -\{\alpha\}; \ f_r, \mu_r \in \mathbb{R}, \mu_r > -1, r = 1, \ldots, k.$$

Let us consider the Cauchy type problem for the inhomogeneous fractional differential equation with quasi-polynomial free term

$$\left(D_{0+}^{\alpha} y\right)(x) = \lambda (x-a)^{\beta} y(x) + \sum_{r=1}^{k} f_r(x-a)^{\mu_r} \ (a < x < b \leq +\infty; \ \lambda \in \mathbb{R}),$$
(8.2.51)

$$\left(D_{0+}^{\alpha-j} y\right)(a+) = b_j \ (b_j \in \mathbb{R}; \ j = 1, \ldots, n).$$
(8.2.52)

This problem has a unique solution in the space $I_{loc}(a, b)$ of locally integrable functions on (a, b) (see [KiSrTr06, pp. 251–252])

$$y(x) = y_0 + \sum_{j=1}^{n} \frac{b_j}{\Gamma(\alpha - j + 1)} (x-a)^{\alpha-j} E_{\alpha, 1+\beta/\alpha, (\beta-j)/\alpha} \left(\lambda(x-a)^{\alpha+\beta}\right),$$
(8.2.53)

where

$$y_0 = \sum_{r=1}^{k} \frac{f_r \Gamma(\mu_r + 1)}{\Gamma(\mu_r + \alpha + 1)} (x-a)^{\alpha+\mu_r} E_{\alpha, 1+\beta/\alpha, (\beta+\mu_r)/\alpha} \left(\lambda(x-a)^{\alpha+\beta}\right). \quad (8.2.54)$$

The Mellin transform method is applied to solve the following fractional differential equations with power-type coefficients (which are sometimes called Euler type fractional equations, see, e.g. [Zhu12, ZhuSit18])

$$\sum_{k=0}^{m} A_k x^{\alpha+k} \left(D_{0+}^{\alpha+k} y\right)(x) = f(x) \ (x > 0, \alpha > 0). \quad (8.2.55)$$

In this case one can use the following property of the Mellin transform

$$\left(\mathcal{M} x^{\alpha+k} \left(D_{0+}^{\alpha+k} y\right)\right)(s) = \frac{\Gamma(1-s)}{\Gamma(1-s-\alpha-k)} (\mathcal{M} y)(s).$$

Hence by applying the Mellin transform to (8.2.55), one obtains the solution in the form of a Mellin convolution (see, e.g. [KiSrTr06, p. 330])

$$y(x) = \int_0^{\infty} G_\alpha(t) f(xt) dt, \quad (8.2.56)$$

where the *Mellin fractional analogue of Green's function* $G_\alpha(x)$ is given by the formula

$$G_\alpha(x) = \left(\mathcal{M}^{-1} \left[\frac{1}{P_\alpha(1-s)}\right]\right), \ P_\alpha(s) = \sum_{k=0}^{m} A_k \frac{\Gamma(s)}{\Gamma(s-\alpha-k)}. \quad (8.2.57)$$

8.2 Fractional Ordinary Differential Equations

In some cases it is possible to determine an explicit representation of the analogue of Green's function (and thus an explicit solution to Eq. (8.2.55)) (see, e.g. [KilZhu08, KilZhu09a, KilZhu09b]). For instance, in the case $m = 1$, the corresponding analogue of Green's function $G_{\alpha,\lambda}(x)$ for the equation

$$x^{\alpha+1}\left(D_{0+}^{\alpha+1}y\right)(x) + \lambda x^{\alpha}\left(D_{0+}^{\alpha}y\right)(x) = f(x) \quad (\alpha > 0, \lambda \in \mathbb{R}) \tag{8.2.58}$$

has the form

$$G_\alpha(x) = G_{\alpha,\lambda}(x) \tag{8.2.59}$$

$$= x^{-\alpha}\left\{\frac{\Gamma(1-\lambda)}{\Gamma(\alpha+1-\lambda)}x^{\lambda-1} - {}_1\Psi_2\left[\begin{matrix}(1-\lambda,1)\\(\alpha,-1),(2-\lambda,1)\end{matrix}\bigg|-x\right]\right\}.$$

8.2.3 Other Types of Ordinary Fractional Differential Equations

In [KiSrTr06] fractional ordinary differential equations with another type of fractional derivative have been considered.

Here we focus on only a few of them which are solved in an explicit form.
Let us consider the following Cauchy type problem:

$$\left(\mathcal{D}_{a+}^\alpha y\right)(x) - \lambda y(x) = f(x) \quad (a < x \leq b, \; \alpha > 0, \; \lambda \in \mathbb{R}), \tag{8.2.60}$$

$$\left(\mathcal{D}_{a+}^{\alpha-k}y\right)(a+) = b_k \quad (b_k \in \mathbb{R}, \; k = 1, \ldots, n, \; n = -[-\alpha]) \tag{8.2.61}$$

where $f \in \mathcal{C}_{\gamma,log}[a,b] := \{g : (a,b] \to \mathbb{R} : \left[\log(x/a)g(x)\right] \in \mathcal{C}[a,b]\}$ and $\mathcal{D}_{a+}^\alpha$ is the Hadamard fractional derivative

$$\left(\mathcal{D}_{a+}^\alpha y\right)(x) = \left(x\frac{d}{dx}\right)^n \frac{1}{\Gamma(n-\alpha)}\int_a^x \left(\log\frac{x}{t}\right)^{n-\alpha-1}\frac{y(t)dt}{t}. \tag{8.2.62}$$

Employing the same techniques as for one-term ordinary fractional differential equations with Riemann–Liouville fractional derivative, one can show that the Cauchy type problem (8.2.60)–(8.2.61) is equivalent in the space $\mathcal{C}_{n-\alpha,log}[a,b]$ to the Volterra integral equation

$$y(x) = \sum_{k=1}^n \frac{b_k}{\Gamma(\alpha-k+1)}\left(\log\frac{x}{a}\right)^{\alpha-k}$$

$$+ \frac{\lambda}{\Gamma(\alpha)}\int_a^x \left(\log\frac{x}{t}\right)^{\alpha-1} y(t)\frac{dt}{t} + \frac{1}{\Gamma(\alpha)}\int_a^x \left(\log\frac{x}{t}\right)^{\alpha-1} f(t)\frac{dt}{t}. \tag{8.2.63}$$

By applying the method of successive approximation we obtain (see [KiSrTr06, p. 234]) the unique solution to the Volterra integral equation (and thus to the Cauchy type problem (8.2.60)–(8.2.61)) in terms of the two-parametric Mittag-Leffler function

$$y(x) = \sum_{k=1}^{n} b_k \left(\log \frac{x}{a}\right)^{\alpha-k} E_{\alpha,\alpha-k+1}\left[\lambda \left(\log \frac{x}{a}\right)^{\alpha}\right]$$

$$+ \int_{a}^{x} \left(\log \frac{x}{t}\right)^{\alpha-1} E_{\alpha,\alpha-k+1}\left[\lambda \left(\log \frac{x}{t}\right)^{\alpha}\right] f(t) dt. \quad (8.2.64)$$

The solution to the fractional ordinary differential equation

$$\sum_{k=1}^{m} A_k \left(\mathbf{D}^{\alpha_k} y\right)(x) + A_0 y(x) = f(x) \quad (8.2.65)$$

$$(0 < \alpha_1 < \ldots < \alpha_m, \ A_k \in \mathbb{R}, k = 1, \ldots, m)$$

with the Riesz fractional derivative

$$(\mathbf{D}^\alpha y)(x) = \frac{1}{d_1(l,\alpha)} \int_{-\infty}^{+\infty} \frac{(\Delta_t^l y)(x)}{|t|^{1+\alpha}} dt \ (l > \alpha) \quad (8.2.66)$$

is given (see, e.g., [KiSrTr06, p. 344–345]) in term of the Fourier convolution

$$y(x) = \int_{-\infty}^{+\infty} G^F_{\alpha_1,\ldots,\alpha_m}(x-t) f(t) dt. \quad (8.2.67)$$

Here $G^F_{\alpha_1,\ldots,\alpha_m}$ is a fractional analogue of Green's function, which has the following form in the case of an equation with constant coefficients:

$$G^F_{\alpha_1,\ldots,\alpha_m}(x) = \frac{1}{\pi} \int_{0}^{+\infty} \frac{1}{\left[\sum_{k=1}^{m} A_k |\tau|^{\alpha_k} + A_0\right]} \cos \tau x \, d\tau. \quad (8.2.68)$$

Finally, we single out an important class of so-called sequential fractional differential equations

$$\left(D_{a+}^{n\alpha} y\right)(x) + \sum_{k=0}^{n-1} a_k(x) \left(D_{a+}^{k\alpha} y\right)(x) = f(x) \quad (8.2.69)$$

8.2 Fractional Ordinary Differential Equations

treated, e.g., in [KiSrTr06, Chap. 7] by different methods. In the case of constant coefficients the solution of the corresponding Cauchy type problem is given in terms of the α-exponential function

$$e_\alpha^{\lambda z} = z^{\alpha-1} E_{\alpha,\alpha}(\lambda z^\alpha).$$

8.3 Optimal Control for Equations with Fractional Derivatives and Integrals

In this section we present some results on the problem of fractional order optimal control, mainly following [Pet19]. Since this subject and related topics (such as stability, observability etc.) have attracted extended interest in the last few years we do not pretend to give a complete presentation, but instead have restricted ourselves to basic formulations and facts. Among the sources for further study we recommend the reader to look at the paper in the 6th volume of the Handbook of Fractional Calculus with Applications [HAND6]. There are several connections between the fractional optimal control problem and the subject of this book. Thus in some special cases the controller can be similar to what we obtain as the Laplace transform of certain generalizations of the Mittag-Leffler function. Moreover, in the time domain the fractional optimal control problem is reduced to a fractional differential equation. For more details we refer to [Pod99, Chap. 9] (see also [Pod99a]).

8.3.1 Linear Fractional-Order Controllers

The fractional-order $PI^\lambda D^\delta$ (known also as $PI^\lambda D^\mu$) controller (FOC) was proposed in [Pod99a] as a generalization of the PID controller with integrator of real order λ and differentiator of real order δ. The transfer function of such a parallel controller in the Laplace domain has the form:

$$C(s) = \frac{U(s)}{E(s)} = K_p + T_i s^{-\lambda} + T_d s^\delta, \quad (\lambda, \delta > 0), \qquad (8.3.1)$$

where K_p is the proportional constant, T_i is the integration constant and T_d is the differentiation constant. The internal structure of the fractional-order controller consists of the parallel connection of the proportional, integration, and derivative part. The transfer function (8.3.1) corresponds in the time domain to a fractional differential equation of the form:

$$u(t) = K_p e(t) + T_i (D^{-\lambda} e)(t) + T_d (D^\delta e)(t). \qquad (8.3.2)$$

Particular cases of the fractional-order controller, which is more flexible, give an opportunity to better adjust the dynamical properties of the fractional-order control system.

A Matlab implementation of the FOC in discrete form is proposed and described in [Pet11a].

8.3.2 Nonlinear Fractional-Order Controllers

In this subsection, a nonlinear fractional-order controller is defined. Following the traditional structure of a nonlinear PID controller, which is well known in the literature, one can write a new formula for a nonlinear fractional-order $PI^\lambda D^\delta$ controller (NFOC) of the following form

$$u(t) = f(e) \left[K_p e(t) + T_i (D^{-\lambda} e)(t) + T_d (D^\delta e)(t) \right], \quad (8.3.3)$$

where $f(e)$ is a function of the variable e.

Below various types of the nonlinear function f used in the definition of nonlinear fractional order controllers are presented (see [Pet19]).

A. A commonly used function in such a definition (see, e.g. [Bob-et-al99]) is

$$f(e) = K_0 + (1 - K_0)|e(t)|, \quad (8.3.4)$$

which leads to the linear controller (8.3.2) if $K_0 = 1$. If $K_0 \neq 1$, then the controller possesses 6 degrees of freedom (6DOF-controller).

B. Generalizing this kind of nonlinearity, various other piecewise linear functions of nonlinear gain can also be used.

C. For desired low e_l and high e_h control error bounds, we obtain a controller with variable gain and the function $f(e)$ is defined, for example, as in [Bob-et-al99]

$$f(e) = \begin{cases} 1 & \text{for } e < e_l, \\ K_0 & \text{for } e \in (e_l, e_h), \ K_0 \geq 0, \\ 1 & \text{for } e > e_h \end{cases} \quad (8.3.5)$$

When $K_0 = 0$ within the interval (e_l, e_h), the output of the controller does not change and, therefore, the actuator behavior is much smoother.

D. It is also possible to consider a scaled error function to be in the form $f(e) = k(e) \cdot e(t)$, where the nonlinear gain $k(e)$ represents any general nonlinear function of the error $e(t)$, which is bounded in a sector. In this case a general form of the nonlinear fractional-order $PI^\lambda D^\delta$ controller is the following:

$$u(t) = \left[K_p + T_i D^{-\lambda} + T_d D^\delta \right] f(e) \quad (8.3.6)$$
$$= K_p(\cdot) e(t) + T_i(\cdot) D^{-\lambda} e(t) + T_d(\cdot) D^\delta e(t),$$

where $K_p(\cdot)$, $T_i(\cdot)$, $T_d(\cdot)$ are time-varying controller parameters, which may depend on the system state, input, control error, or other variables (see, e.g. [SuSuDu05]). It should be taken into account that by using the nonlinear controller parameters as in Eq. (8.3.6), we obtain a controller with variable structure. Controllers of such kind are often called super-twisting controllers. A nonlinear fractional-order $PI^\lambda D^\delta$ controller can also be tuned as a combination of the linear and nonlinear parameters—for instance, pure gain for the proportional part and a nonlinear time constant for integration and derivation constants, and so on.

E. Another general modification of the nonlinear fractional-order PID controller (8.3.3) is a controller with piecewise linear gain which depends on the signal $\nu(t)$. Then the proportional controller gain is $K_p = f(\nu)$, where the piecewise linear function $f(v)$ can be defined as in A. or B.

When we take into account all possible nonlinear functions and all possible particular cases of the fractional-order controllers, we are presented with a new wide class of controllers. This opens a new area of research and many questions are still open. On the other hand, it is possible to turn to the experimental approach and employ well-known tuning methods proposed for linear fractional-order controllers. However, we still have to check the performance of a control loop via simulation before applying the proposed controller to real objects in order to verify and validate requirements on the control system. For Matlab implementation of the NFOC with nonlinearity (8.3.4) in a discrete form, the function NFOC($\cdot$) can be used (see [Pet15]).

8.3.3 Modification of the Control Actions in Fractional-Order PID Controllers

In [Pet19], several modifications of the fractional-order control are described which can be used in order to avoid such problems as wind-up effect, actuator saturation, and derivative action limitation (see also [Pet12]). Let us list them:

(1) *Filtering the desired value $r(t)$.* Here filtering the desired value $r(t)$, known as setpoint tracking, by a first- or second-order filter is a very frequently used trick to avoid problems with derivative action. The first-order prefilter in the discrete form

$$H_p(z) = \frac{k_f}{1 - k_f z^{-1}}$$

is recommended, where k_f is the prefilter constant.

(2) *Using a controlled value in proportional and derivative parts of a controller.* The problem related to step changes of the control signal due to step changes of the desired value $r(t)$ can also be solved by replacing the control error $e(t) = r(t) - y(t)$ by the controlled value $y(t)$.

(3) *Filtering the derivative action.* Due to a noisy signal on the measured controlled value, the differentiation of noise can involve inappropriate changes in the control signal. For the first-order filter in the derivative part and with a genuine integral action, we can write the transfer function of the fractional-order controller (in the Laplace domain) in the following form:

$$C(s) = \frac{U(s)}{E(s)} = K_p + T_i \frac{s^{1-\lambda}}{s} + \frac{T_d s^\delta}{T_f s + 1}, \quad (\lambda, \delta > 0), \tag{8.3.7}$$

where $T_f = \frac{N}{T_d}$ is the filter constant.

(4) *Limitation of integral action.* This limitation (wind-up of the controller) is due to the fact that the actuator also has limitations and, for instance, if the actuator is at the end position and the control error is not zero, the integral part of the controller rapidly grows, the controller calculates an unrealistic value of the control signal and, therefore, the actuator stays at the end position until the sign of control error is changed. There are different ways to avoid wind-up of the controllers.

Filtering can be also obtained automatically if the derivative is implemented by taking the difference between the reference signal and its filtered version. Instead of filtering just the derivative, it is also possible to use the usual controller and filter the measured signal. The transfer function of the fractional order controller with the filter is then

$$C(s) = \frac{U(s)}{E(s)} = \left[K_p + T_i s^{-\lambda} + T_d s^\delta \right] \frac{1}{T_f s + 1}, \quad (\lambda, \delta > 0), \tag{8.3.8}$$

where a σth order filter is used.

8.3.4 Further Possible Modifications of the Fractional-Order PID Controllers

Besides the linear and nonlinear fractional-order PID controllers presented in previous subsections, there are many additional possible modifications. Below we list just a few of them for illustration.

– Fractional-order PID controllers presented in [ValCos12]:

$$C(s) = K_p \left(1 + \frac{1}{T_i s} \right)^\lambda (1 + T_d s)^\delta, \tag{8.3.9}$$

or

$$C(s) = K_p \left(1 + \frac{1}{T_i s^\lambda} \right) (1 + T_d s)^\delta. \tag{8.3.10}$$

- Fractional-order [PI] and [PD] controllers, respectively, introduced in [LuoChe13]:

$$C(s) = \left(K_p + \frac{T_i}{s}\right)^\lambda, \text{ and } C(s) = \left(K_p + T_d s\right)^\delta, \text{ respectively,} \quad (8.3.11)$$

- Fractional-order PID controllers suggested in [PadVis15]:

$$C(s) = K_p \frac{1 + T_i s^\lambda}{T_i s^\lambda} \left(1 + T_d s^\delta\right), \quad (8.3.12)$$

or

$$C(s) = K_p \left(1 + \frac{1}{T_i s^\lambda} + T_d s^\delta\right) \frac{1}{1 + T_f s}, \quad (8.3.13)$$

or

$$C(s) = K_p \frac{1 + T_i s^\lambda}{T_i s^\lambda} \frac{1 + T_d s^\delta}{1 + (T_d/N)s}. \quad (8.3.14)$$

8.4 Differential Equations with Fractional Partial Derivatives

Partial fractional differential equations are of great theoretical and practical importance (see, e.g., [KiSrTr06, Chap. 6], [KilTru02] and the references therein). Recently a number of books presenting results in this area have been published (e.g., [Die10, Mai10, Psk06, Tar10, Uch13a, Uch13b]). One can find there different aspects of the theory and applications of partial fractional differential equations. Indeed, we have to note that this branch of analysis is far from complete. Furthermore, many partial differential equations serve to describe some models in mechanics, physics, chemistry, biology etc. Some of these equations will be discussed in the following two chapters.

In this section we present a few results concerning the simplest fractional partial differential equations. Our main focus will be on the one-dimensional diffusion-wave equation (with Riemann–Liouville fractional derivative, see Sect. 8.4.1, and with Caputo fractional derivative, see Sect. 8.4.2). The main idea is to present to the reader elements of the techniques developed for fractional partial differential equations. Some more specific equations dealing with certain models are presented in the next two chapters. We point out here that the huge variety of fractional partial differential equations and their methods of solution cannot be completely described within the scope of a single chapter.

8.4.1 Cauchy-Type Problems for Differential Equations with Riemann–Liouville Fractional Partial Derivatives

The simplest partial differential equation with Riemann–Liouville fractional derivative is the so-called fractional diffusion equation

$$\left(D_{0+,t}^\alpha u\right)(x,t) = \lambda^2 \frac{\partial^2 u}{\partial x^2} \quad (x \in \mathbb{R}; t > 0; \lambda > 0; \alpha > 0). \tag{8.4.1}$$

Here $D_{0+,t}^\alpha$ is the Riemann–Liouville fractional derivative of order α with respect to time t.

Partial differential equations where the Riemann–Liouville fractional derivative are in time are supplied by initial conditions known as Cauchy-type conditions. Let us consider Eq. (8.4.1) for $0 < \alpha < 2$. The Cauchy-type initial conditions then have the form

$$\left(D_{0+,t}^{\alpha-k} u\right)(x, 0+) = f_k(x), \quad x \in \mathbb{R}, \tag{8.4.2}$$

where $k = 1$ if $0 < \alpha < 1$, and two conditions with $k = 1, 2$ if $1 < \alpha < 2$.[2]

The problem (8.4.1)–(8.4.2) is usually solved by the method of integral transforms. Let us apply to Eq. (8.4.1) successively the Laplace transform with respect to the time variable t

$$(\mathcal{L}_t u)(x,s) = \int_0^\infty u(x,t) e^{-st} dt \quad (x \in \mathbb{R}; s > 0),$$

and the Fourier transform with respect to the spatial variable x

$$(\mathcal{F}_x u)(\sigma, t) = \int_{-\infty}^{+\infty} u(x,t) e^{ix\sigma} dx \quad (\sigma \in \mathbb{R}; t > 0).$$

The Laplace transform of the Riemann–Liouville fractional derivative satisfies the relation

$$\left(\mathcal{L}_t D_{0+,t}^{\alpha-k} u\right)(x,s) = s^\alpha (\mathcal{L}_t u)(x,s) - \sum_{j=1}^l s^{j-1} \left(D_{0+,t}^{\alpha-j} u\right)(x, 0+), \quad (x \in \mathbb{R}), \tag{8.4.3}$$

where $l - 1 < \alpha \leq l, l \in \mathbb{N}$. In the considered case $(0 < \alpha < 2)$ we take into account the initial conditions (8.4.2) we get from (8.4.1)

---

[2] Note that for $\alpha = 1$ Eq. (8.4.1) becomes the standard diffusion equation and the initial condition (8.4.2) becomes the standard Cauchy condition.

8.4 Differential Equations with Fractional Partial Derivatives

$$s^\alpha \left(\mathcal{L}_t u\right)(x, s) = \sum_{j=1}^{k} s^{j-1} f_j(x) + \lambda^2 \left(\frac{\partial^2}{\partial x^2}\mathcal{L}_t u\right). \tag{8.4.4}$$

Here either $k = 1$ or $k = 2$. For the Fourier transform the following relation is known

$$\left(\mathcal{F}_x\left[\frac{\partial^2 u}{\partial x^2}\right]\right)(\sigma, t) = -|\sigma|^2 \left(\mathcal{F}_x u\right)(\sigma, t). \tag{8.4.5}$$

Thus applying the Fourier transform to (8.4.4) we get for $k = 1$ or $k = 2$

$$(\mathcal{F}_x \mathcal{L}_t u)(\sigma, s) = \sum_{j=1}^{k} \frac{s^{j-1}}{s^\alpha + \lambda^2 |\sigma|^2} \quad (\sigma \in \mathbb{R}; s > 0). \tag{8.4.6}$$

In order to obtain an explicit solution to problem (8.4.1)–(8.4.2) we use the inverse Fourier and the inverse Laplace transform and corresponding tables of these transforms. The final result reads (see, e.g., [KiSrTr06, Thm. 6.1]):
Let $0 < \alpha < 2$ and $\lambda > 0$. Then the formal solution of the Cauchy-type problem (8.4.1)–(8.4.2) is represented in the form

$$u(x, t) = \sum_{j=1}^{k} \int_{-\infty}^{+\infty} G_j^\alpha(x - \tau, t) f_j(\tau) d\tau, \tag{8.4.7}$$

where $k = 1$ if $0 < \alpha < 1$, and $k = 2$ if $1 < \alpha < 2$,

$$G_j^\alpha(x, t) = \frac{1}{2\lambda} t^{\alpha/2 - j} \varphi\left(-\frac{\alpha}{2}, \frac{\alpha}{2} - j + 1; -\frac{|x|}{\lambda} t^{-\alpha/2}\right), (j = 1, 2) \tag{8.4.8}$$

where $\varphi(a, b; z)$ is the classical Wright function (see Definition (7.1.1) in Chap. 7). The formal solution (8.4.7) becomes the real one if the integrals on the right-hand side of (8.4.7) converge.

8.4.2 The Cauchy Problem for Differential Equations with Caputo Fractional Partial Derivatives

As an example we consider here the Cauchy problem for the partial fractional differential equation with Caputo derivative

$$\left({}^C D_{0+,t}^\alpha u\right)(x, t) = \lambda^2 \frac{\partial^2 u}{\partial x^2} \quad (x \in \mathbb{R}; t > 0; \lambda > 0; 0 < \alpha < 2). \tag{8.4.9}$$

This equation is a particular case of the so-called fractional diffusion-wave equation

$$\left({}^C D^\alpha_{0+,t} u\right)(\mathbf{x},t) = \lambda^2 \Delta_{\mathbf{x}} u(\mathbf{x},t) \quad (x \in \mathbb{R}^n; t > 0; \lambda > 0; 0 < \alpha < 2). \quad (8.4.10)$$

The Eq. (8.4.9) is supplied by the Cauchy condition(s)

$$\frac{\partial^k u}{\partial x^k}(x,0) = f_k(x) \quad (x \in \mathbb{R}), \quad (8.4.11)$$

where $k = 0$, if $0 < \alpha < 1$, and two conditions with $k = 0, 1$, if $1 < \alpha < 2$; the 0th order derivative means the value of the solution u at the points $(x, 0)$.[3]

To solve the Cauchy problem (8.4.11) for the fractional differential equation (8.4.9) we use the same method as in the previous subsection. We first apply the Laplace integral transform with respect to the time variable t using the relation

$$\left(\mathcal{L}_t{}^C D^\alpha_{0+,t} u\right)(x,s) = s^\alpha \left(\mathcal{L}_t u\right)(x,s) - \sum_{j=0}^{k-1} s^{\alpha-j-1} \frac{\partial^j u}{\partial t^j}(x,0), \quad (8.4.12)$$

and then the Fourier transform with respect to the spatial variable x using the relation (8.4.5). Then in view of the Cauchy initial condition(s) we obtain from equation (8.4.9)

$$(\mathcal{F}_x \mathcal{L}_t u)(\sigma, s) = \sum_{j=0}^{k-1} \frac{s^{\alpha-j-1}}{s^\alpha + \lambda^2 |\sigma|^2} (\mathcal{F}_x f_k)(\sigma). \quad (8.4.13)$$

By using the inverse Fourier and the inverse Laplace transform and corresponding tables of these transforms we get the final result in the form (see, e.g., [KiSrTr06, Thm. 6.3]):

Let $0 < \alpha < 2$ and $\lambda > 0$. Then the formal solution of the Cauchy problem (8.4.9), (8.4.11) is represented in the form

$$u(x,t) = \sum_{j=0}^{k-1} \int_{-\infty}^{+\infty} G^\alpha_j(x-\tau,t) f_j(\tau) d\tau, \quad (8.4.14)$$

where $k = 1$ if $0 < \alpha < 1$, and $k = 2$ if $1 < \alpha < 2$,

$$G^\alpha_j(x,t) = \frac{1}{2\lambda} t^{j-\alpha/2} \varphi\left(-\frac{\alpha}{2}, j+1-\frac{\alpha}{2}; -\frac{|x|}{\lambda} t^{-\alpha/2}\right), (j = 0, 1) \quad (8.4.15)$$

---

[3] Note once again that for $\alpha = 1$ Eq. (8.4.9) becomes the standard diffusion equation and the initial condition (8.4.11) becomes the standard Cauchy condition.

where $\varphi(a, b; z)$ is the classical Wright function (see Definition (7.1.1)). The formal solution (8.4.14) becomes the real one if the integrals on the right-hand side of (8.4.14) converge.

8.5 Numerical Methods for the Solution of Fractional Differential Equations

In previous sections we presented several results on fractional differential equations for which the Mittag-Leffler function either plays the role of the solution or is incorporated into one or another construction related to the solution (for example, in fractional analogs of Green's function, see, e.g. [KiSrTr06]). For such a situation it is of great interest to find an approximate scheme for the calculation of values of the Mittag-Leffler function. In relation to this, we mention the numerical procedures proposed in [GoLoLu02] (where algorithms for numerical evaluation of the Mittag-Leffler function and its derivatives for all $\alpha > 0$, $\beta \in \mathbb{R}$ in the complex plane $\mathbb{C}$ are proposed), in [GarPop13] (for evaluation of the function $e_{\alpha,\beta}(t; \lambda) = t^{\beta-1} E_{\alpha,\beta}(-t^\alpha \lambda)$ on the real line), in [Gar15] (for approximate calculation of two- and three-parametric Mittag-Leffler functions in different domains in the complex plane), and in [GaRoMa17] (for a numerical study of the Le Roy type function). We also have to mention here the paper [ZenChe14] in which the Padé approximation procedure is applied to calculate $E_{\alpha,\beta}(-x)$ and its inverse for certain values of the parameters. Finally, we mention [RaPaZv69], where the first table of values of the Mittag-Leffler function is presented.

Another approach which is frequently used in the study of fractional and differential equations is to approximate the fractional differential or integral operators. This direction goes back to brilliant works by Letnikov and Grünwald. We mention here the papers [Let68a], [Let68b], [Gru67], as well as the book [LetChe11]. However, this approach is discussed more or less completely in any book on Fractional Calculus.

Lastly, there are many approaches to fractional differential equations (mainly nonlinear) by direct or indirect procedures. Some of these procedures are presented in the survey by Diethelm on such numerical methods included in the book [Bal-et-al17] (see also [Die10], [Die-et-al05, Die19]). We briefly outline below the essence of these methods. Note that we apply the numerical methods to specific fractional differential equations. Extension of these methods to other types of equations needs certain extra techniques.

8.5.1 Direct Numerical Methods

Let us consider the initial value problem for the nonlinear fractional differential equation with the Caputo fractional derivative

$$^C D_{0+}^\alpha y(x) = f(x, y(x)), \quad x \in [0, T], \quad y^{(k)}(0) = y_0^{(k)} \quad (k = 0, 1, \ldots, [\alpha] - 1). \tag{8.5.1}$$

As was already mentioned, the first idea to construct a numerical scheme for the solution is to approximate the fractional derivative using a grid of points $0 = x_0 < x_1 < \ldots < x_m = X$

$$^C D_{0+}^\alpha y(x_n) \approx \sum_{k=0}^{n} a_{k,n} y(x_k), \tag{8.5.2}$$

which by replacing unknown values $y(x_k)$ by their approximate values y_k leads to the following system

$$\sum_{k=0}^{n} a_{k,n} y_k = f(x_n, y_n), \quad n = 1, 2, \ldots. \tag{8.5.3}$$

For simplicity we can consider the uniform grid on $[0, X]$

$$x_j = \frac{j}{N} X \quad (j = 0, 1, \ldots, N); \quad h = X/N. \tag{8.5.4}$$

Thus the question is how to determine y_k for $k = 1, 2, \ldots, N$ ($y_0 = y_0^{(0)}$ is already prescribed by the initial conditions in (8.5.1)).

First we mention *quadrature based direct methods*. Since there is an evident connection between the Caputo and the Riemann–Liouville fractional derivatives

$$^C D_{0+}^\alpha y(x) = {}^{RL}D_{0+}^\alpha \left(y - \sum_{k=0}^{n-1} \frac{t^k}{k!} y^{(k)}(0) \right)(x)$$

it is sufficient to discretize the Riemann–Liouville fractional operator. The latter gives

$$h^\alpha \sum_{j=0}^{k} A_{j,k} y_j = f(x_k, y_k), \quad k = 1, 2, \ldots, N. \tag{8.5.5}$$

With $A_{k,k} = \frac{1}{\Gamma(2-\alpha)}$ this leads to the following system of nonlinear equations

$$y_k = \Gamma(2-\alpha) h^\alpha f(x_k, y_k) - \Gamma(2-\alpha) \sum_{j=0}^{k-1} A_{j,k} y_j, \quad k = 1, 2, \ldots, N. \tag{8.5.6}$$

This system has a unique solution if f is continuous, satisfies the Lipschitz condition with respect to y with Lipschitz constant L and $h < (\Gamma(2-\alpha)L)^{-1/\alpha}$. Moreover [Bal-et-al17, Thm. 2.6] the system (8.5.6) generates the solution to (8.5.1), satisfying

$$|y(x_j) - y_j| = O\left(h^{2-\alpha}\right)$$

uniformly for each j whenever $y \in C^2[0, X])$.

8.5.2 Indirect Numerical Methods

The basic idea for such methods is to reduce the initial boundary value problem (8.5.1) to the following nonlinear Volterra equation with weakly singular kernel. It can be obtained by applying the Riemann–Liouville integral operator J^α to the fractional differential equation in (8.5.1) and taking into account the initial conditions:

$$y(x) = \sum_{k=0}^{\lceil \alpha \rceil - 1} \frac{y_0^{(k)}}{k!} x^k + (J^\alpha f(\cdot, y(\cdot)))(x). \tag{8.5.7}$$

It is known (see, e.g. [DieFor02]) that if the function f is continuous then every continuous solution to (8.5.7) is a solution to (8.5.1) and vice versa.

The first indirect method we mention is an *Adams type predictor-corrector method*, which is based on the approximation of the integral operator in (8.5.7) by the product trapezoid method (see [Bal-et-al17, (2.1.8-9)]). By this we arrive at the following system of nonlinear equations

$$y_k = \sum_{j=0}^{\lceil \alpha \rceil - 1} \frac{x_k^j}{k!} y_0^{(j)} + h^\alpha \sum_{j=0}^{k} a_{jk} f(x_j, y_j), \quad k = 1, 2, \ldots, N, \tag{8.5.8}$$

where a_{jk} are coefficients in the approximate formula for the integral operator J^α. Relation (8.5.8) is called in [Bal-et-al17] a *fractional Adams–Moulton formula*. Sufficient conditions for the unique solvability of the system (8.5.8) are given in [Bal-et-al17, Thm. 2.8]. Based on the Adams–Moulton formula, the following iterative procedure is proposed (which is known as the *predictor-corrector method*):

(1) Compute the so-called *predictor*, i.e. the first approximation $y_{k,0}$ of the solution y_k of (8.5.8) by means of the $O(h^{p-1})$-order explicit algorithm.
(2) Determine the so-called *corrector* $y_{k,1}$ by using a one step iteration via the formula

$$y_{k,1} = \sum_{j=0}^{\lceil \alpha \rceil - 1} \frac{x_k^j}{k!} y_0^{(j)} + h^\alpha \sum_{j=0}^{k-1} a_{jk} f(x_j, y_j) + h^\alpha a_{kk} f(x_k, y_{k,0}) \tag{8.5.9}$$

and use this value in place of the true solution y_k in all further steps.

The theory of numerical methods says that $y_k = y_{k,1}$ approximates the solution $y(x_k)$ up to an error of order $O(h^p)$.

A modification of the above approach (known as the *fractional Adams–Bashforth method*) appears when the product trapezoidal formula is replaced by the product rectangle formula for approximation of the integral operator J^α.

Another modification of the above described predictor-corrector method is proposed in [Den07] and is based on the idea of splitting the Riemann–Liouville integral into subintervals and using the product rectangle formula only on the last subinterval ($j = k - 1$) in order to obtain a suitable predictor $y_{k,0}$ and then m-fold corrector.

8.5.3 Other Numerical Methods

Other numerical methods are also described in [Bal-et-al17, Chap. 2]. Among them we point out *linear fractional multistep methods* (which originated in the paper by Lubich [Lub85], see also [Lub86]). We also mention the article [ForCon06] in which a comparison of the performance of the linear multistep methods with other algorithms for the approximate solution to fractional differential equations is provided.

A class of numerical methods is proposed and investigated in [ZayKar14] (which can be called *spectral methods*). These are methods based on the expansion of the solutions in terms of eigenfunctions of certain operators (for details, see [Bal-et-al17, Sect. 2.5]).

From the classical books by Adomian [Ado89], [Ado94] follows the approach known as the *decomposition method*. A comprehensive review of the new iterative method (NIM) (which is an Adomian decomposition method) is presented in [GejKum17], see also [Gej14].

Some other methods are briefly discussed in [Bal-et-al17, Chap. 2] (see also the list of references corresponding to the results presented in the above mentioned chapter). These are the *variational iteration method*, which is initiated as an Adomian decomposition from the study of abstract operator equations, *the method of nonclassical representations of FDO* based on certain identities for fractional derivatives, *collocation methods*, which are extensively used in the theory of nonlinear equations and make use of a presentation of the unknown function via linear combinations of basis functions in a properly chosen functional space, *the method of the terminal value condition*, when one of the initial conditions is replaced by the condition at the end-point of the considered interval.

8.6 Historical and Bibliographical Notes

The most simple integral equations of fractional order, namely the Abel integral equations of the first kind, were investigated by Abel himself [Abe26a]. Abel integral equations of the second kind were studied by Hille and Tamarkin [HilTam30].

8.6 Historical and Bibliographical Notes

In this paper for the first time the solution was represented via the Mittag-Leffler function. The interested reader is referred to [SaKiMa93], [CraBro86], [GorVes91], and [Gor96], [Gor98] for historical notes and detailed analyses with applications.

It is well known that Niels Henrik Abel was led to his famous equation by the mechanical problem of the *tautochrone*, that is by the problem of determining the shape of a curve in the vertical plane such that the time required for a particle to slide down the curve to its lowest point is equal to a given function of its initial height (which is considered as a variable in an interval $[0, H]$). After appropriate changes of variables he obtained his famous integral equation of the first kind with $\alpha = 1/2$. He did, however, solve the general case $0 < \alpha < 1$. As a special case Abel discussed the problem of the *isochrone*, in which it is required that the time taken for the particle to slide down is independent of the initial height. Already in his earlier publication [Abe23] he had recognized the solution as a derivative of non-integer order. We point out that integral equations of Abel type, including the simplest (8.1.1) and (8.1.14), have found so many applications in diverse fields that it is almost impossible to provide an exhaustive list of them.

Abel integral equations occur in many situations where physical measurements are to be evaluated. In many of these the independent variable is the radius of a circle or a sphere and only after a change of variables does the integral operator take the form J^α, usually with $\alpha = 1/2$, and the equation is of the first kind. Applications are, for example, in the evaluation of spectroscopic measurements of cylindrical gas discharges, the study of the solar or a planetary atmosphere, the investigation of star densities in a globular cluster, the inversion of travel times of seismic waves for the determination of terrestrial sub-surface structure, and spherical stereology. Descriptions and analyses of several problems of this kind can be found in the books by Gorenflo and Vessella [GorVes91] and by Craig and Brown [CraBro86], see also [Gor96]. Equations of the first and of the second kind, depending on the arrangement of the measurements, arise in spherical stereology. See [Gor98] where an analysis of the basic problems and many references to the previous literature are given.

Another field in which Abel integral equations or integral equations with more general weakly singular kernels are important is that of *inverse boundary value problems* in partial differential equations, in particular parabolic ones in which the independent variable naturally has the meaning of time.

A number of integral equations similar to the Abel integral equation of the second kind are discussed in the book by Davis [Dav36, Chap. 6].

In this part of the historical overview of solutions of linear and non-linear fractional differential equations we partly follow the survey papers [KilTru01], [KilTru02], and the books [KiSrTr06] and [Die10].

The paper of O'Shaughnessay [O'Sha18] was probably the first where the methods for solving the differential equation of half-order

$$(D^{1/2} y)(x) = \frac{y}{x} \qquad (8.6.1)$$

were considered. Two solutions of such an equation

$$y(x) = x^{-1/2} e^{-1/x} \qquad (8.6.2)$$

and a series solution

$$y(x) = 1 - i\sqrt{\pi} x^{-1/2} e^{-1/x} + x^{-1/2} e^{-1/x} \int_{-\infty}^{x} t^{3/2} e^{1/t} dt$$

$$= 1 - i\sqrt{\pi} x^{-1/2} - 2x^{-1} + i\sqrt{\pi} x^{-3/2} + \cdots \qquad (8.6.3)$$

were suggested by O'Shaughnessay and discussed later by Post [Pos19]. Their arguments were formal and based on an analogy with the Leibnitz rule, which for the Riemann–Liouville fractional derivative has the form

$$\left(D_{0+}^{\alpha}(fg)\right)(x) = \sum_{k=0}^{\infty} \frac{\Gamma(\alpha+1)}{\Gamma(\alpha-k+1)k!} (D_{0+}^{\alpha-k} f)(x) g^{(k)}(x). \qquad (8.6.4)$$

As was proved later (see, e.g. [MilRos93, pp. 195–199]), (8.6.2) is really a solution of the Eq. (8.6.1)

$$(D_{0+}^{1/2} y)(x) = \frac{y}{x} \qquad (8.6.5)$$

with the Riemann–Liouville fractional derivative $D_{0+}^{1/2} y$.

As for (8.6.3), it is not a solution of the equation (8.6.1) with $D^{\alpha} = D_{a+}^{\alpha}$, a being any real constant, because O'Shaughnessay and Post made a mistake while using the relation for the composition of $I^{1/2} D^{1/2} y$. Such a relation for the Riemann–Liouville derivative $D_{a+}^{\alpha} y$ has the form

$$(I_{a+}^{\alpha} D_{a+}^{\alpha} y)(x) = y(x) - \sum_{k=1}^{n} B_k \frac{(x-a)^{\alpha-k}}{\Gamma(\alpha-k+1)}, \qquad (8.6.6)$$

where

$$B_k = y_{n-\alpha}^{(n-k)}(a), \ y_{n-\alpha}(x) = (I_{a+}^{n-\alpha} y)(x), \ (\alpha \in \mathbf{C} \ n = [\text{Re}(\alpha)] + 1), \qquad (8.6.7)$$

in particular,

$$(I_{a+}^{\alpha} D_{a+}^{\alpha} y)(x) = y(x) - B \frac{(x-a)^{\alpha-1}}{\Gamma(\alpha)}, \ B = y_{1-\alpha}(a), \qquad (8.6.8)$$

8.6 Historical and Bibliographical Notes

for $0 < \operatorname{Re}(\alpha) < 1$. O'Shaughnessay and Post applied (8.6.8) for $\alpha = 1/2$ and $a = 0$ by considering the constant B instead of the monomial $Bx^{-1/2}$ (see, e.g., [KilTru01, Sect. 3]).

Mandelbroit [Man25] arrived at a differential equation of fractional order when he investigated an extremum problem for the functional

$$\int_0^1 F[D_{a+}^\alpha y(x); x] dx$$

with the Riemann–Liouville fractional derivative $D_{a+}^\alpha y(x)$. He had assumed that the corresponding variations are equal to zero and obtained the differential equation with Cauchy conditions

$$dF \equiv F[D_{a+}^\alpha y(x); x] = 0, \quad y^{(k)}(a) = b_k \ (k = 1, 2, \cdots, n). \tag{8.6.9}$$

Fujiwara [Fuj33] considered the differential equation of fractional order

$$(\mathcal{D}_+^\alpha y)(x) = \left(\frac{\alpha}{x}\right)^\alpha y(x) \tag{8.6.10}$$

with the Hadamard fractional derivative of order $\alpha > 0$ defined as followed

$$(\mathcal{D}_+^\alpha y)(x) = \left(\frac{d}{dx}\right)^n \frac{1}{\Gamma(n-\alpha)} \int_0^x \frac{y(t) dt}{t(\log(x/t))^{\alpha-n+1}} \quad (n = [\alpha] + 1), \tag{8.6.11}$$

with $n \in \mathbb{N} = \{1, 2, \ldots\}$ and $\alpha \notin \mathbb{N}$. He obtained a formal solution of (8.6.10) in the form of the Mellin–Barnes integral

$$y(x) = \frac{1}{2\pi i} \int_{\gamma-i\infty}^{\gamma+i\infty} [\Gamma(s) x^s]^\alpha ds \ (\gamma > 0), \tag{8.6.12}$$

and proved that $y(x)$ has the following asymptotics at zero

$$y(x) \sim Ax^\lambda e^{-\mu/x}, \tag{8.6.13}$$

with

$$A = \frac{1}{\sqrt{\alpha}} (2\pi)^{(\alpha-1)/2}, \quad \lambda = \frac{\alpha-1}{2}, \quad \mu = \alpha. \tag{8.6.14}$$

Pitcher and Sewell [PitSew38] first considered the non-linear FDE

$$(D_{a+}^\alpha y)(x) = f(x, y(x)) \ (0 < \alpha < 1, \ a \in \mathbb{R}) \tag{8.6.15}$$

with the Riemann–Liouville fractional derivative $D_{a+}^{\alpha} y$ provided that $f(x, y)$ is bounded in the special region G lying in $\mathbb{R} \times \mathbb{R}$ ($\mathbb{R} = (-\infty, \infty)$) and satisfies the Lipschitz condition with respect to y:

$$|f(x, y_1) - f(x, y_2)| \leq A|y_1 - y_2|, \tag{8.6.16}$$

where the constant $A > 0$ does not depend on x. They tried to prove the uniqueness of a continuous solution $y(x)$ of such an equation on the basis of the corresponding result for the non-linear integral equation

$$y(x) - \frac{1}{\Gamma(\alpha)} \int_a^x \frac{f[t, y(t)]dt}{(x-t)^{1-\alpha}} = 0 \ (x > a;\ 0 < \alpha < 1). \tag{8.6.17}$$

But the result of Pitcher and Sewell given in [PitSew38, Theorem 4.2] is not correct since they made the same mistake as O'Shaughnessay [O'Sha18] and Post [Pos19] by using the relation $I_{a+}^{\alpha} D_{a+}^{\alpha} y = y$ instead of (8.6.8).

However, the paper of Pitcher and Sewell [PitSew38] contained the idea of the reduction of the fractional differential equation (8.6.15) to the Volterra integral equation (8.6.17).

Al-Bassam [Al-B65] first considered the following Cauchy-type problem

$$(D_{a+}^{\alpha} y)(x) = f(x, y(x)) \ (0 < \alpha \leq 1), \tag{8.6.18}$$

$$(D_{a+}^{\alpha-1} y)(x)|_{x=a} \equiv (I_{a+}^{1-\alpha} y)(x)|_{x=a} = b_1,\ b_1 \in \mathbb{R}, \tag{8.6.19}$$

in the space of continuous functions $C[a, b]$ provided that $f(x, y)$ is a real-valued, continuous and Lipschitzian function in a domain $G \subset \mathbb{R} \times \mathbb{R}$ such that $\sup_{(x,y) \in G} |f(x, y)| = b_0 < \infty$. Applying the operator I_{a+}^{α} to both sides of (8.6.18), using the relation (8.6.8) and the initial conditions (8.6.19), he reduced (8.6.18)–(8.6.19) to the Volterra non-linear integral equation

$$y(x) = \frac{b_1 (x-a)^{\alpha-1}}{\Gamma(\alpha)} + \frac{1}{\Gamma(\alpha)} \int_a^x \frac{f[t, y(t)]dt}{(x-t)^{1-\alpha}} \ (x > a;\ 0 < \alpha \leq 1). \tag{8.6.20}$$

Using the method of successive approximations he established the existence of a continuous solution $y(x)$ of the Eq. (8.6.20). Furthermore, he was probably the first to indicate that the method of contracting mapping can be applied to prove the uniqueness of this solution $y(x)$ of (8.6.20), and gave such a formal proof. Al-Bassam also indicated—*but did not prove*—the equivalence of the Cauchy-type problem (8.6.18)–(8.6.19) and the integral equation (8.6.20), and therefore his results on the existence and uniqueness of the continuous solution $y(x)$, formulated in [Al-B65, Theorem 1], could be true only for the integral equation (8.6.20). We also note that the conditions suggested by Al-Bassam are not suitable to solve the Cauchy-type problem (8.6.18)–(8.6.19) in the simplest linear case when $f[x, y(x)] = y(x)$.

8.6 Historical and Bibliographical Notes

The same remarks apply to the existence and uniqueness results formulated without proof in [Al-B65, Theorems 2, 4, 5, 6] for a more general Cauchy-type problem of the form (8.6.18)–(8.6.19) with real $\alpha > 0$:

$$(D_{a+}^{\alpha} y)(x) = f(x, y(x)) \ (n - 1 < \alpha \leq n, \ n = -[-\alpha]), \tag{8.6.21}$$

$$(D_{a+}^{\alpha-k} y)(x)|_{x=a} = b_k, \ b_k \in \mathbb{R} \ (k = 1, 2, \cdots, n), \tag{8.6.22}$$

where the corresponding Volterra equation has the form (8.6.20):

$$y(x) = \sum_{k=1}^{n} \frac{b_k (x - a)^{\alpha-k}}{\Gamma(\alpha - k + 1)} + \frac{1}{\Gamma(\alpha)} \int_a^x \frac{f[t, y(t)] dt}{(x - t)^{1-\alpha}} \ (x > a; \ n - 1 < \alpha \leq n) \tag{8.6.23}$$

for the system of equations (8.6.18) and for non-linear fractional equations more general than (8.6.18)

$$(D_{a+}^{n\alpha} y)(x) =$$

$$f\left(x, y(x), (D_{a+}^{\alpha} y)(x), (D_{a+}^{2\alpha} y)(x), \cdots, (D_{a+}^{(n-1)\alpha} y)(x)\right) \ (0 < \alpha \leq 1) \tag{8.6.24}$$

and linear fractional equations

$$\sum_{k=0}^{n} c_k(x)(D_{a+}^{(n-k)\alpha} y)(x) = f(x), \ (0 < \alpha \leq 1) \tag{8.6.25}$$

for continuous $f(x, x_1, x_2, \cdots, x_m)$ and $f(x), p_k(x) \ (0 \leq k \leq m)$ and under the initial conditions

$$(D_{a+}^{k\alpha-1} y)(x)|_{x=a} = b_k \ (k = 1, 2, \cdots, n). \tag{8.6.26}$$

The above and some other results were presented in Al-Bassam [Al-B82]–[Al-B87]. Cauchy-type problems for non-linear ordinary differential equations of fractional order have been studied by many authors (in particular, developing Al-Bassam's method). An extended bibliography on subject is presented in the survey paper [KilTru01], and in the book [KiSrTr06].

Cauchy-type problems for linear ordinary differential equations of fractional order were investigated mainly by using the method of reduction to Volterra integral equations. First, we have to mention the paper by Barrett [Barr54], which first considered the Cauchy-type problem for the linear differential equation with the Riemann–Liouville fractional derivative on a finite interval (a, b) of the real axis

$$(D^\alpha_{a+}y)(x) - \lambda y(x) = f(x) \ (n-1 \le \operatorname{Re}(\alpha) < n; \ \lambda \in \mathbb{C}), \tag{8.6.27}$$

$$(D^{\alpha-k}_{a+}y)(x)|_{x=a+} = b_k \in \mathbb{C} \ (k=1,2,\cdots,n) \tag{8.6.28}$$

$n = [\operatorname{Re}(\alpha)] + 1$, $\alpha \ne n-1$. He proved that if $f(x)$ belongs to $L(a,b)$ or $L(a,b) \cap C(a,b]$, then the problem has a unique solution $y(x)$ in some subspaces of $L(a,b)$ and this solution is given by

$$y(x) = \sum_{k=1}^{n} b_k (x-a)^{\alpha-k} E_{\alpha,\alpha-k+1}(\lambda(x-a)^\alpha)$$

$$+ \int_a^x (x-t)^{\alpha-1} E_{\alpha,\alpha}(\lambda(x-t)^\alpha) f(t) dt \tag{8.6.29}$$

where $E_{\alpha,\beta}(z)$ is the two-parametric Mittag-Leffler function. Barrett's argument was based on the formula (8.6.6) for the product $I^\alpha_{a+} D^\alpha_{a+} f$. From (8.6.29) Barrett obtained the unique solution

$$y(x) = \sum_{k=1}^{n} b_k (x-a)^{\alpha-k} E_{\alpha,\alpha-k+1}(\lambda(x-a)^\alpha) \tag{8.6.30}$$

of the Cauchy-type problem for the homogeneous equation ($f(x) = 0$) corresponding to (8.6.27):

$$(D^\alpha_{a+}y)(x) - \lambda y(x) = 0 \ (n-1 \le \operatorname{Re}(\alpha) < n), \tag{8.6.31}$$

$$(D^{\alpha-k}_{a+}y)(x)|_{x=a+} = b_k \in \mathbf{C} \ (k=1,2,\cdots,n). \tag{8.6.32}$$

He also proved the uniqueness of the solution $y(x)$ of the simplest such Cauchy-type problem (8.6.31)–(8.6.32) with $0 < \alpha < 1$ and $\lambda = -1$. Barrett implicitly used the method of reduction of the Cauchy-type problem (8.6.27)–(8.6.28) to the Volterra integral equation of the second kind and the method of successive approximations.

Dzhrbashian and Nersesyan [DzhNer68] studied the linear differential equation of fractional order

$$(D^\sigma y)(x) \equiv (D^{\sigma_n} y)(x) + \sum_{k=0}^{n-2} a_k(x)(D^{\sigma_{n-k-1}} y)(x) + a_n(x) y(x) = f(x) \tag{8.6.33}$$

with sequential fractional derivatives $(D^\sigma y)(x)$ and $(D^{\sigma_{n-k-1}} y)(x)$ ($k = 0, 1, \cdots$, $n-1$) defined in terms of the Riemann–Liouville fractional derivatives. Here, the term "sequential" means that the orders of the derivatives are related in the following manner:

8.6 Historical and Bibliographical Notes

$$\sigma_k = \sum_{j=0}^{k} \alpha_j - 1 \ (k = 0, 1, \cdots, n); \ 0 < \alpha_j \leq 1 \ (j = 0, 1, \cdots, n), \quad (8.6.34)$$

$$(\alpha_k = \sigma_k - \sigma_{k-1} \ (k = 1, 2, \cdots, n), \ \alpha_0 = \sigma_0 + 1).$$

They proved that for $\alpha_0 > 1 - \alpha_n$ the Cauchy-type problem

$$(D^\sigma y)(x) = f(x), \ (D^{\sigma_k} y)(x)|_{x=0} = b_k \ (k = 0, 1, \cdots, n-1) \quad (8.6.35)$$

has a unique continuous solution $y(x)$ on an interval $[0, d]$ provided that the functions $a_k(x)$ $(0 \leq k \leq n-1)$ and $f(x)$ satisfy some additional conditions. In particular, when $a_k(x) = 0$ $(k = 0, 1, \cdots, n)$, they obtained the explicit solution

$$y(x) = \sum_{k=0}^{n-1} \frac{b_k x^{\sigma_k}}{\Gamma(1+\sigma_k)} + \frac{1}{\Gamma(\sigma_n)} \int_a^x (x-t)^{\sigma_n - 1} f(t) dt \quad (8.6.36)$$

of the Cauchy-type problem

$$(D^{\sigma_n} y)(x) = f(x), \ (D^{\sigma_k} y)(x)|_{x=0} = b_k \ (k = 0, 1, \cdots, n-1). \quad (8.6.37)$$

Laplace transform methods for ordinary fractional differential equations have successfully been used by many authors. Maravall [Mara71] was probably the first who suggested a formal approach based on the Laplace transform to obtain the explicit solution of a particular case of the equation

$$\sum_{k=1}^{m} c_k (D_{a+}^{\alpha_k} y)(x) + c_0 y(x) = f(x) \ (0 < \text{Re}(\alpha_1) < \text{Re}(\alpha_2) < \cdots < \text{Re}(\alpha_m)),$$

$$(8.6.38)$$

where $m \geq 1$, $D_{a+}^{\alpha_k} y$ $(k = 1, 2, \cdots, m)$, are the Riemann–Liouville fractional derivatives, and $c_k \neq 0$, $(k = 0, 2, \cdots, m)$, are real or complex constants. However, since this paper was published in Spanish, it was practically unknown. Later the method of Laplace transforms was used in a different form, based on the main properties of the Laplace transform (see, e.g., [Doe74] and [DitPru65]) and the analytic properties of so-called characteristic quasi-polynomials. Special attention was paid to this approach in recent FDA-Congresses and FDTAs symposiums (see [Adv-07], [NewTr-10] and references therein, as well as [KilTru01], and the books [KiSrTr06], [Cap-et-al10]).

The operational calculus method for ordinary differential equations of fractional order is based on the interpretation of the Laplace convolution

$$(f * g)(x) = \int_0^x f(x-t)g(t)dt \quad (8.6.39)$$

as a multiplication of elements f and g in the ring of continuous functions on the half-axis $\mathbb{R}_+$ (see, e.g., [Mik59]). In [LucSri95], the operational calculus for the Riemann–Liouville fractional derivative $D_{0+}^\alpha y$ was constructed. This method was generalized and developed by several authors (see [KilTru01]). This calculus was applied to the solution of Cauchy-type problems for fractional differential equations of a special kind.

The idea of the composition method for ordinary differential equations of fractional order is based on the known formula for the Riemann–Liouville fractional derivative

$$(D_{a+}^\alpha (t-a)^{\beta-1})(x) = \frac{\Gamma(\beta)}{\Gamma(\beta-\alpha)}(x-a)^{\beta-\alpha-1} \quad (\text{Re}(\beta) > \text{Re}(\alpha) > 0) \quad (8.6.40)$$

(see [SaKiMa93, (2.26) and (2.35)]). These arguments lead us to the conjecture that compositions of fractional derivatives and integrals with elementary functions can give exact solutions of differential and integral equations of fractional order. Moreover, from here we deduce another possibility concerning such results for compositions of fractional calculus operators with special functions. It allows us to find the exact solutions of new classes of differential and integral equations of fractional order. This method was developed by A. Kilbas with co-authors and uses composition of Riemann–Liouville fractional operators with different types of special functions (in particular, the three parametric Mittag-Leffler function, or the Kilbas–Saigo function, see, e.g. [KilSai95a] and [KiSrTr06] and the references therein).

The above investigations were devoted to the solution of the fractional differential equations with the Riemann–Liouville fractional derivative $D_{a+}^\alpha y$ on a finite interval $[a, b]$ of the real axis $\mathbb{R}$. Such equations with the Caputo fractional derivative $^C D_{a+}^\alpha y$ have not been studied extensively. Gorenflo and Mainardi [GorMai96] applied the Laplace transform to solve the fractional differential equation

$$\left(^C D_{0+}^\alpha y\right)(x) - \lambda y(x) = f(x) \quad (x > 0; a > 0; \lambda > 0)$$

with the Caputo fractional derivative of order $\alpha > 0$ and with the initial conditions

$$y^{(k)}(0) = b_k \ (k = 0, 1, \ldots, n-1; n-1 < \alpha \leq n; n \in \mathbb{N}).$$

They discussed the key role of the Mittag-Leffler function for the cases $1 < \alpha < 2$ and $2 < \alpha < 3$. In relation to this, see also the papers by Gorenflo and Mainardi [GorMai97], Gorenflo and Rutman [GorRut94], and Gorenflo et al. [GoMaSr98]. Luchko and Gorenflo [LucGor99] used the operational method to prove that the above Cauchy problem has a unique solution in terms of the Mittag-Leffler functions in a special space of functions on the half-axis $\mathbb{R}_+$. They also obtained the explicit solution to the Cauchy problem for the more general fractional differential equation

8.6 Historical and Bibliographical Notes

$$\left({}^C D_{0+}^\alpha y\right)(x) - \sum_{k=1}^{m} c_k \left({}^C D_{0+}^{\alpha_k} y\right)(x) = f(x) \ (\alpha > \alpha_1 > \ldots > \alpha_m \geq 0)$$

via certain multivariate Mittag-Leffler functions.

It was probably Dzhrbashian [Dzh70] who first considered the Dirichlet-type problem for the integro-differential equations of fractional order. The problem is to find the solution $y(x)$ (in $L(0, T)$ or in $L_2(0, T)$) on a finite interval $(0, T)$ of the following equation

$$(D^\sigma y)(x) - [\lambda + q(x)]y(x) = 0 \ (0 < x < T), \tag{8.6.41}$$

where the operator D^σ is defined in terms of the Riemann–Liouville fractional derivatives and integrals with $a = 0$, which satisfy the initial conditions

$$a(I_{0+}^{1-\alpha_0} y)(x)|_{x=0} + b(I_{0+}^{1-\alpha_1} y)(x)|_{x=0} = 0, \tag{8.6.42}$$

and

$$c(I_{0+}^{1-\alpha_0} y)(x)|_{x=T} + d(I_{0+}^{1-\alpha_1} y)(x)|_{x=T} = 0, \tag{8.6.43}$$

with Lipschitzian $q(x)$ and real a, b, c and d such that $a^2 + b^2 = 1$ and $c^2 + d^2 = 1$. When $\alpha_0 = \alpha_1 = \alpha_2 = 1$ the problem (8.6.41), (8.6.42)–(8.6.43) is reduced to the Sturm–Liouville problem for the ordinary differential equation of second order:

$$y''(x) - [\lambda + q(x)]y(x) = 0 \ (0 < x < T), \ ay(0) + by(0) = 0, \ cy(T) + by(T) = 0. \tag{8.6.44}$$

In particular, the latter problem is reduced to the study of the distribution of zeros for different types of special functions (two- and multi-parametric Mittag-Leffler functions among them). The approach was developed in a series of articles (see, e.g., [Ale82], [Ale84], [Del94], [Djr93], [Nak74], [Nak77], [Veb88]).

The well-known space S of Schwartz test functions, which are infinitely differentiable and rapidly vanish at infinity together with all derivatives, as well as the space $C_0^\infty \subset S$ of infinitely differentiable functions with compact support, is not completely adapted for fractional derivatives and integrals. Although fractional derivatives and integrals of functions from these spaces are infinitely differentiable, they do not have sufficiently good behavior at infinity. Therefore differential equations of fractional order have to be studied in some spaces of test and generalized functions which are invariant with respect to fractional differentiation and integration. A series of results of this type was described in the survey paper [KilTru02]. Some other methods and results for linear and non-linear ordinary differential equations of fractional order are also presented.

In Sect. 8.3 we mainly follow a survey paper by I. Petráš [Pet19] in the 6th volume of the Handbook of Fractional Calculus with Applications (see also [Pet11], as well as the papers in the above mentioned volume [HAND6] and references therein). We mention here some specific results in the area of fractional order control. We also note that there are several connections of the fractional optimal control problem to the subject of this book. Thus in some special cases the controller can be similar to what we obtain as the Laplace transform of certain generalizations of the Mittag-Leffler function. Moreover, in the time domain the fractional optimal control problem is reduced to a fractional differential equation. For more details we refer to [Pod99, Chap. 9] (see also [Pod99a]). Not all cases of fractional order control problems are solved by means of generalizations of the Mittag-Leffler function. We hope that our presentation of some fractional optimal control models will give some perspective on applications of the Mittag-Leffler function in this area.

Close to the above direction is the study of extremal problems (optimal control problems in a fractional context), which involve the optimization of a certain functional on a set of solutions to a fractional order dynamical system. In [BerBou18] attention is paid to a general optimal control problem involving a dynamical system described by a nonlinear Caputo fractional differential equation of order $0 < \alpha \leq 1$, associated to a general Bolza cost written as the sum of a standard Mayer cost and a Lagrange cost given by a Riemann–Liouville fractional integral of order $\beta \geq \alpha$. The thesis [Tep17] is devoted to the study of fractional-order calculus-based modeling and control of dynamic systems with process control applications. In particular, methods for time and frequency domain identification of fractional order models are proposed and discussed. These methods largely form the foundation of model-based control design, which constitutes the next part of the thesis, where new methods dealing with optimization of fractional controllers as well as stabilization of unstable systems are presented. Implementation of fractional-order systems and controllers is also investigated, as it is especially important in real-time control applications. All the methods discussed are then presented in the context of a fractional-order modeling and control framework developed for the MATLAB/Simulink environment.

The classification of linear and non-linear partial differential equations of fractional order is still far from complete. Several results for partial differential equations are described in [KilTru02] (see also [KiSrTr06]). Among these results we mention the pioneering work by Gerasimov [Ger48] and recent books [Die10], [Mai10]. This area is rapidly growing since most of the results are related to different types of applications. Therefore it is impossible to describe all existing results.

We also mention here several contributions by different authors. In addition to the above cited work we indicate the papers by Veber ([Veb74]–[Veb85b]), Malakhovskya and Shikhmanter [MalShi75] (see also [KiSrTr06]), where ordinary fractional differential equations are studied in spaces of generalized functions. In this regard, see Sects. 2 and 3 of the survey paper by Kilbas and Trujillo [KilTru02]. Some authors have constructed formal partial solutions to ordinary differential equations with other fractional derivatives. Nishimoto [Nis84, Volume II, Chap. 6], Nishimoto et al. [NiOwSr84], Srivatsava et al. [SrOwNi84], [SrOwNi85] and Campos [Cam90] constructed explicit solutions of some particular fractional differential equations with the so-called fractional derivatives of complex order (see, for example, Samko et al. [SaKiMa93, Sect. 22.1]). A series of papers by Wiener (see, e.g. [Wie79])

8.6 Historical and Bibliographical Notes

were devoted to the investigation of ordinary linear fractional differential equations and systems of such equations involving the fractional derivatives defined in the Hadamard finite part sense (see, for example, [KilTru02, Sect. 4]). We also note that many authors have applied methods of fractional integro-differentiation to construct solutions of ordinary and partial differential equations, to investigate integro-differential equations, and to obtain a unified theory of special functions. The methods and results in these fields are presented in Samko et al. [SaKiMa93, Chap. 8]) and in Kiryakova [Kir94]. We mention here the papers by Al-Saqabi [Al-S95], by Al-Saqabi and Vu Kim Tuan [Al-STua96] and by Kiryakova and Al-Saqabi [Al-SKir98], [KirAl-S97a], [KirAl-S97b], where solutions in closed form were constructed for certain integro-differential equations with the Riemann–Liouville and Erdelyi–Kober-type fractional integrals (see also [KeScBe02a, Prus93, Sne75, Fuj90a]).

In [Hil02] the infinitesimal generator of time evolution in the standard equation for exponential (Debye) relaxation is replaced with the infinitesimal generator of composite fractional translations. Composite fractional translations are defined as a combination of translation and the fractional time evolution. The fractional differential equation for composite fractional relaxation is solved. The resulting dynamical susceptibility is used to fit broadband dielectric spectroscopy data of glycerol. The composite fractional susceptibility function can exhibit an asymmetric relaxation peak and an excess wing at high frequencies in the imaginary part.

The numerical methods applied are discussed in Sect. 8.5, and are based on [Bal-et-al17, Chap. 2]. Other sources for the complete discussion can be found in the list of references of the above book as well as in the papers mentioned in Sect. 8.5.

Lastly, we refer interested researchers to the second volume of the Handbook of Fractional Calculus with Applications [HAND2], which is devoted to the study of fractional differential equations. In particular, attempts to construct the general theory of fractional partial differential equations are presented in the papers [Koc19b, Koc19c, Koc19d] by A. Kochubei in this volume, see also [Koc12a, Koc12b].

8.7 Exercises

8.7.1 ([Dav36, p. 280]) Show that the integral equation

$$u(x) = \frac{\mu x^{1-\alpha}}{\Gamma(2-\alpha)} - \frac{\lambda}{\Gamma(1-\alpha)} \int_0^x \frac{u(t)\mathrm{d}t}{(x-t)^\alpha}$$

has a solution

$$u(x) = \frac{\mu}{\lambda} - \frac{\mu}{\lambda} E_\beta(-\lambda x^\beta), \quad \beta = 1 - \alpha.$$

8.7.2 ([Dav36, p. 282]) Show that the equation

$$u(x) + \int_0^x \frac{tu(t)\mathrm{d}t}{(x-t)^{1/2}} = f(x)$$

is equivalent to
$$u(x) - \frac{\pi}{2}\int_0^x t(x+t)u(t)dt = F(x),$$
where
$$F(x) = f(x) - \sqrt{\pi}x D_{0+}^{1/2} f(x) + \frac{1}{2}\sqrt{\pi}D_{0+}^{3/2} f(x).$$

8.7.3 ([Dav36, p. 282]) Show that the equation
$$u(x) + \int_0^x \frac{t^2 u(t)dt}{(x-t)^{2/3}} = f(x)$$
is equivalent to
$$u(x) + \frac{\Gamma^3(1/3)}{243}\int_0^x t^2(44x^4 + 40x^3 t + 75x^2 t^2 40xt^3 + 44t^4)u(t)dt = F(x),$$
where
$$F(x) = f(x) + \Gamma(1/3)\left[x^2 D_{0+}^{1/3} f(x) - \tfrac{2}{3}x D_{0+}^{4/3} f(x) + \tfrac{4}{9}x D_{0+}^{7/3} f(x)\right]$$
$$+ \Gamma^2(2/3)\left[x^4 D_{0+}^{2/3} f(x) - 2x^3 D_{0+}^{5/3} f(x) + \tfrac{34}{9}x^2 D_{0+}^{8/3} f(x)\right.$$
$$\left. - \tfrac{16}{3}x D_{0+}^{11/3} f(x) + \tfrac{352}{81} D_{0+}^{14/3} f(x)\right].$$

8.7.4 ([PolMan08, Eq. 3.1–6.44]) Solve the following integral equation
$$\int_0^1 \frac{y(xt)}{\sqrt{1-t}}dt = f(x), \quad 0 < x < 1.$$

Hint. Reduce to the Abel integral equation of the first kind.

8.7.5 ([PolMan08, Eq. 3.1–6.46]) Solve the following integral equation
$$\int_0^1 \frac{t^\mu y(xt)}{(1-t)^\lambda}dt = f(x), \quad 0 < x < 1 \; (\mu \in \mathbb{R}, 0 < \lambda < 1).$$

Hint. Reduce to the Abel integral equation of the first kind.

8.7.6 ([GorVes91, (7.2.8)]) Find an explicit solution to the so-called test Abel integral equation of the second kind

8.7 Exercises

$$u(x) = 1 + \frac{\lambda}{\Gamma(\alpha)} \int_0^x \frac{u(t) dt}{(x-t)^{1-\alpha}} = f(x), \quad 0 < x < 1 \; (0 < \alpha < 1).$$

8.7.7 ([GorVes91, p. 142]) Solve the integral equation of heat conduction in a semi-infinite rod

$$\varphi(t) = \frac{2c}{\sqrt{\pi}} \sqrt{t} - \frac{c}{\sqrt{\pi}} \int_0^t \frac{\varphi(\tau)}{\sqrt{t-\tau}} d\tau, \quad t \geq 0.$$

8.7.8 ([KiSrTr06, p. 281]) Solve the following fractional differential equation using the Laplace transform

$$\left(D_{0+}^{1/2} y \right)(x) = \frac{y}{x} \quad (x > 0).$$

8.7.9 ([KiSrTr06, p. 282, (5.1.20)]) Solve the Cauchy problem for the fractional differential equation with Caputo derivative

$$\left({}^C D_{0+}^{\alpha} y \right)(x) = f(x), \quad (x > 0, \; m-1 < \alpha \leq m; \; m \in \mathbb{N}),$$

$$y(0) = y'(0) = \ldots = y^{(m-1)}(0) = 0.$$

Hint. Use the Laplace transform.

8.7.10 ([KiSrTr06, p. 282, (5.1.21)]) Solve the Cauchy problem for the fractional differential equation with Caputo derivative

$$y'(x) + a \left({}^C D_{0+}^{\alpha} y \right)(x) = f(x), \; y(0) = c_0 \in \mathbb{R} \; (x > 0, \; 0 < \alpha < 1).$$

Hint. Use the Laplace transform.

8.7.11 ([KiSrTr06, p. 282, (5.1.22)]) Solve the Cauchy problem for the fractional differential equation with Caputo derivative

$$y''(x) + a \left({}^C D_{0+}^{\alpha} y \right)(x) = f(x), \quad (x > 0, \; 0 < \alpha < 1)$$

$$y(0) = c_0, \; y'(0) = c_1; \; c_0, c_1 \in \mathbb{R}.$$

Hint. Use the Laplace transform.

8.7.12 ([KiSrTr06, p. 284, (5.2.5-6)]) Show that the family of functions

$$y_j(x) = x^{\alpha-j} E_{\alpha,\alpha+1-j}(\lambda x^{\alpha}) \quad (j = 1, \ldots, l)$$

forms a fundamental system of solutions of the fractional differential equation

$$\left(D_{0+}^{\alpha}y\right)(x) - \lambda y(x) = 0 \quad (x > 0, l - 1 < \alpha \leq l; \; l \in \mathbb{N}; \; \lambda \in \mathbb{R}). \tag{8.7.1}$$

Hint. Use the Laplace transform.

8.7.13 ([KiSrTr06, p. 286, (5.2.30-31)]) Prove that if $\alpha - 1 + j \geq \beta$ then the functions

$$y_j(x) = x^{\alpha-j} E_{\alpha-\beta, \alpha+1-j}(\lambda x^{\alpha-\beta})$$

form a system of linear independent solutions of the fractional differential equations

$$\left(D_{0+}^{\alpha}y\right)(x) - \lambda \left(D_{0+}^{\beta}y\right)(x) = 0 \quad (x > 0, l - 1 < \alpha \leq l; \; l \in \mathbb{N}; \; \lambda \in \mathbb{R}, \alpha > \beta > 0).$$

Hint. Use the Laplace transform.

8.7.14 ([KiSrTr06, p. 295, (5.2.83-84)]) Show that the general solution to the equation

$$\left(D_{0+}^{\alpha}y\right)(x) - \lambda y(x) = f(x) \quad (x > 0, \alpha > 0; \; \lambda \in \mathbb{R})$$

can be represented in the form of the Laplace convolution:

$$y(x) = \int_0^x (x-t)^{\alpha-1} E_{\alpha,\alpha}\left[\lambda(x-t)^{\alpha}\right] f(t) dt.$$

8.7.15 ([KiSrTr06, p. 295, 310, (5.2.83), (5.2.172)]) Solve the Cauchy type problem

$$\left(D_{0+}^{\alpha-k}y\right)(0+) = b_k \quad (b_k \in \mathbb{R}; \; k = 1, \ldots, l; l - 1 < \alpha \leq l)$$

for the fractional differential equation

$$\left(D_{0+}^{\alpha}y\right)(x) - \lambda y(x) = f(x) \quad (x > 0, \alpha > 0; \; \lambda \in \mathbb{R}).$$

Answer.

$$y(x) = \int_0^x (x-t)^{\alpha-1} E_{\alpha,\alpha}\left[\lambda(x-t)^{\alpha}\right] f(t) dt + \sum_{j=1}^l b_j x^{\alpha-j} E_{\alpha,\alpha+1-j}[\lambda x^{\alpha}].$$

8.7.16 ([KiSrTr06, p. 331, (5.4.14–5.4.16–17)]) Show that the particular solution to the inhomogeneous fractional differential equation

$$x^{\alpha+1}\left(D_{0+}^{\alpha+1}y\right)(x) + \lambda x^{\alpha}\left(D_{0+}^{\alpha}y\right)(x) = f(x) \quad (\alpha > 0, \lambda \in \mathbb{R}; \; x > 0)$$

can be represented in the form

8.7 Exercises

$$y(x) = \int_0^1 G_1^{\alpha,\lambda}(t) f(xt) dt,$$

where

$$G_1^{\alpha,\lambda}(x) = x^{-\alpha} \left\{ \frac{\Gamma(1-\lambda)}{\Gamma(\alpha+1-\lambda)} x^{\lambda-1} - {}_1\Psi_2 \left[\begin{matrix} (1-\lambda, 1) \\ (\alpha, -1), (2-\lambda, 1) \end{matrix} \bigg| -x \right] \right\}.$$

Hint. Use the Mellin transform.

8.7.17 ([GejKum17, Ex. 8, p. 12)]) Consider the following nonlinear time-fractional gas dynamics equation

$$D_t^\alpha u(x,t) + \frac{1}{2}(u^2)_x - u(1-u) = 0, \quad t > 0 \ (0 < \alpha \le 1)$$

along with the initial condition

$$u(x, 0) = e^{-x}.$$

Show that this problem has the following solution

$$u(x, t) = e^{-x} E_\alpha(t^\alpha).$$

Hint. Use the NIM method as described in [Gej14], [GejKum17].

8.7.18 ([GejKum17, Ex. 11, p. 14)]) Consider the following time-fractional biological population equation

$$D_t^\alpha u = (u^2)_{xx} + (u^2)_{yy} + hu, \quad t > 0 \ (0 < \alpha \le 1)$$

with the initial condition

$$u(x, y, 0) = \sqrt{xy}.$$

Show that this problem has the following solution

$$u(x, t) = \sqrt{xy} E_\alpha(ht^\alpha).$$

Hint. Use the NIM method as described in [Gej14], [GejKum17].

8.7.19 ([GejKum17, Ex. 15, p. 16)]) Consider the two-dimensional initial-boundary value problem

$$D_t^\alpha u = \frac{1}{12} \left(x^2 u_{xx} + y^2 u_{yy} \right) + hu, \quad 0 < x, y < 1, \ t > 0 \ (1 < \alpha \le 2)$$

subject to the Neumann conditions

$$u_x(0, y, t) = 0, \ u_x(1, y, t) = 4E_\alpha(t),$$
$$u_y(0, y, t) = 0, \ u_y(1, y, t) = 4t E_{\alpha,2}(t),$$

and the initial conditions

$$u(x, y, 0) = x^4, \ u_t(x, y, 0) = y^4.$$

Show that this problem has the following solution

$$u(x, y, t) = x^4 E_\alpha(t) + y^4 t E_{\alpha,2}(t).$$

Hint. Use the NIM method as described in [Gej14], [GejKum17].

8.7.20 ([GejKum17, Ex. 9, p. 13)]) Consider the following time-fractional coupled Burgers equations

$$D_t^\alpha u - u_{xx} - 2uu_x + (uv)_x = 0,$$
$$D_t^\alpha u - v_{xx} - 2vv_x + (uv)_x = 0,$$

with the initial conditions

$$u(x, 0) = e^x, \quad v(x, 0) = e^x.$$

Show that this problem has the following solution

$$u(x, t) = e^x E_\alpha(t^\alpha), \quad v(x, t) = e^x E_\alpha(t^\alpha).$$

Chapter 9
Applications to Deterministic Models

Here we present material illuminating the role of the Mittag-Leffler function and its generalizations in the study of deterministic models. It has already been mentioned that the Mittag-Leffler function is closely related to the Fractional Calculus (being called 'The Queen Function of the Fractional Calculus'). This is why we focus our attention here on fractional (deterministic) models. We start with a technical Sect. 9.1 in which the fractional differential equations, related to the fractional relaxation and oscillation phenomena, are discussed in full detail.

Later we present other physical models involving fractional calculus. Interest in such models is growing rapidly nowadays, and several books on the subject have appeared recently. It would be impossible to give a detailed discussion of fractional models here. Instead, we have chosen some examples related to the above discussed equations (or their simple generalizations) which demonstrate the essential role of the Mittag-Leffler function in fractional modelling.

In the second part of the chapter (Sect. 9.2) some examples of physical and mechanical models involving fractional derivatives are briefly outlined. The main focus is on the problems of fractional visco-elasticity. For other deterministic fractional models we derive only the corresponding fractional differential equation. This section is intended to show how fractional models can appear and which features of fractional objects are useful for such modelling.

9.1 Fractional Relaxation and Oscillations

We now analyze the most simple differential equations of fractional order which have appeared in applications. For this purpose, we choose some examples which, by means of fractional derivatives, generalize the well-known ordinary differential equations related to relaxation and oscillation phenomena.

In the first subsection we treat the simplest types, which we refer to as the *simple fractional relaxation and oscillation equations*. Then, in the next subsection we consider the types, somewhat more cumbersome, which we refer to as the *composite fractional relaxation and oscillation equations*.

9.1.1 Simple Fractional Relaxation and Oscillation

The classical phenomena of relaxation and oscillation in their simplest form are known to be governed by linear ordinary differential equations, of order one and two respectively, that hereafter we recall with the corresponding solutions. Let us denote by $u = u(t)$ the field variable and by $q(t)$ a given continuous function, with $t \geq 0$. The *relaxation* differential equation reads as

$$u'(t) = -u(t) + q(t), \qquad (9.1.1)$$

whose solution, under the initial condition $u(0^+) = c_0$, is

$$u(t) = c_0 \, e^{-t} + \int_0^t q(t-\tau) \, e^{-\tau} \, d\tau. \qquad (9.1.2)$$

The *oscillation* differential equation reads as

$$u''(t) = -u(t) + q(t), \qquad (9.1.3)$$

whose solution, under the initial conditions $u(0^+) = c_0$ and $u'(0^+) = c_1$, is

$$u(t) = c_0 \cos t + c_1 \sin t + \int_0^t q(t-\tau) \sin \tau \, d\tau. \qquad (9.1.4)$$

From the point of view of the fractional calculus a natural generalization of Eqs. (9.1.1) and (9.1.3) is obtained by replacing the ordinary derivative with a fractional one of order α. In order to preserve the type of initial conditions required in the classical phenomena, we agree to replace the first and second derivative in (9.1.1) and (9.1.3) with a Caputo fractional derivative of order α with $0 < \alpha < 1$ and $1 < \alpha < 2$, respectively. We agree to refer to the corresponding equations as the *simple fractional relaxation equation* and the *simple fractional oscillation equation*.

Generally speaking, we consider the following differential equation of fractional order $\alpha > 0$, (see [GorMai97])

$$D_*^\alpha u(t) = D_{0+}^\alpha \left(u(t) - \sum_{k=0}^{m-1} \frac{t^k}{k!} u^{(k)}(0^+) \right) = -u(t) + q(t), \quad t > 0. \qquad (9.1.5)$$

9.1 Fractional Relaxation and Oscillations

Here m is a positive integer uniquely defined by $m - 1 < \alpha \leq m$, which provides the number of the prescribed initial values $u^{(k)}(0^+) = c_k, \ k = 0, 1, 2, \ldots, m - 1$. Implicit in the form of (9.1.5) is our desire to obtain solutions $u(t)$ for which the $u^{(k)}(t)$ are continuous for $t \geq 0, \ k = 0, 1, \ldots, m - 1$. In particular, the cases of *fractional relaxation* and *fractional oscillation* are obtained for $m = 1$ and $m = 2$, respectively. We note that when $\alpha = m$ is an integer, then Eq. (9.1.5) reduces to an ordinary differential equation whose solution can be expressed in terms of m linearly independent solutions of the *homogeneous* equation and of one particular solution of the *inhomogeneous* equation. We summarize this well-known result as follows

$$u(t) = \sum_{k=0}^{m-1} c_k u_k(t) + \int_0^t q(t - \tau) u_\delta(\tau) \, d\tau \, . \tag{9.1.6}$$

$$u_k(t) = J^k u_0(t) \, , \quad u_k^{(h)}(0^+) = \delta_{kh} \, , \quad h, k = 0, 1, \ldots, m - 1 \, , \tag{9.1.7}$$

$$u_\delta(t) = -u_0'(t) \, , \tag{9.1.8}$$

where J^k is a k-times repeated integral, $J^1 u(t) = J u(t) = \int_0^t u(\tau) d\tau$. Thus, the m functions $u_k(t)$ represent the *fundamental solutions* of the differential equation of order m, namely those linearly independent solutions of the *homogeneous* equation which satisfy the initial conditions in (9.1.7). The function $u_\delta(t)$, with which the free term $q(t)$ appears convoluted, represents the so-called *impulse-response solution*, namely the particular solution of the *inhomogeneous* equation with all $c_k \equiv 0, \ k = 0, 1, \ldots, m - 1$, and with $q(t) = \delta(t)$. In the cases of ordinary relaxation and oscillation we recognize that $u_0(t) = e^{-t} = u_\delta(t)$ and $u_0(t) = \cos t$, $u_1(t) = J u_0(t) = \sin t = \cos(t - \pi/2) = u_\delta(t)$, respectively.

Remark 9.1 The more general equation

$$D_{0+}^\alpha \left(u(t) - \sum_{k=0}^{m-1} \frac{t^k}{k!} u^{(k)}(0^+) \right) = -\rho^\alpha u(t) + q(t) \, , \ \rho > 0, \ t > 0, \tag{9.1.9}$$

can be reduced to (9.1.5) by a change of scale $t \to t/\rho$. We prefer, for ease of notation, to discuss the "dimensionless" form (9.1.5).

Let us now solve (9.1.5) by the method of Laplace transforms. For this purpose we can use the Caputo formula directly or, alternatively, reduce (9.1.5) with the prescribed initial conditions to an equivalent (fractional) integral equation and then treat the integral equation by the Laplace transform method. Here we prefer to follow the second approach. Then, applying the operator of fractional integration I^α to both sides of (9.1.5) we obtain

$$u(t) = \sum_{k=0}^{m-1} c_k \frac{t^k}{k!} - I_{0+}^\alpha u(t) + I_{0+}^\alpha q(t). \tag{9.1.10}$$

The application of the Laplace transform yields

$$\tilde{u}(s) = \sum_{k=0}^{m-1} \frac{c_k}{s^{k+1}} - \frac{1}{s^\alpha} \tilde{u}(s) + \frac{1}{s^\alpha} \tilde{q}(s),$$

hence

$$\tilde{u}(s) = \sum_{k=0}^{m-1} c_k \frac{s^{\alpha-k-1}}{s^\alpha + 1} + \frac{1}{s^\alpha + 1} \tilde{q}(s). \tag{9.1.11}$$

Introducing the Mittag-Leffler type functions

$$e_\alpha(t) \equiv e_\alpha(t; 1) := E_\alpha(-t^\alpha) \div \frac{s^{\alpha-1}}{s^\alpha + 1}, \tag{9.1.12}$$

$$u_k(t) := J^k e_\alpha(t) \div \frac{s^{\alpha-k-1}}{s^\alpha + 1}, \quad k = 0, 1, \ldots, m-1, \tag{9.1.13}$$

we find, from inversion of the Laplace transforms in (9.3.10),

$$u(t) = \sum_{k=0}^{m-1} c_k u_k(t) - \int_0^t q(t-\tau) u_0'(\tau)\, d\tau. \tag{9.1.14}$$

To find the last term in the right-hand side of (9.1.14), we have to use the well-known rule for the Laplace transform of the derivative, noting that $u_0(0^+) = e_\alpha(0^+) = 1$, and

$$\frac{1}{s^\alpha + 1} = -\left(s \frac{s^{\alpha-1}}{s^\alpha + 1} - 1\right) \div -u_0'(t) = -e_\alpha'(t). \tag{9.1.15}$$

The formula (9.1.14) encompasses the solutions (9.1.2) and (9.1.4) found for $\alpha = 1, 2$, respectively. When α is not an integer, namely for $m-1 < \alpha < m$, we note that $m-1$ represents the integer part of α (denoted by $[\alpha]$) and m the number of initial conditions necessary and sufficient to ensure the uniqueness of the solution $u(t)$. Thus the m functions $u_k(t) = J^k e_\alpha(t)$ with $k = 0, 1, \ldots, m-1$ represent those particular solutions of the *homogeneous* equation which satisfy the initial conditions

$$u_k^{(h)}(0^+) = \delta_{kh}, \quad h, k = 0, 1, \ldots, m-1, \tag{9.1.16}$$

and therefore they represent the *fundamental solutions* of the fractional equation (9.1.5), in analogy with the case $\alpha = m$. Furthermore, the function $u_\delta(t) = -e_\alpha'(t)$ represents the *impulse-response solution*. Hereafter, we are going to compute and

9.1 Fractional Relaxation and Oscillations

exhibit the *fundamental solutions* and the *impulse-response solution* for the cases (a) $0 < \alpha < 1$ and (b) $1 < \alpha < 2$, pointing out the comparison with the corresponding solutions obtained when $\alpha = 1$ and $\alpha = 2$.

We now infer the relevant properties of the basic functions $e_\alpha(t)$ directly from their representation as a Laplace inverse integral

$$e_\alpha(t) = \frac{1}{2\pi i} \int_{Br} e^{st} \frac{s^{\alpha-1}}{s^\alpha + 1} \, ds, \qquad (9.1.17)$$

in detail for $0 < \alpha \le 2$, without having to make a detour into the general theory of Mittag-Leffler functions in the complex plane. In (9.1.17) Br denotes the Bromwich path, i.e. a line Re $\{s\} = \sigma$ with a value $\sigma \ge 1$, and Im $\{s\}$ running from $-\infty$ to $+\infty$. For reasons of transparency, we separately discuss the cases

$$\text{(a) } 0 < \alpha < 1 \quad \text{and} \quad \text{(b) } 1 < \alpha < 2,$$

recalling that in the limiting cases $\alpha = 1, 2$, we know $e_\alpha(t)$ as an elementary function, namely $e_1(t) = e^{-t}$ and $e_2(t) = \cos t$. For α not an integer the power function s^α is uniquely defined as $s^\alpha = |s|^\alpha e^{i \arg s}$, with $-\pi < \arg s < \pi$, that is, in the complex s-plane cut along the negative real axis. The essential step consists in decomposing $e_\alpha(t)$ into two parts according to $e_\alpha(t) = f_\alpha(t) + g_\alpha(t)$, as indicated below. In case (a) the function $f_\alpha(t)$ and in case (b) the function $-f_\alpha(t)$ is *completely monotone*; in both cases $f_\alpha(t)$ tends to zero as t tends to infinity, from above in case (a) and from below in case (b). The other part, $g_\alpha(t)$, is identically vanishing in case (a), but of *oscillatory* character with exponentially decreasing amplitude in case (b). In order to obtain the desired decomposition of e_α we bend the Bromwich path of integration Br into the equivalent Hankel path Ha(1^+), a loop which starts from $-\infty$ along the lower side of the negative real axis, encircles the circular disk $|s| = 1$ in the positive sense and ends at $-\infty$ along the upper side of the negative real axis. One obtains

$$e_\alpha(t) = f_\alpha(t) + g_\alpha(t), \quad t \ge 0, \qquad (9.1.18)$$

with

$$f_\alpha(t) := \frac{1}{2\pi i} \int_{Ha(\epsilon)} e^{st} \frac{s^{\alpha-1}}{s^\alpha + 1} \, ds, \qquad (9.1.19)$$

where now the Hankel path Ha(ϵ) denotes a loop comprising a small circle $|s| = \epsilon$ with $\epsilon \to 0$ and two sides of the cut negative real semi-axis, and

$$g_\alpha(t) := \sum_h e^{s'_h t} \operatorname{Res}\left[\frac{s^{\alpha-1}}{s^\alpha + 1}\right]_{s'_h} = \frac{1}{\alpha} \sum_h e^{s'_h t}. \qquad (9.1.20)$$

Here s'_h are the relevant poles of $s^{\alpha-1}/(s^\alpha + 1)$. In fact the poles turn out to be $s_h = \exp[i(2h+1)\pi/\alpha]$ with unit modulus; they are all simple but the only relevant ones are those situated in the main Riemann sheet, i.e. the poles s'_h with argument such that $-\pi < \arg s'_h < \pi$. If $0 < \alpha < 1$, there are no such poles, since for all integers h we have $|\arg s_h| = |2h + 1|\pi/\alpha > \pi$. As a consequence,

$$g_\alpha(t) \equiv 0, \quad \text{hence} \quad e_\alpha(t) = f_\alpha(t), \quad \text{if} \quad 0 < \alpha < 1. \tag{9.1.21}$$

If $1 < \alpha < 2$, then there exist precisely two relevant poles, namely $s'_0 = \exp(i\pi/\alpha)$ and $s'_{-1} = \exp(-i\pi/\alpha) = \overline{s'_0}$, which are located in the left half-plane. Then one obtains

$$g_\alpha(t) = \frac{2}{\alpha} e^{t\cos(\pi/\alpha)} \cos\left[t \sin\left(\frac{\pi}{\alpha}\right)\right], \quad \text{if} \quad 1 < \alpha < 2. \tag{9.1.22}$$

We note that this function exhibits oscillations with circular frequency $\omega(\alpha) = \sin(\pi/\alpha)$ and with an exponentially decaying amplitude with rate $\lambda(\alpha) = |\cos(\pi/\alpha)|$.

Remark 9.2 One easily recognizes that (9.1.22) is also valid for $2 \leq \alpha < 3$. In the classical case $\alpha = 2$ the two poles are purely imaginary (coinciding with $\pm i$) so that we recover the sinusoidal behavior with unitary frequency. In the case $2 < \alpha < 3$, however, the two poles are located in the right half-plane, so providing amplified oscillations. This instability, which is common to the case $\alpha = 3$, is the reason why we limit ourselves to consider α in the range $0 < \alpha \leq 2$.

In addition to the basic fundamental solutions $u_0(t) = e_\alpha(t)$ we need to compute the impulse-response solutions $u_\delta(t) = -D^1 e_\alpha(t)$ for cases (a) and (b) and, only in case (b), the second fundamental solution $u_1(t) = J^1 e_\alpha(t)$. For this purpose we note that in general it turns out that

$$J^k f_\alpha(t) = \int_0^\infty e^{-rt} K_{\alpha,k}(r) \, dr, \tag{9.1.23}$$

with

$$K_{\alpha,k}(r) := (-1)^k r^{-k} K_\alpha(r) = \frac{(-1)^k}{\pi} \frac{r^{\alpha-1-k} \sin(\alpha\pi)}{r^{2\alpha} + 2r^\alpha \cos(\alpha\pi) + 1}, \tag{9.1.24}$$

where $K_\alpha(r) = K_{\alpha,0}(r)$, and

$$J^k g_\alpha(t) = \frac{2}{\alpha} e^{t\cos(\pi/\alpha)} \cos\left[t \sin\left(\frac{\pi}{\alpha}\right) - k\frac{\pi}{\alpha}\right]. \tag{9.1.25}$$

This can be done in direct analogy to the computation of the functions $e_\alpha(t)$, the Laplace transform of $J^k e_\alpha(t)$ being given by (9.1.13). For the impulse-response solution we note that the effect of the differential operator D^1 is the same as that of the virtual operator J^{-1}. In conclusion we can resume the solutions for the fractional relaxation and oscillation equations as follows:

(a) $0 < \alpha < 1$,

9.1 Fractional Relaxation and Oscillations

$$u(t) = c_0 \, u_0(t) + \int_0^t q(t-\tau) \, u_\delta(\tau) \, d\tau, \tag{9.1.26}$$

where

$$\begin{cases} u_0(t) = \displaystyle\int_0^\infty e^{-rt} \, K_{\alpha,0}(r) \, dr, \\[2mm] u_\delta(t) = -\displaystyle\int_0^\infty e^{-rt} \, K_{\alpha,-1}(r) \, dr, \end{cases} \tag{9.1.27}$$

with $u_0(0^+) = 1$, $u_\delta(0^+) = \infty$;

(b) $1 < \alpha < 2$,

$$u(t) = c_0 \, u_0(t) + c_1 \, u_1(t) + \int_0^t q(t-\tau) \, u_\delta(\tau) \, d\tau, \tag{9.1.28}$$

where

$$\begin{cases} u_0(t) = \displaystyle\int_0^\infty e^{-rt} \, K_{\alpha,0}(r) \, dr + \frac{2}{\alpha} \, e^{t \cos(\pi/\alpha)} \cos\left[t \sin\left(\frac{\pi}{\alpha}\right)\right], \\[2mm] u_1(t) = \displaystyle\int_0^\infty e^{-rt} \, K_{\alpha,1}(r) \, dr + \frac{2}{\alpha} \, e^{t \cos(\pi/\alpha)} \cos\left[t \sin\left(\frac{\pi}{\alpha}\right) - \frac{\pi}{\alpha}\right], \\[2mm] u_\delta(t) = -\displaystyle\int_0^\infty e^{-rt} \, K_{\alpha,-1}(r) \, dr - \frac{2}{\alpha} \, e^{t \cos(\pi/\alpha)} \cos\left[t \sin\left(\frac{\pi}{\alpha}\right) + \frac{\pi}{\alpha}\right], \end{cases} \tag{9.1.29}$$

with $u_0(0^+) = 1$, $u_0'(0^+) = 0$, $u_1(0^+) = 0$, $u_1'(0^+) = 1$, $u_\delta(0^+) = 0$, $u_\delta'(0^+) = +\infty$.

In Fig. 9.1 we show the plots of the spectral distributions $K_\alpha(r)$ for $\alpha = 0.25$, 0.50, 0.75, 0.9, and $-K\alpha(r)$ for $\alpha = 1.25$, 1.50, 1.75, 1.9.

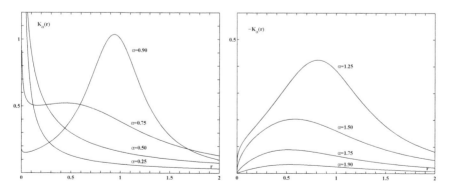

Fig. 9.1 $K_\alpha(r)$ for $0 < \alpha < 1$ (left) and $-K_\alpha(r)$ $1 < \alpha < 2$ (right)

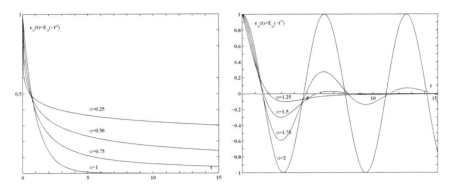

Fig. 9.2 The solution $u_0(t) = e_\alpha(t)$ for $0 < \alpha \le 1$ (left) and $1 < \alpha \le 2$ (right)

In Fig. 9.2 we show the plots of the basic fundamental solution for the following cases: (left) $\alpha = 0.25, 0.50, 0.75, 1$ and (right) $\alpha = 1.25, 1.50, 1.75, 2$, obtained from the first formula in (9.1.27) and (9.1.29), respectively.

We have verified that our present results confirm those obtained by Blank [Bla97] by a numerical treatment and those obtained by Mainardi [Mai96a] by an analytical treatment, valid when α is a rational number. Of particular interest is the case $\alpha = 1/2$, where we recover a well-known formula from the theory of the Laplace transform,

$$e_{1/2}(t) := E_{1/2}(-\sqrt{t}) = e^t \operatorname{erfc}(\sqrt{t}) \div \frac{1}{s^{1/2}(s^{1/2}+1)}, \qquad (9.1.30)$$

where erfc denotes the *complementary error* function.

We now point out that in both cases (a) and (b) (in which α is just non-integer) i.e. for *fractional relaxation* and *fractional oscillation*, all the fundamental and impulse-response solutions exhibit an *algebraic decay* as $t \to \infty$, as discussed below. Let us start with the asymptotic behavior of $u_0(t)$. To this end we first derive an asymptotic series for the function $f_\alpha(t)$, valid for $t \to \infty$. Using the identity

$$\frac{1}{s^\alpha + 1} = 1 - s^\alpha + s^{2\alpha} - s^{3\alpha} + \cdots + (-1)^{N-1} s^{(N-1)\alpha} + (-1)^N \frac{s^{N\alpha}}{s^\alpha + 1},$$

in formula (9.1.19) and the Hankel representation of the reciprocal Gamma function, we (formally) obtain the asymptotic expansion (for non-integer α)

$$f_\alpha(t) = \sum_{n=1}^{N} (-1)^{n-1} \frac{t^{-n\alpha}}{\Gamma(1-n\alpha)} + O\left(t^{-(N+1)\alpha}\right), \quad \text{as } t \to \infty. \qquad (9.1.31)$$

The validity of this asymptotic expansion can be established rigorously using the (generalized) Watson lemma, see [BleHan86]. We can also start from the spectral

9.1 Fractional Relaxation and Oscillations

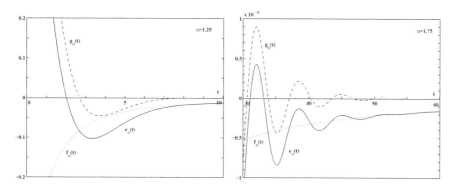

Fig. 9.3 Decay of the solution $u_0(t) = e_\alpha(t)$ for $\alpha = 1.25, 1.75$

representation and expand the spectral function for small r. Then the (ordinary) Watson lemma yields (9.1.31). We note that this asymptotic expansion coincides with that for $u_0(t) = e_\alpha(t)$, having assumed $0 < \alpha < 2$ ($\alpha \neq 1$). In fact the contribution of $g_\alpha(t)$ is identically zero if $0 < \alpha < 1$ and exponentially small as $t \to \infty$ if $1 < \alpha < 2$. The asymptotic expansions of the solutions $u_1(t)$ and $u_\delta(t)$ are obtained from (9.1.31) by integrating or differentiating term-by-term with respect to t. In particular, taking the leading term in (9.1.31), we obtain the asymptotic representations

$$u_0(t) \sim \frac{t^{-\alpha}}{\Gamma(1-\alpha)}, \quad u_1(t) \sim \frac{t^{1-\alpha}}{\Gamma(2-\alpha)}, \quad u_\delta(t) \sim -\frac{t^{-\alpha-1}}{\Gamma(-\alpha)}, \quad \text{as } t \to \infty. \tag{9.1.32}$$

They yield the algebraic decay of the fundamental and impulse-response solutions.

In Fig. 9.3 we show some plots of the *basic fundamental solution* $u_0(t) = e_\alpha(t)$ for $\alpha = 1.25, 1.75$. Here the algebraic decay of the fractional oscillation can be recognized and compared with the two contributions provided by f_α (monotonic behavior) and $g_\alpha(t)$ (exponentially damped oscillation).

The Zeros of the Solutions of the Fractional Oscillation Equation

Now we carry out some investigations concerning the zeros of the basic fundamental solution $u_0(t) = e_\alpha(t)$ in the case (b) of fractional oscillations. For the second fundamental solution and the impulse-response solution the analysis of the zeros can be easily carried out analogously. Recalling the first equation in (9.1.29), the required zeros of $e_\alpha(t)$ are the solutions of the equation

$$e_\alpha(t) = f_\alpha(t) + \frac{2}{\alpha} e^{t \cos(\pi/\alpha)} \cos\left[t \sin\left(\frac{\pi}{\alpha}\right)\right] = 0. \tag{9.1.33}$$

We first note that the function $e_\alpha(t)$ exhibits an *odd* number of zeros, in that $e_\alpha(0) = 1$, and, for sufficiently large t, $e_\alpha(t)$ turns out to be permanently negative, as shown in (9.1.32) by the sign of $\Gamma(1-\alpha)$. The smallest zero lies in the first positivity interval of $\cos[t \sin(\pi/\alpha)]$, hence in the interval $0 < t < \pi/[2 \sin(\pi/\alpha)]$; all other zeros can only lie in the succeeding positivity intervals of $\cos[t\,\sin(\pi/\alpha)]$, in each of these two zeros are present as long as

$$\frac{2}{\alpha} e^{t\,\cos(\pi/\alpha)} \geq |f_\alpha(t)|. \tag{9.1.34}$$

When t is sufficiently large, the zeros are expected to be found approximately from the equation

$$\frac{2}{\alpha} e^{t\,\cos(\pi/\alpha)} \approx \frac{t^{-\alpha}}{|\Gamma(1-\alpha)|}, \tag{9.1.35}$$

obtained from (9.1.33) by ignoring the oscillation factor of $g_\alpha(t)$ (see (9.1.22)) and taking the first term in the asymptotic expansion of $f_\alpha(t)$ (see (9.1.31)–(9.1.32)). This approximation turns out to be useful when $\alpha \to 1^+$ and $\alpha \to 2^-$. For $\alpha \to 1^+$, only one zero is present, which is expected to be very far from the origin in view of the large period of the function $\cos[t\,\sin(\pi/\alpha)]$. In fact, since there is no zero for $\alpha = 1$, and by increasing α more and more zeros arise, we are sure that only one zero exists for α sufficiently close to 1. Putting $\alpha = 1 + \epsilon$, the asymptotic position T_* of this zero can be found from the relation (9.1.35) in the limit $\epsilon \to 0^+$. Assuming in this limit a first-order approximation, we get

$$T_* \sim \log\left(\frac{2}{\epsilon}\right), \tag{9.1.36}$$

which shows that T_* tends to infinity slower than $1/\epsilon$, as $\epsilon \to 0$.

For $\alpha \to 2^-$, there is an increasing number of zeros up to infinity since $e_2(t) = \cos t$ has infinitely many zeros ($t_n^* = (n + 1/2)\pi$, $n = 0, 1, \dots$). Putting now $\alpha = 2 - \delta$ the asymptotic position T_* for the largest zero can be found again from (9.1.35) in the limit $\delta \to 0^+$. Assuming in this limit a first-order approximation, we get

$$T_* \sim \frac{12}{\pi\,\delta} \log\left(\frac{1}{\delta}\right). \tag{9.1.37}$$

Now, for $\delta \to 0^+$ the length of the positivity intervals of $g_\alpha(t)$ tends to π and, as long as $t \leq T_*$, there are two zeros in each positivity interval. Hence, in the limit $\delta \to 0^+$, there is on average one zero per interval of length π, so we expect that $N_* \sim T_*/\pi$.

For the above considerations on the zeros of the oscillating Mittag-Leffler function we were inspired by the paper of Wiman [Wim05b], who at the beginning of the 20th century, after having treated the Mittag-Leffler function in the complex plane, considered the position of the zeros of the function on the negative real semi-axis (without providing any details). Our expressions for T_* differ from those of Wiman in numerical factors; however, the results of our numerical studies confirm and illustrate the validity of our analysis.

9.1 Fractional Relaxation and Oscillations

Table 9.1 $N_* =$ number of zeros, $\alpha =$ fractional order, T_* location of the largest zero

N_*	α	T_*
1 ÷ 3	1.40 ÷ 1.41	1.730 ÷ 5.726
3 ÷ 5	1.56 ÷ 1.57	8.366 ÷ 13.48
5 ÷ 7	1.64 ÷ 1.65	14.61 ÷ 20.00
7 ÷ 9	1.69 ÷ 1.70	20.80 ÷ 26.33
9 ÷ 11	1.72 ÷ 1.73	27.03 ÷ 32.83
11 ÷ 13	1.75 ÷ 1.76	33.11 ÷ 38.81
13 ÷ 15	1.78 ÷ 1.79	39.49 ÷ 45.51
15 ÷ 17	1.79 ÷ 1.80	45.51 ÷ 51.46

Here, we analyse the phenomenon of the transition of the (odd) number of zeros as $1.4 \leq \alpha \leq 1.8$. For this purpose, in Table 9.1 we report the intervals of amplitude $\Delta \alpha = 0.01$ where these transitions occur, and the location T_* of the largest zeros (evaluated within a relative error of 0.1%) found at the two extreme values of the above intervals. We recognize that the transition from 1 to 3 zeros occurs as $1.40 \leq \alpha \leq 1.41$, that the transition from 3 to 5 zeros occurs as $1.56 \leq \alpha \leq 1.57$, and so on. The last transition in the considered range of α is from 15 to 17 zeros, and it occurs as $1.79 \leq \alpha \leq 1.80$.

9.1.2 The Composite Fractional Relaxation and Oscillations

In this subsection we consider the following fractional differential equations for $t \geq 0$, equipped with suitable initial conditions,

$$\frac{du}{dt} + a \frac{d^\alpha u}{dt^\alpha} + u(t) = q(t), \quad u(0^+) = c_0, \quad 0 < \alpha < 1, \quad (9.1.38)$$

$$\frac{d^2v}{dt^2} + a \frac{d^\alpha v}{dt^\alpha} + v(t) = q(t), \quad v(0^+) = c_0, \quad v'(0^+) = c_1, \quad 0 < \alpha < 2, \quad (9.1.39)$$

where a is a positive constant. The unknown functions $u(t)$ and $v(t)$ (the field variables) are required to be sufficiently well behaved to be treated with their derivatives $u'(t)$ and $v'(t)$, $v''(t)$ by the technique of the Laplace transform. The given function $q(t)$ is assumed to be continuous. In the above equations the fractional derivative of order α is assumed to be provided by the operator D_*^α, the *Caputo derivative*, in agreement with our choice in the previous subsection. Note that in (9.1.39) we distinguish the cases (a) $0 < \alpha < 1$, (b) $1 < \alpha < 2$ and $\alpha = 1$. The Eqs. (9.1.38) and (9.1.39) will be referred to as the *composite fractional relaxation equation* and the *composite fractional oscillation equation*, respectively, to be distinguished from the corresponding *simple* fractional equations.

Here we also apply the method of the Laplace transform to solve the fractional differential equations and get some insight into their *fundamental* and *impulse-response solutions*. However, in contrast with the previous subsection, we now find it more convenient to apply the corresponding formula for the Laplace transform of fractional and integer derivatives directly, instead of reducing the equations with the prescribed initial conditions as equivalent (fractional) integral equations to be treated by the Laplace transform.

Let us apply the Laplace transform to the composite fractional relaxation equation (9.1.38). This leads us to the transformed algebraic equation

$$\widetilde{u}(s) = c_0 \frac{1 + a s^{\alpha-1}}{w_1(s)} + \frac{\widetilde{q}(s)}{w_1(s)}, \quad 0 < \alpha < 1, \quad (9.1.40)$$

where

$$w_1(s) := s + a s^{\alpha} + 1, \quad (9.1.41)$$

and $a > 0$. Putting

$$u_0(t) \div \widetilde{u}_0(s) := \frac{1 + a s^{\alpha-1}}{w_1(s)}, \quad u_\delta(t) \div \widetilde{u}_\delta(s) := \frac{1}{w_1(s)}, \quad (9.1.42)$$

and recognizing that

$$u_0(0^+) = \lim_{s \to \infty} s \, \widetilde{u}_0(s) = 1, \quad \widetilde{u}_\delta(s) = -[s \, \widetilde{u}_0(s) - 1], \quad (9.1.43)$$

we conclude that

$$u(t) = c_0 u_0(t) + \int_0^t q(t-\tau) u_\delta(\tau) \, d\tau, \quad u_\delta(t) = -u_0'(t). \quad (9.1.44)$$

Thus $u_0(t)$ and $u_\delta(t)$ are respectively the *fundamental solution* and *impulse-response solution* to Eq. (9.1.38). Let us first consider the problem of finding $u_0(t)$ as the inverse Laplace transform of $\widetilde{u}_0(s)$. We easily see that the function $w_1(s)$ has no zero in the main sheet of the Riemann surface including the sides of the cut (simply show that Im $\{w_1(s)\}$ does not vanish if s is not a real positive number), so that the inversion of the Laplace transform $\widetilde{u}_0(s)$ can be carried out by deforming the original Bromwich path into the Hankel path Ha(ϵ) introduced in the previous subsection, i.e. into the loop constituted by a small circle $|s| = \epsilon$ with $\epsilon \to 0$ and by the two borders of the cut negative real axis. As a consequence we write

$$u_0(t) = \frac{1}{2\pi i} \int_{\text{Ha}(\epsilon)} e^{st} \frac{1 + as^{\alpha-1}}{s + a s^\alpha + 1} \, ds. \quad (9.1.45)$$

9.1 Fractional Relaxation and Oscillations

It is now an exercise in complex analysis to show that the contribution from the Hankel path Ha(ϵ) as $\epsilon \to 0$ is provided by

$$u_0(t) = \int_0^\infty e^{-rt} H_{\alpha,0}^{(1)}(r;a)\, dr\,, \qquad (9.1.46)$$

with

$$H_{\alpha,0}^{(1)}(r;a) = -\frac{1}{\pi} \operatorname{Im} \left\{ \frac{1 + a s^{\alpha-1}}{w_1(s)} \bigg|_{s=r e^{i\pi}} \right\}$$

$$= \frac{1}{\pi} \frac{a\, r^{\alpha-1}\, \sin(\alpha\pi)}{(1-r)^2 + a^2 r^{2\alpha} + 2(1-r)\, a\, r^\alpha \cos(\alpha\pi)}\,. \qquad (9.1.47)$$

For $a > 0$ and $0 < \alpha < 1$ the function $H_{\alpha,0}^{(1)}(r;a)$ is positive for all $r > 0$ since it has the sign of the numerator; in fact in (9.1.47) the denominator is strictly positive, being equal to $|w_1(s)|^2$ as $s = r e^{\pm i\pi}$. Hence, the *fundamental solution* $u_0(t)$ has the peculiar property of being *completely monotone*, and $H_{\alpha,0}^{(1)}(r;a)$ is its *spectral function*. Now the determination of $u_\delta(t) = -u_0'(t)$ is straightforward. We see that the *impulse-response solution* $u_\delta(t)$ is also *completely monotone* since it can be represented by

$$u_\delta(t) = \int_0^\infty e^{-rt} H_{\alpha,-1}^{(1)}(r;a)\, dr\,, \qquad (9.1.48)$$

with *spectral function*

$$H_{\alpha,-1}^{(1)}(r;a) = r\, H_{\alpha,0}^{(1)}(r;a) = \frac{1}{\pi} \frac{a\, r^{\alpha-1}\, \sin(\alpha\pi)}{(1-r)^2 + a^2 r^{2\alpha} + 2(1-r)\, a\, r^\alpha \cos(\alpha\pi)}\,. \qquad (9.1.49)$$

Both solutions $u_0(t)$ and $u_\delta(t)$ turn out to be strictly decreasing from 1 towards 0 as t runs from 0 to ∞. Their behavior as $t \to 0^+$ and $t \to \infty$ can be found by means of a proper asymptotic analysis. The behavior of the solutions as $t \to 0^+$ can be determined from the behavior of their Laplace transforms as $\operatorname{Re}\{s\} \to +\infty$, as is well known from the theory of the Laplace transform, see, e.g. [Doe74]. We obtain as $\operatorname{Re}\{s\} \to +\infty$,

$$\widetilde{u}_0(s) = s^{-1} - s^{-2} + O\left(s^{-3+\alpha}\right),\ \widetilde{u}_\delta(s) = s^{-1} - a s^{-(2-\alpha)} + O\left(s^{-2}\right),\quad (9.1.50)$$

so that

$$u_0(t) = 1 - t + O\left(t^{2-\alpha}\right),\ u_\delta(t) = 1 - a \frac{t^{1-\alpha}}{\Gamma(2-\alpha)} + O(t),\ \text{as } t \to 0^+.$$
$$(9.1.51)$$

The spectral representations (9.1.46) and (9.1.48) are suitable to obtain the asymptotic behavior of $u_0(t)$ and $u_\delta(t)$ as $t \to +\infty$, by using the Watson lemma. In fact, expanding the spectral functions for small r and taking the dominant term in the corresponding asymptotic series, we obtain

$$u_0(t) \sim a\, \frac{t^{-\alpha}}{\Gamma(1-\alpha)}\,,\quad u_\delta(t) \sim -a\, \frac{t^{-\alpha-1}}{\Gamma(-\alpha)}\,,\quad \text{as } t \to \infty. \qquad (9.1.52)$$

We note that the limiting case $\alpha = 1$ can easily be treated by extending the validity of (9.1.40)–(9.1.44) to $\alpha = 1$, as is legitimate. In this case we obtain

$$u_0(t) = \mathrm{e}^{-t/(1+a)}\,,\quad u_\delta(t) = \frac{1}{1+a}\, \mathrm{e}^{-t/(1+a)}\,,\quad \alpha = 1. \qquad (9.1.53)$$

In the case $a \equiv 0$ we recover the standard solutions $u_0(t) = u_\delta(t) = \mathrm{e}^{-t}$.

We conclude with some considerations on the solutions when the order α is a rational number. If we take $\alpha = p/q$, where $p, q \in \mathbb{N}$ are assumed (for convenience) to be relatively prime, a factorization in (9.1.41) is possible by using the procedure indicated by Miller and Ross [MilRos93]. In these cases the solutions can be expressed in terms of a linear combination of q Mittag-Leffler functions of fractional order $1/q$, which, in turn, can be expressed in terms of incomplete gamma functions.

Here we illustrate the factorization in the simplest case $\alpha = 1/2$ and provide the solutions $u_0(t)$ and $u_\delta(t)$ in terms of the functions $e_\alpha(t;\lambda)$ (with $\alpha = 1/2$), introduced in the previous subsection. In this case, in view of the application to the *Basset problem* equation (9.1.38) deserves particular attention.[1] For $\alpha = 1/2$ we can write

$$w_1(s) = s + a\, s^{1/2} + 1 = (s^{1/2} - \lambda_+)(s^{1/2} - \lambda_-)\,,\quad \lambda_\pm = -a/2 \pm (a^2/4 - 1)^{1/2}. \qquad (9.1.54)$$

Here $\lambda_\pm$ denote the two roots (real or conjugate complex) of the second degree polynomial with positive coefficients $z^2 + az + 1$, which, in particular, satisfy the following binary relations

$$\lambda_+ \cdot \lambda_- = 1\,,\ \lambda_+ + \lambda_- = -a\,,\ \lambda_+ - \lambda_- = 2(a^2/4 - 1)^{1/2} = (a^2 - 4)^{1/2}. \qquad (9.1.55)$$

We recognize that we must treat separately the following two cases

$$i)\ 0 < a < 2,\ \text{or}\ a > 2,\quad \text{and}\ ii)\ a = 2,$$

which correspond to two distinct roots ($\lambda_+ \neq \lambda_-$), or two coincident roots ($\lambda_+ \equiv \lambda_- = -1$), respectively.

For this purpose, we write

---

[1] Basset considered in [Bas88] a model of a quiescent fluid which leads in modern language to Eq. (9.1.38) with $\alpha = 1/2$. For arbitrary $0 < \alpha < 1$ the (generalized) Basset problem is discussed in [Mai97].

9.1 Fractional Relaxation and Oscillations

$$\widetilde{M}(s) := \frac{1 + a s^{-1/2}}{s + a s^{-1/2} + 1} = \begin{cases} i) & \dfrac{A_-}{s^{-1/2}(s^{-1/2} - \lambda_+)} + \dfrac{A_+}{s^{-1/2}(s^{-1/2} - \lambda_-)}, \\ ii) & 1\dfrac{1}{(s^{-1/2} + 1)^2} + \dfrac{2}{s^{-1/2}(s^{-1/2} + 1)^2}, \end{cases}$$
(9.1.56)

and

$$\widetilde{N}(s) := \frac{1}{s + a s^{-1/2} + 1} = \begin{cases} i) & \dfrac{A_+}{s^{-1/2}(s^{-1/2} - \lambda_+)} + \dfrac{A_-}{s^{-1/2}(s^{-1/2} - \lambda_-)}, \\ ii) & \dfrac{1}{(s^{-1/2} + 1)^2}, \end{cases}$$
(9.1.57)

where

$$A_\pm = \pm \frac{\lambda_\pm}{\lambda_+ - \lambda_-}.$$
(9.1.58)

Using (9.1.55) we note that

$$A_+ + A_- = 1, \quad A_+ \lambda_- + A_- \lambda_+ = 0, \quad A_+ \lambda_+ + A_- \lambda_- = -a.$$
(9.1.59)

Recalling the Laplace transform pairs we obtain

$$u_0(t) = M(t) := \begin{cases} i) & A_- E_{1/2}(\lambda_+ \sqrt{t}) + A_+ E_{1/2}(\lambda_- \sqrt{t}), \\ ii) & (1 - 2t) E_{1/2}(-\sqrt{t}) + 2\sqrt{t/\pi}, \end{cases}$$
(9.1.60)

and

$$u_\delta(t) = N(t) := \begin{cases} i) & A_+ E_{1/2}(\lambda_+ \sqrt{t}) + A_- E_{1/2}(\lambda_- \sqrt{t}), \\ ii) & (1 + 2t) E_{1/2}(-\sqrt{t}) - 2\sqrt{t/\pi}. \end{cases}$$
(9.1.61)

In (9.1.60)–(9.1.61) the functions $e_{1/2}(t; -\lambda_\pm) = E_{1/2}(\lambda_\pm \sqrt{t})$ and $e_{1/2}(t) = e_{1/2}(t; 1) = E_{1/2}(-\sqrt{t})$ are presented. In particular, the solution of the *Basset problem* can be easily obtained from (9.1.44) with $q(t) = q_0$ by using (9.1.60)–(9.1.61) and noting that $\int_0^t N(\tau) d\tau = 1 - M(t)$. Denoting this solution by $u_B(t)$ we get

$$u_B(t) = q_0 - (q_0 - c_0) M(t).$$
(9.1.62)

When $a \equiv 0$, i.e. in the absence of the term containing the fractional derivative (due to the Basset force), we recover the classical Stokes solution, which we denote by $u_S(t)$,

$$u_S(t) = q_0 - (q_0 - c_0) e^{-t}.$$

In the particular case $q_0 = c_0$, we get the steady-state solution $u_B(t) = u_S(t) \equiv q_0$. For vanishing initial condition $c_0 = 0$, we have the creep-like solutions

$$u_B(t) = q_0 \left[1 - M(t)\right], \quad u_S(t) = q_0 \left[1 - e^{-t}\right].$$

In this case it is instructive to compare the behavior of the two solutions as $t \to 0^+$ and $t \to \infty$. Recalling the general asymptotic expressions of $u_0(t) = M(t)$ in (9.1.51) and (9.1.52) with $\alpha = 1/2$, we recognize that

$$u_B(t) = q_0 \left[t + O\left(t^{3/2}\right)\right], \quad u_S(t) = q_0 \left[t + O\left(t^2\right)\right], \quad \text{as } t \to 0^+,$$

and

$$u_B(t) \sim q_0 \left[1 - a/\sqrt{\pi t}\right], \quad u_S(t) \sim q_0 \left[1 - EST\right], \quad \text{as } t \to \infty,$$

where EST denotes *exponentially small terms*. In particular, we note that the normalized plot of $u_B(t)/q_0$ remains under that of $u_S(t)/q_0$ as t runs from 0 to ∞. The reader is invited to convince himself of the following fact. In the general case $0 < \alpha < 1$ the solution $u(t)$ has the particular property of being equal to 1 for all $t \geq 0$ if $q(t)$ has this property and $u(0^+) = 1$, whereas $q(t) = 1$ for all $t \geq 0$ and $u(0^+) = 0$ implies that $u(t)$ is a creep function tending to 1 as $t \to \infty$.

Let us now apply the Laplace transform to the fractional oscillation equation (9.1.39). This leads us to the transformed algebraic equations

$$\text{(a)} \quad \widetilde{v}(s) = c_0 \frac{s + a s^{\alpha-1}}{w_2(s)} + c_1 \frac{1}{w_2(s)} + \frac{\widetilde{q}(s)}{w_2(s)}, \quad 0 < \alpha < 1, \qquad (9.1.63)$$

or

$$\text{(b)} \quad \widetilde{v}(s) = c_0 \frac{s + a s^{\alpha-1}}{w_2(s)} + c_1 \frac{1 + a s^{\alpha-2}}{w_2(s)} + \frac{\widetilde{q}(s)}{w_2(s)}, \quad 1 < \alpha < 2, \qquad (9.1.64)$$

where

$$w_2(s) := s^2 + a s^\alpha + 1, \qquad (9.1.65)$$

and $a > 0$. Putting

$$\widetilde{v}_0(s) := \frac{s + a s^{\alpha-1}}{w_2(s)}, \quad 0 < \alpha < 2, \qquad (9.1.66)$$

we have

$$v_0(0^+) = \lim_{s \to \infty} s\, \widetilde{v}_0(s) = 1, \quad \frac{1}{w_2(s)} = -[s\, \widetilde{v}_0(s) - 1] \div -v_0'(t), \qquad (9.1.67)$$

and

$$\frac{1 + a s^{\alpha-2}}{w_2(s)} = \frac{\widetilde{v}_0(s)}{s} \div \int_0^t v_0(\tau)\, d\tau. \qquad (9.1.68)$$

Thus we can conclude that

$$\text{(a)} \quad v(t) = c_0 v_0(t) - c_1 v_0'(t) - \int_0^t q(t - \tau)\, v_0'(\tau)\, d\tau, \quad 0 < \alpha < 1, \qquad (9.1.69)$$

9.1 Fractional Relaxation and Oscillations

or

(b) $v(t) = c_0 v_0(t) + c_1 \int_0^t v_0(\tau) d\tau - \int_0^t q(t-\tau) v_0'(\tau) d\tau$, $1 < \alpha < 2$.

(9.1.70)

In both equations the term $-v_0'(t)$ represents the *impulse-response solution* $v_\delta(t)$ for the *composite fractional oscillation equation* (9.1.39), namely the particular solution of the inhomogeneous equation with $c_0 = c_1 = 0$ and with $q(t) = \delta(t)$. For the *fundamental solutions* of (9.1.39) we have two distinct pairs of solutions according to the case (a) and (b) which read

(a) $\{v_0(t), v_{1a}(t) = -v_0'(t)\}$, (b) $\{v_0(t), v_{1b}(t) = \int_0^t v_0(\tau) d\tau\}$. (9.1.71)

We first consider the particular case $\alpha = 1$ for which the fundamental and impulse response solutions are known in terms of elementary functions. This limiting case can also be treated by extending the validity of (9.1.63) and (9.1.69) to $\alpha = 1$, as is legitimate. From

$$\widetilde{v}_0(s) = \frac{s+a}{s^2 + as + 1} = \frac{s + a/2}{(s + a/2)^2 + (1 - a^2/4)} - \frac{a/2}{(s + a/2)^2 + (1 - a^2/4)},$$ (9.1.72)

we obtain the *basic fundamental solution*

$$v_0(t) = \begin{cases} e^{-at/2} \left[\cos(\omega t) + \dfrac{a}{2\omega} \sin(\omega t) \right] & \text{if } 0 < a < 2, \\ e^{-t}(1 - t) & \text{if } a = 2, \\ e^{-at/2} \left[\cosh(\chi t) + \dfrac{a}{2\chi} \sinh(\chi t) \right] & \text{if } a > 2, \end{cases}$$ (9.1.73)

where

$$\omega = \sqrt{1 - a^2/4}, \quad \chi = \sqrt{a^2/4 - 1}.$$ (9.1.74)

By a differentiation of (9.1.73) we easily obtain the *second fundamental solution* $v_{1a}(t)$ and the *impulse-response solution* $v_\delta(t)$ since $v_{1a}(t) = v_\delta(t) = -v_0'(t)$. We point out that all the solutions exhibit an *exponential decay* as $t \to \infty$. Let us now consider the problem of finding $v_0(t)$ as the inverse Laplace transform of $\widetilde{v}_0(s)$,

$$v_0(t) = \frac{1}{2\pi i} \int_{Br} e^{st} \frac{s + a s^{\alpha-1}}{w_2(s)} ds,$$ (9.1.75)

where Br denotes the usual Bromwich path. Using a result of Beyer and Kempfle [BeyKem95] we know that the function $w_2(s)$ (for $a > 0$ and $0 < \alpha < 2$, $\alpha \neq 1$) has exactly *two simple, conjugate complex zeros* on the principal branch in the open left half-plane, cut along the negative real axis, say $s_+ = \rho e^{+i\gamma}$ and $s_- = \rho e^{-i\gamma}$ with $\rho > 0$ and $\pi/2 < \gamma < \pi$. This enables us to repeat the considerations carried out for the simple fractional oscillation equation to decompose the basic fundamental solution $v_0(t)$ into two parts according to $v_0(t) = f_\alpha(t; a) + g_\alpha(t; a)$. In fact, the evaluation of the Bromwich integral (9.1.75) can be achieved by adding the contribution $f_\alpha(t; a)$ from the Hankel path Ha(ϵ), as $\epsilon \to 0$, to the residual contribution $g_\alpha(t; a)$ from the two poles $s_\pm$. As an exercise in complex analysis we obtain

$$f_\alpha(t; a) = \int_0^\infty e^{-rt} H_{\alpha,0}^{(2)}(r; a)\, dr, \qquad (9.1.76)$$

with *spectral function*

$$H_{\alpha,0}^{(2)}(r; a) = -\frac{1}{\pi} \operatorname{Im} \left\{ \left. \frac{s + a s^{\alpha-1}}{w_2(s)} \right|_{s = r e^{i\pi}} \right\}$$

$$= \frac{1}{\pi} \frac{a r^{\alpha-1} \sin(\alpha\pi)}{(r^2 + 1)^2 + a^2 r^{2\alpha} + 2(r^2 + 1) a r^\alpha \cos(\alpha\pi)}. \qquad (9.1.77)$$

Since in (9.1.77) the denominator is strictly positive, being equal to $|w_2(s)|^2$ as $s = r e^{\pm i\pi}$, the *spectral function* $H_{\alpha,0}^{(2)}(r; a)$ turns out to be positive for all $r > 0$ for $0 < \alpha < 1$ and negative for all $r > 0$ for $1 < \alpha < 2$. Hence, in case (a) the function $f_\alpha(t)$ and in case (b) the function $-f_\alpha(t)$ is *completely monotone*; in both cases $f_\alpha(t)$ tends to zero as $t \to \infty$, from above in case (a) and from below in case (b), according to the asymptotic behavior

$$f_\alpha(t; a) \sim a \frac{t^{-\alpha}}{\Gamma(1-\alpha)}, \quad \text{as } t \to \infty,\ 0 < \alpha < 1,\ 1 < \alpha < 2, \qquad (9.1.78)$$

as derived by applying the Watson lemma in (9.1.76) and considering (9.1.77). The other part, $g_\alpha(t; a)$, is obtained as

$$g_\alpha(t; a) = e^{s_+ t} \operatorname{Res}\left[\frac{s + a s^{\alpha-1}}{w_2(s)} \right]_{s_+} + \text{ conjugate complex}$$

$$= 2 \operatorname{Re} \left\{ \frac{s_+ + a s_+^{\alpha-1}}{2 s_+ + a \alpha s_+^{\alpha-1}} e^{s_+ t} \right\}. \qquad (9.1.79)$$

Thus this term exhibits an *oscillatory* character with exponentially decreasing amplitude like $\exp(-\rho t |\cos \gamma|)$. Then we recognize that the basic fundamental solution $v_0(t)$ exhibits a *finite* number of zeros and that, for sufficiently large t, it turns out to be permanently positive if $0 < \alpha < 1$ and permanently negative if $1 < \alpha < 2$ with an *algebraic decay* provided by (9.1.78). For the second fundamental solutions $v_{1a}(t)$, $v_{1b}(t)$ and for the impulse-response solution $v_\delta(t)$, the corresponding analysis is straightforward in view of their connection with $v_0(t)$, pointed out in (9.1.70)–(9.1.71). The *algebraic decay* of all the solutions as $t \to \infty$, for $0 < \alpha < 1$ and

9.1 Fractional Relaxation and Oscillations

$1 < \alpha < 2$, is henceforth resumed in the relations

$$v_0(t) \sim a \frac{t^{-\alpha}}{\Gamma(1-\alpha)}, \quad v_{1a}(t) = v_\delta(t) \sim -a \frac{t^{-\alpha-1}}{\Gamma(-\alpha)}, \quad v_{1b}(t) \sim a \frac{t^{1-\alpha}}{\Gamma(2-\alpha)}. \tag{9.1.80}$$

In conclusion, except in the particular case $\alpha = 1$, all the present solutions of the composite fractional oscillation equation exhibit similar characteristics as the corresponding solutions of the simple fractional oscillation equation, namely a *finite number of damped oscillations* followed by a *monotonic algebraic decay* as $t \to \infty$.

9.2 Examples of Applications of the Fractional Calculus in Physical Models

Here we present a few physical models involving the fractional calculus. An interest in such models is growing rapidly nowadays and several books on the subject have recently appeared. It would be impossible to discuss general fractional models in detail here. Instead we select some models which are related to the above discussed equations (or their simple generalizations) and which demonstrate the essential role of the Mittag-Leffler function in fractional modelling.

9.2.1 Linear Visco-Elasticity

Let us first introduce some notation. We denote the stress by $\sigma = \sigma(x, t)$ and the strain by $\epsilon = \epsilon(x, t)$, where x and t are the space and time variables, respectively. For the sake of convenience, both stress and strain are intended to be normalized, i.e. scaled with respect to a suitable reference state $\{\sigma_*, \epsilon_*\}$.

According to the linear theory of viscoelasticity, assuming the presence of sufficiently small strains, the body may be considered as a linear system with the stress (or strain) as the excitation function (input) and the strain (or stress) as the response function (output).

To formulate general stress-strain relations (or *constitutive equations*), two fundamental hypotheses are required: (i) invariance under time translation and (ii) causality; the former means that a time shift in the input results in an equal shift in the output, the latter that the output for any instant t_1 depends on the values of the input only for $t \leq t_1$. A fundamental role is played by the step response, i.e. the response function, expressed by the Heaviside function $\Theta(t)$.

$$\Theta(t) = \begin{cases} 0 & \text{if } t < 0, \\ 1 & \text{if } t > 0. \end{cases}$$

Two magnitudes are defined in this way:

$$\sigma(t) = \Theta(t) \implies \epsilon(t) = J(t), \qquad (9.2.1)$$
$$\epsilon(t) = \Theta(t) \implies \sigma(t) = G(t). \qquad (9.2.2)$$

The functions $J(t)$ and $G(t)$ are referred to as the *creep compliance* and *relaxation modulus* respectively, or, simply, the *material functions* of the viscoelastic body.

The general stress-strain relation is expressed through a linear hereditary integral of Stieltjes type, namely

$$\epsilon(t) = \int_{-\infty}^{t} J(t-\tau)\,d\sigma(\tau), \qquad (9.2.3)$$

$$\sigma(t) = \int_{-\infty}^{t} G(t-\tau)\,d\epsilon(\tau). \qquad (9.2.4)$$

In the classical *Hook model* for an elastic body we have

$$\sigma(t) = m\epsilon(t), \qquad (9.2.5)$$

and thus $J(t) = 1/m$, $G(t) = m$.

In the classical *Newton model* for an ideal fluid we have

$$\sigma(t) = b_1 \frac{d\epsilon}{dt}, \qquad (9.2.6)$$

and thus $J(t) = t/b_1$, $G(t) = b_1 \delta(t)$.

9.2.2 The Use of Fractional Calculus in Linear Viscoelasticity

Based on certain rheological experiments Scott-Blair [ScoB-Cop39, ScoB-Cop42a] argued that material properties are determined by various states between an elastic solid and a viscous fluid, rather than a combination of an elastic and a viscous element as proposed by Maxwell. The conclusion was that these materials satisfy a law intermediate between Hook's law and Newton's law:

$$\sigma(t) = b_1 \frac{d^\nu \epsilon}{dt^\nu}, \quad 0 < \nu < 1. \qquad (9.2.7)$$

This yields the power-type behavior of the creep function

$$J(t) = \frac{t^\nu}{b_1 \Gamma(1+\nu)} \implies G(t) = \frac{b_1}{\Gamma(1-\nu)} t^{-\nu}. \qquad (9.2.8)$$

9.2 Examples of Applications of the Fractional Calculus in Physical Models

We point out that (9.2.7) is the differential form of the Nutting equation [Nut21] (for more details see [RogMai14]).

By using a power-type law for the creep functions one can rewrite the constitutive relation (9.2.7) in a form involving either a fractional integral (see also the contribution by Rabotnov [Rab80], who presents the constitutive relation in the form of a more cumbersome integral equation)

$$\epsilon(t) = \frac{1}{b_1 \Gamma(1+\nu)} \int_{-\infty}^{t} \frac{\sigma(\tau) d\tau}{(t-\tau)^{1-\nu}} = \frac{1}{b_1} \left({}_{-\infty}I_t^\nu \sigma \right)(t), \qquad (9.2.9)$$

or a fractional derivative (see also the pioneering contribution by Gerasimov [Ger48])

$$\sigma(t) = \frac{b_1}{\Gamma(1-\nu)} \int_{-\infty}^{t} \frac{\dot{\epsilon}(\tau) d\tau}{(t-\tau)^\nu} = b_1 \left({}_{-\infty}D_t^\nu \epsilon \right)(t). \qquad (9.2.10)$$

Here the fractional integrals and derivatives have Liouville (or Liouville–Weyl) form, i.e. with integration from $-\infty$. We note that the fractional derivative in (9.2.10) is similar to the Caputo derivative. Moreover, if we consider causal histories (i.e. starting from $t = 0$), then in (9.2.10) the Liouville fractional derivative will be replaced by the Caputo fractional derivative.

The use of fractional calculus in linear viscoelasticity, started by Scott-Blair, leads us to generalize the classical mechanical models, in that the basic Newton element (dashpot) is substituted by the more general Scott-Blair element (of order ν), sometimes referred to as *pot*. In fact, we can construct the class of these generalized models from Hooke and Scott-Blair elements, disposed singly and in branches of two (in series or in parallel). The material functions are obtained using the combination rule; their determination is made easy if we take into account the following *correspondence principle* between the classical and fractional mechanical models, as introduced in [CapMai71a], which is empirically justified.

Let us also present some constitutive equations and material functions which correspond to the most popular fractional models in visco-elasticity.

$$\text{fractional Newton (Scott-Blair) model: } \sigma(t) = b_1 \frac{d^\nu \epsilon}{dt^\nu}, \qquad (9.2.11)$$

$$\begin{cases} J(t) = \dfrac{t^\nu}{b_1 \Gamma(1+\nu)}, \\ G(t) = b_1 \dfrac{t^{-\nu}}{\Gamma(1-\nu)}; \end{cases}$$

$$\text{fractional Voigt model: } \sigma(t) = m\, \epsilon(t) + b_1 \frac{d^\nu \epsilon}{dt^\nu}, \qquad (9.2.12)$$

$$\begin{cases} J(t) = \dfrac{1}{m}\left\{1 - E_\nu\left[-(t/\tau_\epsilon)^\nu\right]\right\}, \\ G(t) = m + b_1 \dfrac{t^{-\nu}}{\Gamma(1-\nu)}, \end{cases}$$

where $(\tau_\epsilon)^\nu = b_1/m$;

fractional Maxwell model: $\sigma(t) + a_1 \dfrac{\mathrm{d}^\nu \sigma}{\mathrm{d}t^\nu} = b_1 \dfrac{\mathrm{d}^\nu \epsilon}{\mathrm{d}t^\nu}$, (9.2.13)

$$\begin{cases} J(t) = \dfrac{a}{b_1} + \dfrac{1}{b} \dfrac{t^\nu}{\Gamma(1+\nu)}, \\ G(t) = \dfrac{b_1}{a_1} E_\nu\left[-(t/\tau_\sigma)^\nu\right], \end{cases}$$

where $(\tau_\sigma)^\nu = a_1$;

fractional Zener model :

$$\left[1 + a_1 \dfrac{\mathrm{d}^\nu}{\mathrm{d}t^\nu}\right] \sigma(t) = \left[m + b_1 \dfrac{\mathrm{d}^\nu}{\mathrm{d}t^\nu}\right] \epsilon(t),$$ (9.2.14)

$$\begin{cases} J(t) = J_g + J_1\left[1 - E_\nu\left[-(t/\tau_\epsilon)^\nu\right]\right], \\ G(t) = G_e + G_1 E_\nu\left[-(t/\tau_\sigma)^\nu\right], \end{cases}$$

where

$$\begin{cases} J_g = \dfrac{a_1}{b_1}, \ J_1 = \dfrac{1}{m} - \dfrac{a_1}{b_1}, \ \tau_\epsilon = \dfrac{b_1}{m}, \\ G_e = m, \ G_1 = \dfrac{b_1}{a_1} - m, \ \tau_\sigma = a_1 ; \end{cases}$$

fractional anti-Zener model:

$$\left[1 + a_1 \dfrac{\mathrm{d}^\nu}{\mathrm{d}t^\nu}\right] \sigma(t) = \left[b_1 \dfrac{\mathrm{d}^\nu}{\mathrm{d}t^\nu} + b_2 \dfrac{\mathrm{d}^{(\nu+1)}}{\mathrm{d}t^{(\nu+1)}}\right] \epsilon(t),$$ (9.2.15)

$$\begin{cases} J(t) = J_+ \dfrac{t^\nu}{\Gamma(1+\nu)} + J_1\left[1 - E_\nu\left[-(t/\tau_\epsilon)^\nu\right]\right], \\ G(t) = G_- \dfrac{t^{-\nu}}{\Gamma(1-\nu)} + G_1 E_\nu\left[-(t/\tau_\sigma)^\nu\right), \end{cases}$$

where

$$\begin{cases} J_+ = \dfrac{1}{b_1}, \ J_1 = \dfrac{a_1}{b_1} - \dfrac{b_2}{b_1^2}, \ \tau_\epsilon = \dfrac{b_2}{b_1}, \\ G_- = \dfrac{b_2}{a_1}, \ G_1 = \dfrac{b_1}{a_1} - \dfrac{b_2}{a_1^2}, \ \tau_\sigma = a_1 . \end{cases}$$

These models were formerly considered by Capunto and Mainardi in their 1971 review paper and more recently revisited by Mainardi and Spada, who generalized them to include a four-parameter model known as the Burgers model.

9.2.3 The General Fractional Operator Equation

The above equations can be treated in the same way as described in Sect. 9.1.2 for composite type fractional equations, i.e. by the Laplace transform method (for details, see [Mai10]).

Assuming a general network of Hooke and Scott-Blair elements in series and in parallel, as required for the classical networks with Hooke and Newton elements, but requiring the same fractional derivative $\nu \in (0, 1)$, we conjecture that the form of the corresponding operator equation, referred to as the *fractional operator equation*, would read

$$\left[1 + \sum_{k=1}^{p} a_k \frac{d^{\nu_k}}{dt^{\nu_k}}\right] \sigma(t) = \left[m + \sum_{k=1}^{q} b_k \frac{d^{\nu_k}}{dt^{\nu_k}}\right] \epsilon(t), \quad (9.2.16)$$

with $\nu_k = k + \nu - 1$, and

$$\begin{cases} J(t) = J_g + \sum_n J_n \left\{1 - E_\nu \left[-(t/\tau_{\epsilon,n})^\nu\right]\right\} + J_+ \dfrac{t^\nu}{\Gamma(1+\nu)}, \\ G(t) = G_e + \sum_n G_n E_\nu \left[-(t/\tau_{\sigma,n})^\nu\right] + G_- \dfrac{t^{-\nu}}{\Gamma(1-\nu)}, \end{cases} \quad (9.2.17)$$

where all the coefficients are non-negative. The above equations were verified for all of the fractional models considered by Mainardi and Spada [MaiSpa11]. However a general proof for any network of these types is not yet available when the material functions are expressed in terms of pure exponentials (i.e. with $\nu = 1$).

9.3 The Fractional Dielectric Models

The relaxation properties of dielectric materials can be described, in the frequency domain, by several models which have been proposed over the years, of which the most relevant for our applications (i.e. using Mittag-Leffler type functions in a fractional calculus context) are Cole–Cole [ColCol41, ColCol42], Cole–Davidson [DavCol50, DavCol51], and Havriliak–Negami [HavNeg67]. In this section, we survey the main dielectric models and we illustrate the corresponding time-domain functions. In particular, we focus on the completely monotone character of the relaxation and response functions. We also provide a characterization of the models in terms of differential operators of fractional order.

Under the influence of an electric field, matter experiences an electric polarization. The electric displacement effect on free and bound charges is described by the displacement field **D**, which is related to the electric field **E** and to the polarization **P** by

$$\mathbf{D} = \varepsilon_0 \mathbf{E} + \mathbf{P}, \tag{9.3.1}$$

where ε_0 is the permittivity of the free space. For a perfect isotropic dielectric and for harmonic fields of frequency ω, the interdependence between **E** and **P** is described by a constitutive law

$$\mathbf{P} = \varepsilon_0[(\varepsilon_s - \varepsilon_\infty)(\hat{\varepsilon}(i\omega) - 1)]\mathbf{E} = \varepsilon_0[(\varepsilon_s - \varepsilon_\infty)\hat{\chi}(i\omega)]\mathbf{E}, \tag{9.3.2}$$

where ε_s and ε_∞ are the static and infinite dielectric constants. The normalized complex *permittivity* $\hat{\varepsilon}(i\omega)$ and the normalized complex *susceptibility* $\hat{\chi}(i\omega)$ are specific characteristics of the polarized medium and are usually determined by matching experimental data with an appropriate theoretical model.

From a physical point of view, the description of dielectrics, considered as passive and causal linear systems, is also carried out in time by considering two causal functions of time (i.e., vanishing for $t < 0$):

- the *relaxation function* $\Psi(t)$,
- the *response function* $\phi(t)$.

The relaxation function describes the decay of polarization whereas the response function its decay rate (the depolarization current).

Remark 9.3 Note that our notation $\{\Psi(t), \phi(t)\}$ for the relaxation and response functions is in conflict with a notation frequently used in the literature, where the relaxation function is denoted by $\phi(t)$ and the response function by $-d\phi(t)/dt$.

As a matter of fact, the relationship between response and relaxation functions can be better clarified by their probabilistic interpretation investigated in several papers by Karina Weron and her team (e.g., see [Wer91, WerKla00, WerKot96]): interpreting the *relaxation function* as a *survival probability* $\Psi(t)$, the *response function* turns out to be the *probability density function* corresponding to the *cumulative probability function* $\Phi(t) = 1 - \Psi(t)$. Thus the two functions are interrelated as follows

$$\phi(t) = -\frac{d}{dt}\Psi(t) = \frac{d}{dt}\Phi(t), \quad t \geq 0, \tag{9.3.3}$$

and

$$\Psi(t) = 1 - \int_0^t \phi(u)\,du, \quad t \geq 0. \tag{9.3.4}$$

9.3 The Fractional Dielectric Models

In view of their probabilistic meaning, $\phi(t)$ and $\Psi(t)$ are both non-negative and non-increasing functions. In particular, we get the limit $\Psi(0^+) = 1$ whereas $\phi(0^+)$ may be finite or infinite.

The response function $\phi(t)$ is obtained as the inverse Laplace transform of the normalized complex susceptibility by setting the Laplace parameter $s = i\omega$, that is

$$\phi(t) = \mathcal{L}^{-1}\left(\widetilde{\chi}(s); t\right), \qquad (9.3.5)$$

where we have used the superscript $\sim$ to denote a Laplace transform, i.e. $\widetilde{\chi}(s) = \hat{\chi}(i\omega)$; then, for the relaxation function $\Psi(t)$ we have

$$\Psi(t) = 1 - \mathcal{L}^{-1}\left(\frac{1}{s}\widetilde{\chi}(s); t\right) = \mathcal{L}^{-1}\left(\frac{1}{s} - \frac{1}{s}\widetilde{\chi}(s); t\right). \qquad (9.3.6)$$

We can thus outline the basic Laplace transform pairs as follows

$$\phi(t) \div \widetilde{\phi}(s) = \widetilde{\chi}(s), \quad \Psi(t) \div \widetilde{\Psi}(s) = \frac{1 - \widetilde{\phi}(s)}{s}, \qquad (9.3.7)$$

where we have adopted the notation $\div$ to denote the juxtaposition of a function of time $f(t)$ with its Laplace transform $\widetilde{f}(s) = \int_0^\infty e^{-st} f(t)\,dt$.

The standard model in the physics of dielectrics was provided by Debye [Deb12], according to which the normalized complex susceptibility, depending on the frequency of the external field, is provided, unless it is a proper multiplicative constant, such as

$$\hat{\chi}(i\omega) = \frac{1}{1 + i\omega \tau_D}, \qquad (9.3.8)$$

where τ_D is the only expected relaxation time. In this case, both relaxation and response functions turn out to be purely exponential. In fact, recalling standard results on Laplace transforms, we get

$$\Psi_D(t) = e^{-t/\tau_D}, \quad \phi_D(t) = \frac{1}{\tau_D} e^{-t/\tau_D}. \qquad (9.3.9)$$

Even though the Debye relaxation model was first derived on the basis of statistical mechanics, it also finds application in complex systems where it is more reasonable to have a discrete or a continuous distribution of Debye models with different relaxation times, so that the complex susceptibility reads

$$\hat{\chi}(i\omega) = \frac{\rho_1}{1 + i\omega \tau_1} + \frac{\rho_2}{1 + i\omega \tau_2} + \ldots \qquad (9.3.10)$$

with $\rho_1, \rho_2, \ldots$ non-negative constants or, more generally,

$$\hat{\chi}(i\omega) = \int_0^\infty \frac{\rho(\tau)}{1 + i\omega\tau}\,d\tau\,, \qquad (9.3.11)$$

with $\rho(\tau) \geq 0$. In mathematical language the above properties are achieved by requiring relaxation and response to be locally integrable and completely monotone (LICM) functions [HanSer08]. The local integrability is requested to be Laplace transformable in the classical sense. The complete monotonicity means that the functions are non-negative with infinitely many derivatives for $t > 0$ alternating in sign; we provide here its formal definition.

As discussed by Hanyga [Han05b], CM is essential to ensure the monotone decay of the energy in isolated systems (as appears reasonable from physical considerations); thus, restricting to CM functions is essential for the physical acceptability and realizability of the dielectric models (see [AnhMcV]).

For the basic Bernstein theorem for LICM functions [Wid46], $\Psi(t)$ and $\phi(t)$ are represented as real Laplace transforms of non-negative spectral functions (of frequency)

$$\Psi(t) = \int_0^\infty e^{-rt} K^\Psi(r)\,dr\,, \quad \phi(t) = \int_0^\infty e^{-rt} K^\phi(r)\,dr\,. \qquad (9.3.12)$$

Due to the interrelation between $\Psi(t)$ and $\phi(t)$, the corresponding spectral functions are obviously related and indeed

$$K^\Psi(r) = K^\phi(r)/r\,, \quad K^\phi(r) = r\,K^\Psi(r)\,. \qquad (9.3.13)$$

As a matter of fact, the Laplace transform $\widetilde{\Psi}(s)$ of $\Psi(t)$ and $\widetilde{\phi}(s)$ of $\phi(t)$ turn out to be iterated Laplace transforms (that is, Stieltjes transforms) of the corresponding frequency spectral functions $K^\Psi(r)$, $K^\phi(r)$. In fact, by exchanging order in the Laplace integrals, we get

$$\widetilde{\Psi}(s) = \int_0^\infty \frac{K^\Psi(r)}{s+r}\,dr\,, \quad \widetilde{\phi}(s) = \int_0^\infty \frac{K^\phi(r)}{s+r}\,dr\,. \qquad (9.3.14)$$

As a consequence, the frequency spectral functions can be derived from $\widetilde{\Psi}(s)$ and $\widetilde{\phi}(s)$ as their inverse Stieltjes transforms thanks to the Titchmarsh inversion formula [Tit86],

$$K^\Psi(r) = \mp\frac{1}{\pi}\mathrm{Im}\left[\widetilde{\Psi}(s)\big|_{s=re^{\pm i\pi}}\right]\,, \quad K^\phi(r) = \mp\frac{1}{\pi}\mathrm{Im}\left[\widetilde{\phi}(s)\big|_{s=re^{\pm i\pi}}\right]\,. \qquad (9.3.15)$$

For a physical viewpoint, it may be more interesting to deal with spectral functions expressed in terms of relaxation times $\tau = 1/r$ rather than frequencies r. Then we write

$$\Psi(t) = \int_0^\infty e^{-t/\tau} H^\Psi(\tau)\,d\tau\,, \quad \phi(t) = \int_0^\infty e^{-t/\tau} H^\phi(\tau)\,d\tau\,, \qquad (9.3.16)$$

9.3 The Fractional Dielectric Models

so that the time spectral functions are obtained from the corresponding frequency spectral functions by the variable change

$$H^{\Psi,\phi}(\tau) = \frac{K^{\Psi,\phi}(1/\tau)}{\tau^2} \qquad (9.3.17)$$

9.3.1 The Main Models for Anomalous Dielectric Relaxation

The Debye model [Deb12] is one of the first models introduced to describe physical properties of dielectrics and involves relaxation and response functions of exponential type; see Eqs. (9.3.8) and (9.3.9).

As revealed by a number of experiments, a broad variety of dielectric materials exhibit relaxation behaviors which strongly deviate from the exponential Debye law. The observation of "anomalous" phenomena such as broadness, asymmetry and excess in the dielectric dispersion has motivated the proposition of new empirical laws in order to modify the Debye relaxation and match experimental data in a more accurate way.

It is now well established that the relaxation properties of a large variety of materials fit the following models that we are going to briefly discuss

- the Cole–Cole (CC) model,
- the Davidson–Cole (DC) model,
- the Havriliak–Negami (HN) model.

Other models have been discussed in the literature, but most of them can be obtained from one of the above models. In the remainder of this section, we illustrate each model and we discuss their main features from a mathematical perspective.

9.3.2 The Cole–Cole Model

The Cole–Cole model, named after the brothers K.S. Cole and R.H. Cole, was introduced in 1941 [ColCol41] (see also [ColCol42]). As described in [BotBor78], it finds applications in "systems with rather small deviations from a single relaxation time, e.g. many compounds with rigid molecules in the pure liquid state and in solution in non-polar, non-viscous solvents". Nowadays this model is still used to represent impedance of biological tissues, to describe relaxation in polymers, to represent anomalous diffusion in disordered systems and so on [Kal-et-al04, Lin10, MauElw12].

The complex susceptibility of the Cole-Cole model is derived by inserting a real power in the original Debye model, thus to fit data presenting a broader loss peak, and it is given by

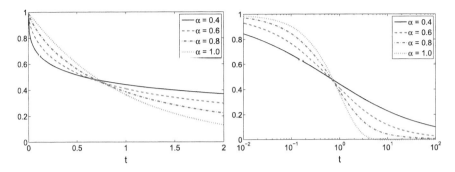

Fig. 9.4 The relaxation function $\Psi_{CC}(t)$ for varying α

$$\hat{\chi}_{CC}(i\omega) = \frac{1}{1 + (i\omega\tau_\star)^\alpha}, \quad 0 < \alpha \leq 1, \tag{9.3.18}$$

where $\tau_\star$ denotes a reference relaxation time. In all the subsequent plots a normalized relaxation time $\tau_\star = 1$ is assumed.

By applying the Laplace inversion of (9.3.18) we get the corresponding response and relaxation functions respectively as

$$\phi_{CC}(t) = \mathcal{L}^{-1}\left(\frac{1}{1 + (s\tau_\star)^\alpha}\right) = \frac{1}{\tau_\star}(t/\tau_\star)^{\alpha-1} E_{\alpha,\alpha}\left(-(t/\tau_\star)^\alpha\right), \tag{9.3.19}$$

and

$$\Psi_{CC}(t) = \mathcal{L}^{-1}\left(\frac{1}{s} - \frac{1}{s\left(1 + (s\tau_\star)^\alpha\right)}\right) \tag{9.3.20}$$
$$= 1 - (t/\tau_\star)^\alpha E_{\alpha,\alpha+1}\left(-(t/\tau_\star)^\alpha\right) = E_{\alpha,1}\left(-(t/\tau_\star)^\alpha\right).$$

Figure 9.4 shows the plots of the relaxation function $\Psi_{CC}(t)$ using linear (left plot) and logarithmic (right plot) scales.

The plots of the relaxation and response functions are also found in Mainardi's book [Mai10] along with their asymptotic representations. Here we just recall that by using standard results on the asymptotic behavior of the Mittag-Leffler function it is possible to verify that

$$\phi_{CC}(t) \sim \begin{cases} \dfrac{1}{\tau_\star \Gamma(\alpha)}(t/\tau_\star)^{\alpha-1}, & \text{for } t \ll \tau_\star, \\ -\dfrac{1}{\tau_\star \Gamma(-\alpha)}(t/\tau_\star)^{-\alpha-1}, & \text{for } t \gg \tau_\star, \end{cases} \tag{9.3.21}$$

and

9.3 The Fractional Dielectric Models

$$\Psi_{CC}(t) \sim \begin{cases} 1 - \dfrac{1}{\Gamma(\alpha+1)}(t/\tau_\star)^\alpha, & \text{for } t \ll \tau_\star, \\ \dfrac{1}{\Gamma(1-\alpha)}(t/\tau_\star)^{-\alpha}, & \text{for } t \gg \tau_\star. \end{cases} \quad (9.3.22)$$

Let us now consider the spectral functions related to the Cole–Cole model, restricting our attention to the relaxation function $\Psi_{CC}(t)$. From (9.3.15) and (9.3.20) the frequency spectral function for $\Psi_{CC}(t)$ turns out to be

$$K^\Psi_{CC}(r) = \frac{\tau_\star}{\pi} \frac{(r\tau_\star)^{\alpha-1} \sin(\alpha\pi)}{(r\tau_\star)^{2\alpha} + 2(r\tau_\star)^\alpha \cos(\alpha\pi) + 1} \geq 0. \quad (9.3.23)$$

With the change of variable $\tau = 1/r$ we get the corresponding spectral representation $H^\Psi_{CC}(\tau) = \tau^{-2} K^\Psi_{CC}(1/\tau)$ in relaxation times, from which it is immediate to evaluate

$$H^\Psi_{CC}(\tau) = \frac{1}{\pi\tau_\star} \frac{(\tau/\tau_\star)^{\alpha-1} \sin(\alpha\pi)}{(\tau/\tau_\star)^{2\alpha} + 2(\tau/\tau_\star)^\alpha \cos(\alpha\pi) + 1} \quad (9.3.24)$$

and thus one easily recognizes the identity $K^\Psi_{CC}(r) = H^\Psi_{CC}(\tau)$ between the two spectral functions when the relaxation time is normalized to $\tau_\star = 1$.

The coincidence between the two spectral functions is a surprising fact pointed out for the Mittag-Leffler function $E_{\alpha,1}(-t^\alpha)$ with $0 < \alpha < 1$ by Mainardi in his 2010 book [Mai10] and his paper [Mai14]. This kind of universal/scaling property seems therefore peculiar for the Cole–Cole relaxation function $\Psi_{CC}(t)$.

For some values of the parameter α and with respect to the relaxation function $\Psi_{CC}(t)$ of the CC model, we show in Fig. 9.5 the time spectral distribution $H^\Psi_{CC}(\tau)$ given by (9.3.24) and its logarithmic representation $L^\Psi_{CC}(u) = e^u H^\Psi_{CC}(e^u)$, i.e.

$$L^\Psi_{CC}(u) = \frac{1}{2\pi} \frac{\sin(\alpha\pi)}{\cosh[\alpha(u - \log\tau_\star)] + \cos(\alpha\pi)}, \quad u = \log(\tau). \quad (9.3.25)$$

Of course, for $\alpha = 1$ the Mittag-Leffler function in (9.3.20) reduces to the exponential function $\exp(-t/\tau_\star)$ and the corresponding spectral distributions are both equal to the Dirac delta generalized function centered, respectively, at $\tau = \tau_\star$ and $u = \log(\tau_\star)$.

Note that both spectral functions were formerly outlined in 1947 by Gross [Gro47] and later revisited, in 1971, by Caputo and Mainardi [CapMai71b, CapMai71a].

The response and relaxation functions of the CC model satisfy some evolution equations expressed by means of fractional differential operators. In particular, for the response $\phi_{CC}(t)$ after applying some basic properties of the Riemann–Liouville derivative $_0D^\alpha_t$ (see [Die10, Sect. 2.2]) it is straightforward to derive

$$_0D^\alpha_t \phi_{CC}(t) = -\frac{1}{\tau_\star^\alpha} \phi_{CC}(t), \quad \lim_{t \to 0^+} {_0J^{1-\alpha}_t} \phi_{CC}(t) = \frac{1}{\tau_\star^\alpha}, \quad (9.3.26)$$

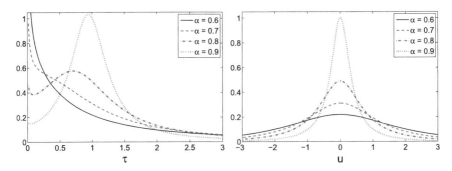

Fig. 9.5 Spectral distributions $H_{CC}^{\Psi}(\tau)$ (left) and $L_{CC}^{\Psi}(u)$ (right)

and, correspondingly, the application of the Caputo fractional derivative ${}_0^C D_t^\alpha$ leads to

$$
{}_0 D_t^\alpha \Psi_{CC}(t) = -\frac{1}{\tau_\star^\alpha} \Psi_{CC}(t), \quad \Psi_{CC}(0) = 1 . \tag{9.3.27}
$$

9.3.3 The Davidson–Cole Model

A decade after the introduction of the CC model, another dielectric model, still depending on one real parameter, was proposed to generalize the standard Debye model. The introduction in 1950–1951 of the new model by D.W. Davidson and R.H. Cole [DavCol50, DavCol51] was motivated by the need to fit the broader range of dispersion observed at high frequencies in some organic compounds such as glycerine, glycerol, propylene glycol, and n-propanol.

This asymmetry is obtained in the Davidson–Cole (DC) model by considering the following complex susceptibility

$$
\hat{\chi}_{DC}(i\omega) = \frac{1}{(1 + i\omega\tau_\star)^\gamma}, \quad 0 < \gamma \le 1 . \tag{9.3.28}
$$

By applying the Laplace transform inversion, we get the corresponding response and relaxation functions

$$
\phi_{DC}(t) = \mathcal{L}^{-1}\left(\frac{1}{(1 + s\tau_\star)^\gamma}\right) = \frac{1}{\tau_\star} (t/\tau_\star)^{\gamma-1} E_{1,\gamma}^\gamma (-t/\tau_\star)
$$
$$
= \frac{1}{\tau_\star} \frac{(t/\tau_\star)^{\gamma-1}}{\Gamma(\gamma)} \exp(-t/\tau_\star) \tag{9.3.29}
$$

and

9.3 The Fractional Dielectric Models

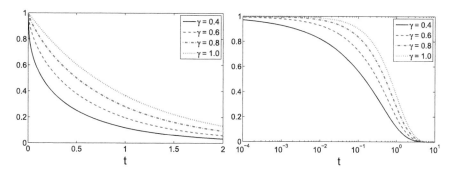

Fig. 9.6 The relaxation function $\Psi_{DC}(t)$ for varying γ

$$\Psi_{DC}(t) = \mathcal{L}^{-1}\left(\frac{1}{s} - \frac{1}{s(1+s\tau_\star)^\gamma}\right) = 1 - (t/\tau_\star)^\gamma E_{1,\gamma+1}^\gamma(-t/\tau_\star)$$
$$= \frac{1}{\Gamma(\gamma)}\Gamma(\gamma, t/\tau_\star) \qquad (9.3.30)$$

where, for $\Re(\gamma) > 0$, $\Gamma(a, z) = \int_z^\infty t^{a-1}e^{-t}\,dt$ is the incomplete gamma function and the last equality for $\Psi_{DC}(t)$ is obtained by integration of the response function $\phi_{DC}(t)$, namely after applying (9.3.4). The plots of the relaxation function $\Psi_{DC}(t)$ using linear and logarithmic scales in the normalized time $\tau_\star = 1$ are shown in Fig. 9.6.

The well-known asymptotic expansion for the incomplete gamma function (e.g., see [AbrSte72, Eq. 6.5.32]) with real and positive argument z

$$\Gamma(a, z) \sim z^{a-1}e^{-z}, \quad z \to \infty, \qquad (9.3.31)$$

allows us to see that, at variance with the CC model, the characteristic functions $\Psi_{DC}(t)$ and $\phi_{DC}(t)$ both decay exponentially for large times, so more rapidly than any power law, namely

$$\phi_{DC}(t) \sim \begin{cases} \dfrac{1}{\tau_\star \Gamma(\gamma)}(t/\tau_\star)^{\gamma-1}, & \text{for } t \ll \tau_\star, \\ \dfrac{1}{\tau_\star \Gamma(\gamma)}(t/\tau_\star)^{\gamma-1}\exp(-t/\tau_\star), & \text{for } t \gg \tau_\star, \end{cases} \qquad (9.3.32)$$

and

$$\Psi_{DC}(t) \sim \begin{cases} 1 - \dfrac{1}{\Gamma(\gamma+1)}(t/\tau_\star)^\gamma, & \text{for } t \ll \tau_\star, \\ \dfrac{1}{\Gamma(\gamma)}(t/\tau_\star)^{\gamma-1}\exp(-t/\tau_\star), & \text{for } t \gg \tau_\star. \end{cases} \qquad (9.3.33)$$

Furthermore, the spectral distribution functions exhibit a cut off, so they vanish at low frequencies $r < 1/\tau_\star$ and indeed we get the following expressions

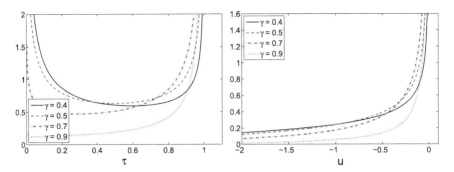

Fig. 9.7 Spectral distributions $H_{DC}^{\Psi}(\tau)$ (left) and $L_{DC}^{\Psi}(u)$ (right)

$$K_{DC}^{\phi}(r) = \begin{cases} 0, & r < 1/\tau_\star, \\ \dfrac{1}{\pi} \dfrac{\sin(\gamma\pi)}{(r\tau_\star - 1)^\gamma}, & r > 1/\tau_\star, \end{cases} \quad (9.3.34)$$

and

$$K_{DC}^{\Psi}(r) = \begin{cases} 0, & r < 1/\tau_\star, \\ \dfrac{1}{\pi} \dfrac{\sin(\gamma\pi)}{r(r\tau_\star - 1)^\gamma}, & r > 1/\tau_\star. \end{cases} \quad (9.3.35)$$

For the plots of spectral distributions, in Fig. 9.7 (as for the CC model) we have limited ourselves to those corresponding to the relaxation function $\Psi_{DC}(t)$, that is

$$H_{DC}^{\Psi}(\tau) = \begin{cases} 0, & \tau > \tau_\star, \\ \dfrac{1}{\pi\tau} \dfrac{\sin(\gamma\pi)}{(\tau_\star/\tau - 1)^\gamma}, & \tau < \tau_\star, \end{cases} \quad (9.3.36)$$

and $L_{DC}^{\Psi}(u) = e^{-u} K_{DC}^{\Psi}(e^{-u})$, where $u = \log(\tau)$.

By a standard derivation it is elementary to see that the response $\phi_{DC}(t)$ satisfies the equation

$$D_t \, \phi_{DC}(t) = -\frac{1}{\tau_\star} \left[1 - (\gamma - 1) \frac{\tau_\star}{t} \right] \phi_{DC}(t) \quad (9.3.37)$$

but by taking into account the composite operator it is also possible to obtain

$$\left(D_t + \tau_\star^{-1}\right)^\gamma \phi_{DC}(t) = e^{-t/\tau_\star} {}_0D_t^\gamma e^{t/\tau_\star} \phi_{DC}(t) = \frac{1}{\tau_\star} e^{-t/\tau_\star} {}_0D_t^\gamma \frac{(t/\tau_\star)^{\gamma-1}}{\Gamma(\gamma)} = 0,$$

where for the last equality we refer to [Die10, Example 2.4]; hence $\phi_{DC}(t)$ satisfies the following equation

$$\left(D_t + \tau_\star^{-1}\right)^\gamma \phi_{DC}(t) = 0, \quad \lim_{t \to 0^+} {}_0J_t^{1-\gamma} \left[e^{t/\tau_\star} \phi_{DC}(t) \right] = \frac{1}{\tau_\star^\gamma}. \quad (9.3.38)$$

9.3 The Fractional Dielectric Models

Observe that, as $t \to 0^+$, the contribution of the exponential $e^{t/\tau_\star}$ in the fractional integral $_0J_t^{1-\gamma}$ is always equal to 1, i.e.

$$\lim_{t \to 0^+} {_0J_t^{1-\gamma}}\left[e^{t/\tau_\star}\phi_{DC}(t)\right] = \lim_{t \to 0^+} {_0J_t^{1-\gamma}}\phi_{DC}(t),$$

thus providing the same initial condition associated to Eq. (9.3.26) for the CC model.

For the relaxation function $\Psi_{DC}(t)$ we use the operator ${}^C(D_t + \lambda)^\gamma$ defined as follows

$$^C(D_t + \lambda)^\gamma = e^{-t\lambda} {_0D_t^\gamma} e^{t\lambda}$$

and standard derivations allow us to compute

$$^C\left(D_t + \tau_\star^{-1}\right)^\gamma \Psi_{DC}(t) = -\frac{1}{\tau_\star^\gamma}\left[1 - (t/\tau_\star)^{1-\gamma} E_{1,2-\gamma}^{1-\gamma}(-t/\tau_\star)\right], \qquad (9.3.39)$$

with the usual initial condition $\Psi_{DC}(0) = 1$. It is worthwhile to note that when $\gamma = 1$ we have $E_{1,2-\gamma}^{1-\gamma}(-t/\tau_\star) \equiv 1$ and, as expected, (9.3.39) returns the standard evolution equation for the relaxation function of the Debye model.

Alternatively, we can consider the particular case, for $\alpha = 1$, of the operator $^C\left({_0D_t^\alpha} + \tau_\star^{-1}\right)^\gamma$ defined as follows

$$^C\left({_0D_t^\alpha} + \lambda\right)^\gamma f(t) \equiv e_{\alpha,1-\alpha\gamma}^{-\gamma}(t;-\lambda) * \frac{d}{dt}f(t) = \int_0^t e_{\alpha,1-\alpha\gamma}^{-\gamma}(t-u;-\lambda)f'(u)\,du.$$

The corresponding equation would read as

$$^C\left(D_t + \tau_\star^{-1}\right)^\gamma \Psi_{DC}(t) = -\frac{1}{\tau_\star^\gamma}, \qquad \Psi_{DC}(0) = 1. \qquad (9.3.40)$$

9.3.4 The Havriliak–Negami Model

In 1967 the American S.J. Havriliak and the Japanese-born S. Negami proposed a new model [HavNeg67] with two real powers to take into account, at the same time, both the asymmetry and the broadness observed in the shape of the permittivity spectrum of some polymers.

The normalized complex susceptibility proposed in the Havriliak–Negami (HN) model is given by

$$\hat{\chi}_{HN}(i\omega) = \frac{1}{\left(1 + (i\tau_\star\omega)^\alpha\right)^\gamma} \qquad (9.3.41)$$

and it is immediate to verify this, since

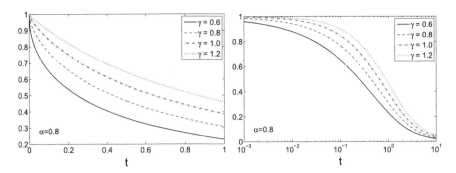

Fig. 9.8 Relaxation functions $\Psi_{HN}(t)$ for $\alpha = 0.8$

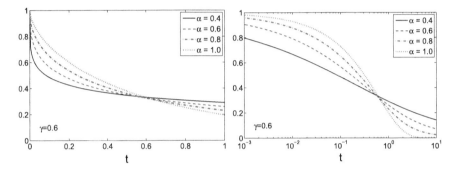

Fig. 9.9 Relaxation functions $\Psi_{HN}(t)$ for $\gamma = 0.6$

$$\hat{\chi}_{HN}(i\omega) \sim (i\tau_\star\omega)^{-\alpha\gamma}, \quad \tau_\star\omega \gg 1,$$
$$\Delta\hat{\chi}_{HN}(i\omega) = \chi_{HN}(0) - \hat{\chi}_{HN}(i\omega) \sim \gamma(i\tau_\star\omega)^\alpha, \quad \tau_\star\omega \ll 1. \tag{9.3.42}$$

The time-domain response and the time-domain relaxation of the HN model are respectively

$$\phi_{HN}(t) = \frac{1}{\tau_\star}(t/\tau_\star)^{\alpha\gamma-1} E_{\alpha,\alpha\gamma}^\gamma\left(-(t/\tau_\star)^\alpha\right) \tag{9.3.43}$$

and

$$\Psi_{HN}(t) = 1 - (t/\tau_\star)^{\alpha\gamma} E_{\alpha,\alpha\gamma+1}^\gamma\left(-(t/\tau_\star)^\alpha\right), \tag{9.3.44}$$

where $E_{\alpha,\beta}^\gamma(z)$ is the Prabhakar function.

Some plots of the relaxation function $\Psi_{HN}(t)$ for varying γ are shown in Fig. 9.8 and for varying α in Fig. 9.9; as usual, the left plots are in normal scale, the right plots in logarithmic scale and $\tau_\star = 1$.

By considering the series definition of the Prabhakar function for small t and its asymptotic expansion for $t \to \infty$, it is possible to verify that the HN response has the following short- and long-time power-law dependencies

9.3 The Fractional Dielectric Models

$$\phi_{HN}(t) \sim \begin{cases} \dfrac{1}{\tau_\star \Gamma(\alpha\gamma)} (t/\tau_\star)^{\alpha\gamma-1}, & \text{for } t \ll \tau_\star, \\ -\dfrac{\gamma}{\tau_\star \Gamma(-\alpha)} (t/\tau_\star)^{-\alpha-1}, & \text{for } t \gg \tau_\star, \end{cases} \quad (9.3.45)$$

which lead, respectively, to the short- and long-time power law dependencies of the HN relaxation for $0 < \alpha < 1$:

$$\Psi_{HN}(t) \sim \begin{cases} 1 - \dfrac{1}{\Gamma(\alpha\gamma+1)} (t/\tau_\star)^{\alpha\gamma}, & \text{for } t \ll \tau_\star, \\ \dfrac{\gamma}{\Gamma(1-\alpha)} (t/\tau_\star)^{-\alpha}, & \text{for } t \gg \tau_\star. \end{cases} \quad (9.3.46)$$

There is a lively debate in the literature about the range of admissibility of the parameters α and γ. Usually it is assumed that $0 < \alpha$ and $\gamma \le 1$ but in [HavHav94], on the basis of the observation of a large amount of experimental data, an extension to $0 < \alpha, \alpha\gamma \le 1$ was proposed. The complete monotonicity of the relaxation and response functions (which is considered an essential feature for the admissibility of the model [Han05b]) has been recently proved in [CdOMai11, MaiGar15] also for this extended range of parameters.

For this purpose we observe that the inversion formulas (9.3.15) lead to

$$K_{HN}^\Psi(r) = \frac{\tau_\star}{\pi} \frac{(\tau_\star r)^{-1} \sin[\gamma \theta_\alpha(r)]}{\left((\tau_\star r)^{2\alpha} + 2(\tau_\star r)^\alpha \cos(\alpha\pi) + 1\right)^{\gamma/2}} \quad (9.3.47)$$

and

$$K_{HN}^\phi(r) = \frac{1}{\pi} \frac{\sin[\gamma \theta_\alpha(r)]}{\left((\tau_\star r)^{2\alpha} + 2(\tau_\star r)^\alpha \cos(\alpha\pi) + 1\right)^{\gamma/2}}, \quad (9.3.48)$$

where

$$\theta_\alpha(r) = \frac{\pi}{2} - \arctan\left[\frac{\cos(\pi\alpha) + (\tau_\star r)^{-\alpha}}{\sin(\pi\alpha)}\right] \in [0, \pi], \quad (9.3.49)$$

and since $(\tau_\star r)^{-\alpha} \ge 0$ the argument of the arctan function is clearly $\ge 1/\tan \pi\alpha$ and hence $\theta_\alpha(r) \le \alpha\pi$, from which it follows that $K_{HN}^\phi(r) \ge 0$ for any $r \ge 0$ and for $0 < \alpha, \alpha\gamma \le 1$.

We can consider for the relaxation function $\Psi_{HN}(t)$ the time spectral distribution

$$H_{HN}^\Psi(\tau) = \frac{1}{\pi\tau} \frac{\sin[\gamma \theta_\alpha(1/\tau)]}{\left((\tau/\tau_\star)^{-2\alpha} + 2(\tau/\tau_\star)^{-\alpha} \cos(\alpha\pi) + 1\right)^{\gamma/2}}, \quad (9.3.50)$$

and its representation on the logarithmic scale $u = \log(\tau)$

$$L_{HN}^\Psi(u) = \frac{1}{\pi} \frac{\sin[\gamma \theta_\alpha(e^{-u})]}{\left(\tau_\star^{2\alpha} e^{-2\alpha u} + 2\tau_\star^\alpha e^{-\alpha u} \cos(\alpha\pi) + 1\right)^{\gamma/2}}. \quad (9.3.51)$$

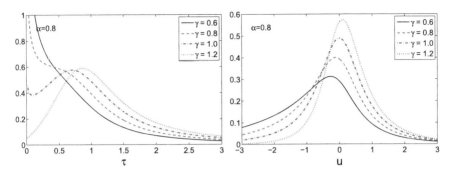

Fig. 9.10 Spectral distributions $H_{\mathrm{HN}}^{\Psi}(\tau)$ (left) and $L_{\mathrm{HN}}^{\Psi}(u)$ (right) for $\alpha = 0.8$

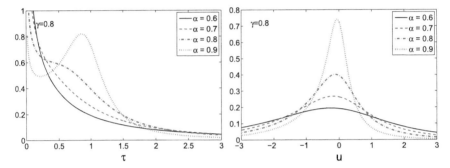

Fig. 9.11 Spectral distributions $H_{\mathrm{HN}}^{\Psi}(\tau)$ (left) and $L_{\mathrm{HN}}^{\Psi}(u)$ (right) for $\gamma = 0.8$

A few instances of the time spectral distributions $H_{\mathrm{HN}}^{\Psi}(\tau)$ and $L_{\mathrm{HN}}^{\Psi}(u)$, for varying γ and α respectively, are presented in Figs. 9.10 and 9.11.

To derive the evolution equations for the HN characteristic functions, we start by recalling that the Laplace transform of the response $\phi_{\mathrm{HN}}(t)$ is

$$\widetilde{\phi}_{\mathrm{HN}}(s) = \frac{1}{\tau_\star^{\alpha\gamma}(s^\alpha + \tau_\star^{-\alpha})^\gamma},$$

and it gives

$$\mathcal{L}\left(\left({}_0D_t^\alpha + \tau_\star^{-\alpha}\right)^\gamma \phi_{\mathrm{HN}}(t);\, s\right) = \frac{1}{\tau_\star^{\alpha\gamma}} - \lim_{t\to 0^+} \mathbf{E}_{\alpha,1-\alpha\gamma,-\tau_\star^{-\alpha},0^+}^{-\gamma} \phi_{\mathrm{HN}}(t). \quad (9.3.52)$$

Moreover, after evaluating the limit and transforming back to the time domain, one easily obtains the equation

$$\left({}_0D_t^\alpha + \tau_\star^{-\alpha}\right)^\gamma \phi_{\mathrm{HN}}(t) = 0, \quad \lim_{t\to 0^+} \mathbf{E}_{\alpha,1-\alpha\gamma,-\tau_\star^{-\alpha},0^+}^{-\gamma} \phi_{\mathrm{HN}}(t) = \frac{1}{\tau_\star^{\alpha\gamma}}. \quad (9.3.53)$$

9.3 The Fractional Dielectric Models

We note that the slight difference with the model by Weron et al. [Wer-et-al05, Eq. (24)] is due to the different way in which the operator $\left({}_0D_t^\alpha + \tau_*^{-\alpha}\right)^\gamma$ is introduced. Indeed, we use the approach proposed in [Gar-et-al14a], published after [Wer-et-al05]. We can easily verify that, if $\gamma = 1$, then the equivalence $\mathbf{E}_{\alpha,1-\alpha\gamma,-\tau_*^{-\alpha},0^+}^{-\gamma} \equiv {}_0J_t^{1-\alpha}$ holds. Hence, as expected, the evolution equation (9.3.26) for the response in the CC model is just the particular case, for $\gamma = 1$, of (9.3.53) in light of the initial condition in (9.3.53).

In a similar way, the equation for the HN relaxation function $\Psi_{\rm HN}(t)$ is derived by first recalling its Laplace transform

$$\widetilde{\Psi}_{\rm HN}(s) = \frac{1}{s} - \frac{1}{\tau_*^{\alpha\gamma} s \left(s^\alpha + \tau_*^{-\alpha}\right)^\gamma} \tag{9.3.54}$$

and by considering the operator ${}^C\left({}_0D_t^\alpha + \tau_*^{-\alpha}\right)^\gamma$ obtained after a regularization (in the Caputo sense) of $\left({}_0D_t^\alpha + \tau_*^{-\alpha}\right)^\gamma$. We have in this case

$$\mathcal{L}\left({}^C\left({}_0D_t^\alpha + \tau_*^{-\alpha}\right)^\gamma \Psi_{\rm HN}(t);\, s\right) = -\frac{1}{\tau_*^{\alpha\gamma} s}, \tag{9.3.55}$$

from which it is an immediate task to obtain

$$^C\left({}_0D_t^\alpha + \tau_*^{-\alpha}\right)^\gamma \Psi_{\rm HN}(t) = -\frac{1}{\tau_*^{\alpha\gamma}}, \quad \Psi_{\rm HN}(0) = 1. \tag{9.3.56}$$

It may be a bit surprising that, with $\gamma = 1$, the above equation slightly differs from the evolution equation (9.3.27) of the CC model. This difference is due to the fact that the operator ${}^C\left({}_0D_t^\alpha + \tau_*^{-\alpha}\right)^\gamma$ is not actually the same as $\left({}_0D_t^\alpha + \tau_*^{-\alpha}\right)^\gamma$, as one might expect.

9.4 The Fractional Calculus in the Basset Problem

In the following we recall the general equation of motion for a spherical particle, in a viscous fluid, pointing out the different force contributions due to effects of inertia, viscous drag and buoyancy. In particular, the so-called Basset force will be interpreted in terms of a fractional derivative of order 1/2 of the particle velocity relative to the fluid. Based on the 1995 work by Mainardi, Pironi and Tampieri [MaPiTa95a], revisited in the CISM chapter by Mainardi in 1997 [Mai97], we shall introduce the *generalized Basset force*, which is expressed in terms of a fractional derivative of any order α ranging in the interval $0 < \alpha < 1$. This generalization, suggested by a mathematical speculation, is expected to provide a phenomenological insight for the experimental data.

We consider the simplified problem, originally investigated by Basset, where the fluid is quiescent and the particle moves under the action of gravity, starting at $t = 0$ with a certain vertical velocity. For the sake of generality, we prefer to consider the problem with the generalized Basset force and will provide the solution for the particle velocity in terms of Mittag-Leffler type functions. The most evident effect of this generalization will be to modify the long-time behavior of the solution, changing its algebraic decay from $t^{-1/2}$ to $t^{-\alpha}$. This effect can be of some interest for a better fit of experimental data.

9.4.1 The Equation of Motion for the Basset Problem

Let us consider a small rigid sphere of radius r_0, mass m_p, density ρ_p, initially centered in $\mathbf{X}(t)$ and moving with velocity $\mathbf{V}(t)$ in a homogeneous fluid, of density ρ_f and kinematic viscosity ν, characterized by a flow field $\mathbf{u}(\mathbf{x}, t)$. In general the equation of motion is required to take into account effects due to inertia, viscous drag and buoyancy, so it can be written as

$$m_p \frac{d\mathbf{V}}{dt} = \mathbf{F}_i + \mathbf{F}_d + \mathbf{F}_g, \qquad (9.4.1)$$

where the forces on the R.H.S. correspond in turn to the above effects. According to Maxey and Riley [MaxRil83] these forces read, adopting our notation,

$$\mathbf{F}_i = m_f \left. \frac{D\mathbf{u}}{Dt} \right|_{\mathbf{X}(t)} - \frac{1}{2} m_f \left(\frac{d\mathbf{V}}{dt} - \left. \frac{D\mathbf{u}}{Dt} \right|_{\mathbf{X}(t)} \right), \qquad (9.4.2)$$

$$\mathbf{F}_d = -\frac{1}{\mu} \left\{ [\mathbf{V}(t) - \mathbf{u}(\mathbf{X}(t), t)] + \sqrt{\frac{\tau_0}{\pi}} \int_{-\infty}^{t} \frac{d\,[\mathbf{V}(\tau) - \mathbf{u}(\mathbf{X}(\tau), \tau)]/d\tau}{\sqrt{t - \tau}} \, d\tau \right\}, \qquad (9.4.3)$$

$$\mathbf{F}_g = (m_p - m_f) \mathbf{g}, \qquad (9.4.4)$$

where $m_f = (4/3)\pi r_0^3 \rho_f$ denotes the mass of the fluid displaced by the spherical particle, and

$$\tau_0 := \frac{r_0^2}{\nu}, \qquad (9.4.5)$$

$$\frac{1}{\mu} := 6\pi r_0 \nu \rho_f = \frac{9}{2} m_f \tau_0^{-1}. \qquad (9.4.6)$$

9.4 The Fractional Calculus in the Basset Problem

The time constant τ_0 represents a sort of time scale induced by viscosity, whereas the constant μ is usually referred to as the *mobility coefficient*.

In (9.4.2) we note two different time derivatives, D/Dt, d/dt, which represent the time derivatives following a fluid element and the moving sphere, respectively, so

$$\left.\frac{D\mathbf{u}}{Dt}\right|_{\mathbf{X}(t)} = \left[\frac{\partial \mathbf{u}}{\partial t} + (\mathbf{u}\cdot\nabla)\mathbf{u}(\mathbf{x},t)\right], \quad \frac{d}{dt}\mathbf{u}[\mathbf{X}(t),t] = \left[\frac{\partial \mathbf{u}}{\partial t} + (\mathbf{V}\cdot\nabla)\mathbf{u}(\mathbf{x},t)\right],$$

where the brackets are computed at $\mathbf{x} = \mathbf{X}(t)$.

The terms on the R.H.S. of (9.4.2) correspond in turn to the effects of pressure gradient of the undisturbed flow and of added mass, whereas those of (9.4.3) represent respectively the well-known viscous Stokes drag, which we shall denote by $\mathbf{F}_S$, and to the augmented viscous Basset drag denoted by $\mathbf{F}_B$. Using the characteristic time τ_0, the Stokes and Basset forces read respectively

$$F_S = -\frac{9}{2}m_f\,\tau_0^{-1}\,[\mathbf{V}(t) - \mathbf{u}(\mathbf{X}(t),t)], \tag{9.4.7}$$

$$F_B = -\frac{9}{2}m_f\,\tau_0^{-1/2}\left\{\frac{1}{\sqrt{\pi}}\int_{-\infty}^t \frac{d[\mathbf{V}(\tau) - \mathbf{u}(\mathbf{X}(\tau),\tau)]/d\tau}{\sqrt{t-\tau}}\,d\tau\right\}. \tag{9.4.8}$$

We thus recognize that the time constant τ_0 provides the natural time scale for the diffusive processes related to the fluid viscosity, and that the integral expression in brackets on the R.H.S. of (9.4.8) just represents the *Caputo fractional derivative* of order $1/2$, with starting point $-\infty$, of the particle velocity relative to the fluid.

Presumably, the first scientist to have pointed out the relationship between the Basset force and the fractional derivative of order $1/2$ was F.B. Tatom in 1988 [Tat88]. However, he limited himself to noting this fact, without treating any related problem by the methods of fractional calculus.

We now introduce the *generalized Basset force* by the definition

$$F_B^\alpha = -\frac{9}{2}m_f\tau_0^{\alpha-1}\frac{d^\alpha}{dt^\alpha}[\mathbf{V}(t) - \mathbf{u}(\mathbf{X}(t),t)], \quad 0 < \alpha < 1, \tag{9.4.9}$$

where the fractional derivative of order α is in Caputo's sense.

Introducing the so-called *effective mass*

$$m_e := m_p + \frac{1}{2}m_f, \tag{9.4.10}$$

and allowing for the *generalized Basset force* in (9.4.3) we can re-write the equation of motion (9.4.1)–(9.4.4) in the more compact and significant form,

$$m_e \frac{d\mathbf{V}}{dt} = \frac{3}{2} m_f \frac{D\mathbf{u}}{Dt} - \frac{9}{2} m_f \left[\frac{1}{\tau_0} + \frac{1}{\tau_0^{1-\alpha}} \frac{d^\alpha}{dt^\alpha} \right] (\mathbf{V} - \mathbf{u}) + (m_p - m_f) \mathbf{g},$$
(9.4.11)

which we refer to as the *generalized equation of motion*. Of course, if in (9.4.11) we put $\alpha = 1/2$, we recover the *basic equation of motion* with the original Basset force.

9.4.2 The Generalized Basset Problem

Let us now assume that the fluid is quiescent, namely $\mathbf{u}(\mathbf{x}, t) = 0$, $\forall \mathbf{x}, t$, and that the particle starts to move under the action of gravity, from a given instant $t_0 = 0$ with a certain velocity $V(0^+) = V_0$, in the vertical direction. This was the problem considered by Basset [Bas88], and which was first solved by Boggio [Bog07] in a cumbersome way, in terms of Gauss and Fresnel integrals.

Introducing the non-dimensional quantities (related to the densities ρ_f, ρ_p of the fluid and particle),

$$\chi := \frac{\rho_p}{\rho_f}, \quad \beta := \frac{9\rho_f}{2\rho_p + \rho_f} = \frac{9}{1 + 2\chi}, \quad (9.4.12)$$

we find it convenient to define a new characteristic time

$$\sigma_e := \mu m_e = \tau_0/\beta, \quad (9.4.13)$$

see (9.4.5), (9.4.10), (9.4.12) and a characteristic velocity (related to the gravity),

$$V_S = (2/9)(\chi - 1) g \tau_0. \quad (9.4.14)$$

Then we can eliminate the mass factors and the gravity acceleration in (9.4.11) and obtain the *equation of motion* in the form

$$\frac{dV}{dt} = -\frac{1}{\sigma_e} \left[1 + \tau_0^\alpha \frac{d^\alpha}{dt^\alpha} \right] V + \frac{1}{\sigma_e} V_S. \quad (9.4.15)$$

If the Basset term were absent, we would obtain the classical Stokes solution

$$V(t) = V_S + (V_0 - V_S) e^{-t/\sigma_e}, \quad (9.4.16)$$

where σ_e represents the characteristic time of the motion, and V_S the final value assumed by the velocity. Later we shall show that in the presence of the Basset term the same final value is still attained by the solution $V(t)$, but with an algebraic rate, which is much slower than the exponential one found in (9.4.16).

In order to investigate the effect of the (generalized) Basset term, we compare the exact solution of (9.4.15) with the Stokes solution (9.4.16); with this aim we

9.4 The Fractional Calculus in the Basset Problem

find it convenient to scale times and velocities in (9.4.15) by $\{\sigma_e, V_S\}$, i.e. to refer to the non-dimensional quantities $t' = t/\sigma_e$, $V' = V/V_S$, $V'_0 = V_0/V_S$. The resulting equation of motion reads (suppressing the indices)

$$\left[\frac{d}{dt} + a\frac{d^\alpha}{dt^\alpha} + 1\right] V(t) = 1, \quad V(0^+) = V_0, \quad a = \beta^\alpha > 0, \quad 0 < \alpha < 1. \tag{9.4.17}$$

This is the *composite fractional relaxation equation* treated in Sect. 9.1.2, precisely Eq. (9.1.38), by using the Laplace transform method. Recalling that in an obvious notation we have

$$V(t) \div \widetilde{V}(s), \quad \frac{d^\alpha}{dt^\alpha} V(t) \div s^\alpha \widetilde{V}(s) - s^{\alpha-1} V_0, \quad 0 < \alpha \leq 1, \tag{9.4.18}$$

the transformed solution of (9.4.17) reads

$$\widetilde{V}(s) = \widetilde{M}(s) V_0 + \frac{1}{s}\widetilde{N}(s), \tag{9.4.19}$$

where

$$\widetilde{M}(s) = \frac{1 + a s^{\alpha-1}}{s + a s^\alpha + 1}, \quad \widetilde{N}(s) = \frac{1}{s + a s^\alpha + 1}. \tag{9.4.20}$$

Noting that

$$\frac{1}{s}\widetilde{N}(s) = \frac{1}{s} - \widetilde{M}(s) \div \int_0^t N(\tau)\, d\tau = 1 - M(t) \iff N(t) = -M'(t), \tag{9.4.21}$$

the actual solution of (9.4.17) turns out to be

$$V(t) = 1 + (V_0 - 1) M(t), \tag{9.4.22}$$

which is "similar" to the Stokes solution (9.4.16) if we consider the replacement of e^{-t} with the function $M(t)$.

We now resume the relevant results from [GorMai97] using the present notation. The integral representation for $M(t)$ turns out to be

$$M(t) = \int_0^\infty e^{-rt} K(r)\, dr, \tag{9.4.23}$$

where

$$K(r) = \frac{1}{\pi} \frac{a r^{\alpha-1} \sin(\alpha\pi)}{(1-r)^2 + a^2 r^{2\alpha} + 2(1-r) a r^\alpha \cos(\alpha\pi)} > 0. \tag{9.4.24}$$

Thus $M(t)$ is a *completely monotone* function [with spectrum $K(r)$], which is decreasing from 1 towards 0 as t runs from 0 to ∞. The behavior of $M(t)$ as $t \to 0^+$ and $t \to \infty$ can be inspected by means of a proper asymptotic analysis, as follows.

The behavior as $t \to 0^+$ can be determined from the behavior of the Laplace transform $\widetilde{M}(s) = s^{-1} - s^{-2} + O\left(s^{-3+\alpha}\right)$, as $\mathrm{Re}\{s\} \to +\infty$. We obtain

$$M(t) = 1 - t + O\left(t^{2-\alpha}\right), \quad \text{as } t \to 0^+. \tag{9.4.25}$$

The spectral representation (9.4.23)–(9.4.24) is suitable to obtain the asymptotic behavior of $M(t)$ as $t \to +\infty$, by using the Watson lemma. In fact, expanding the spectrum $K(r)$ for small r and taking the dominant term in the corresponding asymptotic series, we obtain

$$M(t) \sim a \frac{t^{-\alpha}}{\Gamma(1-\alpha)} = a \frac{\sin(\alpha\pi)}{\pi} \int_0^\infty e^{-rt} r^{\alpha-1}\, dr, \quad \text{as } t \to \infty. \tag{9.4.26}$$

Furthermore, we recognize that $1 > M(t) > e^{-t} > 0$, $0 < t < \infty$, namely, the decreasing plot of $M(t)$ remains above that of the exponential, as t runs from 0 to ∞. Although both functions tend monotonically to 0, the difference between the two plots increases with t: at the initial point $t = 0$, both the curves assume the unitary value and decrease with the same initial rate, but as $t \to \infty$ they exhibit very different decays, algebraic (slow) against exponential (fast).

For the ordinary Basset problem it is convenient to report the result obtained by the factorization method in [MaPiTa95a]. In this case we note that $a = \sqrt{\beta}$, see (9.4.17), ranges from 0 to 3 since from (9.4.12) we recognize that β runs from 0 ($\chi = \infty$, infinitely heavy particle) to 9 ($\chi = 0$, infinitely light particle).

The actual solution is obtained by expanding $\widetilde{M}(s)$ into partial fractions and then inverting. Considering the two roots $\lambda_\pm$ of the polynomial $P(z) \equiv z^2 + az + 1$, with $z = s^{1/2}$, we must treat separately the following two cases

$$i)\ 0 < a < 2\ \text{ or }\ 2 < a < 3, \quad \text{and } ii)\ a = 2,$$

which correspond to two distinct roots ($\lambda_+ \neq \lambda_-$), or two coincident roots ($\lambda_+ \equiv \lambda_- = -1$), respectively. We obtain

(i) $a \neq 2 \iff \beta \neq 4, \chi \neq 5/8$,

$$\widetilde{M}(s) = \frac{1 + a s^{-1/2}}{s + a s^{1/2} + 1} = \frac{A_-}{s^{1/2}(s^{1/2} - \lambda_+)} + \frac{A_+}{s^{1/2}(s^{1/2} - \lambda_-)}, \tag{9.4.27}$$

with

$$\lambda_\pm = \frac{-a \pm (a^2 - 4)^{1/2}}{2} = \frac{1}{\lambda_\mp}, \quad A_\pm = \pm \frac{\lambda_\pm}{\lambda_+ - \lambda_-}; \tag{9.4.28}$$

9.4 The Fractional Calculus in the Basset Problem

(ii) $a = 2 \iff \beta = 4$, $\chi = 5/8$,

$$\widetilde{M}(s) = \frac{1 + 2s^{-1/2}}{s + 2s^{1/2} + 1} = \frac{1}{(s^{1/2} + 1)^2} + \frac{2}{s^{1/2}(s^{1/2} + 1)^2}. \qquad (9.4.29)$$

The Laplace inversion of (9.4.27)–(9.4.29) can be expressed in terms of Mittag-Leffler functions of order $1/2$, $E_{1/2}(\lambda\sqrt{t}) = \exp(\lambda^2 t)\,\mathrm{erfc}(-\lambda\sqrt{t})$, as shown in the Appendix of [GorMai97]. We obtain

$$M(t) = \begin{cases} i) & A_- E_{1/2}(\lambda_+\sqrt{t}) + A_+ E_{1/2}(\lambda_-\sqrt{t}), \\ ii) & (1 - 2t)\, E_{1/2}(-\sqrt{t}) + 2\sqrt{t/\pi}. \end{cases} \qquad (9.4.30)$$

We recall that the analytical solution to the classical Basset problem was formerly provided by Boggio [Bog07] in 1907 by a different (cumbersome) method. One can show that our solution (9.4.30), derived by the tools of the Laplace transform and fractional calculus, coincides with Boggio's solution. Boggio also arrived at the analysis of the two roots $\lambda_\pm$ but his expression of the solution in the case of two conjugate complex roots ($\chi > 5/8$) given as a sum of Fresnel integrals could lead one to forecast un-physical oscillations, in the absence of numerical tables or plots. This disturbed Basset who, when he summarized the state of art about his problem in a later paper of 1910 [Bas10], thought there was some physical deficiency in his own theory. With our integral representation of the solution, see (9.4.23)–(9.4.24), we can prove the monotone character of the solution, even if the arguments of the exponential and error functions are complex.

In order to have some insight about the effects of the two parameters α and a on the (generalized) Basset problem we exhibit some (normalized) plots for the particle velocity $V(t)$, corresponding to the solution of (9.4.17), assuming for simplicity a vanishing initial velocity ($V_0 = 0$).

We consider three cases for α, namely $\alpha = 1/2$ (the ordinary Basset problem) and $\alpha = 1/4$, $3/4$ (the generalized Basset problem), corresponding to Figs. 9.12, 9.13, 9.14, respectively. For each α we consider four values of a corresponding to $\chi := \rho_p/\rho_f = 0.5, 2, 10, 100$. For each couple $\{\alpha, \chi\}$ we compare the Basset solution with its asymptotic expression (dotted line) for large times and the Stokes solution ($a = 0$). From these figures we can recognize the retarding effect of the (generalized) Basset force, which is more relevant for lighter particles, in reaching the final value of the velocity. This effect is of course due to the algebraic decay of the function $M(t)$, see (9.4.26), which is much slower than the exponential decay of the Stokes solution.

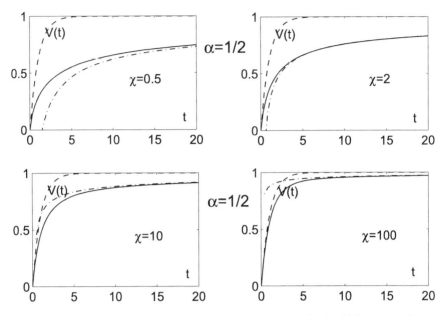

Fig. 9.12 The normalized velocity $V(t)$ for $\alpha = 1/2$ and $\chi = 0.5, 2, 10, 100$ Basset: continuous line; Basset asymptotic: dotted-dashed line; Stokes: dashed line

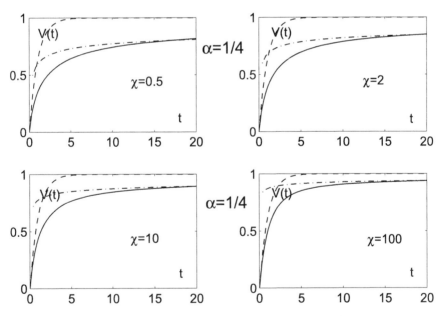

Fig. 9.13 The normalized velocity $V(t)$ for $\alpha = 1/4$ and $\chi = 0.5, 2, 10, 100$ Basset: continuous line; Basset asymptotic: dotted-dashed line; Stokes: dashed line

9.5 Other Deterministic Fractional Models

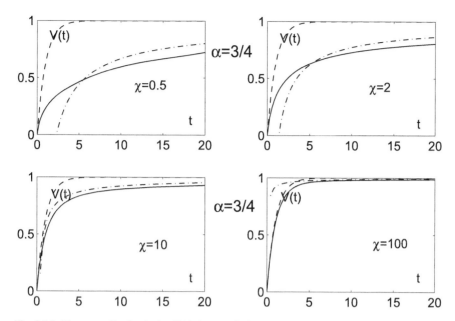

Fig. 9.14 The normalized velocity $V(t)$ for $\alpha = 3/4$ and $\chi = 0.5, 2, 10, 100$ Basset: continuous line; Basset asymptotic: dotted-dashed line; Stokes: dashed line

9.5 Other Deterministic Fractional Models

Here we present a few other formulations of fractional deterministic models. Our choice is determined by two considerations. We try to avoid heavy machinery in order to follow the main idea of this chapter, to give readers an impression of how the fractional models appear and what the advantages are of using the fractional approach. Second, since the subject we are touching is very large and growing rapidly, we understand that any survey text on the applications of the fractional calculus will soon become incomplete after publication. We note also that the material presented here is known and widely spread in the literature.

The Fractional Newton Equation

Let us consider the motion of a body in an incompressible viscous Newtonian fluid. In the absence of external forces, the unsteady flow of such a fluid is governed by the Navier–Stokes system of equations. Following [Uch13b, Sect. 7.3] we write down this (simplified) system in the case of the problem of entrainment of the fluid by a large sized plate moving in the xOz-plane along the x-axis with the given velocity $V(t)$

$$\rho \frac{\partial v}{\partial t} = \nu \frac{\partial^2 v}{\partial z^2}, \quad 0 < t < \infty, \; z \leq 0, \tag{9.5.1}$$

where ρ is the density of the fluid, ν is its viscosity, and $v = v(z, t) = v_x(z, t)$ is the velocity of the fluid in the x-direction at depth z. We suppose here that $v_y = v_z = 0$, and that in the distant past and at a great depth there is no movement of the fluid. Then applying the Fourier transform in the t-variable to (9.5.1) we obtain the ordinary differential equation

$$(-i\omega)\rho\hat{v}(z, \omega) = \nu\frac{d^2\hat{v}}{dz^2}(z, \omega), \qquad (9.5.2)$$

supplied by the boundary conditions

$$\hat{v}(z, \omega) = \hat{V}(\omega), \quad \hat{v}(-\infty, \omega) = 0. \qquad (9.5.3)$$

Its solution has the form

$$\hat{v}(z, \omega) = \hat{V}(\omega)\exp\left\{\sqrt{-\frac{i\omega\rho}{\nu}}z\right\}. \qquad (9.5.4)$$

By introducing the shear stress (the viscous source per unit surface) $\sigma(z, t) = \nu\frac{\partial v}{\partial z}(z, t)$, we get

$$\hat{\sigma}(z, \omega) = (-i\omega)^{1/2}\sqrt{\rho\nu}\hat{v}(z, \omega), \qquad (9.5.5)$$

and thus, by applying the inverse Fourier transform, we obtain the following fractional differential equation

$$\sigma(z, t) = \sqrt{\rho\nu}\left(D_{0+,t}^{1/2}v\right)(z, t), \qquad (9.5.6)$$

where the fractional derivative can be understood either in the Riemann–Liouville or Caputo sense. The physical interpretation of this result is that the shear stress observed at the moment t at the point (x, z) is a result of the contribution of the fluid particles at the points (x', z), where they were located at $t' < t$. This is the simplest mechanism of heredity, the "mechanical" memory.

The Fractional Ohm Law

Here we also follow the presentation given in [Uch13b, Sect. 11.4]. The polarization $P(t)$ of a dielectric, placed in an electric field, consists of two parts, $P_1(t)$ which is proportional to the intensity of the field at time t,

$$P_1(t) = \chi_1 E(t),$$

and the retarded part

$$P_2(t) = \int_{-\infty}^{t} K(t - t')E(t')dt'.$$

9.5 Other Deterministic Fractional Models

Let us denote by $\chi_2 E$ the extreme value of E_2 with constant E and $t \to \infty$. In the classical relaxation theory it is assumed that the rate of the change of P_2 is proportional to the difference between its extreme and current values:

$$\frac{dP_2}{dt} = \frac{1}{\tau}(\chi_2 E - P_2).$$

The simplest macroscopic way to implement a fractional derivative into the relaxation problem is the fractional generalization of the classical law relating the current $i(t)$ and voltage $u(t)$

$$i(t) = K_\alpha \left(D_{a+,t}^\alpha u\right)(t). \tag{9.5.7}$$

This relation describes an element of a chain which, in a sense, possesses an intermediate property between those for an ideal dielectric with capacity $C = K_1$ and a simple conductor with resistance $R = 1/K_0$.

If $u(t')$ varies continuously and rather rapidly approaches 0 as $t' \to -\infty$, then we take $a = -\infty$ in (9.5.7), understanding by $\left(D_{-,t}^\alpha u\right)$ the Liouville form of either the Riemann–Liouville or Caputo fractional derivative.

Taking into account the presence of the active resistance in the chain, we write the Kirchhoff law

$$i(t)R + u(t) = \mathcal{E}(t),$$

where $\mathcal{E}(t)$ is the electromotive force (EMF). Inserting relation (9.5.7) we obtain

$$\left(\left[K_\alpha R_{-\infty}D_t^\alpha + 1\right]u\right)(t) = \mathcal{E}(t). \tag{9.5.8}$$

If we assume $\mathcal{E}(t) = e^{i\omega t}$ then the solution to (9.5.8) coincides with the Cole–Cole response function $\hat{f}_\alpha$ (see [ColCol41, ColCol42]):

$$\left[(i\omega\tau)^\alpha + 1\right]\hat{f}_\alpha(i\omega) = 1, \ 0 < \alpha < 1. \tag{9.5.9}$$

Fractional Equations for Heat Transfer

The classical theory of heat conduction is based on Fourier's law, relating the heat flux $\mathbf{q}$ and the gradient of temperature T

$$\mathbf{q} = -\kappa \nabla T \tag{9.5.10}$$

with conductivity coefficient κ. There exist two fractional generalizations of Fourier's law (see, e.g., [Uch13b, p. 164])

$$\left(1 + \tau_0 \cdot D_{0+,t}^\alpha\right)\mathbf{q} = -\kappa \nabla T, \ 0 < \alpha < 1, \tag{9.5.11}$$

$$\left(1 + \tau \frac{\partial}{\partial t}\right)\mathbf{q} = -\kappa D_{0+,t}^{1-\alpha} \nabla T, \ 1 < \alpha \leq 2. \tag{9.5.12}$$

The standard Fourier law leads to the following heat transfer equation (e.g. in the one-dimensional case, when we consider the heat transfer in a long rod)

$$\frac{\partial T}{\partial t} = \kappa \frac{\partial^2 T}{\partial x^2}. \qquad (9.5.13)$$

This relation can be generalized, taking into account the dependence of the heat flux on the history of the gradient of the temperature (some authors call this generalization the Maxwell–Cattaneo–Lykov equation). The one-dimensional equation of this type has the form:

$$\frac{\partial T}{\partial t} + \tau \frac{\partial^2 T}{\partial t^2} = \kappa \frac{\partial^2 T}{\partial x^2}. \qquad (9.5.14)$$

In the three-dimensional case we have

$$\frac{\partial T}{\partial t} + \tau \frac{\partial^2 T}{\partial t^2} = \kappa \Delta T. \qquad (9.5.15)$$

Considering two extremal possibilities for the relaxation time τ we have

$$\frac{\partial T}{\partial t} + \tau \frac{\partial^2 T}{\partial t^2} = \kappa \Delta T \Rightarrow \begin{cases} \frac{\partial T}{\partial t} = \kappa \Delta T, & \tau \to 0 \text{ (diffusion equation)}; \\ \tau \frac{\partial^2 T}{\partial t^2} = \kappa \Delta T, & \tau \to +\infty \text{ (wave equation)}, \end{cases}$$

which is commonly rewritten as

$$\tau^{\nu-1} \frac{\partial^\nu T}{\partial t^\nu} = \kappa \Delta T, \ \nu = 1, 2.$$

Assuming again the intermediate situation $\nu(0, 2]$, we obtain the time-fractional generalization of the heat transfer equation

$$\tau^{\nu-1} {}_0 D_t^\nu T = \kappa \Delta T, \ \nu(0, 2]. \qquad (9.5.16)$$

The fractional Fourier law allows us to formulate a fractional generalization of the Maxwell–Cattaneo–Lykov equation. Taking the gradient of both sides of Eqs. (9.5.11) and (9.5.12) and applying the heat balance equation

$$\nabla \cdot \mathbf{q} = -\frac{\partial T}{\partial t},$$

we get, respectively,

9.5 Other Deterministic Fractional Models 329

$$\frac{\partial T}{\partial t} + \tau_0 \cdot D_{0+,t}^{1+\alpha} T = \kappa \Delta T, \ 0 < \alpha < 1, \quad (9.5.17)$$

$$\frac{\partial T}{\partial t} + \tau \frac{\partial^2 T}{\partial t^2} = \kappa \cdot {}_0 D_t^{1-\alpha} \Delta T, \ 1 < \alpha < 2. \quad (9.5.18)$$

9.6 Historical and Bibliographical Notes

General Notes

The main part of the material of this chapter is presented in the survey papers by Mainardi, Gorenflo and Luchko in Vols. 4 and 5 of the multivolume Handbook of Fractional Calculus with Applications [HAND4, HAND5].

One of the first investigations of differential equations of fractional order was made by Barrett [Barr54]. He considered differential equations with a fractional derivative of Riemann–Liouville type of arbitrary order α, $\operatorname{Re}\alpha > 0$. n boundary conditions ($n = \operatorname{Re}\alpha + 1$) in the form of the values at the initial points of the fractional derivatives of order $\alpha - k$, $k = 1, 2, \ldots, n$, are posed. It was shown that in a suitable class of functions the solution is unique and is represented via the Mittag-Leffler function.

As former applications in physics we would like to highlight the contributions by K.S. Cole [Col33], quoted by H.T. Davis [Dav36, p. 287] in connection with nerve conduction, and by F.M. de Oliveira Castro [Oli39], K.S. Cole and R.H. Cole (1941–1942) [ColCol41, ColCol42], and B. Gross (1947) [Gro47, Gro48] in connection with dielectric and mechanical relaxation, respectively. Subsequently, in 1971, Caputo and Mainardi [CapMai71a] proved that the Mittag-Leffler function is present whenever derivatives of fractional order are introduced in the constitutive equations of a linear viscoelastic body. Since then, several other authors have pointed out the relevance of the Mittag-Leffler function for fractional viscoelastic models, see Mainardi [Mai97, Mai10]. A number of mechanical problems treated by using the fractional approach are studied by Rossikhin and Shitikova, see, e.g., [RosShi10, Shi19, RosShi19b]. We also mention some other models, see [Hil02b, Nig09, NonGlo91, MetKla02] and the references therein.

Notes on Fractional Differential Equations

The fractional differential equation in (9.1.38) with $\alpha = 1/2$ corresponds to the *Basset problem*, a classical problem in fluid dynamics concerning the unsteady motion of a particle accelerating in a viscous fluid under the action of gravity. The fractional differential equation in (9.1.39) with $0 < \alpha < 2$ models an oscillation process with fractional damping term. It was formerly treated by Caputo, who provided a preliminary analysis using the Laplace transform. The special cases $\alpha = 1/2$ and $\alpha = 3/2$, but with the standard definition D^α for the fractional derivative, were discussed by Bagley [Bag90]. Beyer and Kempfle [BeyKem95] considered (9.1.39) for $-\infty < t < +\infty$, investigating the uniqueness and *causality* of the solutions. As they let t run through $\mathbb{R}$, they used Fourier transforms and characterized the fractional derivative D^α by its properties in frequency space, thereby requiring that for non-integer α the principal branch of $(i\omega)^\alpha$ should be taken. Under the global condition that the solution is square summable, they showed that the system described by (9.1.39) is *causal iff $a > 0$*.

Notes on the Fractional Calculus in Linear Viscoelasticity

1. The First Generation of Pioneers of Fractional Calculus in Viscoelasticity

Linear viscoelasticity is certainly the field in which the most extensive applications of fractional calculus have appeared, in view of its ability to model hereditary phenomena with long memory. During the twentieth century a number of authors have (implicitly or explicitly) used the fractional calculus as an empirical method of describing the properties of viscoelastic materials.

In the first half of that century the early contributors were: Gemant in the UK and the USA, see [Gem36, Gem38], Scott-Blair in the UK, see [ScoB44, ScoB47, ScoB49], and Gerasimov and Rabotnov in the Soviet Union, see [Ger48, Rab48a].

In 1950 Gemant published a series of 16 articles entitled Frictional Phenomena in the Journal of Applied Physics between 1941 and 1943, which were collected in a book of the same title [Gem50]. In his eighth chapter-paper [Gem42, p. 220], he referred to his previous articles [Gem36, Gem38] justifying the necessity of fractional differential operators to compute the shape of relaxation curves for some elasto-viscous fluids. Thus, the words fractional and frictional were coupled, presumably for the first time, by Gemant.

Scott-Blair used the fractional calculus approach to model the observations made in [Nut21, Nut43, Nut46] that the stress relaxation phenomenon could be described by fractional powers of time. He noted that time derivatives of fractional order would simultaneously model the observations of Nutting on stress relaxation and those of Gemant on frequency dependence. It is quite instructive to quote Scott-Blair from [Sti79]:

> *I was working on the assessing of firmness of various materials (e.g. cheese and clay by experts handling them) these systems are of course both elastic and viscous but I felt sure that judgments were made not on an addition of elastic and viscous parts but on something in between the two so I introduced fractional differentials of strain with respect to time.*

Later, in the same letter Scott-Blair added:

> *I gave up the work eventually, mainly because I could not find a definition of a fractional differential that would satisfy the mathematicians.*

More about Scott-Blair's pioneering contribution can be found in the paper by Rogosin and Mainardi [RogMai14].

Later, in (1947–1948) B. Gross [Gro47, Gro48] and, more recently, Garrappa et al. [GarPop13, Gar15] explicitly proposed the Mittag-Leffler function in mechanical relaxation in the framework of linear viscoelasticity. This argument was revisited in 1971 by Caputo and Mainardi [CapMai71b, CapMai71a] in order to propose the so-called fractional Zener model, making use of the time fractional derivative in the Caputo sense. A function strictly related to the Mittag-Leffler function was introduced by Rabotnov in 1948 [Rab48a] and soon afterwards numerical tables of the Rabotnov function appeared by his collaborators. However, the first plots of the Mittag-Leffler function appeared only in the 1971 papers by Caputo and Mainardi [CapMai71b, CapMai71a]. Nowadays, because of the relevance of this function in Fractional Calculus as solutions of differential equations of fractional order, a number of computing routines are available, due to Gorenflo et al. [GoLoLu02], Seybold and Hilfer [HilSey06], and Podlubny et al. [PodKac09].

The 1948 papers by Gerasimov and Rabotnov were published in Russian, so their contents remained unknown to the majority of western scientists until the translation into English of the treatises by Rabotnov, see [Rab69, Rab80]. Whereas Gerasimov explicitly used a fractional derivative to define his model of viscoelasticity (akin to the Scott-Blair model) (see more details in [Nov17]), Rabotnov preferred to use the Volterra integral operators with weakly singular kernels that could be interpreted in terms of fractional integrals and derivatives. Following the appearance of the books by Rabotnov it has become common to speak about Rabotnov's theory of hereditary solid mechanics. The relation between Rabotnov's theory and the models of fractional viscoelasticity is briefly described in [RosShi07]. According to these Russian authors, Rabotnov could express his models in terms of the operators of the fractional calculus, but he considered these operators only as a mathematical abstraction.

2. The Second Generation of Pioneers of Fractional Calculus in Viscoelasticity

In the late sixties, formerly Caputo, see [Cap66, Cap67, Cap69] then Caputo and Mainardi, see [CapMai71a, CapMai71b], explicitly suggested that derivatives of fractional order (of Caputo type) could be successfully used to model dissipation in seismology and in metallurgy.

In relation to this, one of the authors (Mainardi) recalls a correspondence between himself (as a young post-doc student) and the Russian Academician Rabotnov, concerning two courses on Rheology held at CISM (International Centre for Mechanical Sciences, Udine, Italy) in 1973 and 1974, where Rabotnov was an invited speaker but did not participate, see [Rab73, Rab74]. Rabotnov recognized the relevance of the review paper [CapMai71b], writing in his unpublished 1974 CISM Lecture Notes:

> *That's why it was of great interest for me to know the paper of Caputo and Mainardi from the University of Bologna published in 1971. These authors have obtained similar results independently without knowing the corresponding Russian publications...*

Then he added:

> *The paper of Caputo and Mainardi contains a lot of experimental data of different authors in support of their theory. On the other hand a great number of experimental curves obtained by Postnikov and his coworkers and also by foreign authors can be found in numerous papers of Shermergor and Meshkov.*

Unfortunately, the eminent Russian scientist did not cite the 1971 paper by Caputo and Mainardi (presumably for reasons independent from his wishes) in the Russian and English editions of his later book [Rab80].

Nowadays, several articles (originally in Russian) by Shermergor, Meshkov and their associated researchers have been re-printed in English in the Journal of Applied Mechanics and Technical Physics (the English translation of *Zhurnal Prikladnoi Mekhaniki i Tekhnicheskoi Fiziki*), see e.g. [She66, MePaSh66, Mes67, MesRos68, Mes70, Zel-et-al70, GonRos73] available at the URL: http://www.springerlink.com/. In relation to this we cite the recent review papers [Ros10, RosShi07, RosShi19a], where the works of the Russian scientists on fractional viscoelasticity are examined.

The beginning of the modern applications of fractional calculus in linear viscoelasticity is generally attributed to the 1979 Ph.D. thesis by Bagley (under the supervision of Prof. Torvik), see [Bag79], followed by a number of relevant papers, e.g., [BagTor79, BagTor83a, BagTor83b, TorBag84]. However, for the sake of completeness, we should also mention the 1970 Ph.D. thesis of Rossikhin under the supervision of Prof. Meshkov, see [Ros70], and the 1971 Ph.D. thesis of Mainardi under the supervision of Prof. Caputo, summarized in [CapMai71b].

To date, applications of the fractional calculus in linear and non-linear viscoelasticity have been considered by a great and increasing number of authors to whom we have tried to refer in the huge (but not exhaustive) bibliography of the book [Mai10] (see also [GaMaMa16], where a survey of the results on modelling of anomalous dielectric relaxation is presented, as well as the paper [GiuCol18] discussing the fractional Maxwell model with Prabhakar derivatives). Several models of viscoelasticity have been presented in [AchHan09, AdEnOl05, Bag07, Chr82, Mol75, Pip86]. Historical perspectives on the development of this branch of science are discussed in [Mai12].

Finally, let us recall a few novel models related in some way to fractional viscoelasticity involving several special functions and non-local operators, authored by A. Giusti and his collaborators, see e.g. [Col-et-al16, Col-et-al18, Giu17, GiuCol18].

Notes on Fractional Calculus in Dielectrics

The standard and simplest model in the physics of dielectrics was provided by Debye in 1912 [Deb12] based on a relaxation function decaying exponentially in time with a characteristic relaxation time.

However, simple exponential models are often not satisfactory, while advanced non-exponential models (usually referred to as "anomalous relaxation") are commonly required to better explain experimental observations of complex systems. In particular, the relaxation response of many dielectric materials cannot be explained by the standard Debye process and different models have been successively introduced.

Anomalous relaxation and diffusion processes are now recognized in many complex or disordered systems that possess variable structures and parameters and show a time evolution different from the standard exponential pattern [Came09, FePuRy05, Hil02c, KhNiPo14, MetKla00, UchSib13]. Biological tissues are an interesting example of complex systems with anomalous relaxation and diffusion processes [CoKaTi02, Mag-et-al08] and they can be considered as dielectrics with losses.

9.6 Historical and Bibliographical Notes

Since the pioneering work of Kohlrausch in 1854 [Koh54], introducing a stretched exponential relaxation successively rediscovered by Williams and Watts [WilWat70], important models were introduced by Cole and Cole [ColCol41, ColCol42], Davidson and Cole [DavCol50, DavCol51], Havriliak and Negami [HavNeg67] and others.

The challenges are measuring or extrapolating the dielectric properties at high frequencies, fitting the experimental data from various tissues and from different samples of the same tissue, and representing the complex, nonlinear frequency-dependence of the permittivity [FosSch89]. Cole–Cole relaxation models, for instance, are frequently used to model propagation in dispersive biological tissues (Cole–Cole media) because they represent the frequency-dependent relative permittivity better than classical Debye models and over a wide frequency range [MauElw12, Lin10, SaidVar09]. More generally, the universal relaxation response specified by a fractional power-law is used for electromagnetic field propagation [Tar08, Tar09].

Nowadays the aforementioned models, named after their proposers, are considered as the "classical" models for dielectrics, but some other interesting models have been introduced more recently by Jurlewicz, Weron and Stanislavsky [JuTrWe10, StWeTr10] and Hilfer [Hil02, Hil02a] to better fit the experimental data in complex systems.

It is interesting to note that the Cole brothers were not initially interested in expressing relaxation and response in terms of Mittag-Leffler functions in [ColCol41], but, one year later [ColCol42], they made reference to Davis' 1936 treatise [Dav36]. Indirect references to Mittag-Leffler functions in anomalous dielectric relaxation can be found in the works by Gross [Gro37, Gro38, Gro41], but more explicitly in the 1939 papers by his student F.M. de Oliveira Castro [Oli39b, Oli39].

Consequently, the Cole brothers, even though they did not explicitly use fractional derivatives or integrals, can be considered as "indirect" pioneers of this mathematical branch (for a historical perspective see [VaMaKy14]).

Notes on Fractional Calculus in the Basset Problem

The dynamics of a sphere immersed in an incompressible viscous fluid represents a classical problem, which has many applications in flows of geophysical and engineering interest. Usually, the low Reynolds number limit (slow motion approximation) is assumed so that the Navier–Stokes equations describing the fluid motion may be linearized.

The particular but relevant situation of a sphere subjected to gravity was first considered independently by Boussinesq in 1885 [Bouss85] and by Basset in 1888 [Bas88], who introduced a special hydrodynamic force, related to the history of the relative acceleration of the sphere, which is now referred to as *Basset force*. The *Basset problem* refers to the discussion of the topics related to this force and we plan to generalize this problem via fractional calculus.

The relevance of these studies was in that, up to then, only steady motions or small oscillations of bodies in a viscous liquid had been considered, starting from Stokes' celebrated memoir on pendulums in 1851 [Stok51]. The subject matter was considered in more detail in 1907 by Picciati [Picc07] and Boggio [Bog07], and in some notes presented by the great Italian scientist Levi-Civita. The whole was summarized by Basset himself in a later paper of 1910 [Bas10], and, in the 1950s, by Hughes and Gilliand [HugGill52].

Nowadays the dynamics of impurities in unsteady flows is quite relevant as shown by several publications, whose aim is to provide more general expressions for the hydrodynamic forces, including the Basset force, in order to fit experimental data and numerical simulations, see e.g. the papers by Maxey and Riley [MaxRil83] and by Lovalenti and Brady [LovBra95] and the references therein.

Concerning the present situation in the study of the fractional Basset problem we mention the paper by Mainardi, Pironi and Tampieri [MaPiTa95a] in which a factorization method is used to invert $\mathcal{N}(s)$ and henceforth $\mathcal{M}(s)$, applying the procedure indicated by Miller and Ross [MilRos93], which is valid when α is a rational number, say $\alpha = p/q$, where $p, q \in \mathbf{N}$, $p < q$. In this way the actual solution can be finally expressed as a linear combination of certain incomplete gamma functions. This algebraic method is of course convenient for the ordinary Basset problem ($\alpha = 1/2$), but becomes cumbersome for $q > 2$.

Following the analysis by Gorenflo and Mainardi in [GorMai97], we prefer to adopt the general method of inversion based on the complex Bromwich formula. By doing this, we are free from the restriction of α being a rational number and, furthermore, we are able to provide an integral representation of the solution, convenient for numerical computation, which allows us to recognize the monotonicity properties of the solution without need of plotting.

9.7 Exercises

9.7.1 ([DebBha07, p. 313]) Solve the initial-boundary value problem

$$u(x, 0) = f(x), \quad x \in \mathbb{R},$$

$$u(x, t) \to 0, \quad \text{as } |x| \to \infty, \ t > 0,$$

for the linear inhomogeneous fractional Burgers equation

$$\frac{\partial^\alpha u}{\partial t^\alpha} + c\frac{\partial u}{\partial x} - \nu\frac{\partial^2 u}{\partial x^2} = q(x, t), \quad x \in \mathbb{R}, \ t > 0 \quad (0 < \alpha \le 1).$$

Answer.

$$u(x, t) = \frac{1}{\sqrt{2\pi}} \int_{-\infty}^{\infty} \widehat{f}(k) E_{\alpha,1}(-a^2 t^\alpha) e^{ikx} dk$$

9.7 Exercises

$$+\frac{1}{\sqrt{2\pi}}\int_0^t \tau^{\alpha-1}d\tau \int_{-\infty}^{\infty} \widehat{q}(k,t-\tau)E_{\alpha,\alpha}(-a^2\tau^\alpha)e^{ikx}dk,$$

where $a^2 = (ick + \nu k^2)$.

9.7.2 ([DebBha07, p. 314]) Solve the initial-boundary value problem

$$u(x,0) = f(x), \ x \in \mathbb{R}, \ \frac{\partial u}{\partial t}(x,0) = g(x), \ x \in \mathbb{R},$$

$$u(x,t) \to 0, \ \text{as } |x| \to \infty, \ t > 0,$$

for the linear inhomogeneous fractional Klein–Gordon equation

$$\frac{\partial^\alpha u}{\partial t^\alpha} - c^2 \frac{\partial^2 u}{\partial x^2} + d^2 u = q(x,t), \ x \in \mathbb{R}, \ t > 0 \ \ (1 < \alpha \leq 2, \ c = \text{const}, \ d = \text{const}).$$

Answer.

$$u(x,t) = \frac{1}{\sqrt{2\pi}} \int_{-\infty}^{\infty} \widehat{f}(k) E_{\alpha,1}(-a^2 t^\alpha) e^{ikx} dk$$

$$+ \frac{1}{\sqrt{2\pi}} \int_{-\infty}^{\infty} t\widehat{g}(k) E_{\alpha,2}(-a^2 t^\alpha) e^{ikx} dk$$

$$+ \frac{1}{\sqrt{2\pi}} \int_0^t \tau^{\alpha-1} d\tau \int_{-\infty}^{\infty} \widehat{q}(k, t-\tau) E_{\alpha,\alpha}(-a^2 \tau^\alpha) e^{ikx} dk,$$

where $a^2 = (c^2 k^2 + d^2)$.

9.7.3 ([DebBha07, p. 314]) Solve the initial-boundary value problem

$$u(x,0) = f(x), \quad x \in \mathbb{R},$$

$$u(x,t) \to 0 \ \text{as } |x| \to \infty, t > 0,$$

for the linear inhomogeneous fractional KdV equation

$$\frac{\partial^\alpha u}{\partial t^\alpha} + c\frac{\partial u}{\partial x} + b\frac{\partial^3 u}{\partial x^3} = q(x,t), \ x \in \mathbb{R}, \ t > 0,$$

with constant b, c and $0 < \alpha \leq 1$.
  Answer.

$$u(x,t) = \frac{1}{\sqrt{2\pi}} \int_{-\infty}^{+\infty} \widehat{f}(k) E_{\alpha,1}(-a^2 t^\alpha) e^{ikt} dk$$

$$+ \frac{1}{\sqrt{2\pi}} \int_0^t \tau^\alpha d\tau \int_{-\infty}^{+\infty} \widehat{q}(k, t-\tau) E_{\alpha,\alpha}(-a^2 t^\alpha) e^{ikt} dk,$$

where $a^2 = (ick - ik^3 b)$.

9.7.4 ([Uch13b, p. 96–97]) Solve in terms of an H-function the initial-boundary value problem

$$u(y, 0) = 0, \quad y > 0,$$

$$u(0, t) = 1 \quad t > 0,$$

$$u \to 0, \text{ as } y \to \infty,$$

for the Maxwell type fractional equation

$$\frac{\partial u}{\partial t} + \eta^\alpha \cdot {}_0 D_t^{\alpha+1} u = \eta^{\beta-1} \cdot {}_0 D_t^{\beta-1} \left(\frac{\partial^2 u}{\partial y^2} \right), \quad y \in \mathbb{R}_+, \ t > 0.$$

Answer.

$$u(y, t) = 1 + \sum_{n=1}^{\infty} \frac{(-y)^n}{n!} \eta^{n(1+\alpha-\beta)/2} t^{n(\beta-\alpha)/2-n}$$

$$\times H_{1,3}^{1,1} \left[-\frac{t^\alpha}{\eta^\alpha} \left| \begin{matrix} (-n/2, 0) \\ (0, 1) & (-n/2, -1) & (n - n(\beta-\alpha)/2, \alpha) \end{matrix} \right. \right].$$

9.7.5 ([Uch08, p. 303], [Ger48]) Consider the fractional differential model describing the motion of a visco-elastic media between two parallel plates (the lower $x = 0$ is immovable, and the upper $x = a$ is moving in the Oy-direction according to the law $\varphi(t)$, $\varphi(0) = 0$, $\dot\varphi(0) = 0$). Solve the initial-boundary value problem

$$y(x, 0) = 0, \quad \frac{\partial y(x,t)}{\partial t}\bigg|_{t=0} = 0,$$

$$y(0, t) = 0, \quad y(a, t) = \varphi(t),$$

for such a motion described by the equation

$$\rho \frac{\partial^2 y}{\partial t^2} = \frac{\partial \sigma}{\partial x}$$

9.7 Exercises

assuming the visco-elastic media satisfies the fractional constitutive equation (with $\alpha = 1/2$)

$$\sigma(t) = \kappa_\alpha \cdot {}_0 D_t^\alpha \epsilon(t).$$

Find the stress $\sigma(x, t)$ on the upper plane $x = a$.
Answer.

$$y(x, t) = \frac{2}{\pi} \sum_{n=1}^{\infty} \frac{(-1)^n}{n} \sin \frac{n\pi x}{n} \int_0^t \sum_{k=1}^{\infty} \left(\frac{n\pi}{c_{1/2} a}\right)^{2k} \frac{(-1)^k (t-\tau)^{3k/2-1}}{\Gamma(3k/2)} \varphi(\tau) d\tau,$$

$$\sigma(a, t) = \kappa_\alpha c_\alpha \left\{ \frac{t^{-\alpha/2}}{\Gamma(1-\alpha/2)} + 2 \sum_{k=0}^{\infty} \sum_{j=0}^{\infty} \frac{[-(2k+2)c_\alpha a]^j}{j! \Gamma((1-j)(1-\alpha/2))} t^{-j(1-\alpha/2)-\alpha/2} \right\},$$

where $c_\alpha^2 = \rho/\kappa_\alpha^2$.

Chapter 10
Applications to Stochastic Models

This chapter is devoted to the application of the Mittag-Leffler function and related special functions in the study of certain stochastic processes. As this topic is so wide, we restrict our attention to some basic ideas. For more complete presentations of the discussed phenomena we refer to some recent books and original papers which are mentioned in Sect. 10.6.

10.1 Introduction

The structure of the chapter and the notions and phenomena discussed in each part of it are presented in Sect. 10.1.

We start in Sect. 10.2 with a description of an approach to generalizing the Poisson probability distribution due to Pillai [Pil90]. Taking into account the complete monotonicity of the Mittag-Leffler function, Pillai introduced in [Pil90] a probability distribution which he called the *Mittag-Leffler distribution*.

In Sect. 10.3 we present a short introduction to renewal theory and continuous time random walk (CTRW) since these notions (renewal processes and CTRW) are very important for understanding the ideas behind the fractional generalization of stochastic processes. The concept of a *renewal process* has been developed as a stochastic model for describing the class of counting processes for which the times between successive events are independent identically distributed (i.i.d.) non-negative random variables, obeying a given probability law.

Section 10.4 is devoted to a generalization of the standard Poisson process by replacing the exponential function (as waiting time density) by a function of Mittag-Leffler type. Thus the corresponding renewal process can be called the fractional Poisson process or the Mittag-Leffler waiting time process. In this way we discuss how the standard Poisson process is generalized to a fractional process and describe the differences between them. We also discuss here the concept of thinning of a renewal process.

Section 10.5 presents an introduction to the theory of fractional diffusion processes. As a bridge between the simple renewal process and space-time diffusion we consider first the notion of a renewal process with reward. This leads us to the formulation of a fractional master equation which is then reduced to a space-time fraction diffusion equation. We discuss a few properties of the latter, starting with a presentation of the fundamental solution to the space-time fractional diffusion equation. Taking the diffusion limit of the Mittag-Leffler renewal process we derive the space-time fractional diffusion equation. In this connection the rescaling concept is introduced.

Another property mentioned here is the possibility of interpreting the space-time fractional diffusion process as a subordination process. As a by-product of the rescaling-respeeding concept we also obtain the asymptotic universality of the Mittag-Leffler waiting time law.

The role of the Wright function in the fractional stochastic models is illuminated in Sect. 10.6.

We conclude with Sect. 10.7 which presents some historical and bibliographical notes focussing, as in the main text, on those works which concern applications of the Mittag-Leffler function and related special functions. We also point out several notions which have been given different names in recently published papers and books. These have arisen since the discussed theory is not yet complete and has attracted great interest and rapid development because of its applications.

10.2 The Mittag-Leffler Process According to Pillai

We sketch here the theory of a process that has been devised by Pillai [Pil90] as an increasing Lévy process on the spatial half-line $x \geq 0$ happening in natural time $t \geq 0$. Switching notation from t to x, from Φ to F, from ϕ to f, and from β to α, we consider the probability distribution function

$$F_\alpha(x) = 1 - E_\alpha(-x^\alpha), \quad x \geq 0, \quad 0 < \alpha \leq 1 \tag{10.2.1}$$

and its density

$$f_\alpha(x) = -\frac{\mathrm{d}}{\mathrm{d}x} E_\alpha(-x^\alpha), \quad x \geq 0, \quad 0 < \alpha \leq 1. \tag{10.2.2}$$

Their Laplace transforms (denoting by ξ the Laplace parameter corresponding to x) are

$$\widetilde{F}_\alpha(\xi) = \frac{1}{\xi} - \frac{\xi^{\alpha-1}}{1+\xi^\alpha} = \frac{1}{\xi(1+\xi^\alpha)}, \quad \widetilde{f}_\alpha(\xi) = \frac{1}{1+\xi^\alpha}. \tag{10.2.3}$$

10.2 The Mittag-Leffler Process According to Pillai

According to Feller [Fel71], the distribution $F_\alpha(x)$ is infinite divisible if its density can be written as

$$\widetilde{f}_\alpha(\xi) = \exp(-g_\alpha(\xi)), \quad \xi \geq 0$$

where $g_\alpha(\xi)$ is (in a more modern terminology) a Bernstein function, meaning that $g_\alpha(\xi)$ is non-negative and has a completely monotone derivative. Here we have $g_\alpha(\xi) = \log(1 + \xi^\alpha) \geq 0$ so

$$g'_\alpha(\xi) = \frac{\alpha \xi^{\alpha-1}}{1+\xi^\alpha}.$$

As a consequence the derivative $g'_\alpha(\xi)$ is a completely monotone function, being a product of two completely monotone functions, and thus it follows that $g_\alpha(\xi)$ is a Bernstein function.

Now, following Pillai, we can define a stochastic process $x = x(t)$ on the half-line $x \geq 0$ happening in time $t \geq 0$ by its density $f_\alpha(x, t)$ (density in x evolving in t) taking

$$\widetilde{f}_\alpha(\xi, t) = \frac{1}{(1+\xi^\alpha)^t} = (1+\xi^\alpha)^{-t} = (\widetilde{f}_\alpha(\xi))^t. \tag{10.2.4}$$

For the Laplace inversion of $\widetilde{f}_\alpha(\xi, t)$ we write for $\xi > 1$

$$\widetilde{f}_\alpha(\xi, t) = \xi^{-\alpha t} (1 + \xi^{-\alpha})^{-t} = \sum_{k=0}^{\infty} \binom{-t}{k} \xi^{-\alpha(t+k)}. \tag{10.2.5}$$

Then, using the correspondence

$$\frac{x^{\alpha(t+k)-1}}{\Gamma(\alpha(t+k))} \div \xi^{-\alpha(t+k)}, \tag{10.2.6}$$

we get

$$f_\alpha(x, t) = \sum_{k=0}^{\infty} \binom{-t}{k} \frac{x^{\alpha(t+k)-1}}{\Gamma(\alpha(t+k))}. \tag{10.2.7}$$

Hence, by integration

$$F_\alpha(x, t) = \sum_{k=0}^{\infty} \binom{-t}{k} \frac{x^{\alpha(t+k)}}{\Gamma(\alpha(t+k)+1)}. \tag{10.2.8}$$

Manipulation of binomial coefficients yields

$$\binom{-t}{k} = (-1)^k \frac{t(t+1)\ldots(t+k-1)}{k!} = (-1)^k \frac{\Gamma(t+k)}{k!\Gamma(t)}, \tag{10.2.9}$$

so that finally we obtain

$$f_\alpha(x,t) = \sum_{k=0}^{\infty} (-1)^k \frac{\Gamma(t+k)}{k!\,\Gamma(t)\,\Gamma(\alpha(t+k))}\, x^{\alpha(t+k)-1}\,, \qquad (10.2.10)$$

and

$$F_\alpha(x,t) = \sum_{k=0}^{\infty} (-1)^k \frac{\Gamma(t+k)}{k!\,\Gamma(t)\,\Gamma(\alpha(t+k)+1)}\, x^{\alpha(t+k)}\,. \qquad (10.2.11)$$

10.3 Elements of Renewal Theory and Continuous Time Random Walks (CTRWs)

10.3.1 Renewal Processes

The General Renewal Process
For the reader's convenience, we present a brief introduction to renewal theory. For more details see, e.g., the classical treatises by Cox [Cox67], Feller [Fel71], and the more recent book by Ross [Ros97].

By a *renewal process* we mean an infinite sequence $0 = t_0 < t_1 < t_2 < \cdots$ of events separated by i.i.d. (independent and identically distributed) random waiting times $T_j = t_j - t_{j-1}$, whose probability density $\phi(t)$ is given as a function or generalized function in the sense of Gel'fand and Shilov [GelShi64] (interpretable as a measure) with support on the positive real axis $t \geq 0$, non-negative: $\phi(t) \geq 0$, and normalized: $\int_0^\infty \phi(t)\,dt = 1$, but not having a delta peak at the origin $t = 0$. The instant $t_0 = 0$ is not counted as an event. An important global characteristic of a renewal process is its mean waiting time $\langle T \rangle = \int_0^\infty t\,\phi(t)\,dt$. It may be finite or infinite. In any renewal process we can distinguish two processes, namely the *counting number process* and the process inverse to it, that we call the *Erlang process*. The instants $t_1, t_2, t_3, \ldots$ are often called *renewals*. In fact renewal theory is relevant in practice, where it is used to model required exchange of failed parts, e.g., light bulbs.
The Counting Number Process and Its Inverse
We are interested in the *counting number process* $x = N = N(t)$

$$N(t) := \max\{n\,|\,t_n \leq t\} = n \quad \text{for} \quad t_n \leq t < t_{n+1}\,, \quad n = 0, 1, 2, \cdots, \qquad (10.3.1)$$

where in particular $N(0) = 0$. We ask for the counting number probabilities in n, evolving in t,

$$p_n(t) := \mathcal{P}[N(t) = n]\,, \quad n = 0, 1, 2, \cdots\,. \qquad (10.3.2)$$

We denote by $p(x,t)$ the sojourn density for the counting number having the value x. For this process the expectation is

10.3 Elements of Renewal Theory and Continuous Time Random Walks (CTRWs)

$$m(t) := \langle N(t) \rangle = \sum_{n=0}^{\infty} n \, p_n(t) = \int_0^{\infty} x \, p(x,t) \, dx, \qquad (10.3.3)$$

since naturally $p(x,t) = \sum_{n=0}^{\infty} p_n(t) \, \delta(x-n)$. This provides the mean number of events in the half-open interval $(0, t]$, and is called the *renewal function*, see e.g. [Ros97].

We will also look at the process $t = t(N)$, the inverse of the process $N = N(t)$, that we call the *Erlang process*. This gives the time $t = t_N$ of the N-th renewal. We are now looking for the Erlang probability densities

$$q_n(t) = q(t, n), \quad n = 0, 1, 2, \ldots \qquad (10.3.4)$$

For every n the function $q_n(t) = q(t, n)$ is a density in the time variable having value t in the instant of the n-th event. Clearly, this event occurs after n (original) waiting times have passed, so that

$$q_n(t) = \phi^{*n}(t) \quad \text{with Laplace transform} \quad \widetilde{q}_n(s) = (\widetilde{\phi}(s)^n), \qquad (10.3.5)$$

where $\phi^{*n}(t) = [\phi(t)] * \ldots * [\phi(t)]$ is the multiple Laplace convolution in $\mathbb{R}$ with n identical terms. In other words, the function $q_n(t) = q(t, n)$ is a probability density in the variable $t \geq 0$ evolving in the discrete variable $x = n = 0, 1, 2, \ldots$.

10.3.2 Continuous Time Random Walks (CTRWs)

A *continuous time random walk* (CTRW) is given by an infinite sequence of spatial positions $0 = x_0, x_1, x_2, \cdots$, separated by (i.i.d.) random jumps $X_j = x_j - x_{j-1}$, whose probability density function $w(x)$ is given as a non-negative function or generalized function (interpretable as a measure) with support on the real axis $-\infty < x < +\infty$ and normalized: $\int_0^{\infty} w(x) \, dx = 1$, this random walk being subordinated to a renewal process so that we have a random process $x = x(t)$ on the real axis with the property $x(t) = x_n$ for $t_n \leq t < t_{n+1}, n = 0, 1, 2, \cdots$.

We ask for the *sojourn probability density* $u(x, t)$ of a particle wandering according to the random process $x = x(t)$ being at point x at instant t.

Let us define the following cumulative probabilities related to the waiting time density function $\phi(t)$, introduced in Sect. 10.3.1

$$\Phi(t) = \int_0^{t+} \phi(t')\,dt', \quad \Psi(t) = \int_{t+}^{\infty} \phi(t')\,dt' = 1 - \Phi(t). \tag{10.3.6}$$

For definiteness, we take $\Phi(t)$ to be right-continuous and $\Psi(t)$ left-continuous. When the non-negative random variable represents the lifetime of a technical system, it is common to call $\Phi(t) := \mathcal{P}(T \le t)$ the *failure probability* and $\Psi(t) := \mathcal{P}(T > t)$ the *survival probability*, because $\Phi(t)$ and $\Psi(t)$ are the respective probabilities that the system does or does not fail in $(0, t]$. These terms, however, are commonly adopted for any renewal process.

In the Fourier–Laplace domain we have

$$\widetilde{\Psi}(s) = \frac{1 - \widetilde{\phi}(s)}{s}, \tag{10.3.7}$$

and the famous Montroll–Weiss solution formula for a CTRW, see [MonWei65, Wei94],

$$\widehat{\widetilde{u}}(\kappa, s) = \frac{1 - \widetilde{\phi}(s)}{s} \sum_{n=0}^{\infty} \left(\widetilde{\phi}(s)\,\widehat{w}(\kappa)\right)^n = \frac{1 - \widetilde{\phi}(s)}{s} \frac{1}{1 - \widetilde{\phi}(s)\,\widehat{w}(\kappa)}. \tag{10.3.8}$$

In our special situation the jump density has support only on the positive semi-axis $x \ge 0$ and thus, by replacing the Fourier transform by the Laplace transform we obtain the Laplace–Laplace solution

$$\widetilde{\widetilde{u}}(\kappa, s) = \frac{1 - \widetilde{\phi}(s)}{s} \sum_{n=0}^{\infty} \left(\widetilde{\phi}(s)\,\widetilde{w}(\kappa)\right)^n = \frac{1 - \widetilde{\phi}(s)}{s} \frac{1}{1 - \widetilde{\phi}(s)\,\widetilde{w}(\kappa)}. \tag{10.3.9}$$

Recalling the definition of convolutions, in the physical domain we have for the solution $u(x, t)$ the *Cox–Weiss series*, see [Cox67, Wei94],

$$u(x, t) = \left(\Psi * \sum_{n=0}^{\infty} \phi^{*n}\, w^{*n}\right)(x, t). \tag{10.3.10}$$

This formula has an intuitive meaning: Up to and including instant t, there have occurred 0 jumps, or 1 jump, or 2 jumps, or ..., and if the last jump has occurred at instant $t' < t$, the wanderer is resting there for a duration $t - t'$.

The Integral Equation of a CTRW

By natural probabilistic arguments we arrive at the *integral equation* for the probability density $p(x, t)$ (a density with respect to the variable x) of the particle being at point x at instant t,

10.3 Elements of Renewal Theory and Continuous Time Random Walks (CTRWs)

$$p(x,t) = \delta(x)\Psi(t) + \int_0^t \phi(t-t')\left[\int_{-\infty}^{+\infty} w(x-x')\,p(x',t')\,dx'\right]dt'. \tag{10.3.11}$$

Here

$$\Psi(t) = \int_{t+}^{\infty} \phi(t')\,dt' \tag{10.3.12}$$

is the *survival function* (or *survival probability*). It denotes the probability that at instant t the particle is still sitting in its starting position $x = 0$. Clearly, (10.3.11) satisfies the initial condition $p(x, 0^+) = \delta(x)$.

Note that the *special choice*

$$w(x) = \delta(x-1) \tag{10.3.13}$$

gives the *pure renewal process*, with position $x(t) = N(t)$, denoting the *counting function*, and with jumps all of length 1 in the positive direction happening at the renewal instants.

For many purposes the integral equation (10.3.11) of a CTRW can easily be treated by using the Laplace and Fourier transforms. Writing these as

$$\mathcal{L}\{f(t); s\} = \tilde{f}(s) := \int_0^\infty e^{-st} f(t)\,dt,$$

$$\mathcal{F}\{g(x); \kappa\} = \widehat{g}(\kappa) := \int_{-\infty}^{+\infty} e^{+i\kappa x} g(x)\,dx,$$

in the Laplace–Fourier domain Eq. (10.3.11) reads as

$$\widehat{\tilde{p}}(\kappa, s) = \frac{1-\tilde{\phi}(s)}{s} + \tilde{\phi}(s)\,\widehat{w}(\kappa)\,\widehat{\tilde{p}}(\kappa, s). \tag{10.3.14}$$

Formally introducing in the Laplace domain the auxiliary function

$$\tilde{H}(s) = \frac{1-\tilde{\phi}(s)}{s\,\tilde{\phi}(s)} = \frac{\tilde{\Psi}(s)}{\tilde{\phi}(s)}, \quad \text{hence} \quad \tilde{\phi}(s) = \frac{1}{1+s\tilde{H}(s)}, \tag{10.3.15}$$

and assuming that its Laplace inverse $H(t)$ exists, we get, following [Mai-et-al00], in the Laplace–Fourier domain the equation

$$\tilde{H}(s)\left[s\widehat{\tilde{p}}(\kappa, s) - 1\right] = [\widehat{w}(\kappa) - 1]\,\widehat{\tilde{p}}(\kappa, s), \tag{10.3.16}$$

and in the space-time domain the generalized Kolmogorov–Feller equation

$$\int_0^t H(t-t')\frac{\partial}{\partial t'}p(x,t')\,dt' = -p(x,t) + \int_{-\infty}^{+\infty} w(x-x')\,p(x',t)\,dx',$$
(10.3.17)

with $p(x,0) = \delta(x)$.

If the Laplace inverse $H(t)$ of the formally introduced function $\widetilde{H}(s)$ does not exist, we can formally set $\widetilde{K}(s) = 1/\widetilde{H}(s)$ and multiply (10.3.16) by $\widetilde{K}(s)$. Then, if $K(t)$ exists, we get in place of (10.3.17) the alternative form of the generalized Kolmogorov–Feller equation

$$\frac{\partial}{\partial t}p(x,t) = \int_0^t K(t-t')\left[-p(x,t') + \int_{-\infty}^{+\infty} w(x-x')\,p(x',t')\,dx'\right]dt',$$
(10.3.18)

with $p(x,0) = \delta(x)$.

There are some interesting special choices of the memory function $H(t)$. We start the discussion with the following.

(i) $\quad H(t) = \delta(t) \quad \text{corresponding to} \quad \widetilde{H}(s) = 1,$ \hfill (10.3.19)

giving the *exponential waiting time* with

$$\widetilde{\phi}(s) = \frac{1}{1+s}, \quad \phi(t) = -\frac{d}{dt}e^{-t} = e^{-t}, \quad \Psi(t) = e^{-t}.$$
(10.3.20)

In this case we obtain in the Fourier–Laplace domain

$$s\widehat{\widetilde{p}}(\kappa,s) - 1 = [\widehat{w}(\kappa) - 1]\,\widehat{\widetilde{p}}(\kappa,s),$$
(10.3.21)

and in the space-time domain the *classical Kolmogorov–Feller equation*

$$\frac{\partial}{\partial t}p(x,t) = -p(x,t) + \int_{-\infty}^{+\infty} w(x-x')\,p(x',t)\,dx', \quad p(x,0) = \delta(x).$$
(10.3.22)

The other highly relevant choice is

(ii) $\quad H(t) = \dfrac{t^{-\beta}}{\Gamma(1-\beta)}, \quad 0 < \beta < 1 \text{ corresponding to } \widetilde{H}(s) = s^{\beta-1},$ \hfill (10.3.23)

that we will discuss in Sect. 10.4.

10.3.3 The Renewal Process as a Special CTRW

An essential trick in what follows is that we treat renewal processes as continuous time random walks with waiting time density $\phi(t)$ and special jump density $w(x) = \delta(x-1)$ corresponding to the fact that the counting number $N(t)$ increases by 1 at each positive event instant t_n. We then have $\widetilde{w}(\kappa) = \exp(-\kappa)$ and get for the counting number process $N(t)$ the sojourn density in the transform domain ($s \geq 0$, $\kappa \geq 0$),

$$\widetilde{\widetilde{p}}(\kappa, s) = \frac{1-\widetilde{\phi}(s)}{s} \sum_{n=0}^{\infty} \left(\widetilde{\phi}(s)\right)^n e^{-n\kappa} = \frac{1-\widetilde{\phi}(s)}{s} \frac{1}{1-\widetilde{\phi}(s)e^{-\kappa}}. \quad (10.3.24)$$

From this formula we can find formulas for the renewal function $m(t)$ and the probabilities $P_n(t) = \mathcal{P}\{N(t) = n\}$. Because $N(t)$ assumes as values only the non-negative integers, the sojourn density $p(x,t)$ vanishes if x is not equal to one of these, but has a delta peak of height $P_n(t)$ for $x = n$ ($n = 0, 1, 2, 3, \cdots$). Hence

$$p(x, t) = \sum_{n=0}^{\infty} P_n(t) \delta(x-n). \quad (10.3.25)$$

Rewriting Eq. (10.3.24), by inverting with respect to κ, as

$$\sum_{n=0}^{\infty} \left(\Psi * \phi^{*n}\right)(t) \delta(x-n), \quad (10.3.26)$$

we identify

$$P_n(t) = \left(\Psi * \phi^{*n}\right)(t). \quad (10.3.27)$$

According to the theory of the Laplace transform we conclude from Eqs. (10.3.2) and (10.3.25)

$$m(t) = -\frac{\partial}{\partial \kappa} \widetilde{p}(\kappa, t)\big|_{\kappa=0} = \left(\sum_{n=0}^{\infty} n P_n(t) e^{-n\kappa}\right)\bigg|_{\kappa=0} = \sum_{n=0}^{\infty} n P_n(t), \quad (10.3.28)$$

a result naturally expected, and

$$\widetilde{m}(s) = \sum_{n=0}^{\infty} n \widetilde{P}_n(s) = \widetilde{\Psi}(s) \sum_{n=0}^{\infty} n \left(\widetilde{\phi}(s)\right)^n = \frac{\widetilde{\phi}(s)}{s\left(1-\widetilde{\phi}(s)\right)}, \quad (10.3.29)$$

thereby using the identity

$$\sum_{n=0}^{\infty} n z^n = \frac{z}{(1-z)^2}, \quad |z| < 1.$$

Thus we have found in the Laplace domain the reciprocal pair of relationships

$$\widetilde{m}(s) = \frac{\widetilde{\phi}(s)}{s(1-\widetilde{\phi}(s))}, \quad \widetilde{\phi}(s) = \frac{s\,\widetilde{m}(s)}{1+s\,\widetilde{m}(s)}, \quad (10.3.30)$$

telling us that the waiting time density and the renewal function mutually determine each other uniquely. The first formula of Eq. (10.3.30) can also be obtained as the value at $\kappa = 0$ of the negative derivative for $\kappa = 0$ of the last expression in Eq. (10.3.24). Equation (10.3.30) implies the reciprocal pair of relationships in the physical domain

$$m(t) = \int_0^\infty [1 + m(t-t')]\,\phi(t')\,dt',$$

$$m'(t) = \int_0^\infty [1 + m'(t-t')]\,\phi(t')\,dt'.$$
(10.3.31)

The first of these equations is usually called the *renewal equation*.

Considering, formally, the counting number process $N = N(t)$ as a CTRW (with jumps fixed to unit jumps 1), N running increasingly through the non-negative integers $x = 0, 1, 2, \ldots$, happening in natural time $t \in [0, \infty)$, we note that in the Erlang process $t = t(N)$, the roles of N and t are interchanged. The new "waiting time density" is now $w(x) = \delta(x-1)$, the new "jump density" is $\phi(t)$.

It is illuminating to look at the relationships for $t \geq 0, n = 0, 1, 2, \ldots$, between the counting number probabilities $P_n(t)$ and the Erlang densities $q_n(t)$. For Eq. (10.3.5) we have $q_n(t) = \phi^{*n}(t)$, and then by (10.3.27)

$$P_n(t) = (\Psi * q_n)(t) = \int_0^t \left(q_n(t') - q_{n+1}(t')\right) dt'. \quad (10.3.32)$$

We can also express the q_n in another way in terms of the P_n. Introducing the cumulative probabilities $Q_n(t) = \int_0^t q_n(t')\,dt'$, we have

$$Q_n(t) = \mathcal{P}\left(\sum_{k=1}^n T_k \leq t\right) = \mathcal{P}(N(t) \geq n) = \sum_{k=n}^\infty P_k(t), \quad (10.3.33)$$

and finally

$$q_n(t) = \frac{d}{dt} Q(t) = \frac{d}{dt} \sum_{k=n}^\infty P_k(t). \quad (10.3.34)$$

All this is true for $n = 0$ as well, by the empty sum convention $\sum_{k=1}^n T_k = 0$ for $n = 0$.

10.4 The Poisson Process and Its Fractional Generalization (The Renewal Process of Mittag-Leffler Type)

10.4.1 The Mittag-Leffler Waiting Time Density

Returning to the integral equation for the probability density of a CTRW (10.3.11) (see Sect. 10.3.2) we note that besides the classical special case (10.3.19) there exist another one.

(ii) $H(t) = \dfrac{t^{-\beta}}{\Gamma(1-\beta)}$, $0 < \beta < 1$, corresponding to $\widetilde{H}(s) = s^{\beta - 1}$, (10.4.1)

giving the *Mittag-Leffler waiting time density* with

$$\widetilde{\phi}(s) = \frac{1}{1+s^\beta}, \quad \phi(t) = -\frac{d}{dt}E_\beta(-t^\beta) = \phi^{ML}(t), \quad \Psi(t) = E_\beta(-t^\beta). \quad (10.4.2)$$

In this case we obtain in the Fourier–Laplace domain

$$s^{\beta-1}\left[s\widehat{\widetilde{p}}(\kappa, s) - 1\right] = [\widehat{w}(\kappa) - 1]\,\widehat{\widetilde{p}}(\kappa, s), \quad (10.4.3)$$

and in the space-time domain the *time fractional Kolmogorov–Feller equation*

$$_tD_*^\beta\, p(x,t) = -p(x,t) + \int_{-\infty}^{+\infty} w(x-x')\,p(x',t)\,dx', \quad p(x,0^+) = \delta(x),$$
(10.4.4)

where $_tD_*^\beta$ denotes the fractional derivative of order β in the Caputo sense, see Appendix A.

The time fractional Kolmogorov–Feller equation can also be expressed via the Riemann–Liouville fractional derivative $D_{0+,t}^{1-\beta}$, that is

$$\frac{\partial}{\partial t}p(x,t) = D_{0+,t}^{1-\beta}\left[-p(x,t) + \int_{-\infty}^{+\infty} w(x-x')\,p(x',t)\,dx'\right], \quad (10.4.5)$$

with $p(x,0^+) = \delta(x)$. The equivalence of the two forms (10.4.4) and (10.4.5) is easily proved in the Fourier–Laplace domain by multiplying both sides of Eq.(10.4.3) by the factor $s^{1-\beta}$.

We note that the choice (i) may be considered as a limit of the choice (ii) as $\beta = 1$. In fact, in this limit we find $\widetilde{H}(s) \equiv 1$ so $H(t) = t^{-1}/\Gamma(0) \equiv \delta(t)$ (according to a formal representation of the Dirac generalized function [GelShi64]), so that Eqs. (10.3.16)–(10.3.17) reduce to (10.3.21)–(10.3.22), respectively. In this case the order of the Caputo derivative reduces to 1 and that of the R-L derivative to 0, whereas the Mittag-Leffler waiting time law reduces to the exponential.

In the sequel we will formally unite the choices (**i**) and (**ii**) by defining what we call the Mittag-Leffler memory function

$$H^{ML}(t) = \begin{cases} \dfrac{t^{-\beta}}{\Gamma(1-\beta)}, & \text{if } 0 < \beta < 1, \\ \delta(t), & \text{if } \beta = 1, \end{cases} \qquad (10.4.6)$$

whose Laplace transform is

$$\widetilde{H}^{ML}(s) = s^{\beta-1}, \quad 0 < \beta \leq 1. \qquad (10.4.7)$$

Thus we will consider the whole range $0 < \beta \leq 1$ by extending the Mittag-Leffler waiting time law in (10.4.2) to include the exponential law (10.3.20).

10.4.2 The Poisson Process

The most celebrated renewal process is the Poisson process characterized by a waiting time *probability density function (pdf)* of exponential type,

$$\phi(t) = \lambda e^{-\lambda t}, \quad \lambda > 0, \quad t \geq 0. \qquad (10.4.8)$$

The process has *no memory*. Its moments turn out to be

$$\langle T \rangle = \frac{1}{\lambda}, \quad \langle T^2 \rangle = \frac{1}{\lambda^2}, \quad \ldots, \quad \langle T^n \rangle = \frac{1}{\lambda^n}, \quad \ldots, \qquad (10.4.9)$$

and the *survival probability* is

$$\Psi(t) := \mathcal{P}(T > t) = e^{-\lambda t}, \quad t \geq 0. \qquad (10.4.10)$$

We know that the probability that k events occur in the interval of length t is

$$\mathcal{P}(N(t) = k) = \frac{(\lambda t)^k}{k!} e^{-\lambda t}, \quad t \geq 0, \quad k = 0, 1, 2, \ldots. \qquad (10.4.11)$$

The probability distribution related to the sum of k i.i.d. exponential random variables is known to be the so-called *Erlang distribution* (of order k). The corresponding density (the *Erlang pdf*) is thus

$$f_k(t) = \lambda \frac{(\lambda t)^{k-1}}{(k-1)!} e^{-\lambda t}, \quad t \geq 0, \quad k = 1, 2, \ldots, \qquad (10.4.12)$$

so that the Erlang distribution function of order k turns out to be

10.4 The Poisson Process and Its Fractional ...

$$F_k(t) = \int_0^t f_k(t')\,dt' = 1 - \sum_{n=0}^{k-1} \frac{(\lambda t)^n}{n!} e^{-\lambda t} = \sum_{n=k}^{\infty} \frac{(\lambda t)^n}{n!} e^{-\lambda t}, \quad t \geq 0.$$
(10.4.13)

In the limiting case $k = 0$ we recover $f_0(t) = \delta(t)$, $F_0(t) \equiv 1$, $t \geq 0$.

The results (10.4.11)–(10.4.13) can easily be obtained by using the technique of the Laplace transform sketched in the previous section, noting that for the Poisson process we have:

$$\widetilde{\phi}(s) = \frac{\lambda}{\lambda + s}, \quad \widetilde{\Psi}(s) = \frac{1}{\lambda + s},$$
(10.4.14)

and for the Erlang distribution:

$$\widetilde{f}_k(s) = [\widetilde{\phi}(s)]^k = \frac{\lambda^k}{(\lambda + s)^k}, \quad \widetilde{F}_k(s) = \frac{[\widetilde{\phi}(s)]^k}{s} = \frac{\lambda^k}{s(\lambda + s)^k}.$$
(10.4.15)

We also recall that the survival probability for the Poisson renewal process obeys the ordinary differential equation (of relaxation type)

$$\frac{d}{dt}\Psi(t) = -\lambda \Psi(t), \quad t \geq 0; \quad \Psi(0^+) = 1.$$
(10.4.16)

10.4.3 The Renewal Process of Mittag-Leffler Type

A "fractional" generalization of the Poisson renewal process is simply obtained by generalizing the differential equation (10.4.16), replacing there the first derivative with the integro-differential operator $_t D_*^\beta$ that is interpreted as the fractional derivative of order β in the Caputo sense. We write, taking for simplicity $\lambda = 1$,

$$_t D_*^\beta \Psi(t) = -\Psi(t), \quad t > 0, \quad 0 < \beta \leq 1; \quad \Psi(0^+) = 1.$$
(10.4.17)

We also allow the limiting case $\beta = 1$ where all the results of the previous section (with $\lambda = 1$) are expected to be recovered. In fact, taking $\lambda = 1$ is simply a normalized way of scaling the variable t.

We call this renewal process of Mittag-Leffler type the *fractional Poisson process*. To analyze this we work in the Laplace domain where we have

$$\widetilde{\Psi}(s) = \frac{s^{\beta-1}}{1 + s^\beta}, \quad \widetilde{\phi}(s) = \frac{1}{1 + s^\beta}.$$
(10.4.18)

If there is no danger of misunderstanding we will not decorate Ψ and ϕ with the index β. The special choice $\beta = 1$ gives us the standard Poisson process with $\Psi_1(t) = \phi_1(t) = \exp(-t)$.

Whereas the Poisson process has finite mean waiting time (that of its standard version is equal to 1), the *fractional Poisson process* ($0 < \beta < 1$) does not have this property. In fact,

$$\langle T \rangle = \int_0^\infty t\, \phi(t)\, dt = \beta \left. \frac{s^{\beta-1}}{(1+s^\beta)^2} \right|_{s=0} = \begin{cases} 1, & \beta = 1, \\ \infty, & 0 < \beta < 1. \end{cases} \qquad (10.4.19)$$

Let us calculate the renewal function $m(t)$. Inserting $\widetilde{\phi}(s) = 1/(1+s^\beta)$ into Eq. (10.3.24) and taking $w(x) = \delta(x-1)$ as in Sect. 10.3.3, we find for the sojourn density of the counting function $N(t)$ the expressions

$$\widetilde{p}(\kappa, s) = \frac{s^{\beta-1}}{1+s^\beta - e^{-\kappa}} = \frac{s^{\beta-1}}{1+s^\beta} \sum_{n=0}^\infty \frac{e^{-n\kappa}}{(1+s^\beta)^n}, \qquad (10.4.20)$$

and

$$\widetilde{p}(\kappa, t) = E_\beta \left(-(1-e^{-\kappa}) t^\beta \right), \qquad (10.4.21)$$

and then

$$m(t) = -\frac{\partial}{\partial \kappa} \widetilde{p}(\kappa, t) \big|_{\kappa=0} = e^{-\kappa} t^\beta E'_\beta \left(-(1-e^{-\kappa}) t^\beta \right) \big|_{\kappa=0}. \qquad (10.4.22)$$

Using $E'_\beta(0) = 1/\Gamma(1+\beta)$ now yields

$$m(t) = \begin{cases} t, & \beta = 1, \\ \dfrac{t^\beta}{\Gamma(1+\beta)}, & 0 < \beta < 1. \end{cases} \qquad (10.4.23)$$

This result can also be obtained by plugging $\widetilde{\phi}(s) = 1/(1+s^\beta)$ into the first equation in (10.3.30), which yields $\widetilde{m}(s) = 1/s^{\beta+1}$, and then by Laplace inversion Eq. (10.4.23).

Using the general Taylor expansion

$$E_\beta(z) = \sum_{n=0}^\infty \frac{E_\beta^{(n)}}{n!} (z-b)^n, \qquad (10.4.24)$$

in Eq. (10.4.21) with $b = -t^\beta$ we get

$$\widetilde{p}(\kappa, t) = \sum_{n=0}^\infty \frac{t^{n\beta}}{n!} E_\beta^{(n)}(-t^\beta)\, e^{-n\kappa},$$

$$p(x, t) = \sum_{n=0}^\infty \frac{t^{n\beta}}{n!} E_\beta^{(n)}(-t^\beta)\, \delta(x-n), \qquad (10.4.25)$$

10.4 The Poisson Process and Its Fractional ...

and, by comparison with Eq. (10.3.25), the counting number probabilities

$$P_n(t) = \mathcal{P}\{N(t) = n\} = \frac{t^{n\beta}}{n!} E_\beta^{(n)}(-t^\beta). \tag{10.4.26}$$

Observing from Eq. (10.4.20) that

$$\widetilde{p}(\kappa, s) = \frac{s^{\beta-1}}{1 + s^\beta - e^{-\kappa}} = \frac{s^{\beta-1}}{1 + s^\beta} \sum_{n=0}^{\infty} \frac{e^{-n\kappa}}{(1 + s^\beta)^n}, \tag{10.4.27}$$

and inverting with respect to κ,

$$\widetilde{p}(x, s) = \frac{s^{\beta-1}}{1 + s^\beta} \sum_{n=0}^{\infty} \frac{\delta(x - n)}{(1 + s^\beta)^n}, \tag{10.4.28}$$

we finally identify

$$\widetilde{P}_n(s) = \frac{s^{\beta-1}}{(1 + s^\beta)^{n+1}} \div \frac{t^{n\beta}}{n!} E_\beta^{(n)}(-t^\beta) = P_n(t). \tag{10.4.29}$$

En passant we have proved an often cited special case of an inversion formula due to Podlubny (1999) [Pod99, Eq. (1.80)].

For the Poisson process with intensity $\lambda > 0$ we have a well-known infinite system of ordinary differential equations (for $t \geq 0$), see e.g. Khintchine [Khi60],

$$P_0(t) = e^{-\lambda t}, \quad \frac{d}{dt} P_n(t) = \lambda \left(P_{n-1}(t) - P_n(t) \right), \ n \geq 1, \tag{10.4.30}$$

with initial conditions $P_n(0) = 0$, $n = 1, 2, \ldots$, which is sometimes even used to define the Poisson process. We have an analogous system of fractional differential equations for the fractional Poisson process. In fact, from Eq. (10.4.30) we have

$$(1 + s^\beta) \widetilde{P}_n(s) = \frac{s^{\beta-1}}{(1 + s^\beta)^n} = \widetilde{P}_{n-1}(s). \tag{10.4.31}$$

Hence

$$s^\beta \widetilde{P}_n(s) = \widetilde{P}_{n-1}(s) - \widetilde{P}_n(s), \tag{10.4.32}$$

so in the time domain

$$P_0(t) = E_\beta(-t^\beta), \quad {}_*D_t^\beta P_n(t) = P_{n-1}(t) - P_n(t), \; n \geq 1, \quad (10.4.33)$$

with initial conditions $P_n(0) = 0$, $n = 1, 2, \ldots$, where ${}_*D_t^\beta$ denotes the time-fractional derivative of Caputo type of order β. It is also possible to introduce and define the fractional Poisson process by this difference-differential system.

Let us note that by solving the system (10.4.33), Beghin and Orsingher in [BegOrs09] introduce what they call the "first form of the fractional Poisson process", and in [MeNaVe11] Meerschaert et al. show that this process is a renewal process with Mittag-Leffler waiting time density as in (10.4.17), hence is identical to the fractional Poisson process.

Up to now we have investigated the fractional Poisson counting process $N = N(t)$ and found its probabilities $P_n(t)$ in Eq. (10.4.26). To get the corresponding *Erlang probability densities* $q_n(t) = q(t, n)$, densities in t, evolving in $n = 0, 1, 2 \ldots$, we find by Eq. (10.3.34) via telescope summation

$$q_n(t) = \beta \frac{t^{n\beta-1}}{(n-1)!} E_\beta^{(n)}\left(-t^\beta\right), \quad 0 < \beta \leq 1. \quad (10.4.34)$$

We leave it as an exercise to the reader to show that in Eq. (10.4.25) interchange of differentiation and summation is allowed.

Remark With $\beta = 1$ we get the corresponding well-known results for the standard Poisson process. The counting number probabilities are

$$P_n(t) = \frac{t^n}{n!} e^{-t}, \quad n = 0, 1, 2, \ldots \; t \geq 0, \quad (10.4.35)$$

and the Erlang densities

$$q_n(t) = \frac{t^{n-1}}{(n-1)!} e^{-t}, \quad n = 1, 2, 3, \ldots, \; t \geq 0. \quad (10.4.36)$$

By rescaling of time we obtain

$$P_n(t) = \frac{(\lambda t)^n}{n!} e^{-\lambda t}, \quad n = 0, 1, 2, \ldots, \; t \geq 0, \quad (10.4.37)$$

for the classical Poisson process with intensity λ and

$$q_n(t) = \lambda \frac{(\lambda t)^{n-1}}{(n-1)!} e^{-\lambda t}, \quad n = 1, 2, 3, \ldots, \; t \geq 0 \quad (10.4.38)$$

for the corresponding Erlang process.

10.4.4 Thinning of a Renewal Process

We are now going to give an account of the essentials of the thinning of a renewal process with power law waiting times, thereby leaning on the presentation of Gnedenko and Kovalenko [GneKov68] but for reasons of transparency not decorating the power functions by slowly varying functions. Compare also Mainardi, Gorenflo and Scalas [MaGoSc04a] and Gorenflo and Mainardi [GorMai08].

Again with the t_n in strictly increasing order, the time instants of a renewal process, $0 = t_0 < t_1 < t_2 < \ldots$, with i.i.d. waiting times $T_k = t_k - t_{k-1}$ (generically denoted by T), *thinning* (or *rarefaction*) means that for each positive index k a decision is made: the event happening in the instant t_k is deleted with probability p (where $0 < p < 1$) or is maintained with probability $q = 1 - p$. This procedure produces a thinned (or rarefied) renewal process, that is, one with fewer events. Of particular interest for us is the case where q is near zero, which results in very few events in a moderate span of time. To compensate for this loss (wanting to keep a moderate number of events in a moderate span of time) we change the unit of time which amounts to multiplying the (numerical value of) the waiting time by a positive factor τ so that we get waiting times τT_k and instants τt_k in the rescaled process. Loosely speaking, it is our intention to select τ in relation to the rarefaction factor q in such a way that for very small q in some sense the "average" number of events per unit of time remains unchanged. We will make these considerations precise in an asymptotic sense.

Denoting by $F(t) = \mathcal{P}(T \leq t)$ the probability distribution function of the original waiting time T, by $f(t)$ its density (generally this density is a generalized function represented by a measure) so that $F(t) = \int_0^t f(t')\,dt'$, and analogously for the functions $F_k(t)$ and $f_k(t)$, the distribution and density, respectively, of the sum of k waiting times, we have recursively

$$f_1(t) = f(t), \quad f_k(t) = \int_0^t f_{k-1}(t)\,dF(t') \quad \text{for} \quad k \geq 2. \tag{10.4.39}$$

Observing that after a maintained event of the original process the next one is kept with probability q but dropped with probability p in favor of the second-next with probability pq and, generally $n - 1$ events are dropped in favor of the n-th next with probability $p^{n-1}q$, we get for the waiting time density of the thinned process the formula

$$g_q(t) = \sum_{n=1}^{\infty} q\,p^{n-1}\,f_n(t). \tag{10.4.40}$$

With the modified waiting time τT we have $\mathcal{P}(\tau T \leq t) = \mathcal{P}(T \leq t/\tau) = F(t/\tau)$, hence the density $f(t/\tau)/\tau$, and analogously for the density of the sum of n waiting times $f_n(t/\tau)/\tau$. The density of the waiting time of the sum of n waiting times of the rescaled (and thinned) process now turns out as

$$g_{q,\tau}(t) = \sum_{n=1}^{\infty} q\, p^{n-1}\, f_n(t/\tau)/\tau. \tag{10.4.41}$$

In the Laplace domain we have $\widetilde{f}_n(s) = (\widetilde{f}(s))^n$, hence (using $p = 1 - q$)

$$\widetilde{g}_q(s) = \sum_{n=1}^{\infty} q\, p^{n-1}\, (\widetilde{f}(s))^n = \frac{q\, \widetilde{f}(s)}{1 - (1-q)\widetilde{f}(s)}. \tag{10.4.42}$$

By rescaling we get

$$\widetilde{g}_{q,\tau}(s) = \sum_{n=1}^{\infty} q\, p^{n-1}\, (\widetilde{f}(\tau s))^n = \frac{q\, \widetilde{f}(\tau s)}{1 - (1-q)\widetilde{f}(\tau s)}. \tag{10.4.43}$$

Being interested in stronger and stronger thinning (*infinite thinning*) let us consider a scale of processes with the parameters q of *thinning* and τ of *rescaling* tending to zero under a *scaling relation* $q = q(\tau)$ yet to be specified.

Let us consider two cases for the (original) waiting time distribution, namely, case (A) of a finite mean waiting time and case (B) of a power law waiting time. We assume in case (A)

$$\lambda := \int_0^{\infty} t'\, f(t')\, dt' < \infty \quad (A), \quad \text{setting} \quad \beta = 1, \tag{10.4.44}$$

or

$$\Psi(t) = \int_t^{\infty} f(t')\, dt' \sim \frac{c}{\beta} t^{-\beta} \quad \text{for} \quad t \to \infty \quad \text{with} \quad 0 < \beta < 1. \tag{10.4.45}$$

In case (B) we set

$$\lambda = \frac{c\pi}{\Gamma(\beta+1)\, \sin(\beta\pi)}.$$

From Lemma 10.2 in the next Sect. 10.5 we know that $\widetilde{f}(s) = 1 - \lambda s^{\beta} + o(s^{\beta})$ for $0 < s \to 0$.

Passing now to the limit $q \to 0$ of *infinite thinning* under the *scaling relation*

$$q = \lambda \tau^{\beta} \tag{10.4.46}$$

for *fixed* s the Laplace transform (10.4.43) of the rescaled density $g_{q,\tau}(t)$ of the thinned process tends to $\widetilde{g}(s) = 1/(1 + s^{\beta})$ corresponding to the Mittag-Leffler density

$$g(t) = -\frac{d}{dt} E_{\beta}(-t^{\beta}) = \phi_{\beta}^{ML}(t). \tag{10.4.47}$$

Thus, the thinned process converges weakly to the *Mittag-Leffler renewal process* described in Mainardi, Gorenflo and Scalas [MaGoSc04a] (called the *fractional Poisson process* in Laskin [Lai93]) which in the special case $\beta = 1$ reduces to the Poisson process.

10.5 Fractional Diffusion and Subordination Processes

10.5.1 Renewal Process with Reward

The renewal process can be accompanied by a reward, which means that at every renewal instant a space-like variable makes a random jump from its previous position to a new point in "space". Here "space" is used in a very general sense. In the insurance business, for example, the renewal points are instants where the company receives a payment or must give away money to some claim of a customer, so space is money. In such a process occurring in time and in space, also referred to as a *compound renewal process*, the probability distribution of jump widths is as relevant as that of the waiting times.

Let us denote by X_n the jumps occurring at instants t_n, $n = 1, 2, 3, \ldots$. Let us assume that X_n are i.i.d. (real, not necessarily positive) random variables with probability density $w(x)$, independent of the *waiting time* density $\phi(t)$. In a physical context the X_ns represent the jumps of a diffusing particle (the walker), and the resulting random walk model is known as a *continuous time random walk* (abbreviated CTRW) in that the waiting time is assumed to be a *continuous* random variable.

10.5.2 Limit of the Mittag-Leffler Renewal Process

In a CTRW we can, with positive scaling factors h and τ, replace the jumps X by jumps $X_h = h X$ and the waiting times T by waiting times $T_\tau = \tau T$. This leads to the rescaled jump density $w_h(x) = w(x/h)/h$ and the rescaled waiting time density $\phi_\tau(t) = \phi(t/\tau)/\tau$ and correspondingly to the transforms $\widehat{w}_h(\kappa) = \widehat{w}(h\kappa)$, $\widetilde{\phi}_\tau(s) = \widetilde{\phi}(\tau s)$.

For the sojourn density $u_{h,\tau}(x, t)$, the density in x evolving in t, we obtain from Eq. (10.3.8) in the transform domain (the Montroll–Weiss formula)

$$\widehat{\widetilde{u}}_{h,\tau}(\kappa, s) = \frac{1 - \widetilde{\phi}(\tau s)}{s} \frac{1}{1 - \widetilde{\phi}(\tau s)\,\widehat{w}(h\kappa)}, \qquad (10.5.1)$$

where, if $w(x)$ has support on $x \geq 0$, we can work with the Laplace transform instead of the Fourier transform (replace the $\widehat{}$ by $\widetilde{}$). If there exists between h and τ a scaling relation $\mathcal{R}$ (to be introduced later) under which $u(x, t)$ tends as $h \to 0$, $\tau \to 0$ to a meaningful limit $v(x,t) = u_{0,0}(x,t)$, then we call the process $x = x(t)$ with this sojourn density a *diffusion limit*. We find it via

$$\widetilde{v}(\kappa, s) = \lim_{h,\tau \to 0(\mathcal{R})} \widehat{\widetilde{u}}_{h,\tau}(\kappa, s), \qquad (10.5.2)$$

and Fourier–Laplace (or Laplace–Laplace) inversion.

In recent decades power laws in physical (and also economical and other) processes and situations have become increasingly popular for modelling slow (in contrast to fast, mostly exponential) decay at infinity. See Newman [New05] for a general introduction to this concept. For our purpose let us assume that the distribution of jumps is symmetric, and that the distribution of jumps, likewise that of waiting times, either has finite second or first moment, respectively, or decays near infinity like a power with exponent $-\alpha$ or $-\beta$, respectively, $0 < \alpha < 2$, $0 < \beta < 1$. Then we can state two lemmas. These lemmas and more general ones (e.g. with slowly varying decorations of the power laws (a) and (b)) can be distilled from the Gnedenko theorem on the domains of attraction of stable probability laws (see Gnedenko and Kolmogorov [GneKol54]). For wide generalizations (to several space dimensions and to anisotropy) see Meerschaert and Scheffler [MeeSch04]. They can also be modified to cover the special case of smooth densities $w(x)$ and $\phi(t)$ and to the case of fully discrete random walks, see Gorenflo and Abdel-Rehim [GorAbdR05], Gorenflo and Vivoli [GorViv03]. For proofs, see also Gorenflo and Mainardi [GorMai08].

Lemma 10.1 (for the jump distribution) *Assume that $W(x)$ is increasing, $W(-\infty) = 0$, $W(\infty) = 1$, and the symmetry $W(-x) + W(x) = 1$ holds for all continuity points x of $W(x)$, and assume (a) or (b) are valid:*

(a) $\sigma^2 := \int_{-\infty}^{+\infty} x^2 \, dW(x) < \infty$, *labelled as $\alpha = 2$;*

(b) $\int_x^{\infty} dW(x') \sim b\alpha^{-1}x^{-\alpha}$ *for $x \to \infty$, $0 < \alpha < 2$, $b > 0$.*

Then, with $\mu = \sigma^2/2$ in case (a) and $\mu = b\pi/[\Gamma(\alpha+1)\sin(\alpha\pi/2)]$ in case (b) we have the asymptotics

$$1 - \widehat{w}(\kappa) \sim \mu|\kappa|^{\alpha}$$

for $\kappa \to 0$.

Lemma 10.2 (for the waiting time distribution) *Assume $\Phi(t)$ is increasing, $\Phi(0) = 0$, $\Phi(\infty) = 1$, and (A) or (B) is valid:*

(A) $\rho := \int_0^{\infty} t \, d\Phi(t) < \infty$, *labelled as $\beta = 1$,*

(B) $1 - \Phi(t) \sim c\beta^{-1}t^{-\beta}$ *for $t \to \infty$, $0 < \beta < 1$, $c > 0$.*

Then, with $\lambda = \rho$ in case (A) and $\lambda = c\pi/[\Gamma(\beta+1)\sin(\beta\pi)]$ in case (B) we have the asymptotics

$$1 - \widetilde{\phi}(s) \sim \lambda s^{\beta}$$

for $0 < s \to 0$.

10.5 Fractional Diffusion and Subordination Processes

We will now outline the *well-scaled passage to the diffusion limit* by which, via rescaling space and time in a combined way, we will arrive at the Cauchy problem for the space-time fractional diffusion equation. Assuming the conditions of the two lemmata are fulfilled, we carry out this passage in the Fourier–Laplace domain. For rescaling we multiply the jumps and the waiting times by positive factors h and τ and so obtain a random walk $x_n(h) = (X_1 + X_2 + \cdots + X_n)h$ with jump instants $t_n(h) = (T_1 + T_2 + \cdots + T_n)\tau$. We study this rescaled random walk under the intention to send h and τ towards 0. Physically, we change the units of measurement from 1 to $1/h$ in space, from 1 to $1/\tau$ in time, respectively, making intervals of moderate size numerically small, and intervals of large size numerically of moderate size, in this way turning from the microscopic to the macroscopic view. Noting the densities $w_h(x) = w(x/h)/h$ and $\phi_\tau(t/\tau)/\tau$ of the reduced jumps and waiting times, we get the corresponding transforms $\widehat{w}_h(\kappa) = \widehat{w}(\kappa h)$, $\widetilde{\phi}_\tau(s) = \widetilde{\phi}(\tau s)$, and, in analogy with the Montroll–Weiss equation, the result

$$\widehat{\widetilde{p}}_{h,\tau}(\kappa, s) = \frac{1 - \widetilde{\phi}_\tau(s)}{s} \frac{1}{1 - \widehat{w}_h(\kappa)\widetilde{\phi}_\tau(s)} = \frac{1 - \widetilde{\phi}(\tau s)}{s} \frac{1}{1 - \widehat{w}(h\kappa)\widetilde{\phi}(\tau s)}. \tag{10.5.3}$$

Fixing now κ and s, both non-zero, replacing κ by $h\kappa$ and s by τs in the above lemmas, and sending h and τ to zero, we obtain by a trivial calculation the asymptotics

$$\widehat{\widetilde{p}}_{h,\tau}(\kappa, s) \sim \frac{\lambda \tau^\beta s^{\beta-1}}{\mu(h|\kappa|)^\alpha + \lambda(\tau s)^\beta} \tag{10.5.4}$$

that we can rewrite in the form

$$\widehat{\widetilde{p}}_{h,\tau}(\kappa, s) \sim \frac{s^{\beta-1}}{r(h,\tau)|\kappa|^\alpha + s^\beta} \quad \text{with} \quad r(h,\tau) = \frac{\mu h^\alpha}{\lambda \tau^\beta}. \tag{10.5.5}$$

Choosing $r(h,\tau) \equiv 1$ (it suffices to choose $r(h,\tau) \to 1$) we get

$$\widehat{\widetilde{p}}_{h,\tau}(\kappa, s) \to \widehat{\widetilde{p}}_{0,0}(\kappa, s) = \frac{s^{\beta-1}}{|\kappa|^\alpha + s^\beta}. \tag{10.5.6}$$

We will call the condition below the *scaling relation*

$$\frac{\mu h^\alpha}{\lambda \tau^\beta} \equiv 1. \tag{10.5.7}$$

Via $\tau = (\mu/\lambda)h^\alpha)^{1/\beta}$ we can eliminate the parameter τ, apply the inverse Laplace transform to (10.5.4), fix κ and send $h \to 0$. So, by the continuity theorem (for the Fourier transform of a probability distribution, see Feller [Fel71]), we can identify

$$\widehat{\widetilde{p}}_{0,0}(\kappa, s) = \frac{s^{\beta-1}}{|\kappa|^\alpha + s^\beta}$$

as the Fourier–Laplace solution $\widehat{\widetilde{u}}(\kappa, s)$ of the *space-time fractional Cauchy problem* (for $x \in \mathbb{R}, t \geq 0$)

$$^C D^{\beta}_{0+,t} u(x,t) = D^{\alpha}_{0+,x} u(x,t), \quad u(x,0) = \delta(x), \tag{10.5.8}$$

$$0 < \alpha \leq 2, \ 0 < \beta \leq 1. \tag{10.5.9}$$

Here, for $0 < \beta \leq 1$, we denote by $_t D^{\beta}_*$ the regularized fractional differential operator, see Gorenflo and Mainardi [GorMai97], according to

$$^C D^{\beta}_{0+,t} g(t) = D^{\beta}_{0+,t} [g(t) - g(0)] \tag{10.5.10}$$

with the Riemann–Liouville fractional differential operator

$$D^{\beta}_{0+,t} g(t) := \begin{cases} \dfrac{1}{\Gamma(1-\beta)} \dfrac{d}{dt} \displaystyle\int_0^t \dfrac{g(t')\,d\tau}{(t-t')^{\beta}}, & 0 < \beta < 1, \\ \dfrac{d}{dt} g(t), & \beta = 1. \end{cases} \tag{10.5.11}$$

Hence, in longhand:

$$_t D^{\beta}_* g(t) = \begin{cases} \dfrac{1}{\Gamma(1-\beta)} \dfrac{d}{dt} \displaystyle\int_0^t \dfrac{g(t')\,d\tau}{(t-t')^{\beta}} - \dfrac{g(0)t^{-\beta}}{\Gamma(1-\beta)}, & 0 < \beta < 1 \\ \dfrac{d}{dt} g(t), & \beta = 1. \end{cases} \tag{10.5.12}$$

If $g'(t)$ exists we can write

$$^C D^{\beta}_{0+,t} g(t) = \frac{1}{\Gamma(1-\beta)} \int_0^t \frac{g'(t')}{(t-t')^{\beta}} dt', \quad 0 < \beta < 1,$$

and the regularized fractional derivative coincides with the form introduced by Caputo, see Caputo and Mainardi [CapMai71a], Gorenflo and Mainardi [GorMai97], Podlubny [Pod99], henceforth referred to as the Caputo derivative. Observe that in the special case $\beta = 1$ the two fractional derivatives $^C D^{\beta}_{0+,t} g(t)$ and $D^{\beta}_{0+,t} g(t)$ coincide, both then being equal to $g'(t)$.

The Riesz operator is a pseudo-differential operator according to

$$\widehat{D^{\alpha}_{0+,x} f} = -|\kappa|^{\alpha} \widehat{f}(\kappa), \quad \kappa \in \mathbb{R}, \tag{10.5.13}$$

(compare Samko, Kilbas and Marichev [SaKiMa93] and Rubin [Rub96]). It has the Fourier symbol $-|\kappa|^{\alpha}$.

In the transform domain (10.5.8) means

$$s^{\beta} \widehat{\widetilde{u}}(\kappa, s) - s^{\beta-1} = -|\kappa|^{\alpha} \widehat{\widetilde{u}}(\kappa, s)$$

10.5 Fractional Diffusion and Subordination Processes

hence

$$\widehat{\widetilde{u}}(\kappa, s) = \frac{s^{\beta-1}}{|\kappa|^\alpha + s^\beta}, \qquad (10.5.14)$$

and looking back we see: $\widehat{\widetilde{u}}(\kappa, s) = \widehat{\widetilde{p}}_{0,0}(\kappa, s)$. Thus, under the scaling relation (10.5.7), the Fourier–Laplace solution of the CTRW integral equation converges to the Fourier–Laplace solution of the space-time fractional Cauchy problem (10.5.8)–(10.5.9), and we conclude that the sojourn probability of the CTRW converges weakly (or "in law") to the solution of the Cauchy problem for the space-time fractional diffusion equation for every fixed $t > 0$.

It is possible to present another way of passing to the diffusion limit, a way in which by decoupling the transitions in time and in space we circumvent doubts on the correctness of the transition.

For a comprehensive study of integral representations of the solution to the Cauchy problem (10.5.8)–(10.5.9) we recommend the paper by Mainardi, Luchko and Pagnini [MaLuPa01].

The Fundamental Solution to the Space-Time Fractional Diffusion Equation

Let us note that the solution $u(x, t)$ of the Cauchy problem (10.5.8)–(10.5.9), known as the *Green function* or fundamental solution of the space-time fractional diffusion equation, is a probability density in the spatial variable x, evolving in time t. In the case $\alpha = 2$ and $\beta = 1$ we recover the standard diffusion equation for which the fundamental solution is the Gaussian density with variance $\sigma^2 = 2t$.

For completeness, let us consider the more general *space-time fractional diffusion Cauchy problem* (with skewness)

$$^C D_{0+,t}^\beta u(x,t) = D_{0+,x}^\alpha u(x,t), \quad u(x,0) = \delta(x), \qquad (10.5.15)$$

$$0 < \alpha \leq 2, \ \theta \ \text{real}, \ |\theta| \leq \min(\alpha, 2 - \alpha), \ 0 < \beta \leq 1, \qquad (10.5.16)$$

as comprehensively treated by Mainardi, Luchko and Pagnini [MaLuPa01]. Sometimes, to point out the parameters, we may denote the fundamental solution by

$$u(x,t) = G_{\alpha,\beta}^\theta(x,t). \qquad (10.5.17)$$

For our purposes let us confine ourselves here to recalling the representation in the Laplace–Fourier domain of the (fundamental) solution as it results from an application of the Laplace and Fourier transforms to Eq. (10.5.8). Using $\widehat{\delta}(\kappa) \equiv 1$ we have:

$$s^\beta \widehat{\widetilde{u}}(\kappa, s) - s^{\beta-1} = -|\kappa|^\alpha\, i^{\theta \operatorname{sign} \kappa} \widehat{\widetilde{u}}(\kappa, s),$$

hence

$$\widehat{\widetilde{u}}(\kappa, s) = \widehat{\widetilde{G}}_{\alpha,\beta}^\theta(\kappa, s) = \frac{s^{\beta-1}}{s^\beta + |\kappa|^\alpha\, i^{\theta \operatorname{sign} \kappa}}. \qquad (10.5.18)$$

For explicit expressions and plots of the fundamental solution of (10.5.8)–(10.5.9) in the space-time domain we refer the reader to [MaLuPa01]. There, starting from the fact that the Fourier transform of the fundamental solution can be written as a Mittag-Leffler function with complex argument,

$$\widehat{u}(\kappa, t) = \widehat{G^\theta_{\alpha,\beta}}(\kappa, t) = E_\beta \left(-|\kappa|^\alpha i^{\theta \operatorname{sign} \kappa} t^\beta \right), \tag{10.5.19}$$

a Mellin–Barnes integral representation of $u(x, t) = G^\theta_{\alpha,\beta}(x, t)$ is derived, and is used to prove the non-negativity of the solution for values of the parameters $\{\alpha, \theta, \beta\}$ in the range (10.5.9) and analyze the evolution in time of its moments. The representation of $u(x, t)$ in terms of Fox H-functions can be found in Mainardi, Pagnini and Saxena [MaPaSa05], see also Chap. 6 in the recent book [MaSaHa10] by Mathai, Saxena and Haubold.

10.5.3 Subordination in the Space-Time Fractional Diffusion Equation

Now we introduce the analytical and stochastic approaches to subordination in space-time fractional diffusion processes is the fundamental solution of the space-time fractional diffusion equation in the Laplace–Fourier domain given by (10.5.18).

With an integration variable t_* that will play the role of *operational time*, we get the following instructive expression for (10.5.18):

$$\widehat{\widetilde{u}}(\kappa, s) = \int_0^\infty \left[\exp\left(-t_* |\kappa|^\alpha i^{\theta \operatorname{sign} \kappa}\right) \right] \left[s^{\beta-1} \exp\left(-t_* s^\beta\right) \right] dt_*. \tag{10.5.20}$$

We note that the first factor in (10.5.20)

$$\widehat{f}_{\alpha,\theta}(\kappa, t_*) := \exp\left(-t_* |\kappa|^\alpha i^{\theta \operatorname{sign} \kappa}\right) \tag{10.5.21}$$

is the Fourier transform of a skewed stable density in x, evolving in operational time t_*, of a process $x = y(t_*)$ along the real axis x happening in operational time t_*, which we write as

$$f_{\alpha,\theta}(x, t_*) = t_*^{-1/\alpha} L^\theta_\alpha \left(x/t_*^{1/\alpha}\right). \tag{10.5.22}$$

We can interpret the second factor in (10.5.20)

$$\widetilde{q}_\beta(t_*, s) := s^{\beta-1} \exp(-t_* s^\beta) \tag{10.5.23}$$

as the Laplace representation of the probability density in t_* evolving in t of a process $t_* = t_*(t)$, generating the operational time t_* from the physical time t, that is expressed via a fractional integral of a skewed Lévy density as

10.5 Fractional Diffusion and Subordination Processes

$$q_\beta(t_*, t) = t_*^{-1/\beta} {}_t J^{1-\beta} L_\beta^{-\beta}(t/t_*^{1/\beta}) = t^{-\beta} M_\beta(t_*/t^\beta). \tag{10.5.24}$$

We apply to (10.5.23) a second Laplace transformation with respect to t_* with parameter s_* to get

$$\widetilde{\widetilde{q}}_\beta(s_*, s) = \frac{s^{\beta-1}}{s_* + s^\beta}, \tag{10.5.25}$$

so, by inversion with respect to t

$$\widetilde{q}_\beta(s_*, t) = \int_{t_*=0}^\infty e^{-s_* t} q_\beta(t_*, t)\, dt_* = E_\beta(-s_* t^\beta), \tag{10.5.26}$$

and setting $s_* = 0$ we see that the density $q_\beta(t_*, t)$ is normalized:

$$\int_{t_*=0}^\infty q_\beta(t_*, t)\, dt_* = E_\beta(0) = 1. \tag{10.5.27}$$

Weighting the density of $x = y(t_*)$ with the density of $t_* = t_*(t)$ over $0 \le t < \infty$ yields the density $u(x, t)$ in x evolving with time t.

In physical variables $\{x, t\}$, using Eqs. (10.5.20)–(10.5.24), we have the *subordination integral formula*

$$u(x, t) = \int_{t_*=0}^\infty f_{\alpha,\theta}(x, t_*)\, q_\beta(t_*, t)\, dt_*, \tag{10.5.28}$$

where $f_{\alpha,\theta}(x, t_*)$ (density in x evolving in t_*) refers to the process $x = y(t_*)\,(t_* \to x)$ generating in "operational time" t_* the spatial position x, and $q_\beta(t_*, t)$ (density in t_* evolving in t) refers to the process $t_* = t_*(t)\,(t \to t_*)$ generating from physical time t the "operational time" t_*.

Clearly $f_{\alpha,\theta}(x, t_*)$ characterizes a stochastic process describing a trajectory $x = y(t_*)$ in the (t_*, x) plane, that can be visualized as a particle travelling along space x, as operational time t_* is proceeding. Is there also a process $t_* = t_*(t)$, a particle moving along the positive t_* axis, happening in physical time t? Naturally we want $t_*(t)$ to be increasing, at least in the weak sense,

$$t_2 > t_1 \implies t_*(t_2) \ge t_*(t_1).$$

We answer this question in the affirmative by inverting the stable process $t = t(t_*)$ whose probability density (in t, evolving in operational time t_*) is the extremely positively skewed stable density

$$r_\beta(t, t_*) = t_*^{-1/\beta} L_\beta^{-\beta}(t/t_*^{1/\beta}). \tag{10.5.29}$$

In fact, recalling

$$\widetilde{r}_\beta(s, t_*) = \exp(-t_* s^\beta), \tag{10.5.30}$$

there exists the stable process $t = t(t_*)$, weakly increasing, with density in t evolving in t_* given by (10.5.29). We call this process *the leading process*.

Happily, we can invert this process. Inversion of a weakly increasing trajectory means that, in a graphical visualization, horizontal segments are converted to vertical segments and conversely jumps (as vertical segments) are converted to horizontal segments.

Consider a fixed sample trajectory $t = t(t_*)$ and its fixed inversion $t_* = t_*(t)$. Fix an instant T of physical time and an instant T_* of operational time. Then, because $t = t(t_*)$ is increasing, we have the equivalence

$$t_*(T) \leq T_* \iff T \leq t(T_*),$$

which, with the notation slightly changed by

$$t_*(T) \to t'_*, \ T_* \to t_*, \ T \to t, \ t(T_*) \to t',$$

implies

$$\int_0^{t_*} q(t'_*, t) \, dt'_* = \int_t^\infty r_\beta(t', t_*) \, dt', \tag{10.5.31}$$

for the probability density $q(t_*, t)$ in t_* evolving in t. It follows that

$$q(t_*, t) = \frac{\partial}{\partial t_*} \int_t^\infty r_\beta(t', t_*) \, dt' = \int_t^\infty \frac{\partial}{\partial t_*} r_\beta(t', t_*) \, dt'.$$

We continue in the s_*-Laplace domain assuming $t > 0$,

$$\widetilde{q}(s_*, t) = \int_t^\infty \left(s_* \widetilde{r}_\beta(t', s_*) - \delta(t') \right) dt'.$$

It suffices to consider $t > 0$, so that we have $\delta(t') = 0$ in this integral. Observing from (10.5.30)

$$\widetilde{\widetilde{r}}_\beta(s, s_*) = \frac{1}{s_* + s^\beta}, \tag{10.5.32}$$

we find

$$\widetilde{r}_\beta(t, s_*) = \beta t^{\beta-1} E'_\beta(-s_* t^\beta), \tag{10.5.33}$$

so that

$$\widetilde{q}(s_*, t) = \int_t^\infty s_* \beta t'^{\beta-1} E'_\beta(-s_* t'^\beta) \, dt' = E_\beta(-s_* t^\beta), \tag{10.5.34}$$

and finally

$$q(t_*, t) = t^{-\beta} M_\beta(t_*/t^\beta). \tag{10.5.35}$$

10.5 Fractional Diffusion and Subordination Processes

From (10.5.34) we also see that

$$\widetilde{\widetilde{q}}(s_*, s) = \frac{s^{\beta-1}}{s_* + s^\beta} = \widetilde{\widetilde{q}}_\beta(s_*, s), \qquad (10.5.36)$$

implying (10.5.23), see (10.5.25),

$$q(t_*, t) \equiv q_\beta(t_*, t), \qquad (10.5.37)$$

so that the process $t_* = t_*(t)$ is indeed the inverse to the stable process $t = t(t_*)$ and has density $q_\beta(t_*, t)$.

Remark 10.3 For more details on the Gorenflo–Mainardi view of subordination, we draw the reader's attention to [GorMai11, GorMai15], and in particular to the two complementary papers [GorMai12a, GorMai12b]. The highlight of [GorMai12a] is an outline of the way how, by appropriate analytical manipulations from the subordination integral (10.5.28), a CTRW can be derived to produce snapshots of a particle trajectory. In the other paper [GorMai12b], the authors show how, from a generic power law CTRW, by a properly scaled diffusion limit, the subordination integral can be found.

10.5.4 *The Rescaling and Respeeding Concept Revisited. Universality of the Mittag-Leffler Density*

Now we use again the concept of rescaling and respeeding in order to obtain one more property of the Mittag-Leffler waiting time density. First, we generalize the Kolmogorov–Feller equation (10.3.22) by introducing in the Laplace domain the auxiliary function

$$\widetilde{H}(s) = \frac{1 - \widetilde{\phi}(s)}{s\,\widetilde{\phi}(s)} = \frac{\widetilde{\Psi}(s)}{\widetilde{\phi}(s)}, \qquad (10.5.38)$$

which is equivalent to

$$\widetilde{H}(s)\left[s\widehat{\widetilde{p}}(\kappa, s) - 1\right] = [\widehat{w}(\kappa) - 1]\,\widehat{\widetilde{p}}(\kappa, s). \qquad (10.5.39)$$

Then in the space-time domain we get the following *generalized Kolmogorov–Feller equation*

$$\int_0^t H(t - t')\frac{\partial}{\partial t'}p(x, t')\,dt' = -p(x, t) + \int_{-\infty}^{+\infty} w(x - x')\,p(x', t)\,dx', \qquad (10.5.40)$$

with $p(x, 0) = \delta(x)$.

Rescaling time means: with a positive scaling factor τ (intended to be small) we replace the waiting time T by τT. This amounts to replacing the unit 1 of time by $1/\tau$, and if $\tau \ll 1$ then in the rescaled process there will occur very many jumps in a moderate span of time (instead of the original moderate number in a moderate span of time). The rescaled waiting time density and its corresponding Laplace transform are $\phi_\tau(t) = \phi(t/\tau)/\tau$, $\widetilde{\phi}_\tau(s) = \widetilde{\phi}(\tau s)$. Furthermore:

$$\widetilde{H}_\tau(s) = \frac{1 - \widetilde{\phi}_\tau(s)}{s\,\widetilde{\phi}_\tau(s)} = \frac{1 - \widetilde{\phi}(\tau s)}{s\,\widetilde{\phi}(\tau s)}, \quad \text{hence } \widetilde{\phi}_\tau(s) = \frac{1}{1 + s\,\widetilde{H}_\tau(s)}, \qquad (10.5.41)$$

and (10.5.39) goes over into

$$\widetilde{H}_\tau(s)\left[s\widehat{\widetilde{p}}_\tau(\kappa,s) - 1\right] = [\widehat{w}(\kappa) - 1]\,\widehat{\widetilde{p}}_\tau(\kappa,s). \qquad (10.5.42)$$

Respeeding the process means multiplying the left-hand side (actually $\dfrac{\partial}{\partial t'}p(x,t')$) of Eq. (10.5.40) by a positive factor $1/a$, or equivalently its right-hand side by a positive factor a. We honor the number a by the name *respeeding factor*. $a > 1$ means *acceleration* and $a < 1$ *deceleration*. In the Fourier–Laplace domain the rescaled and respeeded CTRW process then assumes the form, analogous to (10.5.39) and (10.5.42),

$$\widetilde{H}_{\tau,a}(s)\left[s\widehat{\widetilde{p}}_{\tau,a}(\kappa,s) - 1\right] = a\,[\widehat{w}(\kappa) - 1]\,\widehat{\widetilde{p}}_{\tau,a}(\kappa,s), \qquad (10.5.43)$$

with

$$\widetilde{H}_{\tau,a}(s) = \frac{\widetilde{H}_\tau(s)}{a} = \frac{1 - \widetilde{\phi}(\tau s)}{a\,s\,\widetilde{\phi}(\tau s)}.$$

What is the effect of such *combined rescaling and respeeding*? We find

$$\widetilde{\phi}_{\tau,a}(s) = \frac{1}{1 + s\,\widetilde{H}_{\tau,a}(s)} = \frac{a\,\widetilde{\phi}(\tau s)}{1 - (1-a)\,\widetilde{\phi}(\tau s)}, \qquad (10.5.44)$$

and are now in a position to address the *asymptotic universality of the Mittag-Leffler waiting time density*.

Using Lemma 10.2 with τs in place of s and taking

$$a = \lambda \tau^\beta, \qquad (10.5.45)$$

fixing s as required by the continuity theorem of probability for Laplace transforms, the asymptotics $\widetilde{\phi}(\tau s) = 1 - \lambda((\tau s)^\beta) + o((\tau s)^\beta)$ for $\tau \to 0$ implies

10.5 Fractional Diffusion and Subordination Processes

$$\widetilde{\phi}_{\tau,\lambda\tau^\beta}(s) = \frac{\lambda\tau^\beta\left[1 - \lambda((\tau s)^\beta) + o((\tau s)^\beta)\right]}{1 - (1 - \lambda\tau^\beta)\left[1 - \lambda((\tau s)^\beta) + o((\tau s)^\beta)\right]} \to \frac{1}{1 + s^\beta} = \widetilde{\phi}_\beta^{ML},$$
(10.5.46)

corresponding to the Mittag-Leffler density

$$\phi_\beta^{ML}(t) = -\frac{d}{dt} E_\beta(-t^\beta).$$

Observe that the parameter λ does not appear in the limit $1/(1 + s^\beta)$. We can make it reappear by choosing the respeeding factor τ^β in place of $\lambda\tau^\beta$. In fact:

$$\widetilde{\phi}_{\tau,\tau^\beta} \to \frac{1}{1 + \lambda s^\beta}.$$

Formula (10.5.46) says that the general density $\phi(t)$ with power law asymptotics as in Lemma 10.2 is gradually deformed into the Mittag-Leffler waiting time density $\phi_\beta^{ML}(t)$. This means that with larger and larger unit of time (by sending $\tau \to 0$) and stronger and stronger deceleration (by $a = \lambda\tau^\beta$) as described our process becomes indistinguishable from one with Mittag-Leffler waiting time (the probability distribution of jumps remaining unchanged). Likewise a pure renewal process with asymptotic power law density becomes indistinguishable from the one with Mittag-Leffler waiting time (the fractional generalization of the Poisson process due to Laskin [Las03] and Mainardi, Gorenflo and Scalas [MaGoSc04a]).

10.6 The Wright M-Functions in Probability

The classical Wright functions of *the second kind* play a role in probability and stochastic processes, so in this section we present some properties of these functions, which we refer to as *auxiliary functions of the Wright type* (introduced by Mainardi), see Chap. 7, Sects. 7.5.3 and 7.6.3. We have already recognized that the Wright M-function with support in $\mathbb{R}^+$ can be interpreted as a probability density function (*pdf*). Consequently, extending the function in a symmetric way to all of $\mathbb{R}$ and dividing by two we have a *symmetric pdf* with support in $\mathbb{R}$. In the former case the variable is usually a time coordinate whereas in the latter the variable is the absolute value of a space coordinate. We now provide more details on these densities in the framework of the theory of probability. As before, we agree to denote by x and $|x|$ the variables in $\mathbb{R}$ and $\mathbb{R}^+$, respectively (see on Fig. 10.1 plots of M-function for certain values of parameters).

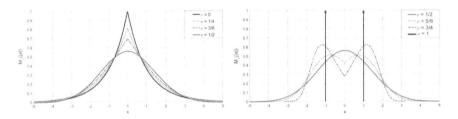

Fig. 10.1 Plots of the functions $M_\nu(|x|)$ for $|x| \leq 5$ at $t = 1$; Left: for $\nu = 0, 1/4, 3/8, 1/2$. Right: for $\nu = 1/2, 5/8, 3/4, 1$

10.6.1 The Absolute Moments of Order δ

The *absolute moments* of order $\delta > -1$ of the Wright M-function in $\mathbb{R}^+$ are finite and turn out to be

$$\int_0^\infty x^\delta M_\nu(x)\,dx = \frac{\Gamma(\delta+1)}{\Gamma(\nu\delta+1)}, \quad \delta > -1, \quad 0 \leq \nu < 1. \tag{10.6.1}$$

In order to derive this fundamental result we proceed as follows, based on the integral representation (F.1.15).

$$\int_0^\infty x^\delta M_\nu(x)\,dx = \int_0^\infty x^\delta \left[\frac{1}{2\pi i}\int_{Ha} e^{\sigma - x\sigma^\nu} \frac{d\sigma}{\sigma^{1-\nu}}\right] dx$$

$$= \frac{1}{2\pi i}\int_{Ha} e^\sigma \left[\int_0^\infty e^{-x\sigma^\nu} x^\delta\,dx\right] \frac{d\sigma}{\sigma^{1-\nu}}$$

$$= \frac{\Gamma(\delta+1)}{2\pi i}\int_{Ha} \frac{e^\sigma}{\sigma^{\nu\delta+1}}\,d\sigma = \frac{\Gamma(\delta+1)}{\Gamma(\nu\delta+1)}.$$

Above we have legitimately changed the order of two integrals and we have used the identity

$$\int_0^\infty e^{-x\sigma^\nu} x^\delta\,dx = \frac{\Gamma(\delta+1)}{(\sigma^\nu)^{\delta+1}},$$

derived from (A.23) along with the Hankel formula (A.19a).

In particular, for $\delta = n \in \mathbb{N}$, the above formula provides the moments of integer order that can also be computed from the Laplace transform pair (7.6.13) as follows:

$$\int_0^{+\infty} x^n M_\nu(x)\,dx = \lim_{s \to 0} (-1)^n \frac{d^n}{ds^n} E_\nu(-s) = \frac{\Gamma(n+1)}{\Gamma(\nu n+1)}.$$

10.6 The Wright M-Functions in Probability

Incidentally, we note that the Laplace transform pair (7.6.13) could be obtained using the fundamental result (10.6.1) by developing in power series the exponential kernel of the Laplace transform and then transforming the series term-by-term.

10.6.2 The Characteristic Function

As is well known in probability theory, the Fourier transform of a density provides the so-called *characteristic function*. In our case we have:

$$\mathcal{F}\left[\frac{1}{2}M_\nu(|x|)\right] := \frac{1}{2}\int_{-\infty}^{+\infty} e^{i\kappa x} M_\nu(|x|)\, dx \qquad (10.6.2)$$
$$= \int_0^\infty \cos(\kappa x)\, M_\nu(x)\, dx = E_{2\nu}(-\kappa^2).$$

To prove this, it is sufficient to develop in series the cosine function and use formula (10.6.1),

$$\int_0^\infty \cos(\kappa x)\, M_\nu(x)\, dx = \sum_{n=0}^\infty (-1)^n \frac{\kappa^{2n}}{(2n)!} \int_0^\infty x^{2n} M_\nu(x)\, dx$$
$$= \sum_{n=0}^\infty (-1)^n \frac{\kappa^{2n}}{\Gamma(2\nu n + 1)} = E_{2\nu}(-\kappa^2).$$

10.6.3 Relations with Lévy Stable Distributions

We find it worthwhile to discuss the relations between the Wright M-functions and the so-called *Lévy stable distributions*. The term stable has been assigned by the French mathematician Paul Lévy, who, in the 1920s, started a systematic study in order to generalize the celebrated *Central Limit Theorem* to probability distributions with infinite variance. For stable distributions we can assume the following

Definition 10.4 If two independent real random variables with the same shape or type of distribution are combined linearly and the distribution of the resulting random variable has the same shape, the common distribution (or, more precisely, its type) is said to be stable.

The restrictive condition of stability enabled Lévy (and then other authors) to derive the *canonical form* for the characteristic function of the densities of these distributions. Here we follow the parametrization in [Fel52, Fel71] revisited in [GorMai98] and in [MaLuPa01]. Denoting by $L_\lambda^\theta(x)$ a generic stable density in $\mathbb{R}$, where λ is the *index of stability* and θ the asymmetry parameter, improperly called *skewness*, its characteristic function reads:

Fig. 10.2 The Feller–Takayasu diamond for Lévy stable densities

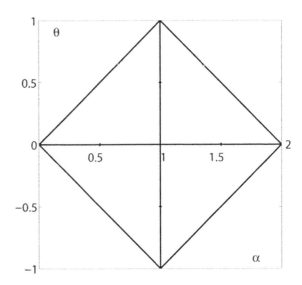

$$L_\lambda^\theta(x) \div \widehat{L}_\lambda^\theta(\kappa) = \exp\left[-\psi_\lambda^\theta(\kappa)\right], \quad \psi_\lambda^\theta(\kappa) = |\kappa|^\lambda \, e^{i(\mathrm{sign}\,\kappa)\theta\pi/2}, \qquad (10.6.3)$$

$$0 < \lambda \le 2, \ |\theta| \le \min\{\lambda, 2-\lambda\}.$$

For $\nu = 1/2$, we get back the diffusion equation and the $M_\nu(x)$ function becomes the Gaussian function (known as the fundamental solution of the diffusion equation for the Cauchy Problem) whereas, for $\nu \to 1$, we get the wave equation and the $M_\nu(x)$ function tends to two Dirac delta functions as fundamental solutions of the Cauchy Problem, centered at $x = \pm 1$.

We note that the allowed region for the parameters λ and θ turns out to be a diamond in the plane $\{\lambda, \theta\}$ with vertices at the points $(0, 0)$, $(1, 1)$, $(1, -1)$, $(2, 0)$, that we call the *Feller–Takayasu diamond*, see Fig. 10.2. For values of θ on the border of the diamond (that is $\theta = \pm\lambda$ if $0 < \lambda < 1$, and $\theta = \pm(2-\lambda)$ if $1 < \lambda < 2$) we obtain the so-called *extremal stable densities*.

We note the *symmetry relation* $L_\lambda^\theta(-x) = L_\lambda^{-\theta}(x)$, so that a stable density with $\theta = 0$ is symmetric.

Stable distributions have noteworthy properties which the interested reader can find in the relevant existing literature. Below we list some of the more peculiar properties:

- *The class of stable distributions possesses its own domain of attraction*, see e.g. [Fel71].
- *Any stable density is unimodal and indeed bell-shaped*, i.e. its n-th derivative has exactly n zeros in $\mathbb{R}$, see [Gaw84].
- *The stable distributions are self-similar and infinitely divisible*.

10.6 The Wright M-Functions in Probability

These properties derive from the canonical form (10.6.3) through the scaling property of the Fourier transform.

Self-similarity means

$$L_\lambda^\theta(x,t) \div \exp\left[-t\psi_\lambda^\theta(\kappa)\right] \iff L_\lambda^\theta(x,t) = t^{-1/\lambda}\left[L_\lambda^\theta(x/t^{1/\lambda})\right], \quad (10.6.4)$$

where t is a positive parameter. If t is time, then $L_\lambda^\theta(x,t)$ is a spatial density evolving in time with self-similarity.

Infinite divisibility means that for every positive integer n, the characteristic function can be expressed as the nth power of some characteristic function, so that any stable distribution can be expressed as the n-fold convolution of a stable distribution of the same type. Indeed, taking in (10.6.3) $\theta = 0$, without loss of generality, we have

$$e^{-t|\kappa|^\lambda} = \left[e^{-(t/n)|\kappa|^\lambda}\right]^n \iff L_\lambda^0(x,t) = \left[L_\lambda^0(x,t/n)\right]^{*n}, \quad (10.6.5)$$

where

$$\left[L_\lambda^0(x,t/n)\right]^{*n} := L_\lambda^0(x,t/n) * L_\lambda^0(x,t/n) * \cdots * L_\lambda^0(x,t/n)$$

is the multiple Fourier convolution in $\mathbb{R}$ with n identical terms.

The inversion of the Fourier transform in (10.6.3) can be carried out only for a few particular cases, using standard tables, and well-known probability distributions are obtained.

For $\lambda = 2$ (so $\theta = 0$), we recover the *Gaussian pdf*, which turns out to be the only stable density with finite variance, and more generally with finite moments of any order $\delta \geq 0$. In fact

$$L_2^0(x) = \frac{1}{2\sqrt{\pi}} e^{-x^2/4}. \quad (10.6.6)$$

All the other stable densities have finite absolute moments of order $\delta \in [-1, \lambda)$, as we will later show.

For $\lambda = 1$ and $|\theta| < 1$, we get

$$L_1^\theta(x) = \frac{1}{\pi} \frac{\cos(\theta\pi/2)}{[x + \sin(\theta\pi/2)]^2 + [\cos(\theta\pi/2)]^2}, \quad (10.6.7)$$

which for $\theta = 0$ includes the *Cauchy–Lorentz pdf*,

$$L_1^0(x) = \frac{1}{\pi} \frac{1}{1+x^2}. \quad (10.6.8)$$

In the limiting cases $\theta = \pm 1$ for $\lambda = 1$ we obtain the *singular Dirac pdfs*

$$L_1^{\pm 1}(x) = \delta(x \pm 1). \quad (10.6.9)$$

In general, we must recall the power series expansions provided in [Fel71]. We restrict our attention to $x > 0$ since the evaluations for $x < 0$ can be obtained using the symmetry relation. The convergent expansions of $L_\lambda^\theta(x)$ ($x > 0$) turn out to be:

for $0 < \lambda < 1$, $|\theta| \leq \lambda$:

$$L_\lambda^\theta(x) = \frac{1}{\pi x} \sum_{n=1}^{\infty} (-x^{-\lambda})^n \frac{\Gamma(1+n\lambda)}{n!} \sin\left[\frac{n\pi}{2}(\theta - \lambda)\right]; \quad (10.6.10)$$

for $1 < \lambda \leq 2$, $|\theta| \leq 2 - \lambda$:

$$L_\lambda^\theta(x) = \frac{1}{\pi x} \sum_{n=1}^{\infty} (-x)^n \frac{\Gamma(1+n/\lambda)}{n!} \sin\left[\frac{n\pi}{2\alpha}(\theta - \alpha)\right]. \quad (10.6.11)$$

From the series in (10.6.10) and the symmetry relation we note that *the extremal stable densities for $0 < \alpha < 1$ are unilateral*, more precisely they vanish for $x > 0$ if $\theta = \alpha$, and for $x < 0$ if $\theta = -\alpha$. In particular, the unilateral extremal densities $L_\alpha^{-\alpha}(x)$ with $0 < \alpha < 1$ have support in $\mathbb{R}^+$ and Laplace transform $\exp(-s^\alpha)$. For $\alpha = 1/2$ we obtain the so-called *Lévy–Smirnov pdf*:

$$L_{1/2}^{-1/2}(x) = \frac{x^{-3/2}}{2\sqrt{\pi}} e^{-1/(4x)}, \quad x \geq 0. \quad (10.6.12)$$

As a consequence of the convergence of the series in (10.6.10)–(10.6.11) and of the symmetry relation we recognize that the stable *pdfs* with $1 < \alpha \leq 2$ are entire functions, whereas with $0 < \alpha < 1$ have the form

$$L_\alpha^\theta(x) = \begin{cases} (1/x)\,\Phi_1(x^{-\alpha}) & \text{for } x > 0, \\ (1/|x|)\,\Phi_2(|x|^{-\alpha}) & \text{for } x < 0, \end{cases} \quad (10.6.13)$$

where $\Phi_1(z)$ and $\Phi_2(z)$ are distinct entire functions. The case $\alpha = 1$ ($|\theta| < 1$) must be considered in the limit for $\alpha \to 1$ of (10.6.10)–(10.6.11), because the corresponding series reduce to power series akin with geometric series in $1/x$ and x, respectively, with a finite radius of convergence. The corresponding stable *pdfs* are no longer represented by entire functions, as can be noted directly from their explicit expressions (10.6.7)–(10.6.8).

We do not provide the asymptotic representations of the stable densities, instead referring the interested reader to [MaLuPa01]. However, based on asymptotic representations, we can state the following: for $0 < \alpha < 2$ the stable *pdfs* exhibit *fat tails* in such a way that their absolute moment of order δ is finite only if $-1 < \delta < \alpha$. More precisely, one can show that for non-Gaussian, non-extremal, stable densities the asymptotic decay of the tails is

$$L_\alpha^\theta(x) = O\left(|x|^{-(\alpha+1)}\right), \quad x \to \pm\infty. \quad (10.6.14)$$

10.6 The Wright M-Functions in Probability

For the extremal densities with $\alpha \neq 1$ this is valid only for one tail (as $|x| \to \infty$), the other (as $|x| \to \infty$) being of exponential order. For $1 < \alpha < 2$ the extremal *pdfs* are two-sided and exhibit an exponential left tail (as $x \to -\infty$) if $\theta = +(2-\alpha)$, or an exponential right tail (as $x \to +\infty$) if $\theta = -(2-\alpha)$. Consequently, the Gaussian *pdf* is the unique stable density with finite variance. Furthermore, when $0 < \alpha \leq 1$, the first absolute moment is infinite so we should use the median instead of the non-existent expected value in order to characterize the corresponding *pdf*.

Let us also recall a relevant identity between stable densities with index α and $1/\alpha$ (a sort of reciprocity relation) pointed out in [Fel71], that is, assuming $x > 0$,

$$\frac{1}{x^{\alpha+1}} L^{\theta}_{1/\alpha}(x^{-\alpha}) = L^{\theta^*}_{\alpha}(x), \quad 1/2 \leq \alpha \leq 1, \quad \theta^* = \alpha(\theta+1) - 1. \quad (10.6.15)$$

The condition $1/2 \leq \alpha \leq 1$ implies $1 \leq 1/\alpha \leq 2$. A check shows that θ^* falls within the prescribed range $|\theta^*| \leq \alpha$ if $|\theta| \leq 2 - 1/\alpha$. We leave as an exercise for the interested reader the verification of this reciprocity relation in the limiting cases $\alpha = 1/2$ and $\alpha = 1$.

From a comparison between the series expansions in (10.6.10)–(10.6.11) and in (7.5.6)–(7.5.7), we recognize that for $x > 0$ our *auxiliary functions of the Wright type are related to the extremal stable densities as follows*, see [MaiTom97],

$$L^{-\alpha}_{\alpha}(x) = \frac{1}{x} F_{\alpha}(x^{-\alpha}) = \frac{\alpha}{x^{\alpha+1}} M_{\alpha}(x^{-\alpha}), \quad 0 < \alpha < 1, \quad (10.6.16)$$

$$L^{\alpha-2}_{\alpha}(x) = \frac{1}{x} F_{1/\alpha}(x) = \frac{1}{\alpha} M_{1/\alpha}(x), \quad 1 < \alpha \leq 2. \quad (10.6.17)$$

In Eqs. (10.6.16)–(10.6.17), for $\alpha = 1$, the skewness parameter turns out to be $\theta = -1$, so we get the singular limit

$$L^{-1}_{1}(x) = M_1(x) = \delta(x-1). \quad (10.6.18)$$

More generally, all (regular) stable densities, given in Eqs. (10.6.10)–(10.6.11), were recognized to belong to the class of Fox H-functions, as formerly shown by [Sch86], see also [GorMai03]. This general class of high transcendental functions is outside the scope of this book.

10.6.4 The Wright $\mathbb{M}$-Function in Two Variables

In view of time-fractional diffusion processes related to time-fractional diffusion equations it is worthwhile to introduce the function in two variables

$$\mathbb{M}_{\nu}(x,t) := t^{-\nu} M_{\nu}(xt^{-\nu}), \quad 0 < \nu < 1, \quad x, t \in \mathbb{R}^+, \quad (10.6.19)$$

which defines a spatial probability density in x evolving in time t with self-similarity exponent $H = \nu$. Of course for $x \in \mathbb{R}$ we have to consider the symmetric version obtained from (10.6.19) by multiplying by $1/2$ and replacing x by $|x|$.

Hereafter we provide a list of the main properties of this function, which can be derived from the Laplace and Fourier transforms for the corresponding Wright M-function in one variable.

From Eq. (7.6.17) we derive the Laplace transform of $\mathbb{M}_\nu(x,t)$ with respect to $t \in \mathbb{R}^+$,

$$\mathcal{L}\{\mathbb{M}_\nu(x,t); t \to s\} = s^{\nu-1}\, \mathrm{e}^{-xs^\nu} . \tag{10.6.20}$$

From Eq. (7.6.13) we derive the Laplace transform of $\mathbb{M}_\nu(x,t)$ with respect to $x \in \mathbb{R}^+$,

$$\mathcal{L}\{\mathbb{M}_\nu(x,t); x \to s\} = E_\nu(-st^\nu) . \tag{10.6.21}$$

From Eq. (10.6.2) we derive the Fourier transform of $\mathbb{M}_\nu(|x|, t)$ with respect to $x \in \mathbb{R}$,

$$\mathcal{F}\{\mathbb{M}_\nu(|x|, t); x \to \kappa\} = 2E_{2\nu}\left(-\kappa^2 t^\nu\right) . \tag{10.6.22}$$

Using the Mellin transform [MaPaGo03] derived the following integral formula,

$$\mathbb{M}_\nu(x,t) = \int_0^\infty \mathbb{M}_\lambda(x,\tau)\, \mathbb{M}_\mu(\tau, t)\, \mathrm{d}\tau, \quad \nu = \lambda\mu . \tag{10.6.23}$$

Special cases of the Wright $\mathbb{M}$-function are easily derived for $\nu = 1/2$ and $\nu = 1/3$ from the corresponding ones in the complex domain, see e.g. [Mai10, App. F, formulas (F.16)–(F.17)]. We devote particular attention to the case $\nu = 1/2$ for which we get from [Mai10, App. F, formula (F.16)] the Gaussian density in $\mathbb{R}$,

$$\mathbb{M}_{1/2}(|x|, t) = \frac{1}{2\sqrt{\pi}t^{1/2}}\, \mathrm{e}^{-x^2/(4t)} . \tag{10.6.24}$$

For the limiting case $\nu = 1$ we obtain

$$\mathcal{M}_1(|x|, t) = \frac{1}{2}\left[\delta(x-t) + \delta(x+t)\right] . \tag{10.6.25}$$

10.7 Historical and Bibliographical Notes

A fractional generalization of the Poisson probability distribution was presented by Pillai in 1990 in his pioneering work [Pil90]. He introduced the probability distribution (which he called the *Mittag-Leffler distribution*) using the complete monotonicity of the Mittag-Leffler function. As has already been mentioned, the complete monotonicity of this function was proved by Pollard in 1948, see [Poll48, Fel49], as well

10.7 Historical and Bibliographical Notes

as in more recent works by Schneider [Sch96], and Miller and Samko [MilSam97]. Nowadays the concept of complete monotonicity is widely investigated in the framework of the Bernstein functions (non-negative functions with a complete monotone first derivative), see the recent book by Schilling et al. [SchSoVo12].

The concept of a *geometrically infinitely divisible* distribution was introduced in 1984 in [KlMaMe84]. Later in 1995 Pillai introduced [PilJay95] (see also [JayPil93, JayPil96, Jay03]) a discrete analogue of such a distribution (the *discrete Mittag-Leffler distribution*). In [CahWoy18], formal estimation procedures for the parameters of the generalized, heavy-tailed three-parameter Linnik and Mittag-Leffler distributions are proposed. The estimators are derived from the moments of the log-transformed random variables and are shown to be asymptotically unbiased. These distributions are used for modeling processes in finance.

Another possible generalization of the Poisson distribution is that introduced by Lamperti in 1958 [Lam58] whose density has the expression

$$f_{X_\alpha}(y) = \frac{\sin \pi\alpha}{\pi} \frac{y^{\alpha-1}}{y^{2\alpha} + 2y^\alpha \cos \pi\alpha + 1},$$

see also [Jam10]. We recognize in it the spectral distribution of the Mittag-Leffler function $E_\alpha(-t^\alpha)$ formerly derived in 1947 by Gross for linear viscoelasticity [Gro47] and then used by Caputo and Mainardi [CapMai71a, CapMai71b]. Lamperti considered a random variable equal to the ratio of two variables $X_\alpha = S_\alpha/S_{\alpha,0}$, $0 < \alpha < 1$, where S_α is a positive stable random variable, with density f_α, and having Laplace transform

$$\mathbb{E}\left(e^{-sS_\alpha}\right) = e^{-s^\alpha},$$

and S_α is a variable independent of $S_{\alpha,\theta}$, $\theta > -\alpha$, whose laws follow a polynomially tilted stable distribution having density proportional to $t^{-\theta} f_\alpha(t)$.

The concept of a *renewal process* has been developed as a stochastic model for describing the class of counting processes for which the times between successive events are independent identically distributed (i.i.d.) non-negative random variables, obeying a given probability law. These times are referred to as waiting times or inter-arrival times. The process of accumulation of waiting times is inverse to the counting number process, and is called the Erlang process in honor of the Danish mathematician and telecommunication engineer A.K. Erlang (see [BrHaJe48]).

For more details see, for example, the classical treatises by Khintchine [Khi37, Khi60], Cox [Cox67], Gnedenko and Kovalenko [GneKov68], Feller [Fel71], and the book by Ross [Ros97], as well as the recent survey paper [MaiRog06] and the book [RogMai11] by Rogosin and Mainardi (see also [RogMai17]).

The Mittag-Leffler function also appears in the solution of the fractional master equation. This equation characterizes the renewal processes with reward modelled by the random walk model known as *continuous time random walks* (abbreviated CTRWs). In this the waiting time is assumed to be a *continuous* random variable. The name CTRW became popular in physics in the 1960s after Montroll, Weiss and Scher (just to cite the pioneers) published a celebrated series of papers on random walks to model diffusion processes on lattices, see e.g. [Wei94] and the references therein. The basic role of the Mittag-Leffler waiting time probability density in time fractional continuous time random walks has become well-known via the fundamental paper of Hilfer and Anton [HilAnt95] (see also [BalR07, Hil03]). Earlier in the theory of thinning (rarefaction) of a renewal process under power law assumptions (see Gnedenko and Kovalenko's book [GneKov68]), this density had been found as a limit density by a combination of thinning followed by rescaling the time and imposing a proper relation between the rescaling factor and the thinning parameter. In 1985 Balakrishnan [BalV85] defined a special class of anomalous random walks where the anomaly appears by growth of the second moment of the sojourn probability density like a power of time with exponent between 0 and 1. This paper appeared a few years before the fundamental paper by Schneider and Wyss [SchWys89], but did not attract much attention (probably because of its style of presentation). However, it should be mentioned that, by a well-scaled passage to the limit, from a CTRW the space-time fractional diffusion equation in the form of an equivalent integro-differential equation was obtained in [BalV85]. Remarkably, Gnedenko and Kovalenko [GneKov68] and Balakrishnan [BalV85] ended their analysis by giving the solution only as a Laplace transform without inverting it.

CTRWs are rather good and general phenomenological models for diffusion, including anomalous diffusion, provided that the resting time of the walker is much greater than the time it takes to make a jump. In fact, in the formalism, jumps are instantaneous. In more recent times, CTRWs have been applied to economics and finance by Hilfer [Hil84], by Gorenflo–Mainardi–Scalas and their co-workers [ScGoMa00, Mai-et-al00, Gor-et-al01, RaScMa02, Sca-et-al03], and, later, by Weiss and co-workers [MaMoWe03]. It should be noted, however, that the idea of combining a stochastic process for waiting times between two consecutive events and another stochastic process which associates a reward or a claim to each event dates back at least to the first half of the twentieth century with the so-called Cramér–Lundberg model for insurance risk, see for a review [EmKlMi01]. In a probabilistic framework, we now find it more appropriate to refer to all these processes as *compound renewal processes*.

Serious studies of the fractional generalization of the Poisson process—replacing the exponential waiting time distribution by a distribution given via a Mittag-Leffler function with modified argument—began around the turn of the millennium, and since then many papers on its various aspects have appeared. There are in the literature many papers on this generalization where the authors have outlined a number of aspects and definitions, see e.g. Repin and Saichev [RepSai00], Wang et al. [AldWai70, WanWen03, WaWeZh06], Laskin [Las03, Las09], Mainardi et al. [MaGoSc04a], Uchaikin et al. [UcCaSi08], Beghin and Orsingher [OrsBeg04, BegOrs09], Cahoy et al. [CaUcWo10], Meerschaert et al. [MeNaVe11, BaeMee01], Politi et al. [PoKaSc11], Kochubei [Koc11], so it would be impossible to list them all exhaustively. However, in effect this generalization had already been used in 1995: Hilfer and Anton [HilAnt95] showed (using different terminology) that the fractional Kolmogorov–Feller equation (replacing the first-order time derivative by a fractional derivative of order between 0 and 1) requires the underlying random walk

10.7 Historical and Bibliographical Notes

to be subordinated to a renewal process with Mittag-Leffler waiting time. Gorenflo and Mainardi [GorMai08, Gor10] mention the asymptotic universality of the Mittag-Leffler waiting time density for the family of power law renewal processes.

An alternative renewal process called the Wright process was investigated by Mainardi et al. [Mai-et-al00, MaGoVi05, MaGoVi07] as a process arising by discretization of the stable subordinator (see the survey of recent results in [Baz18]). This approach is based on the concept of the extremal Lévy stable density (Lévy stable processes are widely discussed in several books on probability theory, see, e.g., [Fel71, Sat99]). For the study of the Wright processes an essential role is played by the so-called M-Wright function (see, e.g., [Mai10]). A scaled version of this process has been used by Barkai [Bark02] to approximate the time-fractional diffusion process directly by a random walk subordinated to it (executing this scaled version in natural time), and he has found rather poor convergence in refinement. In Gorenflo et al. [GoMaVi07] the way of using this discretized stable subordinator has been modified. By appropriate discretization of the relevant spatial stable process we have then obtained a simulation method equivalent to the solution of a pair of Langevin equations, see Fogedby [Fog94] and Kleinhans and Friedrich [KleFri07]. For simulation of space-time fractional diffusion one then obtains a sequence of precise snapshots of a true particle trajectory, see for details Gorenflo et al. [GoMaVi07], and also Gorenflo and Mainardi [GorMai12a, GorMai12b]. Other interesting results in this area can be found in [ChGoSo02, MaPaGo03, MaiPir96].

There are several ways to generalize the classical diffusion equation by introducing space and/or time derivatives of fractional order. We mention here the seminal paper by Schneider and Wyss [SchWys89] for the time fractional diffusion equation and the influential paper by Saichev and Zaslavsky [SaiZas97] for diffusion in time as well as in space. In the recent literature several authors have stressed the viewpoint of subordination, and special attention is being paid to diffusion equations with distributed order of fractional temporal or/and spatial differentiation. The transition from CTRW to such generalized types of diffusion has been investigated by different methods, not only for its purely mathematical interest but also due to its applications in Physics, Chemistry, and other Applied Sciences, including Economics and Finance. As good reference texts on these topics, containing extended lists of relevant works, we refer to the review papers by Metzler and Klafter [MetKla00, MetKla04] (see also [EvaLen18, BouGeo90, MaMuPa10]).

In [Mag-et-al19] a review of the mechanical models for ultraslow diffusion is presented from both a macroscopic and a microscopic perspective, which have been developed in recent decades to depict time evolution of ultraslow diffusion in heterogeneous media.

Lastly, we mention the connection of the above processes to classes of Lévy processes [Bert96] and stable processes, see, e.g. [Khi38, KhiLev36, Levy37] and more recent results in [Fuj90b, GorMai98a, Non90b, SteVH04].

Concerning the Mainardi auxiliary function of the Wright type it should be noted that in the book by Prüss [Prus93] we find a figure quite similar to our figure showing the M-Wright function in the linear scale, namely the Wright function of the second kind *in the transition from diffusion to wave*. It was derived by Prüss by inverting the

Fourier transform expressed in terms of the Mittag-Leffler function, following the approach by Fujita [Fuj90a] for the fundamental solution of the Cauchy problem for the diffusion-wave equation, fractional in time. However, our plot must be considered independent of that of Prüss because Mainardi used the Laplace transform in his former paper presented at the WASCOM conference in Bologna, October 1993 [Mai94a] (and published later in a number of papers and in his 2010 book) so he was aware of the book by Prüss only later.

10.8 Exercises

10.8.1 *A random variable X is said to be gamma distributed (or have gamma density) with parameters (α, β), $\alpha > 0$, $\beta > 0$, if its density of probability has the form*

$$f(x) = \frac{x^{\alpha-1}}{\beta^\alpha \Gamma(\alpha)} e^{-\frac{x}{\beta}}, \text{ for } x \geq 0, \text{ and } f(x) \equiv 0, \; x < 0.$$

Let X_1, X_2 be independently distributed gamma variables with parameters $(\alpha, 1)$ and $(\alpha + \frac{1}{2}, 1)$, respectively.
Let $U = X_1 X_2$. Show that the density $f(u)$ of this distribution is given by

$$f(u) = \frac{2^{2\alpha-1}}{\Gamma(2\alpha)} u^{\alpha-1} e^{-2u^{\frac{1}{2}}}, \; u \geq 0; \text{ and } f(x) \equiv 0, \; x < 0.$$

10.8.2 Let X_1, X_2, X_3 be independently distributed gamma variables with parameters $(\alpha, 1)$, $(\alpha + \frac{1}{3}, 1)$, and $(\alpha + \frac{2}{3}, 1)$, respectively.
Let $U = X_1 X_2 X_3$. Show that the density $f(u)$ of this distribution can be represented via an H-function in the form

$$f(u) = \frac{27}{\Gamma(3\alpha)} H_{0,1}^{1,0} \left[27u \bigg|_{(3\alpha - 3, 3)} \right], \; u \geq 0; \text{ and } f(x) \equiv 0, \; x < 0.$$

10.8.3 ([Pil90, p. 190]) Let us consider the stochastic process $x = x(t)$, $t \geq 0$, with $x \geq 0$ (called in some sources the Mittag-Leffler process) having the density

$$f_\alpha(x,t) = \sum_{k=0}^{\infty} (-1)^k \frac{\Gamma(t+k)}{k! \Gamma(t) \Gamma(\alpha(t+k)+1)} x^{\alpha(t+k)}.$$

Show that this process obeys the following subordination formula

$$f_\alpha(x,t) = \int_0^\infty r_\alpha(x, t_*) \gamma(t_*, t) dt_*,$$

10.8 Exercises

where $r_\alpha(x, t)$ is the distribution of the stable process with the Laplace transform equal to $\exp(-t\xi^\alpha)$ and

$$\gamma(t_*, t) = \frac{1}{\Gamma(t)} t_*^{t-1} e^{-t_*}.$$

10.8.4 Consider the stochastic process $x = x(t)$, $t \geq 0$, with $x \geq 0$ having the density $\gamma(t_*, t)$ with Laplace transform (a so-called γ-process)

$$\tilde{\gamma}(s_*, t) = (1 + s_*)^{-t} = e^{-t \log(1+s_*)}.$$

Find the density $\beta(t, t_*)$ of the inverse process $t = t(t_*)$ happening in $t \geq 0$, running along $t_* \geq 0$.

Answer.

$$\beta(t, t_*) = \frac{d}{dt_*} \mathcal{L}_{s_*}^{-1} \left\{ \frac{s_*(1 + s_*)^{-t}}{\log(1 + s_*)} \right\}.$$

10.8.5 ([GorMai12a]) Represent the fundamental solution $\mathcal{G}_\beta^*(x, t)$ of the rightward time fractional drift equation

$$\left({}_t D_*^\beta u\right)(x, t) = -\frac{\partial u}{\partial x}(x, t), \quad -\infty < x < +\infty, \quad t \geq 0,$$

in terms of the M-Wright function

$$M_\nu(z) := W_{-\nu, 1-\nu}(-z) = \sum_{n=0}^{\infty} \frac{(-z)^n}{n! \Gamma[-\nu n + (1-\nu)]}, \quad 0 < \nu < 1.$$

Answer.

$$\mathcal{G}_\beta^*(x, t) = \begin{cases} t^{-\beta} M_\beta\left(\frac{x}{t^\beta}\right), & x > 0, \\ 0, & x < 0. \end{cases}$$

10.8.6 Rescaling and respeeding.

Show that the Mittag-Leffler waiting time density

$$\phi_\beta^{ML}(t) = -\frac{d}{dt} E_\beta(-t^\beta), \quad 0 < \beta \leq 1,$$

is invariant under combined rescaling and respeeding if $a = \tau^\beta$.

Appendix A
The Eulerian Functions

Here we consider the so-called Eulerian functions, namely the well-known *Gamma function* and *Beta function* together with some special functions that turn out to be related to them, such as the *Psi function* and the *incomplete gamma functions*. We recall not only the main properties and representations of these functions, but we also briefly consider their applications in the evaluation of certain expressions relevant for the fractional calculus.

A.1 The Gamma Function

The *Gamma function* $\Gamma(z)$ is the most widely used of all the special functions: it is usually discussed first because it appears in almost every integral or series representation of other advanced mathematical functions. We take as its definition the *integral formula*

$$\Gamma(z) := \int_0^\infty u^{z-1} \, e^{-u} \, du \, , \quad \operatorname{Re} z > 0 \, . \tag{A.1}$$

This integral representation is the most common for Γ, even if it is valid only in the right half-plane of $\mathbb{C}$.

The analytic continuation to the left half-plane can be done in different ways. As will be shown later, the *domain of analyticity* D_Γ of Γ is

$$D_\Gamma = \mathbb{C} \setminus \{0, -1, -2, \ldots, \} \, . \tag{A.2}$$

Using integration by parts, (A.1) shows that, at least for $\operatorname{Re} z > 0$, Γ satisfies the simple *difference equation*

$$\Gamma(z+1) = z \, \Gamma(z) \, , \tag{A.3}$$

© Springer-Verlag GmbH Germany, part of Springer Nature 2020
R. Gorenflo et al., *Mittag-Leffler Functions, Related Topics and Applications*,
Springer Monographs in Mathematics,
https://doi.org/10.1007/978-3-662-61550-8

which can be iterated to yield

$$\Gamma(z+n) = z(z+1)\ldots(z+n-1)\Gamma(z), \quad n \in \mathbb{N}. \tag{A.4}$$

The recurrence formulas $(A.3-4)$ can be extended to any $z \in D_\Gamma$. In particular, since $\Gamma(1) = 1$, we get for *non-negative integer values*

$$\Gamma(n+1) = n!, \quad n = 0, 1, 2, \ldots. \tag{A.5}$$

As a consequence Γ can be used to define the *Complex Factorial Function*

$$z! := \Gamma(z+1). \tag{A.6}$$

By the substitution $u = v^2$ in (A.1) we get the *Gaussian Integral Representation*

$$\Gamma(z) = 2\int_0^\infty e^{-v^2} v^{2z-1} dv, \quad \operatorname{Re}(z) > 0, \tag{A.7}$$

which can be used to obtain Γ when z assumes *positive semi-integer values*. Starting from

$$\Gamma\left(\frac{1}{2}\right) = \int_{-\infty}^{+\infty} e^{-v^2} dv = \sqrt{\pi} \approx 1.77245, \tag{A.8}$$

we obtain for $n \in \mathbb{N}$,[1]

$$\Gamma\left(\frac{n+1}{2}\right) = \int_{-\infty}^{+\infty} e^{-v^2} v^n dv = \Gamma\left(\frac{1}{2}\right) \frac{(2n-1)!!}{2^n} = \sqrt{\pi}\, \frac{(2n)!}{2^{2n} n!}. \tag{A.9}$$

A.1.1 Analytic Continuation

A common way to derive the domain of analyticity (A.2) is to carry out the analytic continuation by the *mixed representation* due to Mittag-Leffler:

$$\Gamma(z) = \sum_{n=0}^{\infty} \frac{(-1)^n}{n!(z+n)} + \int_1^\infty e^{-u} u^{z-1} du, \quad z \in D_\Gamma. \tag{A.10}$$

This representation can be obtained from the so-called *Prym's decomposition*, namely by splitting the integral in (A.1) into two integrals, one over the interval $0 \le u \le 1$ which is then developed as a series, the other over the interval $1 \le u \le \infty$, which, being uniformly convergent inside $\mathbb{C}$, provides an entire function. The terms of the series (uniformly convergent inside D_Γ) provide the *principal parts* of Γ at the cor-

---

[1] The double factorial $m!!$ means $m!! = 1 \cdot 3 \cdot \ldots m$ if m is odd, and $m!! = 2 \cdot 4 \cdot \ldots m$ if m is even.

Appendix A: The Eulerian Functions

responding poles $z_n = -n$. So we recognize that Γ is analytic in the entire complex plane except at the points $z_n = -n$ ($n = 0, 1, \ldots$), which turn out to be simple poles with residues $R_n = (-1)^n/n!$. The point at infinity, being an accumulation point of poles, is an essential non-isolated singularity. Thus Γ is a transcendental *meromorphic* function.

A formal way to obtain the domain of analyticity D_Γ is to carry out the required analytical continuation via the *Recurrence Formula*

$$\Gamma(z) = \frac{\Gamma(z+n)}{(z+n-1)(z+n-2)\ldots(z+1)z}, \tag{A.11}$$

which is equivalent to (A.4). In this way we can enter the left half-plane step by step. The numerator on the R.H.S. of (A.11) is analytic for $\operatorname{Re} z > -n$; hence, the L.H.S. is analytic for $\operatorname{Re} z > -n$ except for simple poles at $z = 0, -1, \ldots, (-n+2), (-n+1)$. Since n can be arbitrarily large, we deduce the properties discussed above.

Another way to interpret the analytic continuation of the Gamma function is provided by the *Cauchy–Saalschütz representation*, which is obtained by iterated integration by parts in the basic representation (A.1). If $n \geq 0$ denotes any non-negative integer, we have

$$\Gamma(z) = \int_0^\infty u^{z-1} \left[e^{-u} - 1 + u - \frac{1}{2!}u^2 + \cdots + (-1)^{n+1}\frac{1}{n!}u^n \right] du \tag{A.12}$$

in the strip $-(n+1) < \operatorname{Re} z < -n$.

To prove this representation the starting point is provided by the integral

$$\int_0^\infty u^{z-1} \left[e^{-u} - 1 \right] du, \quad -1 < \operatorname{Re} z < 0.$$

Integration by parts gives (the integrated terms vanish at both limits)

$$\int_0^\infty u^{z-1} \left[e^{-u} - 1 \right] du = \frac{1}{z} \int_0^\infty u^z e^{-u} du = \frac{1}{z}\Gamma(z+1) = \Gamma(z).$$

So, by iteration, we get (A.12).

A.1.2 The Graph of the Gamma Function on the Real Axis

Plots of $\Gamma(x)$ (continuous line) and $1/\Gamma(x)$ (dashed line) for $-4 < x \leq 4$ are shown in Fig. A.1 and for $0 < x \leq 3$ in Fig. A.2.

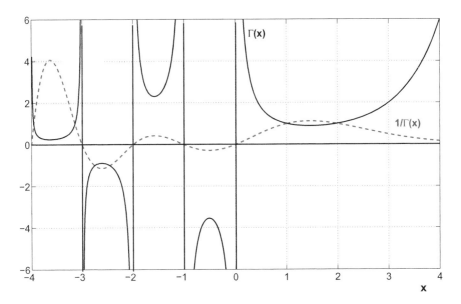

Fig. A.1 Plots of $\Gamma(x)$ (continuous line) and $1/\Gamma(x)$ (dashed line)

Hereafter we provide some analytical arguments that support the plots on the real axis. In fact, one can get an idea of the graph of the Gamma function on the real axis using the formulas

$$\Gamma(x+1) = x\Gamma(x), \quad \Gamma(x-1) = \frac{\Gamma(x)}{x-1},$$

to be iterated starting from the interval $0 < x \le 1$, where $\Gamma(x) \to +\infty$ as $x \to 0^+$ and $\Gamma(1) = 1$.

For $x > 0$ the integral representation (A.1) yields $\Gamma(x) > 0$ and $\Gamma''(x) > 0$ since

$$\Gamma(x) = \int_0^\infty e^{-u} u^{x-1}\, du, \quad \Gamma''(x) = \int_0^\infty e^{-u} u^{x-1} (\log u)^2\, du.$$

As a consequence, on the positive real axis $\Gamma(x)$ turns out to be positive and *convex* so that it first decreases and then increases, exhibiting a minimum value. Since $\Gamma(1) = \Gamma(2) = 1$, we must have a minimum at some x_0, $1 < x_0 < 2$. It turns out that $x_0 = 1.4616\ldots$ and $\Gamma(x_0) = 0.8856\ldots$; hence x_0 is quite close to the point $x = 1.5$ where Γ attains the value $\sqrt{\pi}/2 = 0.8862\ldots$.

On the negative real axis $\Gamma(x)$ exhibits vertical asymptotes at $x = -n$ ($n = 0, 1, 2, \ldots$); it turns out to be positive for $-2 < x < -1$, $-4 < x < -3$, ..., and negative for $-1 < x < 0$, $-3 < x < -2$, ...

Appendix A: The Eulerian Functions

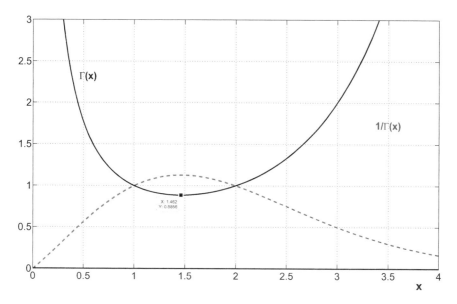

Fig. A.2 Plots of $\Gamma(x)$ (continuous line) and $1/\Gamma(x)$ (dashed line)

A.1.3 The Reflection or Complementary Formula

$$\Gamma(z)\,\Gamma(1-z) = \frac{\pi}{\sin \pi z}\,. \tag{A.13}$$

This formula, which shows the relationship between the Γ function and the trigonometric sin function, is of great importance together with the recurrence formula (A.3). It can be proved in several ways; the simplest proof consists in proving (A.13) for $0 < \mathrm{Re}\,z < 1$ and extending the result by analytic continuation to $\mathbb{C}$ except at the points $0, \pm 1, \pm 2, \ldots$

The reflection formula shows that Γ has no zeros. In fact, the zeros cannot be in $z = 0, \pm 1, \pm 2, \ldots$ and, if Γ vanished at a non-integer z, then because of (A.13), this zero would be a pole of $\Gamma(1-z)$, which cannot be true. This fact implies that $1/\Gamma$ is an entire function. Loosely speaking, $\frac{1}{\Gamma(1-z)}$ collects the positive zeros of $\sin(\pi z)$, while $\frac{1}{\Gamma(z)}$ collects the non-positive zeros.

A.1.4 The Multiplication Formulas

Gauss proved the following *Multiplication Formula*

$$\Gamma(nz) = (2\pi)^{(1-n)/2} \, n^{nz-1/2} \prod_{k=0}^{n-1} \Gamma(z + \frac{k}{n}), \quad n = 2, 3, \ldots, \tag{A.14}$$

which reduces, for $n = 2$, to *Legendre's Duplication Formula*

$$\Gamma(2z) = \frac{1}{\sqrt{2\pi}} \, 2^{2z-1/2} \, \Gamma(z) \, \Gamma(z + \frac{1}{2}), \tag{A.15}$$

and, for $n = 3$, to the *Triplication Formula*

$$\Gamma(3z) = \frac{1}{2\pi} \, 3^{3z-1/2} \, \Gamma(z) \, \Gamma(z + \frac{1}{3}) \, \Gamma(z + \frac{2}{3}). \tag{A.16}$$

A.1.5 Pochhammer's Symbols

Pochhammer's symbols $(z)_n$ are defined for any non-negative integer n as

$$(z)_n := z \, (z+1) \, (z+2) \ldots (z+n-1), \quad n \in \mathbb{N}. \tag{A.17}$$

with $(z)_0 = 1$. If $z \in \mathbb{C} \setminus \{0, -1, -2, \ldots\}$, then $(z)_n = \frac{\Gamma(z+n)}{\Gamma(z)}$. In particular, for $z = 1/2$, we obtain from (A.9)

$$\left(\frac{1}{2}\right)_n := \frac{\Gamma(n + 1/2)}{\Gamma(1/2)} = \frac{(2n-1)!!}{2^n}.$$

We extend the above notation to negative integers, defining

$$(z)_{-n} := z \, (z-1) \, (z-2) \ldots (z-n+1), \quad n \in \mathbb{N}. \tag{A.18}$$

If $z \in \mathbb{C} \setminus \{-1, -2, \ldots\}$, then $(z)_{-n} = \frac{\Gamma(z+1)}{\Gamma(z-n+1)}$.

A.1.6 Hankel Integral Representations

In 1864 Hankel provided a complex integral representation of the function $1/\Gamma(z)$ valid for *unrestricted* z; it reads:

$$\frac{1}{\Gamma(z)} = \frac{1}{2\pi i} \int_{\text{Ha}_-} \frac{e^t}{t^z} \, dt, \quad z \in \mathbb{C}, \tag{A.19a}$$

where Ha$_-$ denotes the *Hankel path* defined as a contour that begins at $t = -\infty - ia$ ($a > 0$), encircles the branch cut that lies along the negative real axis, and ends up

Appendix A: The Eulerian Functions

Fig. A.3 The left Hankel contour Ha$_-$ and the right Hankel contour Ha$_+$

at $t = -\infty + ib$ ($b > 0$). Of course, the branch cut is present when z is non-integer because t^{-z} is a multivalued function; in this case the contour can be chosen as in Fig. A.3 left, where

$$\arg(t) = \begin{cases} +\pi, & \text{above the cut,} \\ -\pi, & \text{below the cut.} \end{cases}$$

When z is an integer, the contour can be taken to be simply a circle around the origin, described in the counterclockwise direction.

An alternative representation is obtained assuming the branch cut along the positive real axis; in this case we get

$$\frac{1}{\Gamma(z)} = -\frac{1}{2\pi i} \int_{\text{Ha}_+} \frac{e^{-t}}{(-t)^z} dt, \quad z \in \mathbb{C}, \tag{A.19b}$$

where Ha$_+$ denotes the Hankel path defined as a contour that begins at $t = +\infty + ib$ ($b > 0$), encircles the branch cut that lies along the positive real axis, and ends up at $t = +\infty - ia$ ($a > 0$). When z is non-integer the contour can be chosen as in Fig. A.3 left, where

$$\arg(t) = \begin{cases} 0, & \text{above the cut,} \\ 2\pi, & \text{below the cut.} \end{cases}$$

When z is an integer, the contour can be taken to be simply a circle around the origin, described in the counterclockwise direction.

We note that

$$\text{Ha}_- \to \text{Ha}_+ \text{ if } t \to t\,e^{-i\pi}, \text{ and } \text{Ha}_+ \to \text{Ha}_- \text{ if } t \to t\,e^{+i\pi}.$$

The advantage of the Hankel representations (A.19a) and (A.19b) compared with the integral representation (A.1) is that they converge for *all* complex z and not just for Re$z > 0$. As a consequence $1/\Gamma$ is a transcendental *entire* function (of maximum exponential type); the point at infinity is an essential isolated singularity, which is an accumulation point of zeros ($z_n = -n$, $n = 0, 1, \ldots$). Since $1/\Gamma$ is entire, Γ does not vanish in $\mathbb{C}$.

The formulas (A.19a) and (A.19b) are very useful for deriving integral representations in the complex plane for several special functions. Furthermore, using the reflection formula (A.13), we can get the integral representations of Γ itself in terms of the Hankel paths (referred to as *Hankel integral representations* for Γ), which turn out to be valid in the whole domain of analyticity D_Γ.

The required Hankel integral representations that provide the analytical continuation of Γ turn out to be:

(a) using the path Ha_-

$$\Gamma(z) = \frac{1}{2i\,\sin\pi z} \int_{\mathrm{Ha}_-} e^t\, t^{z-1}\, dt\,, \quad z \in D_\Gamma\,; \tag{A.20a}$$

(b) using the path Ha_+

$$\Gamma(z) = -\frac{1}{2i\,\sin\pi z} \int_{\mathrm{Ha}_+} e^{-t}\, (-t)^{z-1}\, dt\,, \quad z \in D_\Gamma\,. \tag{A.20b}$$

A.1.7 Notable Integrals via the Gamma Function

$$\int_0^\infty e^{-st}\, t^\alpha\, dt = \frac{\Gamma(\alpha+1)}{s^{\alpha+1}}\,, \quad \mathrm{Re}(s) > 0,\ \mathrm{Re}(\alpha) > -1\,. \tag{A.21}$$

This formula provides the *Laplace transform of the power function* t^α.

$$\int_0^\infty e^{-at^\beta}\, dt = \frac{\Gamma(1+1/\beta)}{a^{1/\beta}}\,, \quad \mathrm{Re}(a) > 0,\ \beta > 0\,. \tag{A.22}$$

This integral for fixed $a > 0$ and $\beta = 2$ attains the well-known value $\sqrt{\pi/a}$ related to the *Gauss integral*. For fixed $a > 0$, the L.H.S. of (A.22) may be referred to as the *generalized Gauss integral*.

The function $I(\beta) := \Gamma(1+1/\beta)$ strongly decreases from infinity at $\beta = 0$ to a positive minimum (less than the unity) attained around $\beta = 2$ and then slowly increases to the asymptotic value 1 as $\beta \to \infty$. The minimum value is attained at $\beta_0 = 2.16638\ldots$ and $I(\beta_0) = 0.8856\ldots$

A more general formula is

$$\int_0^\infty e^{-zt^\mu}\, t^{\nu-1}\, dt = \frac{1}{\mu}\,\frac{\Gamma(\nu/\mu)}{z^{\nu/\mu}} = \frac{1}{\nu}\,\frac{\Gamma(1+\nu/\mu)}{z^{\nu/\mu}}\,, \tag{A.23}$$

where $\mathrm{Re}\,z > 0$, $\mu > 0$, $\mathrm{Re}(\nu) > 0$. This formula includes (A.21)–(A.22); it reduces to (A.21) for $z = s$, $\mu = 1$ and $\nu = \alpha + 1$, and to (A.22) for $z = a$, $\mu = \beta$ and $\nu = 1$.

Appendix A: The Eulerian Functions

A.1.8 Asymptotic Formulas

$$\Gamma(z) \simeq \sqrt{2\pi}\, e^{-z}\, z^{z-1/2} \left[1 + \frac{1}{12\,z} + \frac{1}{288\,z^2} + \dots\right]; \qquad (A.24)$$

as $z \to \infty$ with $|\arg z| < \pi$. This asymptotic expression is usually referred to as *Stirling's formula*, originally given for $n!$. The accuracy of this formula is surprisingly good on the positive real axis and also for moderate values of $z = x > 0$, as can be noted from the following exact formula,

$$x! = \sqrt{2\pi}\, e^{\left(-x + \frac{\theta}{12x}\right)} x^{x+1/2}\,;\ x > 0, \qquad (A.25)$$

where θ is a suitable number in $(0, 1)$.

The two following asymptotic expressions provide a generalization of the Stirling formula.

If a, b denote two positive constants, we have

$$\Gamma(az + b) \simeq \sqrt{2\pi}\, e^{-az} (az)^{az+b-1/2}, \qquad (A.26)$$

as $z \to \infty$ with $|\arg z| < \pi$, and

$$\frac{\Gamma(z+a)}{\Gamma(z+b)} \simeq z^{a-b} \left[1 + \frac{(a-b)(a+b-1)}{2z} + \dots\right], \qquad (A.27)$$

as $z \to \infty$ along any curve joining $z = 0$ and $z = \infty$ provided $z \neq -a, -a-1, \dots$, and $z \neq -b, -b-1, \dots$.

A.1.9 Infinite Products

An alternative approach to the Gamma function is via infinite products, described by Euler in 1729 and Weierstrass in 1856. Let us start with the original formula given by Euler,

$$\Gamma(z) := \frac{1}{z} \prod_{n=1}^{\infty} \frac{\left(1 + \frac{1}{n}\right)^z}{\left(1 + \frac{z}{n}\right)} = \lim_{n \to \infty} \frac{n!\, n^z}{z(z+1)\dots(z+n)}. \qquad (A.28)$$

The above limits exist for all $z \in D_\Gamma \subset \mathbb{C}$.

From Euler's formula (A.28) it is possible to derive Weierstrass' formula

$$\frac{1}{\Gamma(z)} = z\, e^{Cz} \prod_{n=1}^{\infty} \left[\left(1 + \frac{z}{n}\right) e^{-z/n}\right], \qquad (A.29)$$

where C, called the *Euler–Mascheroni constant*, is given by

$$C = 0.5772157\ldots = \begin{cases} \lim_{n\to\infty}\left(\sum_{k=1}^{n}\frac{1}{k} - \log n\right), \\ -\Gamma'(1) = -\int_0^\infty e^{-u}\log u\, du\,. \end{cases} \quad (A.30)$$

A.2 The Beta Function

A.2.1 Euler's Integral Representation

The standard representation of the *Beta function* is

$$B(p,q) = \int_0^1 u^{p-1}(1-u)^{q-1}\,du\,, \quad \begin{cases}\operatorname{Re}(p) > 0,\\ \operatorname{Re}(q) > 0.\end{cases} \quad (A.31)$$

Note that, from a historical viewpoint, this representation is referred to as the *Euler integral of the first kind*, while the integral representation (A.1) for Γ is referred to as the *Euler integral of the second kind*.

The Beta function is a complex function of two complex variables whose analyticity properties will be deduced later, as soon as the relation with the Gamma function has been established.

A.2.2 Symmetry

$$B(p,q) = B(q,p)\,. \quad (A.32)$$

This property is a simple consequence of the definition (A.31).

A.2.3 Trigonometric Integral Representation

$$B(p,q) = 2\int_0^{\pi/2}(\cos\vartheta)^{2p-1}(\sin\vartheta)^{2q-1}\,d\vartheta,\quad \begin{cases}\operatorname{Re}(p) > 0,\\ \operatorname{Re}(q) > 0.\end{cases} \quad (A.33)$$

This noteworthy representation follows from (A.31) by setting $u = (\cos\vartheta)^2$.

Appendix A: The Eulerian Functions

A.2.4 Relation to the Gamma Function

$$B(p,q) = \frac{\Gamma(p)\,\Gamma(q)}{\Gamma(p+q)}. \tag{A.34}$$

This relation is of fundamental importance. Furthermore, it allows us to obtain the analytical continuation of the Beta function.

The proof of (A.34) can easily be obtained by writing the product $\Gamma(p)\,\Gamma(q)$ as a double integral that is to be evaluated introducing polar coordinates. In this respect we must use the Gaussian representation (A.7) for the Gamma function and the trigonometric representation (A.33) for the Beta function. In fact,

$$\begin{aligned}
\Gamma(p)\,\Gamma(q) &= 4 \int_0^\infty \int_0^\infty e^{-(u^2+v^2)}\, u^{2p-1}\, v^{2q-1}\, du\, dv \\
&= 4 \int_0^\infty e^{-\rho^2}\, \rho^{2(p+q)-1}\, d\rho \int_0^{\pi/2} (\cos\vartheta)^{2p-1} (\sin\vartheta)^{2q-1}\, d\vartheta \\
&= \Gamma(p+q)\, B(p,q).
\end{aligned}$$

Henceforth, we shall exhibit other integral representations for $B(p,q)$, all valid for $\operatorname{Re} p > 0$, $\operatorname{Re} q > 0$.

A.2.5 Other Integral Representations

Integral representations on $[0,\infty)$ are

$$B(p,q) = \begin{cases} \displaystyle\int_0^\infty \frac{x^{p-1}}{(1+x)^{p+q}}\, dx, \\[2mm] \displaystyle\int_0^\infty \frac{x^{q-1}}{(1+x)^{p+q}}\, dx, \\[2mm] \displaystyle\frac{1}{2}\int_0^\infty \frac{x^{p-1}+x^{q-1}}{(1+x)^{p+q}}\, dx. \end{cases} \tag{A.35}$$

The first representation follows from (A.31) by setting $u = \dfrac{x}{1+x}$; the other two are easily obtained by using the symmetry property of $B(p,q)$.

A further integral representation on $[0,1]$ is

$$B(p,q) = \int_0^1 \frac{y^{p-1}+y^{q-1}}{(1+y)^{p+q}}\, dy. \tag{A.36}$$

This representation is obtained from the first integral in (A.35) as a sum of the two contributions [0, 1] and [1, ∞).

A.2.6 Notable Integrals via the Beta Function

The Beta function plays a fundamental role in the Laplace convolution of power functions. We recall that the Laplace convolution is the convolution between causal functions (i.e. vanishing for $t < 0$),

$$f(t) * g(t) = \int_{-\infty}^{+\infty} f(\tau) g(t-\tau) \, d\tau = \int_0^t f(\tau) g(t-\tau) \, d\tau.$$

The convolution satisfies both the commutative and associative properties:

$$f(t) * g(t) = g(t) * f(t), \quad f(t) * [g(t) * h(t)] = [f(t) * g(t)] * h(t).$$

It is straightforward to prove the following identity, by setting in (A.31) $u = \tau/t$,

$$t^{p-1} * t^{q-1} = \int_0^t \tau^{p-1} (t-\tau)^{q-1} \, d\tau = t^{p+q-1} \, \mathrm{B}(p,q). \tag{A.37}$$

Introducing the causal Gel'fand–Shilov function

$$\Phi_\lambda(t) := \frac{t_+^{\lambda-1}}{\Gamma(\lambda)}, \quad \lambda \in \mathbb{C},$$

(where the suffix $+$ just denotes the causality property of vanishing for $t < 0$), we can write the previous result in the following interesting form:

$$\Phi_p(t) * \Phi_q(t) = \Phi_{p+q}(t). \tag{A.38}$$

In fact, dividing the L.H.S. of (A.37) by $\Gamma(p)\Gamma(q)$, and using (A.34), we obtain (A.38). As a consequence of (A.38) we get the semigroup property of the Riemann–Liouville fractional integral operator.

In the following we describe other relevant applications of the Beta function. The results (A.37)–(A.38) show that the convolution integral between two (causal) functions, which are absolutely integrable in any interval $[0, t]$ and bounded in every finite interval that does not include the origin, is not necessarily continuous at $t = 0$, even if a theorem ensures that this integral turns out to be continuous for any $t > 0$, see e.g. [Doe74, pp. 47–48]. In fact, considering two arbitrary real numbers α, β greater than -1, we have

$$I_{\alpha,\beta}(t) := t^\alpha * t^\beta = \mathrm{B}(\alpha+1, \beta+1) \, t^{\alpha+\beta+1}, \tag{A.39}$$

Appendix A: The Eulerian Functions

so that
$$\lim_{t \to 0^+} I_{\alpha,\beta}(t) = \begin{cases} +\infty & \text{if } -2 < \alpha + \beta < -1, \\ c(\alpha) & \text{if } \alpha + \beta = -1, \\ 0 & \text{if } \alpha + \beta > -1, \end{cases} \quad (A.40)$$

where $c(\alpha) = B(\alpha + 1, -\alpha) = \Gamma(\alpha + 1)\Gamma(-\alpha) = \pi/\sin(-\alpha\pi)$.

We note that in the case $\alpha + \beta = -1$ the convolution integral attains for any $t > 0$ the constant value $c(\alpha) \geq \pi$. In particular, for $\alpha = \beta = -1/2$, we obtain the minimum value for $c(\alpha)$, i.e.

$$\int_0^t \frac{d\tau}{\sqrt{\tau}\sqrt{t-\tau}} = \pi. \quad (A.41)$$

The Beta function is also used to prove some basic identities for the Gamma function, like the *complementary formula* (A.13) and the *duplication formula* (A.15). For the complementary formula it is sufficient to prove it for a real argument in the interval (0, 1), namely

$$\Gamma(\alpha)\Gamma(1-\alpha) = \frac{\pi}{\sin \pi \alpha}, \quad 0 < \alpha < 1.$$

We note from (A.34)–(A.35) that

$$\Gamma(\alpha)\Gamma(1-\alpha) = B(\alpha, 1-\alpha) = \int_0^\infty \frac{x^{\alpha-1}}{1+x} dx.$$

Then it remains to use well-known formula

$$\int_0^\infty \frac{x^{\alpha-1}}{1+x} dx = \frac{\pi}{\sin \pi \alpha}.$$

To prove the duplication formula we note that it is equivalent to

$$\Gamma(1/2)\Gamma(2z) = 2^{2z-1}\Gamma(z)\Gamma(z+1/2),$$

and hence, after simple manipulations, to

$$B(z, 1/2) = 2^{2z-1} B(z, z). \quad (A.42)$$

This identity is easily verified for $\text{Re}(z) > 0$, using the trigonometric representation (A.33) for the Beta function and noting that

$$\int_0^{\pi/2} (\cos \vartheta)^\alpha \, d\vartheta = \int_0^{\pi/2} (\sin \vartheta)^\alpha \, d\vartheta = 2^\alpha \int_0^{\pi/2} (\cos \vartheta)^\alpha (\sin \vartheta)^\alpha \, d\vartheta,$$

with $\text{Re}(\alpha) > -1$, since $\sin 2\vartheta = 2 \sin \vartheta \cos \vartheta$.

A.3 Historical and Bibliographical Notes

For the historical development of the Gamma function we refer the reader to the notable article [Dav59]. It is surprising that the notation Γ and the name Gamma function were first used by Legendre in 1814 whereas in 1729 Euler had represented his function via an infinite product, see Eq. (A.28). As a matter of fact Legendre introduced the representation (A.1) as a generalization of Euler's integral expression for $n!$,

$$n! = \int_0^1 (-\log t)^n \, dt \, .$$

Indeed, changing the variable $t \to u = -\log t$, we get

$$n! = \int_0^\infty e^{-u} u^n \, du = \Gamma(n+1) \, .$$

Lastly, we mention the well-known Bohr–Mollerup theorem which states that the Γ-function is the only function which satisfies the relation $f(z+1) = z f(z)$ where $\log f(z)$ is convex and $f(1) = 1$. The proof of this fact is presented, e.g., in the book by Artin [Art64].

A.4 Exercises

A.4.1. (see, e.g. [Tem96, p. 72]) The Pochhammer symbol is defined as follows

$$(a)_n := a(a+1)\ldots(a+n-1) = \frac{\Gamma(a+n)}{\Gamma(a)}.$$

Verify that for all $a \in \mathbb{C}$, $m, n = 0, 1, \ldots$ the following identities hold

(a) $(-m)_n = \begin{cases} 0, & \text{if } n > m, \\ (-1)^n \frac{m!}{(m-n)!}, & \text{if } n \leq m; \end{cases}$

(b) $(-a)_n = (-1)^n (a-n+1)_n$;

(c) $(-a)_{2n} = 2^{2n} \left(\frac{a}{2}\right)_n \left(\frac{a+1}{2}\right)_n$;

(d) $(-a)_{2n+1} = 2^{2n+1} \left(\frac{a}{2}\right)_{n+1} \left(\frac{a+1}{2}\right)_n$.

Prove the following formulas for suitable values of parameters:

A.4.2. ([Rai71, p. 103])

Appendix A: The Eulerian Functions

$$\frac{\Gamma(1+a/2)}{\Gamma(1+a)} = \frac{\cos\frac{\pi a}{2}\Gamma(1-a)}{\Gamma(1-a/2)}.$$

A.4.3. ([Rai71, p. 103])

$$\frac{\Gamma(1+a-b)}{\Gamma(1+a/2-b)} = \frac{\sin\pi(b-a/2)\Gamma(b-a/2)}{\sin\pi(b-a)\Gamma(b-a)}.$$

A.4.4. ([Tem96, p. 72]) Prove the formula

$$\int_0^\infty t^{z-1} e^{-\alpha t^x} dt = \frac{1}{x}\Gamma\left(\frac{z}{x}\right) \alpha^{-z/x}, \quad \text{Re}\,\alpha > 0, \text{Re}\,x > 0, \text{Re}\,z > 0.$$

A.4.5. Verify the formulas ([Tem96, p. 73])

(a) $\Gamma\left(-n+\frac{1}{2}\right) = (-1)^n \sqrt{\pi} 2^{2n} \frac{n!}{(2n)!}, \quad n = 0, 1, 2, \ldots$.

(b) $\Gamma\left(n+\frac{1}{2}\right) = (-1)^n \sqrt{\pi} 2^{-2n} \frac{(2n)!}{n!}, \quad n = 0, 1, 2, \ldots$.

A.4.6. Verify the alternative reflection formula for the Gamma function ([Tem96, p. 74])

$$\Gamma\left(\frac{1}{2} - z\right)\Gamma\left(\frac{1}{2} + z\right) = \frac{\pi}{\cos \pi z}, \quad z - \frac{1}{2} \notin \mathbb{Z},$$

in particular

$$\Gamma\left(\frac{1}{2} - iy\right)\Gamma\left(\frac{1}{2} + iy\right) = \frac{\pi}{\cosh \pi y}, \quad y \in \mathbb{R}.$$

A.4.7. Verify the generalized reflection formula for the Gamma function ([Tem96, p. 74])

$$\Gamma(z-n) = (-1)^n \frac{\Gamma(z)\Gamma(1-z)}{\Gamma(n+1-z)} = \frac{(-1)^n \pi}{\sin \pi z \Gamma(n+1-z)}, \quad z \notin \mathbb{Z}, n = 0, 1, 2, \ldots.$$

A.4.8. By using the reflection formula show that

$$|\Gamma(iy)| \sim \sqrt{\frac{2\pi}{|y|}} e^{-\pi|y|}, \quad \text{as } y \to \pm\infty.$$

Calculate the following integrals:

A.4.9. ([Tem96, p. 74])

$$\int_0^{\frac{\pi}{2}} (\cos t)^x \cos ty \, dt, \quad \text{Re}\,x > -1.$$

Answer. $\dfrac{\pi}{2^{x+1}}\dfrac{\Gamma(x+1)}{\Gamma\left[\frac{x+y}{2}+1\right]\Gamma\left[\frac{x-y}{2}+1\right]}.$

A.4.10. ([Tem96, pp. 76–77])

$$\int_0^\infty t^{z-1}\log|1-t|\,dt, \quad -1<\operatorname{Re} z<0.$$

Answer. $\dfrac{\pi\cot\pi z}{z}.$

A.4.11. ([Tem96, pp. 76–77], [Sne56])

(a) $\int_0^\infty t^{z-1}\cos t\,dt, \quad \operatorname{Re} z>1.$

Answer. $\Gamma(z)\cos\dfrac{\pi z}{2}.$

(b) $\int_0^\infty t^{z-1}\sin t\,dt, \quad \operatorname{Re} z>1.$

Answer. $\Gamma(z)\sin\dfrac{\pi z}{2}.$

A.4.12 *. ([WhiWat52, p. 300])
Show that for $a<0$, $a=-\nu+\alpha$, $\nu\in\mathbb{N}$, $\alpha>0$,

$$\frac{\Gamma(x)\Gamma(a)}{\Gamma(x+a)}=\sum_{n=1}^\infty\left\{\frac{R_n}{x+n}+G_n(x)\right\},$$

where

$$R_n+\frac{(-1)^n(a-1)(a-2)\ldots(a-n)}{n!}G(-n),$$

$$G(x)=\left(1+\frac{x}{a-1}\right)\left(1+\frac{x}{a-2}\right)\ldots\left(1+\frac{x}{a-\nu}\right),$$

$$G_n(x)=\frac{G(x)-G(-n)}{x+n}.$$

Appendix B
The Basics of Entire Functions

B.1 Definition and Series Representations

A complex-valued function $F : \mathbb{C} \to \mathbb{C}$ is called an *entire function* (or *integral function*) if it is analytic ($\mathbb{C}$-differentiable) everywhere on the complex plane, i.e. if at each point $z_0 \in \mathbb{C}$ the following limit exists

$$\lim_{z \to z_0} \frac{F(z) - F(z_0)}{z - z_0} \in \mathbb{C}.$$

Typical examples of entire functions are the polynomials, the exponential functions and also sums, products and compositions of these functions, thus trigonometric and hyperbolic functions. Among the special functions we point out the following entire functions: Airy functions $\mathrm{Ai}(z)$, $\mathrm{Bi}(z)$, Bessel functions of the first and second kind $J_\nu(z)$, $Y_\nu(z)$, Fox H-functions $H_{p,q}^{m,n}(z)$ for certain values of parameters, the reciprocal Gamma function $\frac{1}{\Gamma(z)}$, the generalized hypergeometric function ${}_pF_q(z)$, Meijer's G-functions $G_{p,q}^{m,n}(z)$, the Mittag-Leffler function $E_\alpha(z)$ and its different generalizations, and the Wright function $\phi(z; \rho, \beta)$.

According to Liouville's theorem an entire function either has a singularity at infinity or it is a constant. Such a singularity can be either a pole (as is the case for a polynomial), or an essential singularity. In the latter case we speak of *transcendental entire functions*. All of the above-mentioned special functions are transcendental.

Every entire function can be represented in the form of a power series

$$F(z) = \sum_{k=0}^{\infty} c_k z^k, \tag{B.1}$$

converging everywhere on $\mathbb{C}$. Thus, according to the *Cauchy–Hadamard* formula, the coefficients of the series (B.1) satisfy the following condition (the necessary and sufficient condition for the sum of a power series to represent an entire function):

$$\lim_{k\to\infty} |c_k|^{\frac{1}{k}} = 0. \tag{B.2}$$

The absolute value of the coefficients of an entire function necessarily decreases to zero (although not monotonically, in general). One can classify the corresponding function in terms of the speed of this decrease (see below in Sect. B.2.). Thus $|c_k| \to 0$ for $z \to \infty$ is a necessary but not sufficient condition for convergence of a power series.

B.2 Growth of Entire Functions. Order, Type and Indicator Function

The global behavior of entire functions of finite order is characterized by their order and type. Recall (see e.g. [Lev56]) that the *order* ρ of an entire function $F(z)$ is defined as an infimum of those k for which we have the inequality

$$M_F(r) := \max_{|z|=r} |F(z)| < e^{r^k}, \quad \forall r > r(k).$$

Equivalently

$$\rho := \rho_F = \limsup_{r\to\infty} \frac{\log\log M_F(r)}{\log r}. \tag{B.3}$$

A more delicate characteristic of an entire function is its type. Recall (see e.g. [Lev56]) that the *type* σ of an entire function $F(z)$ of finite order ρ is defined as the infimum of those $A > 0$ for which the inequality

$$M_F(r) < e^{Ar^\rho}, \quad \forall r > r(k),$$

holds. Equivalently

$$\sigma := \sigma_F = \limsup_{r\to\infty} \frac{\log M_F(r)}{r^\rho}. \tag{B.4}$$

For an entire function $F(z)$ represented in the form of the series

$$F(z) = \sum_{k=0}^{\infty} c_k z^k$$

its order and type can by found by the following formulae

$$\rho = \limsup_{k\to\infty} \frac{k\log k}{\log \frac{1}{|c_k|}}, \tag{B.5}$$

Appendix B: The Basics of Entire Functions

$$(\sigma e \rho)^{\frac{1}{\rho}} = \limsup_{k \to \infty} \left(k^{\frac{1}{\rho}} \sqrt[k]{|c_k|} \right). \tag{B.6}$$

The asymptotic behavior of an entire function is usually studied via its restriction to rays starting at the origin. In order to describe this we introduce the so-called *indicator function* of an entire function of order ρ:

$$h(\theta) = \limsup_{r \to \infty} \frac{\log |F(re^{i\theta})|}{r^\rho}. \tag{B.7}$$

For instance, the exponential function $e^z = \exp z$ has order $\rho = 1$, type $\sigma = 1$ and indicator function $h(\theta) = \cos\theta$, $-\pi \leq \theta \leq \pi$.

B.3 Weierstrass Canonical Representation. Distribution of Zeros

The Weierstrass canonical representation generalizes the representation of complex polynomials in the form of a product of prime factors. By the fundamental theorem of algebra any complex polynomial of order n can be split into exactly n linear factors

$$P_n(z) = a \prod_{k=1}^{n} (z - a_k) \quad \text{with} \quad a \neq 0. \tag{B.8}$$

Entire functions can have infinitely many zeroes. In this case the finite product in (B.8) has to be replaced by an *infinite product*. Then the main question is to take an infinite product in such form that it can represent an entire function (so, is convergent on the whole complex plane). Another problem is that there are some entire functions which have very few zeros (or have no zeros at all, such as the exponential function). These two ideas were taken into account by Weierstrass.

He took *prime factors* (or *Weierstrass elementary factors*) in the form

$$E_n(z) = \begin{cases} (1-z), & \text{if } n = 0; \\ (1-z)\exp\left\{\frac{z}{1} + \frac{z^2}{2} + \ldots + \frac{z^n}{n}\right\}, & \text{otherwise.} \end{cases} \tag{B.9}$$

Using such factors he proved

Theorem B.1 (Weierstrass) *Let* $(z_j)_{j \in \mathbb{N}_0}$ *be a sequence of complex numbers* $(0 = |z_0| < |z_1| \leq |z_2| \leq \ldots)$, *satisfying the following conditions:*

(i) $z_j \to \infty$ as $j \to \infty$;
(ii) *there exists a sequence of positive integers* $(p_j)_{j \in \mathbb{N}}$ *such that*

$$\sum_{j=1}^{\infty}\left|\frac{z}{z_j}\right|^{1+p_j} < \infty.$$

Then there exists an entire function which has zeroes only at the points $(z_j)_{j\in\mathbb{N}_0}$, *in particular, the following one*

$$F(z) = z^k \prod_{j=1}^{\infty} E_{p_j}\left(\frac{z}{z_j}\right). \tag{B.9a}$$

The following theorem is in a sense the converse of the above.

Theorem B.2 (Weierstrass) *Let* $F(z)$ *be an entire function and let* $(z_j)_{j\in\mathbb{N}_0}$ *be the zeros of* $F(z)$, *then there exists a sequence* $(p_j)_{j\in\mathbb{N}}$ *and an entire function* $g(z)$ *such that the following representation holds*

$$F(z) = Cz^k e^{g(z)} \prod_{j=1}^{\infty} E_{p_j}\left(\frac{z}{z_j}\right), \tag{B.10}$$

where C *is a constant and* $k \in \mathbb{N}_0$ *is the multiplicity of the zero of* F *at the origin.*

For entire functions of finite order ρ the Weierstrass theorems have a more exact form.

Theorem B.3 (Hadamard) *Let* $F(z)$ *be an entire function of finite order* ρ *and let* $(z_j)_{j\in\mathbb{N}_0}$ *be the zeros of* $F(z)$, *listed with multiplicity, then the rank* p *of* $F(z)$ *is defined as the least positive integer such that*

$$\sum_{j=1}^{\infty}\left|\frac{1}{z_j}\right|^{1+p} < \infty. \tag{B.11}$$

Then the canonical Weierstrass product is given by

$$F(z) = Cz^k e^{g(z)} \prod_{j=1}^{\infty} E_p\left(\frac{z}{z_j}\right), \tag{B.12}$$

where $g(z)$ *is a polynomial of degree* $q \leq \rho$. *The genus* μ *of* $F(z)$, *defined as* $\max\{p,q\}$, *is then also finite and*

$$\mu \leq \rho. \tag{B.13}$$

B.4 Entire Functions of Completely Regular Growth

Recall (see, e.g., [GoLeOs91, Lev56, Ron92]) that a ray $\arg z = \theta$ is a *ray of completely regular growth (CRG)* for an entire function F of order ρ if the following weak limit exists

$$\lim_{r \to \infty}{}^{*} \frac{\log |F(re^{i\theta})|}{r^\rho}, \tag{B.14}$$

where the term "weak limit" ($\lim^*$ in (B.14)) means that r tends to infinity, omitting the values of a set $E_\theta \subset \mathbb{R}_+$ which satisfies the condition

$$\lim_{r \to \infty} \frac{\text{meas } E_\theta \cap (0, r)}{r} = 0,$$

i.e. is relatively small. If all rays $\theta \in [0, 2\pi)$ are rays of CRG for an entire function F (with the same exceptional set $E = E_\theta$, $\forall \theta$), then such a function is called an *entire function of completely regular growth*. The main characterization of entire functions of completely regular growth is the following:

An entire function $F(z)$ of order ρ is a function of completely regular growth if and only if its set of zeros $(z_j)_{j \in \mathbb{N}_0}$ has

(a) in the case of a non-integer ρ – an *angular density*:

$$\Delta(\phi, \theta) := \lim_{r \to \infty} \frac{n(r, \phi, \theta)}{r^\rho}; \tag{B.14a}$$

(b) in the case of an integer ρ – an angular density and a finite *angular symmetry*:

$$a_\rho := \lim_{r \to \infty} \left\{ q_\rho + \frac{1}{\rho} \sum_{0 < |z_j| < r} \frac{1}{z_j^\rho} \right\}, \tag{B.14b}$$

where $n(r, \phi, \theta)$ is the number of zeros of the function F in the sector $\{z = se^{i\alpha} \in \mathbb{C} : 0 < s < r, \phi < \alpha < \theta\}$, and q_ρ is the coefficient at the ρ-th power in the exponential factor of the Weierstrass product representation (B.12) for F. The function $\Delta(\theta_0, \theta)$ with an arbitrary but fixed $\theta_0 \in [0, 2\pi]$ is then called the *angular density* and the number a_ρ is called the *coefficient of angular symmetry*.

B.5 Historical and Bibliographical Notes

The period from 1850 to 1950 was the "golden century of the theory of entire functions". This theory was one of the central subjects of complex analysis, it was rapidly developed in connection with several deep problems in mathematics as well as due to the usefulness of the analytic machinery for the solution of a wide range of applied problems. Here we mention only a few results of that time which have great influence on contemporary mathematics and which are related in some way to the subject of our book.

Many of these results were influenced by Weierstrass (see, e.g., [Wei66], and the modern description of Weierstrass's results in [AblFok97, Rud74, Kra04]). He introduced the notion of single-valued analytic functions (*eindeutig analytische Funktionen*), established the method of uniform convergence for families of complex functions which he successfully applied to the proof of his celebrated factorization theorem, introduced and applied the notion of (single-valued) analytic elements which became the core of the analytic continuation method, and developed the theory of elliptic functions illustrated by a special collection of these function now bearing his name and used by him and his successors to create and develop the analytic theory of complex differential equations.

Mittag-Leffler was influenced by the genius of Weierstrass during his stay in Berlin (see Chap. 2 of this book). Among Mittag-Leffler's results we point out the Mittag-Leffler factorization theorem for meromorphic functions which he obtained by developing the ideas of Weierstrass. The final form of this theorem was published in 1884 (see [Mit84]). The evolution of the hypotheses of the Mittag-Leffler theorem is worth studying; there were several varying publications of the theorem between 1876 and 1884, the majority of which were by Mittag-Leffler himself, and there is a noticeable evolution of the ideas which is marked by changes in notation from the original Weierstrassian style and the simplification of proofs. In particular, the later versions of Mittag-Leffler's theorem differ markedly from those of his initial papers. Specifically, in 1882 we see a rather abrupt appearance of Cantor's recently developed theory of transfinite sets in statements and proofs of Mittag-Leffler's theorem, despite the generally negative reception of Cantor's work by prominent mathematicians of that period.

Mittag-Leffler was interested in the solution of the analytic continuation problem as applied to the study of the convergence of divergent series (we discuss the corresponding results in detail in Chap. 2). For this purpose he introduced a new entire function (now called the Mittag-Leffler function) which serves as the simplest generalization of the exponential function and also helped him to get a criterion for analytic continuation generalizing Borel's result.

As for the general theory of entire functions, we point out several results which constitute special directions in mathematics (see, e.g., [Boa54, Rud74]). First of all, from the results of Borel–Picard appeared the modern theory of value distributions for entire functions (see, e.g., [Lev56]) and Nevanlinna's theory of value distributions for meromorphic functions (see, [Nev25, Nev36]).

Levin and Pfluger discovered that the asymptotics of the distribution of zeros of entire functions and the growth of these functions at infinity are closely related for a large class of entire functions, namely the functions of completely regular growth. The theory of entire functions of completely regular growth of one variable, developed in the late 1930s independently by Levin and Pfluger, soon found applications in both mathematics and physics. Later, the theory was extended to functions in the half-plane, subharmonic functions in space, and entire functions of several variables.

Appendix B: The Basics of Entire Functions

The monograph [Ron92] describes this theory and presents recent developments based on the concept of weak convergence. This enables a unified approach and provides a comparatively simple presentation of the classical Levin–Pfluger theory (see [GoLeOs91, Lev56]). This theory has also been generalized to analytic functions in angle domains (see [Gov94]).

Rolf Nevanlinna's most important mathematical achievement is the value distribution theory of meromorphic functions. The roots of the theory go back to the result of Picard in 1879, showing that a non-constant complex-valued function which is analytic in the entire complex plane assumes all complex values except at most one. In the early 1920s Rolf Nevanlinna, partly in collaboration with his brother Frithiof, extended the theory to cover meromorphic functions, i.e. functions analytic in the plane except for isolated points in which the Laurent series of the function has a finite number of terms with a negative power of the variable. Nevanlinna's value distribution theory, or Nevanlinna theory, is crystallized in its two Main Theorems. Qualitatively, the first one states that if a value is assumed less frequently than average, then the function comes close to that value more often than average. The Second Main Theorem, more difficult than the first one, says that there are relatively few values which the function assumes less often than average.

One more classical topic related to the subject of this book is the theory of complex differential equations. Many of the considered special functions (the Mittag-Leffler function and its generalizations are among them) satisfy certain complex differential equations. Thus, for the study of the properties of some functions it is natural to construct the corresponding differential equation and apply the general results of the theory. From another perspective, the Mittag-Leffler function and its generalizations serve as solutions of a new type of equation (in particular integral and differential equations of fractional order). The theory of the latter can be compared with the theory of classical complex differential equations. Among the results on complex differential equations we mention their relations to Nevanlinna Theory (see, e.g., [Lai93]) and those results which are related to the Riemann–Hilbert problem (see, [AnoBol94, Bol90, Bol09]).

In this appendix we collect a number of notions and results on entire functions since for certain values of parameters the Mittag-Leffler function and its generalization are entire functions. Thus we use in the main text approaches describing the asymptotics of entire functions, distribution of zeros, series and integral representations and other analytic properties.

B.6 Exercises

Series.

B.6.1. Represent the following functions in the form of Taylor series [Vol70, p. 65].

a) $\cosh z$, b) $\sinh z$,
c) $\sin^2 z$, d) $\cosh^2 z$

Answers:

a) $\sum_{n=0}^{\infty} \dfrac{z^{2n}}{(2n)!}$;

b) $\sum_{n=0}^{\infty} \dfrac{z^{2n+1}}{(2n+1)!}$;

c) $\sum_{n=0}^{\infty} (-1)^{n+1} \dfrac{2^{2n-1} z^{2n}}{(2n)!}$;

d) $\dfrac{1}{2} + \sum_{n=1}^{\infty} \dfrac{2^{2n-1} z^{2n}}{(2n)!}$.

B.6.2. Prove the following formulas [Evg69, p. 134]

(a)
$$\cos(\lambda \arcsin z) = 1 - \sum_{n=0}^{\infty} \dfrac{\lambda^2(\lambda^2 - 2^2)\cdots(\lambda^2 - 4n^2)}{(2n+2)!}(-1)^n z^{2n+2};$$

(b)
$$\sin(\lambda \arcsin z) = \lambda z - \sum_{n=1}^{\infty} \dfrac{\lambda(\lambda^2 - 1^2)\cdots(\lambda^2 - (2n-1)^2)}{(2n+1)!}(-1)^n z^{2n+1};$$

(c)
$$(\arcsin z)^2 = \sum_{n=1}^{\infty} \dfrac{2^{2n-1}((n-1)!)^2}{(2n)!} z^{2n};$$

(d)
$$\dfrac{\ln(z + \sqrt{1+z^2})}{\sqrt{1+z^2}} = \sum_{n=0}^{\infty} (-1)^n \dfrac{2^{2n}((n)!)^2}{(2n+1)!} z^{2n+1}.$$

Infinite Products

B.6.3. Construct an infinite product (Weierstrass product) for a function having the following collection of zeros [Mark66, p. 253]

(a) $z_k = k$ of multiplicity k, $k \in \mathbf{N}$;
(b) $z_k = k$ of multiplicity $|k|$, $k \in \mathbf{Z}$.

B.6.4. Prove the following formulas [LavSha65, p. 433]

(a)
$$\sin z = z \prod_{n=1}^{\infty} \left(1 - \dfrac{z^2}{n^2 \pi^2}\right);$$

Appendix B: The Basics of Entire Functions

(b)
$$\exp z - 1 = z\exp\frac{z}{2}\prod_{n=1}^{\infty}\left(1 + \frac{4z^2}{4n^2\pi^2}\right);$$

(c)
$$\cos z = \prod_{n=1}^{\infty}\left(1 - \frac{4z^2}{(2n-1)^2\pi^2}\right).$$

B.6.5. Prove the following formulas [Evg69, pp. 266–267]

(a)
$$\cos\frac{\pi z}{2} = \prod_{n=1}^{\infty}\left(1 - \frac{z^2}{(2n+1)^2}\right);$$

(b)
$$\sin(z-a) = \sin a \exp{-z\cot a}\prod_{n=1}^{\infty}\left(1 - \frac{z}{a+n\pi}\right)\exp\frac{z}{a+n\pi};$$

(c)
$$\cos\pi z - \cos\pi a = -\frac{\pi^2}{2}(z^2 - a^2)\prod_{n=1}^{\infty}\left[1 - (\frac{z+a}{2n})^2\right]\left[1 - (\frac{z-a}{2n})^2\right];$$

(d)
$$\exp z - \exp a = \exp\frac{z+a}{2}(z-a)\prod_{n=1}^{\infty}\left[1 + (\frac{z-a}{2\pi n})^2\right];$$

(e)
$$\cosh z - \cos z = z^2\prod_{n=1}^{\infty}\left(1 + \frac{z^4}{4\pi^4 n^4}\right);$$

(f)
$$\sin z - z\cos z = \frac{z^2}{3}\prod_{n=1}^{\infty}\left(1 - \frac{z^2}{\lambda_n^2}\right); \qquad \tan\lambda_n = \lambda_n > 0;$$

(g)
$$\frac{\sin\pi z}{\pi z(1-z)} = \prod_{n=1}^{\infty}\left(1 + \frac{z-z^2}{n+n^2}\right);$$

(h)
$$\frac{1}{\Gamma(z)} = z \exp Cz \prod_{n=1}^{\infty} \left(1 + \frac{z}{n}\right) \exp{-\frac{z}{n}},$$

where C is Euler's constant;

B.6.6. Prove the following formulas [GunKuz58, pp. 200–202]

(a)
$$\sinh z = z \prod_{n=1}^{\infty} \left(1 + \frac{z^2}{n^2 \pi^2}\right);$$

(b)
$$\exp az - \exp bz = (a-b)z \exp 1/2(a+b)z \prod_{n=1}^{\infty} \left(1 + \frac{(a-b)^2 z^2}{4n^2 \pi^2}\right);$$

(c)
$$\sin(z-a) + \sin a = \frac{z(\pi + 2a - z)}{\pi + 2a} \prod_{n \neq 0} \left(1 + \frac{z(\pi + 2a - z)}{2\pi n(\pi(2n-1) - 2a)}\right);$$

(d)
$$\exp z^2 + \exp(2z - 1) = 2 \exp 1/2(z^2 + 2z - 1) \prod_{n=1}^{\infty} \left(1 + \frac{(z-1)^4}{\pi^2 (2n-1)^2}\right).$$

B.6.7. Calculate the values of the following infinite products [Vol70, p. 115]

a) $\prod_{n=2}^{\infty} \left(1 - \frac{1}{n^2}\right) = 1/2$,
b) $\prod_{n=3}^{\infty} \left(\frac{n^2-4}{n^2-1}\right) = 1/4$,

c) $\prod_{n=2}^{\infty} \left(\frac{n^3-1}{n^3+1}\right) = 2/3$,
d) $\prod_{n=1}^{\infty} \left(1 + \frac{1}{n(n+2)}\right) = 2$,

e) $\prod_{n=2}^{\infty} \left(1 - \frac{2}{n(n+1)}\right) = 1/3$,
f) $\prod_{n=1}^{\infty} \left(1 + \frac{(-1)^{n+1}}{n}\right) = 1$,

g) $\dfrac{2}{\sqrt{2}} \dfrac{2}{\sqrt{2+\sqrt{2}}} \dfrac{2}{\sqrt{2+\sqrt{2+\sqrt{2}}}} \cdots = \dfrac{\pi}{2}$,
h) $\prod_{n=1}^{\infty} \left(\frac{2n}{2n-1} \frac{2n}{2n+1}\right) = \frac{\pi}{2}$.

B.6.8. Find domains of convergence for the following infinite products [Vol70, pp. 116–117]

Appendix B: The Basics of Entire Functions

a) $\prod_{n=1}^{\infty}(1-z^n)$,　　b) $\prod_{n=1}^{\infty}\left(1+\frac{z^n}{2^n}\right)$,　　c) $\prod_{n=1}^{\infty}\left(1-\frac{z^2}{n^2}\right)$,

d) $\prod_{n=1}^{\infty}\left(1-(1-\frac{1}{n})^{-n}z^{-n}\right)$,　e) $\prod_{n=1}^{\infty}\left(1+(1+\frac{1}{n})^{n^2}z^n\right)$,　f) $\prod_{n=1}^{\infty}\left(1+\frac{z}{n}\right)e^{(-z/n)}$.

Answers:

a) $|z|<1$,　　b) $|z|<1$,　　c) $|z|<\infty$,
d) $|z|>1$,　　e) $|z|<1/e$,　f) $|z|<\infty$.

Characteristics of Entire Functions

B.6.9. Find the order and type of the following entire functions [Vol70, p. 122]

a) $\exp az^n$,　$a>0, n \in \mathbb{N}$;　　b) $z^n \exp 3z$;　　c) $z^2 \exp 2z - \exp 3z$;
d) $\exp 5z - 3\exp 2z^3$;　　　　　　e) $\exp(2-i)z^2$;　f) $\sin z$;
g) $\cosh z$;　　　　　　　　　　　h) $\exp z \cos z$;　i) $\cos \sqrt{z}$.

Answers:

a) $\rho = n, \sigma = a$　b) $\rho = 1, \sigma = 3$　c) $\rho = 1, \sigma = 3$
d) $\rho = 3, \sigma = 2$　e) $\rho = 2, \sigma = \sqrt{5}$　f) $\rho = 1, \sigma = 1$
g) $\rho = 1, \sigma = 1$　h) $\rho = 1, \sigma = \sqrt{2}$　i) $\rho = 1/2, \sigma = 1$.

B.6.10. Calculate the order and type of the following entire functions, represented in the form of series [Vol70, p. 124]

a) $f(z) = \sum_{n=1}^{\infty}\left(\frac{z}{n}\right)^n$,　　　　　　b) $f(z) = \sum_{n=1}^{\infty}\left(\frac{\ln n}{n}\right)^{n/a} z^n, a > 0$,

c) $f(z) = \sum_{n=2}^{\infty}\left(\frac{1}{n \ln n}\right)^{n/a} z^n, a > 0$,　d) $f(z) = \sum_{n=0}^{\infty} e^{-n^2} z^n$,

e) $f(z) = \sum_{n=1}^{\infty} \frac{z^n}{n^{n^{1+a}}}, a > 0$　　　　f) $f(z) = \sum_{n=1}^{\infty} \frac{\cosh \sqrt{n}}{n!} z^n$,

g) $z^{-\nu} J_\nu(z) = \sum_{n=0}^{\infty} \frac{(-1)^n z^{2n}}{n!\Gamma(n+\nu+1)}$, $(\nu > -1)$ where J_ν is the Bessel function.

Answers:

a) $\rho = 1, \sigma = 1/e$　b) $\rho = a, \sigma = \infty$　c) $\rho = 0, \sigma = 0$
d) $\rho = 0$,　　　　　　e) $\rho = 0$,　　　　　f) $\rho = 1, \sigma = 1$
g) $\rho = 1, \sigma = 2$

B.6.11. Find the order and indicator function of the following entire functions [Vol70, p. 125]

a) $\exp z$, b) $\exp(z) + z^2$, c) $\sin z$,
d) $\cos z$, e) $\cosh z$, f) $\exp(z^n)$,
g) $\dfrac{\sin \sqrt{z}}{\sqrt{z}}$

Answers:

a) $\rho = 1, h(\phi) = \cos \phi$
b) $\rho = 1, h(\phi) = \begin{cases} \cos \phi, & -\pi/2 \leq \phi < \pi/2 \\ 0, & \pi/2 \leq \phi < 3\pi/2, \end{cases}$
c) $\rho = 1, h(\phi) = |\sin \phi|$,
d) $\rho = 1, h(\phi) = |\sin \phi|$,
e) $\rho = 1, h(\phi) = |\cos \phi|$,
f) $\rho = n, h(\phi) = \cos n\phi$,
g) $\rho = 1/2, |\sin \phi/2|$

Zeros of Entire Functions

B.6.12. Find all zeros and their multiplicities for the following entire functions [Vol70, p. 70]

a) $(1 - \exp z)(z^2 - 4)^3$, b) $1 - \cos z$, c) $\dfrac{(z^2 - \pi^2)^2 \sin z}{z}$
d) $\sin^3 z$, e) $\dfrac{\sin^3 z}{z}$, f) $\sin(z^3)$,
g) $\cos^3 z$, h) $\cos z^3$

Answers:
a) $z = \pm 2$, — 3rd order; $z = 2k\pi i \, (k = 0, \pm 1, \cdots)$ — simple;
b) $z = 2k\pi \, (k = 0, \pm 1, \cdots)$ — 2nd order;
c) $z = \pm \pi$, — 3rd order; $z = k\pi \, (k = \pm 2, \pm 3, \cdots)$ — simple;
d) $z = k\pi \, (k = 0, \pm 1, \cdots)$ — 3rd order;
e) $z = 0$, 2nd order, $z = k\pi \, (k = \pm 1, \pm 2, \cdots)$ — 3rd order;
f) $z = \sqrt[3]{k\pi}$, $z = 1/2\sqrt[3]{k\pi}(1 \pm i\sqrt{3})$, $(k = \pm 1, \pm 2, \cdots)$ simple;
g) $z = (2k + 1)\dfrac{\pi}{2}$, $(k = 0, \pm 1, \cdots)$ — 3rd order;
h) $z = \sqrt[3]{(2k + 1)\pi/2}$, $z = 1/2\sqrt[3]{(2k + 1)\pi/2}(1 \pm i\sqrt{3})$, $(k = 0, \pm 1, \pm 2, \cdots)$ simple.

B.6.13. Find the solutions of the following equations [Evg69, p. 74]

a) $\sin z = \dfrac{4i}{3}$, b) $\sin z = \dfrac{5}{3}$, c) $\cos z = \dfrac{3i}{4}$,
d) $\cos z = \dfrac{3+i}{4}$, e) $\sinh z = \dfrac{i}{2}$, f) $\cosh z = \dfrac{1}{2}$

Answers:
a) $z = i(-1)^k \ln 3 + k\pi, \, k = 0, \pm 1, \cdots$,
b) $z = \pm i \ln 3 + \pi/2 + 2k\pi, \, k = 0, \pm 1, \cdots$,
c) $z = \pm(-i \ln 2 + \pi/4) + 2k\pi, \, k = 0, \pm 1, \cdots$,
d) $z = \pm(-i/2 \ln 2 + \pi/4) + 2k\pi, \, k = 0, \pm 1, \cdots$,
e) $z = (-1)^k \dfrac{i\pi}{6} + k\pi, \, k = 0, \pm 1, \cdots$,
f) $z = \pm \dfrac{i\pi}{3} + 2k\pi, \, k = 0, \pm 1, \cdots$.

Appendix C
Integral Transforms

In this appendix we give an outline of the properties of some integral transforms. The main focus is on the properties which are often useful in treating applied problems. It is not our intention to present a complete theory of these transforms. In applications, however, it is advantageous to have at our disposal an arsenal of formal manipulations that should be used with a critical mind. Among the thousands of books on the subject we refer to a few in which the theory is developed with different degrees of rigour.

C.1 Fourier Type Transforms

The most general definition of the Fourier transform is

$$(\mathcal{F}f)(\kappa) = F(\kappa) = A \int_{-\infty}^{+\infty} e^{iB\kappa t} f(t) \mathrm{d}t, \quad A, B \in \mathbb{R}, \ A \neq 0, \ B \neq 0.$$

In this book we use the following definition, which is commonly used in Probability Theory and Stochastic Modelling.

Definition *The Fourier transform of a function $f : \mathbb{R} \to \mathbb{R}(\mathbb{C})$ is denoted by $\mathcal{F}f = F(\kappa)$, $\kappa \in \mathbb{R}$, and defined by the integral*

$$(\mathcal{F}f)(\kappa) = F(\kappa) = \int_{-\infty}^{+\infty} e^{i\kappa t} f(t) \mathrm{d}t, \tag{C.1.1}$$

where $\mathcal{F}$ is called the *Fourier transform operator* or the *Fourier transformation*.[2]

The Fourier image $\mathcal{F}f = F(\kappa)$ is also denoted by $\widehat{f}(\kappa)$.

From the theory of harmonic oscillations comes the following terminology: the pre-image (original) of the Fourier transform is usually called the *signal* or *amplitude function* depending on the time-variable t, while the Fourier image is called the *spectral function* depending on the frequency variable κ.

The simplest class of functions for which the Fourier integral transform exists is the so-called class of rapidly decreasing functions, i.e. real- or complex-valued functions defined for all $x \in \mathbb{R}$ and infinitely differentiable everywhere, such that each derivative tends to zero as $|x| \to +\infty$ faster that any positive power of x:

$$\lim_{|x| \to +\infty} |x|^N f^{(n)}(x) = 0 \qquad (C.1.2)$$

for each positive integer N and n.

A more general sufficient condition for a function f to have a Fourier transform is that f is absolutely integrable on $\mathbb{R}$ (see, e.g., [Doe74, p. 154]), i.e. belongs to $L_1(\mathbb{R})$[3]

$$L_1(\mathbb{R}) := \left\{ f : \mathbb{R} \to \mathbb{R}(\mathbb{C}) : \int_{-\infty}^{+\infty} |f(x)| dx < +\infty \right\}.$$

Theorem C.1 *Let the function f be absolutely integrable, i.e.*

$$\int_{-\infty}^{+\infty} |f(t)| dt < \infty. \qquad (C.1.3)$$

Then at every point t_0, where f is of bounded variation in some (arbitrarily small) neighborhood of t_0, the following inversion formula for the Fourier transform holds:

$$\frac{f(t_0 + 0) + f(t_0 - 0)}{2} = \frac{1}{2\pi} \int_{-\infty}^{+\infty} e^{-i\kappa t} \widehat{f}(\kappa) d\kappa, \qquad (C.1.4)$$

---

[2]Among other definitions of the Fourier transform we mention the symmetric form of the Fourier transform $(\mathcal{F}f)(\kappa) = \frac{1}{\sqrt{2\pi}} \int_{-\infty}^{+\infty} e^{-i\kappa t} f(t) dt$ used in Functional Analysis, and $(\mathcal{F}\varphi)(f) = \int_{-\infty}^{+\infty} e^{-i(2\pi f)t} \varphi(t) dt$ which is commonly used in Signal Processing. In the last definition the variable t is time and f is the frequency of a signal. The function $\mathcal{F}f$ is called the spectrum of the signal $f(t)$.

[3]More precisely the space $L_1(\mathbb{R})$ consists of equivalence classes with respect to the equivalence relation: $f \sim g \iff \int_{-\infty}^{+\infty} |f(x) - g(x)| dx = 0$.

Appendix C: Integral Transforms

where the integral is understood in the sense of the Cauchy principal value.

If t_0 is a point of continuity for f, then the value of the right-hand side of (C.1.4) coincides with the value $f(t_0)$.

Moreover, if $f \in L_1(\mathbb{R})$ then its Fourier transform $\mathcal{F}f = F(\kappa)$ is a (uniformly) continuous and bounded function in $\kappa \in \mathbb{R}$, and $\lim\limits_{|\kappa| \to +\infty} F(\kappa) = 0$.

Theorem C.1 determines the *inverse Fourier operator*

$$\left(\mathcal{F}^{-1} F\right)(t) = \frac{1}{2\pi} \int_{-\infty}^{+\infty} e^{-i\kappa t} \widehat{f}(\kappa) d\kappa. \tag{C.1.5}$$

In particular, if a function is locally integrable and has a compact support (i.e. vanishes outside some interval), then its Fourier transform exists. In this case, the Fourier transform $F(\kappa)$ possesses an analytic continuation into the whole complex plane $\kappa \in \mathbb{C}$.

In order to understand the meaning of the Fourier transform let us recall the *Dirichlet condition*. A real- or complex-valued function defined on the whole real line $\mathbb{R}$ is said to satisfy the Dirichlet condition on $\mathbb{R}$ if:

(a) $f(t)$ has in $\mathbb{R}$ no more than a finite number of finite discontinuity points (jump points) and has no infinite discontinuity points;
(b) $f(t)$ has in $\mathbb{R}$ no more than a finite number of maximum and minimum points.

If $f(t)$ satisfies the Dirichlet condition in $\mathbb{R}$ and is absolutely integrable, then the following *Fourier integral formula* holds:

$$\frac{f(t+0) + f(t-0)}{2} = \frac{1}{2\pi} \int_{-\infty}^{+\infty} e^{-i\kappa t} d\kappa \int_{-\infty}^{+\infty} f(\xi) e^{i\kappa \xi} d\xi \tag{C.1.6}$$

at any finite discontinuity point $t \in (-\infty, +\infty)$. This result is also known as the *Fourier integral theorem*.

In particular, if f is continuous at t, then the Fourier integral formula can be written as

$$f(t) = \frac{1}{2\pi} \int_{-\infty}^{+\infty} e^{-i\kappa t} d\kappa \int_{-\infty}^{+\infty} f(\xi) e^{i\kappa \xi} d\xi. \tag{C.1.7}$$

Let us recall some basic properties of the Fourier transform (for more detailed information we refer to the treatise by Titchmarsh (see [Tit86]).

$$\left(\mathcal{F}\overline{f}\right)(\kappa) = \overline{\widehat{f}(-\kappa)}. \tag{C.1.8}$$

If $f(t) = f(-t)$, then $(\mathcal{F}f)(\kappa) = 2\int_0^\infty f(t)\cos \kappa t\, dt.$ (C.1.9)

If $f(t) = -f(-t)$, then $(\mathcal{F}f)(\kappa) = -2i\int_0^\infty f(t)\sin \kappa t\, dt.$ (C.1.10)

$$\left(\mathcal{F}f(a^{-1}t + b)\right)(\kappa) = ae^{iab\kappa}\widehat{f}(a\kappa),\ a > 0,$$ (C.1.11)

$$\left(\mathcal{F}e^{ibt}f(at)\right)(\kappa) = \frac{1}{a}\widehat{f}\left(\frac{\kappa - b}{a}\right),\ a > 0,$$ (C.1.12)

$$\left(\mathcal{F}t^n f(t)\right)(\kappa) = i^n \frac{d^n \widehat{f}(\kappa)}{d\kappa^n},\ n \in \mathbb{N},$$ (C.1.13)

$$\left(\mathcal{F}f^{(n)}(t)\right)(\kappa) = i^n \kappa^n \widehat{f}(\kappa),\ n \in \mathbb{N},$$ (C.1.14)

$$(\mathcal{F}(f * g)(t))(\kappa) = \widehat{f}(\kappa) \cdot \widehat{g}(\kappa),$$ (C.1.15)

provided that all Fourier images on the left- and right-hand sides exist, where $(f * g)(t)$ is the *Fourier convolution*, i.e.

$$(f * g)(t) = \int_{-\infty}^{+\infty} f(t - \tau)g(\tau)d\tau.$$ (C.1.16)

Sufficient conditions for the fulfillment of equality (C.1.15) read that this equality holds if both functions f and g are integrable and square integrable on the real line:

$$f, g \in L_1(\mathbb{R}) \cap L_2(\mathbb{R}).$$ (C.1.17)

These conditions coincide with the conditions which guarantee the fulfillment of the Cauchy–Schwarz inequality:

$$\int_{-\infty}^{+\infty} |f(t) \cdot g(t)|dt \leq \int_{-\infty}^{+\infty} |f(t)|^2 dt \cdot \int_{-\infty}^{+\infty} |g(t)|^2 dt.$$ (C.1.18)

Analogous conditions give us the so-called *Parseval identity* for the Fourier transform.

Theorem C.2 *Let f be integrable and square integrable on the real line, i.e.*

Appendix C: Integral Transforms

$$f \in L_1(\mathbb{R}) \bigcap L_2(\mathbb{R}).$$

Then for all real x the following formula holds:

$$\int_{-\infty}^{+\infty} f(t)\overline{f(t-x)}dt = \frac{1}{2\pi} \int_{-\infty}^{+\infty} e^{-i\kappa x} \widehat{f}(\kappa)\overline{\widehat{f}(\kappa)}d\kappa. \tag{C.1.19}$$

Moreover for $x = 0$ this yields the Parseval formula

$$\int_{-\infty}^{+\infty} |f(t)|^2 dt = \frac{1}{2\pi} \int_{-\infty}^{+\infty} |\widehat{f}(\kappa)|^2 d\kappa. \tag{C.1.20}$$

The Mittag-Leffler function is one of the most important special functions related to the Fractional Calculus. Thus we recall two composition formulas for the Fourier transform and fractional integrals/derivatives.

$$\left(\mathcal{F}I_{\pm}^{\alpha}\varphi\right)(\kappa) = \frac{\widehat{\varphi}(\kappa)}{(\mp i\kappa)^{\alpha}}, \quad 0 < \operatorname{Re}\alpha < 1, \tag{C.1.21}$$

where $I_{\pm}^{\alpha}$ are fractional integrals of the Liouville type:

$$\left(I_{+}^{\alpha}\varphi\right)(x) = \frac{1}{\Gamma(\alpha)} \int_{-\infty}^{x} \frac{\varphi(t)dt}{(x-t)^{1-\alpha}}, \quad \left(I_{-}^{\alpha}\varphi\right)(x) = \frac{1}{\Gamma(\alpha)} \int_{x}^{+\infty} \frac{\varphi(t)dt}{(t-x)^{1-\alpha}}.$$

Formulas (C.1.21) are valid for any function $\varphi \in L_1(-\infty, +\infty)$.

Corresponding formulas for fractional derivatives of the Liouville type

$$\left(\mathcal{D}_{+}^{\alpha}\varphi\right)(x) = \frac{1}{\Gamma(n-\alpha)} \frac{d^n}{dx^n} \int_{-\infty}^{x} \frac{\varphi(t)dt}{(x-t)^{\alpha-n+1}},$$

$$\left(\mathcal{D}_{-}^{\alpha}\varphi\right)(x) = \frac{(-1)^n}{\Gamma(n-\alpha)} \frac{d^n}{dx^n} \int_{x}^{+\infty} \frac{\varphi(t)dt}{(t-x)^{1-\alpha}}, \quad n = [\operatorname{Re}\alpha] + 1,$$

have the form

$$\left(\mathcal{F}\mathcal{D}_{\pm}^{\alpha}\varphi\right)(\kappa) = (\mp i\kappa)^{\alpha}\widehat{\varphi}(\kappa), \quad \operatorname{Re}\alpha > 0. \tag{C.1.22}$$

Formulas (C.1.22) are valid, in particular, for all functions having derivatives up to the order $n = [\operatorname{Re}\alpha] + 1$, and rapidly decreasing at infinity together with all derivatives (see, e.g. [SaKiMa93]).

In (C.1.21)–(C.1.22) the values of the function $(\mp i\kappa)^\alpha$ are calculated according to the formula

$$(\mp ix)^\alpha = e^{\alpha \log |x| \mp \frac{\alpha \pi i}{2} \operatorname{sgn} x}.$$

We mention two formulas for cosine- and sine-integral transforms of the Riemann–Liouville fractional integral (see, e.g., [SaKiMa93, p. 140])

$$\left(\mathcal{F}_c I_{0+}^\alpha \varphi\right)(\kappa) = \kappa^{-\alpha}\left[\cos\frac{\pi\alpha}{2} (\mathcal{F}_c\varphi)(\kappa) - \sin\frac{\pi\alpha}{2} (\mathcal{F}_s\varphi)(\kappa)\right], \quad \kappa > 0, \quad \text{(C.1.23)}$$

$$\left(\mathcal{F}_s I_{0+}^\alpha \varphi\right)(\kappa) = \kappa^{-\alpha}\left[\sin\frac{\pi\alpha}{2} (\mathcal{F}_c\varphi)(\kappa) + \cos\frac{\pi\alpha}{2} (\mathcal{F}_s\varphi)(\kappa)\right], \quad \kappa > 0. \quad \text{(C.1.24)}$$

Here the cosine- and sine-integral transforms are defined as follows

$$(\mathcal{F}_s(f))(\nu) = \int_{-\infty}^{+\infty} f(t) \sin(2\pi\nu t) dt, \quad (\mathcal{F}_c(f))(\nu) = \int_{-\infty}^{+\infty} f(t) \cos(2\pi\nu t) dt.$$

The above results are considered in the classical case and in the distributional sense (see, e.g., [Bre65]). We use them in the text only in the first sense.

C.2 The Laplace Transform

The classical Laplace transform is defined by the following integral formula

$$(\mathcal{L}f)(s) = \int_0^\infty e^{-st} f(t) dt, \quad \text{(C.2.1)}$$

provided that the function f (the *Laplace original*) is absolutely integrable on the semi-axis $(0, +\infty)$. In this case the image of the Laplace transform (also called the *Laplace image*), i.e. the function

$$F(s) = (\mathcal{L}f)(s) \quad \text{(C.2.2)}$$

(sometimes denoted $F(s) = \tilde{f}(s)$) is defined and analytic in the half-plane $\operatorname{Re} s > 0$.

It may happen that the Laplace image can be analytically continued to the left of the imaginary axes $\operatorname{Re} s = 0$ in a bigger domain, i.e. there exist a non-positive real number σ_s (called the *Laplace abscissa of convergence*) such that $F(s) = \tilde{f}(s)$ is analytic in the half-plane $\operatorname{Re} s \geq \sigma_s$. Then the following inverse Laplace transform can be introduced

Appendix C: Integral Transforms

$$\left(\mathcal{L}^{-1} F\right)(t) = \frac{1}{2\pi i} \int_{\mathcal{L}_{ic}} e^{st} F(s) \, ds, \qquad \text{(C.2.3)}$$

where $\mathcal{L}_{ic} = (c - i\infty, c + i\infty)$, $c > \sigma_s$, and the integral is usually understood in the sense of the Cauchy principal value, i.e.

$$\int_{\mathcal{L}_{ic}} e^{st} F(s) \, ds = \lim_{T \to +\infty} \int_{c-iT}^{c+iT} e^{st} F(s) \, ds. \qquad \text{(C.2.4)}$$

If the integral (C.2.2) converges absolutely on the line $\operatorname{Re} s = c$, then at any continuity point t_0 of the original f the integral (C.2.3) gives the value of f at this point, i.e.

$$\frac{1}{2\pi i} \int_{\mathcal{L}_{ic}} e^{st_0} \widetilde{f}(s) \, ds = f(t_0). \qquad \text{(C.2.5)}$$

Thus under these conditions operators $\mathcal{L}$ and $\mathcal{L}^{-1}$ constitute an inverse pair of operators. Correspondingly, the functions f and $F = \widetilde{f}$ constitute a Laplace transform pair. To indicate this fact the following notation is used

$$f(t) \div \widetilde{f}(s) = \int_0^\infty e^{-st} f(t) \, dt, \quad \operatorname{Re} s > \sigma_c, \qquad \text{(C.2.6)}$$

where σ_c is the abscissa of the convergence. Here the sign $\div$ denotes the juxtaposition of a function (depending on $t \in \mathbb{R}^+$) with its Laplace transform (depending on $s \in \mathbb{C}$). In the following the conjugate variables $\{t, s\}$ may be different, e.g., $\{r, s\}$ and the abscissa of the convergence may sometimes be omitted. Furthermore, throughout our analysis, we assume that the Laplace transforms obtained by our formal manipulations are invertible by using the standard Bromwich formula:

$$f(t) = \left(\mathcal{L}^{-1} \widetilde{f}(s)\right)(t) = \frac{1}{2\pi i} \lim_{T \to +\infty} \int_{\gamma-iT}^{\gamma+iT} e^{st} \widetilde{f}(s) \, ds.$$

Among the rules for the Laplace transform pairs we recall the following one, which turns out to be useful for our purposes,

$$\frac{1}{\sqrt{\pi t}} \int_0^\infty e^{-r^2/(4t)} f(r) \, dr \div \frac{\widetilde{f}(s^{1/2})}{s^{1/2}}. \qquad \text{(C.2.7)}$$

Since an examination of convergence conditions is not always possible (see, e.g., [Wid46]) sometimes the terminology "Laplace transform pair" is used for pairs $f, \widetilde{f}$ not necessarily satisfying the equality (C.2.5) at certain points.

There are several properties of the Laplace transform which make it very useful in the study of a wide class of differential and integral equations. Let us recall a few main properties of the Laplace transform (in the form of Laplace transform pairs) (more information can be found, e.g., [BatErd54a, Wid46, Doe74]).

$$e^{-at} f(t) \div \widetilde{f}(s+a), \ a > 0; \tag{C.2.8}$$

$$t^n f(t) \div (-1)^n \frac{d^n \widetilde{f}}{dt^n}(s), \ n \in \mathbb{N}; \tag{C.2.9}$$

$$t^{-n} f(t) \div \int_s^\infty ds_n \int_{s_n}^\infty ds_{n-1} \ldots \int_{s_2}^\infty \widetilde{f}(s_1) ds_1, \ n \in \mathbb{N}; \tag{C.2.10}$$

$$f^{(n)}(t) \div s^n \widetilde{f}(s) - s^{n-1} f(0) - s^{n-2} f'(0) - \ldots - f^{(n-1)}(0), \ n \in \mathbb{N}; \tag{C.2.11}$$

$$\int_0^t dt_n \int_0^{t_n} dt_{n-1} \ldots \int_0^{t_2} f(t_1) dt_1 \div s^{-n} \widetilde{f}(s), \ n \in \mathbb{N}; \tag{C.2.12}$$

$$\left(t \frac{d}{dt}\right)^n f(t) \div \left(-\frac{d}{ds} s\right)^n \widetilde{f}(s), \ n \in \mathbb{N}; \tag{C.2.13}$$

$$\left(\frac{d}{dt} t\right)^n f(t) \div \left(-s \frac{d}{ds}\right)^n \widetilde{f}(s), \ n \in \mathbb{N}; \tag{C.2.14}$$

$$(f_1 * f_2)(t) \div \widetilde{f}_1(s) \cdot \widetilde{f}_2(s), \tag{C.2.15}$$

where

$$(f_1 * f_2)(t) = \int_0^t f_1(\tau) f_2(t-\tau) d\tau \tag{C.2.16}$$

is the so-called *Laplace convolution*.

Among the simple sufficient existence conditions for the Laplace transform we point out the following:

The Laplace Transform Exists in the Half-Plane $\operatorname{Re} s > a \ (a > 0)$ provided that the original f is locally integrable on $\mathbb{R}_+ = (0, +\infty)$ and has an exponential growth of order a at infinity, i.e. there exists a positive constant $K > 0$ and positive $t_0 > 0$ such that

$$|f(t)| < K e^{at}, \ \forall t \geq t_0.$$

From the formula (C.2.15) follows immediately the Laplace transform of the Riemann–Liouville fractional integral

$$\left(\mathcal{L}I_{0+}^{\alpha}\varphi\right)(s) = s^{-\alpha}\left(\mathcal{L}\varphi\right)(s), \ \operatorname{Re}\alpha > 0. \tag{C.2.17}$$

The Laplace transform of the Riemann–Liouville fractional derivative is given by the formula (see, e.g., [OldSpa74, p. 134])

$$\left(\mathcal{L}D_{0+}^{\alpha}\varphi\right)(s) = s^{\alpha}\left(\mathcal{L}\varphi\right)(s) - \sum_{k=1}^{n} s^{\alpha-k}\left(D_{0+}^{\alpha}\varphi\right)(t)\Big|_{t=0}, \ n-1 < \operatorname{Re} \leq n, \tag{C.2.18}$$

and the Laplace transform of the Caputo fractional derivative, by the formula

$$\left(\mathcal{L}\,{}^{C}D_{0+}^{\alpha}\varphi\right)(s) = s^{\alpha}\left(\mathcal{L}\varphi\right)(s) - \sum_{k=0}^{n-1} s^{\alpha-k-1}\varphi(0+), \ n-1 < \operatorname{Re} \leq n. \tag{C.2.19}$$

Formula (C.2.19) is simplified in the case of the Marchaud fractional derivative $\mathbb{D}_{+}^{\alpha}$ and Grünwald–Letnikov fractional derivative (see Appendix E) ${}_{GL}D_{0+}^{\alpha}$:

$$\left(\mathcal{L}\,\mathbb{D}_{+}^{\alpha}\varphi\right)(s) = s^{\alpha}\left(\mathcal{L}\varphi\right)(s), \ \left(\mathcal{L}\,{}_{GL}D_{0+}^{\alpha}\varphi\right)(s) = s^{\alpha}\left(\mathcal{L}\varphi\right)(s). \tag{C.2.20}$$

C.3 The Mellin Transform

Let

$$\mathcal{M}\{f(r);s\} = f^{*}(s) = \int_{0}^{+\infty} f(r)\,r^{s-1}\,dr, \ \gamma_{1} < \operatorname{Re}(s) < \gamma_{2} \tag{C.3.1}$$

be the Mellin transform of a sufficiently well-behaved function $f(r)$, and let

$$\mathcal{M}^{-1}\{f^{*}(s);r\} = f(r) = \frac{1}{2\pi i}\int_{\gamma-i\infty}^{\gamma+i\infty} f^{*}(s)\,r^{-s}\,ds \tag{C.3.2}$$

be the inverse Mellin transform, where $r > 0$, $\gamma = \operatorname{Re}(s)$, $\gamma_{1} < \gamma < \gamma_{2}$.

For the existence of the Mellin transform and the validity of the inversion formula we need to recall the following theorems, adapted from Marichev's treatise [Mari83], see Theorems 11, 12, on page 39.

Theorem C.3 *Let $f(r) \in L^{c}(\epsilon, E)$, $0 < \epsilon < E < \infty$, be continuous in the intervals $(0, \epsilon]$, $[E, \infty)$, and let $|f(r)| \leq M\,r^{-\gamma_{1}}$ for $0 < r < \epsilon$, $|f(r)| \leq M\,r^{-\gamma_{2}}$ for $r > E$, where M is a constant. Then for the existence of a strip in the s-plane in which $f(r)\,r^{s-1}$ belongs to $L^{c}(0, \infty)$ it is sufficient that $\gamma_{1} < \gamma_{2}$. When this condition holds, the Mellin transform $f^{*}(s)$ exists and is analytic in the vertical strip $\gamma_{1} < \gamma = \operatorname{Re}(s) < \gamma_{2}$.*

Theorem C.4 *If $f(t)$ is piecewise differentiable, and $f(r) r^{\gamma-1} \in L^c(0, \infty)$, then the formula (C.3.2) holds true at all points where $f(r)$ is continuous and the (complex) integral in it must be understood in the sense of the Cauchy principal value.*

We refer to specialized treatises and/or handbooks, see, e.g. [BatErd54a, Mari83, PrBrMa-V3], for more details and tables on the Mellin transform. Here, for our convenience we recall the main rules that are useful when adapting the formulae from the handbooks and which are also relevant in the following.

Denoting by $\overset{\mathcal{M}}{\leftrightarrow}$ the juxtaposition of a function $f(r)$ with its Mellin transform $f^*(s)$, the main rules are:

$$f(ar) \overset{\mathcal{M}}{\leftrightarrow} a^{-s} f^*(s), \quad a > 0, \tag{C.3.3}$$

$$r^a f(r) \overset{\mathcal{M}}{\leftrightarrow} f^*(s+a), \tag{C.3.4}$$

$$f(r^p) \overset{\mathcal{M}}{\leftrightarrow} \frac{1}{|p|} f^*(s/p), \quad p \neq 0, \tag{C.3.5}$$

$$h(r) = \int_0^\infty \frac{1}{\rho} f(\rho) g(r/\rho) \, d\rho \overset{\mathcal{M}}{\leftrightarrow} h^*(s) = f^*(s) g^*(s). \tag{C.3.6}$$

The Mellin convolution formula (C.3.6) is useful in treating integrals of Fourier type for $x = |x| > 0$:

$$I_c(x) = \frac{1}{\pi} \int_0^\infty f(\kappa) \cos(\kappa x) \, d\kappa, \tag{C.3.7}$$

$$I_s(x) = \frac{1}{\pi} \int_0^\infty f(\kappa) \sin(\kappa x) \, d\kappa, \tag{C.3.8}$$

when the Mellin transform $f^*(s)$ of $f(\kappa)$ is known. In fact we recognize that the integrals $I_c(x)$ and $I_s(x)$ can be interpreted as Mellin convolutions (C.3.6) between $f(\kappa)$ and the functions $g_c(\kappa)$, $g_s(\kappa)$, respectively, with $r = 1/|x|$, $\rho = \kappa$, where

$$g_c(\kappa) := \frac{1}{\pi |x| \kappa} \cos\left(\frac{1}{\kappa}\right) \overset{\mathcal{M}}{\leftrightarrow} \frac{\Gamma(1-s)}{\pi |x|} \sin\left(\frac{\pi s}{2}\right) := g_c^*(s), \quad 0 < \operatorname{Re}(s) < 1, \tag{C.3.9}$$

$$g_s(\kappa) := \frac{1}{\pi |x| \kappa} \sin\left(\frac{1}{\kappa}\right) \overset{\mathcal{M}}{\leftrightarrow} \frac{\Gamma(1-s)}{\pi |x|} \cos\left(\frac{\pi s}{2}\right) := g_s^*(s), \quad 0 < \operatorname{Re}(s) < 2. \tag{C.3.10}$$

The Mellin transform pairs (C.3.9)–(C.3.10) have been adapted from the tables in [Mari83] by using (C.3.3)–(C.3.5) and the duplication and reflection formulae for the Gamma function. Finally, the inverse Mellin transform representation (C.3.2) provides the required integrals as

Appendix C: Integral Transforms

$$I_c(x) = \frac{1}{\pi x} \frac{1}{2\pi i} \int_{\gamma-i\infty}^{\gamma+i\infty} f^*(s)\, \Gamma(1-s) \sin\left(\frac{\pi s}{2}\right) x^s\, ds, \quad x > 0,\ 0 < \gamma < 1,$$
(C.3.11)

$$I_s(x) = \frac{1}{\pi x} \frac{1}{2\pi i} \int_{\gamma-i\infty}^{\gamma+i\infty} f^*(s)\, \Gamma(1-s) \cos\left(\frac{\pi s}{2}\right) x^s\, ds, \quad x > 0,\ 0 < \gamma < 2.$$
(C.3.12)

First, we present some known properties of the *Mellin integral transform*.

Theorem C.5 *Let* $x^{k-\frac{1}{2}} f(x) \in L_2(0, +\infty)$. *Then the following four statements hold:*

1. *The functions*

$$\mathcal{M}(s; a) = \int_{1/a}^{a} f(x) x^{s-1} dx, \quad \operatorname{Re} s = k, \qquad (C.3.13)$$

converge in the mean when $a \to +\infty$ *on the line* $(k - i\infty, k + i\infty)$, *that is, there exists a function* $\mathcal{M}(s) \in L_2(k - i\infty, k + i\infty)$ *such that*

$$\lim_{a \to +\infty} \int_{k-i\infty}^{k+i\infty} |\mathcal{M}(s) - \mathcal{M}(s; a)|^2\, |ds| = 0. \qquad (C.3.14)$$

2. *The functions*

$$f(x, a) = \frac{1}{2\pi i} \int_{k-ia}^{k+ia} \mathcal{M}(s) x^{-s} ds, \quad 0 < x < +\infty \qquad (C.3.15)$$

converge in the mean on the semi-axis with the weight function x^{2k-1} *to the function* $f(x)$ *when* $a \to +\infty$, *that is,*

$$\lim_{a \to +\infty} \int_0^{+\infty} |f(x) - f(x, a)|^2 x^{2k-1}\, dx = 0, \qquad (C.3.16)$$

moreover, almost everywhere on the semi-axis $(0, +\infty)$ *we have*

$$f(x) = \frac{1}{2\pi i} \frac{d}{dx} \int_{k-i\infty}^{k+i\infty} \mathcal{M}(s) \frac{x^{1-s}}{1-s}\, ds. \qquad (C.3.17)$$

3. *The Parseval identity*

$$\int_0^{+\infty} |f(x)|^2 x^{2k-1} dx = \frac{1}{2\pi} \int_{-\infty}^{+\infty} |\mathcal{M}(k + it)|^2 dt \qquad (C.3.18)$$

is valid.

4. *Conversely, for any function* $\mathcal{M}(s) \in L_2(k - i\infty, k + i\infty)$ *the functions* (C.3.15) *converge in the mean when* $a \to +\infty$ *in the sense of* (C.3.16) *to some function* $f(x) \in L(0, +\infty)$ *which can be represented in the form* (C.3.17). *The functions*

(C.3.13) converge in the mean in the sense of (C.3.14) to the function $\mathcal{M}(s)$ when $a \to +\infty$ and, moreover, the Parseval identity (C.3.18) holds.

We present a number of important formulas for the Mellin transform.

$$(\mathcal{M} x^\alpha f(x))(p) = f^*(p + \alpha), \tag{C.3.19}$$

$$\left(\mathcal{M} \frac{(1-x)^{\alpha-1} H(1-x)}{\Gamma(\alpha)}\right)(p) = \frac{\Gamma(p)}{\Gamma(p+\alpha)}, \tag{C.3.20}$$

where $H(x)$ is the Heaviside function

$$H(x) = \begin{cases} 1, & 0 \leq x < +\infty, \\ 0, & -\infty < x < 0, \end{cases}$$

$$(\mathcal{M} W^{-\alpha} f(x))(p) = \frac{\Gamma(p)}{\Gamma(p+\alpha)} f^*(p+\alpha), \tag{C.3.21}$$

where $W^{-\alpha}$ is the Weyl fractional integral

$$W^{-\alpha} f(x) = \frac{1}{\Gamma(\alpha)} \int_x^\infty (t-x)^{\alpha-1} f(t) dt, \quad 0 < \operatorname{Re} \alpha < 1, \, x > 0. \tag{C.3.22}$$

The Mellin transform of the Riemann–Liouville fractional integral is given by the formula

$$(\mathcal{M} I_{0+}^\alpha f(x))(p) = \frac{\Gamma(1-\alpha-p)}{\Gamma(1-p)} f^*(p+\alpha), \quad \operatorname{Re}(\alpha+p) < 1. \tag{C.3.23}$$

The Mellin transform of the Riemann–Liouville fractional derivative is represented in the form

$$(\mathcal{M} D_{0+}^\alpha f(x))(p) = \frac{\Gamma(1+\alpha-p)}{\Gamma(1-p)} f^*(p-\alpha)$$

$$+ \sum_{k=0}^{n-1} \frac{\Gamma(1+k-p)}{\Gamma(1-p)} \left[(D_{0+}^{\alpha-k-1} f)(x) x^{p-k-1}\right]_{x=0}^{x=\infty}, \tag{C.3.24}$$

and the formula for the Mellin transform of the Caputo derivative has the form

$$(\mathcal{M}\,{}^C D_{0+}^\alpha f(x))(p) = \frac{\Gamma(1+\alpha-p)}{\Gamma(1-p)} f^*(p-\alpha). \tag{C.3.25}$$

Appendix C: Integral Transforms

C.4 Simple Examples and Tables of Transforms of Basic Elementary and Special Functions

Below the selected values of integral transforms are given (see Tables C.1, C.2 and C.3).

Table C.1 Table of selected values of the Fourier integral transform

n/n	f(t)	$(\mathcal{F}f)(\kappa) = F(\kappa) = \widehat{f}(\kappa) = \int_{-\infty}^{+\infty} e^{i\kappa t} f(t) dt$	Conditions
1.	$\exp(-a\|t\|)$	$\frac{2a}{(a^2+\kappa^2)}$	$a > 0$
2.	$t\exp(-a\|t\|)$	$\frac{4ai}{(a^2+\kappa^2)^2}$	$a > 0$
3.	$\exp(-at^2)$	$\sqrt{\frac{\pi}{a}} \exp\left(-\frac{\kappa^2}{4a}\right)$	$a > 0$
4.	$\frac{1}{t^2+a^2}$	$\frac{\pi}{a} \exp(-a\|\kappa\|)$	$a > 0$
5.	$\frac{t}{t^2+a^2}$	$\frac{-\pi\kappa i}{2a} \exp(-a\|\kappa\|)$	$a > 0$
6.	$\begin{cases} c, & a \le t \le b, \\ 0, & \text{otherwise} \end{cases}$	$-\frac{ic}{\kappa}\left\{e^{ib\kappa} - e^{ia\kappa}\right\}$	$a \ge 0$
7.	$\frac{\sin at}{t}$	$2H(a - \|\kappa\|)$	$a > 0$
8.	$\|t\|^\alpha$	$2\Gamma(\alpha+1) \frac{\cos\left(\frac{\pi}{2}(\alpha+1)\right)}{\|\kappa\|^{1+\alpha}}$	$\alpha \in \mathbb{C}, \alpha \ne 0,$ $\alpha \ne -1 - 2k, k \in \mathbb{N}_0$
9.	$\|t\|^\alpha \operatorname{sgn} t$	$2i\Gamma(\alpha+1) \frac{\cos\left(\frac{\pi\alpha}{2}\right)}{\|\kappa\|^{1+\alpha}} \operatorname{sgn} \kappa$	$\alpha \ne -2k, k \in \mathbb{N}$
10.	$\exp\{-t(a - i\omega)\}H(t)$	$\frac{i}{\omega+\kappa+ia}$	$a > 0$
11.	$\frac{H(a-\|t\|)}{\sqrt{a^2-t^2}}$	$2\pi J_0(-a\kappa)$	$a > 0$
12.	$\exp(-at)H(t)$	$\frac{a+i\kappa}{a^2+\kappa^2}$	$a > 0$
13.	$\frac{1}{t^n}$	$\pi i \left[\frac{(i\kappa)^{n-1}}{(n-1)!}\right] \operatorname{sign} \kappa$	$n \in \mathbb{N}$
14.	$E_\alpha(\|t\|)$	$2\pi\delta(\kappa) - \frac{2}{\kappa^2} {}_2\Psi_1\left[\begin{matrix}(2,2), (1,1) \\ (\alpha+1, 2\alpha)\end{matrix} \bigg| -\frac{1}{\kappa^2}\right]$	$\alpha > 1$
15.	$E_{\alpha,\beta}(\|t\|)$	$\frac{2\pi\delta(\kappa)}{\Gamma(\beta)} - \frac{2}{\kappa^2} {}_2\Psi_1\left[\begin{matrix}(2,2), (1,1) \\ (\alpha+\beta, 2\alpha)\end{matrix} \bigg| -\frac{1}{\kappa^2}\right]$	$\alpha > 1, \beta \in \mathbb{C}$

Table C.2 Table of selected values of the Laplace integral transform

n/n	$f(t)$	$(\mathcal{L}f)(s) = F(s) = \widetilde{f}(s) = \int_0^{+\infty} e^{-st} f(t)\, dt$	Conditions
1.	t^a	$\dfrac{\Gamma(a+1)}{s^{a+1}}$	$a > -1$
2.	$e^{at}\cos bt$	$\dfrac{s-a}{(s-a)^2+b^2}$	
	$E_\alpha(at^\alpha)$	$\dfrac{s^{\alpha-1}}{s^\alpha - a}$	$\operatorname{Re}\alpha > 0,\ s > |a|^{\frac{1}{\operatorname{Re}\alpha}}$
3.	$e^{at}\sin bt$	$\dfrac{b}{(s-a)^2+b^2}$	
4.	$t\cos bt$	$\dfrac{s^2 - b^2}{(s^2+b^2)^2}$	
5.	$t\sin bt$	$\dfrac{2bs}{(s^2+b^2)^2}$	
6.	$\dfrac{1}{\sqrt{t}}\exp\left(-\dfrac{a}{t}\right)$	$\sqrt{\dfrac{\pi}{s}}\exp\left(-2\sqrt{as}\right)$	
7.	$t^{\alpha-1}\,{}_2F_1(a,b;c;t)$	$\dfrac{\Gamma(\alpha)}{s^\alpha}\,{}_3F_1(\alpha,a,b;c;\tfrac{1}{s})$	$a,b\in\mathbb{C},\ c\in\mathbb{C}\setminus\mathbb{Z}_0^-,$ $\operatorname{Re}\alpha > 0;\ \operatorname{Re}s > 0$
8.	$t^{\alpha-1}\Phi(a;c;t)$	$\dfrac{\Gamma(\alpha)}{s^\alpha}\,{}_2F_1(\alpha,a;b;\tfrac{1}{s})$	$a\in\mathbb{C},\ c\in\mathbb{C}\setminus\mathbb{Z}_0^-,$ $\operatorname{Re}\alpha > 0;\ \operatorname{Re}s > 0$
9.	$t^{\alpha-1}\Psi(a;c;t)$	$\dfrac{\Gamma(\alpha)\Gamma(\alpha-c-1)}{\Gamma(a-c+\alpha+1)} \times$ ${}_2F_1(\alpha,\alpha-c+1;$ $a-c+\alpha+1;1-\tfrac{1}{s})$	$a\in\mathbb{C},\ c\in\mathbb{C}\setminus\mathbb{Z}_0^-,$ $\operatorname{Re}\alpha > 0;\ \operatorname{Re}c < 1+\operatorname{Re}\alpha;$ $|1-s| < 1$
10.	$\exp(a^2 t)\,\mathrm{erf}(a\sqrt{t})$	$\dfrac{a}{\sqrt{s}(s-a^2)}$	
11.	$\exp(a^2 t)\,\mathrm{erfc}(a\sqrt{t})$	$\dfrac{a}{\sqrt{s}(\sqrt{s}+a)}$	
12.	$E_\alpha(at^\alpha)$	$\dfrac{s^{\alpha-1}}{s^\alpha-a}$	$\operatorname{Re}\alpha > 0,\ s > |a|^{\frac{1}{\operatorname{Re}\alpha}}$
13.	$E_\alpha(at^\alpha)$	$\dfrac{s^{\alpha-1}}{s^\alpha-a}$	$\operatorname{Re}\alpha > 0,\ s > |a|^{\frac{1}{\operatorname{Re}\alpha}}$
14.	$t^{\beta-1}E_{\alpha,\beta}(at^\alpha)$	$\dfrac{s^{\alpha-\beta}}{s^\alpha-a}$	$\operatorname{Re}\alpha > 0,\ \operatorname{Re}\beta > 0,$ $s > |a|^{\frac{1}{\operatorname{Re}\alpha}}$
15.	$t^{m\alpha+\beta-1}E_{\alpha,\beta}^{(m)}(at)$	$\dfrac{m!\,s^{\alpha-\beta}}{(s^\alpha-a)^{m+1}}$	$\operatorname{Re}\alpha > 0,\ \operatorname{Re}\beta > 0,$ $m\in\mathbb{N},\ s > |a|^{\frac{1}{\operatorname{Re}\alpha}}$
16.	$t^{\rho-1}E_{\alpha,\beta}(at^\gamma)$	$\dfrac{1}{s^\rho}\,{}_2\Psi_1\!\left[\dfrac{a}{s^\alpha}\middle|\begin{array}{c}(1,1),(\rho,\gamma)\\ (\beta,\alpha)\end{array}\right]$	$\operatorname{Re}\alpha > 0,\ \operatorname{Re}\beta > 0,$ $\operatorname{Re}\gamma > 0,\ \operatorname{Re}\rho > 0,$ $s > |a|^{\frac{1}{\operatorname{Re}\alpha}}$
17.	$t^{\beta-1}E_{\alpha,\beta}^\gamma(at^\alpha)$	$s^{-\beta}(1-as^{-\alpha})^{-\gamma}$	$\operatorname{Re}\alpha > 0,\ \operatorname{Re}\beta > 0,$ $\operatorname{Re}\gamma > 0,$ $s > |a|^{\frac{1}{\operatorname{Re}\alpha}}$
18.	$E_{\beta,\gamma}^\delta(t)$	$\dfrac{1}{s}\,{}_2\Psi_1\!\left[\dfrac{1}{s}\middle|\begin{array}{c}(\delta,1),(1,1)\\ (\beta,\gamma)\end{array}\right]$	$\operatorname{Re}\alpha > 0,\ \operatorname{Re}\beta > 0,$ $\operatorname{Re}\gamma > 0,\ \operatorname{Re}\delta > 0,$ $\operatorname{Re}\rho > 0,\ s > |a|^{\frac{1}{\operatorname{Re}\alpha}}$
19.	$E_\rho((\alpha_j,\beta_j)_{1,m};t)$	$\dfrac{1}{s}\,{}_2\Psi_m\!\left[\dfrac{1}{s}\middle|\begin{array}{c}(\rho,1),(1,1)\\ (\alpha_j,\beta_j)_{1,m}\end{array}\right]$	$\operatorname{Re}\alpha_j > 0,\ \operatorname{Re}\beta_j > 0,$ $\operatorname{Re}\rho > 0,\ s > |a|^{\frac{1}{\operatorname{Re}\alpha}}$

Table C.3 Table of selected values of the Mellin integral transform

n/n	$f(t)$	$(\mathcal{M}f)(p) = F(p) = f^*(p) = \int_0^{+\infty} f(t)t^{p-1}dt$	Conditions
1.	$e^{-\lambda t}$	$\frac{\Gamma(p)}{\lambda^p}$	$\operatorname{Re}\lambda > 0$, $\operatorname{Re} p > 0$
2.	$e^{-\lambda t^2}$	$\frac{\Gamma(p/2)}{2\lambda^{p/2}}$	$\operatorname{Re}\lambda > 0$, $\operatorname{Re} p > 0$
3.	$(t+1)^{-\sigma}$	$\frac{\Gamma(p)\Gamma(\sigma-p)}{\Gamma(\sigma)}$	$0 < \operatorname{Re} p < \operatorname{Re}\sigma$
4.	$(t^\alpha+1)^{-\sigma}$	$\frac{\Gamma(p/\alpha)\Gamma(\sigma-p/\alpha)}{\alpha\Gamma(\sigma)}$	$0 < \operatorname{Re}\frac{p}{\alpha} < \operatorname{Re}\sigma$
5.	$\frac{t^a}{(1+t)^b}$	$\frac{\Gamma(a+p)\Gamma(b-a-p)}{\Gamma(\nu-\frac{p}{2}+1)(b)}$	
6.	$\cos at$	$\frac{\Gamma(p)\cos\frac{\pi p}{2}}{a^p}$	$0 < \operatorname{Re} p < 1$
7.	$\sin at$	$\frac{\Gamma(p)\sin\frac{\pi p}{2}}{a^p}$	$0 < \operatorname{Re} p < 1$
8.	$_2F_1(a,b;c;-t)$	$\frac{\Gamma(p)\Gamma(a-p)\Gamma(b-p)\Gamma(c)}{\Gamma(c-p)\Gamma(a)\Gamma(b)}$	$a,b,c \in \mathbb{C}, c \notin \mathbb{Z}_0$; $0 < \operatorname{Re} p < \min\{\operatorname{Re} a, \operatorname{Re} b\}$
9.	$\Phi(a;c;-t)$	$\frac{\Gamma(p)\Gamma(a-p)\Gamma(c)}{\Gamma(c-p)\Gamma(a)}$	$a,c \in \mathbb{C}, c \notin \mathbb{Z}_0$; $0 < \operatorname{Re} p < \operatorname{Re} a$
10.	$t^{1/2}J_\nu(t)$	$\frac{2^{p-1/2}\Gamma\left(\frac{1}{2}\left(p+\nu+\frac{1}{2}\right)\right)}{\Gamma\left(\frac{1}{2}\left(\nu-p+\frac{3}{2}\right)\right)}$	
11.	$t^{-\nu}J_\nu(at)$	$\frac{2^{p-\nu-1}a^{\nu-p}\Gamma(\frac{p}{2})}{\Gamma(\nu-\frac{p}{2}+1)}$	
12.	$\operatorname{erfc}(t)$	$\frac{\Gamma\left(\frac{p+1}{2}\right)}{p\sqrt{\pi}}$	
13.	$E_\alpha(t)$	$\frac{\Gamma(p)\Gamma(1-p)}{\Gamma(1-\alpha p)}$	$0 < \operatorname{Re} p < 1$
14.	$E_{\alpha,\beta}(t)$	$\frac{\Gamma(p)\Gamma(1-p)}{\Gamma(\beta-\alpha p)}$	$0 < \operatorname{Re} p < 1$
15.	$E_{\alpha,\beta}^\gamma(t)$	$\frac{\Gamma(p)\Gamma(\gamma-p)}{\Gamma(\gamma)\Gamma(\beta-\alpha p)}$	$0 < \operatorname{Re} p < \operatorname{Re}\gamma$
16.	$E_\rho((\alpha_j,\beta_j)_{1,m};-t)$	$\frac{\Gamma(p)\Gamma(\rho-p)}{\Gamma(\rho)\prod_{j=1}^m \Gamma(\beta_j-\alpha_j p)}$	$0 < \operatorname{Re} p < \operatorname{Re}\rho$

C.5 Historical and Bibliographical Notes

Here we present a few historical remarks on the early development of the integral transform method. For an extended historical exposition we refer to [DebBha15, Chap. 1].

The Fourier integral theorem was originally stated in J. Fourier's famous treatise entitled *La Théorie Analytique da la Chaleur* (1822), and its deep significance was recognized by mathematicians and mathematical physicists. This theorem is one of the most monumental results of modern mathematical analysis and has widespread physical and engineering applications. Fourier's treatise provided the modern mathematical theory of heat conduction, Fourier series, and Fourier integrals with applications. He gave a series of examples before stating that an arbitrary function defined on a finite interval can be expanded in terms of a trigonometric series which is now universally known as the Fourier series. In an attempt to extend his new ideas to functions defined on an infinite interval, Fourier discovered an integral transform and its inversion formula, which are now well-known as the Fourier transform and the inverse Fourier transform. However, this celebrated idea of Fourier was known to P.S. Laplace and A.L. Cauchy as some of their earlier work involved this transformation. On the other hand, S.D. Poisson also independently used the method of transform in his research on the propagation of water waves.

It was the work of Cauchy that contained the exponential form of the Fourier Integral Theorem. Cauchy also introduced the functions of the operator D:

$$\phi(D)f(x) = \frac{1}{2\pi} \int\limits_{-\infty}^{+\infty} \int\limits_{-\infty}^{+\infty} \phi(i\kappa) e^{i\kappa(x-y)} f(y) \mathrm{d}y \mathrm{d}\kappa,$$

which led to the modern form of operator theory.

The birth of the operational method in its popularization as a powerful method for solving differential equations is probably due to O. Heaviside (see, e.g., [Hea93]). He developed this technique as a purely algebraic one, using and developing the classical ideas of Fourier, Laplace and Cauchy. An extended use of the machinery of complex analytic functions in this theory was proposed by T.J. Bromwich (see, e.g., [Bro09, Bro26]). In this direction several types of integral transform appeared. In particular, elaborating the ideas of B. Riemann, Mellin introduced a new type of integral transform which was later given his name. It turned out that (see [Sla66, Mari83]) the Mellin transform is highly suitable for the study of properties of a wide class of special functions, namely G- and H-functions [MaSaHa10]. In [Mari83], a calculation method for integrals with a ratio of products of Gamma functions was developed. This method is based on the properties of the Mellin transform. Based on this, the handbook of Mellin transforms [BrMaSa19] was established, containing the most recent results for this type of integral transform. Applications of the Mellin transform technique are described in [LucKir19].

Many results were obtained by different authors due to a combination of the complex analytic approach and results from the theory of special functions which rapidly developed in the first part of the 20th century. These results led to the theory of operational calculus in its modern form. We mention here several treaties on integral transform theory [Dav02, DebBha07, DebBha15, DitPru65, Doe74, Mik59, Sne74, Tit86, Wid46]. Applications of the integral transform method are presented in many books on integral and differential equations (see, e.g., [AnKoVa93, Boa83, PolMan08]), in particular those related to the fractional calculus [Die10, KiSrTr06,

SaKiMa93]. Tables of integral transforms (see [BatErd54a, GraRyz00, Mari83, Obe74]) constitute a very useful source for applications.

C.6 Exercises

C.6.1. ([DebBha07, p. 119]) Evaluate the Fourier transform of the functions

$$a)\, f(t) = t\exp\left\{-\frac{at^2}{2}\right\},\, a > 0 \quad b)\, f(t) = t^2 \exp\left\{-\frac{t^2}{2}\right\}.$$

C.6.2. ([DebBha07, p. 121]) Solve the following integral equations with respect to the unknown function $y(t)$

$$a)\, \int_{-\infty}^{+\infty} \phi(x-t) y(t) dt = g(x),$$

$$b)\, \int_{-\infty}^{+\infty} \exp\left(-at^2\right) y(x-t) dt = \exp\left(-ax^2\right),\, a > b > 0.$$

Hint. Use the Fourier transform.

C.6.3. ([DebBha07, p. 128]) Use the Fourier transform to solve the boundary value problem

$$u_{xx} + u_{yy} = x\exp\left(-x^2\right),\, -\infty < x < +\infty,\, 0 < y < +\infty,$$
$$u(x, 0) = 0,\, -\infty < x < +\infty,$$

in the class of continuously differentiable functions such that

$$\lim_{y \to +\infty} u(x, y) = 0,\, \forall x,\, -\infty < x < +\infty.$$

Answer.

$$u(x, y) = \sqrt{2} \int_0^{+\infty} [1 - \exp(-ty)] \frac{\sin tx}{t} \exp\left(-\frac{t^2}{4}\right) dt.$$

C.6.4. ([DebBha07, p. 173]) Find the Laplace transform of the functions:

a) $2t + a\sin at,\, a > 0,$ \qquad b) $(1 - 2t)\exp\{-2t\},$

c) $H(t-a)\exp\{t-a\},\, a > 0,$ \quad d) $(t-a)^k H(t-a),\, a > 0,\, k \in \mathbb{N}.$

C.6.5. ([DebBha07, p. 174]) Evaluate the inverse Laplace transform of the functions

a) $\frac{1}{(s-a)(s-b)^2}$, $a, b > 0$; b) $\frac{1}{s^2(s-a)^2}$, $a > 0$;

c) $\frac{1}{s^2(s^2+a^2)}$, $a > 0$; d) $\frac{s}{(s^2+a^2)(s^2+b^2)}$, $a, b > 0$.

C.6.6. ([DebBha07, p. 179]) Using the change of variables, $s = c + i\omega$, show that the inverse Laplace transformation is a Fourier transformation, that is,

$$f(t) = \left(\mathcal{L}^{-1}\tilde{f}(s)\right)(t) = \frac{e^{ct}}{2\pi}\int_{-\infty}^{+\infty}\tilde{f}(c+i\omega)e^{i\omega t}d\omega.$$

C.6.7. ([DebBha07, p. 365]) Calculate the Mellin transform of the functions

a) $f(t) = H(a-t)$, $a > 0$; b) $f(t) = t^\alpha e^{-\beta t}$, $\alpha, \beta > 0$;

c) $f(t) = \frac{1}{1+t^2}$; d) $f(t) = J_0^2(t)$.

C.6.8. ([DebBha07, p. 365]) Prove that

$$\left(\mathcal{M}\frac{1}{(1+at)^n}\right)(p) = \frac{\Gamma(p)\Gamma(n-p)}{a^p\Gamma(n)}, \quad a > 0;$$

$$\left(\mathcal{M}t^{-n}J_n(at)\right)(p) = \frac{1}{2}\left(\frac{a}{2}\right)^{n-p}\frac{\Gamma\left(\frac{p}{2}\right)}{\Gamma\left(n-\frac{p}{2}+1\right)}, \quad a > 0, \ n > -\frac{1}{2}.$$

C.6.9. ([DebBha07, p. 370]) Prove the following relations of the Mellin transform to the Laplace and the Fourier transforms:

$$(\mathcal{M}f(t))(p) = \left(\mathcal{L}f(e^{-t})\right)(p);$$

$$(\mathcal{M}f(t))(a+i\omega) = \left(\mathcal{F}f(e^{-t})e^{-at}\right)(\omega).$$

C.6.10. ([KiSrTr06, p. 36, (1.7.34)]) Calculate the Laplace transform of the Bessel function of the first kind $J_\nu(t)$ for $\mathrm{Re}\,\nu > -1$, where

$$J_\nu(z) = \sum_{k=0}^{\infty}\frac{(-1)^k(z/2)^{2k+\nu}}{k!\Gamma(\nu+k+1)} \quad (z \in \mathbb{C}\setminus(-\infty, 0]; \ \nu \in \mathbb{C}).$$

Answer.

Appendix C: Integral Transforms

$$(\mathcal{L} J_\nu(t))(s) = \frac{1}{\sqrt{s^2+1}\left[s+\sqrt{s^2+1}\right]^\nu} \quad (\operatorname{Re} s > 0).$$

C.6.11. ([KiSrTr06, p. 36, (1.7.35)]) Calculate the Laplace transform of the Bessel function of the second kind $Y_\nu(t)$ for $|\operatorname{Re} \nu| < 1$, $\nu \neq 0$, where

$$Y_\nu(z) = \frac{\cos(\nu\pi) J_\nu(z) - J_{-\nu}(z)}{\sin \nu\pi} \quad (\nu \in \mathbb{C}\setminus\mathbb{Z}).$$

Answer.

$$(\mathcal{L} Y_\nu(t))(s) = \frac{\cot(\nu\pi) - \csc(\nu\pi)\left[s+\sqrt{s^2+1}\right]^{2\nu}}{\sqrt{s^2+1}\left[s+\sqrt{s^2+1}\right]^\nu} \quad (\operatorname{Re} s > 0).$$

C.6.12. ([KiSrTr06, p. 50, (1.10.10)]) Prove the following relation ($\alpha, \beta, \lambda \in \mathbb{C}$, $\operatorname{Re} \alpha > 0$)

$$\left(\mathcal{L}\left[t^{\alpha n+\beta-1}\left(\frac{\partial}{\partial \lambda}\right)^n E_{\alpha,\beta}(\lambda t^\alpha)\right]\right)(s) = \frac{n! s^{\alpha-\beta}}{(s^\alpha - \lambda)^{n+1}} \quad \left(\left|\frac{\lambda}{s^\alpha}\right| < 1\right).$$

C.6.13. ([KiSrTr06, p. 52, (1.10.27)]) Let $e_\alpha^{\lambda z} = z^{\alpha-1} E_{\alpha,\alpha}(\lambda z^\alpha)$ ($z \in \mathbb{C}$; $\operatorname{Re} \alpha > 0$; $\lambda \in \mathbb{C}$) be the so-called α-Exponential function. Prove the following relation for it:

$$\left(\mathcal{L}\left[\left(\frac{\partial}{\partial \lambda}\right)^n e_\alpha^{\lambda t}\right]\right)(s) = \frac{n!}{(s^\alpha - \lambda)^{n+1}} \quad (n \in \mathbb{N};\ \operatorname{Re} s > 0;\ \left|\frac{\lambda}{s^\alpha}\right| < 1).$$

C.6.14. ([MaSaHa10, p. 52, (2.29)]) Calculate the Laplace transform of the Meyer G-function

$$t^{\rho-1} G_{p,q}^{m,n}\left[at^\sigma \Big| \begin{matrix} a_1,\ldots,a_p \\ b_1,\ldots,b_q \end{matrix}\right]$$

$\rho \in \mathbb{C}$, $\sigma > 0$, $\operatorname{Re}\rho + \sigma \min_{1\leq j\leq m} \operatorname{Re} b_j > 0$,

$$|\arg a| < \frac{\pi c^*}{2},\ c^* = m+n - \frac{p+q}{2} > 0;\ \operatorname{Re} s > 0.$$

C.6.15. ([MaSaHa10, p. 50, (2.19)]) Calculate for all $s \in \mathbb{C}$, $\operatorname{Re} s > 0$, the Laplace transform of the Fox H-function

$$t^{\rho-1} H_{p,q}^{m,n}\left[at^\sigma \Big| \begin{matrix} (a_1,\alpha_1),\ldots,(a_p,\alpha_p) \\ (b_1,\beta_1),\ldots,(b_q,\beta_q) \end{matrix}\right]$$

$\rho, a \in \mathbb{C}$, $\sigma > 0$,

$$\alpha > 0, \ |\arg a| < \frac{\alpha\pi}{2} \quad \text{or } \alpha = 0 \text{ and } \operatorname{Re}\delta < -1$$

$$\operatorname{Re}\rho + \sigma \min_{1 \le j \le m} \frac{\operatorname{Re} b_j}{\beta_j} > 0, \ \text{for } \alpha > 0 \text{ or } \alpha = 0, \mu \ge 0,$$

and

$$\operatorname{Re}\rho + \sigma \min_{1 \le j \le m} \left[\frac{\operatorname{Re} b_j}{\beta_j} + \frac{\operatorname{Re}\delta + 1/2}{\mu} \right] > 0, \ \text{for } \alpha = 0, \mu < 0,$$

where

$$\alpha = \sum_{j=1}^n \alpha_j - \sum_{j=n+1}^p \alpha_j + \sum_{j=1}^m \beta_j - \sum_{j=m+1}^q \beta_j;$$

$$\delta = \sum_{j=1}^q b_j - \sum_{j=1}^p a_j + \frac{p-q}{2}; \ \mu = \sum_{j=1}^q \beta_j - \sum_{j=1}^p \alpha_j.$$

C.6.16. ([KiSrTr06, p. 55, (1.11.6-7)]) Calculate the Laplace and Mellin transforms of the classical Wright function $\phi(\alpha, \beta; t)$, where

$$\phi(\alpha, \beta; z) = \sum_{k=0}^\infty \frac{z^k}{k!\Gamma(\alpha k + \beta)}.$$

Answers.

$$(\mathcal{L}\phi(\alpha, \beta; t))(s) = \frac{1}{s} E_{\alpha,\beta}\left(\frac{1}{s}\right) \quad (\alpha > -1; \ \beta \in \mathbb{C}; \ \operatorname{Re} s > 0);$$

$$(\mathcal{M}\phi(\alpha, \beta; t))(s) = \frac{\Gamma(s)}{\Gamma(\beta - \alpha s)} \quad (\alpha > -1; \ \beta \in \mathbb{C}; \ \operatorname{Re} s > 0).$$

C.6.17. ([MaSaHa10, p. 46, Exercise 2.1]) Find the Mellin transform of the Gauss hypergeometric function $_2F_1(a, b; c; -t)$ $(a, b, c \in \mathbb{C})$.
Answer.

$$(\mathcal{M}\,_2F_1(a, b; c; -t))(s) = \frac{\Gamma(s)\Gamma(a-s)\Gamma(b-s)\Gamma(c)}{\Gamma(c-s)\Gamma(a)\Gamma(b)}$$

$$(\min\{\operatorname{Re} a, \operatorname{Re} b\} > \operatorname{Re} s > 0).$$

C.6.18. ([MaSaHa10, p. 48, (2.9)]) Calculate the Mellin transform of the Meyer G-function

$$G_{p,q}^{m,n}\left[at \left| \begin{matrix} a_1, \ldots, a_p \\ b_1, \ldots, b_q \end{matrix} \right. \right]$$

Appendix C: Integral Transforms

$$a \in \mathbb{C}, \ |\arg a| \frac{\pi c^*}{2}, \ c^* = m + n - \frac{p+q}{2} > 0;$$

$$s \in \mathbb{C}, \ -\min_{1 \le j \le m} \operatorname{Re} b_j < \operatorname{Re} s < 1 - \max_{1 \le j \le n} \operatorname{Re} a_j.$$

Appendix D
The Mellin–Barnes Integral

D.1 Definition. Contour of Integration

In the modern theory of special functions it has become customary to call to an integral of the type (see, e.g., [ParKam01])

$$I(z) = \frac{1}{2\pi i} \int_{\mathcal{L}} f(s) z^{-s} ds \qquad (D.1)$$

a *Mellin–Barnes integral*. Here the density function $f(s)$ is usually a solution to a certain ordinary differential equation with polynomial coefficients. Thus this integral is similar to the Mellin transform applied to special types of originals (see, e.g., [Mari83]). The most crucial part of this definition is the choice of the contour of integration. The contour $\mathcal{L}$ is usually either a loop in the complex s plane, a vertical line indented to avoid certain poles of the integrand, or a curve midway between these two, in the sense of avoiding certain poles of the integrand and tending to infinity in certain fixed directions (see, e.g., [ParKam01]). A short introduction to the theory of Mellin–Barnes integrals is given in [ErdBat-1, pp. 49–50].

In order to be more precise we consider a density function $f(s)$ of the type

$$f(s) = \frac{A(s)B(s)}{C(s)D(s)}, \qquad (D.2)$$

where A, B, C, D are products of Gamma functions depending on parameters. Such integrals appear, in particular, in the representation of the solution to the hypergeometric equation

$$z(1-z)\frac{d^2 u}{dz^2} + [c - (b + a - 1)z]\frac{du}{dz} - abu = 0, \qquad (D.3)$$

i.e. in the representation of the Gauss hypergeometric function

$$F(a,b;c;z) = \sum_{n=0}^{\infty} \frac{(a)_n (b)_n}{(c)_n} z^n,$$

and in the representation of the solution to a more general equation, namely, the generalized hypergeometric equation

$$\left[z \prod_{j=0}^{p} \left(z \frac{d}{dz} + a_j \right) - z \frac{d}{dz} \prod_{k=0}^{q} \left(z \frac{d}{dz} + b_k - 1 \right) \right] u(z) = 0, \qquad (D.4)$$

i.e. in the representation of the generalized hypergeometric function

$$_pF_q(a_1, \ldots, a_p; b_1, \ldots, b_q; z) = \sum_{n=0}^{\infty} \frac{(a_1)_n \cdots (a_p)_n}{(b_1)_n \cdots (b_q)_n} z^n.$$

In [Mei36] more general classes of transcendental functions were introduced via a generalization of the Gauss hypergeometric function presented in the form of a series (commonly known now as Meijer G-functions). Later this definition was replaced by the Mellin–Barnes representation of the G-function

$$G_{p,q}^{m,n} \left[z \,\Big|\, \begin{matrix} a_1, \ldots, a_p \\ b_1, \ldots, b_q \end{matrix} \right] = \frac{1}{2\pi i} \int_{\mathcal{L}} \mathcal{G}_{p,q}^{m,n}(s) z^s \, ds, \qquad (D.5)$$

where $\mathcal{L}$ is a suitably chosen path, $z \neq 0$, $z^s := \exp[s(\ln|z| + i\arg z)]$ with a single valued branch of $\arg z$, and the integrand is defined as

$$\mathcal{G}_{p,q}^{m,n}(s) = \frac{\prod_{k=1}^{m} \Gamma(b_k - s) \prod_{j=1}^{n} \Gamma(1 - a_j + s)}{\prod_{k=m+1}^{q} \Gamma(1 - b_k + s) \prod_{j=n+1}^{p} \Gamma(a_j - s)}. \qquad (D.6)$$

In (D.6) the empty product is assumed to be equal to 1 (the *empty product convention*), the parameters m, n, p, q satisfy the relation $0 \leq n \leq q, 0 \leq n \leq p$, and the complex numbers a_j, b_k are such that no pole of $\Gamma(b_k - s), k = 1, \ldots, m$, coincides with a pole of $\Gamma(1 - a_j + s), j = 1, \ldots, n$.

Let

$$\delta = m + n - \frac{1}{2}(p+q), \quad \nu = \sum_{k=1}^{q} b_k - \sum_{j=1}^{p} a_j.$$

The contour of integration $\mathcal{L}$ in (D.5) can be of the following three types (see [Mari83], [Kir94, p. 313]):

(i) $\mathcal{L} = \mathcal{L}_{i\gamma\infty}$, which starts at $-i\gamma\infty$ and terminates at $+i\gamma\infty$, leaving to the right all poles of Γ-functions $\Gamma(b_k - s), k = 1, \ldots, m$, and leaving to the left all poles of Γ-functions $\Gamma(1 - a_j + s), j = 1, \ldots, n$. Integral (D.5) con-

Appendix D: The Mellin–Barnes Integral

verges for $\delta > 0$, $|\arg z| < \pi\delta$. If $|\arg z| = \pi\delta$, $\delta \geq 0$, then the integral converges absolutely when $p = q$, $\text{Re}\nu < -1$, and when $p \neq q$ if $(q-p)\text{Re}s > \text{Re}\nu + 1 - \frac{1}{2}(q-p)$ as $\text{Im}s \to \pm\infty$.

(ii) $\mathcal{L} = \mathcal{L}_{+\infty}$, which starts at $\varphi_1 + i\infty$, terminates at $\varphi_2 + i\infty$, $-\infty < \varphi_1 < \varphi_2 < +\infty$, encircles once in the negative direction all poles of the Γ-functions $\Gamma(b_k - s)$, $k = 1, \ldots, m$, but no pole of the Γ-functions $\Gamma(1 - a_j + s)$, $j = 1, \ldots, n$. Integral (D5) converges if $q \geq 1$ and either $p < q$ or $p = q$ and $|z| < 1$.

(iii) $\mathcal{L} = \mathcal{L}_{-\infty}$, which starts at $\varphi_1 - i\infty$, terminates at $\varphi_2 - i\infty$, $-\infty < \varphi_1 < \varphi_2 < +\infty$, encircles once in the positive direction all poles of the Γ-functions $\Gamma(1 - a_j + s)$, $j = 1, \ldots, n$, but no pole of the Γ-functions $\Gamma(b_k - s)$, $k = 1, \ldots, m$. Integral (D.5) converges if $p \geq 1$ and either $p > q$ or $p = q$ and $|z| > 1$.

Since the generalized hypergeometric function ${}_pF_q(a_1, \ldots, a_p; b_1, \ldots, b_q; z)$ can be considered as a special case of the Meijer G-functions, the function ${}_pF_q(a_1, \ldots, a_p; b_1, \ldots, b_q; z)$ also possesses a representation via a Mellin–Barnes integral

$${}_pF_q(a_1, \ldots, a_p; b_1, \ldots, b_q; z) = \frac{\prod_{k=1}^{q} \Gamma(b_k)}{\prod_{j=1}^{p} \Gamma(a_j)} G_{p,q+1}^{1,p}\left[-z \left| \begin{array}{c} (1-a_1), \ldots, (1-a_p) \\ 0, (1-b_1), \ldots, (1-b_q) \end{array}\right.\right], \quad (D.7)$$

where

$$G_{p,q+1}^{1,p}\left[-z \left| \begin{array}{c} (1-a_1), \ldots, (1-a_p) \\ 0, (1-b_1), \ldots, (1-b_q) \end{array}\right.\right]$$

$$= \frac{1}{2\pi i} \int_{-i\infty}^{+i\infty} \frac{\Gamma(a_1 + s) \ldots \Gamma(a_p + s)\Gamma(-s)}{\Gamma(b_1 + s) \ldots \Gamma(b_q + s)} (-z)^s ds, \quad (D.8)$$

$$a_j \neq 0, -1, \ldots; j = 1, \ldots, p; |\arg(1 - iz)| < \pi.$$

Although the Meijer G-functions are quite general in nature, there still exist examples of special functions, such as the Mittag-Leffler and the Wright functions, which do not form particular cases of them. A more general class which includes those functions can be obtained by introducing the Fox H-functions [Fox61], whose representation in terms of the Mellin–Barnes integral is a straightforward generalization of that for the G-functions. To introduce it one needs to add to the sets of the complex parameters a_j and b_k the new sets of positive numbers α_j and β_k with $j = 1, \ldots, p$, $k = 1, \ldots, q$, and to replace in the integral of (D.5) the kernel $\mathcal{G}_{m,n}^{p,q}(s)$ by the new one

$$\mathcal{H}_{p,q}^{m,n}(s) = \frac{\prod_{k=1}^{m} \Gamma(b_k - \beta_k s) \prod_{j=1}^{n} \Gamma(1 - a_j + \alpha_j s)}{\prod_{k=m+1}^{q} \Gamma(1 - b_k + \beta_k s) \prod_{j=n+1}^{p} \Gamma(a_j - \alpha_j s)}. \quad (D.9)$$

The Fox H-functions are then defined in the form

$$H_{m,n}^{p,q}(z) = H_{m,n}^{p,q}(z)\left[z \,\Big|\, \begin{matrix}(a_j, \alpha_j)_{j=1}^p \\ (b_k, \beta_k)_{k=1}^q\end{matrix}\right] = \frac{1}{2\pi i}\int_{\mathcal{L}} \mathcal{H}_{p,q}^{m,n}(s) z^s \, ds. \qquad (D.10)$$

A representation of the type (D.10) is usually called a Mellin–Barnes integral representation. The convergence questions for these integrals are completely discussed in [ParKam01, Sect. 2.4]. For further information we refer the reader to the treatises on Fox H-functions by Mathai, Saxena and Haubold [MaSaHa10], Srivastava, Gupta and Goyal [SrGuGo82], Kilbas and Saigo [KilSai04] and the references therein.

D.2 Asymptotic Methods for the Mellin–Barnes Integral

The asymptotics of the Mellin–Barnes integral is based on the following two general lemmas on the expansion of quotients of Gamma functions as inverse factorial expansions (see [Wri40b, Wri40c, Bra62]).

Let us consider the quotient of products of Gamma functions

$$P(s) = \frac{\prod_{j=1}^{p} \Gamma(\alpha_j s + a_j)}{\prod_{k=1}^{q} \Gamma(\beta_k s + b_k)}. \qquad (D.11)$$

Define the parameters

$$\begin{cases} h = \prod_{j=1}^{p} \alpha_j^{\alpha_j} \prod_{k=1}^{q} \beta_k^{-\beta_k}, \\ \vartheta = \sum_{j=1}^{p} a_j - \sum_{k=1}^{q} b_k + \tfrac{1}{2}(q - p + 1), \; \vartheta' = 1 - \vartheta, \\ \kappa = \sum_{k=1}^{q} \beta_k - \sum_{j=1}^{p} \alpha_j. \end{cases} \qquad (D.12)$$

The following lemma presents the inverse factorial expansions for the functions $P(s)$ (see [Wri40c, Bra62]).

Lemma D.1 ([ParKam01, p. 39]) *Let M be a positive integer and suppose $\kappa > 0$. Then there exist numbers A_r ($0 \le r \le M - 1$), independent of s and M, such that the function $P(s)$ in (D.11) possesses the inverse factorial expansion given by*

$$P(s) = (h\kappa^\kappa)^s \left\{ \sum_{r=1}^{M-1} \frac{A_r}{\Gamma(\kappa s + \vartheta' + r)} + \frac{\sigma_M(s)}{\Gamma(\kappa s + \vartheta' + M)} \right\}, \qquad (D.13)$$

Appendix D: The Mellin–Barnes Integral

where the parameters h, κ and ϑ' are defined in (D.12).
In particular, the coefficient A_0 has the value

$$A_0 = (2\pi)^{1/2(p-q+1)} \kappa^{1/2-\vartheta} \prod_{j=1}^{p} \alpha_j^{a_j-1/2} \prod_{k=1}^{q} \beta_k^{1/2-b_k}. \tag{D.14}$$

The remainder function $\sigma_M(s)$ is analytic in s except at the points $s = -(a_j + t)/\alpha_j$, $t = 0, 1, 2, \ldots$ $(1 \leq j \leq p)$, where $P(s)$ has poles, and is such that

$$\sigma_M(s) = O(1)$$

for $|s| \to \infty$ uniformly in $|\arg s| \leq \pi - \varepsilon$, $\varepsilon > 0$.

Let now $Q(s)$ be another type of quotient of products of Gamma functions:

$$Q(s) = \frac{\prod_{k=1}^{q} \Gamma(1 - b_k + \beta_k s)}{\prod_{j=1}^{p} \Gamma(1 - a_j + \alpha_j s)}. \tag{D.15}$$

Let the parameters h, ϑ, ϑ', κ be defined in the same manner as in (D.12).
The following lemma presents the corresponding inverse factorial expansions for the functions $Q(s)$ (see [Wri40c, Bra62]).

Lemma D.2 ([ParKam01, p. 39]) *Let M be a positive integer and suppose $\kappa > 0$. Then there exist numbers A_r $(0 \leq r \leq M-1)$, independent of s and M, such that the function $Q(s)$ in (D.15) possesses the inverse factorial expansion given by*

$$Q(s) = \frac{(h\kappa^{\kappa})^{-s}}{(2\pi)^{p-q+1}} \left\{ \sum_{r=1}^{M-1} (-1)^r A_r \Gamma(\kappa s + \vartheta - r) + \rho_M(s) \Gamma(\kappa s + \vartheta - M) \right\}, \tag{D.16}$$

where the parameters h, κ and ϑ' are defined in (D.12).
The remainder function $\rho_M(s)$ is analytic in s except at the points $s = (b_k - 1 - t)/\beta_k$, $t = 0, 1, 2, \ldots$ $(1 \leq k \leq q)$, where $Q(s)$ has poles, and is such that

$$\rho_M(s) = O(1)$$

for $|s| \to \infty$ uniformly in $|\arg s| \leq \pi - \varepsilon$, $\varepsilon > 0$.

An algebraic method for determining the coefficients A_r in (D.13) and (D.16) is presented in [ParKam01, pp. 46–49].

D.3 Historical and Bibliographical Notes

As a historical note, we point out that "Mellin–Barnes integrals" are named after the two authors (namely Hj. Mellin and E.W. Barnes) who in the early 1910s developed the theory of these integrals, using them for a complete integration of the hypergeometric differential equation. However, these integrals were first used in 1888 by Pincherle, see, e.g., [MaiPag03]. Recent treatises on Mellin–Barnes integrals are [Mari83, ParKam01].

In Vol. 1, p. 49, of Higher Transcendental Functions of the Bateman Project [ErdBat-1] we read "Of all integrals which contain Gamma functions in their integrands, the most important ones are the so-called Mellin–Barnes integrals. Such integrals were first introduced by S. Pincherle, in 1888 [Pin88]; their theory has been developed in 1910 by H. Mellin (where there are references to earlier work) [Mel10] and they were used for a complete integration of the hypergeometric differential equation by E.W. Barnes [Barn08]."

In the classical treatise on Bessel functions by Watson [Wat66, p. 190], we read "By using integrals of a type introduced by Pincherle and Mellin, Barnes has obtained representations of Bessel functions ...".

Salvatore Pincherle (1853–1936) was Professor of Mathematics at the University of Bologna from 1880 to 1928. He retired from the University just after the International Congress of Mathematicians which he had organized in Bologna, following the invitation received at the previous Congress held in Toronto in 1924. He wrote several treatises and lecture notes on Algebra, Geometry, and Real and Complex Analysis. His main book related to his scientific activity is entitled "*Le Operazioni Distributive e loro Applicazioni all'Analisi*"; it was written in collaboration with his assistant, Dr. Ugo Amaldi, and was published in 1901 by Zanichelli, Bologna. Pincherle can be considered one of the most prominent founders of Functional Analysis, and was described as such by J. Hadamard in his review lecture "*Le développement et le rôle scientifique du Calcul fonctionnel*", given at the Congress of Bologna (1928). A description of Pincherle's scientific works requested from him by Mittag-Leffler, who was the Editor of the prestigious journal *Acta Mathematica*, appeared (in French) in 1925 [Pin25]. A collection of selected papers (38 from 247 notes plus 24 treatises) was edited by Unione Matematica Italiana (UMI) on the occasion of the centenary of his birth, and published by Cremonese, Roma 1954. S. Pincherle was the first President of UMI, from 1922 to 1936.

Here we point out that S. Pincherle's 1888 paper (in Italian) on the *Generalized Hypergeometric Functions* led him to introduce what was later called the Mellin–Barnes integral to represent the solution of a generalized hypergeometric differential equation investigated by Goursat in 1883. Pincherle's priority was explicitly recognized by Mellin and Barnes themselves, as reported below.

In 1907 Barnes, see p. 63 in [Barn07b], wrote: "The idea of employing contour integrals involving gamma functions of the variable in the subject of integration appears to be due to Pincherle, whose suggestive paper was the starting point of the investigations of Mellin (1895) though the type of contour and its use can be

Appendix D: The Mellin–Barnes Integral

traced back to Riemann." In 1910 Mellin, see p. 326ff in [Mel10], devoted a section (Sect. 10: Proof of Theorems of Pincherle) to revisit the original work of Pincherle; in particular, he wrote "Before we prove this theorem, which is a special case of a more general theorem of Mr. Pincherle, we want to describe more closely the lines L over which the integration is preferably to be carried out." [free translation from German].

The Mellin–Barnes integrals are the essential tools for treating the two classes of higher transcendental functions known as G-and H-functions, introduced by Meijer (1946) [Mei46] and Fox (1961) [Fox61] respectively, so Pincherle can be considered their precursor. For an exhaustive treatment of the Mellin–Barnes integrals we refer to the recent monograph by Paris and Kaminski [ParKam01].

D.4 Exercises

D.4.1. ([ParKam01, p. 67]) The exponential function can be represented in terms of the Mellin–Barnes integral

$$\mathrm{e}^{-z} = \frac{1}{2\pi i} \int_{c-i\infty}^{c+i\infty} \Gamma(s) z^{-s} \mathrm{d}s \quad (c > 0).$$

Find the domain of convergence of the integral in the right-hand side of this representation.

D.4.2. ([ParKam01, p. 68]) Prove the following Mellin–Barnes integral representation of the hypergeometric function

$$\frac{\Gamma(a)\Gamma(b)}{\Gamma(c)} {}_2F_1(a,b;c;z) = \frac{1}{2\pi i} \int_{\mathcal{L}_{i\gamma\infty}} \Gamma(-s) \frac{\Gamma(s+a)\Gamma(s+b)}{\Gamma(s+c)} (-z)^s \mathrm{d}s,$$

where the contour of integration $\mathcal{L}_{i\gamma\infty}$ is a vertical line, which starts at $-i\gamma\infty$ ends at $+i\gamma\infty$ and separates the poles of the Gamma function $\Gamma(-s)$ from those of the Gamma functions $\Gamma(s+a)$, $\Gamma(s+b)$ for a and $b \neq 0, -1, -2, \ldots$.

Find the domain of convergence of the integral in the above representation. If the vertical line $\mathcal{L}_{i\infty}$ is replaced by the contour $\mathcal{L}_{-\infty}$, then show that the domain of convergence coincides with an open unit disk.

D.4.3. ([ParKam01, p. 68]) Find the domain of convergence of the following Mellin–Barnes integral:

$$\frac{1}{2\pi i} \int_{\mathcal{L}_{+\infty}} \frac{\Gamma(a-s/2)\Gamma(\lambda s+b)\Gamma(c+s/4)}{\Gamma(d+s/4)\Gamma(s+1)} (z)^s \mathrm{d}s,$$

where the parameters a, b, c and λ and the contour of integration $\mathcal{L}_{+\infty}$ are such that the poles of $\Gamma(a - s/2)$ are separated from those of $\Gamma(\lambda s + b)$ and $\Gamma(c + s/4)$.

D.4.4. ([ParKam01, p. 109]) Prove that the following Mellin–Barnes integral representation

$$\frac{\Gamma(a)}{\Gamma(b)} {}_1F_1(a; b; c; z) = \frac{1}{2\pi i} \int_{c-i\infty}^{c+i\infty} \frac{\Gamma(-s)\Gamma(s+a)}{\Gamma(s+b)} (-z)^s ds$$

is valid for all finite values of c provided that the contour of integration can be deformed to separate the poles of $\Gamma(-s)$ and $\Gamma(s + a)$ (which is always possible when a is not a negative integer or zero). Here ${}_1F_1(a; b; c; z)$ is the confluent hypergeometric function

$$_1F_1(a; b; c; z) = \sum_{n=}^{\infty} \frac{(a)_n}{(b)_n} \frac{z^n}{n!}, \quad (|z| < \infty).$$

D.4.5. ([ParKam01, p. 113]) For the incomplete gamma function

$$\gamma(a, z) = \int_0^z t^{a-1} e^{-t} dt \quad (\mathrm{Re}\, a > 0)$$

prove the representation

$$\gamma(a, z) = \Gamma(a) + \frac{1}{2\pi i} \int_{c-i\infty}^{c+i\infty} \frac{\Gamma(-s)}{s+a} z^{s+a} ds$$

where $c < \min\{0, -\mathrm{Re}\, a\}$.

D.4.6. ([ParKam01, p. 113]) For the second incomplete gamma function

$$\Gamma(a, z) = \int_z^{\infty} t^{a-1} e^{-t} dt$$

prove the representation

$$\Gamma(a, z) = \frac{1}{2\pi i} \int_{c-i\infty}^{c+i\infty} \frac{\Gamma(s+a)}{s} z^{-s} ds$$

for all z, $|\arg z| \pi/2$.

Appendix E
Elements of Fractional Calculus

An interest in fractional generalizations of the usual integral and derivatives goes back to the discussion of such a possibility between G. Leibniz and G. L'Hôpital (see, e.g. [TeKiMa11]). Several attempts were made to provide the technical realization of such an approach. Here we mention a few of them.

First of all, based on Leibniz's formula for fractional derivatives of the exponential function, we have

$$\frac{d^\alpha e^{mx}}{dx^\alpha} = m^\alpha e^{mx}, \quad \alpha > 0. \tag{E.1}$$

Using this formula Liouville later introduced the fractional derivative of functions represented in the form of a convergent Dirichlet-type series

$$y(x) = \sum_{k=1}^{\infty} A_k e^{m_k x}. \tag{E.2}$$

Another approach is due to Euler, who introduced the fractional derivative of the power function

$$\frac{d^\alpha x^m}{dx^\alpha} = \frac{\Gamma(m+1)}{\Gamma(m-\alpha+1)} x^{m-\alpha}, \quad \alpha \notin \mathbb{N}. \tag{E.3}$$

The above mentioned Liouville construction has an evident restriction to a particular class of functions. To overcome this difficulty Liouville used a formula similar to Euler's

$$\frac{d^\alpha \frac{1}{x^m}}{dx^\alpha} = \frac{(-1)^\alpha \Gamma(m+\alpha)}{\Gamma(m) x^{m+\alpha}}, \quad \alpha > 0, m + \alpha > 0, \tag{E.4}$$

which follows from the definition of Euler's Gamma function

$$\frac{1}{x^m} = \frac{1}{\Gamma(m)} \int_0^\infty e^{-zx} z^{m-1} dz. \tag{E.5}$$

Since at that time the Legendre–Gauss definition of the Gamma function of a complex argument was not known, Liouville assumed that the fractional derivative of a general power-exponential function is defined only up to an additional function, and showed that such a function must be a polynomial of finite degree. Only in the 1860s and 1870s was it shown by Grünwald and Letnikov that a fractional derivative can be defined on the above mentioned class without any additional functions (see [RogDub18]).

One more approach is due to J. Fourier [Fou22], who noted that his formula for the derivative of the Fourier integral

$$\frac{d^\alpha f(x)}{dx^\alpha} = \frac{1}{2\pi} \int_{-\infty}^{+\infty} f(z)dz \int_{-\infty}^{+\infty} \cos\left(px - pz + \frac{\pi\alpha}{2}\right) dp \qquad (E.6)$$

is valid not only for integer α, but also for all non-integer α.

Fractional integrals are usually defined (see, e.g., [SaKiMa93]) as a generalization of the repeated integral formula

$$J^n f(t) = \int_0^t dt_1 \int_0^{t_1} dt_2 \ldots \int_0^{t_n} f(t_n) dt_n = \frac{1}{(n-1)!} \int_0^t (t-\tau)^{n-1} f(\tau) d\tau, \qquad (E.7)$$

namely

$$J^\alpha f(t) = \frac{1}{\Gamma(\alpha)} \int_0^t (t-\tau)^{\alpha-1} f(\tau) d\tau, \; \alpha > 0. \qquad (E.8)$$

Other motivations as well as a number of known definitions of fractional derivatives and integrals are presented in [ST-TM-CO19].

E.1 The Riemann–Liouville Fractional Calculus

Riemann–Liouville (R–L) fractional integrals and derivatives are one of the most popular fractional constructions. For a function defined on a finite interval (a, b) the *R–L left-sided fractional integral* is given by the formula

$$\left(J_{a+}^\alpha f\right)(t) := \frac{1}{\Gamma(\alpha)} \int_a^t \frac{f(\tau)}{(t-\tau)^{1-\alpha}} d\tau, \quad \alpha > 0. \qquad (E-RL.1)$$

and the *R–L right-sided fractional integral* is given by the formula

$$\left(J_{b-}^\alpha f\right)(t) := \frac{1}{\Gamma(\alpha)} \int_t^b \frac{f(\tau)}{(\tau-t)^{1-\alpha}} d\tau, \quad \alpha > 0. \qquad (E-RL.2)$$

Appendix E: Elements of Fractional Calculus

The fractional integrals $(E-RL.1)$ and $(E-RL.2)$ are defined for functions $f \in L_1(a,b)$ existing almost everywhere.

If a function f is Hölder continuous ($f \in H^{m,\lambda}[a,b]$, i.e. $f^{(m)} \in H^{\lambda}[a,b]$, $0 < \lambda < 1$), then its left-sided fractional integral has the following representation

$$J_{a+}^{\alpha} f(t) = \sum_{k=0}^{m} \frac{f^{(k)}(a)}{\Gamma(\alpha+k+1)} (t-a)^{\alpha+k} + \psi(t), \qquad (E-RL.3)$$

where $\psi \in H^{m+[\lambda+\alpha],\{\lambda+\alpha\}}[a,b]$ in the case of non-integer $\lambda + \alpha$, and $[\cdot]$, $\{\cdot\}$ are the integer and fractional parts of a positive number, respectively. In the case of integer $\lambda + \alpha$ the behavior of $J_{a+}^{\alpha} f(t)$ is more specific (cf. for details [SaKiMa93, Chap. 1]). Similar formulas are valid for the right-sided fractional integrals of Hölder continuous functions.

If $0 < \alpha < 1$ and $1 < p < \frac{1}{\alpha}$ the operator J_{a+}^{α} is bounded from $L_p(a,b)$ into $L_q(a,b)$, $q = \frac{p}{1-\alpha p}$. For bigger p the behavior is even better, namely, if $\alpha > 0$ and $p > \frac{1}{\alpha}$ (but $\alpha - 1/p$ is non-integer), then the operator J_{a+}^{α} is bounded from $L_p(a,b)$ into $H^{[\alpha-1/p],\{\alpha-1/p\}}[a,b]$. Other relations between α and p are carefully discussed in [SaKiMa93, Chap. 1].

One of the most important properties of the fractional integrals is the semigroup relation

$$\left(J_{a+}^{\alpha} J_{a+}^{\beta} f\right)(t) = \left(J_{a+}^{\alpha+\beta} f\right)(t), \qquad (E-RL.4)$$

as well as the composition relation

$$\left(J_{a+}^{\alpha} J_{a+}^{\beta} f\right)(t) = \left(J_{a+}^{\beta} J_{a+}^{\alpha} f\right)(t), \qquad (E-RL.5)$$

which are valid for all functions f for which the integrals exist and all $\alpha > 0$, $\beta > 0$.

Another relation important for applications is the integration by parts formula

$$\int_a^b g(t) \left(J_{a+}^{\alpha} f\right)(t) dt = \int_a^b f(t) \left(J_{b-}^{\alpha} g\right)(t) dt, \qquad (E-RL.6)$$

which holds true, e.g., for $f \in L_p(a,b)$, $g \in L_q(a,b)$, $p \geq 1$, $q \geq 1$, $1/p + 1/q \leq 1 + \alpha$ (and $p \neq 1$, $q \neq 1$, whenever $1/p + 1/q = 1 + \alpha$).

A number of calculations of the fractional integral for elementary and special functions are presented below.

$$\left(J_{a+}^{\alpha}(\tau-a)^{\beta-1}\right)(t) = \frac{\Gamma(\beta)}{\Gamma(\alpha+\beta)} (t-a)^{\alpha+\beta-1}, \quad \beta > 0, \qquad (E-RL.7)$$

$$\left(J_{b-}^{\alpha}(b-\tau)^{\beta-1}\right)(t) = \frac{\Gamma(\beta)}{\Gamma(\alpha+\beta)} (b-t)^{\alpha+\beta-1}, \quad \beta > 0, \qquad (E-RL.8)$$

$$\left(J_{a+}^{\alpha}(\tau-a)^{\beta-1}(b-\tau)^{\gamma-1}\right)(t) \qquad (E-RL.9)$$

$$= \frac{\Gamma(\beta)}{\Gamma(\alpha+\beta)} \frac{(t-a)^{\alpha+\beta-1}}{(b-a)^{1-\gamma}} {}_2F_1(1-\gamma, \beta, \alpha+\beta; \frac{t-a}{b-a}), \ \beta > 0, \gamma \in \mathbb{R},$$

where

$$_2F_1(a, b, c; z) = \sum_{k=0}^{\infty} \frac{(a)_k (b)_k}{(c)_k} z^k$$

is the Gauss hypergeometric function, and $(d)_k = d(d+1)\ldots(d+k-1)$ is the Pochhammer symbol.

$$\left(J_{a+}^{\alpha}(\tau-a)^{\beta-1}\ln(\tau-a)\right)(t) \qquad (E-RL.10)$$

$$= \frac{\Gamma(\beta)}{\Gamma(\alpha+\beta)}(t-a)^{\alpha+\beta-1}(\psi(\beta) - \psi(\alpha+\beta) + \ln(t-a)), \ \beta > 0,$$

where

$$\psi(z) = \frac{\Gamma'(z)}{\Gamma(z)}$$

is the Euler psi function (the logarithmic derivative of the Euler Gamma function).

$$\lambda \left(J_{a+}^{\alpha} E_{\alpha}(\lambda \tau^{\alpha})\right)(t) = E_{\alpha}(\lambda t^{\alpha}) - 1, \ \lambda \in \mathbb{C}, \qquad (E-RL.11)$$

where

$$E_{\alpha}(z) = \sum_{k=0}^{\infty} \frac{z^k}{\Gamma(\alpha k + 1)}$$

is the classical Mittag-Leffler function.

The *Riemann–Liouville fractional derivative* is introduced as the left-inverse operator for the corresponding fractional integrals. To be more precise we assume that the function f satisfies the condition $\left(J_{a+}^{n-\alpha} f\right)(t) \in \mathcal{AC}^n[a, b]$ with $n \in \mathbb{N}, n-1 < \alpha \le n$, and $\mathcal{AC}^n[a, b] := \{g \in C^{n-1}[a, b] : g^{(n-1)} \in \mathcal{AC}[a, b]\}$, $\mathcal{AC}$ denotes a class of absolutely continuous functions. Assuming this condition, the left- and right-sided Riemann–Liouville fractional derivatives are defined by the relations

$$\left({}^{RL}D_{a+}^{\alpha} f\right)(t) := \frac{d^n}{dt^n}\left(J_{a+}^{n-\alpha} f\right)(t), \ t \in (a, b), \qquad (E-RL.12)$$

$$\left({}^{RL}D_{b-}^{\alpha} f\right)(t) := (-1)^n \frac{d^n}{dt^n}\left(J_{a+}^{n-\alpha} f\right)(t), \ t \in (a, b). \qquad (E-RL.13)$$

Note that the condition $\left(J_{a+}^{n-\alpha} f\right)(t) \in \mathcal{AC}^n[a, b]$ follows from $f \in \mathcal{AC}^n[a, b]$. In this case the Riemann–Liouville fractional derivative exists almost everywhere on

Appendix E: Elements of Fractional Calculus

the interval (a, b) and the following equality holds

$$\left(^{RL}D^\alpha_{a+}f\right)(t) = \sum_{k=0}^{n-1} \frac{f^{(k)}(a)(t-a)^{k-\alpha}}{\Gamma(1+k-\alpha)} + \frac{1}{\Gamma(n-\alpha)} \int_a^t \frac{f^{(k)}(\tau)d\tau}{(t-\tau)^{\alpha-n+1}}.$$
$$(E-RL.14)$$

As was already mentioned, the Riemann–Liouville fractional derivative is the left-inverse to the corresponding Riemann–Liouville fractional integral, that is, the following relation

$$\left(^{RL}D^\alpha_{a+} J^\alpha_{a+} f\right)(t) = f(t) \qquad (E-RL.15)$$

holds for any function $f \in L_1(a, b)$. The opposite is not true, i.e. the Riemann–Liouville fractional derivative is not the right-inverse to the Riemann–Liouville fractional integral. Instead, for all $f \in L_1(a, b)$ such that $\left(J^{n-\alpha}_{a+} f\right)(t) \in \mathcal{AC}^n[a, b]$ we have the following relation

$$\left(J^\alpha_{a+}{}^{RL}D^\alpha_{a+} f\right)(t) = f(t) - \sum_{k=0}^{n-1} \frac{(t-a)^{\alpha-k-1}}{\Gamma(\alpha-k)} \left(\frac{d^{n-k-1}}{dt^{n-k-1}} J^{n-\alpha}_{a+} f\right)(a).$$
$$(E-RL.16)$$

The sum in $(E-RL.16)$ converges if the function f can be represented as a value of the fractional integral, i.e. there exists a function $g \in L_1(a, b)$ such that $f(t) = \left(J^\alpha_{a+} g\right)(t)$.

If $\alpha \in \mathbb{R}$, $\beta < 1$, and f is an analytic function, then the semigroup relation

$$\left(^{RL}D^\alpha_{a+}{}^{RL}D^\beta_{a+} f\right)(t) = \left(^{RL}D^{\alpha+\beta}_{a+} f\right)(t) \qquad (E-RL.17)$$

holds true (note that this relation is not valid in general, see, e.g., [SaKiMa93]).

The *Leibniz rule* for the Riemann–Liouville fractional derivative can be written in different forms, e.g.

$$\left(^{RL}D^\alpha_{a+} f \cdot g\right)(t) = \sum_{k=0}^{+\infty} \binom{\alpha}{k} \left(^{RL}D^{\alpha-k}_{a+} f\right)(t) g^{(k)}(t), \quad \alpha \in \mathbb{R}, \qquad (E-RL.18)$$

$$\left(^{RL}D^\alpha_{a+} f \cdot g\right)(t) = \sum_{k=-\infty}^{+\infty} \binom{\alpha}{k+\beta} \left(^{RL}D^{\alpha-\beta-k}_{a+} f\right)(t) \left(^{RL}D^{\beta+k}_{a+} g\right)(t),$$
$$(E-RL.19)$$

$\alpha, \beta \in \mathbb{R}$ ($\alpha \neq -1, -2, \ldots$, if $\beta \in \mathbb{Z}$).

In the case of functions defined on the whole real line, analogs of the fractional integrals and derivatives (sometimes called the *Liouville fractional integrals and derivatives*, respectively) are given by the following formulas

$$\left(J_+^\alpha f\right)(t) = \frac{1}{\Gamma(\alpha)} \int_{-\infty}^{t} \frac{f(\tau) d\tau}{(t-\tau)^{1-\alpha}}, \quad t \in \mathbb{R}, \qquad (E-RL.20)$$

$$\left(J_-^\alpha f\right)(t) = \frac{1}{\Gamma(\alpha)} \int_{t}^{+\infty} \frac{f(\tau) d\tau}{(\tau-t)^{1-\alpha}}, \quad t \in \mathbb{R}, \qquad (E-RL.21)$$

$$\left(^L D_+^\alpha f\right)(t) = \frac{d^n}{dt^n} \left(J_+^{n-\alpha} f\right)(t), \quad t \in \mathbb{R}. \qquad (E-RL.22)$$

$$\left(^L D_-^\alpha f\right)(t) = (-1)^n \frac{d^n}{dt^n} \left(J_-^{n-\alpha} f\right)(t), \quad t \in \mathbb{R}. \qquad (E-RL.23)$$

For applications it is important to know the values of the Fourier, Laplace and Mellin integral transforms of the fractional operators introduced in this section. Let us present some of the most well-known results.

The Fourier transform of the fractional integral of a function $f \in L_1(a,b)$ is given in the case $0 < \operatorname{Re} \alpha < 1$ by the formula

$$\{\mathcal{F}\left(J_\pm^\alpha f\right)(t)\}(\kappa) = \frac{1}{(\mp i\kappa)^\alpha} \{\mathcal{F} f(t)\}(\kappa). \qquad (E-RL.24)$$

Note that if $\operatorname{Re} \alpha \geq 1$, then formula $(E-RL.24)$ cannot be understood in the usual sense (see, e.g., [HAND1]).

A result similar to $(E-RL.24)$ is valid under suitable conditions on the class of functions and on the parameter of the fractional derivative (see, e.g. [SaKiMa93]):

$$\{\mathcal{F}\left(^L D_\pm^\alpha f\right)(t)\}(\kappa) = (\mp i\kappa)^\alpha \{\mathcal{F} f(t)\}(\kappa). \qquad (E-RL.25)$$

The Riemann–Liouville fractional integral can be interpreted as the Laplace convolution of a function and a truncated power monomial:

$$\left(J_{0+}^\alpha f\right)(t) = \left(f * \frac{\tau_+^{\alpha-1}}{\Gamma(\alpha)}\right)(t), \quad \operatorname{Re} \alpha > 0, \ t_+ = \begin{cases} t, & t \geq 0, \\ 0, & t < 0. \end{cases} \qquad (E-RL.26)$$

It follows that for any function $f \in L_1(0,b), \forall b > 0$, with sub-exponential growth (i.e. $|f(t)| \leq Ae^{p_0 t}$, $p_0 \geq 0$), the Laplace transform of the fractional integral satisfies the following formula, valid for all $\operatorname{Re} p > p_0$

$$\{\mathcal{L}\left(J_{0+}^\alpha f\right)(t)\}(p) = p^{-\alpha}\{\mathcal{L} f(t)\}(p), \quad \operatorname{Re} \alpha > 0. \qquad (E-RL.27)$$

Analogously, if $f \in AC^n[a,b]$, $f^{(k)}(0) = 0$, $k = 0, 1, \ldots, n-1$, then

$$\{\mathcal{L}\left(^{RL} D_{0+}^\alpha f\right)(t)\}(p) = p^\alpha \{\mathcal{L} f(t)\}(p), \quad \operatorname{Re} \alpha > 0. \qquad (E-RL.28)$$

Appendix E: Elements of Fractional Calculus

If not all derivatives vanish at 0 the formula $(E - RL.28)$ becomes more cumbersome

$$\left\{ \mathcal{L} \left({}^{RL}D_{0+}^{\alpha} f \right)(\cdot) \right\}(p) = p^{\alpha} \left\{ \mathcal{L}f(\cdot) \right\}(p) - \sum_{k=0}^{n-1} \lim_{t \to +0} \frac{d^k}{dt^k} \left(J_{0+}^{n-\alpha} f \right)(t) p^{n-k-1}.$$

$(E - RL.29)$

Similar results are known for the Mellin transform of the Riemann–Liouville fractional derivative.

$$\left\{ \mathcal{M} \left({}^{RL}D_{0+}^{\alpha} f \right)(\cdot) \right\}(s) = \frac{\Gamma(1+\alpha-s)}{\Gamma(1-s)} \left\{ \mathcal{M}f(\cdot) \right\}(s-\alpha) \qquad (E - RL.30)$$

$$+ \sum_{k=0}^{n-1} \frac{\Gamma(1+k-s)}{\Gamma(1-s)} \left[t^{s-k-1} \left(J_{0+}^{n-\alpha} f \right)(t) \right]_{t=0}^{+\infty},$$

$$\left\{ \mathcal{M} \left({}^{L}D_{-}^{\alpha} f \right)(\cdot) \right\}(s) = \frac{\Gamma(s)}{\Gamma(s-\alpha)} \left\{ \mathcal{M}f(\cdot) \right\}(s-\alpha) \qquad (E - RL.31)$$

$$+ \sum_{k=0}^{n-1} (-1)^{n-k} \frac{\Gamma(s)}{\Gamma(s-k)} \left[t^{s-k-1} \left(J_{-}^{n-\alpha} f \right)(t) \right]_{t=0}^{+\infty}.$$

E.2 The Caputo Fractional Calculus

It should be noted that in Fractional Calculus, the Riemann–Liouville construction of the fractional integral is common in many approaches. This is not the case for fractional derivatives. One of the most attractive and popular approaches is the so-called *Caputo fractional derivative* (known also as the *Caputo–Dzherbashian* or *Caputo–Gerasimov fractional derivative*). It is simply a regularization of the Riemann–Liouville fractional derivative, namely, if $n-1 < \alpha \leq n$, $n \in \mathbb{N}$, and $f \in AC^n[a,b]$ then the Caputo fractional derivatives are defined by the following formulas

$$\left({}^{C}D_{a+}^{\alpha} f \right)(t) = \left({}^{RL}D_{a+}^{\alpha} \left[f(\tau) - \sum_{k=0}^{n-1} \frac{f^{(k)}(a)}{k!} (\tau - a)^k \right] \right)(t), \qquad (E - C.1)$$

$$\left({}^{C}D_{b-}^{\alpha} f \right)(t) = \left({}^{RL}D_{b-}^{\alpha} \left[f(\tau) - \sum_{k=0}^{n-1} \frac{f^{(k)}(b)}{k!} (b - \tau)^k \right] \right)(t). \qquad (E - C.2)$$

If $\alpha \notin \mathbb{N}_0$, then interchanging of the order of integration and differentiation gives another form of the Caputo fractional derivative

$$\left({}^{C}D_{a+}^{\alpha}f\right)(t) = \frac{1}{\Gamma(n-\alpha)} \int_{a}^{t} \frac{f^{(n)}(\tau) d\tau}{(t-\tau)^{\alpha-n+1}} =: \left(J_{a+}^{n-\alpha} f^{(n)}\right)(t), \quad (E-C.3)$$

$$\left({}^{C}D_{b-}^{\alpha}f\right)(t) = \frac{(-1)^{n}}{\Gamma(n-\alpha)} \int_{t}^{b} \frac{f^{(n)}(\tau) d\tau}{(\tau-t)^{\alpha-n+1}} =: (-1)^{n} \left(J_{b-}^{n-\alpha} f^{(n)}\right)(t). \quad (E-C.4)$$

Note that relation $(E - RL.14)$ gives the relation between the Riemann–Liouville and the Caputo derivatives, provided that all components in this formula are well-defined.

Similar to the Riemann–Liouville derivatives we have for all $\beta > n - 1$

$$\left({}^{C}D_{a+}^{\alpha}(\tau-a)^{\beta}\right)(t) = \frac{\Gamma(\beta+1)}{\Gamma(\beta-\alpha+1)} (t-a)^{\beta-\alpha}, \quad (E-C.5)$$

$$\left({}^{C}D_{b-}^{\alpha}(b-\tau)^{\beta}\right)(t) = \frac{\Gamma(\beta+1)}{\Gamma(\beta-\alpha+1)} (b-t)^{\beta-\alpha}, \quad (E-C.6)$$

but the Caputo derivative vanishes on the integer power monomials (which is not the case for the Riemann–Liouville derivative), i.e. for all $k = 0, 1, \ldots, n-1$ we have

$$\left({}^{C}D_{a+}^{\alpha}(\tau-a)^{k}\right)(t) = 0 \text{ and } \left({}^{C}D_{b-}^{\alpha}(b-\tau)^{\beta}\right)(t) = 0. \quad (E-C.7)$$

In particular,

$$\left({}^{C}D_{a+}^{\alpha}1\right)(t) = 0 \text{ and } \left({}^{C}D_{b-}^{\alpha}1\right)(t) = 0. \quad (E-C.8)$$

We also have to mention other formulas for the values of the Caputo fractional derivatives

$$\left({}^{C}D_{+}^{\alpha}e^{\lambda\tau}\right)(t) = \lambda^{\alpha}e^{\lambda t} \text{ and } \left({}^{C}D_{-}^{\alpha}e^{-\lambda\tau}\right)(t) = \lambda^{\alpha}e^{-\lambda t}, \quad (E-C.9)$$

$$\left({}^{C}D_{a+}^{\alpha} E_{\alpha}[\lambda(\tau-a)^{\alpha}]\right)(t) = \lambda E_{\alpha}[\lambda(t-a)^{\alpha}], \quad (E-C.10)$$

$$\left({}^{C}D_{-}^{\alpha} \tau^{\alpha-1} E_{\alpha}[\lambda\tau^{-\alpha}]\right)(t) = \frac{1}{t} E_{\alpha,1-\alpha}[\lambda t^{-\alpha}]. \quad (E-C.11)$$

The Caputo fractional derivative is left inverse to the fractional integral

$$\left({}^{C}D_{a+}^{\alpha} J_{a+}^{\alpha} f\right)(t) = f(t), \quad (E-C.12)$$

but not the right inverse, that is, for all $f \in \mathcal{AC}^{n}[a, b]$ the following relation holds

Appendix E: Elements of Fractional Calculus

$$\left(J_{a+}^{\alpha}\,{}^{C}D_{a+}^{\alpha}f\right)(t) = f(t) - \sum_{k=0}^{n-1}\frac{f^{(k)}(a)}{k!}(t-a)^{k}. \qquad (E-C.13)$$

The Laplace transform of the Caputo fractional derivative is similar to the Laplace transform formula for integer order derivatives

$$\left\{\mathcal{L}\left({}^{C}D_{0+}^{\alpha}f\right)(t)\right\}(p) = p^{\alpha}\left\{\mathcal{L}f(t)\right\}(p) - \sum_{k=0}^{n-1}f^{(k)}(0)p^{\alpha-k-1}. \qquad (E-C.14)$$

The formulas for Mellin transform are similar (existence is discussed, for example, in [KiSrTr06]).

$$\left\{\mathcal{M}\left({}^{C}D_{0+}^{\alpha}f\right)(\cdot)\right\}(s) = \frac{\Gamma(1+\alpha-s)}{\Gamma(1-s)}\left\{\mathcal{M}f(\cdot)\right\}(s-\alpha) \qquad (E-C.15)$$

$$+\sum_{k=0}^{n-1}\frac{\Gamma(1+k+\alpha-n-s)}{\Gamma(1-s)}\left[t^{s+n-\alpha-k-1}f^{(n-k-1)}(t)\right]_{t=0}^{+\infty},$$

$$\left\{\mathcal{M}\left({}^{C}D_{-}^{\alpha}f\right)(\cdot)\right\}(s) = \frac{\Gamma(s)}{\Gamma(s-\alpha)}\left\{\mathcal{M}f(\cdot)\right\}(s-\alpha) \qquad (E-C.16)$$

$$+\sum_{k=0}^{n-1}(-1)^{n-k}\frac{\Gamma(s)}{\Gamma(s+n-\alpha-k)}\left[t^{s+n-\alpha-k-1}f^{(n-k-1)}(t)\right]_{t=0}^{+\infty}.$$

E.3 The Marchaud Fractional Calculus

Marchaud's idea is simple and straightforward (for details, see [RogDub18]). He started from the Riemann–Liouville fractional integral

$$I_{a+}^{\alpha} = \frac{1}{\Gamma(\alpha)}\int_{a}^{x}\frac{f(\tau)d\tau}{(x-\tau)^{1-\alpha}} = \frac{1}{\Gamma(\alpha)}\int_{0}^{x-a}t^{\alpha-1}f(x-t)dt$$

and replaced the positive parameter α by a negative one $-\alpha$. He then arrived at the divergent integral

$$\frac{1}{\Gamma(-\alpha)}\int_{0}^{x-a}t^{-\alpha-1}f(x-t)dt. \qquad (E-M.1)$$

In order to regularize this definition, Marchaud made some transformations to the Riemann–Liouville fractional integral.

For arbitrary values of α, $\operatorname{Re}\alpha > 0$, the definition of the *Marchaud fractional derivative* reads [SaKiMa93, Sect. 5.6]

$$\mathbb{D}_\pm^\alpha f(x) = -\frac{1}{\Gamma(-\alpha)A_l(\alpha)} \int_0^\infty \frac{(\Delta_{\pm t}^l f)(x)}{t^{1+\alpha}} dt, \quad l > \operatorname{Re}\alpha > 0, \qquad (E-M.2)$$

where

$$A_l(\alpha) = \sum_{k=0}^\infty (-1)^{k-1} \binom{l}{k} k^\alpha, \quad (\Delta_{\pm t}^l f)(x) = \sum_{k=0}^\infty (-1)^k \binom{l}{k} f(x \mp kt).$$

If $0 < \alpha < 1$, then the left- and right-hand sides are defined, respectively by (see, e.g., [SaKiMa93, Sect. 5.4])

$$\mathbb{D}_+^\alpha f(x) = \frac{\alpha}{\Gamma(1-\alpha)} \int_0^\infty \frac{f(t) - f(x-t)}{t^{1+\alpha}} dt, \qquad (E-M.3)$$

$$\mathbb{D}_-^\alpha f(x) = \frac{\alpha}{\Gamma(1-\alpha)} \int_0^\infty \frac{f(t) - f(x+t)}{t^{1+\alpha}} dt. \qquad (E-M.4)$$

Since the integral in $(E-M.2)$ is in general divergent, the Marchaud derivative can be defined via the limit of the truncated derivative (if it exists)

$$\mathbb{D}_\pm^\alpha f(x) = \lim_{\varepsilon \to +0} \mathbb{D}_{\pm,\varepsilon}^\alpha f(x) = \lim_{\varepsilon \to +0} -\frac{1}{\Gamma(-\alpha)A_l(\alpha)} \int_\varepsilon^\infty \frac{(\Delta_{\pm t}^l f)(x)}{t^{1+\alpha}} dt. \quad (E-M.4)$$

The definition of the Marchaud fractional derivative on a finite interval is more tricky. Since the integral in $(E-M.1)$ is divergent, this definition needs some transformation. For a function defined on an interval (a, a_1) it was proposed to extend it by zero on $(-\infty, a]$ and to denote the corresponding fractional integral by $f_{(a)}$ and the fractional derivative by $f^{(a)}$.

In order to find a proper transformation Marchaud started by regularizing the fractional integral. In the following formula (valid, for example, for the exponential function)

$$f_{(a)}(x) \int_0^\infty t^{\alpha-1} e^{-t} dt = \int_0^\infty t^{\alpha-1} f(x-t) dt$$

Appendix E: Elements of Fractional Calculus

he replaced t by $k_j t$ ($j = 0, 1, \ldots, p$) and formed a linear combination of the obtained equalities with unknown coefficients C_j, $j = 0, 1, \ldots, p$. This leads to the following formula

$$f_{(\alpha)}(x) \int_0^\infty t^{\alpha-1} \psi(t) dt = \int_0^\infty t^{\alpha-1} \varphi(x, t) dt, \qquad (E-M.5)$$

where

$$\psi(t) = \sum_{j=0}^p C_j e^{-k_j t}, \quad \varphi(x, t) = \sum_{j=0}^p C_j f(x - k_j t). \qquad (E-M.6)$$

Then replacing α by $-\alpha$ we arrive at the following formal relation, which needs some extra conditions for its validity

$$f^{(\alpha)}(x) \int_0^\infty t^{-\alpha-1} \psi(t) dt = \int_0^\infty t^{-\alpha-1} \varphi(x, t) dt. \qquad (E-M.7)$$

If the integral

$$\gamma(\alpha) = \int_0^\infty t^{-\alpha-1} \psi(t) dt$$

is defined for some $\alpha = \alpha_0$ then γ_α is defined and continuous for all $\alpha \leq \alpha_0$.

Denoting $[\alpha]$ by r, one can see that $\gamma(\alpha)$ is well-defined whenever $\psi(t)$ has order of infinitesimality equal to $r + 1$. It is required that

$$\sum_{j=0}^p k_j^s C_j = 0, \quad s = 0, 1, \ldots, r.$$

r-times integration by parts gives

$$\gamma(\alpha) = \Gamma(-\alpha) \sum_{j=0}^p k_j^\alpha C_j \quad (\alpha \neq r); \qquad (E-M.8)$$

$$\gamma(r) = \lim_{\alpha \to r} \gamma(\alpha) = \frac{(-1)^{r+1}}{r!} \sum_{j=0}^p k_j^r C_j \log k_j.$$

The simplest function $\psi(t)$ which has order of infinitesimality equal to p is the following one

$$\psi(t) = e^{-t}\left(1 - e^{-t}\right)^p = e^{-t} - \binom{p}{1}e^{-2t} + \binom{p}{2}e^{-3t} + \dots \qquad (E-M.9)$$

In this case

$$\varphi(x,t) = f(x-t) - \binom{p}{1}f(x-2t) + \binom{p}{2}f(x-3t) + \dots \qquad (E-M.10)$$

The simple properties of the Marchaud derivatives are similar to those of the Liouville derivative, e.g. permutability with the operators of reflection, translation and scaling (see [SaKiMa93, Sect. 2.5]), composition formulas with the singular integral operator S and the relation between $\mathbb{D}_-^\alpha$ and $\mathbb{D}_+^\alpha$:

$$\left(\mathbb{D}_-^\alpha f\right) = \cos\alpha\pi \left(\mathbb{D}_+^\alpha f\right) - \sin\alpha\pi \left(S\mathbb{D}_+^\alpha f\right).$$

Among the characteristic properties we point out the vanishing of the Marchaud derivative on the constant function:

$$\mathbb{D}_\pm^\alpha \text{const} \equiv 0.$$

For all $\alpha > 0$ the Marchaud derivative $\mathbb{D}_\pm^\alpha$ is defined on all bounded functions $f \in \mathcal{C}^{[\alpha]}(\mathbb{R}^1)$ satisfying the following condition:

$$|f^{([\alpha])}(x+h) - f^{([\alpha])}(x)| \le A(x)|h|^\lambda.$$

The Marchaud derivative appeared due to a formal replacement of the positive parameter α for negative one $-\alpha$ in the definition of the Liouville fractional integral. The new object is not well-defined. In order to give a sense to the integral in $(E-M.2)$, one can use Hadamard's finite part (p.f.) (see [SaKiMa93, Lemma 5.2]):

$$\left(\mathbb{D}_\pm^\alpha f\right)(x) = p.f.\left(I_\pm^{(-\alpha)} f\right)(x).$$

An application of this approach gives, in particular, the following result (see [SaKiMa93, Lemma 5.3]): let $f \in C_{loc}^{m,\lambda}(\mathbb{R})$, $0 \le \lambda < 1$, $m = [\alpha]$, $\alpha > 0$, $\alpha \ne 1, 2, \dots$. Then, the following representation holds:

$$p.f. \int_0^\infty \frac{f(x-t)dt}{t^{1+\alpha}} = \int_0^\infty \frac{f(x-t) - \sum_{k=0}^m \frac{(-1)^k f^{(k)}(x)}{k!}}{t^{1+\alpha}} dt.$$

Finally we note that the Marchaud derivative is a suitable object for the representation of fractional powers of operators.

E.4 The Erdélyi–Kober Fractional Calculus

In this section we present the main definitions of the *Erdélyi–Kober fractional integrals and derivatives* (more details can be found in [SaKiMa93, Kir94]).

Let (a, b) be a finite or infinite interval of the real positive semi-axis ($0 \leq a < b \leq +\infty$). We fix parameters α, σ, η such that $\operatorname{Re} \alpha > 0, \sigma > 0, \eta \in \mathbb{C}$. Left- and right-sided Erdélyi–Kober fractional integrals depending on these parameters are defined as follows

$$\left(I^\alpha_{a+;\sigma,\eta} f\right)(t) := \frac{\sigma t^{-\sigma(\alpha+\eta)}}{\Gamma(\alpha)} \int_a^t \frac{\tau^{\sigma\eta+\sigma-1} f(\tau) d\tau}{(t^\sigma - \tau^\sigma)^{1-\alpha}}, \qquad (E-EK.1)$$

$$\left(I^\alpha_{b-;\sigma,\eta} f\right)(t) := \frac{\sigma t^{\sigma\eta}}{\Gamma(\alpha)} \int_t^b \frac{\tau^{\sigma(1-\alpha-\eta)-1} f(\tau) d\tau}{(t^\sigma - \tau^\sigma)^{1-\alpha}}. \qquad (E-EK.2)$$

There exist relationships between Erdélyi–Kober and Riemann–Liouville fractional integrals

$$\left(I^\alpha_{a+;\sigma,\eta} f\right)(t) = \left(N_\sigma M_{-\alpha-\eta} J^\alpha_{a^\sigma+} M_\eta N_{1/\sigma} f\right)(t), \qquad (E-EK.3)$$

$$\left(I^\alpha_{b-;\sigma,\eta} f\right)(t) = \left(N_\sigma M_\eta J^\alpha_{b^\sigma-} M_{-\alpha-\eta} N_{1/\sigma} f\right)(t), \qquad (E-EK.4)$$

where

$$(M_c \varphi)(t) = t^c \varphi(t), \ t \in \mathbb{R}, c \in \mathbb{C},$$
$$(N_c \varphi)(t) = \varphi(t^c), \ t \in \mathbb{R}, c \in \mathbb{R} \setminus \{0\}.$$

Thus, under suitable conditions on function one can prove properties similar to those for the Riemann–Liouville integral, namely, acting properties, the semi-group property, the Leibniz rule, etc.

The corresponding Erdélyi–Kober fractional derivatives are defined by the following formulas ($\operatorname{Re} \alpha \geq 0, \alpha \neq 0, n-1 < \operatorname{Re} \alpha \leq n, \sigma > 0, \eta \in \mathbb{C}$)

$$\left(D^\alpha_{a+;\sigma,\eta} f\right)(t) := t^{-\sigma\eta} \left(\frac{1}{\sigma t^{\sigma-1}} \frac{d}{dt}\right)^n t^{\sigma(n+\eta)} \left(I^{n-\alpha}_{a+;\sigma,\eta+\alpha} f\right)(t), \qquad (E-EK.5)$$

$$\left(D^\alpha_{b-;\sigma,\eta} f\right)(t) := t^{\sigma(\eta+\alpha)} \left(-\frac{1}{\sigma t^{\sigma-1}} \frac{d}{dt}\right)^n t^{\sigma(n-\eta-\alpha)} \left(I^{n-\alpha}_{b-;\sigma,\eta+\alpha-n} f\right)(t). \qquad (E-EK.6)$$

The fractional differential operators $(E-EK.5), (E-EK.6)$ are left inverses for the fractional integral operators $(E-EK.3), (E-EK.4)$, respectively.

In applications, the Mellin transform of the Erdélyi–Kober fractional integrals plays an important role. The corresponding results are the following. Let $\operatorname{Re}\alpha > 0$, $\sigma > 0$, $\eta \in \mathbb{C}$ and $f \in L_p(\mathbb{R}^+)$.

If $\operatorname{Re}(\eta - s/\sigma) > -1$, then

$$\left\{\mathcal{M}\left(I_{0+;\sigma,\eta}^\alpha f(t)\right)\right\}(s) = \frac{\Gamma(1+\eta-s/\sigma)}{\Gamma(1+\eta+\alpha-s/\sigma)}\{\mathcal{M}f\}(s). \qquad (E-EK.7)$$

If $\operatorname{Re}(\eta + s/\sigma) > 0$, then

$$\left\{\mathcal{M}\left(I_{-;\sigma,\eta}^\alpha f(t)\right)\right\}(s) = \frac{\Gamma(\eta+s/\sigma)}{\Gamma(\eta+\alpha+s/\sigma)}\{\mathcal{M}f\}(s). \qquad (E-EK.8)$$

E.5 The Hadamard Fractional Calculus

The *Hadamard fractional integrals* are defined as follows

$$\left(^H\mathcal{J}_{a+}^\alpha f\right)(t) := \frac{1}{\Gamma(\alpha)} \int_a^t \left(\log\frac{t}{\tau}\right)^{\alpha-1} \frac{f(\tau)d\tau}{\tau}, \quad a < t < b, \qquad (E-H.1)$$

$$\left(^H\mathcal{J}_{b-}^\alpha f\right)(t) := \frac{1}{\Gamma(\alpha)} \int_t^b \left(\log\frac{\tau}{t}\right)^{\alpha-1} \frac{f(\tau)d\tau}{\tau}, \quad a < t < b. \qquad (E-H.2)$$

Let $n-1 < \alpha \leq n$, then the *Hadamard fractional derivatives* are defined by the following formulas

$$\left(^H\mathcal{D}_{a+}^\alpha f\right)(t) := \left(t\frac{d}{dt}\right)^n \left(^H\mathcal{J}_{a+}^{n-\alpha} f\right)(t), \qquad (E-H.3)$$

$$\left(^H\mathcal{D}_{b-}^\alpha f\right)(t) := \left(-t\frac{d}{dt}\right)^n \left(^H\mathcal{J}_{b-}^{n-\alpha} f\right)(t). \qquad (E-H.4)$$

These fractional integrals and derivatives are suitable to apply to logarithmic type functions, e.g. for all α, β, $\operatorname{Re}\beta > \operatorname{Re}\alpha > 0$

$$\left(^H\mathcal{J}_{a+}^\alpha \left(\log\frac{\tau}{a}\right)^{\beta-1}\right)(t) = \frac{\Gamma(\beta)}{\Gamma(\beta+\alpha)}\left(\log\frac{t}{a}\right)^{\beta+\alpha-1},$$

$$\left(^H\mathcal{D}_{a+}^\alpha \left(\log\frac{\tau}{a}\right)^{\beta-1}\right)(t) = \frac{\Gamma(\beta)}{\Gamma(\beta-\alpha)}\left(\log\frac{t}{a}\right)^{\beta-\alpha-1}.$$

Appendix E: Elements of Fractional Calculus

In particular, the Hadamard fractional derivatives of the constant are not equal to zero. Indeed, for all α, $0 < \text{Re}\,\alpha < 1$

$$\left({}^H\mathcal{D}_{a+}^\alpha 1\right)(t) = \frac{\left(\log\frac{t}{a}\right)^{-\alpha}}{\Gamma(1-\alpha)}, \quad \left({}^H\mathcal{D}_{b-}^\alpha 1\right)(t) = \frac{\left(\log\frac{b}{t}\right)^{-\alpha}}{\Gamma(1-\alpha)}, \qquad (E-H.5)$$

but on the other hand, for all $j = 1, 2, \ldots, n = [\text{Re}\,\alpha] + 1$

$$\left({}^H\mathcal{D}_{a+}^\alpha \left(\log\frac{\tau}{a}\right)^{\alpha-j}\right)(t) = 0. \qquad (E-H.6)$$

More general forms of fractional integrals and derivatives (called Hadamard-type fractional integrals and derivatives) were introduced in [Kil01] and studied in detail in [BuKiTr02a].

$$\left(\mathcal{J}_{a+,\mu}^\alpha f\right)(t) := \frac{1}{\Gamma(\alpha)} \int_a^t \left(\frac{\tau}{t}\right)^\mu \left(\log\frac{t}{\tau}\right)^{\alpha-1} \frac{f(\tau)d\tau}{\tau}, \quad a < t < b, \qquad (E-H.7)$$

$$\left(\mathcal{J}_{b-}^\alpha f\right)(t) := \frac{1}{\Gamma(\alpha)} \int_t^b \left(\frac{t}{\tau}\right)^\mu \left(\log\frac{\tau}{t}\right)^{\alpha-1} \frac{f(\tau)d\tau}{\tau}, \quad a < t < b. \qquad (E-H.8)$$

$$\left(\mathcal{D}_{a+,\mu}^\alpha f\right)(t) = t^{-\mu}\delta^n t^\mu \, (\mathcal{J}_{a+,\mu}^{n-\alpha} f)(t), \qquad (E-H.9)$$

$$\left(\mathcal{D}_{b-,\mu}^\alpha f\right)(t) = t^{-\mu}(-\delta)^n t^\mu \, (\mathcal{J}_{b-,\mu}^{n-\alpha} f)(t), \qquad (E-H.10)$$

$$\delta = t\frac{d}{dt}, \quad (\alpha > 0; \; n = [\alpha] + 1, \; \mu \in \mathbb{R}).$$

When $\mu = 0$, the Hadamard-type constructions coincide with the standard Hadamard integrals and derivatives.

The properties of the operators $\mathcal{J}_{a+,\mu}^\alpha$ and $\mathcal{D}_{a+,\mu}^\alpha$ with $a > 0$ were investigated in the space $X_c^p(a, b)$ ($c \in \mathbb{R}$, $1 \le p \le \infty$) of Lebesgue measurable functions h, defined on a finite interval $[a, b]$ of the real axis $\mathbb{R}$, for which $\|h\|_{X_c^p} < \infty$, where

$$\|h\|_{X_c^p} = \left(\int_a^b |t^c h(t)|^p \frac{dt}{t}\right)^{1/p} \quad (c \in \mathbb{R}, \; 1 \le p < \infty),$$

$$\|h\|_{X_c^\infty} = \text{esssup}_{a \le t \le b}[t^c |h(t)|] \quad (c \in \mathbb{R}).$$

If $\alpha > 0$, $1 \le p \le \infty$, $0 < a < b < \infty$ and $\mu \ge c$, then the following formula holds for functions $g \in X_c^p(a, b)$:

$$(\mathcal{D}^\alpha_{a+,\mu}\mathcal{J}^\alpha_{a+,\mu}g)(t) = g(t),$$

i.e. the Hadamard-type fractional derivative is left-inverse to the Hadamard-type fractional integral. The opposite is not true, namely, if $\alpha > 0$, $n = -[-\alpha]$, $\mu \in \mathbb{R}$, $0 < a < b < \infty$, and $(\mathcal{J}^{n-\alpha}_{a+,\mu}g)(t)$ is the Hadamard-type fractional integral, then for all $g \in X^1_\mu(a,b)$ and $(\mathcal{J}^{n-\alpha}_{a+,\mu}g)(t) \in \mathcal{AC}^n_{\delta;\mu}[a,b]$, the following holds

$$(\mathcal{J}^\alpha_{a+,\mu}\mathcal{D}^\alpha_{a+,\mu}g)(t) = g(t) - t^{-\mu}\sum_{k=1}^n \frac{g^{n-k}_{n-\alpha}(a)}{\Gamma(\alpha-k+1)}\left(\ln\frac{t}{a}\right)^{\alpha-k}, \quad (E-H.11)$$

where

$$g^k_{n-\alpha}(t) = \delta^k t^\mu(\mathcal{J}^{n-\alpha}_{a+,\mu}g)(t), \quad \delta = t\frac{d}{dt} \quad (k=0,\ldots,n-1).$$

Note that the Hadamard-type fractional derivative of order α, $0 < \alpha < 1$, can be represented in the Marchaud form (see [KilTit07])

$$\left(\mathbf{D}^\alpha_{0+,\mu}f\right)(t) = \frac{\alpha}{\Gamma(1-\alpha)}\int_0^{+\infty} e^{-\mu\tau}\frac{f(t)-f(te^{-\tau})}{\tau^{1+\alpha}}d\tau + \mu^\alpha f(t), \quad (E-H.12)$$

which coincides with $\left(\mathcal{D}^\alpha_{a+,\mu}f\right)(t)$ for all $f \in X^p_c(\mathbb{R}_+)$.

E.6 The Grünwald–Letnikov Fractional Calculus

Let us briefly describe some results which preceded the work of Letnikov. Although the idea of fractional derivatives goes back to the end of the 19th century, the real results in the area were made by Liouville [Lio32a, Lio32b].

Liouville applied his construction to functions represented in the form of the following (convergent!) series

$$y(x) = \sum_{k=1}^\infty A_k e^{m_k x}. \quad (E-GL.1)$$

For such functions he used Leibniz's idea of an arbitrary order differentiation of the exponential function:

$$\frac{d^p y}{dx^p} := \sum_{k=1}^\infty A_k m_k^p e^{m_k x}. \quad (E-GL.2)$$

Liouville considered definition $(E-GL.2)$ as the only possible, stressing (see [LetChe11, p. 14]): "…it is impossible to get an exact and complete understanding of

Appendix E: Elements of Fractional Calculus

the nature of the arbitrary order derivative without taken certain series representation of the function".

Letnikov noted that Liouville's construction has an essential drawback. It follows from the definition $(E-GL.2)$ that it can be used only for functions whose derivatives (of all positive integer orders) vanish at infinity. Liouville himself met this first difficulty when trying to apply his definition to power functions with a positive exponent x^m, $m>0$, which do not satisfy the above condition. To overcome this difficulty he started with the function $y(x)=\frac{1}{x^m}$, $m>0$. It follows from the definition of the Γ-function that the following holds

$$\frac{1}{x^m}=\frac{1}{\Gamma(m)}\int_0^\infty e^{-zx}z^{m-1}dz.$$

The right-hand side of this formula can be considered as an expansion of the type $(E-GL.1)$, namely, $\sum A_n e^{-nx}$ with A_n being sufficiently small as $x\to\infty$. Thus, formula $(E-GL.2)$ applied in this case leads to the following result

$$\frac{d^p\frac{1}{x^m}}{dx^p}=\frac{1}{\Gamma(m)}\int_0^\infty e^{-zx}(-z)^p z^{m-1}dz \qquad (E-GL.3)$$

$$=\frac{(-1)^p}{\Gamma(m)x^{m+p}}\int_0^\infty e^{-t}t^{p+m-1}dz=\frac{(-1)^p\Gamma(m+p)}{\Gamma(m)x^{m+p}},$$

which coincides with the celebrated Euler formula for the arbitrary order derivative of the power function.

Note that Liouville considered only the case when $m>0$ and $m+p>0$.[4] For the remaining cases he used the notion of so-called additional functions in order to correct the above definition. These are the functions whose derivative of order $(-p)$ is equal to zero. Liouville gave a proof that additional functions should have the form

$$A_0+A_1x+\ldots+A_nx^n$$

for a certain finite power n and arbitrary constant coefficients A_j.[5]

During the next 30 years several attempts were made to correct Liouville's construction (see [RogDub18] and the references therein). It was only in 1867–1868 that A.K. Grünwald [Gru67] and A.V. Letnikov [Let68a] proposed a truly general definition of the fractional derivative. Both constructions (which differ only in a few details) are based on the following formula for representing the derivative of an arbitrary positive integer order via finite differences:

---

[4]This predates the Legendre–Gauss definition of the Γ function of a complex argument.
[5]This proof was not considered satisfactory by many mathematicians, even in Liouville's time.

$$\frac{d^p f(x)}{dx^p} = \lim_{h \to +0} \frac{f(x) - \binom{p}{1} f(x-h) + \ldots + (-1)^n \binom{p}{n} f(x-nh)}{h^p},$$

$$(E-GL.4)$$

where n is an arbitrary positive number, $n \geq p$.

Letnikov also considered an expression

$$f^{-p)}(x) := \lim_{h \to 0} h^p \left[f(x) + \binom{p}{1} f(x-h) + \ldots + \binom{p}{n} f(x-nh) \right].$$

If p is a positive integer in the last formula, then $f^{-p)}(x)$ is vanishing whenever n is finite. Hence it is interesting to consider the case when n tends to infinity as $h \to 0$. Assuming $h = (x-x_0)/n$ Letnikov defined the following object[6]

$$\left[D^{-p} f(x) \right]_{x_0}^{x} := \lim_{h \to 0} f^{-p)}(x). \qquad (E-GL.5)$$

It was shown that if p is a positive integer, then the following relation holds whenever the integral in the right-hand side exists

$$\left[D^{-p} f(x) \right]_{x_0}^{x} = \lim_{h \to 0} \sum_{r=0}^{n} h^p \binom{p}{r} f(x-rh) = \frac{1}{(p-1)!} \int_{x_0}^{x} (x-\tau)^{p-1} f(\tau) d\tau.$$

$$(E-GL.6)$$

It follows from $(E-GL.6)$ that if this formula is valid for a positive integer p, then it is valid for $p+1$ too. Moreover, the above introduced object $\left[D^{-p} f(x) \right]_{x_0}^{x}$ is a function whose p-th derivative coincides with $f(x)$:

$$\frac{d^p}{dx^p} \left[D^{-p} f(x) \right]_{x_0}^{x} = f(x),$$

and

$$\frac{d^j}{dx^j} \left[D^{-p} f(x) \right]_{x_0}^{x} \bigg|_{x=x_0} = 0, \quad \forall j = 0, 1, \ldots, p-1.$$

The above definition and its properties constitute the basis for further generalizations. Thus, by using the properties of binomial coefficients, Letnikov proved that for any function f continuous on $[x_0, x]$ there exist the following limits

$$\lim_{n \to \infty} \frac{\sum_{r=0}^{n} (-1)^r \binom{p}{r} f(x-rh)}{h^p}, \quad \lim_{n \to \infty} \sum_{r=0}^{n} h^p \binom{p}{r} f(x-rh),$$

---

[6]Called by him a derivative of negative order in finite limits.

Appendix E: Elements of Fractional Calculus

where $h = (x - x_0)/n$ and $p \in \mathbb{C}$, Re $p > 0$. The values of these limits, denoted by him $[D^p f(x)]_{x_0}^{x}$ and $[D^{-p} f(x)]_{x_0}^{x}$, respectively, are formally equal in this case to

$$[D^p f(x)]_{x_0}^{x} = \frac{1}{\Gamma(-p)} \int_{x_0}^{x} \frac{f(\tau) d\tau}{(x-\tau)^{p+1}}, \qquad (E-GL.7)$$

$$[D^{-p} f(x)]_{x_0}^{x} = \frac{1}{\Gamma(p)} \int_{x_0}^{x} \frac{f(\tau) d\tau}{(x-\tau)^{1-p}}. \qquad (E-GL.8)$$

Formula $(E - GL.7)$ gives the derivative of arbitrary order $p \in \mathbb{C}$, Re $p > 0$, and formula $(E - GL.8)$ gives the integral of arbitrary order $p \in \mathbb{C}$, Re $p > 0$. As has already been mentioned, the integral in $(E - GL.8)$ exists whenever f is continuous, but this is not the case for the integral in $(E - GL.7)$. To overcome this difficulty, Letnikov transformed the right-hand side of (E-GL.7) to the form

$$[D^p f(x)]_{x_0}^{x} = \sum_{k=0}^{m} \frac{f^{(k)}(x_0)(x-x_0)^{-p+k}}{\Gamma(-p+k+1)} + \frac{1}{\Gamma(-p+m+1)} \int_{x_0}^{x} \frac{f^{(m+1)}(\tau) d\tau}{(x-\tau)^{p-m}}. \qquad (E-GL.9)$$

The assumption of existence and continuity of all derivatives up to order $m + 1$ is sufficient for this representation. It is suitable here to take $m = [\text{Re } p]$.

Integration by parts in $(E - GL.8)$ leads to an analogous formula for $[D^{-p} f(x)]_{x_0}^{x}$, which is valid under the same conditions for any integer positive m:

$$[D^{-p} f(x)]_{x_0}^{x} = \sum_{k=0}^{m} \frac{f^{(k)}(x_0)(x-x_0)^{p+k}}{\Gamma(p+k+1)} + \frac{1}{\Gamma(p+m+1)} \int_{x_0}^{x} \frac{f^{(m+1)}(\tau) d\tau}{(x-\tau)^{p-1+m}}. \qquad (E-GL.10)$$

We can also consider the special case when $x_0 \to +\infty$. It follows from $(E - GL.8)$ that

$$[D^{-p} f(x)]_{+\infty}^{x} = \frac{(-1)^p}{\Gamma(p)} \int_{0}^{+\infty} \tau^{p-1} f(x+\tau) d\tau, \qquad (E-GL.11)$$

and from $(E - GL.9)$,

$$[D^p f(x)]_{+\infty}^{x} = \frac{(-1)^{m+1-p}}{\Gamma(-p+m+1)} \int_{0}^{+\infty} \tau^{m-p} f^{(m+1)}(x+\tau) d\tau. \qquad (E-GL.12)$$

Letnikov noted that the integrals in the right-hand side of $(E-GL.12)$ and $(E-GL.11)$ converge in particular if $\lim_{x_0 \to +\infty} f^{(k)}(x_0) = 0$, $k = 0, \ldots, m$. This is exactly the class of functions considered by Liouville and formulas $(E-GL.12)$ and $(E-GL.11)$ coincide with the corresponding formulas presented in [Lio32a, Lio32b]. Further developments of the Grünwald–Letnikov construction are described in [LetChe11] (see also [RogDub18]).

E.7 The Riesz Fractional Calculus

The fractional calculus in multi-dimensional spaces is due to M. Riesz (see [Rie49], cf. [SaKiMa93]). The corresponding fractional integral and derivative are defined by using the multi-dimensional Fourier transform. Formally, such objects are given by the following formulas

$$(-\Delta)^{-\alpha/2} f = \mathcal{F}^{-1} |\mathbf{x}|^{-\alpha} \mathcal{F} f = \begin{cases} \mathbf{I}^\alpha f, & \operatorname{Re} \alpha > 0, \\ \mathbf{D}^{-\alpha} f, & \operatorname{Re} \alpha < 0, \end{cases} \qquad (E-R.1)$$

where $\mathbf{x} = (x_1, x_2, \ldots, x_m) \in \mathbb{R}^m$, $|\mathbf{x}| = \sqrt{x_1^2 + x_2^2 + \ldots + x_m^2}$.

The operators $\mathbf{I}^\alpha$, $\mathbf{D}^{-\alpha}$, defined in $(E-R.1)$, are called the *Riesz fractional integral* and the *Riesz fractional derivative*, respectively.

The Riesz fractional integral $\mathbf{I}^\alpha$, $\operatorname{Re} \alpha > 0$, is known to be represented in the form of a so-called Riesz potential

$$(\mathbf{I}^\alpha f)(\mathbf{x}) = \int_{\mathbb{R}^m} k_\alpha(\mathbf{x} - \mathbf{t}) f(\mathbf{t}) d\mathbf{t}, \qquad (E-R.2)$$

where the Riesz kernel k_α is given by the following formulas:

$$k_\alpha(\mathbf{x}) = \frac{1}{\gamma_m(\alpha)} \begin{cases} |\mathbf{x}|^{\alpha-m}, & \alpha - m \neq 0, 2, 4, \ldots, \\ |\mathbf{x}|^{\alpha-m} \log \frac{1}{|\mathbf{x}|}, & \alpha - m = 0, 2, 4, \ldots, \end{cases} \qquad (E-R.3)$$

with the constant $\gamma_m(\alpha)$ of the form

$$\gamma_m(\alpha) = \begin{cases} 2^\alpha \pi^{\frac{m}{2}} \frac{\Gamma(\frac{\alpha}{2})}{\Gamma(\frac{m-\alpha}{2})}, & \alpha - m \neq 0, 2, 4, \ldots, \\ (-1)^{\frac{m-\alpha}{2}} 2^{\alpha-1} \pi^{\frac{m}{2}} \Gamma\left(1 + \frac{\alpha-m}{2}\right) \Gamma\left(\frac{\alpha}{2}\right), & \alpha - m = 0, 2, 4, \ldots. \end{cases}$$
$$(E-R.4)$$

In other words, for all α, $\operatorname{Re} \alpha > 0$, $\alpha - m \neq 0, 2, 4, \ldots$ the Riesz potential takes the form

$$(\mathbf{I}^\alpha f)(\mathbf{x}) = \frac{1}{\gamma_m(\alpha)} \int_{\mathbb{R}^m} \frac{f(\mathbf{t}) d\mathbf{t}}{|\mathbf{x} - \mathbf{t}|^{m-\alpha}}. \qquad (E-R.5)$$

Appendix E: Elements of Fractional Calculus

Note that if $0 < \alpha < m$ and $1 < p < \frac{m}{\alpha}$, then the Riesz potential $(\mathbf{I}^\alpha f)(\mathbf{x})$ is well defined in the space $L_p(\mathbb{R}^m)$. Moreover, the operator $\mathbf{I}^\alpha$ is bounded from $L_p(\mathbb{R}^m)$ into $L_q(\mathbb{R}^m)$ if, and only if,

$$0 < \alpha < m, \ 1 < p < \frac{m}{\alpha}, \ \text{and} \ \frac{1}{q} = \frac{1}{p} - \frac{\alpha}{m}.$$

Other properties are related to composition of the Riesz potential and certain operators, in particular, the Fourier transform

$$(\mathcal{F}\mathbf{I}^\alpha f)(\mathbf{x}) = \frac{1}{|\mathbf{x}|^\alpha}(\mathcal{F}f)(\mathbf{x}), \qquad (E-R.6)$$

and the Laplace operator

$$(\Delta \mathbf{I}^\alpha f)(\mathbf{x}) = -\left(\mathbf{I}^{\alpha-2}f\right)(\mathbf{x}), \ \operatorname{Re}\alpha > 2. \qquad (E-R.7)$$

A detailed discussion of these and some other properties of the Riesz potential can be found in [SaKiMa93, Sect. 25, 26].

The Riesz fractional derivative is realized in the form of the hypersingular integral and is defined by the following relation

$$(\mathbf{D}^\alpha f)(\mathbf{x}) := \frac{1}{d_m(l,\alpha)} \int_{\mathbb{R}^m} \frac{(\Delta_\mathbf{t}^l f)(\mathbf{x})}{|\mathbf{t}|^\alpha} d\mathbf{t}, \ (l > \alpha), \qquad (E-R.8)$$

where

$$(\Delta_\mathbf{t}^l f)(\mathbf{x}) := \sum_{k=0}^{l}(-1)^k \binom{l}{k} f(\mathbf{x} - k\mathbf{t}), \qquad (E-R.9)$$

and the constant $d_m(l, \alpha)$ is defined by the following formula

$$d_m(l,\alpha) := \frac{2^{-\alpha}\pi^{1+m/2}}{\Gamma\left(1+\frac{\alpha}{2}\right)\Gamma\left(\frac{m+\alpha}{2}\right)} \frac{A_l(\alpha)}{\sin\frac{\alpha\pi}{2}}, \qquad (E-R.10)$$

$$A_l(\alpha) = \sum_{k=0}^{l}(-1)^{k-1}\binom{l}{k}k^\alpha.$$

Among the properties of this form $(\mathbf{D}^\alpha f)$ of the Riesz fractional derivative we point out its Fourier transform

$$(\mathcal{F}\mathbf{D}^\alpha f)(\mathbf{x}) = |\mathbf{x}|^\alpha (\mathcal{F}f)(\mathbf{x}), \qquad (E-R.11)$$

and the composition with the Riesz fractional integral

$$(\mathbf{D}^\alpha \mathbf{I}^\alpha f)(\mathbf{x}) = f(\mathbf{x}). \qquad (E-R.12)$$

The corresponding conditions under which formulas $(E-R.11)$ and $(E-R.12)$ hold true are discussed in detail in [SaKiMa93, Sect. 26]. An extended study of the properties of the hypersingular operators is presented in [Sam02].

E.8 Historical and Bibliographical Notes

Fractional calculus is the field of mathematical analysis which deals with the investigation and application of integrals and derivatives of arbitrary order. The term *fractional* is a misnomer, but it has been retained following the prevailing use.

The fractional calculus may be considered an *old* and yet *novel* topic. It is an *old* topic since, starting from some speculations of G.W. Leibniz (1695, 1697) and L. Euler (1730), it has been developed up to the present day. In fact the idea of generalizing the notion of derivative to non-integer order, in particular to the order 1/2, is contained in the correspondence of Leibniz with Bernoulli, L'Hôpital and Wallis. Euler took the first step by observing that the result of the evaluation of the derivative of the power function has a meaning for non-integer order thanks to his Gamma function.

A list of mathematicians who have provided important contributions up to the middle of the 20th century includes P.S. Laplace (1812), J.B.J. Fourier (1822), N.H. Abel (1823–1826), J. Liouville (1832–1837), B. Riemann (1847), H. Holmgren (1865–67), A.K. Grünwald (1867–1872), A.V. Letnikov (1868–1872), N.Ya. Sonine (1872–1884), H. Laurent (1884), P.A. Nekrassov (1888), A. Krug (1890), J. Hadamard (1892), O. Heaviside (1892–1912), S. Pincherle (1902), G.H. Hardy and J.E. Littlewood (1917–1928), H. Weyl (1917), P. Lévy (1923), A. Marchaud (1927), H.T. Davis (1924–1936), A. Zygmund (1935–1945), E.R. Love (1938–1996), A. Erdélyi (1939–1965), H. Kober (1940), D.V. Widder (1941), M. Riesz (1949), and W. Feller (1952).

In [LetChe11] A.V. Letnikov's main results on fractional calculus are presented, including his dissertations and a long discussion between A.V. Letnikov and N.Ya. Sonine on the foundations of fractional calculus. The modern development of the ideas by Letnikov is given and their applications to underground dynamics and population dynamics are presented.

However, it may be considered a *novel* topic as well, since it has only been the subject of specialized conferences and treatises in the last 30 years. The merit is due to B. Ross for organizing the *First Conference on Fractional Calculus and its Applications* at the University of New Haven in June 1974 and editing the proceedings [Ros75] (see also [Ros77]). For the first monograph the merit is ascribed to K.B. Oldham and J. Spanier [OldSpa74], who, after a joint collaboration starting in 1968, published a book devoted to fractional calculus in 1974.

To our knowledge, the current list of texts in book form with a title explicitly devoted to fractional calculus (and its applications) includes around ten titles, namely

Oldham and Spanier (1974) [OldSpa74], McBride (1979) [McB79], Samko, Kilbas and Marichev (1987–1993) [SaKiMa93], Nishimoto(1984–1996) [Nis84], Miller and Ross (1993) [MilRos93], Kiryakova (1994) [Kir94], Rubin (1996) [Rub96], Podlubny (1999) [Pod99], West, Bologna and Grigolini (2003) [WeBoGr03], Kilbas, Strivastava and Trujillo (2006) [KiSrTr06], Magin (2006) [Mag06], Mainardi (2010) [Mai10], Diethelm (2010) [Die10], Uchaikin (2008-2013) [Uch08, Uch13a, Uch13b], Atanackovic, Pilipovic, Stankovic and Zorica [Ata-et-al14], Baleanu, Diethelm, Scalas and Trujillo (2017) [Bal-et-al17], Evangelista and Lenzi (2018) [EvaLen18], Hermann [Her18], Mathai and Haubold [MatHau17, MatHau18] Sandev and Tomovski (2019) [SanTom19] Capelas (2019) [Cap19].

Furthermore, we draw the reader's attention to the treatises by Davis (1936) [Dav36], Erdélyi (1953–1954) [ErdBat-1, ErdBat-2, ErdBat-3], Gel'fand and Shilov (1959–1964) [GelShi64], Djrbashian (or Dzherbashian) [Dzh66], Caputo [Cap69], Babenko [Bab86], Gorenflo and Vessella [GorVes91], Zaslavsky (2005) [Zas05], which contain a detailed analysis of some mathematical aspects and/or physical applications of fractional calculus, without referring to fractional calculus in the title. See also [Levy23c, BuKiTr03, Sne75].

For more details on the historical development of the fractional calculus we refer the interested reader to Ross' bibliography in [OldSpa74] and to the historical notes generally available in the above quoted texts. In particular, we mention the two posters edited by Machado, Kiryakova and Mainardi on the old and recent (up to 2010) history of Fractional Calculus that are available on Research Gate and at the website of Fractional Calculus and Applied Analysis (FCAA) http://www.math.bas.bg/~fcaa/. The latter includes free access to full length papers from 2004–2010.

In recent years considerable interest in fractional calculus has been stimulated by the applications that it finds in different fields of science, including numerical analysis, economics and finance, engineering, physics, biology, etc., as well as several contributions to the fractional theory [Hil19, Koc19a, KocLuc19] and its applications [Die19, GorMai19, Tar19] in the recent series, edited by Machado (published by De Gruyter), of 8 Handbooks on Fractional Calculus with applications, [HAND1, HAND2, HAND3, HAND4, HAND5, HAND6, HAND7, HAND8]. See also the special issue of Mathematics (MDPI) edited by Mainardi [Mai-spec18].

For economics and finance we quote the collection of articles on the topic of *Fractional Differencing and Long Memory Processes*, edited by Baillie and King (1996), which appeared as a special issue in the Journal of Econometrics [BaiKin96]. For engineering and physics we mention the book edited by Carpinteri and Mainardi (1997) [CarMai97], entitled *Fractals and Fractional Calculus in Continuum Mechanics*, which contains lecture notes of a CISM Course devoted to some applications of related techniques in mechanics, and the book edited by Hilfer (2000) [Hil00], entitled *Applications of Fractional Calculus in Physics*, which provides an introduction to fractional calculus for physicists, and collects review articles written by some of the leading experts. In the above books we recommend the introductory surveys on fractional calculus by Gorenflo and Mainardi [GorMai97] and by Butzer and Westphal [ButWes00], respectively.

In addition to some books containing proceedings of international conferences and workshops on related topics, see e.g. [McBRoa85, Nis90, RuDiKi94, RuDiKi96], we mention regular journals devoted to fractional calculus, i.e. *Journal of Fractional Calculus* (Descartes Press, Tokyo), started in 1992, with Editor-in-Chief Prof. Nishimoto and *Fractional Calculus and Applied Analysis* from 1998, with Editor-in-Chief Prof. Kiryakova (De Gruyter, Berlin). For information on this journal, please visit the WEB site

https://www.degruyter.com/view/j/fca

Furthermore, websites devoted to fractional calculus have also appeared, of which we call attention to www.fracalmo.org, whose name comes from FRActional CALculus MOdelling, and the related links.

Appendix F
Higher Transcendental Functions

F.1 Hypergeometric Functions

F.1.1 Classical Gauss Hypergeometric Functions

The hypergeometric function $F(a; b; c; z) = F\left(\begin{matrix} a, b \\ c \end{matrix}; z\right) = {}_2F_1(a; b; c; z)$ is defined by *the Gauss series*

$$F(a; b; c; z) = \sum_{n=0}^{\infty} \frac{(a)_n (b)_n}{(c)_n n!} z^n = 1 + \frac{ab}{c} z + \frac{a(a+1)b(b+1)}{c(c+1)2!} z^2 + \ldots$$

$$= \frac{\Gamma(c)}{\Gamma(a)\Gamma(b)} \sum_{n=0}^{\infty} \frac{\Gamma(a+n)\Gamma(b+n)}{\Gamma(c+n) n!} z^n \qquad (F.1.1)$$

on the disk $|z| < 1$, and by analytic continuation elsewhere. In general, $F(a; b; c; z)$ does not exist when $c = 0, -1, -2, \ldots$. The branch obtained by introducing a cut from 1 to $+\infty$ on the real z-axis and by fixing the value $F(0) = 1$ is the principal branch (or principal value) of $F(a; b; c; z)$.

For all values of c another type of classical hypergeometric function is defined as

$$\mathbf{F}(a; b; c; z) = \sum_{n=0}^{\infty} \frac{(a)_n (b)_n}{\Gamma(c+n) n!} z^n = \frac{1}{\Gamma(c)} F(a; b; c; z), \quad |z| < 1, \qquad (F.1.2)$$

again with analytic continuation for other values of z, and with the principal branch defined in a similar way. In (F.1.2) it is supposed by definition that if for an index n we have $c + n$ as a non-positive integer, then the corresponding term in the series is identically equal to zero (e.g. if $c = -1$, then we omit the term with $n = 0$ and $n = 1$). This agreement is natural, because $1/\Gamma(-m) = 0$ for $m = 0, -1, -2, \ldots$.

On the circle $|z| = 1$, the Gauss series:

(a) converges absolutely when Re $(c - a - b) > 0$;
(b) converges conditionally when $-1 < \text{Re}(c - a - b) \leq 0$, and $z = 1$ is excluded;
(c) diverges when $\text{Re}(c - a - b) \leq -1$.

The principal branch of $\mathbf{F}(a; b; c; z_0)$ is an entire function of parameters a, b, and c for any fixed values $z_0, |z_0| < 1$ (see [ErdBat-1, p. 68]). The same is true of other branches, since the series in (F.1.2) converges for a fixed $z_0, |z_0| < 1$, in any bounded domain of the complex a, b, c-space. As a multi-valued function of z, $\mathbf{F}(a; b; c; z)$ is analytic everywhere except for possible branch points at $z = 1$, and $+\infty$. The same properties hold for $F(a; b; c; z)$, except that as a function of c, $F(a; b; c; z)$ in general has poles at $c = 0, -1, -2, \ldots$. Because of the analytic properties with respect to a, b, and c, it is usually legitimate to take limits in formulas involving functions that are undefined for certain values of the parameters.

For special values of parameters the hypergeometric function coincides with elementary functions

(a) $F(1; 1; 2; z)$ $= -\dfrac{\ln(1-z)}{z}$,

(b) $F(1/2; 1; 3/2; z^2)$ $= \dfrac{1}{2z} \ln\left(\dfrac{1+z}{1-z}\right)$,

(c) $F(1/2; 1; 3/2; -z^2)$ $= \dfrac{\arctan z}{z}$,

(d) $F(1/2; 1/2; 3/2; z^2)$ $= \dfrac{\arcsin z}{z}$,

(e) $F(1/2; 1/2; 3/2; -z^2)$ $= \dfrac{\ln\left(z + \sqrt{1+z^2}\right)}{z}$,

(f) $F(a; b; b; z)$ $= (1-z)^{-a}$,

(g) $F(a; 1/2 + a; 1/2; z^2) = \dfrac{1}{2}\left((1+z)^{-2a} + (1-z)^{-2a}\right)$

(see also the additional formulas below in Sect. F.5, cf. [NIST]).

We have the following asymptotic formulas for the hypergeometric function near the branch point $z = 1$:

(i) $\lim\limits_{z \to 1-0} -\dfrac{F(a; b; a+b; z)}{\ln(1-z)} = \dfrac{\Gamma(a+b)}{\Gamma(a)\Gamma(b)}$;

(ii) $\lim\limits_{z \to 1-0} (1-z)^{a+b-c} \left(F(a; b; c; z) - \dfrac{\Gamma(c)\Gamma(c-a-b)}{\Gamma(c-a)\Gamma(c-b)}\right)$

Appendix F: Higher Transcendental Functions

$$= \frac{\Gamma(c)\Gamma(a+b-c)}{\Gamma(a)\Gamma(b)}, \quad \text{where } \operatorname{Re}(c-a-b)=0, \; c \neq a+b;$$

(iii) $\lim_{z \to 1-0} (1-z)^{a+b-c} F(a;b;c;z)$

$$= \frac{\Gamma(c)\Gamma(a+b-c)}{\Gamma(a)\Gamma(b)}, \quad \text{where } \operatorname{Re}(c-a-b)<0.$$

All these formulas can be immediately verified, they follow from the definition (F.1.1).

F.1.2 Euler Integral Representation. Mellin–Barnes Integral Representation

If $\operatorname{Re} c > \operatorname{Re} b > 0$, the hypergeometric function satisfies the Euler integral representation (cf., [ErdBat-1, p. 59])

$$F(a;b;c;z) = \frac{\Gamma(c)}{\Gamma(b)\Gamma(c-b)} \int_0^1 \frac{t^{b-1}(1-t)^{c-b-1}}{(1-zt)^a} dt. \qquad (\text{F.1.3})$$

Here the right-hand side is a single-valued analytic function of z in the domain $|\arg(1-z)| < \pi$. Therefore, this formula determines an analytic continuation of the hypergeometric function $F(a;b;c;z)$ into this domain too.

In order to prove (F.1.3) in the unit disk $|z| < 1$ it is sufficient to expand the function $(1-zt)^{-a}$ into a binomial series and calculate the series term-by-term by using standard formulas for the Beta function.

The Euler integral representation can be rewritten as

$$F(a;b;c;z) = \frac{i\Gamma(c)e^{i\pi(b-c)}}{\Gamma(b)\Gamma(c-b)2\sin\pi(c-b)} \int_0^{(1+)} \frac{t^{b-1}(1-t)^{c-b-1}}{(1-zt)^a} dt,$$

$$\operatorname{Re} b > 0, \; |\arg(1-z)| < \pi, \; c-b \neq 1,2,\ldots;$$

$$F(a;b;c;z) = \frac{-i\Gamma(c)e^{-i\pi b}}{\Gamma(b)\Gamma(c-b)2\sin\pi b} \int_{(0+)}^1 \frac{t^{b-1}(1-t)^{c-b-1}}{(1-zt)^a} dt,$$

$$\operatorname{Re} c > \operatorname{Re} b > 0, \; |\arg(-z)| < \pi, \; b \neq 1,2,\ldots;$$

$$F(a;b;c;z) = \frac{-i\Gamma(c)e^{-i\pi c}}{\Gamma(b)\Gamma(c-b)4\sin\pi b \sin\pi(c-b)} \int_{(1+,0+,1-,0-)} \frac{t^{b-1}(1-t)^{c-b-1}}{(1-zt)^a} dt,$$

$$|\arg(-z)| < \pi; \quad b, 1-c, b-c \neq 1, 2, \ldots.$$

In each case we suppose that the path of integration starts and ends at corresponding points on the Riemann surface of the function

$$t^{b-1}(1-t)^{c-b-1}(1-zt)^{-a}$$

with real t, $0 \leq t \leq 1$, where t^b, $(1-t)^{c-b}$ mean the principal branches of these functions and $(1-zt)^{-a}$ is defined in such a way that $(1-zt)^{-a} \to 1$ whenever $z \to 0$.

The second type of integral representation is the so-called Mellin–Barnes representation (see Appendix D)

$$\frac{\Gamma(a)\Gamma(b)}{\Gamma(c)} F(a; b; c; z) = \frac{1}{2\pi i} \int_{-i\infty}^{+i\infty} \frac{\Gamma(a+s)\Gamma(b+s)\Gamma(-s)}{\Gamma(c+s)} (-z)^s ds, \quad \text{(F.1.4)}$$

where $|\arg(-z)| < \pi$ and the integration contour separates the poles of the functions $\Gamma(a+s)$ and $\Gamma(b+s)$ from those of the function $\Gamma(-s)$, and $(-z)^s$ assumes its principal values.

F.1.3 Basic Properties of Hypergeometric Functions

In the domain of definition the hypergeometric function satisfies some differential relations

$$\frac{d}{dz} F(a; b; c; z) = \frac{ab}{c} F(a+1; b+1; c+1; z),$$

$$\frac{d^n}{dz^n} F(a; b; c; z) = \frac{(a)_n (b)_n}{(c)_n} F(a+n; b+n; c+n; z),$$

$$\frac{d^n}{dz^n} \left(\frac{z^{c-1}}{(1-z)^{c-a-b}} F(a; b; c; z) \right) = \frac{(c-n)_n z^{c-n-1}}{(1-z)^{c+n-a-b}} F(a-n; b-n; c-n; z),$$

$$\left(z\frac{d}{dz} z \right)^n \left(z^{a-1} F(a; b; c; z) \right) = (a)_n z^{a+n-1} F(a+n; b; c; z).$$

(F.1.5)

Here the operator $\left(z\frac{d}{dz}z\right)^n$ is defined by the operator identity $\left(z\frac{d}{dz}z\right)^n (\cdot) = z^n \frac{d^n}{dz^n} z^n (\cdot)$, $n = 1, 2, \ldots$.

F.1.4 The Hypergeometric Differential Equation

If the second-order homogeneous differential equation has at most three singular points, we can assume that these are the points $0, 1, \infty$. If all these points are regular (see [Bol90]), then the equation can be reduced to the form

$$z(z-1)\frac{d^2w}{dz^2} + [c - (a+b+1)z]\frac{dw}{dz} - abw = 0. \tag{F.1.6}$$

This is *the hypergeometric differential equation*. It has regular singularities at $z = 0, 1, \infty$, with corresponding exponent pairs $\{0, 1-c\}$, $\{0, c-a-b\}$, $\{a, b\}$, respectively.

Here we use the standard terminology for the complex ordinary differential equation (see, e.g., [NIST, Sect. 2.7(i)])

$$\frac{d^2w}{dz^2} + f(z)\frac{dw}{dz} + g(z)w = 0.$$

A point z_0 is an ordinary point for this equation if the coefficients f and g are analytic in a neighborhood of z_0. In this case all solutions of the equation are analytic in a neighborhood of z_0. All other points z_0 are called singular points (or simply, singularities) for the differential equation. If both $(z - z_0)f(z)$ and $(z - z_0)^2 g(z)$ are analytic in a neighborhood of z_0, then this point is a regular singularity. All other singularities are called irregular. An irregular singularity z_0 is of rank $l - 1$ if l is the least integer such that $(z - z_0)^l f(z)$ and $(z - z_0)^{2l} g(z)$ are analytic in a neighborhood of z_0. The most common type of irregular singularity for special functions has rank 1 and is located at infinity. In this case the coefficients f, g have series representations

$$f(z) = \sum_{s=0}^{\infty} \frac{f_s}{z^s}, \quad g(z) = \sum_{s=0}^{\infty} \frac{g_s}{z^s},$$

where at least one of f_0, g_0, g_1 is non-zero.

Regular singularities are characterized by a pair of exponents (or indices) that are roots α_1, α_2 of the indicial equation

$$Q(\alpha) \equiv \alpha(\alpha - 1) + f_0\alpha + g_0 = 0,$$

where $f_0 = \lim_{z \to z_0}(z - z_0)f(z)$ and $g_0 = \lim_{z \to z_0}(z - z_0)^2 g(z)$. Provided that $(\alpha_1 - \alpha_2) \notin \mathbb{Z}$ the differential equation has two linear independent solutions

$$w_j = (z - z_0)^{\alpha_j} \sum_{s=0}^{\infty} a_{s,j}(z - z_0)^s, \quad j = 1, 2.$$

In the case of the hypergeometric differential equation, when none of $c, c - a - b, a - b$ is an integer, we have the pair $f_1(z)$, $f_2(z)$ of fundamental solutions. They are also numerically satisfactory ([NIST, Sect. 2.7(iv)]) in a neighborhood of the corresponding singularity.

F.1.5 Kummer's and Tricomi's Confluent Hypergeometric Functions

Kummer's differential equation

$$z \frac{d^2 w}{dz^2} + (c - z) \frac{dw}{dz} - aw = 0 \tag{F.1.7}$$

has a regular singularity at the origin with indices 0 and $1 - b$, and an irregular singularity at infinity of rank one. It can be regarded as the limiting form of the hypergeometric differential equation (F.1.6) that is obtained on replacing z by z/b, letting $b \to \infty$, and subsequently replacing the symbol c by b. In effect, the regular singularities of the hypergeometric differential equation at b and ∞ coalesce into an irregular singularity at ∞. Equation (F.1.7) is called *the confluent hypergeometric equation*, and its solutions are called *confluent hypergeometric functions*.

Two standard solutions to equation (F.1.7) are the following:

$$M(a; c; z) = \sum_{n=0}^{\infty} \frac{(a)_n}{(c)_n n!} z^n = 1 + \frac{a}{c} z + \frac{a(a+1)}{c(c+1)2!} z^2 + \ldots \tag{F.1.8}$$

and

$$\mathbf{M}(a; c; z) = \sum_{n=0}^{\infty} \frac{(a)_n}{\Gamma(c+n) n!} z^n = \frac{1}{\Gamma(c)} + \frac{a}{\Gamma(c+1)} z + \frac{a(a+1)}{\Gamma(c+2)2!} z^2 + \ldots. \tag{F.1.9}$$

The first of these functions $M(a; c; z)$ does not exist if c is a non-positive integer. For all other values of parameters the following identity holds:

$$M(a; c; z) = \Gamma(c) \mathbf{M}(a; c; z).$$

The series (F.1.8) and (F.1.9) converge for all $z \in \mathbb{C}$. $M(a; c; z)$ is an entire function in z and a, and is meromorphic in c. $\mathbf{M}(a; c; z)$ is an entire function in z, a, and c.

The function $M(a; c; z)$ is known as the *Kummer confluent hypergeometric function*. Sometimes the notation

$$M(a; c; z) = {}_1 F_1(a; c; z)$$

Appendix F: Higher Transcendental Functions

is used, which is also Humbert's symbol [ErdBat-1, p. 248]

$$M(a; c; z) = \Phi(a; c; z).$$

Another standard solution to the confluent hypergeometric equation (F.1.7) is the function $U(a; c; z)$ which is determined uniquely by the property

$$U(a; c; z) \sim z^{-a}, \quad z \to \infty, \quad |\arg z| < \pi - \delta. \qquad \text{(F.1.10)}$$

Here δ is an arbitrary small positive constant. In general, $U(a; c; z)$ has a branch point at $z = 0$. The principal branch corresponds to the principal value of z^{-a} in (F.9), and has a cut in the z-plane along the interval $(-\infty; 0]$. The function $U(a; c; z)$ was introduced by Tricomi (see, e.g., [ErdBat-1, p. 257]). Sometimes it is called *the Tricomi confluent hypergeometric function*. It is related to the Erdélyi function $_2F_0(a; c; z)$

$$_2F_0(a; c; -1/z) = z^a U(a; a - c + 1; z)$$

and the Kummer confluent hypergeometric function

$$U(a; c; z) = \frac{\Gamma(1-c)}{\Gamma(a-c+1)} M(a; c; z) + \frac{\Gamma(c-1)}{\Gamma(a)} z^{1-a} M(a - c + 1; 2 - c; z).$$

The following notation (Humbert's symbol) is used too (see, e.g., [ErdBat-1, p. 255]):

$$U(a; c; z) = \Psi(a; c; z).$$

Integral Representations
Two main types of integral representations for confluent hypergeometric functions can be mentioned here. First of all these are representations analogous to the Euler integral representation of the classical hypergeometric function.

The integral representation of Kummer's confluent hypergeometric function

$$M(a; c; z) = \frac{\Gamma(c)}{\Gamma(a)\Gamma(c-a)} \int_0^1 e^{tz} t^{a-1}(1-t)^{c-a-1} dt, \quad \operatorname{Re} c > \operatorname{Re} a > 0, \qquad \text{(F.1.11)}$$

can be immediately verified by expanding the exponential function e^{tz}.

The formula

$$\frac{1}{\Gamma(a)} \int_0^\infty e^{-tz} t^{a-1}(1+t)^{c-a-1} dt, \quad \operatorname{Re} a > 0, \qquad \text{(F.1.12)}$$

gives the solution of the differential equation (F.1.7) in the right half-plane $\operatorname{Re} z > 0$ and coincides in this domain with $U(a; c; z)$. The analytic continuation of (F.1.12) yields the integral representation of Tricomi's confluent hypergeometric function $U(a; c; z)$.

Another type of integral representation uses Mellin–Barnes integrals. For confluent hypergeometric functions this type of integral representation has the form

$$M(a; c; z) = \frac{1}{2\pi i} \frac{\Gamma(c)}{\Gamma(a)} \int_{\gamma-i\infty}^{\gamma+i\infty} \frac{\Gamma(-s)\Gamma(a+s)}{\Gamma(c+s)} (-z)^s ds, \qquad \text{(F.1.13)}$$

$|\arg(-z)| < \pi/2$, $-\operatorname{Re} a < \gamma < 0$, $c \neq 0, 1, 2, \ldots$;

$$U(a; c; z) = \frac{1}{2\pi i} \int_{\gamma-i\infty}^{\gamma+i\infty} \frac{\Gamma(-s)\Gamma(a+s)\Gamma(1-c-s)}{\Gamma(c+s)\Gamma(a-c+1)} (z)^s ds, \qquad \text{(F.1.14)}$$

$|\arg(z)| < 3\pi/2$, $-\operatorname{Re} a < \gamma < \min\{0, 1 - \operatorname{Re} c\}$.

The conditions on the parameters in (F.1.13) and in (F.1.14) can be relaxed by suitable deformation of the contour of integration. Thus, (F.1.13) is valid with any γ whenever a is not a non-negative integer, provided that the contour of integration separates the poles of $\Gamma(-s)$ from the poles of $\Gamma(a+s)$. Similarly, (F.1.14) is valid with any γ whenever neither a nor $a - c + 1$ is a non-negative integer, provided that the contour of integration separates the poles of $\Gamma(-s)\Gamma(1-c-s)$ from the poles of $\Gamma(a+s)$. The conditions on $\arg z$ cannot be relaxed.

Asymptotics

As $z \to \infty$ then (see, e.g., [NIST, p. 328])

$$\mathbf{M}(a; c; z) \sim \frac{e^z z^{a-c}}{\Gamma(a)} \sum_{n=0}^{\infty} \frac{(1-a)_n (c-a)_n}{n!} z^{-n}$$

$$+ \frac{e^{\pm \pi i a} z^{-a}}{\Gamma(c-a)} \sum_{n=0}^{\infty} \frac{(a)_n (a-c+1)_n}{n!} (-z)^{-n}, \qquad \text{(F.1.15)}$$

$-\pi/2 + \delta < \pm \arg z < 3\pi/2 - \delta$, $a \neq 0, -1, -2, \ldots$; $c - a \neq 0, -1, -2, \ldots$;

$$U(a; c; z) \sim z^{-a} \sum_{n=0}^{\infty} \frac{(a)_n (a-c+1)_n}{n!} (-z)^{-n}, \qquad \text{(F.1.16)}$$

$|\arg z| < 3\pi/2 - \delta$, with δ being sufficiently small positive.

Other asymptotic formulas with respect to large values of the variable z and/or with respect to large values of the parameters a and c can be found in [NIST, pp. 330–331].

Appendix F: Higher Transcendental Functions

Relation to Elementary and Other Special Functions
In the following special cases the confluent hypergeometric functions coincide with elementary functions:

$$M(a; a; z) = e^z;$$

$$M(1; 2; 2z) = \frac{e^z}{z} \sinh z;$$

$$M(0; c; z) = U(0; c; z) = 1;$$

$$U(a; a+1; z) = z^{-a};$$

with incomplete gamma function $\gamma(a, z)$ (in the case when $a - c$ is an integer or a is a positive integer):

$$M(a; a+1; -z) = e^{-z} M(1; a+1; z) = az^{-a}\gamma(a, z);$$

with the error functions erf and erfc:

$$M(1/2; 3/2; -z^2) = \frac{\sqrt{\pi}}{2z} \operatorname{erf}(z);$$

$$U(1/2; 1/2; z^2) = \sqrt{\pi} e^{z^2} \operatorname{erfc}(z);$$

with the orthogonal polynomials:
- the Hermite polynomials $H_\nu(z)$:

$$M(-n; 1/2; z^2) = (-1)^n \frac{n!}{(2n)!} H_{2n}(z);$$

$$M(-n; 3/2; z^2) = (-1)^n \frac{n!}{(2n+1)!2z} H_{2n+1}(z);$$

$$U(1/2 - n/2; 3/2; z^2) = \frac{2^{-n}}{z} H_n(z);$$

- the Laguerre polynomials $L_\nu^{(\alpha)}(z)$:

$$U(-n; \alpha+1; z) = (-1)^n (\alpha+1)_n M(-n; \alpha+1; z) = (-1)^n n! L L_n^{(\alpha)}(z);$$

with the modified Bessel functions I_ν, K_ν (in the case when $c = 2b$):

$$M(\nu + 1/2; 2\nu + 1; z) = \Gamma(\nu + 1) e^z (z/2)^{-\nu} I_\nu(z);$$

$$U(\nu + 1/2; 2\nu + 1; z) = \frac{1}{\sqrt{\pi}} e^z (2z)^{-\nu} K_\nu(z).$$

F.1.6 Generalized Hypergeometric Functions and their Properties

Generalized hypergeometric functions $_pF_q$ (or $_pF_q\left(\begin{matrix}\mathbf{a}\\\mathbf{b}\end{matrix};z\right) = {_pF_q}(\mathbf{a},\mathbf{b};z) = {_pF_q}(a_1,\ldots,a_p;b_1,\ldots,b_q;z))$ are defined by the following series (where none of the parameters b_j is a non-positive integer):

$$_pF_q\left(\begin{matrix}a_1,a_2,\ldots a_p\\ b_1,b_2,\ldots b_q\end{matrix};z\right) = \sum_{n=0}^{\infty} \frac{(a_1)_n (a_2)_n \ldots (a_p)_n}{(b_1)_n (b_2)_n \ldots (b_q)_n} \frac{z^n}{n!}. \tag{F.1.17}$$

This series converges for all z whenever $p < q$ and thus represents an entire function of z.

If $p = q + 1$, then under the assumption that none of the a_j is a non-positive integer the radius of convergence of the series (F.1.17) is equal to 1, and outside the open disk $|z| < 1$ the generalized hypergeometric function is defined by analytic continuation with respect to z. The branch obtained by introducing a cut from 1 to $+\infty$ on the real axis, that is, the branch in the sector $|\arg(1-z)| < \pi$, is *the principal branch* (or *principal value*) of $_{q+1}F_q(\mathbf{a};\mathbf{b};z)$. Elsewhere the generalized hypergeometric function is a multi-valued function that is analytic except for possible branch points at $z = 0; 1, \infty$. On the circle $|z| = 1$ the series (F.1.17) is absolutely convergent if $\mathrm{Re}\,\gamma_q > 0$, convergent except at $z = 1$ if $-1 < \mathrm{Re}\,\gamma_q \le 0$, and divergent if $\mathrm{Re}\,\gamma_q < -1$, where $\gamma_q = (b_1 + \ldots + b_q) - (a_1 + \ldots + a_{q+1})$.

In general the series (F.1.17) diverges for all non-zero values of z whenever $p > q + 1$. However, when one or more of the top parameters a_j is a non-positive integer the series terminates and the generalized hypergeometric function is a polynomial in z. Note that if $m = \max\{-a_1,\ldots,-a_q\}$ is a positive integer, then the following identity holds:

$$_{p+1}F_q\left(\begin{matrix}-m,\mathbf{a}\\ \mathbf{b}\end{matrix};z\right) = \frac{(\mathbf{a})_m(-z)^m}{(\mathbf{b})_m} {_{q+1}F_p}\left(\begin{matrix}-m,1-m-\mathbf{b}\\ 1-m-\mathbf{a}\end{matrix};\frac{(-1)^{p+q+1}}{z}\right),$$

which can be used to interchange $\mathbf{a}$ and $\mathbf{b}$.

If $p \le q = 1$ and z is not a branch point of the generalized hypergeometric function $_pF_q$, then the function

$$_p\mathbf{F}_q\left(\begin{matrix}\mathbf{a}\\ \mathbf{b}\end{matrix};z\right) = \frac{1}{\Gamma(b_1)\ldots\Gamma(b_q)} {_pF_q}\left(\begin{matrix}\mathbf{a}\\ \mathbf{b}\end{matrix};z\right) = \sum_{n=0}^{\infty} \frac{(a_1)_n \ldots (a_p)_n}{\Gamma(b_1+n)\ldots\Gamma(b_q+n)} \frac{z^n}{n!}$$

is an entire function of each parameter $a_1,\ldots,a_p,b_1,\ldots,b_q$.

F.2 Wright Functions

F.2.1 The Classical Wright Function

The simplest Wright function $\phi(\alpha, \beta; z)$ is defined for $z, \alpha, \beta \in \mathbb{C}$ by the series

$$\phi(\alpha, \beta; z) = {}_0\Psi_1 \left[\begin{array}{c} \text{---} \\ (\beta, \alpha) \end{array} \bigg| z \right] := \sum_{k=0}^{\infty} \frac{1}{\Gamma(\alpha k + \beta)} \frac{z^k}{k!}. \qquad (\text{F.2.1})$$

If $\alpha > -1$, this series is absolutely convergent for all $z \in \mathbb{C}$, while for $\alpha = -1$ it is absolutely convergent for $|z| < 1$ and for $|z| = 1$ and $\text{Re}(\beta) > -1$; see [KiSrTr06, Sect. 1.11]. Moreover, for $\alpha > -1$, $\phi(\alpha, \beta; z)$ is an entire function of z. Using formulas for the order (B.5) and the type (B.6) of an entire function and Stirling's asymptotic formula for the Gamma function (A.24) one can deduce that for $\alpha > -1$ the Wright function $\phi(\alpha, \beta; z)$ has order $\rho = \dfrac{1}{\alpha + 1}$ and type $\sigma = (\alpha + 1)\alpha^{\frac{1}{(\alpha+1)}} = \dfrac{1}{\rho}\alpha^{\rho}$.

The classical Wright function is closely related to the multiparametric Mittag-Leffler function. It is discussed in detail in Chap. 7 in this book.

F.2.2 The Bessel–Wright Function. Generalized Wright Functions and Fox–Wright Functions

When $\alpha = \mu$, $\beta = \nu + 1$, and z is replaced by $-z$, the function $\phi(\alpha, \beta; z)$ is denoted by $J_\nu^\mu(z)$:

$$J_\nu^\mu(z) := \psi(\mu, \nu + 1; -z) = \sum_{k=0}^{\infty} \frac{1}{\Gamma(\mu k + \nu + 1)} \frac{(-z)^k}{k!}, \qquad (\text{F.2.2})$$

and this function is known as the *Bessel–Wright* function, or the *Wright generalized Bessel function*,[7] see [PrBrMa-V2] and [Kir94, p. 352]. When $\mu = 1$, the Bessel function of the first kind is connected with (F.2.2) by

$$J_\nu(z) = \left(\frac{z}{2}\right)^\nu J_\nu^1\left(\frac{z^2}{4}\right). \qquad (\text{F.2.3})$$

If $\mu = \dfrac{p}{q}$ is a rational number then [Kir94, p. 352] the Bessel–Wright function satisfies the following differential equation of order $(p + q)$

---

[7] Also misnamed the Bessel–Maitland function after E.M. Wright's second name: Maitland.

$$\left[\frac{(-z)^q}{q^q p^p} - \prod_{j=1}^{p+q}\left(\frac{1}{q}z\frac{d}{dz} - d_j\right)\right] J_\nu^\mu(z) = 0, \qquad (F.2.4)$$

where

$$d_j = \begin{cases} \frac{j}{q}, & 1 \le j \le q-1, \\ 1 - \frac{j+\nu+1-q}{q}, & q \le j \le q+p. \end{cases}$$

The original differential equation for $J_\nu^\mu(z)$, equivalent to (F.2.4), was formulated by Wright [Wri33]:

$$(-1)^q z^{\nu/\mu} J_\nu^\mu(z) = \left(\mu z^{1-1/\mu}\frac{d}{dz}\right)^p z^{\frac{\nu+p}{\mu}}\left(\frac{d}{dz}\right)^q J_\nu^\mu(z).$$

There exists a further generalization of the Bessel–Wright function (the so-called generalized Bessel–Wright function) given in [Pat66, Pat67]:

$$J_{\nu,\lambda}^\mu(z) = \left(\frac{z}{2}\right)^{\nu+2\lambda}\sum_{k=0}^{\infty}\frac{(-1)^k \left(\frac{z}{2}\right)^{2k}}{\Gamma(\nu+k\mu+\lambda+1)\Gamma(\lambda+k+1)}, \quad \mu > 0. \qquad (F.2.5)$$

This function generates the Lommel function, and is a special case of the four-parametric Mittag-Leffler function.

The more general function ${}_p\Psi_q(z)$ is defined for $z \in \mathbb{C}$, complex $a_l, b_j \in \mathbb{C}$, and real $\alpha_l, \beta_j \in \mathbb{R}$ $(l = 1, \cdots, p;\ j = 1, \cdots, q)$ by the series

$${}_p\Psi_q(z) = {}_p\Psi_q\left[\begin{array}{c}(a_l, \alpha_l)_{1,p}\\(b_l, \beta_l)_{1,q}\end{array}\bigg| z\right] := \sum_{k=0}^{\infty}\frac{\prod_{l=1}^{p}\Gamma(a_l+\alpha_l k)}{\prod_{j=1}^{q}\Gamma(b_j+\beta_j k)}\frac{z^k}{k!} \qquad (F.2.6)$$

$(z, a_l, b_j \in \mathbb{C};\ \alpha_l, \beta_j \in \mathbb{R};\ l = 1, \cdots, p;\ j = 1, \cdots, q).$

This general (*Wright* or, more appropriately, *Fox–Wright*) function was investigated by Fox [Fox28] and Wright [Wri35b, Wri40b, Wri40c], who presented its asymptotic expansion for large values of the argument z under the condition

$$\sum_{j=1}^{q}\beta_j - \sum_{l=1}^{p}\alpha_l > -1. \qquad (F.2.7)$$

If these conditions are satisfied, the series in (F.2.6) is convergent for any $z \in \mathbb{C}$. This result follows from the assertion [KiSrTr06, Theorem 1.]:

Theorem F.1 *Let $a_l, b_j \in \mathbb{C}$ and $\alpha_l, \beta_j \in \mathbb{R}$ $(l = 1, \cdots p;\ j = 1, \cdots, q)$ and let*

Appendix F: Higher Transcendental Functions

$$\Delta = \sum_{j=1}^{q} \beta_j - \sum_{l=1}^{p} \alpha_l, \tag{F.2.8}$$

$$\delta = \prod_{l=1}^{p} |\alpha_l|^{-\alpha_l} \prod_{j=1}^{q} |\beta_j|^{\beta_j}, \tag{F.2.9}$$

and

$$\mu = \sum_{j=1}^{q} b_j - \sum_{l=1}^{p} a_l + \frac{p-q}{2}. \tag{F.2.10}$$

(a) If $\Delta > -1$, then the series in (F.2.6) is absolutely convergent for all $z \in \mathbb{C}$.
(b) If $\Delta = -1$, then the series in (F.2.6) is absolutely convergent for $|z| < \delta$ and for $|z| = \delta$ and $\text{Re}(\mu) > 1/2$.

When $\alpha_l, \beta_j \in \mathbb{R}$ ($l = 1, \cdots, p$; $j = 1, \cdots, q$), the generalized Wright function $_p\Psi_q(z)$ has the following integral representation as a Mellin–Barnes contour integral:

$$_p\Psi_q \left[\begin{array}{c} (a_l, \alpha_l)_{1,p} \\ (b_l, \beta_l)_{1,q} \end{array} \Big| z \right] = \frac{1}{2\pi i} \int_{\mathcal{C}} \frac{\Gamma(s) \prod_{l=1}^{p} \Gamma(a_l - \alpha_l s)}{\prod_{j=1}^{q} \Gamma(b_j - \beta_j s)} (-z)^{-s} ds, \tag{F.2.11}$$

where the path of integration $\mathcal{C}$ separates all the poles at $s = -k$ ($k \in \mathbb{N}_0$) to the left and all the poles $s = (a_l + n_l)/\alpha_l$ ($l = 1, \cdots, p$; $n_l \in \mathbb{N}$ to the right. If $\mathcal{C} = (\gamma - i\infty, \gamma + i\infty)$ ($\gamma \in \mathbb{R}$), then representation (F.2.11) is valid if either of the following conditions holds:

$$\Delta < 1, \ |\arg(-z)| < \frac{(1-\Delta)\pi}{2}, \ z \neq 0 \tag{F.2.12}$$

or

$$\Delta = 1, \ (\Delta + 1)\gamma + \frac{1}{2} < \text{Re}(\mu), \ \arg(-z) = 0, \ z \neq 0. \tag{F.2.13}$$

Conditions for the representation (F.2.11) were also given for the case when $\mathcal{C} = \mathcal{L}_{-\infty}$ ($\mathcal{L}_{\infty}$) is a loop situated in a horizontal strip starting at the point $-\infty + i\varphi_1$ ($\infty + i\varphi_1$) and terminating at the point $-\infty + i\varphi_2$ ($\infty + i\varphi_2$) with $-\infty < \varphi_1 < \varphi_2 < \infty$.

If we put $\alpha_l = 1$, $1 \leq l \leq p$ and $\beta_l = 1$, $1 \leq l \leq q$ in representation (F.2.6) then we get the following relation of the generalized Wright function $_p\Psi_q$ with the generalized hypergeometric function $_pF_q$:

$$_p\Psi_q \left[\begin{array}{c} (a_l, 1)_{1,p} \\ (b_l, 1)_{1,q} \end{array} \Big| z \right] = \frac{\prod_{l=1}^{p} \Gamma(a_l)}{\prod_{l=1}^{q} \Gamma(b_l)} {}_pF_q(a_1, \ldots, a_p; b_1, \ldots, b_q; z). \tag{F.2.14}$$

F.3 Meijer G-Functions

F.3.1 Definition via Integrals. Existence

In [Mei36] more general classes of transcendental functions were introduced via generalization of the Gauss hypergeometric functions presented in the form of series (commonly known now as Meijer G-functions):

$$G_{p,q}^{m,n}\left[z \left|\begin{array}{c} a_1,\ldots,a_p \\ b_1,\ldots,b_q \end{array}\right.\right]. \qquad (F.3.1)$$

This definition is due to the relation between the generalized hypergeometric function and the Meijer G-functions (see, e.g., [Sla66, p. 42]):

$$G_{p,q}^{m,n}\left[-z \left|\begin{array}{c} 1-a_1,\ldots,1-a_p \\ 0, 1-b_1,\ldots,1-b_q \end{array}\right.\right] = \frac{\Gamma(a_1)\ldots\Gamma(a_p)}{\Gamma(b_1)\ldots\Gamma(b_q)} {}_pF_q(\mathbf{a};\mathbf{b};z). \qquad (F.3.2)$$

Later this definition was replaced by the Mellin–Barnes representation of the G-function

$$G_{p,q}^{m,n}\left[z \left|\begin{array}{c} a_1,\ldots,a_p \\ b_1,\ldots,b_q \end{array}\right.\right] = \frac{1}{2\pi i} \int_{\mathcal{L}} \mathcal{G}_{p,q}^{m,n}(s) z^s \, \mathrm{d}s, \qquad (F.3.3)$$

where $\mathcal{L}$ is a suitably chosen path, $z \neq 0$, $z^s := \exp[s(\ln|z| + i\arg z)]$ with a single valued branch of $\arg z$, and the integrand is defined as

$$\mathcal{G}_{p,q}^{m,n}(s) = \frac{\prod_{k=1}^{m} \Gamma(b_k - s) \prod_{j=1}^{n} \Gamma(1 - a_j + s)}{\prod_{k=m+1}^{q} \Gamma(1 - b_k + s) \prod_{j=n+1}^{p} \Gamma(a_j - s)}. \qquad (F.3.4)$$

In (F.3.4) the empty product is assumed to be equal to 1, parameters m, n, p, q satisfy the relation $0 \leq n \leq q$, $0 \leq n \leq p$, and the complex numbers a_j, b_k are such that no pole of $\Gamma(b_k - s)$, $k = 1,\ldots,m$, coincides with a pole of $\Gamma(1 - a_j + s)$, $j = 1,\ldots,n$.

F.3.2 Basic Properties of the Meijer G-Functions

1. Symmetry
The G-functions are symmetric with respect to their parameters in the following sense: if the value of a parameter from the group $(a_1,\ldots,a_n)$ (respectively, from the group $(a_{n+1},\ldots,a_p)$) is equal to the value of a parameter from the group $(b_{m+1},\ldots,b_q)$ (respectively, to the value of a parameter from the group $(b_1,\ldots,b_m)$), then these parameters can be excluded from the definition of the G-

Appendix F: Higher Transcendental Functions

function, i.e. the "order" of the G-function decreases. For instance, if $a_1 = b_q$, then

$$G_{p,q}^{m,n}\left[z \Bigg| \begin{matrix} a_1,\ldots,a_p \\ b_1,\ldots,b_q \end{matrix}\right] = G_{p-1,q-1}^{m,n-1}\left[z \Bigg| \begin{matrix} a_2,\ldots,a_p \\ b_1,\ldots,b_{q-1} \end{matrix}\right]. \quad (F.3.5)$$

2. *Shift of Parameters*
The two properties below indicate how to "shift" parameters. They follow from the change of variable in the definition of the G-functions (F.3.3).

$$z^\sigma G_{p,q}^{m,n}\left[z \Bigg| \begin{matrix} a_1,\ldots,a_p \\ b_1,\ldots,b_q \end{matrix}\right] = G_{p,q}^{m,n}\left[z \Bigg| \begin{matrix} a_1+\sigma,\ldots,a_p+\sigma \\ b_1+\sigma,\ldots,b_q+\sigma \end{matrix}\right]; \quad (F.3.6)$$

$$G_{p,q}^{m,n}\left[z \Bigg| \begin{matrix} a_1,\ldots,a_p \\ b_1,\ldots,b_q \end{matrix}\right] = G_{p,q}^{m,n}\left[\frac{1}{z} \Bigg| \begin{matrix} 1-b_1,\ldots,1-b_q \\ 1-a_1,\ldots,1-a_p \end{matrix}\right]. \quad (F.3.7)$$

3. *Multiplication Formula*
For any natural number $r \in \mathbb{N}$ the following formula holds:

$$G_{p,q}^{m,n}\left[z \Bigg| \begin{matrix} a_1,\ldots,a_p \\ b_1,\ldots,b_q \end{matrix}\right] =$$

$$(2\pi)^u r^v G_{pr,qr}^{mr,nr}\left[\frac{z^r}{r^{r(q-p)}} \Bigg| \begin{matrix} c_{1,1},\ldots,c_{1,r},\ldots,c_{p,1},\ldots,c_{p,r} \\ d_{1,1},\ldots,d_{1,r},\ldots,d_{q,1},\ldots,d_{q,r} \end{matrix}\right], \quad (F.3.8)$$

where

$$c_{j,l} = \frac{a_j+l-1}{r}, \ (l=1,\ldots,r); \ d_{k,l} = \frac{b_k+l-1}{r}, \ (l=1,\ldots,r).$$

F.3.3 Special Cases

The basic elementary functions can be represented as G-function with special values of parameters. They mostly follow from the relation between the generalized hypergeometric function ${}_pF_q$ and the Meijer G-function (F.3.2).

Let us recall some of these representations.

$$\exp z = G_{1,0}^{0,1}[z \,|\, 0]; \quad (F.3.9)$$

$$\sin z = \sqrt{\pi} G_{0,2}^{1,0}\left[\frac{z^2}{4} \Bigg| \frac{1}{2},0\right]; \quad (F.3.10)$$

$$\cos z = \sqrt{\pi} G_{0,2}^{1,0}\left[\frac{z^2}{4} \Bigg| 0,\frac{1}{2}\right]; \quad (F.3.11)$$

$$\log(1+z) = G_{2,2}^{1,2}\left[z \left|\begin{array}{c}1,1\\1,0\end{array}\right.\right] \quad (|z|<1); \tag{F.3.12}$$

$$\arcsin z = \frac{1}{2\sqrt{\pi}} G_{2,2}^{1,2}\left[-z^2 \left|\begin{array}{c}1,1\\ \frac{1}{2},0\end{array}\right.\right] \quad (|z|<1); \tag{F.3.13}$$

$$\arctan z = \frac{1}{2} G_{2,2}^{1,2}\left[z^2 \left|\begin{array}{c}1,\frac{1}{2}\\ \frac{1}{2},0\end{array}\right.\right] \quad (|z|<1); \tag{F.3.14}$$

$$z^\gamma = G_{1,1}^{1,0}\left[z \left|\begin{array}{c}\gamma+1\\ \gamma\end{array}\right.\right] \quad (|z|<1); \tag{F.3.15}$$

$$(1\pm z)^{-\alpha} = \frac{1}{\Gamma(\alpha)} G_{1,1}^{1,1}\left[\mp z \left|\begin{array}{c}1-\alpha\\ 0\end{array}\right.\right] \quad (|z|<1, \operatorname{Re}\alpha > 0); \tag{F.3.16}$$

$$(1\pm z)^{-\alpha} = \Gamma(1-\alpha) G_{1,1}^{1,0}\left[\mp z \left|\begin{array}{c}1-\alpha\\ 0\end{array}\right.\right] \quad (|z|<1, \operatorname{Re}\alpha < -1). \tag{F.3.17}$$

F.3.4 Relations to Fractional Calculus

Let us consider the Riemann–Liouville fractional integral (see (E-RL.1))

$$J_{a+}^\alpha \phi(x) := \frac{1}{\Gamma(\alpha)} \int_a^x (x-\xi)^{\alpha-1} \phi(\xi)\,d\xi, \quad a < x < b, \quad \alpha > 0.$$

It is well-known (see, e.g. [SaKiMa93, Chap. 4]) that this integral can be extended into the complex domain. For this we fix a point $z \in \mathbb{C}$ and introduce the multi-valued function

$$(z-\zeta)^{\alpha-1}.$$

By fixing an arbitrary single-valued branch of this function in the complex plane cut along the line L starting at $\zeta = a$ and ending at $\zeta = \infty$ and containing the point $\zeta = z$ we define the Riemann–Liouville fractional integral in the complex domain

$$J_{a+}^\alpha f(z) = \frac{1}{\Gamma(\alpha)} \int_a^z (z-\zeta)^{\alpha-1} f(\zeta)\,d\zeta, \tag{F.3.18}$$

where the integration is performed along a part of the above-described contour (cut) L starting at $\zeta = a$ and ending at $\zeta = z$. Analogously, the so-called "right-sided" Riemann–Liouville fractional integral (or Weyl type fractional integral) in the complex domain is defined by

Appendix F: Higher Transcendental Functions

$$J_{-}^{\alpha} f(z) = \frac{1}{\Gamma(\alpha)} \int_{z}^{\infty} (z - \zeta)^{\alpha-1} f(\zeta) \, d\zeta. \quad \text{(F.3.19)}$$

The density f in both formulas (F.3.18) and (F.3.19) is assumed to be defined in a neighborhood of the contour L. If this function is analytic in the whole plane, then by the Cauchy Integral Theorem one can replace the contour L by the straight interval $[a, z]$ and by the ray $[z, \infty)$, respectively.

Let us present the formulas which show that the Riemann–Liouville fractional integration of the Meijer G-function yields the Meijer G-function with other values of parameters (see, e.g. [Kir94, p. 318], [PrBrMa-V3]).

$$J_{a+}^{\alpha} G_{p,q}^{m,n}(\eta z) = \frac{1}{\Gamma(\alpha)} \int_{a}^{z} (z - \zeta)^{\alpha-1} G_{p,q}^{m,n}\left[\eta \zeta \,\bigg|\, \begin{matrix} a_1, \ldots, a_p \\ b_1, \ldots, b_q \end{matrix}\right] d\zeta$$

$$= z^{\alpha} \int_{0}^{1} \frac{(1-\sigma)^{\alpha-1}}{\Gamma(\alpha)} G_{p,q}^{m,n}\left[\eta z\sigma \,\bigg|\, \begin{matrix} a_1, \ldots, a_p \\ b_1, \ldots, b_q \end{matrix}\right] d\sigma$$

$$= z^{\alpha} G_{p+1,q+1}^{m,n+1}\left[\eta \zeta \,\bigg|\, \begin{matrix} 0, a_1, \ldots, a_p \\ b_1, \ldots, b_q, -\alpha \end{matrix}\right]. \quad \text{(F.3.20)}$$

This formula is valid when $\operatorname{Re} b_k > -1$, $k = 1, \ldots, m$, and $p \leq q$. If $p = q$, then it is valid for $|\eta z| < 1$. For $p + q < 2(m + n)$ we require additionally that $|\arg \eta z| < \left(m + n - \frac{p+q}{2}\right) \pi$.

$$J_{-}^{\alpha} G_{p,q}^{m,n}(\eta z) = \frac{1}{\Gamma(\alpha)} \int_{z}^{\infty} (z - \zeta)^{\alpha-1} G_{p,q}^{m,n}\left[\eta \zeta \,\bigg|\, \begin{matrix} a_1, \ldots, a_p \\ b_1, \ldots, b_q \end{matrix}\right] d\zeta$$

$$= z^{\alpha} \int_{1}^{\infty} \frac{(\sigma-1)^{\alpha-1}}{\Gamma(\alpha)} G_{p,q}^{m,n}\left[\eta z\sigma \,\bigg|\, \begin{matrix} a_1, \ldots, a_p \\ b_1, \ldots, b_q \end{matrix}\right] d\sigma$$

$$= z^{\alpha} G_{p+1,q+1}^{m+1,n}\left[\eta \zeta \,\bigg|\, \begin{matrix} a_1, \ldots, a_p, 0 \\ -\alpha, b_1, \ldots, b_q \end{matrix}\right]. \quad \text{(F.3.21)}$$

This formula is valid when $0 < \operatorname{Re} \alpha < 1 - \operatorname{Re} a_j$, $j = 1, \ldots, n$, and $p \geq q$. If $p = q$, then it is valid for $|\eta z| > 1$. For $p + q < 2(m + n)$ we require additionally that $|\arg \eta z| < \left(m + n - \frac{p+q}{2}\right) \pi$.

F.3.5 Integral Transforms of G-Functions

The structure of the Meijer G function is very similar to that of the Mellin integral transform. Hence, the Mellin transform of the Meijer G-function is up to a power factor equal to the ratio of the products of Γ-functions (see, e.g., [BatErd54b, p. 301], [Kir94, p. 318]).

$$\mathcal{M}\left(G_{p,q}^{m,n}(\eta t)\right)(s) = \int_0^\infty t^{s-1} G_{p,q}^{m,n}\left[\eta t \left|\begin{array}{c} a_1,\ldots,a_p \\ b_1,\ldots,b_q \end{array}\right.\right] dt$$

$$= \eta^s \mathcal{G}_{p,q}^{m,n}(-s) = \eta^s \frac{\prod\limits_{k=1}^m \Gamma(b_k + s) \prod\limits_{j=1}^n \Gamma(1 - a_j - s)}{\prod\limits_{k=m+1}^q \Gamma(1 - b_k - s) \prod\limits_{j=n+1}^p \Gamma(a_j + s)}, \qquad \text{(F.3.22)}$$

where, e.g., $p + q < 2(m + n)$, $|\arg \eta| < \left(m + n - \tfrac{1}{2}(p + q)\right)\pi$, $-\min\limits_{1\leq k \leq m} \mathrm{Re}\, b_k < \mathrm{Re}\, \eta < 1 - \max\limits_{1\leq j \leq n} \mathrm{Re}\, a_j$. Other conditions under which formula (F.3.22) is valid can be found in [Luk1, pp. 157–159] and [BatErd54b, pp. 300–301]. We also mention several formulas for the Mellin integral transform presented in [BatErd54a, pp. 295–296].

The Laplace transform of the Meijer G-function can be determined by the relation:

$$\mathcal{L}\left(G_{p,q}^{m,n}(\eta x)\right)(t) = \int_0^\infty e^{-tx} G_{p,q}^{m,n}\left[\eta x \left|\begin{array}{c} a_1,\ldots,a_p \\ b_1,\ldots,b_q \end{array}\right.\right] dx$$

$$= t^{-1} G_{p+1,q}^{m,n+1}\left[\frac{\eta}{t} \left|\begin{array}{c} 0, a_1,\ldots,a_p \\ b_1,\ldots,b_q \end{array}\right.\right], \qquad \text{(F.3.23)}$$

where, e.g., $p + q < 2(m + n)$, $|\arg \eta| < \left(m + n - \tfrac{1}{2}(p + q)\right)\pi$, $|\arg t| < \pi/2$, $\mathrm{Re}\, b_k - 1$, $k = 1,\ldots,m$.

More general formulas for the Laplace transform of the Meijer G-function with different weights can also be found in [Luk1, pp. 166–169] and [BatErd54b, p. 302].

F.4 Fox H-Functions

F.4.1 Definition via Integrals. Existence

A straightforward generalization of the Meijer G-functions are the so-called Fox H-functions, introduced and studied by Fox in [Fox61]. According to a standard notation the Fox H-functions are defined by

$$H_{p,q}^{m,n}(z) = H_{p,q}^{m,n}\left[z \left| \begin{array}{c} (a_1,\alpha_1),\ldots,(a_p,\alpha_p) \\ (b_1,\beta_1),\ldots,(b_q,\beta_q) \end{array}\right.\right] = \frac{1}{2\pi i}\int_{\mathcal{L}} \mathcal{H}_{p,q}^{m,n}(s)\, z^s\, ds\,, \quad (F.4.1)$$

where $\mathcal{L}$ is a suitable path in the complex plane $\mathbb{C}$ to be found later on, $z^s = \exp\{s(\log|z| + i\,\mathrm{arg}z)\}$, and

$$\mathcal{H}_{p,q}^{m,n}(s) = \frac{A(s)\,B(s)}{C(s)\,D(s)}\,, \quad (F.4.2)$$

$$A(s) = \prod_{j=1}^{m} \Gamma(b_j - \beta_j s)\,,\quad B(s) = \prod_{j=1}^{n} \Gamma(1 - a_j + \alpha_j s)\,, \quad (F.4.3)$$

$$C(s) = \prod_{j=m+1}^{q} \Gamma(1 - b_j + \beta_j s)\,,\quad D(s) = \prod_{j=n+1}^{p} \Gamma(a_j - \alpha_j s)\,, \quad (F.4.4)$$

with $0 \leq n \leq p$, $1 \leq m \leq q$, $\{a_j, b_j\} \in \mathbb{C}$, $\{\alpha_j, \beta_j\} \in \mathbb{R}^+$. As usual, an empty product, when it occurs, is taken to be equal to 1. Hence

$$n = 0 \Leftrightarrow B(s) = 1\,,\ m = q \Leftrightarrow C(s) = 1\,,\ n = p \Leftrightarrow D(s) = 1\,.$$

Due to the occurrence of the factor z^s in the integrand of (D.1), the H-function is, in general, multi-valued, but it can be made one-valued on the Riemann surface of $\log z$ by choosing a proper branch. We also note that when the αs and βs are equal to 1, we obtain the Meijer G-functions $G_{p,q}^{m,n}(z)$, thus the Meijer G-functions can be considered as special cases of the Fox H-functions:

$$H_{p,q}^{m,n}\left[z \left|\begin{array}{c}(a_1,1),\ldots,(a_p,1) \\ (b_1,1),\ldots,(b_q,1)\end{array}\right.\right] = G_{p,q}^{m,n}\left[z \left|\begin{array}{c}a_1,\ldots,a_p \\ b_1,\ldots,b_q\end{array}\right.\right].$$

The above integral representation of H-functions in terms of products and ratios of Gamma functions is known to be of *Mellin–Barnes integral* type. A compact notation is usually adopted for (F.4.1):

$$H_{p,q}^{m,n}(z) = H_{p,q}^{m,n}\left[z\left|\begin{array}{c}(a_j,\alpha_j)_{j=1,\ldots,p}\\(b_j,\beta_j)_{j=1,\ldots,q}\end{array}\right.\right]. \quad (F.4.5)$$

The singular points of the kernel $\mathcal{H}$ are the poles of the Gamma functions appearing in the expressions of $A(s)$ and $B(s)$, that we assume do not coincide. Denoting by $\mathcal{P}(A)$ and $\mathcal{P}(B)$ the sets of these poles, we write $\mathcal{P}(A) \cap \mathcal{P}(B) = \emptyset$. The conditions for the existence of the H-functions can be determined by inspecting the convergence of the integral (2.1), which can depend on the selection of the contour $\mathcal{L}$ and on certain relations between the parameters $\{a_i, \alpha_i\}$ ($i = 1, \ldots, p$) and $\{b_j, \beta_j\}$ ($j = 1, \ldots, q$). For the analysis of the general case we refer to the specialized treatises on H-functions, e.g., [MatSax73, MaSaHa10, SrGuGo82] and, in particular to the paper by Braaksma [Bra62], where an exhaustive discussion on the asymptotic expansions and analytical continuation of these functions can be found, see also [KilSai99] and the book [KilSai04].

In the following we limit ourselves to recalling the essential properties of the H-functions, preferring to analyze later in detail those functions related to *fractional diffusion*. As will be shown later, this phenomenon depends on one real independent variable and three parameters; in this case we shall have $z = x \in \mathbb{R}$ and $m \le 2$, $n \le 2$, $p \le 3$, $q \le 3$.

The contour $\mathcal{L}$ in (F.4.1) can be chosen as follows:

(i) $\mathcal{L} = \mathcal{L}_{i\gamma\infty}$ chosen to go from $-i\gamma\infty$ to $+i\gamma\infty$ leaving to the right all the poles of $\mathcal{P}(A)$, namely the poles $s_{j,k} = (b_j + k)/\beta_j$; $j = 1, 2, \ldots, m$; $k = 0, 1, \ldots$ of the functions Γ appearing in $A(s)$, and to left all the poles of $\mathcal{P}(B)$, namely the poles $s_{j,l} = (a_j - 1 - l)/\beta_j$; $j = 1, 2, \ldots, n$; $l = 0, 1, \ldots$ of the functions Γ appearing in $B(s)$.

(ii) $\mathcal{L} = \mathcal{L}_{+\infty}$ is a loop beginning and ending at $+\infty$ and encircling once in the negative direction all the poles of $\mathcal{P}(A)$, but none of the poles of $\mathcal{P}(B)$.

(iii) $\mathcal{L} = \mathcal{L}_{-\infty}$ is a loop beginning and ending at $-\infty$ and encircling once in the positive direction all the poles of $\mathcal{P}(B)$, but none of the poles of $\mathcal{P}(A)$.

Braaksma has shown that, independently of the choice of $\mathcal{L}$, the Mellin–Barnes integral makes sense and defines an analytic function of z in the following two cases

$$\mu > 0, \ 0 < |z| < \infty, \ \text{where } \mu = \sum_{j=1}^{q} \beta_j - \sum_{j=1}^{p} \alpha_j, \tag{F.4.6}$$

$$\mu = 0, \ 0 < |z| < \delta, \ \text{where } \delta = \prod_{j=1}^{p} \alpha_j^{-\alpha_j} \prod_{j=1}^{q} \beta_j^{\beta_j}. \tag{F.4.7}$$

Via the following useful and important formula for the H-function

$$H_{p,q}^{m,n}\left[z \left| \begin{array}{l} (a_j, \alpha_j)_{1,p} \\ (b_j, \beta_j)_{1,q} \end{array} \right. \right] = H_{q,p}^{n,m}\left[\frac{1}{z} \left| \begin{array}{l} (1 - b_j, \beta_j)_{1,q} \\ (1 - a_j, \alpha_j)_{1,p} \end{array} \right. \right] \tag{F.4.8}$$

we can transform the H-function with $\mu < 0$ and argument z to one with $\mu > 0$ and argument $1/z$. This property is useful when comparing the results of the theory of

Appendix F: Higher Transcendental Functions

H-functions based on (F.4.1) using z^s with the theory that uses z^{-s}, which is often found in the literature.

More detailed information on the existence of the H-function is presented, e.g., in [KilSai04] (see also [PrBrMa-V3]). In order to formulate this result we introduce a set of auxiliary parameters.

Let $m, n, p, q, \alpha_j, a_j, \beta_k, b_k$ ($j = 1, \ldots, p$, $k = 1, \ldots, q$). Define

$$\begin{cases} a^* = \sum_{j=1}^{n} \alpha_j - \sum_{j=n+1}^{p} \alpha_j + \sum_{k=1}^{m} \beta_k - \sum_{k=m+1}^{q} \beta_k; \\ a_1^* = \sum_{k=1}^{m} \beta_k - \sum_{j=n+1}^{p} \alpha_j; \\ a_2^* = \sum_{j=1}^{n} \alpha_j - \sum_{k=m+1}^{q} \beta_k; \\ \Delta = \sum_{k=1}^{q} \beta_k - \sum_{j=1}^{p} \alpha_j; \\ \delta = \prod_{j=1}^{p} \alpha_j^{-\alpha_j} \prod_{k=1}^{q} \beta_k^{\beta_k}; \\ \mu = \sum_{k=1}^{q} b_k - \sum_{j=1}^{p} a_j + \tfrac{1}{2}(p - q); \\ \xi = \sum_{k=1}^{m} b_k - \sum_{k=m+1}^{q} b_k + \sum_{j=1}^{n} a_j - \sum_{j=n+1}^{p} a_j; \\ c^* = m + n - \tfrac{1}{2}(p + q). \end{cases} \quad \text{(F.4.9)}$$

Theorem F.2 ([KilSai04, pp. 4–5]) *Let the parameters a^*, Δ, δ, μ be defined as in (F.4.9). Then the Fox H-function $H_{p,q}^{m,n}(z)$ (as defined in (F.4.1)–(F.4.4)) exists in the following cases:*

If $\mathcal{L} = \mathcal{L}_{-\infty}, \Delta > 0$, then $H_{p,q}^{m,n}(z)$ exists for all $z : z \neq 0$. (F.4.10)

If $\mathcal{L} = \mathcal{L}_{-\infty}, \Delta = 0$, then $H_{p,q}^{m,n}(z)$ exists for all $z : 0 < |z| < \delta$. (F.4.11)

If $\mathcal{L} = \mathcal{L}_{-\infty}, \Delta = 0, \operatorname{Re} \mu < -1$, then $H_{p,q}^{m,n}(z)$ exists for $z : |z| = \delta$. (F.4.12)

If $\mathcal{L} = \mathcal{L}_{+\infty}, \Delta < 0$, then $H_{p,q}^{m,n}(z)$ exists for all $z : z \neq 0$. (F.4.13)

If $\mathcal{L} = \mathcal{L}_{+\infty}, \Delta = 0$, then $H_{p,q}^{m,n}(z)$ exists for all $z : |z| > \delta$. (F.4.14)

If $\mathcal{L} = \mathcal{L}_{+\infty}, \Delta = 0, \operatorname{Re} \mu < -1$, then $H_{p,q}^{m,n}(z)$ exists for $z : |z| = \delta$. (F.4.15)

If $\mathcal{L} = \mathcal{L}_{i\gamma\infty}, a^* > 0$, then $H_{p,q}^{m,n}(z)$ exists for all $z : |\arg z| < \dfrac{a^* \pi}{2}$, $z \neq 0$. (F.4.16)

If $\mathcal{L} = \mathcal{L}_{i\gamma\infty}, a^* = 0, \Delta\gamma + \operatorname{Re} \mu < -1$, then $H_{p,q}^{m,n}(z)$ exists for all $z : \arg z = 0, z \neq 0$. (F.4.17)

Other important properties of the Fox H-functions, that can easily be derived from their definition, are included in the list below.

(i) The H-function is symmetric in the set of pairs
$(a_1, \alpha_1), \ldots, (a_n, \alpha_n), (a_{n+1}, \alpha_{n+1}), \ldots, (a_p, \alpha_p)$ and
$(b_1, \beta_1), \ldots, (b_m, \beta_m), (b_{m+1}, \beta_{m+1}), \ldots, (b_q, \beta_q)$.

(ii) If one of the (a_j, α_j), $j = 1, \ldots, n$, is equal to one of the (b_k, β_k), $k = m + 1, \ldots, q$; [or one of the pairs (a_j, α_j), $j = n+1, \ldots, p$, is equal to one of the (b_k, β_k), $k = 1, \ldots, m$], then the H-function reduces to one of the lower order, that is, p, q and n [or m] decrease by unity. Provided $n \geq 1$ and $q > m$, we have

$$H_{p,q}^{m,n}\left[z \left| \begin{array}{c} (a_j, \alpha_j)_{1,p} \\ (b_k, \beta_k)_{1,q-1} (a_1, \alpha_1) \end{array}\right.\right] = H_{p-1,q-1}^{m,n-1}\left[z \left| \begin{array}{c} (a_j, \alpha_j)_{2,p} \\ (b_k, \beta_k)_{1,q-1} \end{array}\right.\right], \quad (\text{F.4.18})$$

$$H_{p,q}^{m,n}\left[z \left| \begin{array}{c} (a_j, \alpha_j)_{1,p-1} (b_1, \beta_1) \\ (b_1, \beta_1) (b_k, \beta_k)_{2,q} \end{array}\right.\right] = H_{p-1,q-1}^{m-1,n}\left[z \left| \begin{array}{c} (a_j, \alpha_j)_{1,p-1} \\ (b_k, \beta_k)_{2,q} \end{array}\right.\right]. \quad (\text{F.4.19})$$

(iii)
$$z^\sigma H_{p,q}^{m,n}\left[z \left| \begin{array}{c} (a_j, \alpha_j)_{1,p} \\ (b_k, \beta_k)_{1,q} \end{array}\right.\right] = H_{p,q}^{m,n}\left[z \left| \begin{array}{c} (a_j + \sigma\alpha_j, \alpha_j)_{1,p} \\ (b_k + \sigma\beta_k, \beta_k)_{1,q} \end{array}\right.\right]. \quad (\text{F.4.20})$$

(iv)
$$\frac{1}{c} H_{p,q}^{m,n}\left[z \left| \begin{array}{c} (a_j, \alpha_j)_{1,p} \\ (b_k, \beta_k)_{1,q} \end{array}\right.\right] = H_{p,q}^{m,n}\left[z^c \left| \begin{array}{c} (a_j, c\alpha_j)_{1,p} \\ (b_k, c\beta_k)_{1,q} \end{array}\right.\right], \quad c > 0. \quad (\text{F.4.21})$$

The convergent and asymptotic expansions (for $z \to 0$ or $z \to \infty$) are mostly obtained by applying the residue theorem in the poles (assumed to be simple) of the Gamma functions appearing in $A(s)$ or $B(s)$ that are found inside the specially chosen path. In some cases (in particular if $n = 0 \Leftrightarrow B(s) = 1$) we find an exponential asymptotic behavior.

In the presence of a multiple pole s_0 of order N the treatment becomes more cumbersome because we need to expand in a power series at the pole the product of the involved functions, including z^s, and take the first N terms up to $(s - s_0)^{N-1}$ inclusive. Then the coefficient of $(s - s_0)^{N-1}$ is the required residue.

Let us consider the case $N = 2$ (double pole) of interest for the fractional diffusion. Then the expansions for the Gamma functions are of the form

$$\Gamma(s) = \Gamma(s_0)\left[1 + \psi(s_0)(s - s_0) + O\left((s - s_0)^2\right)\right], \quad s \to s_0, \quad s_0 \neq 0, -1, -2, \ldots$$

$$\Gamma(s) = \frac{(-1)^k}{\Gamma(k+1)(s+k)}\left[1 + \psi(k+1)(s+k) + O\left((s+k)^2\right)\right], \quad s \to -k,$$

where $k = 0, 1, 2, \ldots$ and $\psi(z)$ denotes the logarithmic derivative of the Γ function,

Appendix F: Higher Transcendental Functions

$$\psi(z) = \frac{d}{dz} \log \Gamma(z) = \frac{\Gamma'(z)}{\Gamma(z)},$$

whereas the expansion of z^s yields the logarithmic term

$$z^s = z^{s_0} \left[1 + \log z \, (s - s_0) + O((s - s_0)^2)\right], \quad s \to s_0.$$

F.4.2 Series Representations and Asymptotics. Recurrence Relations

Series representations of the Fox H-functions can be found by applying the Residue Theory to calculate the corresponding Mellin–Barnes integral. These calculations critically depend on the choice of the contour of integration $\mathcal{L}$ in (F.4.1) and the distribution of poles of the integrand there.

For simplicity, let us suppose that the poles of the Gamma functions $\Gamma(b_k + \beta_k s)$ and the poles of the Gamma functions $\Gamma(1 - a_j - \alpha_j s)$ do not coincide, i.e.

$$\alpha_j(b_k + r) \neq \beta_k(a_i - t - 1), \quad j = 1, \ldots, n, \; k = 1, \ldots, m, \; r, t = 0, 1, 2, \ldots. \tag{F.4.22}$$

This assumption allows us to choose one of the above described contours of integration $\mathcal{L} = \mathcal{L}_{-\infty}$, or $\mathcal{L} = \mathcal{L}_{+\infty}$, or $\mathcal{L} = \mathcal{L}_{i\gamma\infty}$. The series representations of Fox H-functions are based on the following theorem:

Theorem F.3 ([KilSai04, p. 5]) *Let the conditions (F.4.22) be satisfied. Then the following assertions hold true.*

(i) *In the cases (F.4.10) and (F.4.11) the H-function (F.4.1) is analytic in z and can be calculated via the formula*

$$H_{p,q}^{m,n}(z) = \sum_{k=1}^{m} \sum_{r=0}^{\infty} \mathrm{Res}_{s=b_{kr}} \left[\mathcal{H}_{p,q}^{m,n}(s) z^{-s}\right], \tag{F.4.23}$$

where the sum is calculated with respect to all poles b_{kr} of the Gamma functions $\Gamma(b_k + \beta_k s)$.

(ii) *In the cases (F.4.13) and (F.4.14) the H-function (F.4.1) is analytic in z and can be calculated via the formula*

$$H_{p,q}^{m,n}(z) = -\sum_{k=1}^{m} \sum_{r=0}^{\infty} \mathrm{Res}_{s=a_{jt}} \left[\mathcal{H}_{p,q}^{m,n}(s) z^{-s}\right], \tag{F.4.24}$$

where the sum is calculated with respect to all poles a_{jt} of the Gamma functions $\Gamma(1 - a_j - \alpha_j s)$.

(iii) In the case (F.4.16) the H-function (F.4.1) is analytic in z in the sector $|\arg z| < \frac{a^*\pi}{2}$.

Theorem F.4 ([KilSai04, p. 6]) *Suppose the poles of the Gamma functions $\Gamma(b_k + \beta_k s)$ and the poles of the Gamma functions $\Gamma(1 - a_j - \alpha_j s)$ do not coincide.*

(i) *If all the poles of the Gamma functions $\Gamma(b_k + \beta_k s)$ are simple, and either $\Delta > 0, z \neq 0$, or $\Delta = 0, 0 < |z| < \delta$, then the Fox H-function $H_{p,q}^{m,n}(z)$ has the power series expansion*

$$H_{p,q}^{m,n}(z) = \sum_{k=1}^{m} \sum_{r=0}^{\infty} h_{kr}^* z^{(b_k+r)/\beta_k}, \quad (F.4.25)$$

where

$$h_{kr}^* = \lim_{s \to b_{kr}} \left[(s - b_{kr}) \mathcal{H}_{p,q}^{m,n}(z)\right]$$

$$= \frac{(-1)^r}{r!\beta_k} \frac{\prod_{i=1,i\neq k}^{m} \Gamma\left(b_i - [b_k + r]\frac{\beta_i}{\beta_k}\right) \prod_{i=1}^{n} \Gamma\left(1 - a_i + [b_k + r]\frac{\alpha_i}{\beta_k}\right)}{\prod_{i=n+1}^{p} \Gamma\left(a_i - [b_k + r]\frac{\alpha_i}{\beta_k}\right) \prod_{i=m+1}^{q} \Gamma\left(1 - b_i + [b_k + r]\frac{\beta_i}{\beta_k}\right)}.$$

(F.4.26)

(ii) *If all the poles of the Gamma functions $\Gamma(1 - a_j - \alpha_j s)$ are simple, and either $\Delta < 0, z \neq 0$, or $\Delta = 0, |z| > \delta$, then the Fox H-function $H_{p,q}^{m,n}(z)$ has the power series expansion*

$$H_{p,q}^{m,n}(z) = \sum_{k=1}^{m} \sum_{r=0}^{\infty} h_{jt} z^{(a_j-1-t)/\alpha_j}, \quad (F.4.27)$$

where

$$h_{jt} = \lim_{s \to a_{jt}} \left[-(s - a_{jt}) \mathcal{H}_{p,q}^{m,n}(z)\right]$$

$$= \frac{(-1)^t}{t!\alpha_j} \frac{\prod_{i=1}^{m} \Gamma\left(b_i + [1 - a_j + t]\frac{\beta_i}{\alpha_j}\right) \prod_{i=1,i\neq j}^{n} \Gamma\left(1 - a_i + [1 - a_j + t]\frac{\alpha_i}{\alpha_j}\right)}{\prod_{i=n+1}^{p} \Gamma\left(a_i + [1 - a_j + t]\frac{\alpha_i}{\alpha_j}\right) \prod_{i=m+1}^{q} \Gamma\left(1 - b_i - [1 - a_j + t]\frac{\beta_i}{\alpha_j}\right)}.$$

(F.4.28)

Theorem F.5 ([KilSai04, p. 9]) *Suppose the poles of the Gamma functions $\Gamma(b_k + \beta_k s)$ and the poles of the Gamma functions $\Gamma(1 - a_j - \alpha_j s)$ do not coincide. Let either $\Delta < 0, z \neq 0$ or $\Delta = 0, |z| > \delta$.*

Then the Fox H-function $H_{p,q}^{m,n}(z)$ has the following power-logarithmic series expansion

Appendix F: Higher Transcendental Functions

$$H_{p,q}^{m,n}(z) = \sideset{}{'}\sum_{j,t} h_{jt} z^{(a_j-1-t)/\alpha_j} + \sideset{}{''}\sum_{j,t} \sum_{l=0}^{N_{jt}-1} H_{jtl} z^{(a_j-1-t)/\alpha_j} \left[\log z\right]^l, \quad (F.4.29)$$

where the summation in $\sum'$ is performed over all j, t, $j = 1, \ldots, n$, $t = 0, 1, 2, \ldots$, for which the poles of $\Gamma(1 - a_j - \alpha_j s)$ are simple, and the summation in $\sum''$ is performed over all values of parameters j, t for which the poles of $\Gamma(1 - a_j - \alpha_j s)$ have order N_{jt}, the constants h_{jt} are given by the formulas (F.4.28), and the constants H_{jtl} can be explicitly calculated (see [KilSai04, p. 8]).

From Theorems F.3, F.4 and F.5 follow corresponding asymptotic power- and power-logarithmic type expansions at infinity of the Fox H-functions $H_{p,q}^{m,n}(z)$ (see, e.g., [KilSai04]). Exponential asymptotic expansions in the case $\Delta > 0$, $a^* = 0$ and in the case $n = 0$ are presented, for example, in [KilSai04, Sect. 1.6, 1.7] (see also [Bra62, MaSaHa10, SrGuGo82]).

More detailed information on the asymptotics of the Fox H-functions at infinity can be found in [KilSai04, Chap. 1]. The asymptotic behavior of the Fox H-functions at zero is also discussed there.

The following two three-term recurrence formulas are linear combinations of the H-function with the same values of parameters m, n, p, q in which some a_j and b_k are replaced by $a_j \pm 1$ and by $b_k \pm 1$, respectively. Such formulas are called *contiguous relations* in [SrGuGo82].

Let $m \geq 1$ and $1 \leq n \leq p - 1$, then the following recurrence relation holds:

$$(b_1 \alpha_p - a_p \beta_1 + \beta_1) H_{p,q}^{m,n}\left[z^\sigma \left| \begin{array}{c} (a_j, \alpha_j)_{j=1,\ldots,p} \\ (b_j, \beta_j)_{j=1,\ldots,q} \end{array} \right. \right]$$

$$= \alpha_p H_{p,q}^{m,n}\left[z^\sigma \left| \begin{array}{c} (a_j, \alpha_j)_{j=1,\ldots,p} \\ (b_1+1, \beta_1), (b_j, \beta_j)_{j=2,\ldots,q} \end{array} \right. \right]$$

$$- \beta_1 H_{p,q}^{m,n}\left[z^\sigma \left| \begin{array}{c} (a_j, \alpha_j)_{j=1,\ldots,p-1}, (a_p-1, \alpha_p) \\ (b_j, \beta_j)_{j=1,\ldots,q} \end{array} \right. \right]. \quad (F.4.30)$$

Let $n \geq 1$ and $1 \leq m \leq q - 1$, then the following recurrence relation holds:

$$(b_q \alpha_1 - a_1 \beta_q + \beta_q) H_{p,q}^{m,n}\left[z^\sigma \left| \begin{array}{c} (a_j, \alpha_j)_{j=1,\ldots,p} \\ (b_j, \beta_j)_{j=1,\ldots,q} \end{array} \right. \right]$$

$$= \beta_p H_{p,q}^{m,n}\left[z^\sigma \left| \begin{array}{c} (a_1-1, \alpha_1), (a_j, \alpha_j)_{j=2,\ldots,p} \\ (b_j, \beta_j)_{j=1,\ldots,q} \end{array} \right. \right]$$

$$- \alpha_1 H_{p,q}^{m,n}\left[z^\sigma \left| \begin{array}{c} (a_j, \alpha_j)_{j=1,\ldots,p} \\ (b_j, \beta_j)_{j=1,\ldots,q-1}, (b_q+1, \beta_q) \end{array} \right. \right]. \quad (F.4.31)$$

A complete list of contiguous relations for H-functions can be found in [Bus72].

F.4.3 Special Cases

Most elementary functions can be represented as special cases of the Fox H-function. Let us present a list of such representations.

$$H_{0,1}^{1,0}\left[z \left| \begin{array}{c} -- \\ (b,\beta) \end{array} \right.\right] = \frac{1}{\beta} z^{b/\beta} \exp\left(-z^{\frac{1}{\beta}}\right); \tag{F.4.32}$$

$$H_{1,1}^{1,1}\left[z \left| \begin{array}{c} (1-a,1) \\ (0,1) \end{array} \right.\right] = \Gamma(a)(1+z)^a = \Gamma(a) {}_1F_0(a;-z); \tag{F.4.33}$$

$$H_{1,1}^{1,0}\left[z \left| \begin{array}{c} (\alpha+\beta+1,1) \\ (\alpha,1) \end{array} \right.\right] = z^\alpha (1-z)^\beta; \tag{F.4.34}$$

$$H_{0,2}^{1,0}\left[\frac{z^2}{4} \left| \begin{array}{c} ---- \\ \left(\frac{1}{2},1\right),(0,1) \end{array} \right.\right] = \frac{1}{\sqrt{\pi}} \sin z; \tag{F.4.35}$$

$$H_{0,2}^{1,0}\left[\frac{z^2}{4} \left| \begin{array}{c} ---- \\ (0,1),\left(\frac{1}{2},1\right) \end{array} \right.\right] = \frac{1}{\sqrt{\pi}} \cos z; \tag{F.4.36}$$

$$H_{0,2}^{1,0}\left[-\frac{z^2}{4} \left| \begin{array}{c} ---- \\ \left(\frac{1}{2},1\right),(0,1) \end{array} \right.\right] = \frac{i}{\sqrt{\pi}} \sinh z; \tag{F.4.37}$$

$$H_{0,2}^{1,0}\left[-\frac{z^2}{4} \left| \begin{array}{c} ---- \\ (0,1),\left(\frac{1}{2},1\right) \end{array} \right.\right] = \frac{i}{\sqrt{\pi}} \cosh z; \tag{F.4.38}$$

$$\pm H_{2,2}^{1,0}\left[z \left| \begin{array}{c} (1,1),(1,1) \\ (1,1),(0,1) \end{array} \right.\right] = \log(1 \pm z); \tag{F.4.39}$$

$$H_{2,2}^{1,2}\left[-z^2 \left| \begin{array}{c} \left(\frac{1}{2},1\right),\left(\frac{1}{2},1\right) \\ (0,1),\left(-\frac{1}{2},1\right) \end{array} \right.\right] = 2 \arcsin z; \tag{F.4.40}$$

$$H_{2,2}^{1,2}\left[z^2 \left| \begin{array}{c} \left(\frac{1}{2},1\right),(1,1) \\ \left(\frac{1}{2},1\right),(0,1) \end{array} \right.\right] = 2 \arctan z. \tag{F.4.41}$$

A number of representation formulas relating some special functions to the Fox H-function is presented in [KilSai04, Sect. 2.9]. Two of these formulas appear below.

$$H_{1,2}^{1,1}\left[-z \left| \begin{array}{c} (0,1) \\ (0,1),(1-\beta,\alpha) \end{array} \right.\right] = E_{\alpha,\beta}(z), \tag{F.4.42}$$

where $E_{\alpha,\beta}$ is the Mittag-Leffler function

Appendix F: Higher Transcendental Functions

$$E_{\alpha,\beta}(z) = \sum_{k=0}^{\infty} \frac{z^k}{\Gamma(\alpha k + \beta)}.$$

$$\frac{1}{\Gamma(\gamma)} H_{1,2}^{1,1}\left[-z \left|\begin{array}{c}(1-\gamma, 1) \\ (0,1), (1-\beta, \alpha)\end{array}\right.\right] = E_{\alpha,\beta}^{\gamma}(z), \ \mathrm{Re}\,\gamma > 0, \qquad (\mathrm{F}.4.43)$$

where $E_{\alpha,\beta}^{\gamma}$ is the three-parametric (Prabhakar) Mittag-Leffler function

$$E_{\alpha,\beta}^{\gamma}(z) = \sum_{k=0}^{\infty} \frac{(\gamma)_k z^k}{\Gamma(\alpha k + \beta)}.$$

$$H_{0,2}^{1,0}\left[-z \left|\begin{array}{c}-- \\ (0,1), (-\nu, \mu)\end{array}\right.\right] = J_{\nu}^{\mu}(z), \qquad (\mathrm{F}.4.44)$$

where J_{ν}^{μ} is the Bessel–Maitland (or the Bessel–Wright) function

$$J_{\nu}^{\mu}(z) = \sum_{k=0}^{\infty} \frac{(-z)^k}{\Gamma(\nu + k\mu + 1) k!}.$$

$$H_{1,3}^{1,1}\left[\frac{z^2}{4} \left|\begin{array}{c}(\lambda + \frac{\nu}{2}, 1) \\ (\lambda + \frac{\nu}{2}, 1), (\frac{\nu}{2}, 1), (\mu(\lambda + \frac{\nu}{2}) - \lambda - \nu, \mu)\end{array}\right.\right] = J_{\nu,\lambda}^{\mu}(z), \qquad (\mathrm{F}.4.45)$$

where $J_{\nu,\lambda}^{\mu}$ is the generalized Bessel–Maitland (or the generalized Bessel–Wright) function

$$J_{\nu,\lambda}^{\mu}(z) = \sum_{k=0}^{\infty} \frac{(-1)^k \left(\frac{z}{2}\right)^{\nu+2\lambda+2k}}{\Gamma(\nu+k\mu+\lambda+1)\Gamma(k+\lambda+1)}.$$

$$H_{p,q+1}^{1,p}\left[z \left|\begin{array}{c}(1-a_1, \alpha_1), \ldots, (1-a_p, \alpha_p) \\ (0,1), (1-b_1, \beta_1), \ldots, (1-b_q, \beta_q)\end{array}\right.\right] = {}_p\Psi_q\left[\begin{array}{c}(a_l, \alpha_l)_{1,p} \\ (b_l, \beta_l)_{1,q}\end{array}\bigg| z\right], \qquad (\mathrm{F}.4.46)$$

where ${}_p\Psi_q$ is the generalized Wright function

$${}_p\Psi_q\left[\begin{array}{c}(a_l, \alpha_l)_{1,p} \\ (b_l, \beta_l)_{1,q}\end{array}\bigg| z\right] = \sum_{k=0}^{\infty} \frac{\prod_{l=1}^{p} \Gamma(a_l + \alpha_l k)}{\prod_{j=1}^{q} \Gamma(b_j + \beta_j k)} \frac{z^k}{k!}.$$

F.4.4 Relations to Fractional Calculus

Following [KilSai04, Sect. 2.7] we present here two theorems describing the relationship of the Fox H-function to fractional calculus.

Theorem F.6 ([KilSai04, pp. 52–53]) *Let $\alpha \in \mathbb{C}$ (Re $\alpha > 0$), $\omega \in \mathbb{C}$, and $\sigma > 0$. Let us assume that either $a^* > 0$ or $a^* = 0$ and Re $\mu < -1$. Then the following statements hold:*

(i) If

$$\sigma \min_{1 \le k \le m} \left[\frac{\operatorname{Re} b_k}{\beta_k} \right] + \operatorname{Re} \omega > -1, \qquad (F.4.47)$$

for $a^* > 0$ or $a^* = 0$ and $\Delta \ge 0$, while

$$\sigma \min_{1 \le k \le m} \left[\frac{\operatorname{Re} b_k}{\beta_k}, \frac{\operatorname{Re} \mu + 1/2}{\Delta} \right] + \operatorname{Re} \omega > -1, \qquad (F.4.48)$$

for $a^* = 0$ and $\Delta < 0$, then the Riemann–Liouville fractional integral I_{0+}^{α} of the Fox H-function exists and the following relation holds:

$$\left(I_{0+}^{\alpha} t^{\omega} H_{p,q}^{m,n} \left[t^{\sigma} \;\middle|\; \begin{matrix} (a_1, \alpha_1), \ldots, (a_p, \alpha_p) \\ (0, 1), (b_1, \beta_1), \ldots, (b_q, \beta_q) \end{matrix} \right] \right)(x)$$

$$= x^{\omega + \alpha} H_{p+1, q+1}^{m, n+1} \left[x^{\sigma} \;\middle|\; \begin{matrix} (-\omega, \sigma), (a_1, \alpha_1), \ldots, (a_p, \alpha_p) \\ (b_1, \beta_1), \ldots, (b_q, \beta_q), (-\omega - \alpha, \sigma) \end{matrix} \right]. \qquad (F.4.49)$$

(ii) If

$$\sigma \min_{1 \le j \le n} \left[\frac{\operatorname{Re} a_j - 1}{\alpha_k} \right] + \operatorname{Re} \omega + \operatorname{Re} \alpha < 0, \qquad (F.4.50)$$

for $a^* > 0$ or $a^* = 0$ and $\Delta \le 0$, while

$$\sigma \min_{1 \le k \le m} \left[\frac{\operatorname{Re} a_j - 1}{\alpha_j}, \frac{\operatorname{Re} \mu + 1/2}{\Delta} \right] + \operatorname{Re} \omega + \operatorname{Re} \alpha < 0, \qquad (F.4.51)$$

for $a^* = 0$ and $\Delta > 0$, then the Riemann–Liouville fractional integral I_-^{α} of the Fox H-function exists and the following relation holds:

$$\left(I_-^{\alpha} t^{\omega} H_{p,q}^{m,n} \left[t^{\sigma} \;\middle|\; \begin{matrix} (a_1, \alpha_1), \ldots, (a_p, \alpha_p) \\ (0, 1), (b_1, \beta_1), \ldots, (b_q, \beta_q) \end{matrix} \right] \right)(x)$$

$$= x^{\omega + \alpha} H_{p+1, q+1}^{m+1, n} \left[x^{\sigma} \;\middle|\; \begin{matrix} (a_1, \alpha_1), \ldots, (a_p, \alpha_p), (-\omega, \sigma) \\ (-\omega - \alpha, \sigma), (b_1, \beta_1), \ldots, (b_q, \beta_q) \end{matrix} \right]. \qquad (F.4.52)$$

Appendix F: Higher Transcendental Functions

Theorem F.7 ([KilSai04, p. 55]) *Let $\alpha \in \mathbb{C}$ (Re $\alpha > 0$), $\omega \in \mathbb{C}$, and $\sigma > 0$. Let us assume that either $a^* > 0$ or $a^* = 0$ and Re $\mu < -1$. Then the following statements hold:*

(i) If either the condition in (F.4.44) is satisfied for $a^* > 0$ or $a^* = 0$ and $\Delta \geq 0$, or the condition in (F.4.45) is satisfied for $a^* = 0$ and $\Delta < 0$, then the Riemann–Liouville fractional derivative D_{0+}^α of the Fox H-function exists and the following relation holds:

$$\left(D_{0+}^\alpha t^\omega H_{p,q}^{m,n} \left[t^\sigma \left| \begin{array}{c} (a_1, \alpha_1), \ldots, (a_p, \alpha_p) \\ (0,1), (b_1, \beta_1), \ldots, (b_q, \beta_q) \end{array} \right. \right] \right)(x)$$

$$= x^{\omega - \alpha} H_{p+1, q+1}^{m, n+1} \left[x^\sigma \left| \begin{array}{c} (-\omega, \sigma), (a_1, \alpha_1), \ldots, (a_p, \alpha_p) \\ (b_1, \beta_1), \ldots, (b_q, \beta_q), (-\omega + \alpha, \sigma) \end{array} \right. \right]. \quad \text{(F.4.53)}$$

(ii) If

$$\sigma \min_{1 \leq j \leq n} \left[\frac{\operatorname{Re} a_j - 1}{\alpha_k} \right] + \operatorname{Re} \omega + 1 - \{\operatorname{Re} \alpha\} < 0, \quad \text{(F.4.54)}$$

for $a^* > 0$ or $a^* = 0$ and $\Delta \leq 0$, while

$$\sigma \min_{1 \leq k \leq m} \left[\frac{\operatorname{Re} a_j - 1}{\alpha_j}, \frac{\operatorname{Re} \mu + 1/2}{\Delta} \right] + \operatorname{Re} \omega + 1 - \{\operatorname{Re} \alpha\} < 0, \quad \text{(F.4.55)}$$

for $a^* = 0$ and $\Delta > 0$, where $\{\operatorname{Re} \alpha\}$ denotes the fractional part of the number Re α, then the Riemann–Liouville fractional derivative D_-^α of the Fox H-function exists and the following relation holds:

$$\left(D_-^\alpha t^\omega H_{p,q}^{m,n} \left[t^\sigma \left| \begin{array}{c} (a_1, \alpha_1), \ldots, (a_p, \alpha_p) \\ (0,1), (b_1, \beta_1), \ldots, (b_q, \beta_q) \end{array} \right. \right] \right)(x)$$

$$= x^{\omega - \alpha} H_{p+1, q+1}^{m+1, n} \left[x^\sigma \left| \begin{array}{c} (a_1, \alpha_1), \ldots, (a_p, \alpha_p), (-\omega, \sigma) \\ (-\omega + \alpha, \sigma), (b_1, \beta_1), \ldots, (b_q, \beta_q) \end{array} \right. \right]. \quad \text{(F.4.56)}$$

F.4.5 Integral Transforms of H-Functions

Here we consider the Fox H-function where, in the definition (F.4.1)–(F.4.4), the poles of the Gamma functions $\Gamma(b_k + \beta_k s)$ and the poles of the Gamma functions $\Gamma(1 - a_j - \alpha_j s)$ do not coincide. The notation introduced in (F.4.9) will be useful for us in this subsection too.

The first result is related to the Mellin transform of the Fox H-function. This follows from Theorem F.2. and the Mellin inversion theorem (see, e.g., [Tit86, Sect. 1.5]).

Theorem F.8 ([KilSai04, p. 43]) *Let $a^* \geq 0$, and $s \in \mathbb{C}$ be such that*

$$-\min_{1\leq k\leq m}\left[\frac{\operatorname{Re} b_k}{\beta_k}\right] < \operatorname{Re} s < \min_{1\leq j\leq n}\left[\frac{1-\operatorname{Re} a_j}{\alpha_j}\right] \tag{F.4.57}$$

when $a^* > 0$, and, additionally

$$\Delta \operatorname{Re} s + \operatorname{Re} \mu < -1,$$

when $a^* = 0$.
Then the Mellin transform of the Fox H-function exists and the following relation holds:

$$\left(\mathcal{M} H_{p,q}^{m,n}\left[x\,\bigg|\,\begin{matrix}(a_j,\alpha_j)_{1,p}\\(b_j,\beta_j)_{1,q}\end{matrix}\right]\right)(s) = \mathcal{H}_{p,q}^{m,n}\left[\begin{matrix}(a_j,\alpha_j)_{1,p}\\(b_j,\beta_j)_{1,q}\end{matrix}\,\bigg|\,s\right], \tag{F.4.58}$$

where $\mathcal{H}_{p,q}^{m,n}$ is the kernel in the Mellin–Barnes integral representation of the H-function.

A number of more general formulas for the Mellin transforms of the Fox H-function are presented in [KilSai04, p. 44], [MaSaHa10, pp. 39–40].

The next theorem gives the formula for the Laplace transform of the Fox H-function.

Theorem F.9 ([KilSai04, p. 45]) *Let either $a^* > 0$, or $a^* = 0$, $\operatorname{Re}\mu < -1$ be such that*

$$\min_{1\leq k\leq m}\left[\frac{\operatorname{Re} b_k}{\beta_k}\right] > -1, \tag{F.4.59}$$

when $a^* > 0$, or $a^* = 0$, $\Delta \geq 0$, and

$$\min_{1\leq k\leq m}\left[\frac{\operatorname{Re} b_k}{\beta_k},\frac{\operatorname{Re}\mu+\frac{1}{2}}{\Delta}\right] > -1, \tag{F.4.60}$$

when $a^* = 0$, $\Delta < 0$.
Then the Laplace transform of the Fox H-function exists and the following relation holds for all $t \in \mathbb{C}$, $\operatorname{Re} t > 0$:

$$\left(\mathcal{L} H_{p,q}^{m,n}\left[x\,\bigg|\,\begin{matrix}(a_j,\alpha_j)_{1,p}\\(b_j,\beta_j)_{1,q}\end{matrix}\right]\right)(t) = \frac{1}{t} H_{p+1,q}^{m,n+1}\left[\frac{1}{t}\,\bigg|\,\begin{matrix}(0,1),(a_j,\alpha_j)_{1,p}\\(b_j,\beta_j)_{1,q}\end{matrix}\right]. \tag{F.4.61}$$

A number of more general formulas for the Laplace transform of the Fox H-function are presented in [KilSai04, pp. 46–48].

F.5 Historical and Bibliographical Notes

The (classical) hypergeometric function $_2F_1$ is commonly defined by the following *Gauss series* representation (see, e.g., [NIST, p. 384])

$$_2F_1(a,b;c;z) = \sum_{k=0}^{\infty} \frac{(a)_k(b)_k}{(c)_k k!} z^k = \frac{\Gamma(c)}{\Gamma(a)\Gamma(b)} \sum_{k=0}^{\infty} \frac{\Gamma(a+k)\Gamma(b+k)}{\Gamma(c+k)k!} z^k.$$

The term *hypergeometric series* was proposed by J. Wallis in his book *Arithmetica Infinitorum* (1655). Hypergeometric series were studied by L. Euler, and a systematic analysis of their properties was presented in C.-F. Gauss's 1812 paper (see the reprint in the collection of Gauss's works [Gauss, pp. 123–162]). Studies in the nineteenth century included those of E. Kummer [Kum36], and the fundamental characterization by Bernhard Riemann of the hypergeometric function by means of the differential equation it satisfies. Riemann showed that the second-order differential equation for $_2F_1$, examined in the complex plane, could be characterized (on the Riemann sphere) by its three regular singularities.

Historically, *confluent hypergeometric functions* were introduced as solutions of a degenerate form of the hypergeometric differential equation. Kummer's confluent hypergeometric function $M(a;b;z)$ (known also as the Φ-function, see [ErdBat-1]) was introduced by Kummer in 1837 ([Kum37]) as a solution to (Kummer's) differential equation

$$z\frac{d^2 w}{dz^2} + (b-z)\frac{dw}{dz} + aw = 0.$$

The function $M(a,b;z)$ can be represented in the form of a series too

$$M(a,b;z) = \sum_{k=0}^{\infty} \frac{(a)_k}{(b)_k k!} z^k = {}_1F_1(a;b;z).$$

Another (linearly independent) solution to Kummer's differential equation $U(a;b;z)$ (known also as the Ψ-function, see [ErdBat-1]) was found by Tricomi in 1947 ([Tri47]).

The function $\phi(\alpha,\beta;z)$, called the Wright function, was introduced by Wright in 1933 (see [Wri33]) in relation to the asymptotic theory of partitions. An extended discussion of its properties and applications is given in [GoLuMa99]. Special attention is given to the key role of the Wright function in the theory of fractional partial differential equations.

The generalized hypergeometric functions introduced by Pochhammer [Poc70] and Goursat [Gou83a, Gou83b] are solutions of linear differential equations of order n with polynomial coefficients. These functions were considered later by Pincherle. Thus, Pincherle's paper [Pin88] is based on what he called the "duality principle", which relates linear differential equations with rational coefficients to linear difference equations with rational coefficients. Let us recall that the phrase "rational coefficients" means that the coefficients are in general rational functions (i.e. a ratio of two polynomials) of the independent variable and, in particular, polynomials (for more details see [MaiPag03]).

These integrals were used by Meijer in 1946 to introduce the G-function into mathematical analysis [Mei46]. From 1956 to 1970 a lot of work was done on this function, which can be seen from the bibliography of the book by Mathai and Saxena [MatSax73].

The H-functions, introduced by Fox [Fox61] in 1961 as symmetrical Fourier kernels, can be regarded as the extreme generalization of the generalized hypergeometric functions $_pF_q$ beyond the Meijer G functions (see, e.g. [Sax09]). The importance of this function is appreciated by scientists, engineers and statisticians due to its vast potential of applications in diverse fields. These functions include, among others, the functions considered by Boersma [Boe62], Mittag-Leffler [ML1, ML2, ML3, ML4], the generalized Bessel function due to Wright [Wri35a], the generalization of the hypergeometric functions studied by Fox (1928), and Wright [Wri35b, Wri40c], the Krätzel function [Kra79], the generalized Mittag-Leffler function due to Dzherbashyan [Dzh60], the generalized Mittag-Leffler function due to Prabhakar [Pra71] and to Kilbas and Saigo [KilSai95a], the multi-index Mittag-Leffler function due to Kiryakova [Kir99, Kir00], and Luchko [Luc99] (see also [KilSai96]), etc. Except for the functions of Boersma [Boe62], the aforementioned functions cannot be obtained as special cases of the G-function of Meijer [Mei46], hence a study of the H-function will cover a wider range than the G-function and gives general, deeper, and useful results directly applicable in various problems of a physical, biological, engineering and earth sciences nature, such as fluid flow, rheology, diffusion in porous media, kinematics in viscoelastic media, relaxation and diffusion processes in complex systems, propagation of seismic waves, anomalous diffusion and turbulence, etc. See, Caputo [Cap69], Glöckle and Nonnenmacher [GloNon93], Mainardi et al. [MaLuPa01], Saichev and Zaslavsky [SaiZas97], Hilfer [Hil00], Metzler and Klafter [MetKla00], Podlubny [Pod99], Schneider [Sch86] and Schneider and Wyss [SchWys89] and others.

A major contribution of Fox involves a systematic investigation of the asymptotic expansion of the generalized hypergeometric function (now called *Wright functions*, or *generalized Wright functions*, or *Fox–Wright functions*):

$$_pF_q((a_1,\alpha_1),\ldots,(a_p,\alpha_p);(b_1,\beta_1),\ldots,(b_q,\beta_q);z) = \sum_{l=0}^{\infty} \frac{\prod_{j=1}^{p} \Gamma(a_j+\alpha_j l)}{\prod_{k=1}^{q} \Gamma(b_k+\beta_k l)} \frac{z^l}{l!},$$

where $z \in \mathbb{C}, a_j, b_k \in \mathbb{C}, \alpha_j, \beta_k \in \mathbb{R}, j=1,\ldots,p; k=1,\ldots,q; \sum_{k=1}^{q}\beta_k - \sum_{j=1}^{p}\alpha_j \geq -1$. His method is an improvement of the approach by Barnes (see, e.g., [Barn07b]) who found an asymptotic expansion of the ordinary generalized hypergeometric function $_pF_q(z)$. In 1961 Fox introduced [Fox61] the H-function in the theory of special functions, which generalized the MacRobert's E-function (see [Mac-R38], [Sla66, p. 42]), the generalized Wright hypergeometric function, and the Meijer G-function. In the mentioned paper he investigated the far-most generalized Fourier (or Mellin) kernel associated with the H-function and established many properties and special cases of this kernel.

Like the Meijer G-functions, the Fox H-functions turn out to be related to the Mellin–Barnes integrals and to the Mellin transforms, but in a more general way. After Fox, the H-functions were carefully investigated by Braaksma [Bra62], who provided their convergent and asymptotic expansions in the complex plane, based on their Mellin–Barnes integral representation.

More recently, the H-functions, being related to the Mellin transforms (see [Mari83]), have been recognized to play a fundamental role in probability theory and statistics (see e.g. [MaSaHa10, Sch86, SaxNon04, UchZol99, Uch03]), in fractional calculus [KilSai99, KiSrTr06, Kir94], and its applications [AnhLeo01, AnhLeo03, AnLeSa03, Hil00, MatSax73], including phenomena of non-standard (anomalous) relaxation and diffusion [GorMai98, Koc90]. Several books specially devoted to H-functions and their applications have been published recently. Among them are the books by Kilbas *and* Saigo [KilSai04] and by Mathai, Saxena *and* Haubold [MaSaHa10].

In [MaPaSa05] the fundamental solutions of the Cauchy problem for the space-time fractional diffusion equation are expressed in terms of proper Fox H-functions, based on their Mellin–Barnes integral representations.

The asymptotic properties of special functions are one of the most important questions to be solved. Several technical approaches are described in [Olv74] and in some chapters of [NIST] (see also [Evg78], where asymptotic analysis is developed in the framework of the theory of general entire functions, and the survey paper [Par19]).

Different properties of special functions are discussed in [Mat93, Mik59a, WongZh99a, WongZh99b]

F.6 Exercises

F.6.1. ([NIST, p. 386]) Prove that the following equalities hold for all z, $|z| < \pi/4$:

(i) $F(a; 1/2 + a; 1/2; -\tan^2 z) = (\cos z)^{2a} \cos(2az);$

(ii) $F(a; 1/2 + a; 3/2; -\tan^2 z) = (\cos z)^{2a} \dfrac{\sin((1-2a)z)}{(1-2a)\sin z}.$

F.6.2. ([NIST, p. 386]) Prove that the following equalities hold for all z, $|z| < \pi/2$:

(i) $F(-a; a; 1/2; \sin^2 z) = \cos(2az);$

(ii) $F(a; 1-a; 1/2; \sin^2 z) = \dfrac{\cos((2a-1)z)}{\cos z};$

(iii) $F(a; 1 - a; 3/2; \sin^2 z) = \dfrac{\sin((2a-1)z)}{(2a-1)\sin z}.$

F.6.3. ([KilSai01, p. 63]) Prove the following representations of the Kummer confluent hypergeometric function $_1F_1$ and the Gauss hypergeometric function $_2F_1$ in terms of the Fox H-function:

(i) $H_{1,2}^{1,1}\left[z \left|\begin{array}{c}(1-a,1)\\(0,1),(1-c,1)\end{array}\right.\right] = \dfrac{\Gamma(a)}{\Gamma(c)}\,_1F_1(a;c;-z);$

(ii) $H_{2,2}^{1,2}\left[z \left|\begin{array}{c}(1-a,1),(1-b,1)\\(0,1),(1-c,1)\end{array}\right.\right] = \dfrac{\Gamma(a)\Gamma(b)}{\Gamma(c)}\,_2F_1(a;b;c;-z).$

F.6.4. ([Kir94, p. 334]) Prove the following relations for the incomplete gamma- and Beta-functions:

(i) $\gamma(\alpha, z) := \displaystyle\int_0^z e^{-t} t^{\alpha-1} dt = \,_1F_1(\alpha; \alpha+1; -z);$

(ii) $B_z(p, q) := \displaystyle\int_0^z t^{p-1}(1-t)^{q-1} dt = \,_2F_1(p, 1-q; p+1; z).$

F.6.5. ([Kir94, p. 334]) Prove the following representation of the error function in terms of confluent hypergeometric function:

$\mathrm{erf}(z) := \dfrac{2}{\sqrt{\pi}} \displaystyle\int_0^z e^{-t^2} dt = \dfrac{2z}{\sqrt{\pi}}\,_1F_1(\tfrac{1}{2}; \tfrac{3}{2}; -z^2).$

F.6.6. Prove the following representation of the classical polynomials in terms of special cases of the hypergeometric functions $_pF_q$:

(i) Laguerre polynomials ([MaSaHa10, p. 29])

$L_n^{(\alpha)}(z) := \dfrac{e^z}{z^\alpha n!} \dfrac{d^n}{dz^n}\left\{e^{-z} z^{n+\alpha}\right\} = \dfrac{(1+\alpha)_n}{n!}\,_1F_1(-n; \alpha+1; z);$

(ii) Jacobi polynomials ([MaSaHa10, p. 28])

$P_n^{(\alpha,\beta)}(z) := \dfrac{(-1)^n}{2^n n!}(1-z)^{-\alpha}(1+z)^{-\beta} \dfrac{d^n}{dz^n}\left\{(z^2-1)^n\right\}$

$= \dfrac{(\alpha+1)_n}{n!}\,_2F_1\left(-n, n+\alpha+\beta+1; \beta+1; \dfrac{1-z}{2}\right);$

(iii) Legendre polynomials ([Kir94, p. 333]) (for $|z| < 1$)

$$P_n(z) := \frac{1}{2^n n!} \frac{d^n}{dz^n} \left\{ (1-z)^{\alpha+n}(1+z)^{\beta+n} \right\}$$

$$= (-1)^n {}_2F_1\left(-n, n+1; 1; \frac{1-z}{2}\right), \quad |z| < 1;$$

(iv) Tchebyshev polynomials ([Kir94, p. 333])

$$T_n(z) := \cos(\arccos z) = \frac{n!\sqrt{\pi}}{\Gamma(n+\frac{1}{2})} P_n^{(-\frac{1}{2}, -\frac{1}{2})}(z) = {}_2F_1\left(-n, n; \frac{1}{2}; \frac{1-z}{2}\right);$$

(v) Bessel polynomials ([Kir94, p. 333])

$$Y_n(z, a, b) := \sum_{k=0}^{n} \frac{(-n)_k (a+n-1)_k}{k!} \left(-\frac{z}{b}\right)^k = {}_2F_0\left(-n, -a+n-1; -; -\frac{z}{b}\right);$$

(vi) Hermite polynomials ([NIST, p. 443])

$$H_n(z) := n! \sum_{k=0}^{[n/2]} \frac{(-1)^k (2z)^{n-2k}}{k!(n-2k)!} = (2z)^n {}_2F_0\left(-\frac{n}{2}, -\frac{n}{2}+\frac{1}{2}; -; -\frac{1}{z^2}\right).$$

F.6.7. ([Kir94, pp. 331–332]) Prove the following representations of some elementary functions in terms of the Meijer G-function:

$$(i) \quad \frac{z^\gamma}{1+az^\alpha} = a^{-\frac{\gamma}{\alpha}} G_{1,1}^{1,1}\left[az^\alpha \,\middle|\, \begin{array}{c} \frac{\gamma}{\alpha} \\ \frac{\gamma}{\alpha} \end{array}\right];$$

$$(ii) \quad z^\beta e^{-\eta z^\alpha} = \eta^{-\frac{\beta}{\alpha}} G_{0,1}^{1,0}\left[\eta z^\alpha \,\middle|\, \begin{array}{c} \beta \\ \alpha \end{array}\right];$$

$$(iii) \quad \log\left(\frac{1+z}{1-z}\right) = G_{1,1}^{1,2}\left[-z^2 \,\middle|\, \begin{array}{c} 1, \frac{1}{2} \\ \frac{1}{2}, 0 \end{array}\right], \quad |z| < 1;$$

$$(iv) \quad \left[\frac{1}{2}(1+\sqrt{1-z})\right]^{1-2a} = \frac{2a-1}{\sqrt{\pi} 2^{2-2a}} G_{2,2}^{1,2}\left[-z \,\middle|\, \begin{array}{c} 1-a, \frac{3}{2}-a \\ 0, 1-2a \end{array}\right].$$

F.6.8. ([Bra62, p. 279]) Prove the following series representation for the Fox H-function:

$$\beta_1 z^{-b_1} H_{p,q}^{1,n}\left[z^{\beta_1} \,\middle|\, \begin{array}{c} (a_1, \alpha_1), \ldots, (a_p, \alpha_p) \\ (b_1, \beta_1), \ldots, (b_q, \beta_q) \end{array}\right]$$

$$= \sum_{\nu}^{\infty} \frac{(-z)^{\nu}}{\nu!} \frac{\prod_{j=1}^{n} \Gamma\left(1 - a_j + \alpha_j \left(\frac{b_1+\nu}{\beta_1}\right)\right)}{\prod_{k=2}^{q} \Gamma\left(1 - b_k + \beta_k \left(\frac{b_1+\nu}{\beta_1}\right)\right) \prod_{j=n+1}^{p} \Gamma\left(a_j - \alpha_j \left(\frac{b_1+\nu}{\beta_1}\right)\right)}.$$

References

[Abe23] Abel, N.H.: Oplösning af et Par Opgaver ved Hjelp af bestemie Integraler Magazin for Naturvidenskaberne, Aargang 1, Bind 2, 11–27 (1823), in Norwegian. French translation "Solution de quelques problèmes à l'aide d'intégrales définies". In: Sylov, L., Lie, S. (eds.) Oeuvres Complètes de Niels Henrik Abel (Deuxième Edition), I, Christiania, 11–27 (1881). Reprinted by Éditions Jacques Gabay, Sceaux (1992)

[Abe26a] Abel, N.H.: Untersuchungen über die Reihe: $1+\frac{m}{1}x+\frac{m\cdot(m-1)}{1\cdot 2}x^2+\frac{m\cdot(m-1)\cdot(m-2)}{1\cdot 2\cdot 3}x^3+\cdots\cdot$ u. s. w. J. Reine Angew. Math., **1**, 311–339 (1826), translation into English in: Researches on the series $1+\frac{m}{1}x+\frac{m(m-1)}{1.2}x^2+\frac{m(m-1)(m-2)}{1.2.3}x^3+\cdots$. Tokio Math. Ges. **IV**, 52–86 (1891)

[AblFok97] Ablowitz, M.J., Fokas, A.S.: Complex Variables. Cambridge University Press, Cambridge (1997)

[AbrSte72] Abramovitz, M., Stegun, I.A.: Handbook of Mathematical Functions. National Bureau of Standards, Washington D.C. (1972) (10th printing)

[AchHan09] Achar, B.N.N., Hanneken, J.W.: Microscopic formulation of fractional calculus theory of viscoelasticity based on lattice dynamics. Phys. Scripta **T136**, 014011/1–7 (2009)

[AdEnOl05] Adolfsson, K., Enelund, M., Olsson, P.: On the fractional order model of viscoelasticity. Mech. Time-Depend. Mater. **9**(1), 15–34 (2005)

[Ado89] Adomian, G.: Nonlinear Stochastic Systems Theory and Applications to Physics. Kluwer, Dordrecht (1989)

[Ado94] Adomian, G.: Solving Frontier Problems of Physics: Decomposition Method. Kluwer, Dordrecht (1994)

[Adv-07] Sabatier, J., Agrawal, O.P., Tenreiro Machado, J.A. (eds.): Advances in Fractional Calculus: Theoretical Developments and Applications in Physics and Engineering. Springer, Berlin (2007)

[AgaAli19] Agahi, H., Alipour, M.: Mittag-Leffler-Gaussian distribution: theory and application to real data. Math. Comp. Simul. **156**, 227–235 (2019)

[AgMiNi15] Agarwal, P., Milovanović, G.V., Nisar, K.S.: A fractional integral operator involving the Mittag-Leffler type function with four parameters, Facta Universitatis (NIŠ) Ser. Math. Inform. **30**(5), 597–605 (2015)

[Aga53] Agarwal, R.P.: A propos d'une note de M. Pierre Humbert. C. R. Acad. Sci. Paris. **236**, 2031–2032 (1953)

© Springer-Verlag GmbH Germany, part of Springer Nature 2020
R. Gorenflo et al., *Mittag-Leffler Functions, Related Topics and Applications*,
Springer Monographs in Mathematics,
https://doi.org/10.1007/978-3-662-61550-8

[AldWai70] Alder, B.J., Wainwright, T.E.: Decay of velocity autocorrelation function. Phys. Rev. A **1**, 18–21 (1970)
[Ale82] Aleroev, T.S.: The Sturm–Liouville problem for a second-order differential equation with fractional derivatives in the lower terms. (Russian) Differentsial'nye Uravneniya, **18**(2), 341–342 (1982)
[Ale84] Aleroev, T.S.: Special analysis of a class of nonselfadjoint operators. (Russian) Differentsial'nye Uravneniya, **20**(1), 171–172 (1984)
[Al-B65] Al-Bassam, M.A.: Some existence theorems on differential equations of generalized order. J. Reine Angew. Math. **218**, 70–78 (1965)
[Al-B82] Al-Bassam, M.A.: On fractional calculus and its applications to the theory of ordinary differential equations of generalized order. Nonlinear analysis and applications (St. Johns, Nfld., 1981), pp. 305–331, Lecture Notes in Pure and Applied Mathematics, vol. 80. Marcel Dekker, New York (1982)
[Al-B86] Al-Bassam, M.A.: On Fractional Analysis and Its Applications, Modern Analysis and Its Applications (New Delhi, 1983), 269–307. Prentice-Hall of India, New Delhi (1986)
[Al-B87] Al-Bassam, M.A.: On Generalized Power Series and Generalized Operational Calculus and Its Applications, Nonlinear Analysis, pp. 51–88. World Scientific Publishing, Singapore (1987)
[Al-BLuc95] Al-Bassam, M.-A., Luchko, Y.F.: On generalized fractional calculus and it application to the solution of integro-differential equations. J. Fract. Calc. **7**, 69–88 (1995)
[AlKiKa02] Ali, I., Kiryakova, V., Kalla, S.L.: Solutions of fractional multi-order integral and differential equations using a Poisson-type transform. J. Math. Anal. Appl. **269**, 172–199 (2002)
[Al-MKiVu02] Al-Musallam, F., Kiryakova, V., Kim Tuan, V.: A multi-index Borel–Dzrbashjan transform. Rocky Mt. J. Math. **32**(2), 409–428 (2002)
[Al-S95] Al Saqabi, B.N.: Solution of a class of differintegral equations by means of the Riemann-Liouville operator. J. Fract. Calc. **8**, 95–102 (1995)
[Al-SKir98] Al Saqabi, B.N., Kiryakova, V.S.: Explicit solutions of fractional integral and differential equations involving Erdelyi-Kober operators. Appl. Math. Comp. **95**(1), 1–13 (1998)
[Al-STua96] Al Saqabi, B.N., Tuan, V.K.: Solution of a fractional differintegral equation. Integr. Transform. Spec. Funct. **4**, 321–326 (1996)
[AnVHBa12] An, J., Van Hese, E., Baes, M.: Phase-space consistency of stellar dynamical models determined by separable augmented densities. Mon. Not. R. Astron. Soc. **422**(1), 652–664 (2012)
[AnhLeo01] Anh, V.V., Leonenko, N.N.: Spectral analysis of fractional kinetic equations with random data. J. Stat. Phys. **104**(5–6), 1349–1387 (2001)
[AnhLeo03] Anh, V.V., Leonenko, N.N.: Harmonic analysis of random fractional diffusion-wave equations. Appl. Math. Comput. **141**(1), 77–85 (2003)
[AnLeSa03] Anh, V.V., Leonenko, N.N., Sakhno, L.M.: Higher-order spectral densities of fractional random fields. J. Stat. Phys. **111**(3–4), 789–814 (2003)
[AnhMcV] Anh, V.V., McVinish, R.: Completely monotone property of fractional Green functions. Fract. Calc. Appl. Anal. **6**(2), 157–173 (2003)
[AnoBol94] Anosov, D.V., Bolibruch, A.A.: The Riemann-Hilbert Problem, A Publication from the Steklov Institute of Mathematics. Aspects Math. Friedr. Vieweg & Sohn, Braunschweg (1994)
[AnsShe14] Ansari, A., Sheikhani, A.R.: New identities for the Wright and the Mittag-Leffler functions using the Laplace transform. Asian-Eur. J. Math., **7**(3), 1450038 (8 pages) (2014)
[AnKoVa93] Antimirov, M.Ya., Kolyshkin, A.A., Vaillancourt, R.: Applied Integral Transforms, American Mathematical Society, Providence (1993)
[Ape08] Apelblat, A.: Volterra Functions. Nova Science Publishers Inc, New York (2008)

[Art64]	Artin, E.: The Gamma Function. Holt, Rinehart and Winston, New York (1964) [first published by B.G. Teubner, Leipzig (1931)]
[AskAns16]	Askari, H., Ansari, A.: Fractional calculus of variations with a generalized fractional derivative. Fract. Differ. Calc. **6**(1), 57–72 (2016)
[Ata-et-al14]	Atanackovic, T.M., Pilipovic, S., Stankovic, B., Zorica, D.: Fractional Calculus with Applications in Mechanics: Vibrations and Diffusion Processes. Wiley, London (2014)
[Bab86]	Babenko, Y.I.: Heat and Mass Transfer. Chimia, Leningrad (1986). (in Russian)
[BaeMee01]	Baeumer, B., Meerschaert, M.M.: Stochastic solutions for fractional Cauchy problems. Fract. Calc. Appl. Anal. **4**, 481–500 (2001)
[Bag79]	Bagley, R.L.: Applications of generalized derivatives to viscoelasticity. Ph.D. Dissertation, Air Force Institute of Technology (1979)
[Bag90]	Bagley, R.L.: On the fractional order initial value problem and its engineering applications. In: Nishimoto, K. (ed.) Fracional Calculus and Its Application, pp. 12–20. Nihon University, College of Engineering (1990)
[Bag07]	Bagley, R.L.: On the equivalence of the Riemann-Liouville and the Caputo fractional order derivatives in modeling of linear viscoelastic materials. Fract. Calc. Appl. Anal. **10**(2), 123–126 (2007)
[BagTor79]	Bagley, R.L., Torvik, P.J.: A generalized derivative model for an elastomer damper. Shock Vib. Bull. **49**, 135–143 (1979)
[BagTor83a]	Bagley, R.L., Torvik, P.J.: A theoretical basis for the application of fractional calculus. J. Rheology **27**, 201–210 (1983)
[BagTor83b]	Bagley, R.L., Torvik, P.J.: Fractional calculus - a different approach to the finite element analysis of viscoelastically damped structures. AIAA J. **21**, 741–748 (1983)
[BaiKin96]	Baillie, R.T., King, M.L.: Fractional differencing and long memory processes. J. Econom. **73**, 1–3 (1996)
[BalV85]	Balakrishnan, V.: Anomalous diffusion in one dimension. Phys. A. **132**, 569–580 (1985)
[Bal-et-al17]	Baleanu, D., Diethelm, K., Scalas, E., Trujillo, J.J.: Fractional Calculus: Models and Numerical Methods. Series on Complexity, Nonlinearity and Chaos, 2nd edn. vol. 5. World Scientific, Singapore (2017)
[BalR07]	Balescu, R.: V-Langevin equations, continuous time random walks and fractional diffusion. Chaos, Solitons Fractals **34**, 62–80 (2007)
[BanPra16]	Bansal, D., Prajapat, J.K.: Certain geometric properties of the Mittag-Leffler functions. Complex Var. Elliptic Equ. **61**(3), 338–350 (2016)
[Bark02]	Barkai, E.: CTRW pathways to the fractional diffusion equation. Chem. Phys. **284**, 13–27 (2002)
[Barn02]	Barnes, E.W.: A memoir of integral functions. Lond. Phil. Trans. **199**(A), 411–500; Lond. Royal Soc. Proc. **69**, 121–125 (1902)
[Barn06]	Barnes, E.W.: The asymptotic expansion of integral functions defined by Taylor's series. Phil. Trans. Roy. Soc. Lon. A **206**, 249–297 (1906)
[Barn08]	Barnes, E.W.: A new development of the theory of the hypergeometric functions. Proc. London Math. Soc., ser. 2. **6**, 141–177 (1908)
[Barn07b]	Barnes, E.W.: The asymptotic expansion of integral functions defined by generalized hypergeometric series, *Proc. London Math. Soc., Ser 2*. **5**, 59–116 (1907)
[Barr54]	Barrett, J.H.: Differential equations of non-integer order. Can. J. Math. **6**, 529–541 (1954)
[Bas88]	Basset, A.B.: A Treatise on Hydrodynamics, vol. 2. Deighton Bell, Cambridge (1888), Chap. 22
[Bas10]	Basset, A.B.: On the descent of a sphere in a viscous liquid. Quart. J. Math **41**, 369–381 (1910)
[BatErd54a]	Bateman, H., Erdelyi, A.: with the participation of Magnus, W., Oberhettinger, F. and Tricomi, F.G., Tables of the Integrals Transforms, vol. 1. McGraw-Hill, New York (1954)

[BatErd54b]	Bateman, H., Erdelyi, A.: with the participation of Magnus, W., Oberhettinger, F. and Tricomi, F.G., Tables of the Integrals Transforms, vol. 2. McGraw-Hill, New York (1954)
[Baz18]	Bazhlekova, E.. Subordination in a class of generalized time-fractional diffusion-wave equations. Fract. Calc. Appl. Anal. **21**(4), 869–900 (2018)
[BazDim13]	Bazhlekova, E., Dimovski, I.: Time-fractional Thornley's problem. J. Inequal. Spec. Funct. **4**(1), 21–35 (2013)
[BazDim14]	Bazhlekova, E., Dimovski, I.: Exact solution of two-term time-fractional Thornley's problem by operational method. Integr. Trans. Spec. Funct. **25**(1), 61–74 (2014)
[BegOrs09]	Beghin, L., Orsingher, E.: Iterated elastic Brownian motions and fractional diffusion equations. Stoch. Proc. Appl. **119**, 1975–2003 (2009)
[BenOrs87]	Bender, C.M., Orszag, S.A.: Advanced Mathematical Methods for Scientists and Engineers. McGraw-Hill, Singapore (1987)
[Ber-S05a]	Berberan-Santos, M.N.: Analytic inversion of the Laplace transform without contour integration: application to luminescence decay laws and other relaxation functions. J. Math. Chem. **38**, 165–173 (2005)
[Ber-S05b]	Berberan-Santos, M.N.: Relation between the inverse Laplace transforms of $I(t^\beta)$ and $I(t)$: Application of the Mittag-Leffler and asymptotic power law relaxation functions. J. Math. Chem. **38**, 265–270 (2005)
[Ber-S05c]	Berberan-Santos, M.N.: Properties of the Mittag-Leffler relaxation function. J. Math. Chem. **38**, 629–635 (2005)
[BerBou18]	Bergounioux, M., Bourdin, L.: Filippov's existence theorem and Pontryagin maximum principle for general Caputo fractional optimal control problems. Preprint, 39 p. (2018)
[Bern28]	Bernstein, S.: Sur les fonctions absolutment monotones. Acta Math. **51**, 1–66 (1928)
[Bert96]	Bertoin, J.: Lévy Processes, Cambridge University Press [Cambridge Tracts in Mathematics, vol. 121], Cambridge (1996)
[BeyKem95]	Beyer, H., Kempfle, S.: Definition of physically consistent damping laws with fractional derivatives. Z. angew. Math. Mech. (ZAMM) **75**(8), 623–635 (1995)
[Bie31]	Bieberbach, L.: Lehrbuch der Funktionentheorie. Bd. vol. II: Moderne Funktionentheorie. 2. Aufl. (German), Leipzig, B.G. Teubner (1931)
[Bla97]	Blank, L.: Numerical treatment of differential equations of fractional order. Nonlinear World. **4**(4), 473–491 (1997)
[BleHan86]	Bleistein, N., Handelsman, R.A.: Asymptotic Expansions of Integrals. Dover, New York (1986)
[Boa83]	Boas, M.: Mathematical Methods in the Physical Sciences. Wiley, New York (1983)
[Boa54]	Boas, R.P.: Entire Functions. Academic, New York (1954)
[Bob-et-al99]	Bobal, V., Bohm, J., Prokop, R., Fessl, J.: Practical Aspects of Self-Tuning Controllers: Algorithms and Implementation. VUT, Brno (1999). (in Czech)
[Boc37]	Bochner, S.: Completely monotone functions of the Laplace operator for torus and sphere. Duke Math. J. **3**, 488–502 (1937)
[Boe62]	Boersma, J.: On a function which is a special case of Meijer's G-function. Comp. Math. **15**, 34–63 (1962)
[Bog07]	Boggio, T.: Integrazione dell'equazione funzionale che regge la caduta di una sfera in un liquido viscoso. Rend. R. Acc. Naz. Lincei (ser. 5), **16**, 613–620, 730–737 (1907)
[Bol90]	Bolibrukh, A.A.: Riemann–Hilbert problem. Uspekhi mat. nauk. **45**(2) (272), 3–47 (1990) (Russian)
[Bol09]	Bolibruch, A.A.: Inverse Monodromy Problems in the Analytic Theory of Differential Equations. MTsNMO, Moscow (2009). (in Russian)

References

[Bon-et-al02]	Bonilla, B., Rivero, M., Rodriguez-Germa, L., Trujillo, J.J., Kilbas, A.A., Klimets, N.G.: Mittag-Leffler integral transform on $\mathcal{L}_{\nu,r}$-spaces. Rev. Acad. Canar. Cienc. **14**(1–2), 65–77 (2002)
[Bor01]	Borel, E.: Lecons sur la series divergentes. Gauthier-Villars, Paris (1901)
[BotBor78]	Böttcher, C.J.F., Borderwijk, P.: Theory of Electric Polarization, vol. 2. Dielectrics in Time-dependent Fields. Elsevier, New York (1978)
[BouGeo90]	Bouchaud, J.-P., Georges, A.: Anomalous diffusion in disordered media: statistical mechanisms, models and physical applications. Phys. Rep. **195**, 127–293 (1990)
[BoSiVa19]	Boudabsa, L., Simon, T., Vallois, P.: Fractional extreme distribution, arXiv:1908.00584v1 [math.PR] 1 Aug 2019
[Bouss85]	Boussinsesq, J.: Sur la résistance qu'oppose un liquid indéfini en repos, san pesanteur, au mouvement varié d'une sphère solide qu'il mouille sur toute sa surface, quand les vitesses restent bien continues et assez faibles pour que leurs carrés et produits soient négligeables. C.R. Acad. Paris, **100**, 935–937 (1885)
[Bra62]	Braaksma, B.L.J.: Asymptotic expansions and analytical continuations for a class of Barnes-integrals. Compositio Mathematica, **15**, 239–341 (1962–1964)
[Bra96]	Brankov, J.G.: Introduction to Finite-Size Scaling. Leuven University Press, Leuven (1996)
[BraTon92]	Brankov, J.G., Tonchev, N.S.: Finite-size scaling for systems with long-range interactions. Phys. A. **189**, 583–610 (1992)
[Bre65]	Bremermann, H.: Distributions, Complex Variables and Fourier Transforms. Addison-Wesley, Reading (1965)
[BrHaJe48]	Brockmeyer, E., Halstrøm, H.L., Jensen, A.: The Life and Works of A.K. Erlang, Transactions of the Danish Academy of Technical Sciences (The Copenhagen Telephone Company) No 2, Copenhagen (1948)
[Bro09]	Bromwich, T.J.: An asymptotic formula for generalized hypergeometric series. Proc. Lon. Math. Soc. **72**(2), 101–106 (1909)
[Bro26]	Bromwich, T.J.I.: An Introduction to the Theory of Infinite Series. American Mathematical Society Chelsea Publish (1926)
[BrMaSa19]	Brychkov, YuA., Marichev, O.I., Savischenko, N.V.: Handbook of Mellin Transforms. Chapman and Hall/CRC Press, Boca Raton (2019)
[BucMai75]	Buchen, P.W., Mainardi, F.: Asymptotic expansions for transient viscoelastic waves. J. de Mécanique **14**, 597–608 (1975)
[BucLuc98]	Buckwar, E., Luchko, Yu.: Invariance of a partial differential equation of fractional order under the Lie group of scaling transformations. J. Math. Anal. Appl. **227**, 81–97 (1998)
[Bus72]	Buschman, R.G.: Contiguous relations and related formulas for H-function of Fox. Jñānābha Sect. A **2**, 39–47 (1972)
[Buh25a]	Buhl, A.: Sommabilité et fonction $E_\alpha(x)$. Enseignment **24**, 69–76 (1925)
[Buh25b]	Buhl, A.: Séries Analytiques. Acad. Sci. Paris, Fasc. VII, Gauthier-Villars, Paris, Sommabilité, Mémorial des Sciences Mathématiques (1925)
[BuKiTr02a]	Butzer, P.L., Kilbas, A.A., Trujillo, J.J.: Fractional calculus in the Mellin settings and Hadamard-type fractional integrals. J. Math. Anal. Appl. **269**, 1–27 (2002)
[BuKiTr03]	Butzer, P.L., Kilbas, A.A., Trujillo, J.J.: Generalized Stirling functions of second kind and representations of fractional order differences via derivatives. J. Diff. Equat. Appl. **9**(5), 503–533 (2003)
[ButWes75]	Butzer, P., Westphal, U.: An access to fractional differentiation via fractional difference quotients. In: Ross, B. (ed.) Fractional Calculus and Its Applications. Lecture Notes in Mathematics, vol. 457, pp. 116–145. Springer, Berlin (1975)
[ButWes00]	Butzer, P., Westphal, U.: Introduction to fractional calculus. In: Hilfer, R. (ed.) Fractional Calculus, pp. 1–85. Applications in Physics. World Scientific, Singapore (2000)
[CaUcWo10]	Cahoy, D.O., Uchaikin, V.V., Woyczyński, W.A.: Parameter estimation for fractional Poisson processes. J. Stat. Plann. Infer. **140**, 3106–3120 (2010)

[CahWoy18] Cahoy, D.O., Woyczynski, W.A.: Log-moment estimators for the generalized linnik and mittag-leffler distributions with applications to financial modeling. J. Math. Stat. **14**, 156–166 (2018)

[Came09] Cametti, C.: Dielectric and conductometric properties of highly heterogeneous colloidal systems. Rivista del Nuovo Cimento **32**(5), 185–260 (2009)

[Cam90] Campos, L.M.: On the solution of some simple fractional differential equations. Int. J. Math. Math. Sci. **13**(3), 481–496 (1990)

[Cap04] Capelas de Oliveira, E.: Special Functions with Applications, Editora Livraria da Fisica, Sao Paulo, Brazil (2004) [in Portuguese]

[Cap13] Capelas de Oliveira, E.: Capelas's relations. Personal Communication (2013)

[Cap19] Capelas de Oliveira, E.: Solved Exercises in Fractional Calculus. Springer Nature, Cham (2019)

[CdOMai11] Capelas de Oliveira, E., Mainardi, F., Vaz, J. Jr.: Models based on Mittag-Leffler functions for anomalous relaxation in dielectrics. Eur. Phys. J.-Special Topics **193**, 161–171 (2011) [Revised Version as E-print arXiv:1106.1761v2]

[CdOMai14] Capelas de Oliveira, E., Mainardi, F., Vaz, J. Jr.: Fractional models of anomalous relaxation based on the Kilbas and Saigo function. Meccanica, **49**(9), 2049–2060 (2014)

[Cap-et-al10] Caponetto, R., Dongola, G., Fortuna, L., Petráš, I.: Fractional Order Systems: Modeling and Control Applications. World Scientific, Singapore (2010)

[Cap66] Caputo, M.: Linear models of dissipation whose Q is almost frequency independent. Annali di Geofisica **19**, 383–393 (1966)

[Cap67] Caputo, M.: Linear models of dissipation whose Q is almost frequency independent: Part II. Geophys. J. R. Astr. Soc. **13**, 529–539 (1967)

[Cap69] Caputo, M.: Elasticità e Dissipazione. Zanichelli, Bologna (1969)

[CapMai71a] Caputo, M., Mainardi, F.: Linear models of dissipation in anelastic solids. Riv. Nuovo Cimento (Ser. II). **1**, 161–198 (1971)

[CapMai71b] Caputo, M., Mainardi, F.: A new dissipation model based on memory mechanism, Pure Appl. Geophys. (PAGEOPH), **91**, 134–147 (1971) [Reprinted in Fract. Calc. Appl. Anal. **10**(3), 309–324 (2007)]

[CarMai97] Carpinteri, A., Mainardi, F. (Eds): Fractals and Fractional Calculus in Continuum Mechanics. Springer, Wien and New York (1997) (vol. 378, Series CISM Courses and Lecture Notes.)

[CeLuDo18] Cerutti, R.A., Luque, L.L., Dorrego, G.A.: On the p-k-Mittag-Leffler function. Appl. Math. Sci. **11**(51), 2541–2560 (2017)

[ChaTon06] Chamati, H., Tonchev, N.S.: Generalized Mittag-Leffler functions in the theory of finite-size scaling for systems with strong anisotropy and/or long-range interaction. J. Phys. A: Math. Gen. **39**(3), 469–478 (2006)

[ChGoSo02] Chechkin, A.V., Gorenflo, R. and Sokolov, I.M.: Retarding subdiffusion and accelerating superdiffusion governed by distributed-order fractional diffusion equations. Phys. Rev. E, **66**, 046129/1–6 (2002)

[Chr82] Christensen, R.M.: Theory of Viscoelasticity. Academic, New York (1982) [1-st ed. 1971, 2-nd ed. 1982]

[ClShNe09] Clauset, A., Shalizi, C.R., Newman, M.E.J.: Power-Law distributions in empirical data. SIAM Rev. **51**(4), 661–703 (2009)

[CoKaTi02] Coffey, W.T., Kalmykov, YuP, Titov, S.V.: Anomalous dielectric relaxation in the context of the Debye model of noninertial rotational diffusion. J. Chem. Phys. **116**(15), 6422–6426 (2002)

[Col33] Cole, K.S.: Electrical conductance of biological systems. Electrical excitation in nerves. In: Proceedings Symposium on Quantitative Biololgy, Cold Spring Harbor, New York, vol. 1, pp. 107–116 (1933)

[ColCol41] Cole, K.S., Cole, R.H.: Dispersion and absorption in dielectrics. I. Alternating current characteristics. J. Chem. Phys. **9**, 341–351 (1941)

References

[ColCol42]	Cole, K.S., Cole, R.H.: Dispersion and absorption in dielectrics. II. Direct current characteristics. J. Chem. Phys. **10**, 98–105 (1942)
[Col-et-al16]	Colombaro, I., Giusti, A., Mainardi, F.: A class of linear viscoelastic models based on Bessel functions. Meccanica **52**(4–5), 825–832 (2016)
[Col-et-al18]	Colombaro, I., Garra, R., Giusti, A., Mainardi, F.: Scott-Blair models with time-varying viscosity. Appl. Math. Lett. **86**, 57–63 (2018)
[ConMax62]	Conway, R.W., Maxwell, W.L.: A queueing model with state dependent service rate. J. Industr. Engn. **XII**(2), 132–136 (1962)
[Cox67]	Cox, D.R.: Renewal Theory, 2nd edn. Methuen, London (1967)
[CraBro86]	Craig, J.D., Brown, J.C.: Inverse Problems in Astronomy. Adam Hilger Ltd, Bristol (1986)
[Dau80]	Dauben, J.W.: Mathematicians and World War I: The international diplomacy of G.H. Hardy and Gösta Mittag-Leffler as reflected in their personal correspondence. Historia Math. **7**, 261–288 (1980)
[DavCol50]	Davidson, D.W., Cole, R.H.: Dielectric relaxation in glycerine. J. Chem. Phys. **18**(10), 1417, Letter to the Editor (1950)
[DavCol51]	Davidson, D.W., Cole, R.H.: Dielectric relaxation in glycerol, propylene glycol, and n-propanol. J. Chem. Phys. **19**(12), 1484–1490 (1951)
[Dav02]	Davies, B.J.: Integral Transforms and Their Applications, 3rd edn. Springer, Berlin (2002)
[DavHig03]	Davies, P.I., Higham, N.J.: A Schur–Parlett algorithm for computing matrix functions. SIAM J. Matrix Anal. Appl. **25**(2), 464–485 (electronic) (2003)
[Dav59]	Davis, P.J.: Leonard Euler's integral: a historical profile of the Gamma function. Amer. Math. Mon. **66**, 849–869 (1959)
[Dav36]	Davis, H.T.: The Theory of Linear Operators. The Principia Press, Bloomington (1936)
[Deb12]	Debye, P.: Zur theorie der spezifischen Wärme. Annalen der Physik **39**, 789–839 (1912)
[DebBha07]	Debnath, L., Bhatta, D.: Integral Transforms and Their Applications, 2nd edn. Chapman & Hall/CRC, Boca Raton (2007)
[DebBha15]	Debnath, L., Bhatta, D.: Integral Transforms and Their Applications, 3rd edn. Chapman & Hall/CRC, Boca Raton (2015)
[Del94]	Delbosco, D.: Fractional calculus and function spaces. J. Fract. Calc. **6**, 45–53 (1994)
[Den07]	Deng, W.H.: Numerical algorithm for the time fractional Fokker-Planck equation. J. Comput. Phys. **227**, 1510–1522 (2007)
[Die10]	Diethelm, K.: The Analysis of Differential Equations of Fractional Order: An Application-Oriented Exposition Using Differential Operators of Caputo Type. Lecture Notes in Mathematics, vol. 2004. Springer, Berlin (2010)
[Die19]	Diethelm, K.: General theory of Caputo-type fractional differential equations, In: *Handbook of Fractional Calculus with Applications* (J. Tenreiro Machado Ed.), **2** Fractional Differential equations. (A. Kochubei, Yu. Luchko eds.), Berlin-Boston, De Gruyter, 1–20 (2019)
[DieFor02]	Diethelm, K., Ford, N.J.: Analysis of fractional differential equations. J. Math. Anal. Appl. **265**, 229–248 (2002)
[Die-et-al05]	Diethelm, K., Ford, N.J., Freed, A.D., Luchko, Yu.: Algorithms for the fractional calculus: a selection of numerical methods. Comput. Methods Appl. Mech. Engrg. **194**, 743–773 (2005)
[DitPru65]	Ditkin, V.A., Prudnikov, A.P.: Integral Transforms and Operational Calculus. Pergamon Press, Oxford (1965)
[Doe74]	Doetsch, G.: Introduction to the Theory and Applications of the Laplace Transformation. Springer, New York (1974)
[DokMac09]	Dokoumetzidis, A., Macheras, P.: Fractional kinetics in drug absorption and disposition processes. J Pharmacokinet. Pharmacodyn. **36**, 165–178 (2009)

[Dua18]	Junsheng, D.: A generalization of the Mittag-Leffler function and solution of system of fractional differential equations. Advances in Difference Equations, 2018:239, 12 p. (2018)
[Dur83]	Duren, P.L.: Univalent functions, vol. 259. Grundlehren der Mathematischen Wissenschaften. Springer, New York (1983)
[Dzh54a]	Dzherbashian [=Djrbashian], M.M.: On integral representation of functions continuous on given rays (generalization of the Fourier integrals). Izvestija Akad. Nauk SSSR, Ser. mat. **18**, 427–448 (1954) (in Russian)
[Dzh54b]	Dzherbashian [=Djrbashian], M.M.: On the asymptotic expansion of a function of Mittag-Leffler type. Akad. Nauk Armjan. SSR, Doklady. **19**, 65–72 (1954) (in Russian)
[Dzh54c]	Dzherbashian [=Djrbashian], M.M.: On Abelian summation of the generalized integral transform. Akad. Nauk Armjan. SSR, Izvestija, fiz-mat. estest. techn. nauki. **7**(6, 1–26 (1954) (in Russian)
[Dzh60]	Dzherbashian [=Djrbashian], M.M.: On integral transforms generated by the generalized Mittag-Leffler function. Izv. Akad. Nauk Armjan. SSR. **13**(3), 21–63 (1960) (in Russian)
[Dzh66]	Dzherbashian [=Djrbashian], M.M.: Integral Transforms and Representation of Functions in the Complex Domain. Nauka, Moscow (1966) (in Russian)
[Dzh70]	Dzherbashian [=Djrbashian], M.M.: A boundary value problem for a Sturm–Liouville type differential operator of fractional order. (Russian) Izv. Akad. Nauk Armyan. SSR, Ser. Mat. **5**(2), 71–96 (1970)
[Dzh84]	Dzherbashian [=Djrbashian], M.M.: Interpolation and spectral expansions associated with differential operators of fractional order. (English. Russian original) Sov. J. Contemp. Math. Anal., Arm. Acad. Sci. **19**(2), 116 p. (1984), translation from Izv. Akad. Nauk Arm. SSR, Mat. **19**(2), 81–181 (1984)
[Djr93]	Djrbashian, M.M.: Harmonic Analysis and Boundary Value Problems in the Complex Domain. Birkhäuser Verlag, Basel (1993)
[DjrBag75]	Djrbashian, M.M., Bagian, R.A.: On integral representations and measures associated with Mittag-Leffler type functions. Izv. Akad. Nauk Armjanskoy SSR, Matematika **10**, 483–508 (1975) (in Russian)
[DzhNer68]	Dzherbashian [=Djrbashian], M.M., Nersesian, A.B.: Fractional derivatives and the Cauchy problem for differential equations of fractional order. Izv. Acad. Nauk Armjanskvy SSR, Matematika **3**(1), 3–29 (1968) (in Russian)
[DorCer12]	Dorrego, G., Cerutti, R.: The k-Mittag-Leffler Function. J. Contemp. Math. Sci. **7**, 705–716 (2012)
[D'OPol17]	D'Ovidio, M., Polito, F.: Fractional diffusion-telegraph equations and their associated stochastic solutions. Teor. Veroyatnost. i Primenen. **62**(4), 692–718 (2017)
[EmKlMi01]	Embrechts, P., Klüppelberg, C., Mikosch, T.: Modelling Extreme Events for Insurance and Finance. Springer, Berlin (2001)
[ErdBat-1]	Erdelyi, A., Magnus, W., Oberhettinger, F., Tricomi, F.G.: Higher Transcendental Functions, vol. 1. McGraw-Hill, New York (1953)
[ErdBat-2]	Erdelyi, A., Magnus, W., Oberhettinger, F., Tricomi, F.G.: Higher Transcendental Functions, vol. 2. McGraw Hill, New York (1954)
[ErdBat-3]	Erdelyi, A., Magnus, W., Oberhettinger, F., Tricomi, F.G.: Higher Transcendental Functions, vol. 3. McGraw-Hill, New York (1955)
[EshAns16]	Eshaghi, S., Ansari, A.: Lyapunov inequality for fractional differential equations with Prabhakar derivative. Math. Inequal. Appl. **19**(1), 349–358 (2016)
[EvaLen18]	Evangelista, L.R., Lenzi, E.K.: Fractional Diffusion Equations and Anomalous Diffusion. Cambridge University Press, Cambridge (2018)
[Evg69]	Evgrafov, M.A., et al.: Collection of Exercises on the Theory of Analytic Functions. Nauka, Moscow (1969). (in Russian)
[Evg78]	Evgrafov, M.A.: Asymptotic Estimates and Entire Functions. Nauka, Moscow (1978) (2nd edition) (in Russian)

[Fej36]	Féjer, L.: Untersuchungen ber Potenzreihen mit mehrfach monotoner Koeffizientenfolge. Acta Literarum Sci. **8**, 89–115 (1936)
[FePuRy05]	Feldman, Y., Puzenko, A., Ryabov, Y.: Dielectric relaxation phenomena in complex materials. In: Coffey, W.T., Kalmykov, Y.P. (eds.) Fractals, Diffusion, and Relaxation in Disordered Complex Systems. Special Volume of Advances in Chemical Physics, vol. 133, pp. 1–125. Part A. Wiley, New York (2005)
[Fel49]	Feller, W.: Fluctuation theory of recurrent events. Trans. Amer. Math. Soc. **67**, 98–119 (1949)
[Fel52]	Feller, W.: On a generalization of Marcel Riesz' potentials and the semi-groups generated by them. Meddelanden Lunds Universitets Matematiska Seminarium (Comm. Sém. Mathém. Université de Lund), Tome suppl. dédié à M. Riesz, Lund, pp. 73–81 (1952)
[Fel71]	Feller, W.: An Introduction to Probability Theory and its Applications, vol. **2**, 2nd edn. Wiley, New York (1971)
[FiCaVa12]	Figueiredo Camargo, R., Capelas de Oliveira, E., Vaz J. Jr.: On the generalized Mittag-Leffler function and its application in a fractional telegraph equation. Math. Phys. Anal. Geom. **15**(1), 1–16 (2012)
[Fog94]	Fogedby, H.C.: Langevin equations for continuous time Lévy flights. Phys. Rev. E **50**, 1657–1660 (1994)
[ForCon06]	Ford, N.J., Connolly, J.A.: Comparison of numerical methods for fractional differential equations. Commun. Pure Appl. Anal. **5**, 289–307 (2006)
[FosSch89]	Foster, K.R., Schwan, H.P.: Dielectric properties of tissues and biological materials: a critical review. Crit. Rev. Biomed. Eng. **17**(1), 25–104 (1989)
[Fou22]	Fourier, J.B.J.: Théorie Analitique de la Chaleure. Didot, Paris (1822)
[Fox28]	Fox, C.: The asymptotic expansion of generalized hypergeometric functions. Proc. London Math. Soc., ser. 2. **27**, 389–400 (1928)
[Fox61]	Fox, C.: The G and H functions as symmetrical Fourier kernels. Trans. Amer. Math. Soc. **98**, 395–429 (1961)
[FrDiLu02]	Freed, A., Diethelm, K., Luchko, Y.: Fractional-order viscoelasticity (FOV): constitutive development using the fractional calculus. First Annual Report, NASA/TM-2002-211914, Gleen Research Center , pp. I-XIV, 1–121 (2002)
[Fuj90a]	Fujita, Y.: Integrodifferential equation which interpolates the heat and the wave equations. I. II. Osaka J. Math. **27**(309–321), 797–804 (1990)
[Fuj90b]	Fujita, Y.: Cauchy problems of fractional order and stable processes. Japan J. Appl. Math. **7**(3), 459–476 (1990)
[Fuj33]	Fujiwara, M.: On the integration and differentiation of an arbitrary order. Tohoku Math. J. **37**, 110–121 (1933)
[GajSta76]	Gajić, L.J., Stanković, B.: Some properties of Wright's function. Publ. de l'Institut Mathèmatique, Beograd, Nouvelle Sèr. **20**(34), 91–98 (1976)
[GaMaKa13]	Garg, M., Manohar, P., Kalla, S.L.: A Mittag-Leffler function of two variables. Integr. Transf. Spec. Funct. **24**(11), 934–944 (2013)
[GaShMa15]	Garg, M., Sharma, A., Manohar, P.: A Generalized Mittag-Leffler type function with four parameters. Thai J. Math. **14**(3), 637–649 (2016)
[GarGar18]	Garra, R., Garrappa, R.: The Prabhakar or three parameter Mittag-Leffler function: theory and application. Commun. Nonlinear Sci. Numer. Simulat. **56**, 314–329 (2018)
[Gar-et-al14a]	Garra, R., Gorenflo, R., Polito, F., Tomovski, Ž.: Hilfer-Prabhakar derivatives and some applications. Appl. Math. Comput. **242**, 576–589 (2014)
[GarPol13]	Garra, R., Polito, F.: On some operators involving Hadamard derivatives. Int. Trans. Spec. Funct. **24**(10), 773–782 (2013)
[Gar15]	Garrappa, R.: Numerical evaluation of two and three parameter Mittag-Leffler functions. SIAM J. Numer. Anal. **53**(3), 1350–1369 (2015)
[GarMai16]	Garrappa, R., Mainardi, F.: On Volterra functions and Ramanujan integrals. Analysis **36**(2), 89–105 (2016)

[GaMaMa16] Garrappa, R., Mainardi, F., Maione, G.: Models of dielectric relaxation based on completely monotone functions. Fract. Calc. Appl. Anal. **19**(5), 1105–1160 (2016)

[GarPop13] Garrappa, R., Popolizio, M.: Evaluation of generalized Mittag-Leffler functions on the real line. Adv. Comput. Math. **39**(1), 205–225 (2013)

[GarPop18] Garrappa, R., Popolizio, M.: Computing the matrix Mittag-Leffler function with applications to fractional calculus. J. Sci. Comput. **77**(1), 129–153 (2018)

[GaRoMa17] Garrappa, R., Rogosin, S., Mainardi, F.: On a generalized three-parameter Wright function of Le Roy type. Fract. Calc. Appl. Anal. **20**(5), 1196–1215 (2017)

[Gat73] Gatteschi, L.: Funzioni Speciali. UTET, Torino (1973)

[Gauss] Gauss, C.F.: Werke, Bd. III. Analysis (various texts, in Latin and German, orig. publ. between 1799–1851, or found in the "Nachlass", annotated by E.J. Schering). Dieterichschen Universitäts-Druckerei W.Fr. Kaestner (1866)

[Gaw84] Gawronski, W.: On the bell-shape of stable distributions. Ann. Probab. **12**, 230–242 (1984)

[Gej14] Gejji, V.D. (ed.): Fractional Calculus: Theory and Applications. Narosa Publishing House, New Delhi (2014)

[GejKum17] Gejji, V.D., Kumar, M.: New Iterative Method: A Review. Chapter 9 in Frontiers in Fractional Calculus, 2017 (S. Bhalekar (ed.)). Bentham Science Publishers (2017)

[GelShi64] Gel'fand, I.M., Shilov, G.E.: Generalized Functions, vol. **1**. Academic, New York (1964) [English translation from the Russian (Nauka, Moscow, 1959)]

[Gem36] Gemant, A.: A method of analyzing experimental results obtained from elastiviscous bodies. Physics **7**, 311–317 (1936)

[Gem38] Gemant, A.: On fractional differentials. Phil. Mag. (Ser. 7) **25**, 540–549 (1938)

[Gem42] Gemant, A.: Frictional phenomena: VIII. J. Appl. Phys. **13**, 210–221 (1942)

[Gem50] Gemant, A.: *Frictional Phenomena*. Chemical Public Co, Brooklyn (1950)

[Ger48] Gerasimov, A.N.: A generalization of linear laws of deformation and its application to internal friction problem. Akad. Nauk SSSR. Prikl. Mat. Mekh. **12**, 251–260 (1948). (in Russian)

[Ger12] Gerhold, S.: Asymptotics for a variant of the Mittag-Leffler function. Int. Trans. Spec. Func. **23**(6), 397–403 (2012)

[Giu17] Giusti, A.: On infinite order differential operators in fractional viscoelasticity. Fract. Calc. Appl. Anal. **20**(4), 854–867 (2017)

[GiuCol18] Giusti, A., Colombaro, I.: Prabhakar-like fractional viscoelasticity. Commun. Nonlinear Sci. Numer. Simul. **56**, 138–143 (2018)

[Giu-et-al20] Giusti, A., Colombaro, I., Garra, R., Garrappa, R., Polito, F., Popolizio, M., Mainardi, F.: A practical guide to Prabhakar fractional calculus. Fract. Calc. Appl. Anal. **23**(1), 9–54 (2020)

[GloNon93] Glöckle, W.G., Nonnenmacher, T.F.: Fox function representation of non-Debye relaxation processes. J. Stat. Phys. **71**, 741–757 (1993)

[GneKol54] Gnedenko, B.V., Kolmogorov A.N.: Limit Distributions for Sums of Independent Random Variables. Addison-Wesley, Cambridge (1954). [Translated from the Russian edition, Moscow 1949, with notes by K.L. Chung, revised (1968)]

[GneKov68] Gnedenko, B.V., Kovalenko, I.N.: *Introduction to Queueing Theory*, Israel Program for Scientific Translations, Jerusalem (1968) [Translated from the 1966 Russian edition]

[GoLeOs91] Gol'dberg, A.A., Levin, B.Ya., Ostrovskij, I.V.: Entire and meromorphic functions. *Itogi Nauki Tekh., Ser. Sovrem. Probl. Mat., Fundam. Napravleniya*, vol. 85, pp 5–186 (1991) (in Russian)

[GonRos73] Gonsovski, V.L.: Rossikhin, YuA: Stress waves in a viscoelastic medium with a singular hereditary kernel. J. Appl. Mech. Tech. Physics **14**(4), 595–597 (1973)

[Goo83] Goodman, A.W.: Univalent functions, vols. 1–2. Mariner, Tampa (1983)

[Gor96] Gorenflo, R.: Abel Integral Equations with Special Emphasis on Applications, Lectures in Mathematical Sciences, vol. 13. The University of Tokyo, Graduate School of Mathematical Sciences (1996)

[Gor98]	Gorenflo, R.: The tomato salad problem in spherical stereology. In: Rusev, P., Dimovski, I., Kiryakova, V. (eds.) Transform Methods and Special Functions, Varna 1996, pp. 132–149. Science Culture Technology, Singapore (1998)
[Gor02]	Gorenflo, R.: The tomato salad problem in spherical stereology. In: Kabanikhin, S.L., Romanov, V.G. (eds.) Ill-Posed Problems and Inverse Problems, pp. 117–134. AH Zeist, VSP (2002)
[Gor10]	Gorenflo, R.: Mittag-Leffler waiting time, power laws, rarefaction, continuous time random walk, diffusion limit. In: Pai, S.S., Sebastian, N., Nair, S.S., Joseph, D.P., Kumar, D. (eds.), Proceedings of the (Indian) National Workshop on Fractional Calculus and Statistical Distributions (November 25–27, 2009), pp. 1–22, Publ. No 41, CMS Pala/Kerala, India (2010). arXiv:1004.4413
[GorAbdR05]	Gorenflo, R., Abdel-Rehim, E.: Discrete models of time-fractional diffusion in a potential well. Fract. Calc. Appl. Anal. **8**(2), 173–200 (2005)
[GKMR]	Gorenflo, R., Kilbas, A.A., Mainardi, F., Rogosin, S.V.: Mittag-Leffler Functions. Related Topics and Applications. Springer, Berlin (2014)
[GoKiRo98]	Gorenflo, R., Kilbas, A.A., Rogosin, S.V.: On generalized Mittag-Leffler type functions. Integr. Trans. Special Funct. **7**(3–4), 215–224 (1998)
[GoLoLu02]	Gorenflo, R., Loutchko, J. and Luchko, Yu.: Computation of the Mittag-Leffler function $E_{\alpha,\beta}(z)$ and its derivative. Fract. Calc. Appl. Anal. **5**(4), 491–518 (2002). Corrections in: Fract. Calc. Appl. Anal. **6**(1), 111–112 (2003)
[GoLuMa99]	Gorenflo, R., Luchko, Yu., Mainardi, F.: Analytical properties and applications of the Wright function. Frac. Cal. Appl. Anal. **2**(4), 383–414 (1999)
[GoLuMa00]	Gorenflo, R., Luchko, Yu., Mainardi, F.: Wright functions as scale-invariant solutions of the diffusion-wave equation.J. Comput. Appl. Math. **118**(1–2), 175–191 (2000)
[GoLuRo97]	Gorenflo, R., Luchko, Yu., Rogosin, S.V.: Mittag-Leffler type functions: notes on growth properties and distribution of zeros, Preprint No. A04-97, Freie Universität Berlin. Serie A. Mathematik (1997)
[GorMai96]	Gorenflo, R., Mainardi, F.: Fractional oscillations and Mittag-Leffler functions, In: University Kuwait, D.M.C.S. (ed.) International Workshop on the Recent Advances in Applied Mathematics (Kuwait, RAAM'96), Kuwait, pp. 193–208 (1996)
[GorMai97]	Gorenflo, R., Mainardi, F.: Fractional calculus: integral and differential equations of fractional order. In: Carpinteri, A., Mainardi, F. (eds.) Fractals and Fractional Calculus in Continuum Mechanics, pp. 223–276. Springer, Wien (1997)
[GorMai98]	Gorenflo, R., Mainardi, F.: Random walk models for space-fractional diffusion processes. Fract. Calc. Appl. Anal. **1**, 167–191 (1998)
[GorMai98a]	Gorenflo, R., Mainardi, F.: Fractional calculus and stable probability distributions. Arch. Mech. **50**, 377–388 (1998)
[GorMai03]	Gorenflo, R., Mainardi, F.: Fractional diffusion processes: probability distributions and continuous time random walk. In: Rangarajan, G., Ding, M. (eds.) Processes with Long Range Correlations, pp. 148–166. Springer, Berlin (2003) [Lecture Notes in Physics, No. 621]
[GorMai08]	Gorenflo, R., Mainardi, F.: Continuous time random walk, Mittag-Leffler waiting time and fractional diffusion: mathematical aspects. In: Klages, R., Radons, G., Sokolov, I.M. (eds.) Anomalous Transport, Foundations and Applications, pp. 93–127. Wiley-VCH Verlag Gmbh & Co., KGaA, Weinheim, Germany (2008)
[GorMai11]	Gorenflo, R., Mainardi, F.: Subordination pathways to fractional diffusion. Eur. Phys. J. Special Topics **193**, 119–132 (2011). arXiv:1104.4041
[GorMai12a]	Gorenflo, R., Mainardi, F.: Parametric subordination in fractional diffusion processes. In: Klafter, J., Lim, S.C., Metzler, R. (eds.) Fractional Dynamics. World Scientific, Singapore (2012), Chapter 10, pp. 229–263. arXiv:1210.8414
[GorMai12b]	Gorenflo, R., Mainardi, F.: Laplace-Laplace analysis of the fractional Poisson process. In: Rogosin, S. (ed.) Analytical Methods of Analysis and Differential

	Equations. AMADE 2011, pp. 43–58. Belarusan State University, Minsk (2012). [Kilbas Memorial Volume]
[GorMai15]	Gorenflo, R., Mainardi, F.: On the fractional Poisson process and the discretized stable subordinator. Axioms **4**, 321–344 (2015). arXiv:1305.3074
[GorMai19]	Gorenflo, R., Mainardi, F.: Fractional Relaxation-Oscillation Phenomena. In: Tarasov, V.E. (ed.) Handbook of Fractional Calculus with Applications, Applications in Physics, Part A, vol. 4, pp. 45–74. De Gruyter, Berlin (2019)
[GoMaRo19]	Gorenflo, R., Mainardi, F., Rogosin, S.: Mittag-Leffler function: properties and applications. In: Kochubei, A., Luchko, Yu. (eds.) Handbook of Fractional Calculus with Applications, vol. 1, pp. 269–296. De Gruyter, Berlin (2019)
[Gor-et-al01]	Gorenflo, R., Mainardi, F., Scalas, E., Raberto, M.: Fractional calculus and continuous-time finance III: the diffusion limit. In: Kohlmann, M., Tang, S. (eds.) Trends in Mathematics - Mathematical Finance, pp. 171–180. Birkhäuser, Basel (2001)
[GoMaSr98]	Gorenflo, R., Mainardi, F., Srivastava, H.M.: Special functions in fractional relaxation-oscillation and fractional diffusion-wave phenomena. In: The Eighth International Colloquium on Differential Equations (Plovdiv, 1997), pp. 195–202. VSP Publishing Company, Utrecht (1998)
[GoMaVi07]	Gorenflo, R., Mainardi, F., Vivoli, A.: Continuous-time random walk and parametric suboredination in fractional diffusion. Chaos Solitons Fractals **34**, 87–103 (2007)
[GorRut94]	Gorenflo, R., Rutman, R.: On ultraslow and intermediate processes. In: Rusev, P., Dimovski, I., Kiryakova, V. (eds.) Transform Methods and Special Functions, Sofia 1994, pp. 171–183. Science Culture Technology, Singapore (1994)
[GorVes91]	Gorenflo, R., Vessella, S.: Abel Integral Equations: Analysis and Applications. Springer, Berlin (1991)
[GorViv03]	Gorenflo, R., Vivoli, A.: Fully discrete random walks for space-time fractional diffusion equations. Signal Process. **83**, 2411–2420 (2003)
[Go-et-al16]	Górska, K., Horzela, A., Bratek, L., Penson, K.A., Dattoli, G.: The probability density function for the Havriliak-Negami relaxation (2016). arXiv:1611.06433. Accessed 19 Nov 2016
[GoHoGa19]	Górska, K., Horzela, A., Garrappa, R.: Some results on the complete monotonicity of the Mittag-Leffler functions of Le Roy type. Fract. Calc. Appl. Anal. **22**(5), 1284–1306 (2019)
[Gov94]	Govorov, N.V.: Riemann's Boundary Value Problem with Infinite Index. Birkhäuser Verlag, Basel (1994)
[Gou83a]	Goursat, E.: Sur les fonctions hypergéométriques d'ordre supérieu. Comptes Rendus. Académie des Sciences, Paris **96**, 185–188 (1883) [Séance du 15 Janvier 1883]
[Gou83b]	Goursat, E.: Mémoire sur les fonctions hypergéométriques d'ordre supérieur, Ann. Sci. École Norm. Sup. (Ser 2) **12**, 261–286, 395–430 (1883)
[GraRyz00]	Gradshteyn, I.S., Ryzhik, I.M. (ed. A. Jeffrey): Table of Integrals, Series, and Products, Academic, New York (2000)
[GraCso06]	Graven, T., Csordas, G.: The Fox's Wright functions and Laguerre multiplier sequences. J. Math. Anal. Appl. **314**, 109–125 (2006)
[GrLoSt90]	Gripenberg, G., Londen, S.O., Staffans, O.J.: Volterra Integral and Functional Equations. Cambridge University Press, Cambridge (1990)
[Gro37]	Gross, B.: Über die anomalien der festen dielektrika. Zeitschrift für Physik A: Hadrons and Nuclei **107**, 217–234 (1937)
[Gro38]	Gross, B.: Zum verlauf des einsatzstromes im anomalen dielektrikum. Zeitschrift für Physik A: Hadrons and Nuclei **108**, 598–608 (1938)
[Gro41]	Gross, B.: On the theory of dielectric loss. Phys. Rev. (Series I) **59**, 748–750 (1941)
[Gro47]	Gross, B.: On creep and relaxation. J. Appl. Phys. **18**, 212–221 (1947)
[Gro48]	Gross, B.: On creep and relaxation II. J. Appl. Phys. **19**, 257–264 (1948)

References

[Gro53] Gross, B.: Mathematical Structure of the Theories of Viscoelasticity. Hermann, Paris (1953)

[Gru67] Grünwald, A.K.: Über "begrenzte" Derivation und deren Anwendung. Z. angew. Math. und Phys. **12**, 441–480 (1867)

[GunKuz58] Günter, N.M., Kuz'min, R.O.: Collection of Exercises on Higher Mathematics. Phyzmatgiz, Moscow (1958). (in Russian)

[GupDeb07] Gupta, I.S., Debnath, L.: Some properties of the Mittag-Leffler functions. Integr. Trans. Special Func. **18**(5), 329–336 (2007)

[HAND1] Tenreiro Machado J.A. (ed.): Handbook of Fractional Calculus with Applications, vol. 8. In: Kochubei, A., Luchko , Yu. (eds.) Basic Theory, vol. 1. De Gruyter, Berlin (2019)

[HAND2] Tenreiro Machado, J.A. (ed.): Handbook of Fractional Calculus with Applications, vol. 8. In: Kochubei, A., Luchko, Yu. (eds.) Fractional Differential Equations, vol. 2. De Gruyter, Berlin/Boston (2019)

[HAND3] Tenreiro Machado, J.A. (ed.): Handbook of Fractional Calculus with Applications, vol. 8. In: Karniadakis, G. (ed.) Numerical Methods, vol. 3. De Gruyter, Berlin (2019)

[HAND4] Tenreiro Machado, J.A. (ed.): Handbook of Fractional Calculus with Applications in 8 Volumes. In: Tarasov, V. (ed.) Fractional Differential Equations. De Gruyter, Berlin (2019)

[HAND5] Tenreiro Machado, J.A. (ed.): Handbook of Fractional Calculus with Applications, vol. 8. In: Tarasov, V. (ed.) Fractional Differential Equations, vol. 5. De Gruyter, Berlin/Boston (2019)

[HAND6] Tenreiro Machado, J.A. (ed.): Handbook of Fractional Calculus with Applications, vol. 8. In: Petráš, I. (ed.) Applications in Control, vol. 6. De Gruyter, Berlin (2019)

[HAND7] Tenreiro Machado, J.A. (ed.): Handbook of Fractional Calculus with Applications, vol. 8. In: Baleanu, D., Lopez, A.N. (eds.) Applications in Engineering, Life and Social Sciences. Part A, vol. 7. De Gruyter, Berlin (2019)

[HAND8] Tenreiro Machado, J.A. (ed.): Handbook of Fractional Calculus with Applications, vol. 8. In: Baleanu, D., Lopez, A.N. (eds.) Applications in Engineering, Life and Social Sciences. Part B, vol. 8. De Gruyter, Berlin/Boston (2019)

[Han-et-al09] Hanneken, J.W., Achar, B.N.N., Puzio, R. and Vaught, D.M.: Properties of the Mittag-Leffler function for negative α. Phys. Scripta. **T136**, 014037/1–5 (2009)

[Han05b] Hanyga, A.: Physically acceptable viscoelastic models. In: Hutter, K., Wang, Y. (eds.) Trends in Applications of Mathematics to Mechanics, Ber. Math., pp. 125–136. Shaker Verlag, Aachen (2005)

[HanSer08] Hanyga, A., Seredynska, M.: On a mathematical framework for the constitutive equations of anisotropic dielectric relaxation. J. Stat. Phys. **131**, 269–303 (2008)

[HanSer12] Hanyga, A., Seredynska, M.: Anisotropy in high-resolution diffusion-weighted MRI and anomalous diffusion. J. Magnetic Resonance **220**, 85–93 (2012)

[Har28a] Hardy, G.H.: Gösta Mittag-Leffler, 1846–1927. Proc. R. Soc. Lond. (A) **119**, V-VIII (1928)

[Har28b] Hardy, G.H.: Gösta Mittag-Leffler. J. Lond. Math. Soc. **3**, 156–160 (1928)

[Har92] Hardy, G.H.: Divergent series. AMS Chelsea, Rhode Island (1992). [reprinted from the 1949 edition]

[HaMaSa11] Haubold, H.J., Mathai, A.M., Saxena, R.K.: Mittag-Leffler Functions and Their Applications, Hindawi Publishing Corporation. J. Appl. Math. Vol. **2011**, Article ID 298628, 51 pp

[HavNeg67] Havriliak, S., Negami, S.: A complex plane representation of dielectric and mechanical relaxation processes in some polymers. Polymer **8**, 161–210 (1967)

[HavHav94] Havriliak, S. Jr., Havriliak, S.J: Results from an unbiased analysis of nearly 1000 sets of relaxation data. J. Non-Cryst. Solids 172–174, PART 1, 297–310 (1994)

[Hea93] Heaviside, O.: Operators in mathematical physics. Proc. Roy. Soc. Lond., Ser. A **52**, 504–529 (1893)

[Her18]	Hermann, R.: Fractional Calculus: An Introduction for Physicists, 3rd edn. World Scientific, Singapore (2018)
[Hig02]	Higham, N.J.: Accuracy and Stability of Numerical Algorithms, 2nd edn. Society for Industrial and Applied Mathematics (SIAM), Philadelphia (2002)
[Hig08]	Higham, N.J.: Functions of Matrices. Society for Industrial and Applied Mathematics (SIAM), Philadelphia (2008)
[Hil84]	Hilfer, R.: Stochastische Modelle für die betriebliche Planung. GBI-Verlag, Münich (1984)
[Hil00]	Hilfer, R. (ed.): Applications of Fractional Calculus in Physics. World Scientific, Singapore (2000)
[Hil02]	Hilfer, R.: Experimental evidence for fractional time evolution in glass materials. Chem. Phys. **284**(1), 399–408 (2002)
[Hil02a]	Hilfer, R.: Fitting the excess wing in the dielectric α-relaxation of propylene carbonate. J. Phys. Condens. Matter **14**(9), 2297–2301 (2002)
[Hil02b]	Hilfer, R.: H-function representations of stretched exponential relaxation and non-Debye susceptibilities in glass systems. Phys. Rev. E **65**, 061510/1–5 (2002)
[Hil02c]	Hilfer, R.: Fitting the excess wing in the dielectric α-relaxation of propylene carbonate. J. Phys. Condens. Matter **14**(9), 2297–2301 (2002)
[Hil03]	Hilfer, R.: On fractional diffusion and continuous time random walks. Phys. A **329**(1–2), 35–40 (2003)
[Hil16]	Hilfer, R.: Mathematical analysis of time flow. Analysis **36**(1), 49–64 (2016)
[Hil17]	Hilfer, R.: Composite continuous time random walks. Eur. Phys. J. B **90**, 233, 4 pages (2017) https://doi.org/10.1140/epjb/e2017-80369-y
[Hil19]	Hilfer, R.: Mathematical and physical interpretations of fractional derivatives and integrals, In: Tenreiro Machado, J. (ed.) Handbook of Fractional Calculus with Applications, vol. 1. In: Kochubei, A., Luchko, Yu. (eds.) Basic Theory, pp. 47–86. De Gruyter, Berlin (2019)
[HilAnt95]	Hilfer, H., Anton, L.: Fractional master equations and fractal time random walks. Phys. Rev. E **51**, R848–R851 (1995)
[HiLuTo09]	Hilfer, R., Luchko, Yu., Tomovski, Ž.: Operational method for the solution of fractional differential equations with generalized Riemann-Liouville fractional derivatives. Fract. Calc. Appl. Anal. **12**(3), 299–318 (2009)
[HilSey06]	Hilfer, R., Seybold, H.J.: Computation of the generalized Mittag-Leffler function and its inverse in the complex plane. Integr. Trans. Special Funct. **17**(9), 637–652 (2006)
[HilTam30]	Hille, E., Tamarkin, J.D.: On the theory of linear integral equations. Ann. Math. **31**, 479–528 (1930)
[HoJaLa18]	Ho, M.-W., James, L.F., Lau, J.W.: Gibbs partitions, Riemann–Liouville fractional operators, Mittag-Leffler functions, and fragmentations derived from stable subordinators. arXiv:1802.05352v2 [math.PR]. Accessed 19 Feb 2018
[HugGill52]	Hughes, R.R., Gilliand, E.R.: The mechanics of drops. Chem. Eng. Progress **48**, 497–504 (1952)
[Hui16]	Huillet, T.E.: On Mittag-Leffler distributions and related stochastic processes. J. Comp. Appl. Math. **296**, 181–211 (2016)
[Hum45]	Humbert, P.: Nouvelles correspondances symboliques. Bull. Sci. Mathém. (Paris, II ser.), **69**, 121–129 (1945)
[Hum53]	Humbert, P.: Quelques résultats relatifs à la fonction de Mittag-Leffler. C. R. Acad. Sci. Paris **236**, 1467–1468 (1953)
[HumAga53]	Humbert, P., Agarwal, R.P.: Sur la fonction de Mittag-Leffler et quelquenes de ses généralisationes. Bull. Sci. Math. (Ser. II). **77**, 180–185 (1953)
[HumDel53]	Humbert, P., Delerue, P.: Sur une extension à deux variables de la fonction de Mittag-Leffler. C. R. Acad. Sci. Paris **237**, 1059–1060 (1953)
[JaAgKi17]	Jain, S., Agarwal, P., Kilicman, A.: Pathway fractional integral operator associated with $3m$-parametric Mittag-Leffler functions. Int. J. Appl. Comput. Math. **4**, 115 (2018). https://doi.org/10.1007/s40819-018-0549-z

[Jam10]	James, L.F.: Lamperti-type laws. Ann. Appl. Probab. **20**(4), 1303–1340 (2010)
[Jay03]	Jayakamur, K.: Mittag-Leffler process. Math. Comput. Modell. **37**, 1427–1434 (2003)
[JayPil93]	Jayakamur, K., Pillai, R.N.: The first order autoregressive Mittag-Leffler process. J. Appl. Probab. **30**, 462–466 (1993)
[JayPil96]	Jayakamur, K., Pillai, R.N.: Characterization of Mittag-Leffler distribution. J. Appl. Stat. Sci. **4**(1), 77–82 (1996)
[JaySur03]	Jaykumar, K., Suresh, R.P.: Mittag-Leffler distribution. J. Ind. Soc. Probab. Statist. **7**, 51–71 (2003)
[JuTrWe10]	Jurlewicz, A., Trzmiel, J., Weron, K.: Two-power-law relaxation processes in complex materials. Acta Phys. Pol. B **41**(5), 1001–1008 (2010)
[Kal-et-al04]	Kalmykov, Y.P., Coffey, W.T., Crothers, D.S.F., Titov, S.V.: Microscopic models for dielectric relaxation in disordered systems. Phys. Rev. E **70**(41), 041103/1–11 (2004)
[KarTik17]	Karev, A.V., Tikhonov, I.V.: Zeros distribution of an entire function of the Mittag-Leffler type with applications in the theory of inverse problems. Chelyabinsk Phys.-Math. J. **2**(4), 430–446 (2017) (in Russian)
[KeScBe02a]	Kempfle, S., Schäfer, I., Beyer, H.: Functional calculus and a link to fractional calculus. Fract. Calc. Appl. Anal. **5**(4), 411–426 (2002)
[KhNiPo14]	Khamzin, A.A., Nigmatullin, R.R., Popov, I.I.: Justification of the empirical laws of the anomalous dielectric relaxation in the framework of the memory function formalism. Fract. Calc. Appl. Anal. **17**(1), 247–258 (2014)
[Kha-et-al18]	Khawaja, U.A., Al-Refai, M., Shchedrin, G., Carr, L.D.: High-accuracy power series solutions with arbitrarily large radius of convergence for the fractional non-linear Schrödinger-type equations. J. Phys. A: Math. Theor. **51**(23), 235201 (2018)
[Khi37]	Khintchine, A.Ya.: Zur Theorie der unbeschränkt teilbaren Verteilungsgesetze. Mat. Sbornik [*Rec. Mat. [Mat. Sbornik]*, New Series] **2**(44), No. 1, 79–119 (1937)
[Khi38]	Khintchine, A.Ya.: Limit Distributions for the Sum of Independent Random Variables. O.N.T.I., Moscow (1938), 115 pp. (in Russian)
[Khi60]	Khintchine, A.Ya.: Mathematical Methods in the Theory of Queueing. Charles Griffin, London (1960)
[KhiLev36]	Khintchine, A., Lévy, P.: Sur le lois stables. C. R. Acad. Sci Paris **202**, 374–376 (1936)
[Kil01]	Kilbas, A.A.: Hadamard-type fractional calculus. J. Korean Math. Soc. **38**(1), 1191–1204 (2001)
[Kil05]	Kilbas, A.A.: Fractional calculus of the generalized Wright function. Fract. Calc. Appl. Anal. **8**(2), 113–126 (2005)
[KilKor05]	Kilbas, A.A., Koroleva, A.A.: Generalized Mittag-Leffler function and its extension. Tr. Inst. Mat., Minsk **13**(1), 23–32 (2005) (in Russian)
[KilKor06a]	Kilbas, A.A., Koroleva, A.A.: Inversion of the integral transform with an extended generalized Mittag-Leffler function. Doklady Math. **74**(3), 805–808 (2006)
[KilKor06b]	Kilbas, A.A., Koroleva, A.A.: Integral transform with the extended generalized Mittag-Leffler function. Math. Modell. Anal. **11**(1), 161–174 (2006)
[KilKor06c]	Kilbas, A.A., Koroleva, A.A.: Extended generalized Mittag-Leffler functions as H-functions, generalized Wright functions and differentiation formulas. Vestnik of Belruasian State University, Ser. **1**(2), 53–60 (2006) (in Russian)
[KiKoRo13]	Kilbas, A.A., Koroleva, A.A., Rogosin, S.V.: Multi-parametric Mittag-Leffler functions and their extension. Fract. Calc. Appl. Anal. **16**(2), 378–404 (2013)
[KilRep10]	Kilbas, A.A., Repin, O.A.: An analog of the Tricomi problem for a mixed type equation with partial fractional derivative. Fract. Calc. Appl. Anal. **13**, 69–84 (2010)
[KilSai95a]	Kilbas, A.A., Saigo, M.: On solution of integral equations of Abel-Volterra type. Diff. Integr. Equ. **8**(5), 993–1011 (1995)

[KilSai95b]   Kilbas, A.A., Saigo, M.: Fractional integral and derivatives of Mittag-Leffler type function. Dokl. Akad. Nauk Belarusi **39**(4), 22–26 (1995) (in Russian)

[KilSai96]   Kilbas, A.A., Saigo, M.: On Mittag-Leffler type functions, fractional calculus operators and solution of integral equations. Integr. Trans. Special Funct. **4**, 355–370 (1996)

[KilSai99]   Kilbas, A.A., Saigo, M.: On the H functions. J. Appl. Math. Stoch. Anal. **12**, 191–204 (1999)

[KilSai00]   Kilbas, A.A., Saigo, M.: The solution of a class of linear differential equations via functions of the Mittag-Leffler type. Diff. Equ. **36**(2), 193–202 (2000)

[KilSai04]   Kilbas, A.A., Saigo, M.: H-Transform. Theory and Applications. Chapman and Hall/CRC, Boca Raton (2004)

[KiSaSa02]   Kilbas, A.A., Saigo, M., Saxena, R.K.: Solution of Volterra integrodifferential equations with generalized Mittag-Leffler function in the kernel. J. Integr. Equ. Appl. **14**, 377–396 (2002)

[KiSaSa04]   Kilbas, A.A., Saigo, M., Saxena, R.K.: Generalized Mittag-Leffler function and generalized fractional calculus operators. Integr. Trans. Special Funct. **15**(1), 31–49 (2004)

[KiSaTr02]   Kilbas, A., Saigo, M., Trujillo, J.J.: On the generalized Wright function. Fract. Calc. Appl. Anal. **5**(4), 437–460 (2002)

[Kil-et-al12]   Kilbas, A., Saxena, R.K., Saigo, M., Trujillo, J.J.: Series representations and asymptotic expansions of extended generalized hypergeometric function. In: Rogosin, S.V. (ed.) Analytic Methods of Analysis and Differential Equations: AMADE-2009, pp. 31–59. Cambridge Scientific Publishers, Cambridge (2012)

[KiSrTr06]   Kilbas, A.A., Srivastava, H.M., Trujillo, J.J.: Theory and Applications of Fractional Differential Equations, North-Holland Mathematics Studies 204. Elsevier, Amsterdam (2006)

[KilTit07]   Kilbas, A.A., Titioura, A.A.: Nonlinear differential equations with Marchaud-Hadamard-type fractional derivative in the weighted spaces of summable functions. Math. Model. Anal. **12**(3), 343–356 (2007)

[KilTru01]   Kilbas, A.A., Trujillo, J.J.: Differential equations of fractional order: methods, results and problems I. Appl. Anal. **78**(1–2), 153–192 (2001)

[KilTru02]   Kilbas, A.A., Trujillo, J.J.: Differential equations of fractional order: methods, results and problems II. Appl. Anal. **81**(2), 435–494 (2002)

[KilZhu08]   Kilbas, A.A., Zhukovskaya, N.V.: Solution of Euler-type non-homogeneous differential equations with three fractional derivatives. In: Kilbas, A.A., Rogosin, S.V. (eds.) Analytic Methods of Analysis and Differential Equations: AMADE-2006, pp. 111–137. Cambridge Scientific Publisher, Cottenham, Cambridge (2008)

[KilZhu09a]   Kilbas, A.A., Zhukovskaya, N.V.: Euler-type non-homogeneous differential equations with three Liouville fractional derivatives. Fract. Calc. Appl. Anal. **12**(2), 205–234 (2009)

[KilZhu09b]   Kilbas, A.A., Zhukovskaya, N.V.: Solution in closed form linear inhomogeneous Euler type equations with fractional derivatives. Doklady NAS Belarus **53**(4), 30–36 (2009) (in Russian)

[Kir94]   Kiryakova, V.: Generalized Fractional Calculus and Applications. Harlow, Longman (1994) [Pitman Research Notes in Mathematics, vol. 301]

[Kir99]   Kiryakova, V.: Multiindex Mittag-Leffler functions, related Gelfond-Leontiev operators and Laplace type integral transforms. Fract. Calc. Appl. Anal. **2**(4), 445–462 (1999)

[Kir00]   Kiryakova, V.: Multiple (multiindex) Mittag-Leffler functions and relations to generalized fractional calculus. J. Comput. Appl. Math. **118**, 241–259 (2000)

[Kir08]   Kiryakova, V.: Some special functions related to fractional calculus and fractional (non-integer) order control systems and equations. Facta Universitatis (Sci. J. of University of Nis), Series: Automatic Control and Robotics **7**(1), 79–98 (2008)

References

[Kir10a]	Kiryakova, V.: The special functions of fractional calculus as generalized fractional calculus operators of some basic functions. Comp. Math. Appl. **59**(5), 1128–1141 (2010)
[Kir10b]	Kiryakova, V.: The multi-index Mittag-Leffler functions as important class of special functions of fractional calculus. Comp. Math. Appl. **59**(5), 1885–1895 (2010)
[KirAl-S97a]	Kiryakova, V., Al-Saqabi, B.N.: Solutions of Erdelyi-Kober fractional integral, differential and differintegral equations of second type. C. R. Acad. Bulgare Sci. **50**(1), 27–30 (1997)
[KirAl-S97b]	Kiryakova, V., Al-Saqabi, B.N.: Transmutation method for solving Erdelyi-Kober fractional differintegral equations. J. Math. Anal. Appl. **211**(1), 347–364 (1997)
[KirLuc10]	Kiryakova, V., Luchko, Yu.: The multi-index Mittag-Leffler functions and their applications for solving fractional order problems in applied analysis. AIP Conf. Proc. **1301**, 597–613 (2010)
[KlMaMe84]	Klebanov, L.B., Maniya, G.M., Melamed, I.A.: Zolotarev's problem and analogous infinitely divisible and stable distributions. Theory Probab. Appl. **29**, 791–794 (1984)
[KleFri07]	Kleinhans, D., Friedrich, R.: Continuous-time random walks: simulations of continuous trajectories. Phys. Rev E **76**, 061102/1–6 (2007)
[Koc90]	Kochubei, A.N.: Fractional order diffusion. J. Diff. Equ. **26**, 485–492 (1990) [English transl. from Russian Differenzial'nye Uravnenija]
[Koc11]	Kochubei, A.N.: General fractional calculus, evolution equations, and renewal processes. Integr. Equ. Oper. Theory **71**, 583–600 (2011)
[Koc12a]	Kochubei, A.N.: Fractional-Hyperbolic Systems. Fract. Calc. Appl. Anal. **13**(4), 860–873 (2013)
[Koc12b]	Kochubei, A.N.: Fractional-parabolic systems. Potential Anal. **37**, 1–30 (2012)
[Koc19a]	Kochubei, A.N.: General fractional calculus. In: Tenreiro Machado, J. (ed.) Handbook of Fractional Calculus with Applications, vol. 1. In: Kochubei, A., Luchko, Yu. (eds.) Basic Theory, pp. 111–126. De Gruyter, Berlin (2019)
[Koc19b]	Kochubei, A.: Fractional-parabolic equations and systems. Cauchy problem, In: J. Tenreiro Machado (ed.) Handbook of Fractional Calculus with Applications, vol. 2. In: Kochubei, A., Luchko, Yu. (eds.) Fractional Differential Equations, pp. 145–158. De Gruyter, Berlin-Boston (2019)
[Koc19c]	Kochubei, A.: Fractional-hyperbolic equations and systems. Cauchy problem. In: Kochubei, A., Luchko, Yu. (eds.) Handbook of Fractional Calculus with Applications. In: Tenreiro Machado, J. (ed.) Fractional Differential Equations, vol. 2, pp. 197–222. De Gruyter, Berlin (2019)
[Koc19d]	Kochubei, A.: Equations with general fractional time derivatives – Cauchy problem, In: Tenreiro Machado, J. (ed.) Handbook of Fractional Calculus with Applications, vol. 2. In: Kochubei, A., Luchko, Yu. (eds.) Fractional Differential Equations, pp. 223–234. De Gruyter, Berlin (2019)
[KoKoSi18]	Kochubei, A.N., Kondratiev, Yu., da Silva, J.L.: From random times to fractional kinetics. arXiv:1811.10531v1 [math.PR]. Accessed 15 Nov 2018
[KocLuc19]	Kochubei, A.N., Luchko, Yu.: Basic FC operators and their properties, In: Tenreiro Machado, J. (ed.) Handbook of Fractional Calculus with Applications, vol. 1. In: Kochubei, A., Luchko, Yu. (eds.) Basic Theory, pp. 23–46 . De Gruyter, Berlin (2019)
[Koh54]	Kohlrausch, R.: Theorie des elektrischen rückstandes in der leidner flasche. Annalen der Physik und Chemie **91**(56–82), 179–213 (1854)
[KolYus96]	Kolmogorov, A.N., Yushkevich, A.P. (eds.): Mathematics of the 19th Century: Geometry, Analytic Function Theory, Birkäuser Verlag , Basel (1996) [first published in Russian by Nauka, Moscow (1981)]
[Kon67]	Konhauser, J.D.E.: Biorthogonal polynomials suggested by the Laguerre polynomials. Pacific J. Math. **21**, 303–314 (1967)

[KoGoZa18]	Korolev, V.Yu., Gorshenin, A.K., Zeifman, A.I.: On mixture representations for the generalized Linnik distribution and their applications in limit theorems. arXiv:1810.06389v1 [math.PR]. Accessed 12 Oct 2018
[Kra04]	Krantz, S.: Complex Analysis: The Geometric Viewpoint, 2nd edn. Mathematical Association of America Inc, Washington (2004)
[Kra79]	Krätzel, E.: Integral transformations of Bessel type. In: Generalized functions & operational calculus, (Proc. Conf. Varna, 1975), Bulg. Acad. Sci., Sofia, 148–165 (1979)
[Kum36]	Kummer, E.E.: Über die hypergeometrische Reihe $1 + \frac{\alpha \cdot \beta}{1 \cdot \gamma} x + \frac{\alpha(\alpha+1)\beta(\beta+1)}{1 \cdot 2 \cdot \gamma(\gamma+1)} x^2 + \frac{\alpha(\alpha+1)(\alpha+2)\beta(\beta+1)(\beta+2)}{1 \cdot 2 \cdot 3 \cdot \gamma(\gamma+1)(\gamma+2)} x^3 +$ etc. J. für die reine und angewandte Mathematik **15**, 39–83, 127–172 (1836) (in German)
[Kum37]	Kummer, E.E.: De integralibus quibusdam definitis et seriebus infinitis. J. für die reine und angewandte Mathematik **17**, 228–242 (1837) (in Latin)
[Lai93]	Laine, I.: Nevanlinna Theory and Complex Differential Equations. de Gruyter, Basel (1993)
[Lam58]	Lamperti, J.: An occupation time theorem for a class of stochastic processes. Trans. Amer. Math. Soc. **88**, 380–387 (1958)
[Las03]	Laskin, N.: Fractional Poisson processes. Comm. Nonlinear Sci. Num. Sim. **8**, 201–213 (2003)
[Las09]	Laskin, N.: Some applications of the fractional Poisson probability distribution, *Journal of Mathematical Physics* **50**, 113513/1–12 (2009)
[Lav18]	Lavault, Ch.: Integral representations and asymptotic behaviours of Mittag-Leffler type functions of two variables. Adv. Oper. Theory **3**(2), 40–48 (2018)
[LavSha65]	Lavrent'ev, M.A., Shabat, B.V.: Methods of the Theory of Functions of Complex Variable. Nauka, Moscow (1965). (in Russian)
[LeR99]	Roy, Le: É Valeurs asymptotiques de certaines séries procdant suivant les puissances entères et positives d'une variable réelle (French). Darboux Bull. **2**(24), 245–268 (1899)
[LeR00]	Le Roy, É.: Sur les séries divergentes et les fonctions définies par un développement de Taylor. Toulouse Ann. **2**(2), 317–430 (1900)
[Let68a]	Letnikov, A.V.: Theory of differentiation with an arbitrary index. Mat. Sb. **3**, 1–66 (1868). (in Russian)
[Let68b]	Letnikov, A.V.: On historical development of differentiation theory with an arbitrary index. Mat. Sb. **3**, 85–112 (1868). (in Russian)
[LetChe11]	Letnikov, A.V., Chernykh, V.A.: The Foundation of Fractional Calculus. Neftegaz, Moscow (2011). (in Russian)
[Lev56]	Levin, B.Ya.: Distribution of Zeros of Entire Functions. AMS, Rhode Island (1980) (2nd printing) [first published in Russian by Nauka, Moscow (1956)]
[Levy23c]	Lévy, P.: Sur une opération fonctionnelle généralisant la dérivation d'ordre non entier. Compt. Rendus Acad. Sci. Paris **176**, 1441–1444 (1923)
[Levy37]	Lévy, P.: *Théorie de l'Addition des Variables Aléatoires*, 2nd edn. Gauthier-Villars, Paris (1954) [1st edn, 1937],
[LiSaKa17]	Liemert, A., Sandev, T., Kantz, H.: Generalized Langevin equation with tempered memory kernel. Phys. A **466**, 356–369 (2017)
[Lin10]	Lin, Z.: On the FDTD formulations for biological tissues with Cole-Cole dispersion. IEEE Microw. Compon. Lett. **20**(5), 244–246 (2010)
[Lio32a]	Liouville, J.: Mémoire sur quelques questiones de géométrie et de mécanique, et nouveau gendre de calcul pour résoudre ces questiones. J. l'Ecole Roy. Polytéchn. **13**(21), 1–69 (1832)
[Lio32b]	Liouville, J.: Mémoire sur le calcul des différentielles a indices quelconques. J. l'Ecole Roy. Polytéchn. **13**(21), 71–162 (1832)
[Lin03]	Lindelöf, E.: Sur la détermination de la croissance des fonctions entières définies par un developpment de Taylor. Bull. des Sciences Mathématigues, (ser. II). **27**, 213–226 (1903)

References

[LovBra95] Lovalenti, P.M., Brady, J.F.: The temporal behaviour of the hydrodynamic force on a body in response to an abrupt change in velocity at small but finite Reynolds number. J. Fluid Mech. **293**, 35–46 (1995)

[Lub85] Lubich, C.: Fractional linear multistep methods for Abel-Volterra integral equations of the second kind. Math. Comput. **45**, 463–469 (1985)

[Lub86] Lubich, C.: Discretized fractional calculus. SIAM J. Math. Anal. **17**, 704–719 (1986)

[Luc99] Luchko, Yu.: Operational method in fractional calculus. Fract. Calc. Appl. Anal. **2**, 463–488 (1999)

[Luc00] Luchko, Yu.: Asymptotics of zeros of the Wright function. J. Anal. Appl. (ZAA) **19**, 583–596 (2000)

[Luc01] Luchko, Yu.: On the distribution of zeros of the Wright function. Integr. Trans. Special Funct. **11**, 195–200 (2001)

[Luc08] Luchko, Yu.: Algorithms for evaluation of the Wright function for the real arguments' values. Fract. Calc. Appl. Anal. **11**, 57–75 (2008)

[Luc19] Luchko, Yu.: The Wright function and its applications, In: Tenreiro Machado, J. (ed.) Handbook of Fractional Calculus with Applications, vol. 1. In: Kochubei, A., Luchco,Yu. (eds.) Basic Theory, pp. 241–268 (2019)

[LucGor98] Luchko, Yu., Gorenflo, R.: Scale-invariant solutions of a partial differential equation of fractional order. Fract. Calc. Appl. Anal. **1**, 63–78 (1998)

[LucGor99] Luchko, Yu., Gorenflo, R.: An operational method for solving fractional differential equations with the Caputo derivatives. Acta Math. Vietnamica **24**, 207–233 (1999)

[LucKir19] Luchko, Yu., Kiryakova, V.: Applications of the Mellin integral transform technique in fractional calculus, In: Tenreiro Machado, J. (ed.) Handbook of Fractional Calculus with Applications, vol. 1. In: Kochubei, A., Luchko, Yu. (eds.) Basic Theory, pp. 195–224. De Gruyter, Berlin (2019)

[LucSri95] Luchko, Yu., Srivastava, H.M.: The exact solution of certain differential equations of fractional order by using operational calculus. Comput. Math. Appl. **29**, 73–85 (1995)

[Luk1] Luke, Y.L.: The Special Functions and Their Approximations, vol. 1. Academic, New York (1969)

[LuoChe13] Luo, Y., Chen, Y.Q.: Fractional Order Motion Control. Wiley, New York (2013)

[Mac-R38] Mac Robert, T.M.: Proofs of some formulae for generalized hypergeometric function and certain related functions. Phil. Mag. **26**, 82–93 (1938)

[Mag06] Magin, R.L.: Fractional Calculus in Bioengineering. Begell House Publishers, Connecticut (2006)

[Mag19] Magin, R.: Kilbas and Saigo Function (https://www.mathworks.com/matlabcentral/fileexchange/70999-kilbas-and-saigo-function), MATLAB Central File Exchange. Accessed 8 July 2019

[Mag-et-al08] Magin, R.L., Abdullah, O., Baleanu, D., Zhou, X.J.: Anomalous diffusion expressed through fractional order differential operators in the Bloch-Torrey equation. J. Magn. Reson. **190**(2), 255–270 (2008)

[Mag-et-al19] Liang, Y., Wang, S., Chen, W., Zhou, Z., Magin, R.L.: A survey of models of ultraslow diffusion in heterogeneous materials. Appl. Mech. Rev., in print (2019)

[Mai94a] Mainardi, F.: On the initial value problem for the fractional diffusion-wave equation. In: Rionero, S., Ruggeri, T. (eds.) Waves and Stability in Continuous Media, pp. 246–251. World Scientific, Singapore (1994)

[Mai96a] Mainardi, F.: Fundamental solutions for the fractional diffusion-wave equation. Appl. Math. Lett. **9**(6), 23–28 (1996)

[Mai96b] Mainardi, F.: Fractional relaxation-oscillation and fractional diffusion-wave phenomena. Chaos Solitons Fractals **7**, 1461–1477 (1996)

[Mai97] Mainardi, F.: Fractional calculus: some basic problems in continuum and statistical mechanics. In: Carpinteri, A., Mainardi, F. (eds.) Fractals and Fractional Calculus in Continuum Mechanics, pp. 291–348. Springer, Wien (1997)

[Mai10] Mainardi, F.: Fractional Calculus and Waves in Linear Viscoelasticity. Imperial College Press, London (2010)

[Mai12] Mainardi, F.: An historical perspective on fractional calculus in linear viscoelasticity. Fract. Calc. Appl. Anal. **15**(4), 712–717 (2012)

[Mai14] Mainardi, F.: On some properties of the Mittag-Leffler function $E_\alpha(-t^\alpha)$, completely monotone for $t>0$ with $0<\alpha<1$. Discr. and Cont. Dyn. Syst. Ser. B, **19**(7), 2267–2278 (2014)

[Mai-spec18] Mainardi, F. (ed.): Special issue "Fractional Calculus: Theory and Applications". Mathematics, **3** (2018)

[MaiGar15] Mainardi, F., Garrappa, R.: On complete monotonicity of the Prabhakar function and non-Debye relaxation in dielectrics. J. Comput. Phys. **293**, 70–80 (2015)

[MaiGor00] Mainardi, F., Gorenflo, R.: Fractional calculus: Special functions and applications. In: Cocolicchio, D., et al. (ed.) Advanced special functions and applications, Proceedings of the workshop, Melfi, Italy, May 9–12, 1999. Rome: Aracne Editrice. Proc. Melfi Sch. Adv. Top. Math. Phys. **1**, 165–188 (2000)

[MaiGor07] Mainardi, F., Gorenflo, R.: Time-fractional derivatives in relaxation processes: a tutorial survey. Fract. Calc. Appl. Anal. **10**, 269–308 (2007)

[MaGoSc04a] Mainardi, F., Gorenflo, R., Scalas, E.: A fractional generalization of the Poisson processes. Vietnam J. Math., **32**(Spec. Iss.), 53–64 (2004)

[MaGoVi05] Mainardi, F., Gorenflo, R., Vivoli, A.: Renewal processes of Mittag-Leffler and Wright type. Fract. Calc. Appl. Anal. **8**, 7–38 (2005)

[MaGoVi07] Mainardi, F., Gorenflo, R., Vivoli, A.: Beyond the Poisson renewal process: a tutorial survey. J. Comp. Appl. Math. **205**, 725–735 (2007)

[MaLuPa01] Mainardi, F., Luchko, Yu., Pagnini, G.: The fundamental solution of the space-time fractional diffusion equation. Fract. Calc. Appl. Anal. **4**, 153–192 (2001)

[MaMuPa10] Mainardi, F., Mura, A., Pagnini, G.: The M-Wright function in time-fractional diffusion processes: a tutorial survey. Int. J. Diff. Equ. **2010**, Article ID 104505, 29 pages (2010)

[MaiPag03] Mainardi, F., Pagnini, G.: Salvatore Pincherle: the pioneer of the Mellin-Barnes integrals. J. Comp. Appl. Math. **153**, 331–342 (2003)

[MaiPag03a] Mainardi, F., Pagnini, G.: The Wright functions as solutions of the time-fractional diffusion equations. Appl. Math. Comput. **141**, 51–66 (2003)

[MaPaGo03] Mainardi, F., Pagnini, G., Gorenflo, R.: Mellin transform and subordination laws in fractional diffusion processes. Fract. Calc. Appl. Anal. **6**(4), 441–459 (2003)

[MaPaSa05] Mainardi, F., Pagnini, G., Saxena, R.K.: Fox H-functions in fractional diffusion. J. Comput. Appl. Math. **178**, 321–331 (2005)

[MaiPir96] Mainardi, F., Pironi, P.: The fractional Langevin equation: the Brownian motion revisited. Extracta Math. **11**, 140–154 (1996)

[MaPiTa95a] Mainardi, F., Pironi, P., Tampieri, F.: On a generalization of the Basset problem via fractional calculus. In: Tabarrok, B., Dost, S. (eds.) Proceedings CANCAM 95, vol. **II**, pp. 836–837 (1995) [15th Canadian Congress of Applied Mechanics, Victoria, British Columbia, Canada, 28 May–2 June 1995]

[Mai-et-al00] Mainardi, F., Raberto, M., Gorenflo, R., Scalas, E.: Fractional calculus and continuous-time finance II: the waiting-time distribution. Phys. A **287**, 468–481 (2000)

[MaiRog06] Mainardi, F., Rogosin, S.: The origin of infinitely divisible distributions: from de Finetti's problem to Lévy-Khintchine formula. Math. Methods Econ. Financ. **1**, 37–55 (2006)

[MaiSpa11] Mainardi, F., Spada, G.: Creep, relaxation and viscosity properties for basic fractional models in rheology. Eur. Phys. J. (Special Topics) **193**, 119–132 (2011)

[MaiTom95] Mainardi, F., Tomirotti, M.: On a special function arising in the time fractional diffusion-wave equation. In: Rusev, P., Dimovski, I., Kiryakova, V. (eds.) Transform Methods and Special Functions, Sofia 1994, pp. 171–183. Science Culture Technology, Singapore (1995)

[MaiTom97]	Mainardi, F., Tomirotti, M.: Seismic pulse propagation with constant Q and stable probability distributions. Annali di Geofisica **40**, 1311–1328 (1997)
[MalShi75]	Malakhovskaya, R.M., Shikhmanter, E.D.: Certain formulae for the realization of operators, and their applications to the solution of integro-diferential equations. Trudy Tomsk. Gos. Univ. **220**, 46–56 (1975)
[Mal03]	Malmquist, J.: Sur le calcul des integrales d'un système d'équations différentielles par la methodé de Cauchy-Lipschitz. Arkiv för Mat. Astr. och Fysik **1**, 149–156 (1903)
[Mal05]	Malmquist, J.: Étude d'une fonction entière. Acta Math. **29**, 203–215 (1905)
[Mam18]	Mamchuev, M.O.: Boundary value problem for a multidimensional system of equation with Riemann–Liouville derivatives, p. 12. arXiv:1806.08521v1. Accessed 22 June 2018
[Man25]	Mandelbroit, S.: Sulla generalizzazione del calcolo delle variazione. Atti Reale Accad. Naz. Lincei. Rend. Cl. sci. fis. mat. s natur., Ser. 6 **1**, 151–156 (1925)
[Mara71]	Maravall, D.: Linear differential equations of non-integer order and fractional oscilations (Spanish). Rev. Ac. Ci. Madrid **65**, 245–258 (1971)
[Mari83]	Marichev, O.I.: Handbook of Integral Transforms of Higher Transcendental Functions. Ellis Horwood, Theory and Algorithmic Tables. Chichester (1983)
[Mark66]	Markushevich, A.I.: Entire Functions. Elsevier, New York (1966)
[MaMoWe03]	Masoliver, J., Montero, M., Weiss, G.H.: Continuous-time random-walk model for financial distributions. Phys. Rev. E **67**, 021112/1–9 (2003)
[Mat93]	Mathai, A.M.: A Handbook of Generalized Special Functions for Statistical and Physical Sciences. Oxford University Press, Oxford (1993)
[MatHau08]	Mathai, A.M., Haubold, H.J.: Special Functions for Applied Scientists. Springer, New York (2008)
[MatHau17]	Mathai, A.M., Haubold, H.J.: Fractional and Multivariable Calculus. Model Building and Optimization Problems. Springer Nature, Cham (2017)
[MatHau18]	Mathai, A.M., Haubold, H.J.: Erdélyi-Kober Fractional Calculus. Inspired by Solar Neutrino Physics. From a Statistical Perspective. Springer Nature, Singapore (2018)
[MatSax73]	Mathai, A.M., Saxena, R.K.: Generalized Hypergeometric Functions with Applications in Statistics and Physical Sciences. Lecture Notes in Mathematics, vol. 348. Springer, Berlin (1973)
[MatSax78]	Mathai, A.M., Saxena, R.K.: The H-Function with Applications in Statistics and Other Disciplines. Wiley Eastern Ltd, New Delhi (1978)
[MaSaHa10]	Mathai, A.M., Saxena, R.K., Haubold, H.J.: The H-Function. Theory and Applications. Springer, Dordrecht (2010)
[MauElw12]	Maundy, B., Elwakil, A.S.: Extracting single dispersion Cole-Cole impedance model parameters using an integrator setup. Analog Integr. Circuits Signal Process. **71**(1), 107–110 (2012)
[MaxRil83]	Maxey, M.R., Riley, J.J.: Equation of motion for a small rigid sphere in a nonuniform flow. Phys. Fluids **26**, 883–889 (1983)
[McB79]	McBride, A.C.: Fractional Calculus and Integral Transforms of Generalized Functions. Pitman, London (1979)
[McBRoa85]	McBride, A.C., Roach, G.F. (eds.): Fractional Calculus. Pitman Research Notes in Mathematics, No. 132. Pitman, London (1985)
[MeeSch04]	Meerschaert, M.M., Scheffler, H.P.: Limit theorems for continuous-time random walks with infinite mean waiting times. J. Appl. Probab. **41**, 623–638 (2004)
[MeNaVe11]	Meerschaert, M.M., Nane, E., Vellaisamy, P.: The fractional Poisson process and the inverse stable subordinator. Electronic J. Prob. **16**, 1600–1620 (2011)
[Meh17]	Mehrez, Kh.: Monotonicity patterns and functional inequalities for classical and generalized Wright functions. arXiv:1708.00461v1 [math.CA]. Accessed 24 Jul 2017

[Mei36]	Meijer, C.S.: Über Whittakersche bzw. Besselsche Funktionen und deren Produkte. Nieuw Arch. Wiskd. **18**, 10–39 (1936) (in German)
[Mei46]	Meijer, C.S.: On the G-function. Proc. Nederl. Akad. Wetensch. **49**, 227–236, 344–356, 457–469, 632–641, 765–772, 936–943, 1062–1072, 1165–1175 (1946)
[Mel02]	Mellin, Hj.: Eine Formel für den Logarithmus transzendenter Funktionen von endlichem Geschlecht. Helsingfors. Acta Soc. Fenn., 50 S. 4° (1902)
[Mel10]	Mellin, Hj.: Abriss einer einheitlichen Theorie der Gamma- und der hypergeometrischen Funktionen. Math. Ann. **68**, 305–337 (1910)
[Mes67]	Meshkov, S.I.: Description of internal friction in the memory theory of elasticity using kernels with a weak singularity. J. Appl. Mech. Tech. Phys. **8**(4), 100–102 (1967)
[Mes70]	Meshkov, S.I.: Integral representation of fractional exponential functions and their application to dynamic problems in linear viscoelasticity. J. Appl. Mech. Tech. Phys. **11**(1), 100–107 (1970). Transl. from Zh. Prikl. Mekh. i Techn. Fiz. **11**(1), 103–110 (1970)
[MePaSh66]	Meshkov, S.I., Pachevskaya, G.N., Shermergor, T.D.: Internal friction described with the aid of fractionally-exponential kernels. J. Appl. Mech. Tech. Phys. **7**(3), 63–65 (1966)
[MesRos68]	Meshkov, S.I.: Rossikhin, YuA: Propagation of acoustic waves in a hereditary elastic medium. J. Appl. Mech. Tech. Phys. **9**(5), 589–592 (1968)
[MetKla00]	Metzler, R., Klafter, J.: The random walk's guide to anomalous diffusion: a fractional dynamics approach. Phys. Rep. **339**, 1–77 (2000)
[MetKla02]	Metzler, R., Klafter, J.: From stretched exponential to inverse power-law: fractional dynamics, Cole-Cole relaxation processes, and beyond. J. Non-Crystalline Solids **305**, 81–87 (2002)
[MetKla04]	Metzler, R., Klafter, J.: The restaurant at the end of the random walk: Recent developments in the description of anomalous transport by fractional dynamics. J. Phys. A. Math. Gen. **37**, R161–R208 (2004)
[Mik59]	Mikusiński, J.: Operational Calculus. Pergamon Press, New York (1959)
[Mik59a]	Mikusiński, J.: On the function whose Laplace transform is $\exp(-s^\alpha \lambda)$. Studia Math. **18**, 191–198 (1959)
[Mil93]	Miller, K.S.: The Mittag-Leffler and related functions. Integr. Trans. Special Funct. **1**, 41–49 (1993)
[MilRos93]	Miller, K.S., Ross, B.: An Introduction to the Fractional Calculus and Fractional Differential Equations. Wiley, New York (1993)
[MilSam97]	Miller, K.S., Samko, S.G.: A note on the complete monotonicity of the generalized Mittag-Leffler function. Real. Anal. Exchange **23**, 753–755 (1997)
[MilSam01]	Miller, K.S., Samko, S.G.: Completely monotonic functions. Integr. Transf. Spec. Funct. **12**(4), 389–402 (2001)
[Mis09]	Miskinis, P.: The Havriliak-Negami susceptibility as a nonlinear and nonlocal process. Phys. Scripta **2009**(T136), 014019 (2009)
[Mit84]	Mittag-Leffler, M.G.: Sur la representation analytique des fonctions monogène uniformes d'une variable independante. Acta Math. **4**, 1–79 (1884)
[ML5-1]	Mittag-Leffler, M.G.: Sur la représentation analytique d'une branche uniforme d'une fonction monogène (première note). Acta Math. **23**, 43–62 (1899)
[ML5-2]	Mittag-Leffler, M.G.: Sur la représentation analytique d'une branche uniforme d'une fonction monogène (seconde note). Acta Math. **24**, 183–204 (1900)
[ML5-3]	Mittag-Leffler, M.G.: Sur la représentation analytique d'une branche uniforme d'une fonction monogène (troisème note). Acta Math. **24**, 205–244 (1900)
[ML5-4]	Mittag-Leffler, M.G.: Sur la représentation analytique d'une branche uniforme d'une fonction monogène (quatrième note). Acta Math. **26**, 353–392 (1902)
[ML1]	Mittag-Leffler, M.G.: Sur l'intégrale de Laplace-Abel. Comp. Rend. Acad. Sci. Paris **135**, 937–939 (1902)

[ML2]	Mittag-Leffler, M.G.: Une généralization de l'intégrale de Laplace-Abel. Comp. Rend. Acad. Sci. Paris **136**, 537–539 (1903)
[ML3]	Mittag-Leffler, M.G.: Sur la nouvelle fonction $E_\alpha(x)$. Comp. Rend. Acad. Sci. Paris **137**, 554–558 (1903)
[ML4]	Mittag-Leffler, M.G.: Sopra la funzione $E_\alpha(x)$. Rend. R. Acc. Lincei, (Ser. 5) **13**, 3–5 (1904)
[ML5-5]	Mittag-Leffler, M.G.: Sur la representation analytique d'une branche uniforme d'une fonction monogène (cinquième note). Acta Math. **29**, 101–181 (1905)
[ML5-6]	Mittag-Leffler, M.G.: Sur la representation analytique d'une branche uniforme d'une fonction monogène (sixième note). Acta Math. **42**, 285–308 (1920)
[ML08]	Mittag-Leffler, M.G.: Sur la représentation arithmétique des fonctions analytiques d'une variable complexe, Atti del IV Congresso Internazionale dei Matematici (Roma, 6–11 Aprile 1908), vol. I, pp. 67–86 (1908)
[MiPaJo16]	Mittal, E., Pandey, R.M., Joshi, S.: On extension of Mittag-Leffler function. Appl. Appl. Math. **11**(1), 307–316 (2016)
[MolvLoa78]	Moler, C., Van Loan, Ch.: Nineteen dubious ways to compute the exponential of a matrix. SIAM Rev. **20**(4), 801–836 (1978)
[MolvLoa03]	Moler, C., Van Loan, Ch.: Nineteen dubious ways to compute the exponential of a matrix, twenty-five years later. SIAM Rev. **45**(1), 3–49 (2003)
[Mol75]	Molinari, A.: Viscoélasticité linéaire et function complètement monotones. J. de Mécanique **12**, 541–553 (1975)
[MonWei65]	Montroll, E.W., Weiss, G.H.: Random walks in lattices II. J. Math. Phys. **6**(2), 167–181 (1965)
[NagMen12]	Nagar, H., Menaria, N.: Certain families of generalized Mittag-Leffler functions and their integral representation. J. Comp. Math. Sci. **3**(5), 498–556 (2012)
[Nak74]	Nakhushev, A.M.: Inverse problems for degenerate equations, and Volterra integral equations of the third kind. Differentsial'nye Uravnenija **10**(1), 100–111 (1974) (in Russian)
[Nak77]	Nakhushev, A.M.: The Sturm–Liouville problem for a second order ordinary differential equation with fractional derivatives in the lower terms. Dokl. Akad. Nauk SSSR **234**(2), 308–311 (1977) (in Russian)
[Nak03]	Nakhushev, A.M.: Fractional calculus and its applications, (Drobnoe ischislenie i ego primenenie). Fizmatlit, Moskva (2003). (in Russian)
[Nev25]	Nevanlinna, R.: Zur Theorie der Meromorphen Funktionen. Acta Math. (Springer, Netherlands) **46**, 1–99 (1925)
[Nev36]	Nevanlinna, R.: Analytic functions, Die Grundlehren der mathematischen Wissenschaften, vol. 162. Springer, Berlin (1970) [reprinted of 1936 edition]
[New05]	Newman, M.E.J.: Power law, Pareto distributions and Zip's law. Contemp. Phys. **46**(5), 323–351 (2005)
[NewTr-10]	Baleanu, D., Güvenç, Z.B., Tenreiro Machado, J.A. (eds.): New Trends in Nanotechnology and Fractional Calculus Applications. Springer, Dordrecht (2010)
[Nig09]	Nigmatullin, R.R.: Dielectric relaxation phenomenon based on the fractional kinetics: theory and its experimental confirmation. Phys. Scripta **T136**, 014001/1–6 (2009)
[Nis84]	Nishimoto, K.: Fractional Calculus: Integration and Differentiation of Arbitrary Order, Vols. I–V. Descartes Press, Koriyama (1984) [2nd edn, (1989); 3rd edn, (1991); 4th edn, (1996)]
[Nis90]	Nishimoto, K. (ed.): Fractional Calculus and its Applications. Nihon University, Tokyo (1990)
[NiOwSr84]	Nishimoto, K., Owa, S., Srivastava, H.M.: Solutions to a new class of fractional differintegral equations. J. College Engrg. Nihon Univ. Ser. B **25**, 75–78 (1984)
[NIST]	Olver, F.W.J. (editor-in-chief), Lozier, D.W., Boisvert, R.F., Clark, C.W. (eds.) NIST Handbook of Mathematical Functions. Gaithersburg, Maryland, National Institute of Standards and Technology, and New York, Cambridge University Press, 951 + xv pages and a CD (2010)

[Non90b]	Nonnenmacher, T.F.: Fractional integral and differential equations for a class of Lévy-type probability densities. J. Phys A: Math. Gen. **23**, L697–L700 (1990)
[NonGlo91]	Nonnenmacher, T.F., Glöckle, W.G.: A fractional model for mechanical stress relaxation. Phil. Mag. Lett. **64**(2), 89–93 (1991)
[Noe27]	Nörlund, N.E.: G. Mittag-Leffler. Acta Math. **50**, pp. I–XXIII (1927)
[Nov17]	Novozhenova, O.G.: Life and science of Alexey Gerasimov, one of the pioneers of fractional calculus in Soviet Union. Fract. Calc. Appl. Anal. **20**(3), 790–809 (2017)
[Nut21]	Nutting, P.G.: A new general law of deformation. J. Frankline Inst. **191**, 679–685 (1921)
[Nut43]	Nutting, P.G.: A general stress-strain-time formula. J. Frankline Inst. **235**, 513–524 (1943)
[Nut46]	Nutting, P.G.: Deformation in relation to time, pressure and temperature. J. Frankline Inst. **242**, 449–458 (1946)
[Obe74]	Oberhettinger, F.: Tables of Mellin Transforms. Springer, New York (1974)
[Obr16]	Obrechkoff, N.: On the summation of Taylor's series on the contour of the domain of summability (Archive Paper). Fract. Calc. Appl. Anal. **19**(5), 1316–1346 (2016)
[OldSpa74]	Oldham, K.B., Spanier, J.: The Fractional Calculus. Academic, New York (1974)
[Oli39]	de Oliveira Castro, F.M.: Zur Theorie der dielektrischen Nachwirkung. Zeits. für Physik **114**, 116–126 (1939)
[Oli39b]	de Oliveira Castro, F.M.: Nota sobra uma equacao integro-diffrencial que intressa a elelectrotecnica. Ann. Acad. Brasilieria de Sciencias **11**, 151–163 (1939)
[Olv74]	Olver, F.W.J.: Asymptotics and Special Functions. Academic, New York (1974)
[OrsBeg04]	Orsingher, E., Beghin, L.: Time-fractional telegraph equations and telegraph processes with Brownian time. Probab. Theory Relat. Fields **128**(1), 141–160 (2004)
[OrsPol09]	Orsingher, E., Polito, F.: Some results on time-varying fractional partial differential equations and birth-death processes. In: Proceedings of the XIII International EM Conference on Eventological Mathematics and Related Fields, Krasnoyarsk, pp. 23–27 (2009)
[Ost01]	Ostrovskii, I.V.: On zero distribution of sections and tails of power series. Israel Math. Conf. Proc. **15**, 297–310 (2001)
[OstPer97]	Ostrovski, I.V., Peresyolkova, I.N.: Nonasymptotic results on distribution of zeros of the function $E_\rho(z;\mu)$. Anal. Math. **23**, 283–296 (1997)
[OzaYlm14]	Ozarslan, M.A., Ylmaz, B.: The extended Mittag-Leffler function and its properties. J. Inequal. Appl. **2014**, 85 (2014)
[PadVis15]	Padula, F., Visioli, A.: Advances in Robust Fractional Control. Springer, Berlin (2015)
[Pan18]	Pandey, S.C.: The Lorenzo-Hartley's function for fractional calculus and its applications pertaining to fractional order modelling of anomalous relaxation in dielectrics. Comput. Appl. Math. **37**(3), 2648–2666 (2018)
[Pan-K11]	Paneva-Konovska, J.: Multi-index ($3m$-parametric) Mittag-Leffler functions and fractional calculus. Comptes rendus de l'Academie bulgare des Sciences **64**(8), 1089–1098 (2011)
[Pan-K12]	Paneva-Konovska, J.: Three-multi-index Mittag-Leffler functions, series and convergence theorems. In: Proceedings of the FDA'12, Nanjing, China, May 2012 (2012)
[Pan-K13]	Paneva-Konovska, J.: On the multi-index ($3m$-parametric) Mittag-Leffler functions, fractional calculus relations and series convergence. Cent. Eur. J. Phys. **11**(10), 1164–1177 (2013). https://doi.org/10.2478/s11534-013-0263-8
[Pan-K16]	Paneva-Konovska, J.: From Bessel to Multi-Index Mittag-Leffler Functions: Enumerable Families. Series in them and Convergence. World Scientific Publication, London (2016)
[Pari02]	Paris, R.B.: Exponential asymptotics of the Mittag-Leffler function. Proc. R. Soc. Lond., Ser. A, Math. Phys. Eng. Sci. **458**(2028), 3041–3052 (2002)

[Par10]	Paris, R.B.: Exponentially small expansions in the asymptotics of the Wright function. J. Comp. Appl. Math. **234**, 488–504 (2010)
[Par17]	Paris, R.B.: Some remarks on the theorems of Wright and Braaksma on the Wright function ${}_p\Psi_q(z)$. arXiv:1708.04824v1 [math.CA]. Accessed 16 Aug 2017
[Par19]	Paris, R.B.: Asymptotics of the special functions of fractional calculus. In: Kochubei, A., Luchco, Yu. (eds.) Handbook of Fractional Calculus with Applications, vol. 1. Basic Theory, pp. 297–326 (2019)
[ParKam01]	Paris, R.B., Kaminski, D.: Asymptotic and Mellin-Barnes Integrals. Cambridge University Press, Cambridge (2001)
[ParVin16]	Paris, R.B., Vinogradov, V.: Asymptotic and structural properties of special cases of the Wright function arising in probability theory. Lithuanian Math. J. **56**(3), 377–409 (2016)
[PaLuRa16]	Parmar, R.K., Luo, M., Raina, R.K.: On a multivariable class of Mittag-Leffler type functions. J. Appl. Anal. Comput. **6**, 981–999 (2016)
[Pat66]	Pathak, R.S.: Certain convergence theorems and asymptotic properties of a generalization of Lommel and Maitland transformations. Proc. Nat. Acad. Sci. India **A-36**(1), 81–86 (1966)
[Pat67]	Pathak, R.S.: An inversion formula for a generalization of the Lommel and Maitland transformations. J. Sci. Banaras Hindu Univ. **17**(1), 65–69 (1966–67)
[Wright-ob]	Pears, A.R.: Edward Maitland Wright 1906–2005 (obituary). Bull. Lond. Math. Soc. **39**, 857–865 (2007)
[PengLi10]	Peng, Jigen: Li, Kexue: A note on property of the Mittag-Leffler function. J. Math. Anal. Appl. **370**, 635–638 (2010)
[Per00]	Peresyolkova, I.N.: On the distribution of zeros of generalized functions of Mittag-Leffler type. Mat. Stud. **13**(2), 157–164 (2000)
[Pet11]	Petráš, I.: Fractional-Order Nonlinear Systems. Springer, Berlin (2011)
[Pet11a]	Petráš, I.: Discrete fractional-order PID controller. Matlab Central File Exchange, MathWorks, Inc. (2011). http://www.mathworks.com/matlabcentral/fileexchange/33761
[Pet12]	Petráš, I.: Tuning and implementation methods for fractional-order controllers. Fract. Calc. Appl. Anal. **15**, 282–303 (2012)
[Pet15]	Petráš, I.: Discrete fractional-order PID controller, Matlab Central File Exchange, MathWorks, Inc. (2015). http://www.mathworks.com/matlabcentral/fileexchange/51190
[Pet19]	Petráš, I.: Modified versions of the fractional-order PID controller. In: Petráš, I (ed.) Handbook of Fractional Calculus with Applications, Applications in Control, vol. 6, pp. 57–72 (2019)
[Phr04]	Phragmén, E.: Sur une extension d'un théorème classique de la théorie des fonctions. Acta Math. **28**, 351–368 (1904)
[PhrLin08]	Phragmén, E., Lindelöf, E.: Sur une extension d'un principle classique de l'analyse. Acta Math. **31**, 381–406 (1908)
[Picc07]	Picciati, G.: Sul moto di una sfera in un liquido viscoso. Rend. R. Acc. Naz. Lincei (ser. 5) **16**, 943–951 (1907)
[Pil90]	Pillai, R.N.: On Mittag-Leffler functions and related distributions. Ann. Inst. Stat. Math. **42**(1), 157–161 (1990)
[PilJay95]	Pillai, R.N., Jayakumar, K.: Discrete Mittag-Leffler distributions. Stat. Probab. Lett. **23**(3), 271–274 (1995)
[Pin88]	Pincherle, S.: Sulle funzioni ipergeometriche generalizzate. Atti R. Accad. Lincei, Rend. Cl. Sci. Fis. Mat. Natur. **4**(4), 694–700, 792–799 (1888)
[Pin25]	Pincherle, S.: Notices sur les travaux. Acta Math. **46**, 341–362 (1925)
[Pip86]	Pipkin, A.C.: Lectures on Viscoelastic Theory. Springer, New York (1986) [1st edn, (1972), 2nd edn, (1986)]
[PitSew38]	Pitcher, E., Sewell, W.E.: Existence theorem for solutions of differential equations of non-integral order. Bull. Amer. Math. Soc. **44**(2), 100–107 (1938) (A correction in No. 12, p. 888)

[Poc70] Pochhammer, L.: Über hypergeometrische Functionen n-ter Ordnung. Journal für die reine und angewandte Mathematik (Crelle) **71**, 316–352 (1870)

[Pod99] Podlubny, I.: Fractional Differential Equations. Academic, New York (1999)

[Pod99a] Podlubny, I.: Fractional order systems and $PI^\lambda D^\mu$-controllers. IEEE Trans. Autom. Control **44**(1), 208–214 (1999)

[Pod06] Podlubny, I.: Mittag-Leffler function, WEB Site of MATLAB Central File exchange (2006). http://www.mathworks.com/matlabcentral/fileexchange

[Pod11] Podlubny, I.: Fitting data using the Mittag-Leffler function. Matlab Central File Exchange, (11 Jul 2011 (Updated 02 Apr 2012)). http://www.mathworks.com/matlabcentral/fileexchange

[PodKac09] Podlubny, I., Kacenak, M.: The Matlab mlf code. MATLAB Central, File Exchange (2001–2009). File ID: 8738

[Pog16] Pogány, T.: Integral form of the COM-Poisson renormalization constant. Stat. Probab. Lett. **119**, 144–145 (2016)

[PogTom16] Pogány, T.K., Tomovski, Ž.: Probability distribution built by Prabhakar function. Related Turán and Laguerre inequalities. Integr. Trans. Special Funct. **27**(10), 783–793 (2016)

[PoKaSc11] Politi, M., Kaizoji, T., Scalas, E.: Full characterization of the fractional Poisson process. Eur. Phys. Lett. (EPL) **96**, 20004/1-6 (2011)

[PolSca16] Polito, F., Scalas, E.: A generalization of the space-fractional Poisson process and its connection to some Lévy processes. Electron Commun. Probab. **21**, Paper No. 20, 14 (2016)

[Poll46] Pollard, H.: The representation of e^{-x^λ} as Laplace integral. Bull. Amer. Math. Soc. **52**, 908–910 (1946)

[Poll48] Pollard, H.: The completely monotonic character of the Mittag-Leffler function $e_\alpha(-x)$. Bull. Amer. Math. Soc. **54**, 1115–1116 (1948)

[Poly21] Polya, G.: Bemerkung über die Mittag-Lefflerschen Funktionen $E_{1/\alpha}(x)$. Tôhoku Math. J. **19**, 241–248 (1921)

[PolSze76] Pólya, G., Szegö, G.: Problems and Theorems in Analysis, vol. II. Theory of Functions, Zeros, Polynomials, Determinants, Number Theory, Geometry. Springer, New York (1976)

[PolMan08] Polyanin, A.D., Manzhirov, A.V.: Handbook of Integral Equations, 2nd edn. CRC Press, Boca Raton (2008)

[Pop02] Popov, A.Yu.: The spectral values of a boundary value problem and the zeros of Mittag-Leffler functions. (English. Russian original) Differ. Equ. **38**(5), 642–653 (2002). Translation from Differ. Uravn. **38**(5), 611–621 (2002)

[Pop06] Popov, A.Yu.: On a zeroes of Mittag-Leffler functions with parameter $\rho<1/2$. Anal. Math. **32**(3), 207–246 (2006) (in Russian)

[PopSed03] Popov, A.Yu., Sedletskii, A.M.: Zeros distribution of Mittag-Leffler functions. Dokl. Math. **67**(3), 336–339 (2003). Translation from Dokl. Akad. Nauk. Ross. Akad. Nauk. **390**(2), 165–168 (2003)

[PopSed11] Popov, A.Yu., Sedletskii, A.M.: Zeros distribution of Mittag-Leffler functions. Contemp. Math. Fund. Dir. **40**, 3–171 (2011) (in Russian), transl. in *J. Math. Sci.* **190**, 209–409 (2013)

[Pos19] Post, E.L.: Discussion of the solution of $(d/dx)^{1/2} y = y/x$ (problem ♯ 433). Amer. Math. Month. **26**, 37–39 (1919)

[Pra71] Prabhakar, T.R.: A singular integral equation with a generalized Mittag-Leffler function in the kernel. Yokohama Math. J. **19**, 7–15 (1971)

[Pra15] Prajapat, J.K.: Certain geometric properties of the Wright function. Int. Transf. Spec. Funct. **26**(3), 203–212 (2015)

[PrMaBa18] Prajapat, J.K., Maharana, S., Bansal, D.: Radius of Starlikeness and Hardy Space of Mittag-Leffler Functions. Filomat **32**(18), 6475–6486 (2018)

[Pre16] Preda, L.: Splitting and accelerating Gaussian beam modulated by Mittag Leffler function. Optik **127**, 1066–1070 (2016)

[PrOrVa00]	Prilepko, A.I., Orlovsky, D.G., Vasin, I.A.: Methods for Solving Inverse Problems in Mathematical Physics. Marcel Dekker, Basel (2000)
[PrBrMa-V2]	Prudnikov, A.P., Brychkov, Yu.A., Marichev, O.I.: Integral and Series. Special Functions, vol. 2. Gordon and Breach, New York (1986)
[PrBrMa-V3]	Prudnikov, A.P., Brychkov, Yu.A., Marichev, O.I.: Integral and Series. More Special Functions, vol. 3. Gordon and Breach, New York (1990)
[Prus93]	Prüss, J.: Evolutionary Integral Equations and Applications. Birkhäuser, Basel (1993). Reprinted (2011)
[Psk05]	Pskhu, A.V.: On real zeros of a Mittag-Leffler-type function. Math. Notes **77**(4), 592–599 (2005) [translation from Russian *Matematicheskie Zametki*]
[Psk06]	Pskhu, A.V.: Partial Differential Equations of Fractional Order. Nauka, Moscow (2006). (in Russian)
[RaScMa02]	Raberto, M., Scalas, E., Mainardi, F.: Waiting-times and returns in high-frequency financial data: an empirical study. Phys. A **314**, 749–755 (2002)
[Rab48a]	Rabotnov, YuN: The equilibrium of an elastic medium with after-effect. Prikl. Mat. Mekh. **12**, 53–62 (1948). (in Russian)
[Rab69]	Rabotnov, Yu.N.: Creep Problems in Structural Members. North-Holland, Amsterdam (1969) [English translation of the 1966 Russian edition]
[Rab73]	Rabotnov, Yu.N.: On the Use of Singular Operators in the Theory of Viscoelasticity, p. 50, Moscow, (1973). Unpublished Lecture Notes for the CISM course on Rheology held in Udine, October 1973. http://www.cism.it
[Rab74]	Rabotnov, Yu.N.: Experimental Evidence of the Principle of Hereditary in Mechanics of Solids, pp. 80, Moscow, (1974). Unpublished Lecture Notes for the CISM course on Experimental Methods in Mechanics, A) Rheology, held in Udine. Accessed 24–29 Oct 1974. http://www.cism.it
[Rab80]	Rabotnov, Yu.N.: Elements of Hereditary Solid Mechanics. MIR, Moscow (1980) [English translation, revised from the 1977 Russian edition]
[RaPaZv69]	Rabotnov, YuN., Papernik, L.H., Zvonov, E.N.: Tables of a Fractional-Exponential Function of Negative Parameter and Its Integral. Nauka, Moscow (1969). (in Russian)
[Rad16]	Răducanu, D.: On partial sums of normalized Mittag-Leffler functions. arXiv:1606.04690v1 [math.CV]. Accessed 15 Jun 2016
[Rah-et-al17]	Rahman, G., Ghaffar, A., Mubeen, S., Arshad, M., Khan, S.U.: The composition of extended Mittag-Leffler functions with pathway integral operator. Adv. Diff. Equ. **2017**, 176 (2017). https://doi.org/10.1186/s13662-017-1237-8
[Rai71]	Rainville, E.D.: Special Functions. Chelsea Publishing Comp, Bronx (1971)
[ReeSim78]	Reed, M., Simon, B.: Methods of Modern Mathematical Physics, vol. IV. Analysis of Operators. Academic, New York (1978)
[RepSai00]	Repin, O.N., Saichev, A.I.: Fractional Poisson law. Radiophys. Quantum Electron. **43**, 738–741 (2000)
[Rie49]	Riesz, M.: L'integral de Riemann-Liouville et le probleme de Cauchy. Acta Math. **81**, 1–223 (1949)
[RogDub18]	Rogosin, S.V., Dubatovskaya, M.V.: Letnikov vs Marchaud. A survey on two prominent constructions of fractional derivatives. Mathematics **6**(1), 3 (2018). https://doi.org/10.3390/math6010003
[RogKor10]	Rogosin, S., Koroleva, A.: Integral representation of the four-parametric generalized Mittag-Leffler function. Lithuanian Math. J. **50**(3), 337–343 (2010)
[RogMai11]	Rogosin, S., Mainardi, F.: The Legacy of A. Ya. Khintchine's Work in Probability Theory. Cambridge Scientific Publishers, Cottenham (2011)
[RogMai17]	Rogosin, S., Mainardi, F.: A. Ya. Khintchine's Work in Probability Theory. Notice ICCM **5**(2), 60–75 (2017)
[RogMai14]	Rogosin, S., Mainardi, F.: George William Scott Blair – the pioneer of factional calculus in rheology. Comm. Appl. Indust. Math. **6**(1), e481 (2014). arXiv:1404.3295.v1

[Ron92]   Ronkin, L.I.: Functions Of Completely Regular Growth. Kluwer Academic Publishers, Amsterdam (1992)
[Ros75]   Ross, B. (ed.): Fractional Calculus and its Applications. Lecture Notes in Mathematics No. 457. Springer, Berlin (1975)
[Ros77]   Ross, B.: The development of fractional calculus 1695–1900. Historia Math. **4**, 75–89 (1977)
[Ros97]   Ross, S.M.: Introduction to Probability Models, 6th edn. Academic, New York (1997)
[RosDix17]   Rosenfeld, J.A., Dixon, W.E.: Approximating the Caputo fractional derivative through the Mittag-Leffler reproducing kernel Hilbert space and the kernelized Adams-Bashforth-Moulton method. SIAM J. Numer. Anal. **55**(3), 1201–1217 (2017)
[RoRuDi18]   Rosenfeld, J.A., Russo, B., Dixon, W.E.: The Mittag-Leffler reproducing kernel Hilbert space of entire and analytic functions. J. Math. Anal. Appl. **463**, 576–592 (2018)
[Ros70]   Rossikhin, Yu.A.: Dynamic Problems of Linear Viscoelasticity connected with the Investigation of Retardation and Relaxation Spectra. Ph.D. Dissertation, Voronezh Polytechnic Institute, Voronezh (1970) (in Russian)
[Ros10]   Rossikhin, Yu.A.: Reflections on two parallel ways in the progress of fractional calculus in mechanics of solids. Appl. Mech. Review **63**, 010701/1–12 (2010)
[RosShi07]   Rossikhin, YuA, Shitikova, M.V.: Comparative analysis of viscoelastic models involving fractional derivatives of different orders. Fract. Calc. Appl. Anal. **10**(2), 111–121 (2007)
[RosShi10]   Rossikhin, Yu.A., Shitikova, M.V.: Applications of fractional calculus to dynamic problems of solid mechanics: novel trends and recent results. Appl. Mech. Rev. **63**, 010801/1–52 (2010)
[RosShi19a] Rossikhin, Yu.A., Shitikova, M.V.: Fractional calculus models in dynamic problems of viscoelasticity. In: Tenreiro Machado, J. (ed.) Handbook of Fractional Calculus with Applications, vol. 7. In: Baleanu, D., Lopez, A.N. (eds.) Applications in Engineering, Life and Social Sciences. Part A, pp. 139–158. De Gruyter, Berlin (2019)
[RosShi19b] Rossikhin, Yu.A., Shitikova, M.V.: Fractional calculus in structural mechanics. In: Tenreiro Machado, J. (ed.) Handbook of Fractional Calculus with Applications. In: Baleanu, D., Lopez, A.N. (eds.) Applications in Engineering, Life and Social Sciences. Part A, pp. 159–192 . De Gruyter, Berlin-Boston (2019)
[Rub96]   Rubin, B.: Fractional Integrals and Potentials. Addison-Wesley & Longman, Harlow (1996)
[Rud74]   Rudin, W.: Real and Complex Analysis. McGraw-Hill, New York (1974)
[RuDiKi94]   Rusev, P., Dimovski, I., Kiryakova, V. (eds.): Transform Methods and Special Functions, Sofia 1994. Science Culture Technology, Singapore (1995)
[RuDiKi96]   Rusev, P., Dimovski, I., Kiryakova, V. (eds.).: Transform Methods and Special Functions, Varna 1996. Inst. Maths & Informatics, Bulg. Acad. Sci. Sofia (1998). [Proc. 2nd Int. Workshop TMSF, Varna, Bulgaria, 23–29 August 1996]
[SaiZas97]   Saichev, A., Zaslavsky, G.: Fractional kinetic equations: solutions and applications. Chaos **7**, 753–764 (1997)
[SaidVar09] Said, T., Varadan, V.V.: Variation of Cole-Cole model parameters with the complex permittivity of biological tissues. In: IEEE MTT-S International Microwave Symposium Digest, pp. 1445–1448 (2009)
[SaiKil98]   Saigo, M., Kilbas, A.A.: On Mittag-Leffler type function and applications. Integr. Trans. Special Funct. **7**(1–2), 97–112 (1998)
[SaiKil00]   Saigo, M., Kilbas, A.A.: Solution of a class of linear differential equations in terms of functions of Mittag-Leffler type. Differ. Equ. **36**(2), 193–200 (2000)
[ST-TM-CO19] Sales Teodoro, G., Tenreiro Machado, J.A., Capelas de Oliveira, E.: A review of definitions of fractional derivatives and other operators. J. Comput. Phys. **388**, 195–208 (2019)

[Sam02]	Samko, S.G.: Hypersingular Integrals and Their Applications. Taylor & Francis, London (2002)
[SaKiMa93]	Samko, S.G., Kilbas, A.A., Marichev, O.I.: Fractional Integrals and Derivatives: Theory and Applications. Gordon and Breach Science Publishers, New York (1993) [Extended edition of the Russian original, Nauka i Tekhnika, Minsk (1987)]
[SanTom19]	Sandev, T., Tomovski, Ž.: Fractional Equations and Models: Theory and Applications. Springer Nature, Switzerland (2019)
[SanGer60]	Sansone, G., Gerretsen, J.: Lectures on the Theory of Functions of a Complex Variable, vol. I. Holomorphic Functions. Noordhoff, Groningen (1960)
[SLSPL]	Santarelli, M.F., Della Latta, D., Scipioni, M., Positano, V., Landini, L.: A Conway-Maxwell-Poisson (CMP) model to address data dispersion on positron emission tomography. Comput. Biol. Med. **77**, 90–101 (2016)
[San88]	Sanz-Serna, J.M.: A numerical method for a partial integro-differential equation. SIAM J. Numer. Anal. **25**(2), 319–327 (1988)
[Sat99]	Sato, K.: Lévy Processes and Infinitely Divisible Distributions. Cambridge University Press, Cambridge (1999)
[Sax02]	Saxena, R.K.: Certain properties of generalized Mittag-Leffler function, In: Proceedings of the 3rd Annual Conference of the Society for Special Functions and Their Applications, Chennai, India, pp. 77–81 (2002)
[Sax09]	Saxena, R.K.: In memorium of Charles Fox. Fract. Calc. Appl. Anal. **12**(3), 337–344 (2009)
[SaKaKi03]	Saxena, R.K., Kalla, S.L., Kiryakova, V.S.: Relations connecting multiindex Mittag-Leffler functions and Riemann-Liouville fractional calculus. Algebras Groups Geom. **20**(4), 363–386 (2003)
[SaKaSa11]	Saxena, R.K., Kalla, S.L., Saxena, R.: On a multivariate analogue of generalized Mittag-Leffler function. Integr. Transf. Special Funct. **22**(7), 533–548 (2011)
[SaxNis10]	Saxena, R.K., Nishimoto, K.: N-Fractional Calculus of Generalized Mittag-Leffler functions. J. Fract. Calc. **37**, 43–52 (2010)
[SaxNon04]	Saxena, R.K., Nonnenmacher, T.F.: Application of the H-function in Markovian and non-Markovian chain models. Fract. Calc. Appl. Anal. **7**(2), 135–148 (2004)
[Sax-et-al10]	Saxena, R.K., Pogány, T.K., Ram, J., Daiya, J.: Dirichlet averages of generalized multi-index Mittag-Leffler functions. Amer. J. Math. **3**(4), 174–187 (2010)
[SaxSai05]	Saxena, R.K., Saigo, M.: Certain properties of fractional calculus operators associated with generalized Wright function. Fract. Calc. Appl. Anal. **6**, 141–154 (2005)
[ScGoMa00]	Scalas, E., Gorenflo, R., Mainardi, F.: Fractional calculus and continuous-time finance. Phys. A **284**, 376–384 (2000)
[Sca-et-al03]	Scalas, E., Gorenflo, R., Mainardi, F., Raberto, M.: Revisiting the derivation of the fractional diffusion equation. Fractals 11. Suppl. S 281–289 (2003)
[SchSoVo12]	Schilling, R.L., Song, F., Vondraček, Z.: Bernstein Functions. Theory and Applications, 2nd edn. de Gruyter, Berlin (2012)
[Sch86]	Schneider, W.R.: Stable distributions: fox function representation and generalization. In: Albeverio, S., Casati, G., Merlini, D. (eds.) Stochastic Processes in Classical and Quantum Systems, pp. 497–511. Springer, Berlin (1986)
[Sch90]	Schneider, W.R.: Fractional diffusion. In: Lima, R., Streit, L., Vilela Mendes, D. (eds.) Dynamics and Stochastic Processes, Theory and Applications. Lecture Notes in Physics, vol. 355, pp. 276–286. Springer, Heidelberg (1990)
[Sch96]	Schneider, W.R.: Completely monotone generalized Mittag-Leffler functions. Expositiones Mathematicae **14**, 3–16 (1996)
[SchWys89]	Schneider, W.R., Wyss, W.: Fractional diffusion and wave equations. J. Math. Phys. **30**, 134–144 (1989)
[ScoB44]	Scott-Blair, G.W.: Analytical and integrative aspects of the stress-strain-time problem. J. Sci. Inst. **21**, 80–84 (1944)
[ScoB47]	Scott-Blair, G.W.: The role of psychophysics in rheology. J. Colloid Sci. **2**, 21–32 (1947)

[ScoB49] Scott-Blair, G.W.: Survey of General and Applied Rheology. Pitman, London (1949)

[ScoB-Cop39] Scott Blair, G.W., Coppen, F.M.V.: The subjective judgement of the elastic and plastic properties of soft bodies: the "differential thresholds" for viscosities and compression moduli. Proc. R. Soc. B **128**, 109–125 (1939)

[ScoB-Cop42a] Scott Blair, G.W., Coppen, F.M.V.: The classification of rheological properties of industrial materials in the light of power-law relations between stress, strain, and time. J. Sci. Instr. **19**, 88–93 (1942)

[Sed94] Sedletskii, A.M.: Asymptotic formulas for zeros of a function of Mittag-Leffler type. Anal. Math. **20**, 117–132 (1994). (in Russian)

[Sed98] Sedletskii, A.M.: Approximation properties of systems of exponentials on a line and a half-line. Mat. Sb. [Russian Acad. Sci. Sb. Math.] **189**, 125–140 (1998)

[Sed00] Sedletskii, A.M.: On zeros of a function of Mittag-Leffler type. Math. Notes. **68**(5), 117–132 (2000) [translation from Russian Matematicheskie Zametki]

[Sed04] Sedletskii, A.M.: Non asymptotic properties of roots of a Mittag-Leffler type function. Math. Notes **75**(3), 372–386 (2004) [translation from Russian Matematicheskie Zametki]

[Sed07] Sedletskii, A.M.: Asymptotics of zeros of the Mittag-Leffler type function of order 1/2. (Russian, English) Vestn. Mosk. Univ., Ser. I(1), 22–28 (2007), translation in Mosc. Univ. Math. Bull. **62**(1), 22–28 (2007)

[SeyHil05] Seybold, H.J., Hilfer, R.: Numerical results for the generalized Mittag-Leffler function. Fract. Calc. Appl. Anal. **8**, 127–139 (2005)

[SeyHil08] Seybold, H.J., Hilfer, R.: Numerical algorithm for calculating the generalized Mittag-Leffler function. SIAM J. Numer. Anal. **47**(1), 69–88 (2008)

[O'Sha18] O'Shaughnessy, L.: Problem No. 433. Amer. Math. Month. **25**, 172–173 (1918)

[She66] Shermergor, T.D.: On application of operators of fractional differentiation for the description of hereditary properties of materials. Prikl. Mat. Tekh. Fiz. **6**, 118–121 (1966)

[Shi19] Shitikova, M.V.: The fractional derivative expansion method in nonlinear dynamic analysis of structures. Nonlinear Dyn. **1–14**, (2019)

[ShuPra07] Shukla, A.K., Prajapati, J.C.: On a generalization of Mittag-Leffler function and its properties. J. Math. Anal. Appl. **336**, 797–811 (2007)

[Shu01] Shushin, A.I.: Anomalous two-state model for anomalous diffusion. Phys. Rev. E. **64**, 051108 (2001)

[Sim14] Simon, T.: Comparing Fréchet and positive stable laws. Electron. J. Probab. **19**, 1–25 (2014)

[Sla66] Slater, L.J.: Generalized Hypergeometric Functions. Cambridge University Press, London (1966)

[Sne56] Sneddon, I.N.: Special Functions in Mathematical Physics and Chemistry. Oliver & Boyd, Edinburgh (1956)

[Sne74] Sneddon, I.N.: The Use of Integral Transforms. TATA McGraw-Hill, New Dehli (1974) [first published by McGraw-Hill in 1972]

[Sne75] Sneddon, I.N.: The use in Mathematical Physics of Erdélyi-Kober operators and some of their applications. In: Ross, B. (ed.) Fractional Calculus and its Applications, pp. 37–79. Springer, Berlin (1975)

[Sne95] Sneddon, I.N.: Fourier Transforms. Dover, New York (1995) [first published by McGraw-Hill in 1951]

[Sri63] Srivastav, R.P.: A note on certain integral equations of Abel type. Proc. Edinburgh Math. Soc. (Ser. II) **13**, 271–272 (1963)

[SrGuGo82] Srivastava, H.M., Gupta, K.C., Goyal, S.P.: The H-Functions of One and Two Variables with Applications. South Asian Publishers, New Delhi (1982)

[SrOwNi84] Srivastava, H.M., Owa, S., Nishimoto, K.: A note on a certain class of fractional differintegral equations. J. College Engrg. Nihon Univ. Ser. B **25**, 69–73 (1984)

[SrOwNi85]	Srivastava, H.M., Owa, S., Nishimoto, K.: Some fractional differintegral equations. J. Math. Anal. Appl. **106**(2), 360–366 (1985)
[SriTom09]	Srivastava, H.M., Tomovski, I.: Fractional claculus with an integral operator containing generalized Mittag-Leffler function in the kernel. Appl. Math. Comput. **211**(1), 198–210 (2009)
[StaWer16]	Stanislavsky, A., Weron, K.: Atypical case of the dielectric relaxation responses and its fractional kinetic equation. Fract. Calc. Appl. Anal. **19**(1), 212–228 (2016)
[StWeTr10]	Stanislavsky, A., Weron, K., Trzmiel, J.: Subordination model of anomalous diffusion leading to the two-power-law relaxation responses. EPL **91**, 40003/1–5 (2010)
[Sta70]	Stankovic, B.: On the function of E.M. Wright. Publ. de l'Institut Mathèmatique, Beograd, Nouvelle Sèr. **10**, 113–124 (1970)
[SteVH04]	Steutel, F.W., van Harn, K.: Infinite Divisibility of Probability Distributions on the Real Line. Marcel Dekker, New York (2004) [No. 259, Series on Pure and Applied Mathematics]
[Sti79]	Stiassnie, M.: On the application of fractional calculus on the formulation of viscoelastic models. Appl. Math. Modelling **3**, 300–302 (1979)
[Stok51]	Stokes, G.G.: On the effect of the internal friction of fluids on the motion of pendulums. Cambridge Phil. Trans., **9** [p 8], 1–86 (1851), reprinted with additional notes by the author in *Mathematical and Physical Papers*, Vol. III , 1–141, Cambridge University Press, Cambridge (1901)
[SuSuDu05]	Su, Y.X., Sun, D., Duan, B.Y.: Design of an enhanced nonlinear PID controller. Mechatronics **15**(8), 1005–1024 (2005)
[Tar08]	Tarasov, V.E.: Universal electromagnetic waves in dielectric. J. Phys. Condens. Matter **20**(17), 175223/1–7 (2008)
[Tar09]	Tarasov, V.E.: Fractional integro-differential equations for electromagnetic waves in dielectric media. Theoret. Math. Phys. **158**(3), 355–359 (2009)
[Tar10]	Tarasov, V.E.: Fractional Dynamics: Applications of Fractional Calculus to Dynamics of Particles. Fields and Media. Springer, New York (2010)
[Tar13]	Tarasov, V.E.: Review of some promising fractional physical models. Int. J. Mod. Phys. B **27**(9), 1330005 (32 pages) (2013)
[Tar19]	Tarasov, V.E.: Economic models with power-law memory. In: Tenreiro Machado, J. (ed.) Handbook of Fractional Calculus with Applications. In: Baleanu, D., Lopez, A.N. (eds.) Applications in Engineering, Life and Social Sciences. Part B, vol. 8, pp. 1–32. De Gruyter, Berlin (2019)
[TarTar18]	Tarasov, V.E., Tarasova, V.V.: Macroeconomic models with long dynamic memory: fractional calculus approach. Appl. Math. Comp. **338**, 466–486 (2018)
[TaraTar16]	Tarasova, V.V., Tarasov, V.E.: Elasticity for economic processes with memory: fractional differential calculus approach. Frac. Diff. Calc. **6**(2), 219–232 (2016)
[Tat88]	Tatom, F.B.: The Basset term as a semiderivative. Appl. Sci. Res. **45**, 283–285 (1988)
[Tem96]	Temme, N.M.: Special Functions: An Introduction to the Classical Functions of Mathematical Physics. Wiley, New York (1996)
[TeKiMa11]	Tenreiro Machado, J.A., Kiryakova, V., Mainardi, F.: History of fractional calculus. Commun. Nonlinear Sci. Numer. Simulat **6**, 1140–1153 (2011)
[Tep17]	Tepljakov, A.: Fractional-order Modeling and Control of Dynamic Systems. Springer Theses, Springer International, Cham (2017)
[TikEid94]	Tikhonov, I.V.: Éidel'man, YuS: The questions of correctness of the direct and inverse problems for the special form of evolution equation. Matematicheskie Zametki **56**(2), 99–113 (1994)
[TikEid02]	Tikhonov, I.V.: Éidel'man, YuS: Inverse scattering transform for differential equations in Banach space and the distribution of zeros of an entire Mittag-Leffler type function. Differentsial'nye Uravneniya [Differential Equations]. **38**(5), 637–644 (2002)

[TikEid05]	Tikhonov, I.V., Éidel'man, Yu.S.: Uniqueness criterion in an inverse problem for an abstract differential equation with nonstationary inhomogeneous term. Math. Notes **77**(2), 246–262 (2005) [Translated from Matematicheskie Zametki **77**(2), 273–290 (2005)]
[Tit86]	Titchmarsh, E.C.: Introduction to the Theory of Fourier Transforms. Chelsea, New York (1986) [First edition Oxford University Press, Oxford (1937)]
[Ton07]	Tonchev, N.S.: Finite-size scaling in anisotropic systems. Phys. Rev. E **75**, 031110 (2007)
[TorBag84]	Torvik, P.J., Bagley, R.L.: On the appearance of the fractional derivatives in the behavior of real materials. ASME J. Appl. Mech. **51**, 294–298 (1984)
[Tri47]	Tricomi, F.G.: Sulle funzioni ipergeometriche confluenti. Annali di Matematica Pura ed Applicata, Serie Quarta **26**, 141–175 (1947). (in Italian)
[Tri60]	Tricomi, F.G.: Fonctions Hypergéometriques Confluentes. Mém. Sci. Math. No. 140, Gauthier-Villars, Paris (1960)
[Tua17]	Tuan, H.T.: On some special properties of Mittag-Leffler functions. arXiv:1708.02277v1 [math.CA]. Accessed 7 Aug 2017
[Uch03]	Uchaikin, V.V.: Relaxation processes and fractional differential equations. Int. J. Theor. Phys. **42**(1), 121–134 (2003)
[Uch08]	Uchaikin, V.V.: Method of Fractional Derivatives. Ul'yanovsk, Artishok (2008). (in Russian)
[Uch13a]	Uchaikin, V.V.: Fractional Derivatives for Physicists and Engineers, vol. I. Background and Theory. Springer, Berlin; Higher Education Press, Beijing (2013)
[Uch13b]	Uchaikin, V.V.: Fractional Derivatives for Physicists and Engineers. vol. II. Applications. Springer, Berlin; Higher Education Press, Beijing (2013)
[UcCaSi08]	Uchaikin, V.V., Cahoy, D.O., Sibatov, R.T.: Fractional processes: from Poisson to branching one. Int. J. Bifurcation Chaos **18**, 1–9 (2008)
[UchSib12]	Uchaikin, V.V., Sibatov, R.T.: Anomalous kinetics of charge carriers in disordered solids: fractional derivative approach. Int. J. Modern Physics B **26** (31) paper N 1230016 (34 pages) (2012). https://doi.org/10.1142/S0217979212300162
[UchSib13]	Uchaikin, V.V., Sibatov, R.T.: Fractional Kinetics in Solids: Anomalous Charge Transport in Semiconductors. Dielectrics and Nanosystems. World Scientific Publishing, Singapore (2013)
[UchZol99]	Uchaikin, V.V., Zolotarev, V.M.: Chance and Stability. Stable Distributions and their Applications. VSP, Utrecht (1999)
[ValCos12]	Valério, D., da Costa, J.S.: An Introduction to Fractional Control. IET, London (2012)
[VaMaKy14]	Valério, D., Tenreiro Machado, J., Kiryakova, V.: Some pioneers of the applications of fractional calculus. Fract. Calc. Appl. Anal. **17**(2), 552–578 (2014)
[Veb74]	Veber, V.K.: A space of fundamental functions in the theory of Liouville differentiation. Trudy Kirgiz. Gos. University, Ser. Mat. Nauk. **9**, 164–168 (1974) (in Russian)
[Veb76]	Veber, V.K.: The structure of general solution of the system $y^{(\alpha)} = Ay$; $0<a<1$. Trudy Kirgiz. Gos. University Ser. Mat. Nauk, Vyp. **11**, 26–32 (1976) (in Russian)
[Veb83a]	Veber, V.K.: Asymptotic behavior of solutions of a linear system of differential equations of fractional order. Stud. IntegroDiff. Equ. **16**, 19–25 (1983). (in Russian)
[Veb83b]	Veber, V.K.: Passivity of linear systems of differential equations with fractional derivative and quasiasymptotic behaviour of solutions. Stud. IntegroDiff. Equ. **16**, 349–356 (1983). (in Russian)
[Veb85a]	Veber, V.K.: Linear equations with fractional derivatives and constant coefficients in spaces of generalized functions. Stud. IntegroDiff. Equ. **18**, 306–312 (1985). (in Russian)
[Veb85b]	Veber, V.K.: On the general theory of linear systems with fractional derivatives. Stud. IntegroDiff. Equ. **18**, 301–305 (1985). (in Russian)

References

[Veb88]	Veber, V.K.: Solution of problem of Sturm-Liouville type. Stud. Integro Diff. Equ. **21**, 245–249 (1988). (in Russian)
[Vol70]	Volkovyskii, L.I., et al.: Collection of exercises on the theory of functions of complex variable. Nauka, Moscow (1970). (in Russian)
[Volt28]	Volterra, V.: Sur la théorie mathématique des phénomènes héréditaires. J. Math. Pures Appl. **7**, 249–298 (1928)
[Volt59]	Volterra, V.: Theory of Functionals and of Integral and Integro - Differential Equations. Dover, New York (1959) [first publ. in 1930]
[WanWen03]	Wang, X., Wen, Z.: Poisson fractional processes. Chaos Solitons Fractals **18**, 169–177 (2003)
[WaWeZh06]	Wang, X., Wen, Z., Zhang, S.: Fractional Poisson process (II). Chaos Solitons Fractals **28**, 143–147 (2006)
[WaZhOR18]	Wang, J.R., Zhou, Y., O'Regan, D.: A note on asymptotic behaviour of Mittag-Leffler functions. Int. Transf. Spec. Funct. **29**(2), 81–94 (2018)
[Wat66]	Watson, G.N.: A Treatise on the Theory of Bessel Functions, 3rd edn. Cambridge University Press, Cambridge (1966) [1st edn, (1922); 2nd edn, (1944)]
[Wei66]	Weierstrass, K.: Abhandlungen aus der Funktionenlehre. Springer, Berlin (1866)
[Weil82]	Weil, A.: Mittag-Leffler as I remember him. Acta Math. **148**(1), 9–13 (1982)
[Wei94]	Weiss, G.H.: Aspects and Applications of Random Walks. North-Holland, Amsterdam (1994)
[Wer91]	Weron, K.: A probabilistic mechanism hidden behind the universal power law for dielectric relaxation: general relaxation equation. J. Phys. Condens. Matter **3**(46), 9151–9162 (1991)
[Wer-et-al05]	Weron, K., Jurlewicz, A., Magdziarz, M.: Havriliak-Negami response in the framework of the continuous-time random walk. Acta Phys. Pol. B **36**(5), 1855–1868 (2005)
[Wer-et-al10]	Weron, K., Jurlewicz, A., Magdziarz, M., Weron, A., Trzmiel, J.: Overshooting and undershooting subordination scenario for fractional two-power-law relaxation responses. Phys. Rev. E **81**(4), 041123/1–7 (2010)
[WerKla00]	Weron, K., Klauzer, A.: Probabilistic basis for the Cole-Cole relaxation law. Ferroelectrics **236**(1), 59–69 (2000)
[WerKot96]	Weron, K., Kotulski, M.: On the Cole-Cole relaxation function and related Mittag-Leffler distribution. Phys. A **232**(1–2), 180–188 (1996)
[WeBoGr03]	West, B.J., Bologna, M., Grigolini, P.: Physics of Fractal Operators. Springer, New York (2003)
[WhiWat52]	Whittaker, E.T., Watson, G.N.: A Course of Modern Analysis, 4th edn. Cambridge University Press, Cambridge (1952) [reprinted 1990]
[Wie79]	Wiener, K.: On solutions of boundary value problems for differential equations of noninteger order. Math. Nachr. **88**, 181–190 (1979)
[Wim05a]	Wiman, A.: Über den Fundamentalsatz der Theorie der Funkntionen $E_\alpha(x)$. Acta Math. **29**, 191–201 (1905)
[Wim05b]	Wiman, A.: Über die Nullstellen der Funkntionen $E_\alpha(x)$. Acta Math. **29**, 217–234 (1905)
[Wid46]	Widder, D.V.: The Laplace Transform. Princeton University Press, Princeton (1946)
[WilWat70]	Williams, G., Watts, D.C.: Non-symmetrical dielectric relaxation behaviour arising from a simple empirical decay function. Trans. Faraday Soc. **66**, 80–85 (1970)
[Wong89]	Wong, R.: Asymptotic Approximations of Integrals. Academic, Boston (1989)
[WongZh99a]	Wong, R., Zhao, Y.-Q.: Smoothing of Stokes' discontinuity for the generalized Bessel function. Proc. R. Soc. Lond. A **455**, 1381–1400 (1999)
[WongZh99b]	Wong, R., Zhao, Y.-Q.: Smoothing of Stokes' discontinuity for the generalized Bessel function II. Proc. R. Soc. Lond. A **455**, 3065–3084 (1999)
[WongZh02]	Wong, R., Zhao, Y.-Q.: Exponential asymptotics of the Mittag-Leffler function. Constr. Approx. **18**, 355–385 (2002)

[Wri33] Wright, E.M.: On the coefficients of power series having exponential singularities. J. Lond. Math. Soc. **8**, 71–79 (1933)

[Wri35a] Wright, E.M.: The asymptotic expansion of the generalized Bessel function. Proc. Lond. Math. Soc. (Ser. II) **38**, 257–270 (1935)

[Wri35b] Wright, E.M.: The asymptotic expansion of the generalized hypergeometric function. J. Lond. Math. Soc. **10**, 287–293 (1935)

[Wri40a] Wright, E.M.: The asymptotic expansion of the generalized Bessel function. Proc. Lond. Math. Soc. (Ser. II) **46**, 389–408 (1940)

[Wri40b] Wright, E.M.: The generalized Bessel function of order greater than one. Quart. J. Math. Oxford Ser. **11**, 36–48 (1940)

[Wri40c] Wright, E.M.: The asymptotic expansion of integral functions defined by Taylor series. Philos. Trans. R. Soc. Lond. A **238**, 423–451 (1940)

[Xu17] Xu, J.: Time-fractional particle deposition in porous media. J. Phys. A: Math. Theor. **50**, 195002 (2017)

[YuZha06] Yu, R., Zhang, H.: New function of Mittag-Leffler type and its application in the fractional diffusion-wave equation. Chaos Solitons Fractals **30**, 946–955 (2006)

[Zas05] Zaslavsky, G.M.: Hamiltonian Chaos and Fractional Dynamics. Oxford University Press, Oxford (2005)

[ZayKar14] Zayenouri, M., Karniadakis, G.E.: Exponentially accurate spectral and spectral element methods for fractional ODEs. J. Comput. Phys. **47**, 2108–2131 (2014)

[Zhe02] Zheltukhina, N.: Asymptotic zero distribution of sections and tails of Mittag-Leffler functions. C. R. Acad. Sci. Paris, Ser. I. **335**, 133–138 (2002)

[Zel-et-al70] Zelenev, V.M., Meshkov, S.I.: Rossikhin, YuA: Damped vibrations of hereditary - elastic systems with weakly singular kernels. J. Appl. Mech. Tech. Phys. **11**(2), 290–293 (1970)

[ZenChe14] Zeng, C., Chen, Y.Q.: Global Padé approximation of the generalized Mittag-Leffler function and its inverse. Fract. Calc. Appl. Anal. **18**(6), 1492–1506 (2015)

[ZhK12] Zhu, K.: Analysis on Fock Spaces. Springer, New York (2012)

[Zhu12] Zhukovskaya, N.V.: Solutions of Euler-type homogeneous differential equations with finite number of fractional derivatives. Integr. Transf. Special Funct. **23**(3), 161–175 (2012)

[ZhuSit18] Zhukovskaya, N.V., Sitnik, S.M.: Euler-type differential equations of fractional order, Mat. zametki SVFU **25**(2), 29–39 (2018) (in Russian)

Suggested Readings. Recently Published Papers/Books

[Ape20] Apelblat, A.: Differentiation of the Mittag-Leffler functions with respect to parameters in the Laplace transform approach. Mathematics **8**, Art 657, 22 p. (2020). https://doi.org/10.3390/math8050657

[Ape20a] Apelblat, A.: Bessel and Related Functions, Vol **1** Theoretical Aspects, Vol **2** Numerical Results W. De Gruyter Gmbh. Berlin (2020)

[ApeMai20] Apelblat, A., Mainardi, F.: Differentiation of the Wright functions with respect to parameters and other results [E-print arXiv:2009.08803 (2020), 20 pp.]

[Baz19] Bazhlekova, E.: Subordination principle for space-time fractional evolution equations and some applications. Integr. Transf. Spec. Funct. **30**, 431–452 (2019)

[CoLuMa19] Consiglio, A., Luchko, Yu., Mainardi, F.: Some notes on the Wright functions in probability theory. WSEAS Trans. Math. **18**, 389–393 (2019)

[ConMai19] Consiglio, A., Mainardi, F.: On the evolution of fractional diffusive waves, *Ricerche di Matematica*, Published on line 06 December 2019, 13 p. https://doi.org/10.1007/s11587-019-00476-6 E-print arXiv:1910.12595 [math.GM]

[ConMai20]	Consiglio, A., Mainardi, F.: Fractional diffusive waves in the Cauchy and signalling problems. In: Beghin, L., Garrappa, R. and Mainardi, F. (eds.) Nonlocal and Fractional Operators: Theory and Applications to Physics, Probability and Numerical Analysis, p. 16. Springer, Berlin (2020)
[Die-et-al20]	Diethelm, K., Garrappa, R., Giusti, A., Stynes, M.: Why fractional derivatives with nonsingular kernel should not be used. Fract. Calc. Appl. Anal. **23**(3), 610–634 (2020) [https://doi.org/10.1515/fca-2020-0032] [E-print arXiv:2006.15237v1, 19 pp]
[FedAvi20]	Fedorov, V.E., Avilovich, A.: Semilinear equations in Banach spaces with Riemann-Liouville derivative and sectorial operators. In: S.V. Rogosin, M.V. Dubatovskaya (eds.) Analytic Methods of Analysis and Differential Equations: AMADE–2018, pp. 187–205. Cambridge Scientific Publishers, (2020)
[GaGiMa19]	Garra, R., Giraldi, F., Mainardi, F.: Wright-type generalized coherent states, WSEAS Transactions on Mathematics. **18**, Art. 52, 428–431 (2019)
[GarMai20]	Garra, R., Mainardi, F.: Some aspects of Wright functions in fractional differential equations. Rep. Math. Phys. (2020) (accepted) E-print arXiv:2007.13340 [math.CA]
[GaKaPo19]	Garrappa, R. Kaslik, E. and Popolizio, M.: Evaluation of fractional integrals and derivatives of elementary functions: overview and tutorial. Mathematics **7**(407), p. 21 (2019) [https://doi.org/10.3390/math7050407]
[Giu20]	Giusti, A.: General fractional calculus and Prabhakar theory, Comm. Nonlinear Sci. Numer. Simul. **83**, Art. 105114, 7 (2020) [https://doi.org/10.1016/j.cnsns.2019.105114] [Eprint arXiv:1911.06695]
[Giu-et-al20]	Giusti, A., Colombaro, I., Garra, R., Garrappa, R., Polito, F., Popolizio, M., Mainardi, F.: A Guide to Prabhakar functions and operators. Fract. Calc. Appl. Anal. **23**(1), 9–54 (2020) https://doi.org/10.1515/fca-2020-0002 E-print arXiv:2002.10978 [math.CA]
[GiuMai20]	Giusti, A., Mainardi, F.: Editorial on Special Issue "Advanced Mathematical Methods: Theory and Applications". Mathematics **8**, 107. https://doi.org/10.3390/math8010107
[GorHor20]	Gorska, K., Horzela, A.: The Volterra type equations related to the non-Debye relaxation. Comm. Nonlinear Sci. Numer. Simul. **85**, 105246/1–14 (2020)
[Gor-et-al20]	Gorska, K., Horzela, A., Lattanzi, A., Pogany, T.K.: On complete monotonicity of three parameter Mittag-Leffler function. Appl. Anal. Discrete Math. in press (2020)
[Han20a]	Hanyga, A.: A comment on a controversial issue: a generalized fractional derivative cannot have a regular kernel. Fract. Calc. Appl. Anal. **23** (1), 211–223 (2020). https://doi.org/10.1515/fca-2020-0008 [E-print arXiv:2003.04385]
[Han20b]	Hanyga, A.: Effects of Newtonian viscosity and relaxation on linear viscoelastic wave propagation. Arch. Appl. Mech. **90**, 467–474 (2020). https://doi.org/10.1007/s00419-019-01620-2 [E-print arXiv:1903.03814]
[Kar20]	Karp, D.B.: A Note on Fox's H Function in the Light of Braaksma's Results. In: P. Agarwal, R.P. Agarwal, M. Ruzhanski (eds.): Special Functions and Analysis of Differential Equations, pp. 243–254, CRC Press, (2020)
[Kir19]	Kiryakova, V.: Commentary: "A remark on the fractional integral operators and the image formulas of generalized Lommel-Wright function". Frontiers in Phys. **7**(145), 4 (2019). https://doi.org/10.3389/fphy.2019.0014
[Kir20]	Kiryakova, V.: A Guide to Special Functions in Fractional Calculus, *Mathematics*, Special Issue "Special Functions with Applications to Mathematical Physics II" (2020) (submitted)
[KocKon20]	Kochubei, A.N., Kondratiev, Yu. G.: Random time change and related evolution equations. Time Asymptot. Behav. Stoch. S Dyn. 2050034, 24 (2020) [https://doi.org/10.1142/S0219493720500343]

[KoKoDa20]	Kochubei, A.N., Kondratiev, Yu.G., Da Silva, J.L.: From random times to fractional kinetics. Interdiscipl. Stud. Complex Systems **16**(2020), 5–32 (2020)
[KraSit20]	Kravchenko, V.V., Sitnik, S.M.: Transmutation operators and applications Birhhäuser. Springer Nature, Switzerland (2020)
[Luc19]	Luchko, Yu.: Some schema for applications of the integral transforms of Mathematical Physics. Mathematics **7**(254), 18 (2019) [https://doi.org/10.3390/math7030254]
[Luc20a]	Luchko, Yu.: Fractional derivatives and the fundamental theorem of fractional calculus. Fract. Calc. Appl. Anal. **23**(4), 939–966 (2020) E-print arXiv:2006.14383 [math.CA]
[Luc20b]	Luchko, Yu.: The four-parameters Wright function of the second-kind and its applications in Fractional Calculus. Mathematics **8**(970) 16 (2020) https://doi.org/10.3390/math8060970
[Luc20c]	Luchko, Yu.: On complete monotonicity of solution to the fractional relaxation equation with the nth level fractional derivative. Mathematics **8**(1561) 14 (2020) https://doi.org/10.3390/8091561. E-print arXiv:2007.11992 [math.CA]
[Mai20]	Mainardi, F.: A Tutorial on the Basic Special Functions of Fractional Calculus, WSEAS Transactions on Mathematics. **19**, 74–98 (2020). https://doi.org/10.37394/23206.2020.19.8 E-print arXiv:2003.12385 [math.GM]
[MaiCon20a]	Mainardi, F., Consiglio, A.: The Wright functions of the second kind in Mathematical Physics. Mathematics **8**(6), Art 884/1-26 (2020). https://doi.org/10.3390/MATH8060884 E-print arXiv:2007.02098 [math.GM]
[MaiCon20b]	Mainardi, F., Consiglio, A.: The Pioneers of the Mittag-Leffler Functions in Dielectrical and Mechanical Relaxation Processes. WSEAS Trans., **19**, 289–300 (2020). https://doi.org/10.37394/23206.2020.19.29 E-print arXiv:2006.07653 [math.GM]
[P-KKir20]	Paneva-Konowska, J., Kiryakova, V.: On the multi-index Mittag-Leffler function. Int. J. Appl. Math. **33**(4), 549–571 (2020)
[Par20a]	Paris, R.B.: Asymptotics expansions of the of the modified exponential integral involving the Mittag-Leffler function. Mathematics **6**(428), 13 (2020) [https://doi.org/10.3390/math8030428]
[Par20ab]	Paris, R.B.: Asymptotics of the Mittag-Leffler function $E_\alpha(z)$ on the negative real axis when $\alpha \to 1$. 9 p (2020). E-print: arXiv:2005.05737
[PaCoMa20]	Paris, R.B., Consiglio, A., Mainardi, F.: On the asymptotics of Wright functions of the second kind. Fract. Calc. Appl. Anal. (2020) (submitted)
[Pop19]	Popolizio, M.: On the matrix Mittag-Leffler function: theoretical properties and numerical computation. Mathematics **7**, Art. 1140, 12 p. (2019) [https://doi.org/10.3390/math7121140]
[Pov19]	Povstenko, Y.: Generalized theory of diffusive stresses associated with the time-fractional diffusion equation and nonlocal constitutive equations for the stress tensor. Comput. Math. Appl. **78**, 1819–1825 (2019)
[Psk20]	Pskhu, A.: The Stankovich Integral Transform and Its Applications. In: P. Agarwal, R.P. Agarwal, M. Ruzhanski (eds.): Special Functions and Analysis of Differential Equations, CRC Press, 197–212 (2020)
[RogDub19]	Rogosin, S.V., Dubatovskaya, M.V.: Special Functions Method for Fractional Analysis and Fractional Modeling. In: S. Rogosin, A.O. Celebi (eds.) Analysis as a Life, Birkhäuser, 261–278 (2019)
[RogDub20a]	Rogosin, S., Dubatovskaya, M.: A short survey on basic elements of Fractional Calculus. In: S.V. Rogosin, M.V. Dubatovskaya (eds.) Analytic Methods of Analysis and Differential Equations: AMADE–2018, 207–237, Cambridge Scientific Publishers (2020)
[RogDub20b]	Rogosin, S., Dubatovskaya, M.: Mkhitar Djrbashian and his contribution to Fractional Calculus. Fract. Calc. Appl. Anal. (2020) (submitted)

References

[RuzTor20a]	Ruzhansky, M., Torebek, B.T.: Van der Corput lemmas for Mittag-Leffler functions. E-print arXiv:2002.07492, 32 (2020)
[RuzTor20b]	Ruzhansky, M., Torebek, B.T.: Van der Corput lemmas for Mittag-Leffler functions, II. α-directions. E-print arXiv:2005.04546, 19 (2020)
[Sae20a]	Saenko, V.V.: Singular points of the integral representation of the Mittag-Leffler function. E-print arXiv:2004.08164, 11 (2020)
[Sae20b]	Saenko, V.V.: Two forms of the integral representation of the Mittag-Leffler function. E-print arXiv:2005.11745, 33 (2020)
[Sae20c]	Saenko, V.V.: The calculation of the Mittag-Leffler function. E-print arXiv:2006.14916, 33 (2020)
[SaMeCh20]	Sandev, T., Metzler, R., Chechkin, A.: Generalized diffusion and wave equation: recent advances. In: S.V. Rogosin, M.V. Dubatovskaya (eds.) Analytic Methods of Analysis and Differential Equations: AMADE–2018, Cambridge Scientific Publishers 187–205 (2020)
[ShiSit20]	Shishkina, E., Sitnik, S.: Singular and Fractional Differential Equations with Applications to Mathematical Physics, Transmutations. Elsevier and Academic Press, London (2020)
[VanM20]	Van Mieghem, P.: The Mittag-Leffler function. E-print arXiv:2005.1330, 69 (2020)
[VuBoDi19]	Vu Kim Tuan, Boumenir, A., Dinh Thanh Duc.: Real Variable Inverse Laplace Transform. In: S. Rogosin, A.O. Celebi (eds.): Analysis as a Life, Birkhäuser, 303–318 (2019)
[Yam20]	Yamamoto, M.: Uniqueness in determining the orders of time and spatial fractional derivatives. E-print arXiv:2006.15046, 11 (2020)

Index

A
Abel equation of the first kind, 235
Abel equation of the second kind, 237
Abstract differential equation, 84
Acta Mathematica, 8
Adams type predictor-corrector method, 263
α-exponential function, 94
Angular density, 401
Angular symmetry, 401
Asymptotic universality of Mittag-Leffler waiting time density, 366
Auxiliary functions of the Wright type, 222

B
Basset problem, 294
Bernstein function, 341
Bessel polynomials, 497
Bessel–Wright function, 473
Beta function, 390
Bromwich formula, 47
Bromwich inversion formula, 48
Bromwich path, 229

C
Caputo fractional derivative, 445
Caputo–Dzherbashian fractional derivative, 445
Caputo–Gerasimov fractional derivative, 445
Cauchy–Lorentz probability density, 371
Cauchy problem for OFDEs, 243
Cauchy type initial conditions, 241
Cauchy type problem for OFDEs, 242
Central limit theorem, 369

Characteristic function, 369
Classical Kolmogorov–Feller equation, 346
Close-to-convex function, 95
Cole–Cole response function, 327
Completely monotonic function, 53, 230
Complex factorial function, 382
Composite fractional oscillation equation, 292
Composite fractional relaxation and oscillation equations, 291
Composite fractional relaxation equation, 291, 321
Compound renewal processes, 376
Constitutive equations, 299
Contiguous relations, 487
Continuous time random walk (CTRW), 343
Correspondence principle, 301
Counting number process, 342
Creep compliance, 300

D
Diffusion limit, 358
Dirichlet condition, 411
Discrete Mittag-Leffler distribution, 375
Djrbashian summation formula, 66

E
Empty product convention, 128
Entire function, 397
Entire function of completely regular growth, 401
Erdélyi–Kober fractional integrals and derivatives, 451
Erlang process, 343

Error function, 22
Euler integral of the second kind, 390
Euler–Mascheroni constant, 390
Euler transform of the Mittag-Leffler function, 37
Extended generalized Mittag-Leffler function, 168

F
Failure probability, 344
Feller–Takayasu diamond, 370
Four-parametric generalization of the Prabhakar function, 179
Four-parametric Mittag-Leffler function, 163
Fourier convolution, 412
Fourier integral formula, 411
Fourier integral theorem, 411
Fourier transform, 369, 371, 374
Fourier transformation, 410
Fourier transform operator, 410
Fox H-function, 15, 373, 481
Fractional anti-Zener model, 303
Fractional equations for heat transfer, 328
Fractional KdV equation, 335
Fractional Klein–Gordon equation, 335
Fractional Maxwell model, 302
Fractional Newton equation, 326
Fractional Newton (Scott-Blair) model, 301
Fractional Ohm law, 327
Fractional operator equation, 303
Fractional-order controller, 253
Fractional oscillation, 288
Fractional Poisson process, 351
Fractional relaxation, 288
Fractional Voigt model, 301
Fractional Zener model, 302
Function starlike with respect to origin, 95

G
Gamma function, 381
Gauss probability density, 373
Generalized Bessel–Wright function, 474
Generalized hypergeometric functions, 472
Generalized Kolmogorov–Feller equation, 346
Grünwald–Letnikov fractional derivative, 455

H
Hadamard fractional derivatives, 452

Hadamard fractional integrals, 452
Hankel integral representation, 25, 388
Hankel integration path, 12
Hankel loop, 88
Hankel path, 27, 229, 386
Heaviside function, 299
Hermite polynomials, 471, 497
Hypergeometric differential equation, 467
Hypergeometric function, 463

I
Incomplete gamma function, 23
Indicator function of an entire function, 399
Indirect numerical methods, 263
Infinite divisibility, 371
Infinite product, 399
Integral representations of auxiliary functions of the Wright type, 222
Inverse Fourier operator, 411
Inverse Mellin transform, 417
Inverse problem for the abstract differential equation, 84

J
Jacobi polynomials, 496
Jordan canonical form, 200

K
Kilbas–Saigo function, 128
Kummer confluent hypergeometric function, 468

L
Lévy stable distribution, 373
Laguerre polynomials, 471, 496
Lambert function, 84
Laplace–Abel integral, 10
Laplace abscissa of convergence, 47, 414
Laplace convolution, 416
Laplace image, 46, 414
Laplace original, 46, 414
Laplace transform, 369, 374
Laplace transform method for FDEs, 247
Le Roy function, 16, 147
Le Roy type function, 147
Legendre polynomials, 497
Legendre's duplication formula, 386
Leibniz rule, 443
Lévy stable distribution, 369
Liouville fractional integrals and derivatives, 443

M

Marchaud fractional derivative, 448
Material functions, 300
Maxwell type fractional equation, 336
Meijer G-functions, 14, 476
Mellin–Barnes integral, 431
Mellin–Barnes integral formula, 14
Mellin convolution formula, 418
Mellin fractional analogue of Green's function, 250
Mellin integral transform, 417
Mellin transform, 374
Miller–Ross function, 66
Mittag-Leffler distribution, 374
Mittag-Leffler function, 11, 228, 230
Mittag-Leffler function of several variables, 194
Mittag-Leffler functions with $2n$ parameters, 181
Mittag-Leffler, Gösta Magnus, 7
Mittag-Leffler integral representation, 27
Mittag-Leffler reproducing kernel Hilbert space, 44
Mittag-Leffler star, 43
Mittag-Leffler sum, 41
Multiplication formula for Γ-function, 385

N

Nonlinear fractional-order controller, 254

O

Operational calculus for FDEs, 244
Order of an entire function, 398
Ordinary fractional differential equation, 241

P

Parseval identity, 412
Phragmén–Lindelöf theorem, 13
Pochhammer's symbol, 386
Pot, 301
Prabhakar function, 115
Prabhakar summation formula, 67
Predictor-corrector method, 263
Probability density function, 350

Q

Quadrature based direct numerical methods, 262
Quasi-polynomial, 247

R

Rabotnov function, 66
Relaxation differential equation, 282
Relaxation modulus, 300
Renewal function, 343
Renewal process, 342
Rescaling concept, 366
Respeeding concept, 366
Riemann–Liouville fractional derivative, 442
Riemann–Liouville left-sided fractional integral, 440
Riemann–Liouville right-sided fractional integral, 440
Riesz fractional derivative, 458
Riesz fractional integral, 458

S

Self-similarity, 371
Simple fractional oscillation equation, 282
Simple fractional relaxation equation, 282
Small Prabhakar function, 123
Spectral function, 230
starlike function of order η, 95
Stirling's formula, 389
Stress–strain relation, 299
Survival probability, 344

T

Tchebyshev polynomials, 497
Thinning, 355
Three-parametric Mittag-Leffler function, 15
Time fractional Kolmogorov–Feller equation, 349
Titchmarsh inversion formula, 124
Transcendental entire functions, 397
Tricomi confluent hypergeometric function, 469
Triplication formula, 386
Two-parametric Mittag-Leffler function, 15
Type of an entire function, 398

U

Univalent function, 95

W

Weierstrass elementary factors, 399
Whittaker function, 122
Wright function, 230, 231, 473

Wright function of the first kind, 210
Wright function of the second kind, 210

Wright M-function, 230, 368, 374

Printed by Books on Demand, Germany